EMULSIONS AND EMULSION STABILITY

Second Edition

SURFACTANT SCIENCE SERIES

1. Nonionic Surfactants, *edited by Martin J. Schick* (see also Volumes 19, 23, and 60)
2. Solvent Properties of Surfactant Solutions, *edited by Kozo Shinoda* (see Volume 55)
3. Surfactant Biodegradation, *R. D. Swisher* (see Volume 18)
4. Cationic Surfactants, *edited by Eric Jungermann* (see also Volumes 34, 37, and 53)
5. Detergency: Theory and Test Methods (in three parts), *edited by W. G. Cutler and R. C. Davis* (see also Volume 20)
6. Emulsions and Emulsion Technology (in three parts), *edited by Kenneth J. Lissant*
7. Anionic Surfactants (in two parts), *edited by Warner M. Linfield* (see Volume 56)
8. Anionic Surfactants: Chemical Analysis, *edited by John Cross*
9. Stabilization of Colloidal Dispersions by Polymer Adsorption, *Tatsuo Sato and Richard Ruch*
10. Anionic Surfactants: Biochemistry, Toxicology, Dermatology, *edited by Christian Gloxhuber* (see Volume 43)
11. Anionic Surfactants: Physical Chemistry of Surfactant Action, *edited by E. H. Lucassen-Reynders*
12. Amphoteric Surfactants, *edited by B. R. Bluestein and Clifford L. Hilton* (see Volume 59)
13. Demulsification: Industrial Applications, *Kenneth J. Lissant*
14. Surfactants in Textile Processing, *Arved Datyner*
15. Electrical Phenomena at Interfaces: Fundamentals, Measurements, and Applications, *edited by Ayao Kitahara and Akira Watanabe*
16. Surfactants in Cosmetics, edited by Martin M. Rieger (see Volume 68)
17. Interfacial Phenomena: Equilibrium and Dynamic Effects, *Clarence A. Miller and P. Neogi*
18. Surfactant Biodegradation: Second Edition, Revised and Expanded, *R. D. Swisher*
19. Nonionic Surfactants: Chemical Analysis, *edited by John Cross*
20. Detergency: Theory and Technology, *edited by W. Gale Cutler and Erik Kissa*
21. Interfacial Phenomena in Apolar Media, *edited by Hans-Friedrich Eicke and Geoffrey D. Parfitt*
22. Surfactant Solutions: New Methods of Investigation, *edited by Raoul Zana*
23. Nonionic Surfactants: Physical Chemistry, *edited by Martin J. Schick*
24. Microemulsion Systems, *edited by Henri L. Rosano and Marc Clausse*
25. Biosurfactants and Biotechnology, *edited by Naim Kosaric, W. L. Cairns, and Neil C. C. Gray*
26. Surfactants in Emerging Technologies, *edited by Milton J. Rosen*
27. Reagents in Mineral Technology, *edited by P. Somasundaran and Brij M. Moudgil*
28. Surfactants in Chemical/Process Engineering, *edited by Darsh T. Wasan, Martin E. Ginn, and Dinesh O. Shah*
29. Thin Liquid Films, *edited by I. B. Ivanov*
30. Microemulsions and Related Systems: Formulation, Solvency, and Physical Properties, *edited by Maurice Bourrel and Robert S. Schechter*
31. Crystallization and Polymorphism of Fats and Fatty Acids, *edited by Nissim Garti and Kiyotaka Sato*
32. Interfacial Phenomena in Coal Technology, *edited by Gregory D. Botsaris and Yuli M. Glazman*
33. Surfactant-Based Separation Processes, *edited by John F. Scamehorn and Jeffrey H. Harwell*
34. Cationic Surfactants: Organic Chemistry, *edited by James M. Richmond*

35. Alkylene Oxides and Their Polymers, *F. E. Bailey, Jr., and Joseph V. Koleske*
36. Interfacial Phenomena in Petroleum Recovery, *edited by Norman R. Morrow*
37. Cationic Surfactants: Physical Chemistry, *edited by Donn N. Rubingh and Paul M. Holland*
38. Kinetics and Catalysis in Microheterogeneous Systems, *edited by M. Grätzel and K. Kalyanasundaram*
39. Interfacial Phenomena in Biological Systems, *edited by Max Bender*
40. Analysis of Surfactants, *Thomas M. Schmitt* (see Volume 96)
41. Light Scattering by Liquid Surfaces and Complementary Techniques, *edited by Dominique Langevin*
42. Polymeric Surfactants, *Irja Piirma*
43. Anionic Surfactants: Biochemistry, Toxicology, Dermatology. Second Edition, Revised and Expanded, *edited by Christian Gloxhuber and Klaus Künstler*
44. Organized Solutions: Surfactants in Science and Technology, *edited by Stig E. Friberg and Björn Lindman*
45. Defoaming: Theory and Industrial Applications, *edited by P. R. Garrett*
46. Mixed Surfactant Systems, *edited by Keizo Ogino and Masahiko Abe*
47. Coagulation and Flocculation: Theory and Applications, *edited by Bohuslav Dobiás*
48. Biosurfactants: Production Properties Applications, edited by Naim Kosaric
49. Wettability, *edited by John C. Berg*
50. Fluorinated Surfactants: Synthesis Properties Applications, *Erik Kissa*
51. Surface and Colloid Chemistry in Advanced Ceramics Processing, *edited by Robert J. Pugh and Lennart Bergström*
52. Technological Applications of Dispersions, *edited by Robert B. McKay*
53. Cationic Surfactants: Analytical and Biological Evaluation, *edited by John Cross and Edward J. Singer*
54. Surfactants in Agrochemicals, *Tharwat F. Tadros*
55. Solubilization in Surfactant Aggregates, *edited by Sherril D. Christian and John F. Scamehorn*
56. Anionic Surfactants: Organic Chemistry, *edited by Helmut W. Stache*
57. Foams: Theory, Measurements, and Applications, *edited by Robert K. Prud'homme and Saad A. Khan*
58. The Preparation of Dispersions in Liquids, *H. N. Stein*
59. Amphoteric Surfactants: Second Edition, *edited by Eric G. Lomax*
60. Nonionic Surfactants: Polyoxyalkylene Block Copolymers, *edited by Vaughn M. Nace*
61. Emulsions and Emulsion Stability, *edited by Johan Sjöblom*
62. Vesicles, *edited by Morton Rosoff*
63. Applied Surface Thermodynamics, *edited by A. W. Neumann and Jan K. Spelt*
64. Surfactants in Solution, *edited by Arun K. Chattopadhyay and K. L. Mittal*
65. Detergents in the Environment, *edited by Milan Johann Schwuger*
66. Industrial Applications of Microemulsions, *edited by Conxita Solans and Hironobu Kunieda*
67. Liquid Detergents, *edited by Kuo-Yann Lai*
68. Surfactants in Cosmetics: Second Edition, Revised and Expanded, *edited by Martin M. Rieger and Linda D. Rhein*
69. Enzymes in Detergency, *edited by Jan H. van Ee, Onno Misset, and Erik J. Baas*
70. Structure-Performance Relationships in Surfactants, *edited by Kunio Esumi and Minoru Ueno*

71. Powdered Detergents, *edited by Michael S. Showell*
72. Nonionic Surfactants: Organic Chemistry, *edited by Nico M. van Os*
73. Anionic Surfactants: Analytical Chemistry, Second Edition, Revised and Expanded, *edited by John Cross*
74. Novel Surfactants: Preparation, Applications, and Biodegradability, *edited by Krister Holmberg*
75. Biopolymers at Interfaces, *edited by Martin Malmsten*
76. Electrical Phenomena at Interfaces: Fundamentals, Measurements, and Applications, Second Edition, Revised and Expanded, *edited by Hiroyuki Ohshima and Kunio Furusawa*
77. Polymer-Surfactant Systems, *edited by Jan C. T. Kwak*
78. Surfaces of Nanoparticles and Porous Materials, *edited by James A. Schwarz and Cristian I. Contescu*
79. Surface Chemistry and Electrochemistry of Membranes, *edited by Torben Smith Sørensen*
80. Interfacial Phenomena in Chromatography, *edited by Emile Pefferkorn*
81. Solid–Liquid Dispersions, *Bohuslav Dobiás, Xueping Qiu, and Wolfgang von Rybinski*
82. Handbook of Detergents, *editor in chief: Uri Zoller* Part A: Properties, *edited by Guy Broze*
83. Modern Characterization Methods of Surfactant Systems, *edited by Bernard P. Binks*
84. Dispersions: Characterization, Testing, and Measurement, *Erik Kissa*
85. Interfacial Forces and Fields: Theory and Applications, *edited by Jyh-Ping Hsu*
86. Silicone Surfactants, *edited by Randal M. Hill*
87. Surface Characterization Methods: Principles, Techniques, and Applications, *edited by Andrew J. Milling*
88. Interfacial Dynamics, *edited by Nikola Kallay*
89. Computational Methods in Surface and Colloid Science, *edited by Malgorzata Borówko*
90. Adsorption on Silica Surfaces, *edited by Eugène Papirer*
91. Nonionic Surfactants: Alkyl Polyglucosides, *edited by Dieter Balzer and Harald Lüders*
92. Fine Particles: Synthesis, Characterization, and Mechanisms of Growth, *edited by Tadao Sugimoto*
93. Thermal Behavior of Dispersed Systems, *edited by Nissim Garti*
94. Surface Characteristics of Fibers and Textiles, *edited by Christopher M. Pastore and Paul Kiekens*
95. Liquid Interfaces in Chemical, Biological, and Pharmaceutical Applications, *edited by Alexander G. Volkov*
96. Analysis of Surfactants: Second Edition, Revised and Expanded, *Thomas M. Schmitt*
97. Fluorinated Surfactants and Repellents: Second Edition, Revised and Expanded, *Erik Kissa*
98. Detergency of Specialty Surfactants, *edited by Floyd E. Friedli*
99. Physical Chemistry of Polyelectrolytes, *edited by Tsetska Radeva*
100. Reactions and Synthesis in Surfactant Systems, *edited by John Texter*
101. Protein-Based Surfactants: Synthesis, Physicochemical Properties, and Applications, *edited by Ifendu A. Nnanna and Jiding Xia*
102. Chemical Properties of Material Surfaces, *Marek Kosmulski*

103. Oxide Surfaces, *edited by James A. Wingrave*
104. Polymers in Particulate Systems: Properties and Applications, *edited by Vincent A. Hackley, P. Somasundaran, and Jennifer A. Lewis*
105. Colloid and Surface Properties of Clays and Related Minerals, *Rossman F. Giese and Carel J. van Oss*
106. Interfacial Electrokinetics and Electrophoresis, *edited by Ángel V. Delgado*
107. Adsorption: Theory, Modeling, and Analysis, *edited by József Tóth*
108. Interfacial Applications in Environmental Engineering, *edited by Mark A. Keane*
109. Adsorption and Aggregation of Surfactants in Solution, *edited by K. L. Mittal and Dinesh O. Shah*
110. Biopolymers at Interfaces: Second Edition, Revised and Expanded, *edited by Martin Malmsten*
111. Biomolecular Films: Design, Function, and Applications, *edited by James F. Rusling*
112. Structure–Performance Relationships in Surfactants: Second Edition, Revised and Expanded, *edited by Kunio Esumi and Minoru Ueno*
113. Liquid Interfacial Systems: Oscillations and Instability, *Rudolph V. Birikh, Vladimir A. Briskman, Manuel G. Velarde, and Jean-Claude Legros*
114. Novel Surfactants: Preparation, Applications, and Biodegradability: Second Edition, Revised and Expanded, *edited by Krister Holmberg*
115. Colloidal Polymers: Synthesis and Characterization, *edited by Abdelhamid Elaissari*
116. Colloidal Biomolecules, Biomaterials, and Biomedical Applications, *edited by Abdelhamid Elaissari*
117. Gemini Surfactants: Synthesis, Interfacial and Solution-Phase Behavior, and Applications, *edited by Raoul Zana and Jiding Xia*
118. Colloidal Science of Flotation, *Anh V. Nguyen and Hans Joachim Schulze*
119. Surface and Interfacial Tension: Measurement, Theory, and Applications, *edited by Stanley Hartland*
120. Microporous Media: Synthesis, Properties, and Modeling, *Freddy Romm*
121. Handbook of Detergents, *editor in chief: Uri Zoller* Part B: Environmental Impact, *edited by Uri Zoller*
122. Luminous Chemical Vapor Deposition and Interface Engineering, *HirotsuguYasuda*
123. Handbook of Detergents, editor in chief: Uri Zoller Part C: Analysis, *edited by Heinrich Waldhoff and Rüdiger Spilker*
124. Mixed Surfactant Systems: Second Edition, Revised and Expanded, *edited by Masahiko Abe and John F. Scamehorn*
125. Dynamics of Surfactant Self-Assemblies: Micelles, Microemulsions, Vesicles and Lyotropic Phases, *edited by Raoul Zana*
126. Coagulation and Flocculation: Second Edition, *edited by Hansjoachim Stechemesser and Bohulav Dobiás*
127. Bicontinuous Liquid Crystals, *edited by Matthew L. Lynch and Patrick T. Spicer*
128. Handbook of Detergents, editor in chief: Uri Zoller Part D: Formulation, *edited by Michael S. Showell*
129. Liquid Detergents: Second Edition, *edited by Kuo-Yann Lai*
130. Finely Dispersed Particles: Micro-, Nano-, and Atto-Engineering, *edited by Aleksandar M. Spasic and Jyh-Ping Hsu*
131. Colloidal Silica: Fundamentals and Applications, *edited by Horacio E. Bergna and William O. Roberts*
132. Emulsions and Emulsion Stability, Second Edition, *edited by Johan Sjöblom*

EMULSIONS AND EMULSION STABILITY

Second Edition

Edited by

Johann Sjöblom
Universitetet i Bergen
Bergen, Norway

CRC Press is an imprint of the
Taylor & Francis Group, an **informa** business
A TAYLOR & FRANCIS BOOK

Published in 2006 by
CRC Press
Taylor & Francis Group
6000 Broken Sound Parkway NW, Suite 300
Boca Raton, FL 33487-2742

First issued in paperback 2020

ISBN 13: 978-0-367-57795-7 (pbk)
ISBN 13: 978-0-8247-2695-9 (hbk)

Visit the Taylor & Francis Web site at
http://www.taylorandfrancis.com

and the CRC Press Web site at
http://www.crcpress.com

Library of Congress Cataloging-in-Publication Data

Catalog record is available from the Library of Congress

Preface

The revised and expanded edition of *Emulsions and Emulsion Stability* is a comprehensive treatise of different aspects of emulsions, both from a theoretical as well as a practical point of view. The book contains 19 chapters written by leading scientists in the field of emulsions. In more detail the chapters cover the following:

Dukhin's chapter covers fundamental processes, i.e. flocculation and coalescence in emulsified systems. A central aspect is to cover the time evolution of the droplet size distribution by means of the population balance equation (PBE) and based on subprocesses like droplet aggregation (both reversible and irreversible), aggregate fragmentation, droplet flocculation and droplet coalescence. In order to get an overall picture an attempt to couple these processes is undertaken.

Spontaneous emulsification is covered by Miller in his chapter. He relates the concept of self-emulsification to inversion processes and local supersaturation of oil, and a subsequent nucleation of oil droplets. The author also describes the role of a lamellar liquid crystalline phase in coating the small oil droplets and hence protecting them from coalescence. Also the formation of "nanoemulsions" is brought forward by the author.

The importance of thin liquid films in colloidal systems is presented by Platikanov and Exerowa from the famous Bulgarian school. The chapter provides a valuable insight into theory of thin films, essential properties and experimental techniques to characterize them.

Salager's chapter deals with the central question of inversion in emulsion systems. The phenomenon is presented as depending of different system variables, like formulation variables related to the equilibrium of the system, composition variables related to the quantity in proportions of the compounds (generally surfactant/cosurfactant, oil and water) and protocol variables describing the preparation of the emulsion.

Johansson and Voets elaborate further the phase inversion concept as a tool for optimization of surfactant mixtures in water/oil systems. The chapter by Johansson and Voets also deals with several technical applications of the optimized surfactant mixtures. It is by now well-known that unconcentrated emulsions have a polyheder structure instead of the spherical structure for low internal phase volume fractions.

The chapter of Esquena and Solans takes up the topic of highly concentrated emulsions as templates for solid foams. The idea is to prepare low density foams (< 0.1 g/ml) by polymerizing the continuous phase. These macroporous foams can be based on either organic or inorganic components in the continuous phase.

Colloidal structures like emulsions and microemulsions can serve as templates for chemical reactions. One of the leading scientists in this area, Krister Holmberg gives a comprehensive survey of different organic reactions in dispersed systems. The advantage in using emulsions/microemulsions is in enhanced solubility of reactants and in huge interfacial areas.

The first chapter in the characterization session of the emulsions deals with NMR. Peña and Hirasaki give an excellent review and update of different NMR applications in this field. The chapter covers basic NMR principles, relaxation measurements based on CPMG experiments and diffusion measurements via PGSE and PGSTE experiments. The applications for emulsified systems cover a determination of droplet sizes and stability against coalescence. The chapter also gives a comprehensive analysis of the pros and cons of the technique.

The emulsion characterization continues with Andrei Dukhin's chapter about ultrasound characterization of emulsions and microemulsions. The author gives a brilliant introduction to the topic. After this he gives numerous examples on practical applications on the determination of droplet sizes on a variety of systems, in addressing dairy systems like milk and butter. The author also gives a detailed comparison of the ultrasound technique and other techniques, like scattering measurements.

The book continues with some technical emulsions. The first chapter by Moldestad et al. deals with environmental emulsions, where the authors present in detail problems with oil spills. In an oil spill there is a natural uptake of water, and the wave energy contributes to the emulsification. As a result a stable w/o emulsion will emerge. A contributing factor is also the continuous evaporation of volatile components from the crude oil. Sintef in Trondheim has significantly contributed to a better understanding of the characterization and stability of such emulsions.

Per Redelius from Nynäs gives a comprehensive introduction to the topic of bitumen emulsions. These emulsions are in wide use in what we know as asphalt on our roads. In these formulations the road surface is a result of mixing the bitumen formulation to crushed rock, combined with a binding agent giving the final colour. Alternatively, for grey surfaces, a Portland concrete is used.

In the chapter by Sjöblom et al. the focus is on experimental techniques for characterization of crude oils. There seems to be a need for better and systematic needs to characterize crude oils and predict flow assurance related problems on a sound basis. Problems in relation to production and processing are water-in-oil and oil-in-water emulsions, foams and deposits. These problems (often resulting in shut-in periods) can be related to the properties of the crude oils and hence a detailed characterization is needed. In addition there is also a need to extend the characterization to elevated pressures and temperatures to cover subsea and downhole conditions.

Acid crude oils are covered by C. Hurtevent et al. in their chapter. These crude oils cause production and maintenance problems due to both formation of water-in-oil emulsions and deposits in the form of calcium naphthenates. Obviously different naphthenic acids are responsible for emulsion formation and precipitation, respectively. This is the background to intense research in the field of isolation, separation and characterization of individual acids from the family occurring in crude oils worldwide.

Li et al. present the topic of chemical flooding and related to this the problem of w/o emulsions. Central themes in the chapter are the connection between characterization of Chinese crude oils and the prediction of emulsion stability.

Electrocoalescence as a unit process is presented in detail by Lundgaard et al. The treatment is very thorough and built on a fundamental understanding of droplet-droplet interactions in different electrical fields. Also some practical aspects of the installment of electrocoalescence are discussed by the authors.

Separation technology is also the topic of the next chapter by Ernst Hansen. He deals with numerical simulation of fluids in offshore gravity separations. The chapter deals with both batch and continuous separation, together with the effect of different inlet configurations. It is obvious that CFD (computational fluid dynamics) can provide valuable information about the flow conditions in a separator tank.

Friedemann continues with the same topic of fluid behavior in confined offshore separators. The author also extends the discussion to subsea solutions. This chapter also emphasizes the need for a detailed fluid characterization in order to gain a better understanding of the underlying mechanisms of the emulsion problems and how to treat these. The chapter presents a variety of models for the separation performance.

The problems with production water are emerging and there are huge quantities that are produced and need to be cleaned. In this context droplet sizes and droplet size distributions are essential in the treatment.

Hemmingsen et al. present video microscopy as a technique to monitor on-line the DSD in product lines from the separator. In this context one needs a technique that can be adopted to higher pressures. The DVM technique gives representable DSDs and seems to fulfill the essential requirements for on-line instrumentation.

A central question for processing offshore is to develop on-line instrumentation to follow for instance the demulsification process. The last chapter in the book deals with conductivity measurements as an alternative to such instrumentation. The chapter is written by van Dijk et al. The authors provide convincing data for the DEMCOM technique.

The revised and expanded edition of *Emulsions and Emulsion Stability* serves as a comprehensive treatise of the emulsion phenomenon. We have tried to cover fundamental and basic aspects, experimental characterization, technical systems and, finally, separation technology and examples on on-line instrumentation. I hope that the readers will enjoy this potpourri and that they gain new insight and knowledge that can be implemental both in academia and industry.

To my co-authors, I want to express my deepest gratitude for their contributions. I am very touched by the scientific level of all the chapters and also how the chapters are updated with modern references.

Finally, I do hope that the scientific community will appreciate our efforts to move the frontiers forward in the field of emulsion science.

Enjoy the reading!

Johan Sjöblom
Professor and
Head of Ugelstad Laboratory
NTNU, Trondheim
Norway

The Editor

Johan Sjöblom has been professor in surface and colloid chemistry, and head of the Ugelstad Laboratory at the Norwegian University of Technology and Sciences (NTNU), Trondheim, Norway, since January 2002. Previously, he was a professor at the University of Bergen between 1988 and 1999, and chief researcher at Statoil's R&D Center between 1998 and 2001. Dr Sjöblom has also served as adjunct professor of colloid chemistry and materials science at Clarkson University, Potsdam, New York, and at the Technical University of Helsinki from 1992 to 1999.

Dr Sjöblom is the editor of nine books and the author or co-author of more than 260 professional papers. He has also supervised 28 PhD students and 35 MSc students. He is editor-in-chief for *Journal of Dispersion Science and Technology* and a member of the advisory board of *Micro- and Macroporous Materials*, and was a board member of *Colloids and Surfaces* between 1993 and 2003.

Dr Sjöblom is a member of the American Chemical Society, the Norwegian Chemical Society, the Finnish Chemical Society, and an elected member of the Norwegian Academy of Technological Sciences and the Swedish Academy of Engineering Sciences in Finland. He received his PhD degree (1982) in physical chemistry from Åbo Akademi, Finland, and has been an industrial consultant in colloid chemistry since 1980.

Contributors

Pierre Atten
CNRS/LEMD
Grenoble, France

Inge H. Auflem
Statoil ASA
Stjørdal, Norway

Gunnar Berg
SINTEF Energy Research
Trondheim, Norway

Maurice Bourrel
Total Petrochemicals France
Lacq, France

Øystein Brandal
Norwegian University of
Science and Technology
Trondheim, Norway

Benjamin Brocart
Total Petrochemicals France
Lacq, France

Per S. Daling
Marine Environmental
Technology Department
SINTEF Materials and Chemistry
Trondheim, Norway

Jan van Dijk
Champion Technologies
Delden, Netherlands

Zhaoxia Dong
University of Petroleum
Beijing, China

A. S. Dukhin
Dispersion Technology Inc.
Bedford Hills,
New York, USA

Stanislav S. Dukhin
New Jersey Institute of Technology
Newark, New Jersey, USA

Marit-Helen Ese
Norwegian University of
Science and Technology
Trondheim, Norway

Jordi Esquena
IIQAB
Barcelona, Spain

Dotchi Exerowa
Institute of Physical Chemistry
Bulgarian Academy of Sciences
Sofia, Bulgaria

John D. Friedemann
Vetco Gray
Stavanger, Norway

Wilhelm R. Glomm
Department of Chemical Engineering
Norwegian University of
Science and Technology
Trondheim, Norway

P.J. Goetz
Dispersion Technology Inc.
Bedford Hills, New York, USA

Jixiang Guo
University of Petroleum
Beijing, China

Andreas Hannisdal
Department of Chemical Engineering
Norwegian University of
Science and Technology
Trondheim, Norway

Ernst W.M. Hansen
Complex Flow Design AS
Trondheim, Norway

Trond E. Havre
Champion Technologies
Delden, Netherlands

Pål V. Hemmingsen
Department of Chemical Engineering
Norwegian University of Science and Technology
Trondheim, Norway

George J. Hirasaki
Department of Chemical Engineering
Rice University
Houston, Texas, USA

Krister Holmberg
Applied Surface Chemistry
Chalmers University of Technology
Göteborg, Sweden

Christian Hurtevent
Total Exploration & Production
Pau, France

Stian Ingebrigtsen
SINTEF Energy Research
Trondheim, Norway

Øistein Johansen
Marine Environmental Technology Department
SINTEF Materials and Chemistry
Trondheim, Norway

Ingegärd Johansson
Akzo Nobel Surface Chemistry
Stenungsund, Sweden

Harald Kallevik
Statoil R&D Center
Trondheim, Norway

Magne Knag
Norwegian University of Science and Technology
Trondheim, Norway

Frode Leirvik
Marine Environmental Technology Department Chemistry
SINTEF Materials and Chemistry
Trondheim, Norway

Alun Lewis
Oil Spill Consultant
Staines, United Kingdom

Mingyuan Li
University of Petroleum
Beijing, China

Meiqin Lin
University of Petroleum
Beijing, China

Lars E. Lundgaard
SINTEF Energy Research
Trondheim, Norway

Clarence A. Miller
Department of Chemical Engineering
Rice University
Houston, Texas, USA

Merete Ø. Moldestad
Marine Environmental Technology Department
SINTEF Materials and Chemistry
Trondheim, Norway

Hans-Jörg Oschmann
Champion Technologies
Delden, Netherlands

Gisle Øye
Department of Chemical Engineering
Norwegian University of
Science and Technology
Trondheim, Norway

Alejandro A. Peña
Department of Oilfield Chemical Products
Schlumberger Technology Corp.
Sugar Land, Texas, USA

Bo Peng
University of Petroleum
Beijing, China

Dimo Platikanov
Department of Physical Chemistry
University of Sofia, Bulgaria

Per Redelius
Nynas Bitumen
Nynäshamn, Sweden

Geir J. Rørtveit
Complex Flow Design AS
Trondheim, Norway

Guy Rousseau
Total Exploration & Production
Pau, France

Øystein Sæther
Det Norske Veritas
Section of Materials Technology
and Condition Management
Bergen, Norway

Jean-Louis Salager
FIRP
Universidad de Los Andes
Mérida, Venezuela

Johan Sjöblom
Department of Chemical
Engineering
Norwegian University of
Science and Technology
Trondheim, Norway

Conxita Solans
Consejo Superior de
Investigaciones Científicas
Barcelona, Spain

Ilja Voets
Laboratory of Physical
Chemistry and Colloid Science
Wageningen University
Wageningen, Netherlands

Jean Walter
Nynas Bitumen
South Wirral, UK

Arild Westvik
Statoil R&D Center
Trondheim, Norway

Zhaoliang Wu
University of Petroleum
Beijing, China

Table of Contents

1 An Experimental and Theoretical Approach to the Dynamic Behavior of Emulsions

Stanislav S. Dukhin, Johan Sjöblom, and Øystein Sæther

CONTENTS

Abstract

For emulsion stability the surface chemistry, the dynamic of adsorption, the surface rheology, and the physicochemical kinetics are important. In contrast to the large success in industrial application of the emulsion surface chemistry the physicochemical kinetics is almost not used in emulsion technology. Meanwhile, the population balance equation (PBE) enables prediction of the evolution in time for droplet size distribution if the family of subprocesses including droplet aggregation, aggregate fragmentation, droplet coalescence, and droplet (floccula) creaming are quantified. These subprocesses are characterized in PBE by means of kinetic coefficients (kernels). The coupling of these four subprocesses, droplet poly-dispersity, and immense variety in droplet aggregate configurations cause the extreme difficulty in emulsion dynamics modeling (EDM). Three subprocesses, namely aggregation, fragmentation, and creaming, can be quantified. The systematic consideration of these three subprocesses with account for both Brownian and gravitational aggregation is accomplished in this chapter as it is necessary for EDM. In contrast to those three subprocesses, the experimental approach only is effective now concerning the emulsion film stability and coalescence kernel quantification for EDM. Accordingly, an experimental theoretical approach for the coalescence time determination, based on a dilute emulsion characterization in its simplest state, namely at the singlet doublet quasi-equilibrium, is elaborated. Information about elementary acts of coalescence and fragmentation obtained in experiments with dilute emulsion preserves its significance for concentrated emulsion as well. For modeling of nondiluted emulsions the combining of experimental investigation of the simplest emulsion model system with computer simulation accounting for the characteristics of a concentrated emulsion is proposed. Emulsion dynamics modeling combined with experiments using dilute emulsions at singlet doublet equilibrium may result in: (1) the quantification of emulsion film stability, namely the establishment of the coalescence time dependence on the physicochemical specificity of the adsorption layer of a surfactant (polymer), its structure and the droplet dimensions. This quantification can form a base for the optimization of emulsifier and demulsifier selection and their synthesis for emulsion technology applications, instead of the current empirical level applied in the area; (2) the elaboration of a commercial device for coalescence time measurement, which in combination with EDM will represent a useful approach to the optimization of emulsion technology with respect to stabilization and destabilization.

1.1 GENERAL

1.1.1 Stability Mechanisms

In emulsions, the water droplets in oil, or oil droplets in water, are large (of the order of microns). Because of the large surface area per drop, the excess Gibbs energy per drop is high and cannot be compensated for by entropy contributions as in microemulsions.

The stability of a disperse system is characterized by a constant behavior in time of its basis parameters, namely, the dispersity and the uniform distribution of the dispersed phase in the medium. The problem of stability is one that is most important and complicated in colloid chemistry.

Notwithstanding their thermodynamic instability, many emulsions are kinetically stable and do not change appreciably for a prolonged period (sometimes for decades). These systems exist in the metastable state, that is, the potential barrier preventing aggregation of the particles is sufficiently high.

To understand the reasons for the relative stability of such systems, it is necessary to first determine the stability [1] and mechanisms of destabilization. *Sedimentation stability* distinguishes the stability of the disperse phase with respect to the force of gravity. Phase separation due to sedimentation is a typical phenomenon for droplets in coarsely dispersed emulsions resulting in settling (or floating) of the drops. Highly dispersed emulsions are *kinetically* stable. They are characterized by diffusion–sedimentation subsequent equilibrium. Isothermal distillation of fine drops in coarser ones causes subsequent sedimentation. The loss of *aggregative* stability is due to a combination of drops.

Forces of molecular attraction may result in the formation of a continuously structured system with *phase* stability. *Coagulation* or *flocculation* is a process of particle cohesion, the formation of larger aggregates with a loss of sedimentation and phase stability and the subsequent phase separation, that is, a destruction of the emulsion. Hence, *aggregative stability* can be defined as the ability of emulsion to retain the dispersity and individuality of drops. In aggregates, notwithstanding the change of their mobility, the drops still remain as such for a certain time (the "lifetime") after which they can merge spontaneously with diminishing phase interface. The merging of droplets is called *coalescence.*

The aggregative *stability* or *instability* is most characteristic for colloid systems. This chapter is devoted to this type of instability.

Two types of the aggregative stability can be discriminated: a very low coagulation rate or an equilibrium in the aggregation and disaggregation processes (reversible coagulation).

The theoretical and experimental investigations of flocculation are focused on monodisperse systems because it simplifies to quantify the flocculation kinetics. Polydisperse emulsions are not usually chosen as a model system for the flocculation kinetics. Another reason for not choosing emulsion is the deformation of droplet surfaces upon interaction, which further complicates a quantitative treatment. In addition, the mentioned coupling of flocculation and coalescence creates difficulties in the modeling of the flocculation kinetics.

1.1.2 Hydrodynamics of Flocculation: Main Notions

When two particles approach each other, several types of interaction patterns may arise, affecting the flocculation process. There are two different but related ways in which colloid interactions influence flocculation. First, they have a direct effect on the *collision efficiency*, which is the probability that a pair of colliding particles will form a permanent aggregate. A strong long-range repulsion between the particles will reduce the chance of aggregate formation, and flocculation will occur very slowly, if at all. The other aspect of colloid interaction is their effect in the *strength* of aggregates, which is much less well understood but of great practical importance [2].

Flocculation occurs only if particles collide with each other and adhere when brought together by collision (i.e., the particles have low colloidal stability). To a large extent, these two processes, which could be termed *transport* and *attachment* steps, may be regarded as independent and can be treated separately.

Practically all colloidal interactions are of a short range, almost never extending over distances greater than the size of the particles. Hence, they have little influence over the transport of the particles, although they are crucial in determining the collision efficiency. To a large extent, this justifies the treatment of transport and attachment as separate steps. There is an important type of interaction to which this conclusion does not apply, the so-called *viscous* or *hydrodynamic* interaction, which arises during the approach of particles in a viscous fluid. The effect reduces the rate of approach and gives a lower collision efficiency.

The main qualitative difference in flocculation kinetics is caused by the sign of the surface forces. Naturally, the attractive force enhances the flocculation and repulsive forces retard it. This difference is reflected (expressed) in the notions of rapid and slow flocculation. If attractive forces predominate, the flocculation is called "rapid," whereas "slow" coagulation relates to the systems in which the repulsive interaction between particles predominates.

In addition, the surface force dependence on the interparticle distance is important. As is known, the van der Waals attractive forces can predominate at small and large distances, whereas the repulsive forces cause the electrostatic barrier at intermediate distances. The flocculation/coagulation in the primary minimum is slow because it is retarded by the electrostatic barrier. It is not valid for the flocculation/coagulation in secondary minimum, which is rapid.

The formation of the drop doublet is the basic process in the flocculation kinetics. In a polydisperse emulsion, a statistical ensemble of the pairs of the drops with the radius a_1 and a_2 is considered [3,4]. One of the drops is chosen as the reference sphere, and the flux of all drops with another radius to the surface of the reference drop is considered. Thus, the flux represents the result of the averaging over all drop pairs of the statistical ensemble. The flux determines the loss of the particle drop number during the flocculation and the rate of the doublet formation.

The drop loss rates can be computed from the distribution of drops around the reference sphere. In the case of coagulation, the steady-state capture rate is given by the net inward flux of drops through an arbitrary surface enclosing the test sphere. The net flux depends on the drop distribution as well as the relative sphere velocities, which are known functions of the relative drop positions. Once the drop loss rates are known, the overall stability of an emulsion can be determined by solving the governing population balance equations which incorporate these drop loss rates. The solution of this system of coupled nonlinear partial differential equations gives the drop size distribution as a function of time and position.

1.1.3 Importance of Flocculation Kinetics

Tadros and Vincent [5] classified emulsion-breakdown processes. The transport processes (Brownian diffusion and differential settling) can manifest themselves in any kind of breakdown processes. For instance, it is clear that the first step in Ostwald ripening [6] is due to the molecular diffusion. However, the mutual approach of the droplets caused by the differential settling or Brownian movement influences the molecular diffusion and kinetics of the Ostwald ripening. Thus, one concludes that the role of the transport processes can be understood more widely than in this chapter.

In the case of flocculation, two final states have to be distinguished: creaming and emulsion gelling [7,8] which occupy the emulsion volume as a whole. The larger the difference in liquid densities, the larger the droplet dimensions and the lower the volume fraction, which is favorable for creaming.

The quantification of the conditions of the two final states in flocculated emulsions is, in principle, possible on the basis of the flocculation theory.

Flocculation kinetics is important because it is accompanied by the change of emulsion properties and finally leads to the creaming and phase separation. The coalescence process usually follows the flocculation; that is, the coalescence kinetics are coupled with flocculation kinetics. Thus, the flocculated state and the flocculated kinetics have independent significance [9–11].

1.1.4 Scope of the Chapter

In monographs and reviews, much more attention is paid to surface forces [12,13] and stable colloids than to flocculation kinetics. Exceptions are Refs. 14 and 15 and literature on aerosols [16,17] and flotation [18–20].

In emulsion science, the main attention is focused on interparticle forces or colloid interactions and, correspondingly, to emulsion stability or instability, in terms of demulsification. In this chapter we turn to the question of transport mechanism, the coagulation kinetics, and to a lesser degree to the floc properties.

Now, due to the success of the investigations of flocculation in disperse model systems, the prerequisites are created for the investigation of flocculation in emulsions. In this review, we try to couple the studies of coagulation in disperse systems with flocculation in emulsions.

In a manner similar to Tadros and Vincent [5] and Melik and Fogler [15], we are concerned solely with "dilute" emulsions and suspensions in which the particle behavior is dominated by two-body interactions. Consequently, for emulsions, the present analysis is valid only for conditions predominated by singlets and doublets. However, the true application of this chapter is with emulsion systems which are characterized by coalescence times much faster than the corresponding flocculation times. The present analysis remains rigorous for predicting the flocculation and creaming behavior of a new larger particle, which is the result of a coalescence process, in relation to the remaining particles in the system. At faster coalescence times, an emulsion consists of the singlet and temporary doublet. At faster flocculation times, the emulsion consists of flocs. The structure of these flocs will not be considered in this chapter.

Dilute emulsions play an important role in, for instance, oil-dewatering processes [21]. At high initial volume fractions, the dewatering rate is high as well. The smaller the volume fraction, the lower the dewatering rate. Thus, the efficiency of oil-dewatering technology is connected to the flocculation kinetics. Obviously, the same is true with respect to produced water purification in environmental protection [22]. These two examples prove the importance of the flocculation process in dilute emulsions.

Tadros and Vincent [5] classified a wide range of emulsion systems. The classification subdivides the different types into (1) the nature of the "stabilizing moieties" and (2) the basic "structure" of the system. In both cases, the list represents a hierarchy of increasing complexity. In this hierarchy, the simplest systems will be considered here, which is justified by the limited knowledge of emulsion flocculation. Systems with steric stabilization [23], polymer bridging [24], and depletion stabilization [25] are beyond the scope of this chapter. This obviously does not mean that the Brownian and gravitational coagulations described earlier are not important for the more complicated emulsion systems.

A general mathematical formulation of the problem of evolution of liquid drops spectrum is given in Ref. 1. A deductive method is used for the presentation. Equations of doublet formation kinetics and population balance equations (PBEs) are written in the more general form. Then the stability problem is being considered as a purely mathematical one. In principle, such an approach is very efficient if reliable information is available about coagulation kinetics. Unfortunately, available information about surface forces in emulsions is incomplete and often inaccurate with a clear discrepancy between theory and experiment.

Along with a general approach characterized in Refs. 15, 26, and 27, it is very important at the present stage to discriminate between facts established reliably and the links in the complex process of coagulation in emulsions which need a systematic investigation. As applied to definite systems and conditions, the general model of coagulation kinetics in emulsions described in Ref. 15 can be used at present; however, without guarantee of the required accuracy. Generally, the use of potentially very wide possibilities of the general coagulation theory can lead to unreliable results because of incomplete information about surface forces and subprocesses. There is a possibility to provide efficiency of the general theory with regard to a wider range of problems and systems which can result in a qualitatively new level of the emulsions science. The objective of this review is to provide some movement in this direction. This has also determined the accepted inductive method of presentation from a simple to a complicated matter, from reliable established facts to those insufficiently studied.

During the last years, the influence of droplet surface mobility on collision efficiency and droplet deformation on colloidal interaction in emulsion have been quantified [28], a fact which stimulated the preparation of this paper.

1.2 SURFACE FORCES

Surface forces are described in all monographs and reviews concerning emulsion stability [5,15,29–31]. In the following section we deal with some general notions and some new results.

1.2.1 van der Waals Interaction

1.2.1.1 Macroscopic Approach

The universal attractive forces between atoms and molecules, known as van der Waals forces, also operate between macroscopic objects and play a very important role in the interaction of colloidal particles. Indeed, without these forces, aggregation of particles would usually be prevented by the hydrodynamic interaction.

The interaction between macroscopic bodies arises from spontaneous electric and magnetic polarizations, giving a fluctuating electromagnetic field within the media and in the gap between them. In order to calculate the force, the variation in electromagnetic wave energy with the separation distance has to be determined. Lifshitz derived an expression for the force between two semi-infinite media separated by a plane-parallel gap. His treatment was later extended by Dzyaloshinskii et al. to deal with the case of two bodies separated by a third medium. The expressions can be found in Refs. 12 and 29. In principle, these methods should enable the interaction between systems of interest to be calculated, and direct measurements of van der Waals forces between mica sheets [32] have confirmed the essential correctness of the Lifshitz or *macroscopic* approach. However, proper application of this approach requires detailed knowledge of the dielectric responses of the interacting media over a very wide frequency range.

1.2.1.2 Hamaker Expressions

Because of difficulties in applying the macroscopic theory, an older approach, due mainly to Hamaker [33], is still widely used. This is based on the assumption of pairwise additivity of intermolecular forces. The interaction between two particles is calculated simply by summing the interactions of all molecules in one particle with all of the molecules in the other particle. Hamaker replaced the summation by a double integration procedure, which leads to very simple expressions, especially when the separation distance is small. For two spheres, radii a_1 and a_2, separated by a distance h, the interaction energy at close approach ($h \ll a$) is given by

$$V_A = -\frac{A_{12}}{6h}\frac{a_1 a_2}{a_1 + a_2} \tag{1.1}$$

where V_A is the attraction energy between the two spheres and A_{12} is the *Hamaker constant* for media 1 and 2, of which the spheres are composed. The results given above apply to the interaction of media across a vacuum.

A useful approximation for Hamaker constants of different media is the geometric mean assumption

$$A_{12} \cong (A_{11}A_{22})^{1/2} \tag{1.2}$$

For similar materials, medium 1 interacting across medium 3,

$$A_{131} \cong \left(A_{11}^{1/2} - A_{33}^{1/2}\right)^2 \tag{1.3}$$

Equation 1.3 led Hamaker to the conclusion that the van der Waals interaction between similar materials in a liquid would always be attractive (positive Hamaker constant) whatever the values of A_{11} and A_{33}.

Oil phases are characterized by fairly low dielectric constants ranging between 2 and 5, which makes the pure entropic term fairly constant. The nonretarded dispersion contribution is the major cause of differences in Hamaker constants between different emulsions. The Hamaker constant can be expected to vary between 3×10^{-21} and 10×10^{-21} J in most food emulsions [31].

1.2.1.3 Retardation

Because dispersion forces are electromagnetic in character, they are subject to a *retardation* effect. The finite time of propagation causes a reduced correlation between oscillations in the interacting bodies and a smaller interaction. Pailtorpe and Russel [34] considered the effect of retardation and found that the total Hamaker constant decreases by approximately 70% for $a = 0.1$ μm and 85% for $a = 0.25$ μm over the range $2.001 < S - 2 < 2.5$. Melik and Fogler [15] conclude that the equation proposed in Ref. 35 cannot be considered a reasonable quantitative approximation to the retarded interparticle potential. However, in light of the tedious calculations required to account rigorously for any spatial variation of the Hamaker constant, they used this equation in their calculations of flocculation in the secondary minimum. Its distance to the surface can be equal to 5–7 Debye radius, that is, 10–100 nm (Section 1.5.1.1). According to Lyklema [36] the van der Waals energy dependence on distance can always be expressed through the Hamaker function $A(h)$. At a small separation, $A(h)$ becomes independent of h and identical to the *Hamaker constant*. At large h,

$$A(h) \sim h^{-1} \tag{1.4}$$

However, the systematic investigation of Rabinovich and Churaev [32,37,38] led them to the conclusion that for systems including water and a dielectric, the asymptotic equation [Equation 1.4] is not valid at any distance. In Ref. 32, they carried out a numerical calculation of $A(h)$ for 36 systems.

1.2.2 Electrical Interaction

1.2.2.1 Electrical Double Layer

Most particles in aqueous media are charged due to various reasons, such as the ionization of surface groups, specific adsorption of ions, and so forth. In an electrolyte solution, the distribution of ions around a charged particle is not uniform and gives rise to an *electric double layer* (DL). This topic has been the subject of several reviews [39,40] and details are not given here. The essential point is that the charge on a particle surface is balanced by an equivalent number of oppositely charged *counterions* in solution. These counterions are subject to two opposing effects: electrostatic attraction tending to localize the counterions close to the particles and the tendency of ions to diffuse randomly throughout the solution due to their thermal energy. The surface charge on a particle and the associated counterion charge together constitute the electrical double layer. A widely accepted model for the double layer is that of Stern, later modified by Graham (see Hunter [39]), in which part of the counterion charge is located close to the particle surface (the so-called *Stern layer*) and the remainder is distributed more broadly in the *diffuse layer*.

The interaction between charged particles is governed predominantly by the overlap of diffuse layers, so the potential most relevant to the interaction is that at the boundary between the Stern and diffuse layers (the Stern potential, Ψ_δ), rather than the potential at the particle surface. This boundary (the *Stern plane*) is generally considered to be at a distance of about 0.3–0.5 nm from the particle surface, corresponding to the diameter of a hydrated counterion.

There is no direct experimental method for determining the Stern potential. The two major influences on electrical interaction between particles are the magnitude of the effective "surface potential" (generally assumed to be Ψ_δ or ζ) and the extent of the diffuse layer because the latter governs the range of the interactions. Surface potentials can be modified in two distinct ways. If the ionic strength is raised, then a greater proportion of the potential drop occurs across the Stern layer, giving a smaller Stern potential. This effect can be produced by adding salt, and those which act in this way only are known as *indifferent electrolytes*. A more dramatic effect can be produced by the addition of salts with *specifically adsorbing counterions*. These adsorb on the particles because of some specific, nonelectrostatic affinity and can be regarded as located in the Stern layer. In many cases, such ions can adsorb to such an extent that they reverse the sign of the Stern potential.

The extent of the diffuse layer is also dependent on the ionic strength and is best seen through the variation of the potential as a function of distance from the Stern plane, for which the Poisson–Boltzmann approach is most commonly employed [39,40]. For fairly low potentials the linear form of the Poisson–Boltzmann expression is appropriate and a very simple result is obtained:

$$\Psi = \Psi_\delta \exp(-\kappa x) \tag{1.5}$$

where Ψ is the potential at a distance x from the Stern plane and κ is the *Debye–Hückel* reciprocal length [39,40], which is of great importance in colloid stability. For aqueous solution at 25 °C, κ is given by

$$\kappa = 2.3 \times 10^9 \left(\sum c_i z_i^2\right)^{1/2} \ (m^{-1}) \tag{1.6}$$

where c_i is the molar concentration and z_i is the valence of ion i. The sum is made over all ions in solution.

The Debye–Hückel parameter has the dimensions of reciprocal length and $1/\kappa$ is a characteristic length which determines the extent of the diffuse layer.

1.2.2.2 Double-Layer Interaction

When two charged particles approach each other in an electrolyte solution, their diffuse layers overlap and, in the case of identical particles, a repulsion is experienced between them. The precise way in which the double layers respond to each other depends on a number of factors which cannot be treated in detail here. One distinction is between interaction at constant surface potential or constant surface charge.

The repulsion energy of identical spherical particles at constant potential is [39], according to DLVO theory [1],

$$V_R = \frac{4\pi a_1 a_2}{a_1 + a_2} \varepsilon \Psi^2 \ln(1 + e^{-\kappa h}) \tag{1.7}$$

When deriving Equation 1.7, it was assumed that the Stern potential is low, that is,

$$\tilde{\Psi} = \frac{e\Psi}{kT} < 1 \tag{1.8}$$

where e is the electron charge, k is the Boltzmann constant, T is the absolute temperature, and the double layer is thin, that is,

$$\kappa a \gg 1 \tag{1.9}$$

For larger potentials, the analytic equations for large separation and thin layers [41,42],

$$\kappa a > 2 \tag{1.10}$$

is

$$V_R = 32\pi\varepsilon \left(\frac{kT}{e}\right)^2 \tanh^2\left(\frac{\tilde{\Psi}}{4}\right) a e^{-\kappa h} \tag{1.11}$$

1.2.2.3 The Problem of the Determination of the Stern Potential

The electrostatic interaction is very sensitive to the value of the surface potential. Its identification with the ζ potential is correct for surfaces which can be described by the so-called standard electrokinetic model [39,40]. In this model, the surfaces are described as molecularly smooth, impermeable for ions, and under no conditions porous and rough. The surface of an emulsion droplet is usually covered by the adsorption layer of an organic substance and does not satisfy this condition. In Ref. 31, values concerning the size and penetration of the different polar groups of the adsorbed organic molecular into water are given as 0.3–1.6 nm. The ions of the diffuse layer are distributed partially outside the adsorption layer and partially inside it. The latter part is not small if the electrolyte concentration is not low. The counterions inside the adsorption layer participate in the electrostatic interaction of particles also. The ζ potential yields information about the counterions outside the adsorption layer. Thus, using the ζ potential instead of the Stern potential leads to an underestimation of the electrostatic interaction.

The ζ potential manifests itself in Bickerman surface conductivity. The counterions distributed between the slipping plane and the interface are mobile and produce the additional surface conductivity. Measurements of the surface conductivity enable the evaluation of the total charge of the diffuse layer and the efficient Stern potential [43]. Low-frequency dielectric dispersion is the most exact method for those measurements [44]. High-frequency conductometry is also useful for the Stern potential evaluation [45].

1.2.3 Hydrophobic Interaction

In some cases, surfaces may have significant areas with a hydrophobic (nonwettable) character, such as polymer latex particles with a low density of surface ionic groups, or negative particles with adsorbed cationic surfactant. The possibility then arises of another type of interaction which can give appreciable attraction – the so-called *hydrophobic interaction*. The extensive hydrogen bonding present in ordinary water is responsible for a considerable degree of association between water molecules and a significant "structuring" effect. A hydrophobic surface offers no possibility of hydrogen bonding or ion hydration, so there is no inherent affinity for water. However, the presence of such a surface tends to limit the contact with such surfaces as far as possible. In aqueous solutions, molecules with hydrophobic segments can associate with each other in such a way that contact between water and the hydrophobic regions is minimized. This is known as *hydrophobic bonding*. The best known examples of this are surfactant and protein solutions.

It has recently been found [46] that the same type of interaction occurs between macroscopic hydrophobic surfaces. The resulting attractive force can be surprisingly large and of quite long range [47].

With increasing hydrophilicity with the introduction of charged headgroups onto the surface, the measurable range of the hydrophobic attraction decreases rapidly [48]. No long-range hydrophobic interaction is observed for surfaces having contact angles below about 40–70° (water/air). It, thus, appears that a low density of hydrophilic groups on a hydrophobic surface is sufficient to considerably reduce the magnitude and range of the hydrophobic force. The hydrophobic interactions between macroscopic surfaces seem to be important only in rather pure systems [31] and not in technical emulsions.

1.2.4 Hydration Effects

The nature of water close to a particle surface is usually very different from bulk water. The major consequence of hydration at a particle surface is an increased repulsion between approaching particles because of the need for ions to lose their water of hydration upon contact between particles. This involves work and, hence, an increase in free energy of the system.

The most direct evidence of hydration effects comes from measurements of the force between mica sheets in various electrolyte solutions [49]. At low ionic strengths, the repulsion follows the expected exponential form for the double-layer interaction (see Section 1.2.2.2). At salt concentrations above 1 mM, a monotonic short-range force is apparent, in addition to the double-layer repulsion, which is due to adsorbed hydrated cations. This extra force increases with the degree of hydration ($Li^+ \cong Na^+ > K^+ > Cs^+$) and is roughly exponential over the range 1.5–4 nm, with a decay length of the order of 1 nm. In their earlier studies, Derjaguin and Zorin [50] introduced the concept of a "structural component of disjoining pressure" to account for the anomalous behavior of thin films.

The range of these hydration forces is quite appreciable in relation to the range of double-layer repulsion, and they may be expected to have an effect on colloid stability, especially at high ionic strengths. Similarly, the way some flocculated colloids can be redispersed ("repetized") simply by washing away the electrolyte [51,52] is a clear indication that aggregation is not occurring in a true primary minimum, but in a "hydration minimum," where the particles are prevented from coming into true contact by the presence of hydrated ions. In these cases, the van der Waals attraction is not sufficiently strong to prevent separation of particles when the salt is diluted and double-layer repulsion is re-established.

Another type of hydration repulsion arises when adsorbed layers of hydrophilic material are present, but this is usually considered as a "steric" interaction. In Ref. 31, it is shown that the hydration forces acting between bilayers of different phosphatidylcholines have a comparable magnitude out to a bilayer of about 2–3 nm. The authors suggest the important role of the hydration forces in emulsion stabilization.

1.3 RAPID BROWNIAN COAGULATION

1.3.1 von Smoluchowski Theory

The classical understanding of coagulation kinetics is given by von Smoluchowski theory [3,4], which follows from the assumption that the collisions are binary and that fluctuations in density are sufficiently small. Computer simulations [53–56] serve as a means to test the validity of the mean field approach.

An aggregate formed from i identical particles is called an i-mer. The average number of i-mers per unit volume is the particle concentration z_i.

The coagulation of two clusters of the kind i and j is given by the following relation:

$$i\text{-}mer + j\text{-}mer \xrightarrow{k_{ij}} (i+j)\text{-}mer = k\text{-}mer$$

where k_{ij} is the concentration-independent coagulation constant or kernel. Physically, it means that the coagulation rate between all kinds of i-mers and j-mers is the same.

For dilute dispersions with volume fractions less than 1%, only two-particle collisions need to be considered because the probability of three-particle collisions is small.

The equation describing the temporal evolution of the cluster of kind k is as follows:

$$\frac{dz_k}{dt} = \frac{1}{2} \sum_{i=1, j=k-1} k_{ij} z_i z_j - z_k \sum_{i=1}^{\infty} k_{ik} z_i \tag{1.12}$$

k_{ij} depends on details of the collision process between *i*-mers and *j*-mers. This kernel embodies the dependence on i and j of the meeting of an *i*-mer and a *j*-mer, including effects such as the volume dependence of the collision cross-section and the diffusion constant.

The first term in Equation 1.12 describes the increase in z_k owing to coagulation of an *i*-mer and a *j*-mer and the second term describes the decrease of z_k owing to the coagulation of a *k*-mer with other aggregates.

It is important to note that Equation 1.12 is for *irreversible* aggregation – no account is taken of the breakup of aggregates, which would require a third term on the right-hand side. Also, it is assumed, for the present, that each collision results in the formation of an aggregate (i.e., the collision efficiency and stability ratio are both unity).

In principle, it would be possible to use Equation 1.12 to derive the concentrations of all aggregate types at any time, but there are great difficulties, notably in assigning values to the rate coefficients, which depend not only on aggregate size and shape but also on the particle transport mechanism. By converting Equation 1.12 into an integral expression and considering continuous aggregate size distributions rather than discrete numbers, it is possible to derive solutions in certain cases but only for specific forms of the i-ate coefficients (or kernels), which may not be physically realistic. Nevertheless, such approaches can give some insight into the way aggregate size distributions may evolve during flocculation processes.

Particles in suspensions are subject to Brownian motion [57,58]. As a result, particles will collide. The Smoluchowski approach is to imagine a stationary central particle and to calculate the number of particles colliding with it in unit time. Allowance can then be made for the fact that the central particle itself is one of many similar particles undergoing Brownian motion and so the appropriate collision frequency can be derived:

$$J_{ij} = 4\pi R_{ij}(D_i + D_j)n_i n_j \tag{1.13}$$

J_{ij} is the total number of collisions occurring between particles of types i and j in unit volume per unit time. D_i and D_j, and n_i and n_j, are the diffusion coefficients and number concentrations, respectively. The term R_{ij} is the collision radius for the pair of particles and represents the center-to-center distance at which the particles may be assumed to be in contact. In many cases, the collision radius can be taken simply as the sum of the particle radii, but if there is long-range attraction between the particles, the effective collision radius will be somewhat larger. In what follows, we will assume spherical particles, radii a_i and a_j, and that $R_{ij} = a_i + a_j$. Also, for the diffusion coefficients, the Stokes–Einstein expression is used:

$$D_i = \frac{kT}{6\pi a_i \mu} \tag{1.14}$$

where μ is the viscosity of the fluid, k is the Boltzmann constant, and T is the temperature.

Equation 1.13 then becomes

$$J_{ij} = \frac{(2kT/3\mu)\, n_i n_j \left(a_i + a_j\right)^2}{a_i a_j} \tag{1.15}$$

and so, by comparison with Equation 1.12, the rate constant can be written as

$$k_{ij} = \frac{(2kT/3\mu)\left(a_i + a_j\right)^2}{a_i a_j} \tag{1.16}$$

For an initially monodisperse suspension of particles, radius a_1, the initial collision rate can be calculated easily from Equation 1.12 because only one type of collision (1-1) is involved. Also, the initial rate of decrease of the *total* particle concentration, n_T, follows directly from the collision rate because each collision reduces the number of particles by one (two primary particles lost, one aggregate gained). The result is

$$-\frac{dn_T}{dt} = \left(\frac{4kT}{3\mu}\right) n_T^2 = k_F n_T^2 \tag{1.17}$$

where k_F is known as the flocculation rate constant and has a value of 6.13×10^{-18} m^3 s^{-1} for aqueous dispersions at 25 °C.

The most noteworthy feature of Equation 1.17 is that it does not include the particle size because the size terms cancels from Equation 1.16 when $a_i = a_j$. As the particle size increases, the diffusion coefficients decreases, but the collision radius increases. These have opposing effects on the collision rate and, for equal particles, they balance exactly. Even for spheres of different size, the term $(a_i + a_j)^2 / a_i a_j \cong 4$, provided that the sizes do not differ too greatly (say, by no more than a factor of 3, in which case the size term has a value of about 5).

The simplest second-order rate expression of Equation 1.17 can be integrated to give the total number of particles as a function of time:

$$n_T = \frac{n_0}{1 + k_F n_0 t} \tag{1.18}$$

where n_0 is the initial concentration of primary particles.

There is a characteristic flocculation time t_F, in which the number of particles is reduced to half of the initial value ($n_T = n_0/2$) and this follows immediately from Equation 1.18:

$$t_F = \frac{1}{k_F n_0} \tag{1.19}$$

The flocculation time can also be thought of as the average time in which a particle experiences one collision. From the value of k_F quoted above, t_F turns out to be about $1.6 \times 10^{17}/n_0$ s and so for an initial particle concentration of 10^{16} m^{-3}, the flocculation time would be about 16 s. Particle concentrations are often much lower and the corresponding flocculation time would be greater.

As well as the total particle concentration, it is also possible to calculate the concentration of single particles and aggregates, assuming the rate constant is the same for all collisions. For single particles,

$$n_1 = \frac{n_0}{(1 + t/t_F)^2} \tag{1.20a}$$

and for doublets,

$$n_2 = \frac{n_0 t/t_F}{(1 + t/t_F)^3} \tag{1.20b}$$

The general Smoluchowski expression for k-fold aggregates is

$$n_k = \frac{n_0 (t/t_F)^{k-1}}{(1 + t/t_F)^{k+1}} \tag{1.21}$$

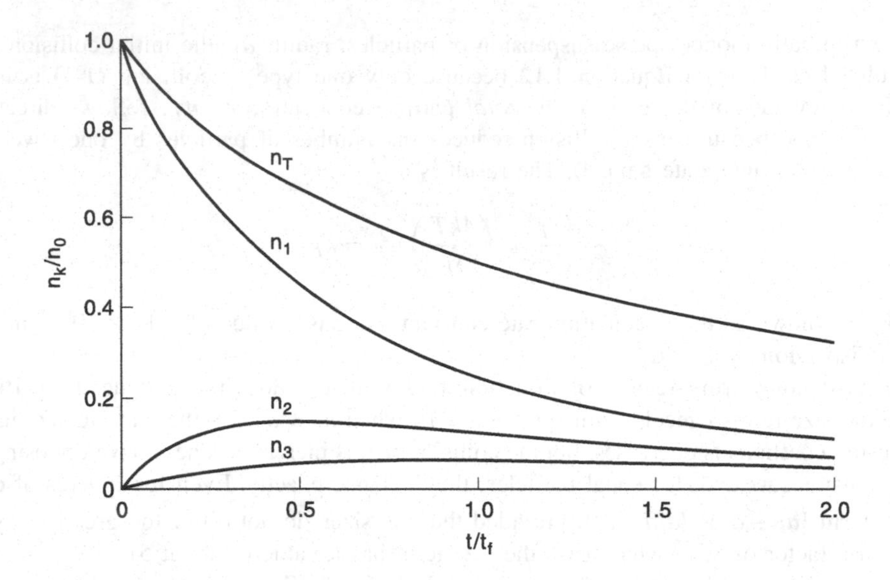

FIGURE 1.1 Reduced concentration of *k*-fold aggregates plotted against reduced time for $k = 1, 2, 3$ and for the total concentration n_T.

Results from these expressions are shown in the dimensionless form (n_k/n_0 versus t/t_F) in Figure 1.1, for aggregates up to three-fold. For all aggregates, the concentration rises to a maximum at a characteristic time and then declines slowly. Note that at all times the concentration of singlets exceeds that of any other aggregate.

Experimental data for rapid flocculation show reasonable agreement with Smoluchowski predictions, at least in the early stages of the process (see Section 1.3.4).

Rate constants determined for the rapid flocculation of latex particles are about half of the value given by Equation 1.17 and this is now known to be a result of hydrodynamic interaction between approaching particles. This effect will be considered briefly in Sections 1.3.2 and 1.3.3.

1.3.2 Incorporation of Hydrodynamic Interaction and Attractive Surface Forces in the Theory of Rapid Coagulation

Theory of perikinetic coagulation continued to develop along the way as proposed by Smoluchowski. Oversimplifications were not allowed, which led step by step to an improvement in theory. In the Smoluchowski model, every collision is accompanied by coagulation. The possibility of disaggregation is neglected. In other words, Smoluchowski proposed the theory of the irreversible coagulation, which corresponds to the case of strong attractive forces. However, the introduction of the notion of long-range surface forces necessitates the generalization of the model of the mutual free diffusion of the particles. At small interparticle distances, the additional flux caused by the particle attraction to each other has to be incorporated in the expression for the particles flux to the reference particle. This important generalization was made by Fuchs [59] with respect to repulsive forces. However, the attractive surface forces can be used in Fuchs expression for particle flux too. Fuchs has not paid attention to another shortcoming in the Smoluchowski model which was noticed by Derjaguin [60].

The particle approach leads to the necessity of a liquid flow out from the gap between them. The thinner the liquid interlayer dividing the particles, the higher the hydrodynamic resistance to its thinning. It leads to the decrease of the coefficient of the particle mutual diffusion. Instead of the constant particle diffusivity in Smoluchowski–Fuchs theory, Derjaguin and Krotova [61] introduced in the Fuchs equation the particle diffusivity which is proportional to h at a small interparticle distance and consequently equals zero at particle contact.

The Einstein equation for interconnection of diffusivity and friction coefficient f,

$$fD = kT \tag{1.22}$$

is valid for any interparticle distance. It allows one to obtain the dependence $D(h)$ using the dependence $f(h)$. The asymptotic expression for $f(h)$ at small h is given by Taylor [62] without derivation. The derivation of this linear dependence is given by Derjaguin [60]. The diffusivity decreases to zero at $h \rightarrow 0$ and does not lead to the disappearance of particle flux because the attractive surface force causes the opposite influence. However, the neglect of the long-range surface forces leads to an erroneous conclusion about when the coagulation takes place. The elimination of this paradoxical result is the main consequence of the incorporation [61] of the attractive surface forces in the theory of perikinetic coagulation. This phenomenon has been considered in some articles [63,64]. To distinguish this from the analytical result of Derjaguin and Krotova [61], the numerical calculation was used [63,64]. The results obtained in Refs. 63 and 64 are less convenient for application and confirm the original investigation [61].

The collision of droplets of intermediate dimensions are controlled by the joint action of the Brownian movement and sedimentation. In this respect, the Fuchs equation was generalized by Batchelor [65]. The Batchelor equation was used by many investigators and led to noticeable progress in the kinetic theory. Taking this into account, we shall represent the Fuchs equation in the next section in the notation used by Batchelor and his successors.

1.3.3 Brownian Collisions in Emulsions

There are few studies available of the coalescence of fluid drops, presumably because of the more complex interactions which involve fluid flow both inside and outside of the drops. Further, there is a possibility that the drops will deform as they collide. A notable exception is that by Zinchenko [66–68] who has calculated the rate of gravity-induced coalescence of spherical drops of different sizes numerically, using a trajectory analysis for pairs of drops, without considering the effects of interparticle forces. His results confirm that, in contrast to rigid spheres, drop collision is possible at finite rates under the action of a finite external force only. He also showed how the rate of drop collisions decreases with an increasing ratio of the drop fluid viscosity and the surrounding fluid viscosity.

Under conditions of low Reynolds numbers, the relative motion of two spherical drops may be decomposed into motions along and normal to their line of centers. Hetsroni and Haber [69] used the method of reflections to describe the hydrodynamic interaction for widely separated drops in both configurations. More accurate image techniques, which are valid unless the separation distance is small relative to the radius of the smaller drop, have been developed by Fuentes et al. [70,71]. Exact solutions based on bispherical coordinate methods have been developed by Haber et al. [72] and Rushton and Davies [73] for axisymmetric motion along the line of centers, and by Zinchenko [67] for asymmetric motion normal to the line of centers. Each of these solution methods yields an infinite series for the hydrodynamic force between the drops, which diverges when the distance between the drops tends to zero. Because coalescence phenomena depend critically on the near-contact interaction, these earlier solutions may be matched with the recent lubrication theories of Davis et al. [74]. In these, the nature of the hydrodynamic force resisting the near-contact relative motion of two spherical drops in the direction along their line of centers has been analyzed in detail.

1.3.3.1 Specification of the Fuchs Equation for Emulsions

Zhang and Davis [75] employ the above solutions for the hydrodynamic interactions of two spherical drops and calculate collision rates by extending the previous work by several authors for rigid particles and by Zinchenko [68] for spherical drops. According to Zhang and Davis [75], for creeping flow, the external driving forces on each drop balance the hydrodynamic forces, and the velocity $\mathbf{V}_{12}$ of drop 1 to drop 2 is linearly related to the sum of the external forces and depends only on the relative position of the two drops. An expression for this relative force has been presented by Batchelor [65] for rigid spheres.

Zhang and Davis [75] use Batchelor's expression for $\mathbf{V}_{12}$ and modify it for spherical drops:

$$\mathbf{V}_{12} = -\frac{D_{12}^{(0)}}{kT} G\left(\frac{dU_{12}}{dr} + kT\frac{d\ln p_{12}}{dr}\right) \tag{1.23}$$

where $\mathbf{r}$ is the vector from the center of drop 2 to the center of drop 1. Similarly, the relative diffusivity due to Brownian motion for two widely separated drops is

$$D_{12}^{(0)} = \frac{kT(\hat{\mu}+1)(1+\lambda^{-1})}{2\pi\mu(3\hat{\mu}+2)a_1} \tag{1.24}$$

The pair-distribution function, p_{12}, represents the probability that drop 1 is at position $\mathbf{r}$ relative to drop 2, normalized such that $p_{12} \to 1$ as $r \to \infty$. The interparticle force is described by the potential function $U_{12}(\mathbf{r})$.

The relative mobility function for motion along the line of centers G describes the effects of hydrodynamic interactions between the two drops. The function depends on the size ratio of the two drops

$$\lambda = \frac{a_1}{a_2} \tag{1.25}$$

the viscosity ratio of the drop fluid and the surrounding fluid

$$\hat{\mu} = \frac{\mu_1}{\mu_2} \tag{1.26}$$

and the dimensionless distance between the drops, $s = 2r/(a_1 + a_2)$. It is unchanged when λ is replaced with λ^{-1}.

In the development by Davis et al. [74], the dimensionless lubrication force between two spherical drops in near contact is shown to depend on a single dimensionless parameter,

$$m = \hat{\mu}\left(\frac{a}{h_0}\right)^{1/2} \equiv \frac{a_1 a_2}{a_1 + a_2} \tag{1.27}$$

where $a_1 a_2/(a_1 + a_2)$ is the reduced radius of the two drops and $h_0 = r - (a_1 + a_2)$ is the closest separation between two drop surfaces. This parameter describes the mobility of the interfaces: When $m \ll 1$, the drops behave as rigid spheres, whereas when $m \gg 1$, the drops have fully mobile interfaces and offer relatively little resistance to the drop relative motion. Note that the interface mobility, m, is not a property of the interfaces themselves but instead represents the viscous resistance of the fluid inside the drops to the flow exerted on their interfaces by the external fluid as it is squeezed out of the gap between the drops. Using this interface mobility, the lubrication forces acting on the drops in the direction along their line of centers can be simply expressed as

$$-F_{1,1} = F_{1,2} = 6\pi\mu a^2 \frac{V_{12}}{h_0} f(m) \tag{1.28}$$

where V_{12} is the component of $\mathbf{V}_{12}$ in the direction along the line of centers and $f(m)$ is a dimensionless function which is approximated by Davis et al. [74] using the following Pade-type expression:

$$f(m) = \frac{1 + 0.402m}{1 + 1.711m + 0.461m^2} \tag{1.29}$$

Note that the lubrication force for drops with mobile interfaces ($m \gg 1$) is inversely proportional to $(h_0/a)^{1/2}$, indicating that spherical drops can come into contact in a finite time under the action of a finite force, in contrast to that for immobile interfaces ($m \ll 1$) for which the lubrication force is inversely proportional to h_0/a. This lubrication force dominates the hydrodynamic resistance unless the droop viscosity is very small [$\hat{\mu} < O(h_0/a)^{1/2}$], in which case the fluid slips out of the gap with little resistance.

When the drops are close to one another ($\zeta \to 0$), the lubrication force dominates the hydrodynamic force and directly balances the external force on each of the drops, that is,

$$G(\xi) = \frac{2 + 3\hat{\mu}}{3 + 3\hat{\mu}} \frac{(1+\lambda)^2}{2\lambda} \frac{\xi}{f(m)} \tag{1.30}$$

1.3.3.2 Expression for the Drop Collision Rate

The rate at which drops of radius a_1 collide with drops of radius a_2 per unit volume is equal to the flux of pairs into the contact surface $r = a_1 + a_2$ and is expressed in terms of the pair-distribution function $p_{12}(\mathbf{r})$ and the drop relative velocity V_{12} by

$$J_{12} = -n_1 n_2 \int_{r=a_1+a_2} p_{12}\mathbf{V}_{12}\mathbf{n}\, dA \tag{1.31}$$

where $\mathbf{n} = \mathbf{r}/r$ is the outward unit normal to the spherical surface represented by $r = a_1 + a_2$, and n_1 and n_2 are the number of drops at the given time in the size categories characterized by radius a_1 and radius a_2, respectively, per unit volume of the dispersion.

For the dilute dispersion, the pair-distribution function is governed by a quasi-steady mass conservation equation for regions of space outside the contact surface:

$$\nabla \bullet (p_{12}\mathbf{V}_{12}) = 0 \tag{1.32}$$

As the colliding drops come into contact, they are assumed to coalesce, and so $p_{12} = 0$ for $r = a_1 + a_2$:

$$p_{12}\big|_{r=a_1+a_2} = 0 \tag{1.33}$$

Provided that all of the drop–drop encounters originate at wide separations in a homogeneous dispersion, the other boundary condition is

$$p_{12} \to 1 \text{ as } r \to \infty \tag{1.34}$$

Due to the spherical symmetry and quasi-steady mass conservation equation, the expression for rate can be simplified:

$$J_{12} = 4\pi r^2 n_1 n_2 p_{12}(\mathbf{r}) V_{12}(\mathbf{r}) \tag{1.35}$$

It does not depend on r. The substitution in it $V_{12}(\mathbf{r})$ according Equation 1.23 yields the linear differential equation of first order where J_{12} is an unknown constant. The integration of this equation and using the boundary conditions for p_{12}, Equations 1.33 and 1.34 yield the equation for the collision rate:

$$J_{12} = 4\pi n_1 n_2 D_{12}^{(0)} \left(\int_{a_1+a_2}^{\infty} \frac{\exp(U_{12}/kT)}{r^2 G} dr \right)^{-1} \tag{1.36}$$

If the drops are assumed to move independently, that is, without any hydrodynamic interactions ($G = 1$) or interparticle forces ($U_{12} = 0$), other than a sticking force on contact, the collision rate is that obtained by Smoluchowski (Equation 1.13). From Equation 1.13, Equation 1.12 for the total number of collisions occurring between particles of types i and j follows. We define the collision efficiency,

$$E_{12} \equiv \frac{J_{12}}{J_{12}^{(0)}} \tag{1.37}$$

as the ratio of the predicted collision rate with hydrodynamic and interparticle interactions to that obtained in their absence. Using the dimensionless center-to-center distance, s, this is then

$$E_{12} = \left(2 \int_{2}^{\infty} \frac{\exp\left[U_{12}(s)/kT\right]}{s^2 G(s)} ds \right)^{-1} \tag{1.38}$$

Note that the inverse of the collision efficiency is often called the "stability ratio." Of considerable interest is the influence of the viscosity ratio on the collision efficiency. For viscous drops, the relative mobility function G for the near-contact relative motion is inversely proportional to the square root of the distance between the drops when the interface mobility is large. It leads to the integration in Equation 1.38 being finite instead of being infinite as for rigid spheres. Furthermore, because G decreases with increasing viscosity ratio, Equation 1.38 indicates that the collision efficiency will decrease monotonically as the ratio of the drop phase viscosity to the suspending phase viscosity is increased. On the other hand, comparing the effects of the hydrodynamic interactions and interparticle force which are represented by G and U_{12}, respectively, on E_{12} through Equation 1.38, it is seen that the hydrodynamic interactions appear in a pre-exponential factor and, therefore, are subordinate, for moderate values of A/kT, to the interparticle forces, which appear in the argument of the exponential.

1.3.3.3 Brownian Collisions without Interparticle Forces

Figure 1.2 shows the results for E_{12} as a function of λ for $\hat{\mu} = 0, 0.1, 1.0, 10, 100$, and 1000. As expected, E_{12} decreases as $\hat{\mu}$ increases because this corresponds to decreasing the interface mobility and internal drop flow, which leads to a higher hydrodynamic resistance to the close approach. In the limit as $\hat{\mu} \to \infty$, corresponding to that of rigid spheres, $E_{12} \to 0$, although this limit is approached only slowly.

As λ decreases from unity, E_{12} increases and it tends to unity when λ tends to zero. One reason for this is that when λ decreases, the influence of the smaller drop on the Brownian diffusion of the larger one is decreased. More important is that the contribution of the smaller drop to the relative Brownian diffusivity increases as λ decreases. As $\lambda \to 0$, the hydrodynamic interactions become important only within an increasingly small boundary layer around the larger drop, and so $E_{12} \to 0$ as $\lambda \to 0$.

1.3.3.4 Brownian Collisions with van der Waals Forces

In addition to λ and $\hat{\mu}$, the collision efficiency depends on A/kT. This dimensionless parameter is called the Hamaker group, and it provides a measure of the strength of the van der Waals forces relative to the Brownian motion. Typically, it attains a value of unity or less.

The effects of van der Waals attraction on the collision efficiency of Brownian drops are shown as a function of $A/6kT$ in Figure 1.3 for different viscosity ratios with $\lambda = 1$. As expected, the attractive force increases the collision rate. In fact, the collision efficiency becomes higher than unity for $A \gg 6kT$, but attractive forces of this magnitude are not usually encountered in practice.

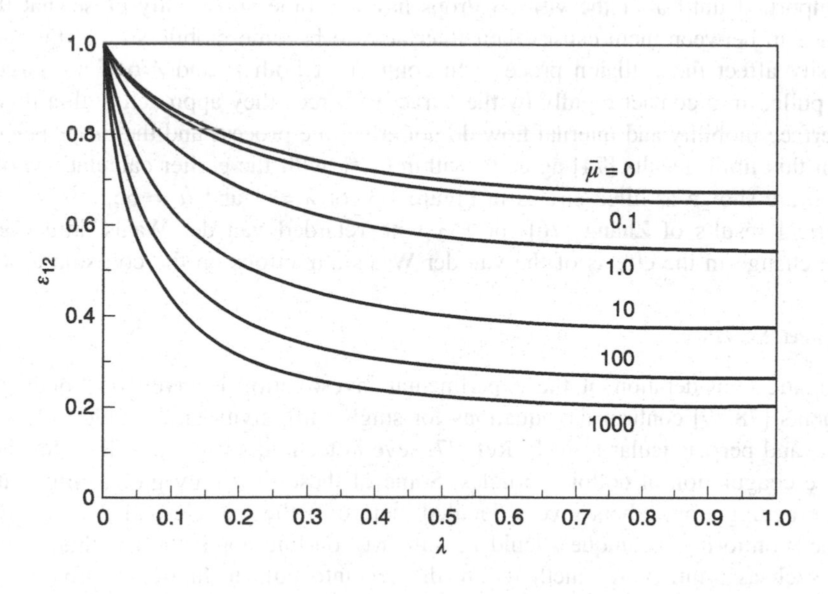

FIGURE 1.2 The collision efficiency for Brownian drops as a function of the size ratio for various viscosity ratios without interparticle forces. (Redrawn from X. Zhang and R.H. Davis, *J. Fluid Mech.* 230: 479 (1991).)

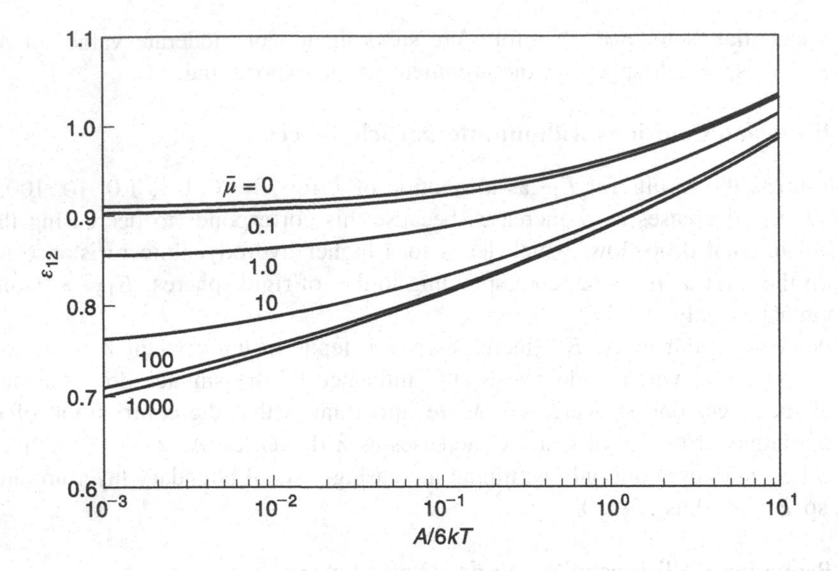

FIGURE 1.3 The collision efficiency for Brownian drops as a function of the Hamaker group for $\lambda = 1.0$ and various ratios with unretarded van der Waals attraction. (Redrawn from X. Zhang and R.H. Davis, *J. Fluid Mech.* 230: 479 (1991).)

Moreover, the van der Waals attraction plays an increasingly important role as μ increases. In particular, the collision efficiency for $\hat{\mu} \gg 1$ is independent of $\hat{\mu}$ at large values of $A/6kT$ but not for small values of $A/6kT$. This is because van der Waals forces are too weak when $A \ll 6kT$ to become important until after the viscous drops have become sufficiently close that the squeeze flow in the gap between them causes their interfaces to become mobile so that the internal flow and viscosity affect the collision process. In contrast, if both $\hat{\mu}$ and $A/6kT$ are large, then the drops are pulled into contact rapidly by the attractive forces they approach within this range and so the interface mobility and internal flow do not affect the process and the drops behave as rigid spheres. In this limit, results [74] agree to within 0.2% with the earlier calculations of Spielman [63], which are shown as filled circles in Figure 1.3 for $\lambda = 1$ and $\hat{\mu} = \infty$.

Numerical results of Zhang [76] for E_{12} with retarded van der Waals attraction show no qualitative change in the effects of the van der Waals attractions on the collision efficiency.

1.3.4 Experiments

The systematic consideration of the experimental investigation is given by Sonntag [77]. The measurements [78,79] confirm the equations for singlet diffusivity and doublet diffusivity in the line of axis and perpendicular to it. In Ref. 77, several techniques are described for the measurement of the coagulation of colloid particles. Some of these give only global information on the state of the aggregation; others give a detailed picture of the particle and floc size distribution. Ideally, the monitoring technique should be suited to on-line application without a sample pretreatment such as dilution. All methods are divided into bulk techniques (turbidity, static light scattering) and single-particle techniques (Coulter counter, flow ultramicroscopy). First, quantitative experiments on rapid coagulation were performed with gold particles by slit ultramicroscopy and individual counting [80–82]. The average value of the coagulation constant of gold particles was found to be $k_{ij} = 12 \times 10^{-18} \pm 1\ \mathrm{m^3\,s^{-1}}$ at 298 K. These are the only values described in the literature that confirm the theoretical value.

The coagulation of gold particles in different electrolytes was also investigated later with streaming ultramicroscopy with visual counting by Derjaguin and Kudravtseva [83]. For sodium chloride, $k_{ij} = 8.2 \times 10^{-18} \pm 1\ \mathrm{m^3\,s^{-1}}$; for magnesium sulfate, $k_{ij} = 8.9 \times 10^{-18} \pm 1\ \mathrm{m^3\,s^{-1}}$; and for lanthanum nitrate $k_{ij} = 6.5 \times 10^{-18} \pm 1\ \mathrm{m^3\,s^{-1}}$ was obtained. Selenium sols were investigated spectrophotometrically by Watillon et al. [84]. The rapid rate constant was $4 \times 10^{-18} \pm 1\ \mathrm{m^3\,s^{-1}}$.

Silver iodide sols were coagulated with barium nitrate and lanthanum nitrate by Ottewill and Rastogi [85]. The coagulation constants were determined to $8.78 \times 10^{-18} \pm 1$ and $10.2 \times 10^{-18} \pm 1\ \mathrm{m^3\,s^{-1}}$, respectively.

The influence of the particle size on rapid coagulation was investigated with hematite particles by Penners and Koopal [86].

In recent years, many experiments have been carried out using polystyrene lattices because of their monodispersity and their ideal spherical shape. The coagulation rate was lower than the theoretical value, even when the hydrodynamic interaction was taken into consideration.

Sonntag [77] considers two ways to explain this behavior. The first one is the introduction of reversibility into the coagulation process. This was suggested by Frens and Overbeek [51]. Under this assumption the coagulation kernel can remain constant or depend on the aggregate size.

The Smoluchowski treatment is based on the collision of spheres and this assumption is questionable in the case of aggregates. Except when coalescence occurs, aggregates cannot be truly spherical. Two colliding solid spheres must form an aggregate in the form of a dumbbell, and with higher aggregates, many different shapes become possible, as illustrated in Figure 1.4. The collision rates of such aggregates are likely to differ from those for spheres. In particular, the effects of size on the collision radius and the diffusion coefficient cannot be expected to balance each other as for spheres, which led to the very simple, size-independent form of the collision rate constant in Equation 1.16. As the size of an irregular aggregate increases, it is likely that the increase in collision radius more than compensates for the decrease in diffusion coefficient and the results of Cahill et al. [87] tend to support this conclusion. Gedan et al. [88], from their measurements of aggregate size distribution, assigned different rate constants to the various types of collision. For the collision of primary particles, k_{11} was found to be $3 \times 10^{-18} \pm 1\ \mathrm{m^3\,s^{-1}}$. For other collisions, they found $k_{12} \cong 2k_{11}$ and $k_{13}, k_{22}, k_{12}, k_{33} \cong 4k_{11}$. These values show a more pronounced effect of size on collision rate than expected. However, the value found for k_{11} is only about 25% of the Smoluchowski value and there may be some doubt about the accuracy of the value.

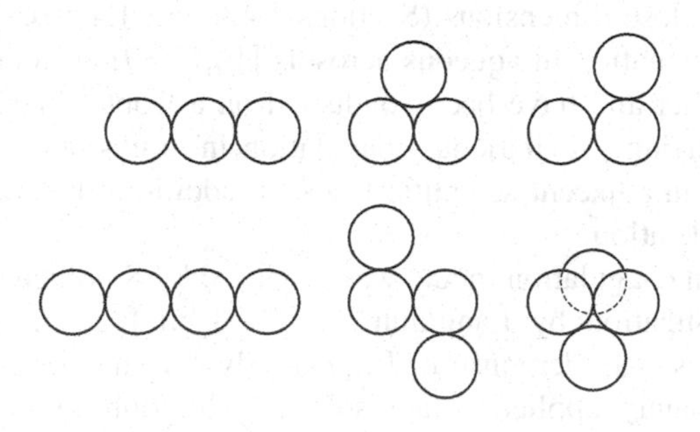

FIGURE 1.4 Possible shapes of aggregates up to four-fold.

1.4 KINETICS OF RAPID GRAVITATIONAL COAGULATION IN THE ABSENCE OF REPULSIVE SURFACE FORCES

1.4.1 ORTHOKINETIC COAGULATION AND PARTICLE CAPTURE FROM FLOW

The role of Brownian diffusion in the coagulation process decreases with increasing particle size, and other factors begin to dominate. The difference in particle dimension and, therefore, in their sedimentation rate play the determining role in the particle approaching a quiescent media. The "orthokinetic coagulation" term used in the case when a directed motion of particles is favorable for coagulation was initially connected just with this mechanism.

Particles of similar size also approach in a inhomogeneous hydrodynamic field. The term "orthokinetic coagulation" is now more frequently used as applied to coagulation in an inhomogeneous hydrodynamic field. Coagulation in stationary medium is often called the "gravitational coagulation," "gravity-induced coagulation," or "differential settling."

The theory is mainly developed to describe an approach of particles with largely differing dimensions. The coordinate system is related to the larger particle, and the trajectory of the smaller particle is considered in the hydrodynamic field of the larger one. The hydrodynamic field arises at the cost of the motion of the larger particle with respect to the medium, which in the laboratory coordinate system is considered to be stationary. But in the coordinate system related to the particle a liquid flow runs on the latter, which is homogeneous at infinity and carries particles of small size. In this problem, the large particle is called the collector because small particles settling from the flow accumulate in it. Only the relative velocity of the medium and collector is substantial for the settling mechanism.

The situation is fully identical to the following: A liquid comprised of small particles is at rest as a whole (in the laboratory coordinate system) and the collector moves with respect to it, for example, under the effect of gravity, electric, or centrifugal field; the collector rests (in the laboratory coordinate system) and liquid flow is present. It is precisely the setting of the experiment aimed at the study of the elementary orthokinetic coagulation act. Similar to that is the elementary act of particle capture from the flow in filtering with the use of porous media, for example of granular filters. A particle from the flow is settled on individual elements of the filtering medium (on a granule of a granular filter or on fibers of a fibrous filter). Thus, considering the process of particle sedimentation from the flow on collector, we obtain, at the same time, information about the elementary act of orthokinetic coagulation.

Processes controlling the formation of cloud drop spectrum or emulsion drop spectrum are more complex, as it is much more difficult to describe coagulation of drops of almost equal size.

It is natural to first consider the gravitational coagulation of drops strongly differing in size (Sections 1.4.1–1.4.8), which will facilitate the understanding of the more complex process of coagulation of drops of close dimensions (Sections 1.4.9 and 1.4.10).

Investigations of coagulation in aqueous aerosols [16,17,89] at flotation [18,19] and filtration [90,91] were started earlier and have been conducted on a front wider than in the case of emulsions. Therefore, in studying gravitational coagulation in emulsions, one should not ignore the experience accumulated in adjacent scientific leads. In addition, the specific nature of emulsions can be taken into consideration.

The first gravitational coagulation model was proposed by Smoluchowski [4]. The next important step was the investigation by Langmuir and Blodgett [92] in which the role of inertial and hydrodynamic forces was demonstrated apparently for the first time. Then investigations have been conducted mainly applied to aerosols and flotation. A systematic consideration of orthokinetic coagulation was given in Refs. 15 and 91.

1.4.2 Primitive Model of Gravitational Flocculation and its Specification for Emulsions

Particles of different size or density will settle at different rates and the resulting relative motion can cause particle collisions and, hence, flocculation. The first attempts to estimate the rate of gravity-induced coagulation of a dispersion were made by Smoluchowski [4].

The collision frequency can be calculated very simply, assuming that the Stokes or the Hadamard–Rybczynski law [93,94] applies and that particle motion is linear up to contact with another particle. The resulting single-particle collision rate was found to be

$$J^0_{Gr} = \pi(u_{02} - u_{01})(a_1 + a_{02})^2 n_1 \tag{1.39}$$

where u_{01} is Stokes creaming velocity for particles of radius a_1. Because all resistances are ignored in this analysis, the Smoluchowski flocculation rate provides a useful scale on which other flocculation rates can be compared. The ratio of Smoluchowski flocculation rate to the actual flocculation rate in a system undergoing gravity-induced flocculation, G_{Gr}, is known as the stability ratio,

$$W_{Gr} = \frac{J^0_{Gr}}{J_{Gr}} \tag{1.40}$$

The gravity-induced capture efficiency is defined as the reciprocal of the stability ratio:

$$E_{Gr} = \frac{1}{W_{Gr}} \tag{1.41}$$

Some authors assume erroneously that this first primitive model of gravity-induced coagulation was proposed by Saffman and Turner [95].

Zhang and Davis [75] specified the Smoluchowski model for emulsions, taking into account that the drag coefficient of a drop is described by the Hadamard–Rybczynski equation [93,94].

In distinguishing this from the Stokes equation, the sedimentation velocity of a droplet expressed according to the Hadamard–Rybczynski theory is given by

$$\bar{u}_0 = \frac{2}{3}\frac{\Delta\rho g a^2}{\mu}\frac{1+\hat{\mu}}{2+3\hat{\mu}} \tag{1.42}$$

where g is the gravitational acceleration vector and $\Delta\rho = \rho' - \rho$, where ρ' and ρ are densities of oil and water, respectively. The internal circulation in a drop and the mobility of its surface reduces the media hydrodynamic resistance to a drop movement that causes the increase in its creaming velocity in comparison with a solid sphere. The relative velocity due to gravity for widely separated drops is given by

$$u^{(0)}_{12} = u^1_0(\hat{\mu}) - u^2_0(\mu) = \frac{2(\hat{\mu}+1)\Delta\rho a_1^2(1-\lambda^2)g}{3(3\hat{\mu}+2)\mu} \tag{1.43}$$

It is seen that when $\hat{\mu} \to \infty$, Equation 1.42 transforms into the equation for the Stokes settling velocity, and when $\hat{\mu} \ll 1$, the settling velocity increases 3/2 times.

In Ref. 96, the experimental data concerning the droplet settling velocity are analyzed. Some experimental data confirm the Stokes equation, whereas others are in agreement with the Hadamard–Rybczynski theory. The systematic experimental theoretical investigations of Frumkin and Levich described in Ref. 97 proved that the transition from the Hadamard–Rybczynski regime to Stokesian behavior of a droplet occurs due to the retardation coefficient χ_r. The Hadamard–Rybczynski equation (1.42) was generalized by Levich and Frumkin by incorporation of the retardation coefficient χ_r:

$$u_0(\hat{\mu}) = \frac{2}{3}\frac{\Delta g a^2}{\mu}\frac{\mu + \mu' + \chi_r}{2\mu + 3\mu' + 3\chi_r} \tag{1.44}$$

where

$$\chi_r = \frac{2}{3}\frac{RT}{D_i}\frac{\Gamma_0^2}{c_0} \tag{1.45}$$

D_i is a surfactant molecule diffusion coefficient and c_0 and Γ_0 are its bulk and surface equilibrium concentration, respectively.

At a very low surfactant concentration, χ_r is small and Equation 1.44 transforms into the Hadamard–Rybczynski equation (1.42). The surfactant concentration growth leads to the surface retardation and the decrease of settling velocity. For $\chi_r \to \infty$, Equation 1.44 transforms into the Stokes equation.

The different equations for χ_r corresponding to the different types of the adsorption kinetics are given in Refs. 96 and 98. The level of impurities in water and their adsorption usually is sufficiently high to cause a strong or even complete retardation of droplet surface mobility in emulsions. The larger the droplet, the lower the retardation. The surface movement is possible for sufficiently large drops and by the special purification of water and oil. Thus, at Re $\ll$ 1, the deviation from the Stokesian drag coefficient is weak even under special experimental conditions. At the same dimension of a drop and a bubble the gravity force and correspondingly the viscous stresses which cause the surface movement are weaker in emulsions than in foams. Thus, neglecting the droplet surface mobility in the differential settling in emulsions is justified even in higher degrees than for bubbles in foam.

One underlying restriction is that the flow in and around the drops is sufficiently slow that inertia is small relative to viscous forces. This requires that the Reynolds number

$$\mathrm{Re} = \frac{\rho \mu_0 a_2}{\mu} \tag{1.46}$$

is small compared to unity for all phases, where a_2 is the larger drop radius and μ_0 is its sedimentation velocity. Typical conditions are $\mu = 0.01\ \mathrm{g\,cm^{-1}\,s^{-1}}$, $\rho = 1\ \mathrm{g\,cm^{-3}}$, $\Delta\rho = 0.1\ \mathrm{g\,cm^{-3}}$, and $g = 10^3\ \mathrm{cm\,s^{-2}}$; this requires that $a_2 < 50\ \mu\mathrm{m}$.

As seen from Equations 1.44 and 1.46, Re is proportional to a_2^3. It means that condition 1.46 is strongly violated for $a_2 > 50\ \mu\mathrm{m}$. In addition, the experimental investigations of the collision efficiency are simplified for large drops.

The droplet velocity in centrifugal field can increase up to hundred times, which leads to a similar growth of Reynolds number. Meanwhile, the ultracentrifugation is important in the investigations of emulsion stability [99–101]. Hence, the case Re $\gg$ 1 becomes interesting as well. For this condition, the empirical dependence of the settling velocity on the radius of a solid particle or a droplet with the retarded surface is known. It follows from the expression for the resistance coefficient for the spherical particle which is a function of the Reynolds number [102,103].

The relative motion of different sized particles can be induced in other ways. In this respect, one important method is the acoustic method, in which ultrasonic waves induce vibrations of suspended particles. The smaller particles are better able to respond to the inducing frequency and to vibrate with greater amplitude than larger particles, facilitating collisions between particles of different size [104].

1.4.3 Long-Range and Short-Range Hydrodynamic Interaction

The process when two droplets of different size approach each other undergoes qualitative changes as the distance between their surfaces diminishes. At larger distances, this process is determined by two parameters: forces of inertia and the long-range hydrodynamic interaction (LRHI).

A sufficiently large drop moves linearly under the effect of the forces of interia until it collides with the bigger one, which takes place if the target distance $b < a_2 + a_1$ (Figure 1.5), where R is the radius of the bubble.

The liquid flow envelops the surface of the big drop, and the small drops are entrained to a greater or a lesser extent by the liquid. The smaller the drops and their difference in density relative to the medium, the weaker the inertial forces acting on them and the more closely the

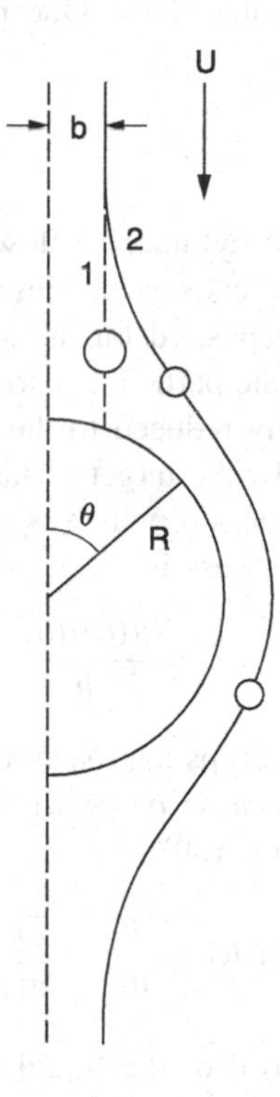

FIGURE 1.5 The influence of the inertia of particles on their trajectory in the vicinity of the floating bubble. Trajectories of the great (inertial) (line 1) and the small (inertia-free) (line 2) particles at the same target distance b.

drop trajectory coincides with the liquid streamlines. Thus, at the same target distance, fairly large drops move almost linearly (Figure 1.5, line 1), whereas fairly small drops move essentially along the corresponding liquid flow line (line 2). The trajectories of drops of intermediate size are distributed within lines 1 and 2; as the size of drops decreases, the trajectories shift from line 1 to line 2 and the probability of collision thus decreases.

The deviation of trajectory of small droplets from the rectilinear path to the surface of the biggest drop at distances of the same order as the biggest drop is caused by LRHI. The biggest drop causes a curving of liquid streamlines and thereby bends the trajectory of small drops (i.e., acts on them hydrodynamically due to the liquid velocity field). In the case of large drops, the forces of inertia considerably exceed the LRHI which is, therefore, not clearly manifested. In the case of small drops, the forces of inertia are small compared with the LRHI.

Thus, the process of the approach of large drops to the biggest drop is ensured by forces of inertia, whereas in the case of small drops, this process occurs in an inertia-free manner and is strongly hindered by the LRHI [105,106].

In addition, the hydrodynamic interaction at distances comparable to the drop radius has to be taken into account; the latter causes the drop's trajectory to deviate from the liquid flow line and should naturally be called the short-range hydrodynamic interaction (SRHI). Using Taylor's solution of the hydrodynamic problem involving the squeezing out of liquid from the gap as spherical particles approach the flat surface, Derjaguin and Dukhin [106] have shown that the SRHI may prevent drops from coming into contact.

The process of the approach of particles to the bubble surface can be described quantitatively by taking into account both LRHI and the SRHI. One introduces a dimensionless parameter of the collision efficiency,

$$E = \frac{b^2}{a^2} \tag{1.47}$$

where b is the maximum radius of the cylinder of flow around the bigger drop encompassing all particles deposited on the small droplets surface (Figure 1.6). The particles moving along the streamline at a target distance b are deposited on the surface of a bigger drop (Figure 1.6, as indicated by a dashed line). Otherwise the particle is carried off by the flow. From Figure 1.6 it is evident that the calculation is essentially reduced to the so-called "limiting (grazing) trajectory" (continuous curve) and, correspondingly, the target distance.

According to Taylor, at a gap thickness h, which is much smaller than a_1, the hydrodynamic resistance of the film to the thinning process is

$$F_h = \frac{u(h)\mu a_1^2}{h} \tag{1.48}$$

where $u(h)$ is the velocity at which the drops approach a certain surface area of the biggest drop, which may be considered to be flat because $a_2 \gg a_1$. If a constant force F_h is applied to the smaller drop, then according to Equation 1.48,

$$u(h) \equiv \frac{dh}{dt} = \frac{F_h h}{a_1^2 \mu}$$

It may be inferred that complete removal of the liquid from the gap requires an infinitely long time:

$$t \sim -\int_h \frac{\mu a_1^2 dy}{F_h y} \sim -\frac{a_1^2 \mu}{F_h} \ln y \,|_h \to \infty$$

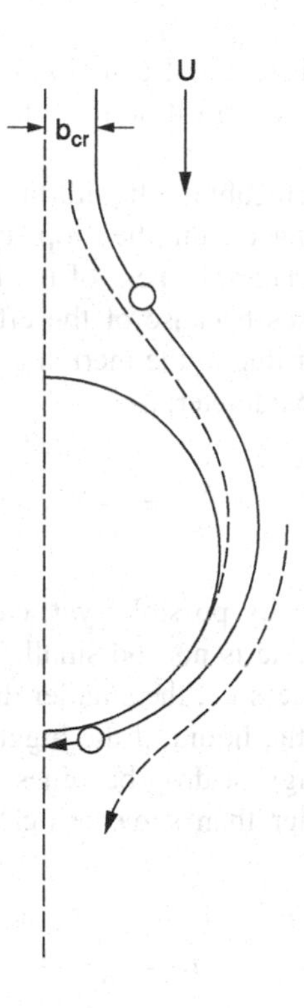

FIGURE 1.6 Continuous line illustrating the concept of the limiting trajectory of particles. Dashed line indicates the trajectories of the particle at $b < b_{cr}$ and $b > b_{cr}$.

In the area above the equatorial plane, the liquid flow lines approach the surface of the biggest drop, which means that the radial component of the liquid velocity is directed toward the surface of the biggest drop. Because the motion of the particle toward the surface is obstructed within the zone of the SRHI, the radial velocity of the liquid is higher than that of the small drop. Thus, at a small gap thickness where the viscous resistance is high, the radial velocity of the liquid will be even higher. The radial flow of liquid envelops the small drop whose approach to the biggest drop has been retarded and presses it against the latter. As a first approximation, this hydrodynamic force can be estimated from the Stokes formula by substituting into it the radius of the small drop and the difference in the local values between the velocity of the liquid and that of the small droplet.

1.4.4 The Effect of Droplet Inertia

In order to understand the mechanism of inertial deposition of smaller drops on the biggest drop, one must introduce the notion of the particle inertial path, l. The latter is defined as the distance the particle with an initial velocity u_0 is able to cover in the presence of the viscous resistance of the liquid due to the initial velocity

$$l = \frac{2}{9}\frac{u_0 a_1^2 \rho'}{\mu} \tag{1.49}$$

This result follows from the linear differential equation of the particle hydrodynamic relaxation which consists of two terms. The first term describes the inertial force and the second one the hydrodynamic resistance.

Because the drop surface is impermeable to liquid, the normal component of the liquid velocity on the surface is zero. As the distance from the drop surface increases, the normal component of the liquid velocity also increases. The thickness of the liquid layer, in which the normal component of the liquid velocity decreases because of the effect of the drop, is of the order of a_2. The particle traverses this liquid layer due to the inertial path whereby the deposition of a smaller drop depends on the dimensionless parameter:

$$\tilde{l} = \frac{l}{a_2} \tag{1.50}$$

When $\tilde{l} > 1$, deposition is obviously possible; yet, calculations have shown that it can also take place at $\tilde{l} < 1$ as long as this value is not too small. This conclusion becomes apparent if it is considered that in a layer of thickness a_2, the smaller drop moves toward the surface not only due to inertia but also together with the liquid of the biggest drop. The motion component of the latter normal to the surface of the biggest drop becomes zero at its surface. Inertial depositions prove to be impossible if $\tilde{l}$ is smaller than some critical value $\tilde{l}_{cr}$. Neglecting the size of the smaller drop, Levin [107] obtained

$$\tilde{l}_{cr} = \frac{1}{24} \tag{1.51}$$

This is taken into account in the equation describing the hydrodynamic relaxation in the hydrodynamic field if the biggest droplet. The one-dimensional equation describes the small droplet movement along the symmetry axis crossing the centers of both droplets:

$$\mathrm{St}\frac{d^2x}{dt^2} + \left(\frac{dx}{dt} - u(x)\right) = 0 \tag{1.52}$$

where the Stokes number $\mathrm{St} = \tilde{l}$ characterizes the ratio of inertial and viscous forces when the Stokes mobility is valid for smaller droplets. It means its Reynolds number is less than 1. x is the distance to the surface of biggest droplet and $u(x)$ is the liquid velocity distribution along the symmetry axis. It is described by the Stokes equation for the hydrodynamic field if the Reynolds number for the biggest droplet is less than 1 also. The characteristic equation $\mathrm{St}m^2 + m + 2 = 0$ has one real (negative) root and separation between the droplet surface decreases exponentially with time. However, the (point) particle never reaches the surface because the particle velocity and fluid velocity coincide. This conclusion holds for nonzero St values if St is smaller than a critical value,

$$\mathrm{St} < \mathrm{St}_{cr} \tag{1.53}$$

Considerable attention was paid to the calculation of St_{cr}. The numerical calculations of Natanson [108] are in good agreement with the original Langmuir data [92] for E. Later, the general solution of the problem for St_{cr} was given by Levin [107] who derived the equation for St_{cr} for any hydrodynamic field. It is important because condition 1.53 is often violated for the biggest drop.

The dependence $E(\mathrm{St})$ can be obtained by the solution of a set of two differential equations for two components of particle velocity. The results of the numerical calculation of Fonda and Herne [109] for the potential and the Stokes flow are given in Ref. 110. These results and the analytical results of Natanson [108] and Levine [107] are in good agreement [89].

The differential equation (1.53) describes the movement of a spherical particle center. Inertial deposition proves to be possible even at $\mathrm{St} < \mathrm{St}_{\mathrm{cr}}$ if it is considered that at a distance equal to $a_1 + a_2$ between the smaller droplet center and the biggest droplet surface, the surfaces touch each other.

The possibility of the inertialess collision due to interception and the importance of the particle dimension was proved by Sutherland [105]. During some decades, there was no coupling between two directions in the theory of the collision efficiency. At $\mathrm{St} < \mathrm{St}_{\mathrm{cr}}$, the collision efficiency was considered to be caused by the interception. The inertial forces were neglected and the dimension of a particle was important. At $\mathrm{St} > \mathrm{St}_{\mathrm{cr}}$, the inertial forces were taken into account. However, the role of a particle dimension was neglected. These simplifications can be verified at extreme cases of

$$\mathrm{St} \ll \mathrm{St}_{\mathrm{cr}} \text{ and } \mathrm{St} \gg \mathrm{St}_{\mathrm{cr}} \tag{1.54}$$

In Ref. 111, it is emphasized that in the intermediate range of Stokes values [i.e., at the violation of conditions 1.54], both the interception mechanism and inertial forces are important.

Condition 1.46 is violated at the conditions which are necessary for the manifestation of inertial forces. It is violated even at smaller droplet dimensions when the inertial forces can be neglected. Thus, the extension of the droplet range in the theory of the collision efficiency can be achieved by the refusal of condition 1.46. The next and more difficult task is the incorporation of inertial forces in the theory of macroemulsions. The inertialess collision at the intermediate Reynolds number will be considered in the next section. Some additional remarks concerning the inertial collision in emulsion will be given in Section 1.4.6.

1.4.5 Collision Efficiency Caused by Long-Range Hydrodynamic Interaction

1.4.5.1 The Grazing Trajectory Method

Collision efficiency calculations were presented for the first time in Ref. 105. They were based on the consideration of liquid streamlines and of the finite size of spherical particles. Functions characterizing liquid streamline are related to the velocity field by well-known differential relations, and in the Stokes case

$$\Psi = a_2^2 U \left(\frac{1}{2}\rho^2 + \frac{1}{4}\rho^{-1} \right) \sin^2\theta \tag{1.55}$$

where ρ denotes the radial position scaled on a_2 and θ denotes the angle from the front stagnation point. As can be seen from Equation 1.55, streamline $\rho(\theta)$ is symmetrical with respect to equatorial plane of bubble: $\theta = \pi/2$. This means that the droplet moving along the liquid streamline approaches the drop surface most closely at $\theta = \pi/2$. Because the center of a droplet moves along the liquid streamline, it will touch the surface of bigger drops only in the case when the distance between the streamline and the droplet is equal to the droplet radius. According to Ref. 105, liquid streamline passing through a point with coordinates

$$r = a_2 + a_1 \text{ and } \theta = \frac{\pi}{2} \tag{1.56}$$

represents the limiting trajectory of the droplet ($\Delta\rho = 0$). Under conditions 1.56, the droplet touches the drop. But if we increase the tangent distance, then even the nearest distance between the streamline and the drop surface ($\theta = \pi/2$) proves to be larger than the droplet radius. By substituting values of r and θ according to Equations 1.56 into Equation 1.55, Ψ_c which characterizes grazing streamline giving the basis for the collision efficiency E_{0c} can be determined:

$$\lim_{\rho\to\infty}(\rho^2\sin^2\theta) = \frac{1}{a_2^2}\lim_{r\to\infty}(r^2\sin^2\theta) = \frac{b^2}{a_2^2} = E_{0c} = \Psi_p$$

After substitution of the expression for Ψ_p in which at $a_1/a_2 \ll 1$, only the term linear in small parameters is considered, we obtain

$$E_{0s} = \frac{3}{2}\frac{a_1^2}{a_2^2} \tag{1.57}$$

We introduced the subscript zero to denote that SHRI and surface forces are not taken into account. According to it, when the size of particles decreases, LRHI retards coagulation kinetics by a factor hundreds and even thousands.

1.4.5.2 Equation for the Smaller Drop Flux on the Surface of a Larger Drop

In the case of non-inertia collisions and without regard for SHRI, the velocity of a smaller drop is well known at any distance from the bigger one:

$$\mathbf{u} = \mathbf{u}(r,\theta) + \mathbf{u}_g \tag{1.58}$$

so that it is advisable to derive an expression for the radial component of smaller drop flux density

$$J_r = n(r,\theta)u_r \tag{1.59}$$

where $n(r, \theta)$ is the smaller drop number concentration next to the surface of a bigger drop. Integrating the flux density over that part of the surface on which sedimentation takes place, we obtain the number of smaller drops colliding with the bigger one in unit time N, which relates to E_0 as

$$N = 4\pi a_2^2 E_0 u_0 \tag{1.60}$$

Earlier [112], a theorem was proven that the particle concentration remains constant if the velocity field is solenoidal. From this, it follows that

$$E_0 = \frac{1}{\pi a_1^2 u_0 n_1}\int_0^{\theta_c} n_1(u_r + u_g)2\pi a_1^2 \sin\theta d\theta \tag{1.61}$$

where θ_c characterizes the boundary of the region of deposition of smaller drops at the surface of the bigger drop. To consider the effect of a finite size of a smaller drop, integration over a concentric sphere of radius $a_1 + a_2$ has to be performed. From this condition, θ_c can be derived:

$$u_r(a_1 + a_2, \theta_c) = 0 \tag{1.62}$$

Substituting the Stokes velocity field into Equation 1.61 and using Equation 1.44 at $\hat{\mu} \to \infty$, we obtain [112]

$$E_{0s} = \frac{3}{2}\left(\frac{a_1}{a_2}\right)^2 - \frac{\Delta\rho}{\rho}\left(\frac{a_1}{a_2}\right)^2 \tag{1.63}$$

This method of calculation was provided almost a quarter of a century later by Weber [113], who, probably, was not aware of the work of Dukhin and Derjaguin [112]. The usefulness of the method was well demonstrated, although the discussion was restricted by the consideration of systems in which the particle density differs only slightly from water density and allows one to ignore the role of sedimentation. A more general approach was demonstrated in the recent work of Nguen Van and Kmet [114].

It is important that Equation 1.61 is applicable not only under Stokes and potential flow but also at any solenoidal hydrodynamic field, which ensures its wide application.

1.4.5.3 Long-Range Hydrodynamic Interaction at Intermediate Reynolds Numbers

The important equation (1.63) is valid for condition 1.46. At higher Reynolds numbers, the hydrodynamic flow pattern around a drop changes and causes the change in the equation for collision efficiency. Flow conditions representative of very high Reynolds numbers can be approximated by the potential flow. It is not valid in the surface vicinity within the so-called hydrodynamic boundary layer. This case of very high Reynolds numbers will not be considered here because it corresponds to very big drop dimensions and the deviation of shape from sphericity.

The hydrodynamic at intermediate Reynolds numbers with respect to spherical particles is considered in Ref. 115. There is a large qualitative distinction in flow pattern around a drop, depending on the ratio $\hat{\mu}$. If an oil viscosity strongly exceeds water viscosity, the flow pattern is similar to that of a solid sphere.

According to experimental investigations and numerical solutions (cf. Clift et al. [115]) qualitatively different hydrodynamic regimes exist at different Reynolds numbers. At $7 < \mathrm{Re} < 20$, a so-called unseparated flow is observed. Flow separation is indicated by a change in the sign of the vorticity and first occurs at the rear stagnation point, approximately at $\mathrm{Re} = 20$.

If Re increases beyond 20, the separation ring moves forward so that the attached recirculation wake widens and lengthens. A steady wake region appears at 20. The onset of wake instability corresponds to separation angle

$$Q_s = 180 - 42.5(\ln \mathrm{Re}/20)^{0.48}$$

To generalize Equation 1.63, the theories of flow pattern around the bubble at intermediate Re can be used [114,116,117].

If an oil viscosity is less than water viscosity, the theory in Ref. 118 predicts the disappearance of flow separation. For this case, the solution of the Navier–Stokes equation can be accomplished according to numerical methods only [119].

Investigations of droplet collisions in emulsions at intermediate Reynolds numbers seem to be lacking. However, the emulsion specificity disappears in the presence of even traces of a surfactant with respect to long-range hydrodynamic interactions. This statement cannot be valid with respect to short-range hydrodynamic interaction, which will be discussed in the next section.

1.4.5.3.1 Complete Surface Retardation

As discussed in Section 1.4.1, a droplet surface can be substantially retarded even in distilled water due to the presence of surfactants. Weber and Paddock [120] applied a curve-fitting technique to the numerical solutions of the Navier–Stokes equations obtained by Masliyah [121] and Woo [122] and derived the expression for the interceptional collision efficiency:

$$E_0 = \frac{3}{2}\lambda^2\left(1 + \frac{(3/16)\mathrm{Re}}{1 + 0.25\mathrm{Re}^{0.56}}\right) \tag{1.64}$$

Yoon and Luttrel [116] empirically developed an analytical expression for the streamline function at intermediate Reynolds number by combination of the stream function for creeping and potential flow regimes and derived the equation for the interceptional collision efficiency. As can be seen from Figure 1.4 of their article [116], the difference of their results from Equation 1.64 is small.

The complete Navier–Stokes equation was solved numerically again by Nguen Van and Kmet [114] who obtained the interceptional collision efficiency. Its first term coincides with Equation 1.64; the second one characterizes the particle sedimentation on the bubble surface. It is important in the condition of mineral flotation and can be neglected in emulsions. Thus, the theoretical results of Refs. 114 and 116 do not contradict Equation 1.64.

1.4.5.3.2 Incomplete Surface Retardation

The term "emulsion" usually relates to the dispersion with drops less than 100 μm in diameter. At these dimensions, the droplet surface mobility is retarded completely or very strongly by the traces of surfactants. At strong retardation, a small residual surface mobility is possible. However, under some special conditions, the participation of the droplet of bigger dimensions in collisions can be of interest. In this extreme case, the interceptional collision efficiency can be described by means of the theory [123] elaborated for the description of the flotation kinetics at intermediate Reynolds number and unretarded bubble surface. As surface retardation decreases, both the tangential liquid velocity and its normal component at the distance of order a_1 from the surface increased. Thus, the smaller drop flux increases, and this increase can be described by the Sutherland equation as the surface retardation decreases. Thus, the interceptional collision efficiency for free surface E_{0c} is less sensitive to the small value of the radius ratio λ than that for immobile surface E_{0r}. For the intermediate Reynolds number, Rulyov and Leshchov [123] used Hamieloe theory [124] and obtained

$$E_{0c} \sim \frac{a_1}{a_2^{0.9}},\ E_{0r} \sim \frac{a_1^2}{a_2^{1.8}} \tag{1.65}$$

For small values of the radius ratio surface retardation reduces the interceptional efficiency by magnitudes of hundred or thousand. Experimental investigation is easier for big drops and the discussed theory [123] can be useful for this purpose.

1.4.5.3.3 Experiments

Experimental investigations of the droplet collisions in emulsions at intermediate Reynolds numbers are not reported. However, some information can be extracted from the experiments accomplished for the investigation of particle collisions with a small bubble at an intermediate Reynolds number. In experiments [114], a single bubble is generated and captured on a needle. Afterwards, the small particles' trajectories in the vicinity of the captive bubble were observed and the grazing trajectory was determined. In these experiments [114], the size of a captive bubble was in the range 0.5 to 2 mm, the Reynolds number in the range 30 to 300, and the particle size in the range 10 to 35 μm. In other experiments [116], the bubble size was in range 100 to 500 μm and the particle size in the range 1 to 40 μm. The experiments [114,116] confirmed the theoretical prediction of collision efficiency dependence on particle and bubble dimensions.

In these experiments, the influence of solid-particle sedimentation on the bubble surface predominated due to the large difference of densities of solid particles and water. It means that the manifestation of a second component of the particle transport to the bubble surface, caused by the normal component of liquid velocity, was weak. Thus, the exactness of the verification of the theory of the convective transport in these experiments was low. Meanwhile, the convective transport predominates in emulsions due to small differences in densities of water and oil.

Thus, Equation 1.64 can be used for the description of the interceptional collision efficiency in emulsions to a first approximation. However, the exactness of its experimental verification with respect to emulsions is limited.

The experimental methods and the experimental installations, described in Ref. 114, can be applied for the experimental investigations of the collision efficiency in emulsions. It is sufficient to replace the captive bubble by a big captive oil drop.

If a particle shape is isometric, its behavior in the long-range hydrodynamic interaction is similar to the spherical particle with the same dimension. However, their behavior in short-range hydrodynamic interaction will be different. The drainage in the gap between a bubble (drop) and particle (small drop) is very sensitive to the microrelief on the surface. The resistance to thinning of the interfacial film can be drastically reduced if the angles between the facets are sufficiently sharp so that "intrusion" of the interface film by a sharpened section of a particle surface may take place. Anfruns and Kitchener [125] determined the efficiency of the accumulation of glass spheres and broken quartz particles and established that the manifestation of SRHI is weakened in the case of quartz particles.

Thus, the discrimination between LRHI and SRHI contributes to measured particle accumulation by a big bubble (drop) is possible. The broken particle accumulation yields information concerning LRHI. The decrease of small droplet accumulation yields information concerning SRHI. The exactness of this procedure can be decreased if SRHI preserves in some degree for the broken particle also.

1.4.6 Flocculation in a Centrifugal Field and Dynamic Adsorption Layer

Centrifugal fields are used for demulsification [126] and in the investigation of emulsion stability [99,100]. The increase of a droplet velocity in a centrifugal field (for example, 100-fold) led to the corresponding increase of Reynolds and Stokes numbers. Thus, the movement of not very small droplets in the centrifugal field occurs in the intermediate range of Reynolds number (Section 1.4.5.3) and their flocculation can be complicated by the action of the inertial force (Section 1.4.4).

Larger drop velocities create stronger viscous stresses in the vicinity of its surface. Thus, the residual surface mobility increases as the Reynolds number increases and has to be taken into account in the theory of orthokinetic coagulation. Probably this question was not considered with respect to drops. Thus, the experience of the flotation theory can be used [18,19].

Steady-state motion of a drop induces adsorption–desorption exchange with the subsurface, with the amount of substance adsorbed on one part of its surface being equal to the amount desorbed from another part. Obviously, the surface concentration is lower on the part where adsorption takes place and it is higher on the part where desorption occurs. Thus, surface concentration varies along the surface of a moving drop, taking a maximum value at the rear pole and a minimum one at the front pole [97]. The adsorption difference between the poles of a drop causes the surface tension drop and Marangoni–Gibbs phenomena; that is, the surface movement in the direction of the surface tension decreases. This secondary surface flow is directed opposite to the primary surface and flow; consequently, the surface velocity retards.

Therefore, the state of adsorption layer on a moving drop surface is qualitatively different from that on a resting one. Such an adsorption layer was called a dynamic adsorption layer [127].

With respect to the large Reynolds numbers, the retardation coefficient can be evaluated at the strong surface retardation by using the concept of the hydrodynamic and diffusion layers [97,127] having a thickness δ_G and δ_D, respectively, independent of angle θ. It follows from the boundary condition expressing the interconnection between adsorption–desorption flux and the surface divergence of the surface convective flux of the adsorbates [127]

$$\chi_b = \frac{RT\Gamma_0^2}{Dc_0}\frac{\delta_D}{a}\frac{\delta_G}{a} = \frac{RT\Gamma_0^2}{Dc_0}\frac{D}{\nu \mathrm{Re}} \tag{1.66}$$

because

$$\frac{\delta_c}{a} \sim Re^{-1/2}, \ \frac{\delta_\delta}{a} \sim \mathrm{Pe}^{-1/2} \tag{1.67}$$

and the Peclet number $\mathrm{Pe} = \frac{ua}{D}$,

$$\mathrm{Pe} = \frac{\nu}{D}\mathrm{Re} \tag{1.68}$$

Comparing Equations 1.66 and 1.45, one concludes that the decrease of the retardation coefficient is

$$\frac{\chi_B|_{\mathrm{Re}\gg 1}}{\chi_B|_{\mathrm{Re}\ll 1}} = \frac{D}{\nu \mathrm{Re}} \tag{1.69}$$

Taking into account that $\nu/D \sim 10^3$, one concludes that the transition from a small Reynolds number to a larger one is accompanied by a reduction in the retardation coefficient by a factor of 10^4–10^5. Thus, for big drops or under the action of centrifugal force, the state of the mobile surface is possible even at noticeable surfactant concentrations.

The dynamic state of the adsorption layer causes the change in all stages of the flocculation. It influences the degree of the drop surface retardation, its velocity and the relative velocity for two widely separated drops, the hydrodynamic velocity distribution around a drop, and, consequently, the LRHI. It modifies surface forces because they depend on the surface concentration. Thus, it is important at the last stage of flocculation too.

1.4.7 Incorporation of Short-Range Hydrodynamic Interaction and Attractive Surface Forces in the Theory of Inertialess Collision

In the process involving an inertialess approach of a smaller drop to a bigger one, the size of the latter plays an important role. It is in the equatorial plane that the closest approach of a streamline of the surface of a bigger drop is attained. In Figure 1.7, the broken line (curve 1) represents the liquid streamline whose distance from the surface of the bigger drop in the equatorial plane is equal to the radius of the smaller drop. Some authors erroneously believe that this liquid streamline is limiting for the drops of that radius. The error consists in that the SRHI is disregarded in this case. Under the influence of the SRHI, the drop is displaced from liquid streamline 1 so that its trajectory (curve 2) in the equatorial plane is shifted from the surface by a separation larger than its radius. Therefore, no contact with the surface occurs and, correspondingly, $b(a_1)$ is not a critical target distance.

Due to the SRHI, the distance from the smaller drop to the surface in the equatorial plane is larger than the distance from the surface to the liquid streamline with which the trajectory of the smaller drop coincides at large distances from the bigger drop. It may thus be concluded that $b_{cr} < b(a_1)$. The limiting liquid streamline (curve 3) is characterized by the drop trajectory (curve 4),

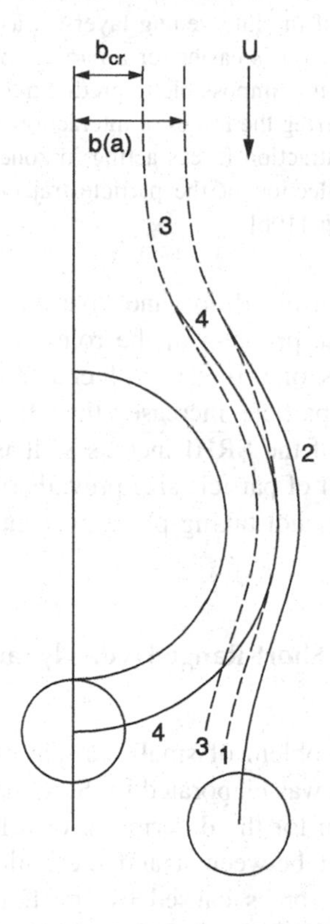

FIGURE 1.7 The influence of the finite dimension of particles in the inertial-free flotation on their trajectory in the vicinity of the floating bubble. The liquid flow lines corresponding to target distances $b(a)$ and b_{cr} are indicated by dashed lines. The continuous lines are characteristic for the deviation of the trajectory of particles from the liquid flow lines under the influence of short-range hydrodynamic interaction.

which, branching off under the influence of the SRHI, runs in the equatorial plane at a distance a_1 from the surface of the bigger drop.

The value of b_{cr} decreases, first due to the deflection of the liquid streamlines under the influence of the LRHI and, second, due to the deflection of the small drop trajectory from the liquid streamline under the influence of SRHI. Therefore, the collision efficiency is expressed as the product of two factors, E_0 and f, both smaller than unity. The first one represents the influence of the LRHI and the second, the influence of the SRHI.

It is evident that SRHI, like LRHI, manifests itself in the gravitational coagulation of drops, solid particles of suspensions, bubbles, and in their sedimentation from a flow on large-size surfaces, with the SRHI mechanism being common in these systems. In essence, identical SRHI theories are created as applied to flotation, aerosols, and suspension purification from particles by filtration. Only lately has the problem received further development as applied to emulsions [26,27].

The mechanism of SRHI is the same under conditions of perikinetic and orthokinetic coagulation. Derjaguin and Dukhin [106] formulated the problem of SRHI as applied to orthokinetic coagulation or, more properly, to the elementary act of flotation of a small spherical particle by a bubble.

> When the mineral particle comes closer than its own radius to the bubble viscous resistance forces should develop retarding thinning of the intervening layer ... assuming the trajectory of the particle center to be independent of its radius it is easier for a particle of larger radius to contact the bubble surface. It can be easily shown that it is impossible to predict the outcome of the competition between two opposite effects without considering the forces of interaction between particle and bubble surfaces. When the radius of action of the attraction forces acting in zone 3 is much smaller than the particle radius the influence of viscous deflection of the particle trajectory will always outweigh the other effect and contact ... will not occur [106].

If we replace "bubble" with "a big drop" and "particle" with "small drops" in this quote, then we obtain the statement of the problem of the role of SRHI in emulsions. As can be seen from the quote, the main features of this stage of coagulation process were already clear at that time. When the size of the particle increase, the effect of LRHI on collision efficiency decreases and the negative effect of the SRHI increases. It is shown in the further development of quantitative theory that the effect of particle size prevails of LRHI. Along with this, the remark about flocculation of large particles not taking place even in the presence of attractive forces is not always correct, as seen below.

1.4.7.1 Quantitative Theory of Short-Range Hydrodynamic Interaction Given Attractive Surface Forces

A more general approach to the problem of small-particle deposition from laminar flow on the surface of a big particle (collector) was elaborated by Spielman [128] and Goren [129,130]. They introduce a local coordinate system for the description of a local hydrodynamic field around the small particle and in the interlayer between it and the collector. They presented the equations for the short-range hydrodynamic forces caused by the field. Their results form an important component of the Derjaguin–Dukhin–Rulyov (DDR) theory of SRHI in flotation [19]. The particles move along a bubble surface and the condition for particle–bubble interactions changes (i.e., the short-range interaction is unsteady). However, in the local coordinate system which is bound to the particle and moves with it, the interaction can be considered as a quasi-steady one.

The cylindrical system of coordinates is introduced. Its center is lying on the bubble surface with the z-axis crossing the center of the moving particle.

Cylindrical symmetry in the chosen coordinate system simplifies the description of hydrodynamic interaction. Separate equations for the normal component $u_r = dh/dt$ and tangential component $u_\theta = a_2(d\theta/dt)$ of a particle velocity are written respectively

$$\frac{dh}{dt} = \frac{F_n f_1(h)}{6\pi \mu a_1} \tag{1.70}$$

$$a_2 \frac{d\theta}{dt} = u_\theta f_2(h) \tag{1.71}$$

The equations describe the stage of a particle movement when the liquid interlayer is thin and the distance between the centers of a particle and the bubble equal to $a_1 + a_2$.

The rates of interlayer thinning and particle movement to the bubble surface coincide and are determined by the action of pressing force F_n and the resistance source which is characterized by the product of the Stokes drag coefficient and the dimensionless function $f_1(H)$.

The particle trajectory equation follows from the set of Equations 1.70 and 1.71 after the introduction of H instead of h and the time exclusion

$$\frac{dH}{d\theta} = \frac{a_2 F_n(\theta, H) f_1(H)}{6\pi \eta a_1^2 u_\theta(\theta, H) f_2(H)} \tag{1.72}$$

In the general case, the pressing force is a superposition of many forces:

$$F_n = F_A + F_\Psi + F_V + F_g \tag{1.73}$$

where F_A, F_Ψ, F_V, and F_g are the attractive surface force, the repulsive surface forces, the hydrodynamic pressing force, and the gravity force, respectively. The semianalytical solution of Equation 1.72 is possible if the radius of action of surface forces is small in comparison with the particle dimension.

The hydrodynamic pressing force is proportional to the local value of the normal component of the hydrodynamic velocity:

$$F_n = F_H = 6\pi a \eta u_r(H, \theta) f_3(H) \tag{1.74}$$

The function $f_3(H)$ yields its dependence of the film thickness. After the substitution of Equation 1.74 into Equation 1.73, one obtains

$$\frac{dH}{d\theta} = \frac{R u_r(\theta, H) f_1(H) f_3(H)}{a u_\theta(\theta, H) f_2(H)} \tag{1.75}$$

The functions f_1, f_2, and f_3 were determined in the work of Goren [130], Goren and O'Neil [129], Goldman et al. [131], and Spielman and Fitzpatrick [90]. At a distance which is comparable with the dimension (i.e., at $H \geq 1$), a simplification occurs:

$$\frac{f_1 f_2}{f_3} \sim 1 \tag{1.76}$$

It means that at this distance, Equation 1.72 transforms into an equation for the liquid streamline. Consequently, at this distance, the particle trajectory coincides with the specific streamline.

The theory specifies how surfaces of the two models influence the drainage [19].

1. A critical thickness h_{cr} exists. Naturally, there is a correlation between h_{cr} and DL thickness. This is not taken into account and the surface forces action beyond h_{cr} is neglected. In this model, a particle is attached after the film thinning is culminated, satisfying the boundary condition

$$H\left(\frac{\pi}{2}\right) = H_{cr}\left(H_{cr} = \frac{h_{cr}}{a_1}\right) \tag{1.77}$$

Naturally, a particle is attached when $h(\theta) = h_{cr}$ at any θ. However, condition 1.77 separates the grazing trajectory. The result for creeping flow is

$$f_s = 1.5(1 - 0.3 \ln H_{cr})^{-2} \tag{1.78}$$

2. The notion of h_{cr} is neglected. The attachment is caused due to a predominating attraction [132]. In the case of the first model, neglecting the molecular attractive force we could not consider the particle attachment on the lower surface of a bubble because any particle departs there from the bubble surface under the gravity action. Neglecting h_{cr}, we considered the particle almost on the bubble surface under the action of the predominating attractive force which can exceed the gravity. Thus, the model enables one to consider the particle attachment on the lower surface of a bubble. The grazing trajectory finishes at the rear stagnant pole of the bubble. Due to cylindrical symmetry,

$$\left(\frac{dH}{d\theta}\right)_{\theta=\pi} = 0 \tag{1.79}$$

Comparing condition 1.39 and Equation 1.72, one concludes that the final coordinates of the grazing trajectory satisfy conditions

$$\theta_0 = \pi,\ F_n(\pi, H_0) = 0 \tag{1.80}$$

1.4.7.2 Theory Specification for Different Models

First, the function (1.78) was calculated. At all values under consideration, this function is smaller than unity; it decreases with H_{cr} and becomes zero at $H_{cr} = 0$. This confirms the aforementioned representation of the mechanism based on the influence of the SRHI on the particle deposition process.

As H_{cr} decreases from 10^{-1} to 10^{-3}, f decreases from 0.5 to 0.15; that is, the dependence of f (and of E) on the absolute value of h_{cr} is weak. Thus, the inclusion of the SRHI is important not only in considering the problem of flotation. This effect reduces the number of collisions by several times.

Rulyov [132] has developed the SRHI theory without considering the phenomenological parameter, h_{cr}, by directly taking into account the dependence of molecular forces on h, and obtained

$$E = E_0 f(W), \quad W = \frac{Aa_2^2}{27u_s\pi\mu a_1^4} \tag{1.81}$$

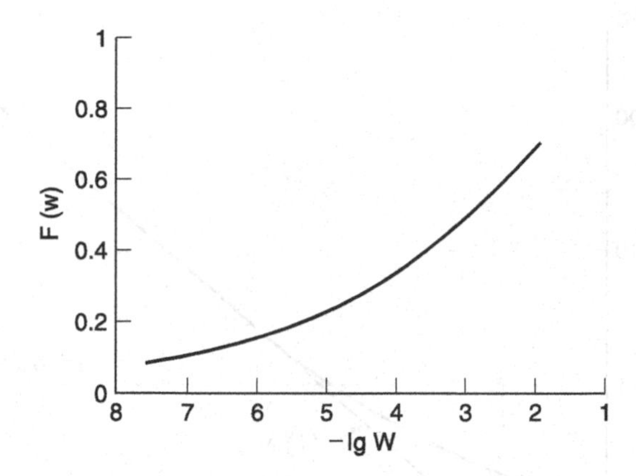

FIGURE 1.8 The multiplier in Equation 1.81 expressing the influence of hydrodynamic and molecular interaction between particle and bubble at inertialess flotation. (After V. Shubin and P. Kekicheff, *J. Colloid Interface Sci.* 155: 108 (1993).)

The function $f(W)$ is plotted in Figure 1.8 and can be approximated [133] to

$$f = 1.2W^{0.15} = \frac{1.2A^{0.15}}{(2\Delta\rho g/9)^{1/6}a_1^{0.7}} \tag{1.82}$$

The first model can be applied to unstable emulsions. The interdroplet lamellae is unstable and its critical thickness is h_{cr}. The second model corresponds to emulsions which are unstable with respect to flocculation and stable with respect to coalescence. The drop doublet forms under the action of attractive molecular forces. However, the coalescence is prevented due to the presence of a stable adsorption layer.

The second model was used by Spielman and Fitzpatric [90,91] who earlier considered the same problem as Rulyov [132]. It is interesting to compare the results of these investigations. The dimensionless number N_A introduced in Ref. 90 differs from W by a factor of 4/3 only. In Ref. 90, the dependence E/E_0 is presented as a function of N_A (Figure 1.9). Taking into account that $E/E_0 = f$, one concludes that the ordinates in Figures 1.8 and 1.9 coincide. Neglecting the difference in abscissas caused by the multiplier 4/3, one concludes that the functions $f(W)$ and $f(N_A)$ practically coincide. In Ref. 90, N_A is called the attraction number. Its physical sense is explained in Ref. 91.

Different velocity and time scales are appropriately close to the particle surface where viscous and colloidal forces become important. The characteristic fluid velocity is smaller [i.e., $O(a_1u_0/a_2)$] due to the proximity of the surface, so the time spent in the neighborhood of collector, the dwell time, is $O(a_2^2/u_0a_1)$. The importance of interparticle attraction is assessed by calculating the time required to capture a particle once it has been moved close to the surface and comparing this to the dwell time. A particle mobility of $O((\mu a_1)^{-1})$ and a dispersion force of A/a (see Section 1.2.1) produce a velocity toward the surface of $O(A/\mu a_1^2)$. Therefore, a representative capture time, the time required to move a distance $O(a_1)$, is $O(\mu a_1^3/A)$. The ratio of the dwell time to the capture time is called the attraction or adhesion number, N_A.

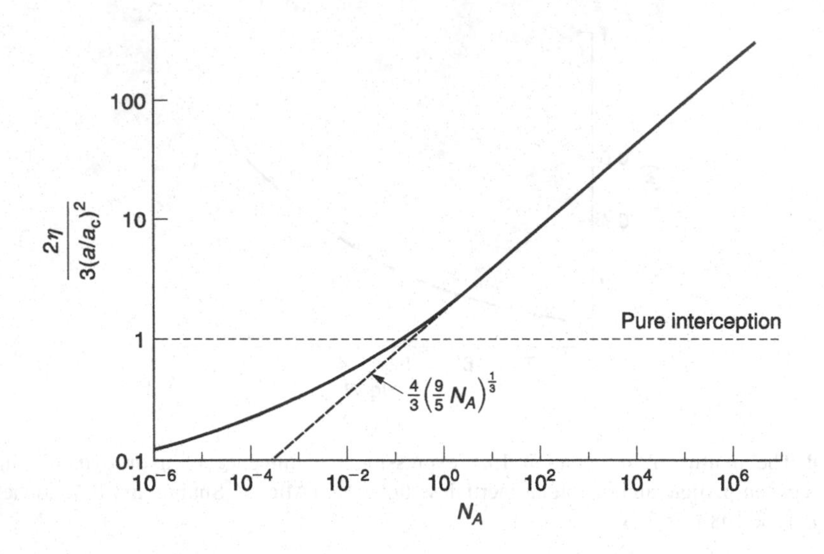

FIGURE 1.9 Capture efficiencies for a spherical collector. Solid line from a numerical solution. (From L.A. Spielman and J.A. Fitzpatrick, *J. Colloid Interface Sci.* 42: 607 (1973).)

As the dwell time exceeds the capture time, that is,

$$N_A > 1, \tag{1.83}$$

the smaller drop moves along the surface of the bigger drop during a sufficiently long time and can be attracted to the surface from a distance of the order of its dimensions. It means that in condition 1.83, the collision efficiency can increase due to attraction forces in comparison with its value caused by the interception. It explains why at condition 1.83, $f > 1$; as can be seen from Figure 1.9, the smaller the N_A, the shorter the distance from the grazing trajectory to the surface of the bigger drop; that is, the smaller the collision efficiency and f.

1.4.7.3 Peculiarities of Gravitational Coagulation at Small Aggregation Numbers

An interesting peculiarity of short-range hydrodynamic interaction is the fact that according to Equation 1.82, it depends to a larger degree on the size of the smaller particle (power 0.7) than on the Hamaker constant (power 1.6). As can be seen from this formula and from Figure 1.8, a decrease in the Hamaker constant is accompanied by a decrease in collision efficiency. But this dependence is very weak; and so, the impression arises that coagulation is possible at any small (but finite) value of the Hamaker constant.

This impression is most likely incorrect. With an increasing size of the smaller drop, the stability of the doublet is attained at the cost of the closest separation between two droplet surfaces, h. In this case, it is assumed that the attractive force infinitely increases. The infinite growth of attractive force with decreasing h cannot be proved due to difficulties in measuring surface forces at very small distances [134].

The initial model of surface forces with an infinitely deep primary energy minimum in DLVO theory was replaced by a model with the potential well of a finite depth in the works by Frens and Overbeek [51] and Martinov and Muller [135]. This has resulted in the abandonment of the notion of completely irreversible coagulation and in the development of important notions of reversible coagulation. In this context, the paradoxical result can be discussed, which points to the possibility of gravitational coagulation at an arbitrarily small but finite value of the Hamaker constant. Hydration effects and/or structural components of disjoining pressure (Section 1.2.4) which will further restrict the possibility of gravitational coagulation should be taken into consideration in the future.

The mentioned factors restricting gravitational coagulation manifest themselves when we consider conditions 1.73 and 1.80, which in the absence of the repulsive force can take the form

$$\frac{aa_1}{6}h_m^2 = 6\pi\mu a_1\left(4.8V_2\frac{a_a^2}{a_2^2} - V_1\right) = (3.8)6\pi\eta a_1 V_2\frac{a_1^2}{a_2^2} = 2.2\pi\Delta\rho g a_1^3 \tag{1.84}$$

The first term on the right-hand side of the equation is the detaching hydrodynamic force and the second one is the gravity force.

Equations for the absolute values of the detaching force and the pressing force are identical. The normal component of liquid flow has different directions in the vicinity of the leading and rear hemispheres and causes the pressing force on the leading hemisphere and the detaching force on the rear hemisphere. Consequences of Equation 1.84 on emulsions will be discussed in the next section.

1.4.7.4 Estimation of the Size of Emulsion Drops Stable Against Flocculation with Larger Drops

As was pointed out in Section 1.4.7.2, the value of the Hamaker constant in oil–water (o/w) emulsions is approximately known. In the case of emulsion-stabilized surfactants, we obtain the lower bound of h also, i.e., h_m. In this case, h_m exceeds the double thickness of the adsorption layer h_a. Then for the critical radius act of the smaller of the drops it follows from Equation 1.84 that

$$a_{1cr} \leq \left(\frac{A}{2.2\pi\Delta\rho g}\right)^{1/2}(2h_a)^{-2} \tag{1.85}$$

Drops with a radius exceeding a_{1cr} cannot flocculate with drops of a much larger radius, $a_2 \gg a_{1cr}$. Substituting the values $A = (3\times10^{-21} \div 10^{-20})$, $\Delta\rho = 0.1\ \mathrm{g\,cm^{-3}}$, $2h_a \cong 2\times10^{-7}$ cm into Equation 1.85, we obtain $a_{1cr} \leq 100$ to 300 μm.

It is that h_a can be both larger and smaller than the estimated value when using emulsifiers of different nature. The condition of a small Reynolds number is poorly fulfilled at this value of a_{1cr} and is completely violated for a larger drop with a radius $a_2 \gg a_{1cr}$. This limits the accuracy to estimate according to Equation 1.85. Nevertheless, it is clear that the detaching force cannot prevent gravitational coagulation in emulsions for drops of Stokesian size if additional repulsion forces are absent because of the small value of $\Delta\rho$. In suspensions, $\Delta\rho$ can be 10 to 70 times higher and a_{1cr} can be 3 to 8 times smaller. Thus, the statement [106] cited in Section 1.4.7.1 that molecular attractive forces cannot result in coagulation of particles of a larger size is to some extent confirmed. But at a radius $a_1 < a_{1cr}$ even very small molecular forces provide coagulation which was underestimated in Ref. 106.

1.4.8 Experiments

There is a lack of systematic experimental investigations of coagulation kinetics in emulsions. This drawback can be compensated in some degree by the consideration of the capture efficiency in adhesion experiments with packed beds and in flotation, because the mechanism of the collision is the same.

1.4.8.1 Experiments with Packed Beds

Experiments with beds of solid spheres have been carried out to establish the applicability of theories based on single collector-to-collector arrays where several transport processes operate and the flow field is not known in detail. Here, filter coefficients are measured instead of single-particle capture efficiencies and a flow structure parameter is used.

An extensive study of nonBrownian particles was carried out by Spielman and Fitzpatrick [90,91] to test theories for attraction-dominated capture. Several hundred experiments were done wherein latex particles were filtered from aqueous suspensions using beds of glass particles. Particle diameters ranged from 0.7 μm to 21 μm, bead diameters from 0.1 mm to 4 mm, and velocities between 0.01 and 0.1 cm s^{-1}. Electrolyte type and ionic strength were varied in order to control electrostatic repulsion and to avoid interference from particle deposits. Filter coefficients were calculated from particle concentrations measured with a Coulter counter.

In Ref. 91, data are shown for conditions in which the electrostatic repulsion and sedimentation are negligible. Although there is a fair amount of uncertainty in the data, the data cluster about the line derived from the theory given by Spielman and Fitzpatrick [90].

At low ionic strengths, where repulsion is strongest, agreement between theory and experiment was poor. This was attributed to the acute sensitivity of the capture rate to small adjustments (~1 mV) to the values of the ζ-potential of the film used in theory; the agreement has improved.

1.4.8.2 Experimental Investigation of Collision Efficiency at a Small Radius Ratio

Equation 1.63, which can be recommended for the collision efficiency in emulsions, was confirmed in systematic investigations devoted to microflotation kinetics. The monodisperse latex or glass spherical particles were used in the experiments. The radius ratio was very small. Thus, in Equations 1.63 and 1.82, a_1 corresponds to the latex particle dimension and a_2 to the bubble dimension. The dependence $E \sim a_2^{-2}$ was confirmed in Refs. 136 and 137. The dependence $E \sim a_1^{1.5}$ was established in Refs. 138 and 139. The combination of Equation 1.63 describing LRHI and Equation 1.82 describing the SRHI influence leads to the dependence $E \sim a_1^{1.4}$. Thus, the theory of short-range hydrodynamic interaction is confirmed by the investigations [134,136–139].

Recently, the direct observation of the grazing trajectory of latex particles (0.9 μm) in the vicinity of the rising bubble with radius 15 μm was accomplished [140,141]. The cell was attached to the microscope stage which is capable of moving vertically in the same velocity as that of the rising bubbles in the cell.

The similar experiments in emulsion can be very important. The predicted particle trajectory agreed well with those obtained experimentally. It was found that the collision efficiency values obtained experimentally were at maximum when the absolute values of ζ-potentials of both the bubbles and particles were at minimum.

1.4.9 Long- and Short-Range Hydrodynamic Interaction and Collision Efficiency at Any Radius Ratio

If the difference between particle dimensions is not large, the dynamics of the interacting spheres have to be represented in terms of the relative and center of mass motion by defining $dr/dt = u_2 - u_1$ and $d\bar{x}/dt = u_{12} = (u_1 + u_2)/2$, where u_1 and u_2 (see Figure 1.10) are the velocities of particles 1 and 2. Every particle moves under the action of the external field and the interaction with the neighboring particle. Thus, its velocity can be represented by the sum

$$u_i = \sum_{i=1}^{2} \omega_{ij} F_j \tag{1.86}$$

with $i = 1,2$. The mobility tensors ω_{ij} express the response of the ith sphere to a force acting on the jth and depend on the separation. The superposition (1.86) is justified by the linearity of the Stokes equation. The set of equations (1.86) can be transformed in the expression for the relative velocity of two sphere centers. Batchelor and colleagues [65,142,143] accomplished this transformation:

$$\mathbf{u}_{12}(\mathbf{r}) = \mathbf{u}_{12}^{(0)}\left[\frac{\mathbf{rr}}{\mathbf{r}^2}L(s) + \left(\mathbf{I} - \frac{\mathbf{rr}}{\mathbf{r}^2}\right)M(s)\right] \tag{1.87}$$

The relative mobility functions for motion along the line of centers (L) and motion normal to the line of centers (M) describe the effects of hydrodynamic interactions between the two drops. These functions depend on λ, $\hat{\mu}$, and s.

When the spheres are very close ($\xi = s - 2 \ll \lambda, \xi \ll 1$) the mobility functions have the asymptotic forms given by Jeffrey and Onishi [144]. Using these asymptotic forms, Davis [145] incorporated them in the equation for the particles' trajectories. Its solution with the boundary condition 1.80 enabled one to determine the grazing trajectory and accomplished the numerical calculations of the collision efficiency. At $\lambda \ll 1$, his results agree with Rulyov's results (Section 1.4.7). Davis' data are plotted in Figure 1.13 for $\hat{\mu} = 1000$.

The kinetics of the doublet formation were also analyzed in Ref. 142 by another method. Authors note the agreements with Ref. 145.

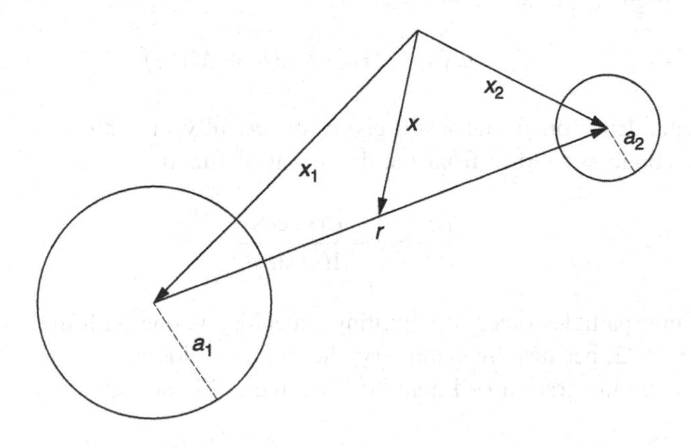

FIGURE 1.10 Two interacting spheres.

1.4.10 Collision Efficiency for Interacting Drops

Zhang and Davis [28] use the Batchelor expression (Equation 1.87) for u_{12} and modify it for spherical drops.

1.4.10.1 Mobility Functions for the Relative Motion of Two Drops

Their derivation [28] of expressions for the function L is based on the equation for the lubrication force on drops in the direction along their line of centers. The force balance enables one to obtain the functions L(s) and M(s) in Equation 1.87. They consider two unequal drops which are nearly touching and which move together as a pair due to gravity. Superimposed on this is a small relative velocity of the larger drop approaching the smaller one. A force balance on each drop yields

$$F_{g,i} + F_{d,i} + F_{l,i} = 0 \tag{1.88}$$

where $F_{g,i}$ is the gravity force acting on the drop i, which can be described by the Hadamard–Rybczynski result, Equation 1.42:

$$F_{g,i} = 6\pi \mu a_i U_i^{(0)} \frac{3\hat{\mu} + 2}{3\hat{\mu} + 3} \tag{1.89}$$

$F_{d,i}$ is the drag force exerted on the drop i by the surrounding fluid and is defined as the total hydrodynamic force minus the lubrication force. According to the analysis of Reed and Morrison [146] for two touching drops, it can be expressed as

$$F_{d,i} = -6\pi \mu a_i U_p \frac{3\hat{\mu} + 2}{3\hat{\mu} + 3} \beta_i \tag{1.90}$$

where β_i is a correction factor of the Hadamard–Rybczynski formula for drop i to account for the presence of the second drop.

Equations 1.88 to 1.90 for $i = 1$ and $i = 2$ may be solved for the pair velocity U_p and the lubrication force, $F_{11} = -F_{12}$. It enables the determination of the coefficients L_1, M_0, and M_1 in asymptotic equations:

$$L\xi = L_1\xi, \quad M(\xi) = M_0 + M_1(\xi) \tag{1.91}$$

The coefficient dependence on $\hat{\mu}$ and λ are given numerically in Table 1 of Ref. 28.

The trajectory equation follows from the definition of functions L and M,

$$\frac{ds}{d\theta} = s\frac{-L(s)\cos\theta}{\mathrm{M(s)}\sin\theta} \tag{1.92}$$

In the absence of interparticle forces, the limiting trajectory is one with the final condition $s = 2$ (contact) when $\theta = \pi/2$, because by symmetry, the point of closest approach occurs at $\theta = \pi/2$. Using the result of the integration of Equation 1.92, the collision efficiency is [28]

$$E_{12} = \exp\left(-2\int_2^1 \frac{M - L}{sL} ds\right) \tag{1.93}$$

With interparticle forces considered, the trajectory equation can no longer solve L analytically for an explicit formula for the dimensionless critical impact parameter, and hence the collision efficiency. Instead, the determination of the collision efficiency has to be performed by integrating Equation 1.92 numerically along the limiting trajectory from the infinite separation of two drops to the termination point.

The dimensionless critical impact parameter may be determined by integrating Equation 1.92 backward along the limiting trajectory from the termination point $\theta = \pi$ and $\xi = \delta$, to a position $s = s_1$ and $\theta = \theta_f$, beyond which the van der Waals forces are negligible. This numerical solution may be matched with the solution in the outer region ($s > s_f$). Setting $s > s_f$ and $\theta = \theta_f$ as the matching condition reveals that

$$E_{12} = \left\{4\left[s_f \sin\theta_f \exp\left(\int_{s_f}^{\chi} \frac{L-M}{sL} ds\right)\right]^2\right\}^{-1} \tag{1.94}$$

Figure 1.11 shows the results for E changing with λ for several different $\hat{\mu}$ [28]. The results for the collision efficiencies predicted by Zinchenko [68] are presented as solid circles for comparison. There is a very good agreement between the present results and Zinchenko's, with the relative difference between them being smaller than 3%. The *collision efficiency* approaches a finite value as the drops become equi-sized ($\lambda \to 1$). However, the *collision rate* goes to zero in this limit because the relative velocity of the two drops approaches zero. The collision rate may be nondimensionalized with a quantity not involving the size ratio: $J_{12}/J_{12}/(n_1 n_2 u_{01} \cdot \pi a_1^2) J_{12}/(n_1 n_2 u_{01} \pi a_1^2) = E_{12}(1-\lambda^2)(1+\lambda)^2$. This quantity is shown in Figure 1.12.

The collision rate is small for small size ratios because of the reduced collision cross section and collision efficiency (as discussed previously) achieves a maximum at moderate size ratios, and then decreases as the size ratio approaches unity because of the reduced relative velocity.

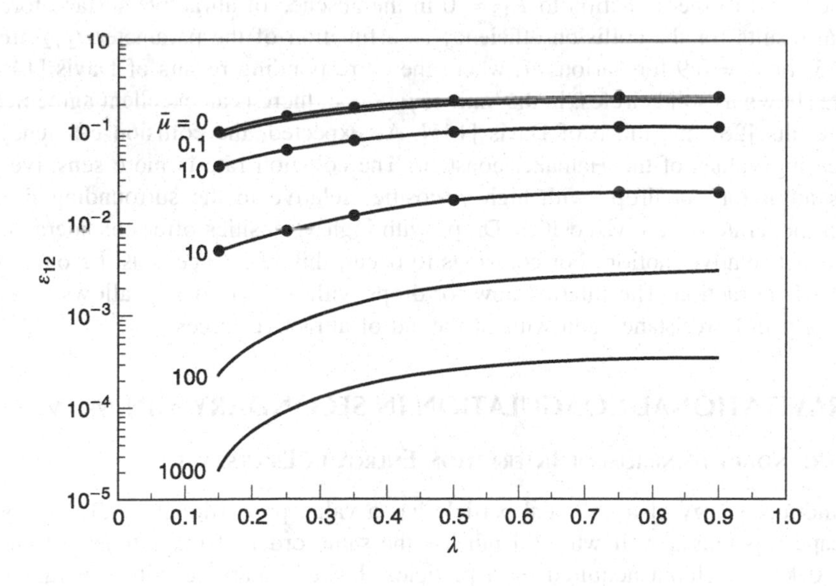

FIGURE 1.11 The dimensionless collision rate, $J_{12}/(n_1 n_2 u_{01} \pi a_1^2) = E_{12}(1-\lambda^2)(1+\lambda^2)$, for gravity sedimentation of drops as a function of the size ratio for various viscosity ratios without interparticle forces. (After X. Zhang and R. Davis, *J. Fluid Mech.* 230: 479 (1991).)

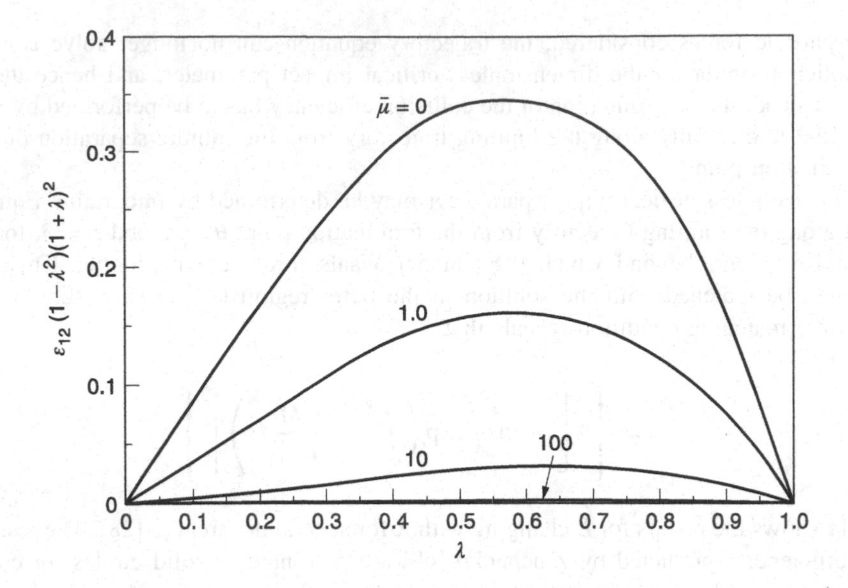

FIGURE 1.12 The collision efficiency for gravity sedimentation of drops as a function of the size ratio of various viscosity ratios without interparticle forces. (After X. Zhang and R. Davis, *J. Fluid Mech.* 230: 479 (1991).)

The unique peculiarity of results characterized by Figure 1.12 is that the nonzero collision efficiency occurs in the absence of attractive molecular forces. This is possible due to the mobility of droplet surfaces. The mobility decrease caused by liquid viscosity increase leads to a drastic decrease in the collision efficiency. As $\hat{\mu}$ increases, E decreases and its value is unusually low for $\hat{\mu} = 100$ and 1000. The increase in the unrestricted $\hat{\mu}$ corresponds to the transition to the solid particle and to the transition to $E_{12} = 0$ in the absence of attractive surface forces.

Typical results for the collision efficiency as a function of the parameter Q_{12} are shown in Figure 1.13 for $\lambda = 0.9$ for various $\hat{\mu}$, where the corresponding results of Davis [145] for rigid spheres are shown as solid circles. In the limit of $\hat{\mu} \to \infty$, there is an excellent agreement between our new results [28] and those of Davis [145]. As expected, the collision efficiency increases with increasing values of the Hamaker constant. The collision rate is more sensitive to the van der Waals attraction for drops with high viscosities relative to the surrounding fluid than for drops with moderate or low viscosities. Drops with high viscosities offer considerable resistance to near-contact relative motion. For collisions to occur, this resistance must be overcome by the van der Waals attraction. The internal flow for drops with low viscosities allows them to collide with relatively little resistance and without the aid of attractive forces.

1.5 GRAVITATIONAL COAGULATION IN SECONDARY MINIMUM

1.5.1 Secondary Minimum of Interaction Energy in Emulsions

Thermal motion energy of a colloidal particle has a value proportional to kT. The particle will easily escape a potential well whose depth has the same order of magnitude. An energy of the order of 10 kT is seldom acquired by a particle. Hence, a particle will stay rather long in a potential well of this magnitude. The depth of the secondary minimum can be fairly large (i.e., of the order of 10 kT and more) if two conditions are fulfilled simultaneously. The thickness of

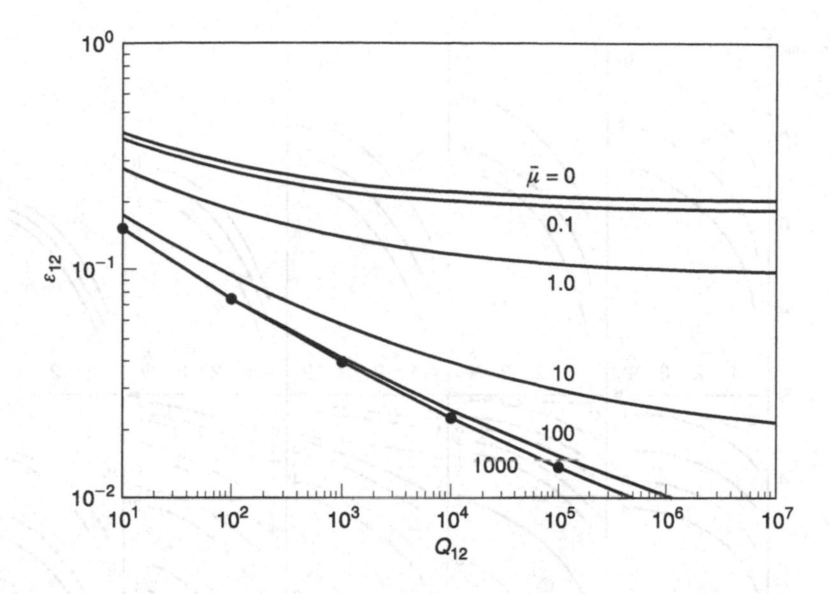

FIGURE 1.13 The collision efficiency for gravity sedimentation of drops as a function of the interparticle force parameter for $\lambda = 0.90$ and various viscosity ratios with unretarded van der Waals attractions. The solid circles are the rigid sphere results of Davis (1984). (After L. Spielman, *J Colloid Interface Sci.* 33: 562 (1970).)

the double layer is small (of the order of nanometers) and the size of the particles is rather large. The necessity of these conditions is clear without further calculations. The secondary minimum appears at the periphery of DL where electrostatic repulsive forces are weakened and molecular attractive forces prevail. Therefore, the thinner the DL, the closer the secondary minimum is to the surface; but the closer the secondary minimum to the surface, the higher the molecular attractive forces. A calculation shows that the secondary minimum is localized at a distance of several Debye radii (Figure 1.14). Then its distance to the surface is of the order of 10 nm as applied to conditions of interest to us. These conditions are realized rather often in natural colloidal and technological processes. Indeed, electrolyte concentration usually is fairly high under technological conditions and it can be equal to several centimols in 1 liter, which corresponds to a DL thickness of 3 nm and more.

The depth of the minimum strongly depends on the value of the Hamaker constant and on the size of particle. The energy due to attractive forces and electrostatic energy (and therefore energy as a whole) is proportional to particle size. As can be seen from Figure 1.15, the depth of the secondary minimum expressed in kilotesla units quickly decreases with the size of particles. Long-range aggregation is possible for particles in colloidal dispersions only for very high values of the Hamaker constant (i.e., for metal particles). On the other hand, long-range aggregation for large particles in microheterogeneous systems ($a > 1$ μm) is carried out over a wide range of Hamaker constants.

When the charge increases, the electrostatic energy of repulsive forces increases, the primary minimum disappears, and the height of the repulsion barrier grows. The electrical contribution to primary coagulation is very substantial. It is practically impossible to prevent long-range coagulation by increasing electrostatic repulsion (at real high electrolyte concentrations). The electrostatic factor of stability is strong with respect to short-range aggregation and is weak with respect to the long-range one. In colloidal systems, it dominates over other factors because

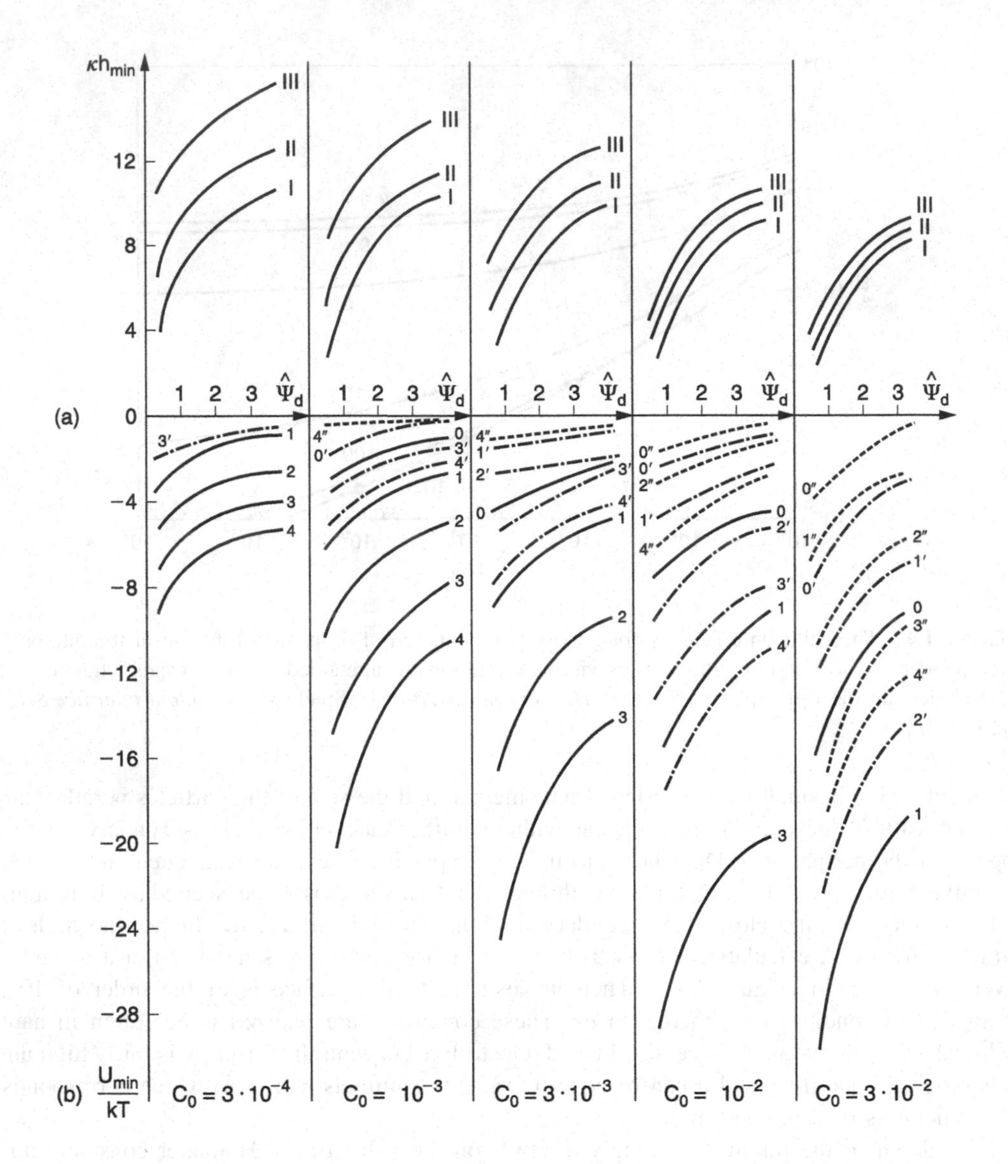

FIGURE 1.14 Dependence on the Stern potential for the coordinate (a) and the depth (b) of the secondary minimum: I, 0–4 without the retardation of screening; II, 0′–4′ with account for retardation; III, 0″–4″ with account for retardation and screening (0-a = 0.5 μm; 1-a = 1 μm; 2-a = 2 μm; 3-a = 3 μm; 4-a = 4 μm). The electrolyte concentrations [M] are shown at the bottom.

the depth of the secondary minimum is small and long-range aggregation is impossible. The electrostatic factor appears as a weak one in microheterogeneous systems when the depth of the secondary minimum is large and slightly depends on particle charge so that it does not prevent the long-range aggregation.

The depth of the secondary minimum was calculated from the expression for the total interaction energy between a spherical particle and a planar surface or (what is the same) with a

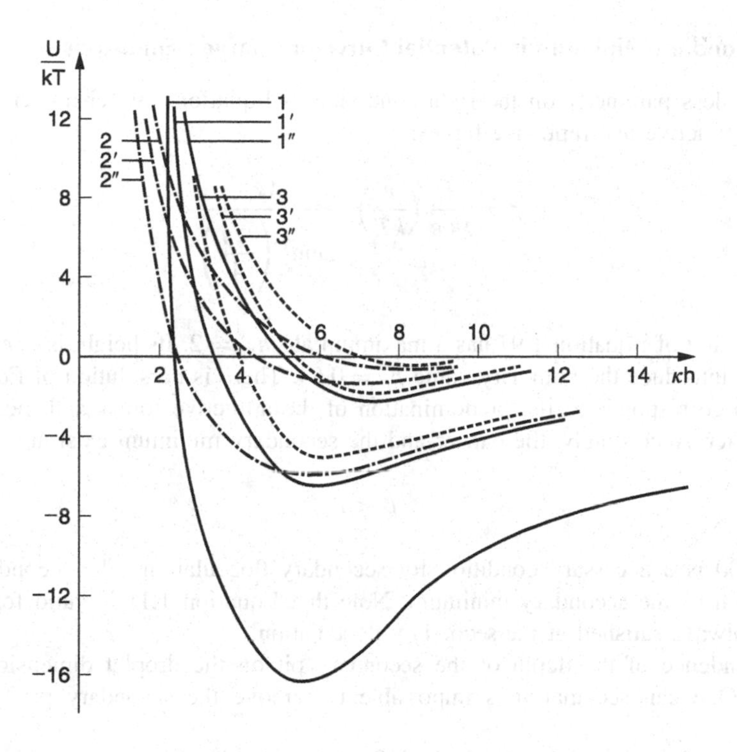

FIGURE 1.15 Pair interaction energy versus κh: 1, 1′, 1″ — $\tilde{\Psi}_d = 1$; $c_0 = 10^{-2}$ mol l^{-1}, 2, 2′, 2″ — $\tilde{\Psi}_d = 0.5$; $c_0 = 10^{-3}$ mol l^{-1}, 3, 3′, 3″ — $\tilde{\Psi}_d = 1$; $c_0 = 10^{-3}$ mol l^{-1}, 1, 2, 3: without retardation and screening; 1′, 2′, 3′: with account of retardation; 1″, 2″, 3″: both screening and retardation are taken into account.

particle of a much larger size. With regard to Equations 1.1 and 1.7, this expression has the form

$$y = de^{-\kappa h} - B\frac{\kappa a}{\kappa h} \tag{1.95}$$

where

$$y = \frac{U}{kT}, \quad U = U_A + U_\Psi, \quad B = \frac{A}{6kT}, \quad d = \frac{\varepsilon\Psi^2 a}{kT} \tag{1.96}$$

From $y = 0$ we can find the equation for κh_m:

$$(\kappa h_{\min})^2 \exp(-\kappa h_{\min})^2 = \frac{B\kappa a}{d} = \beta \tag{1.97}$$

This equation has two solutions, one of which corresponds to the "bottom" of the distant well and the other to the "top" of the barrier. The expression for the depth $U_{\min}$ of the secondary minimum follows from Equations 1.95 and 1.96:

$$-\frac{U_{\min}}{kT} = \frac{A\kappa a(\kappa h_{\min} - 1)}{6kT(\kappa h_{\min})^2} \tag{1.98}$$

1.5.1.1 Secondary Minimum in Potential Curve of Charged Emulsions

The dimensionless parameter on the right-hand side of Equation 1.97 characterizes the balance between the attractive and repulsive forces:

$$\beta = \frac{1}{96\varepsilon}\left(\frac{e}{kT}\right)^2 \frac{A\kappa}{\tanh^2\left(\frac{\tilde{\Psi}_d}{4}\right)} \tag{1.99}$$

The left-hand side of Equation 1.97 has a maximum at $\kappa h_m = 2$. Its height is $2^2 e^{-2} = 0.54$. It is convenient to introduce the boundary value $\beta_b = 0.54$. There is no solution of Equation 1.97 for $\beta > \beta_b$. This corresponds to the predomination of the attractive forces and the absence of the energetic barrier. Accordingly, the barrier and the secondary minimum exist at

$$\beta < \beta_b \tag{1.100}$$

Equation 1.100 is a necessary condition for secondary flocculation. The second condition is a sufficient depth of the secondary minimum. Note that Equation 1.11 is valid for $\kappa h > 2$. This condition is always satisfied at the secondary flocculation.

The dependence of the depth of the secondary pit on the droplet dimension is plotted in Figure 1.14. One can see that it is impossible to remove the secondary pit by the potential increase.

According to Equations 1.1 and 1.7, the influence of the droplet dimensions is identical. Thus, Equation 1.95 can be generalized by multiplying by $\lambda/(1+\lambda)$. The localization of the secondary minimum does not depend on λ. However, the equation for its depth of minimum has to be generalized:

$$\tilde{U}(\lambda) = \frac{\lambda}{1+\lambda} U|_{\lambda=0} \tag{1.101}$$

where $U|_{\lambda=0}$ is given by Equation 1.95. Note that the depth of the potential pit is a factor of 2 smaller for identical droplets in comparison with the case $\lambda \ll 1$.

At lower electrolyte concentrations and higher potentials, the distance of the secondary minimum from the surface is larger by 5 to 10 nm. Thus, the retardation of the dispersion forces (Section 1.2) influences the localization of the secondary minimum and the depth of the potential pit (Figures 1.14 and 1.15).

1.5.1.2 Repulsive Hydration Forces and the Secondary Minimum

Similar to the electrostatic energy, hydration forces decline (Section 1.2) exponentially with the distance. Thus, the conditions for the formation of the secondary potential pit are identical. This identity exists even on the quantitative level because the same function describes the dependence on the distance for electrostatic and hydration forces. Due to this similarity, the introduction of the dimensionless ratio h/h_s, instead of κh, enables one to derive an equation similar to Equation 1.98:

$$\tilde{U}_{\min} = A\frac{a}{h_s}\frac{h_{\min}/h_s - 1}{6kT(h_{\min}/h_s)^2} \tag{1.102}$$

where h_{min} satisfies the equation similar to Equation 1.97:

$$\left(\frac{h_{min}}{h_s}\right)^2 e^{-(h_{min}/h_s)} = \beta_s = \frac{A}{6\pi h_s^3 K_s} \quad (1.103)$$

β_s characterizes the balance between the attractive and the hydration repulsive forces. Because Equations 1.97 and 1.103 are identical, we can conclude that the hydration forces can produce the barrier and the secondary minimum at the condition

$$\beta_s < \beta_{sb} \quad (1.104)$$

which is identical with condition 1.100.

If K_s and h_s are not sufficiently large, the hydration forces cannot give rise to the barrier and the secondary minimum. The manifestation of the hydration repulsive forces was found in the investigation of emulsion stability in Ref. 147. The substitution of the values of K_s and A determined in Ref. 147 and $h_s = 1$ nm, assumed by the authors, into Equation 1.103 yields β_s value exceeding β_{sb}. However, Equation 1.103 for β_s is written for an interaction at small λ, whereas the equation for any λ is used in Ref. 147. For $\lambda \ll 1$, the multiplier 0.5 has to be introduced into Equation 1.103. After this correction, it turns out that the experimental data in Ref. 147 agree with condition 1.104. The information concerning K_s and h_s values in Ref. 147 agrees with condition 1.104. The data cited in Ref. 31 confirm the importance of repulsive hydration forces in the formation of the energy barrier and secondary potential pit. These forces were introduced to explain the forces observed between bilayers in lamellar phases [148,149]. In Ref. 31, the data concerning h_s for different phospholipids, lipids, and a few surfactants are summarized in Table 14 of Ref. 31. The table shows that $h_s = 2$ to 3 nm for phosphatidylcholins [150], about 2 nm for phosphatidylethanolamine, and approximately 1.5 nm for glycerol groups on alkoxyglycerol surfaces. The K_s values change in broad range.

In Ref. 31, the authors emphasize that the surface charge in food emulsions is low, electrolyte concentration is high, and, hence, the DL is not responsible for the emulsion stability. The stabilization can be caused by the repulsive hydration forces. However, the secondary flocculation preserves.

1.5.2 Estimation of the Size of Drops Not Coagulating in the Secondary Minimum. Sedimentation–Hydrodynamic Stability Mechanism of Microheterogeneous Dispersed Systems

The important question of a critical size of drops not undergoing a coagulation in the secondary minimum can be considered to be very simply based on the balance of forces at the rear stagnant pole of a larger drop (Section 1.4.7.4). Most likely, this effect was considered for the first time when applied to flotation [19].

It was shown in Ref. 19 that only particles of a rather small size can float, at the cost of coagulation in the secondary minimum. The larger the particle, the deeper the secondary minimum and the larger the gravity force. Gravity forces increase with increasing particle size faster than the depth of the secondary minimum. Therefore, coagulation of rather large particles in the secondary minimum is impossible. The failure of long-range aggregation for rather large

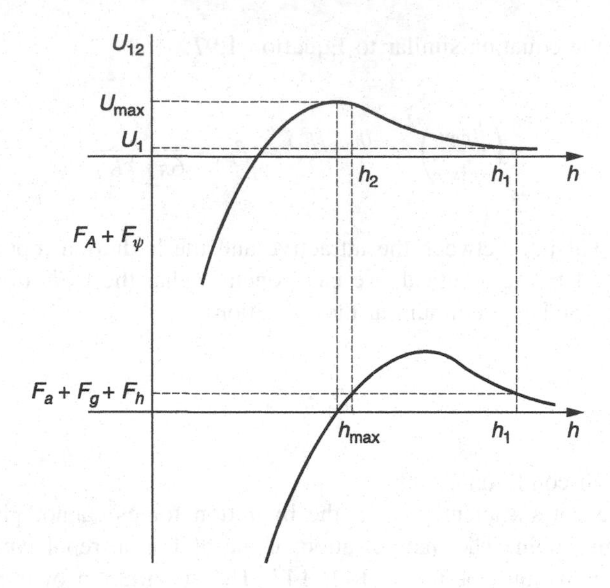

FIGURE 1.16 Illustrative curves representing the dependencies of surface forces $F_A + F_\Psi$ and corresponding energy U_{12} on h. The points h_1 and h_2 are characterized by the equality of the surface forces and the bulk forces $F_a + F_h + F_g$. (After A.S. Dukhin, *Kolloidn. Zh.* 50: 441 (1988).)

particles (microheterogeneous disperse systems) is an experimentally established fact considered by Kruyt [151], although he provided no explanation for the phenomenon.

The notion of a sedimentation–hydrodynamic mechanism of aggregate stability (SHAS) of microheterogeneous systems was introduced in Ref. 152. The importance of this phenomenon for classification of powders is shown in Ref. 153.

The dependence of particle interaction force on h is obtained by differentiating the known curve of total interaction energy (Figure 1.16). The decrease in the curve to the right of the minimum of total interaction force F is caused by the decrease of each of the forces with distance. The decrease at the left is caused by the faster increase in repulsive forces than in the attractive forces. The minimum size of the particle is determined from the condition of equilibrium between the pull force and the maximum force of attraction between particles in the floc:

$$F_A(h_{f\,\min}) + F_\Psi(h_{f\,\min}) = F_h + F_g + F_a \tag{1.105}$$

As mentioned in Section 1.5.1, energy due to molecular attractive forces predominates in the secondary energy minimum. This is true to a still larger extent in the secondary force minimum because $\kappa h_{f\,\min} > \kappa h_{\min}$ (see Figure 1.16). Therefore, the electrostatic repulsive force in Equation 1.95 can be omitted. In view of a large distance from the surface, the molecular attractive force in the region of the secondary force minimum should be calculated with regard to electromagnetic delay, that is, by Equation 1.4. F_g and F_a in Equation 1.105 are the gravity and the Archimed force, respectively, for the smaller of the particles; F_h is the detaching hydrodynamic force. Taking into consideration the aforementioned, the formula follows from

Equation 1.105 [152]:

$$a_{1cr} = \sqrt{\frac{B}{2\Delta\rho g (h_{f\min})^3} F(\lambda)},$$
$$F(\lambda) = \sqrt{\frac{\lambda^2 \left[\lambda\Lambda(\lambda) + \Lambda(\lambda^{-1})\right]}{(1+\lambda)\left[\Lambda(\lambda) - \lambda^2\Lambda(\lambda^{-1})\right]}} \tag{1.106}$$

where $\kappa h_{f\min}$ is the greater of two possible positive roots of the analogue to Equation 1.97,

$$(\kappa h_{0\min})^4 e^{-\kappa h_{0\min}} = \frac{2\pi\kappa^2 B}{\Psi^2 \varepsilon} \tag{1.107}$$

and $B = \lim(A(h))_{h\to\infty}$.

A very complicated function $\Lambda(\lambda)$ was obtained in Ref. 154. Its graphical and asymptotic representations are given in Ref. 155. The plot of the function $F(\lambda)$ is given in Ref. 155. The calculation is performed for the values $B = 2.5 \times 10^{-20}$ J m, $\Psi = 25$ mV, $c = 10^{-3}$ M, and $\lambda = 0$ (i.e., approximately at $a_2 \gg a_1$; $a_{1cr} = 1$ μm corresponds to this case). If we increase λ to 0.9 and make it very close to a_2, $a_{1cr} = 4.2$ μm at $a_2 = 4.2$ μm. These calculations have been performed for titanium carbide powder.

Using formula 1.106, the graphical representation of function $F(\lambda)$, and the calculated value of $a_{1cr}|_{a_{1cr} \ll a_2}$ we obtain curve $a_{1cr}(a_2)$ as presented in Figure 1.17. At any fixed a_2 particles with radius $a_1 > a_{1cr}(a_2)$ cannot aggregate with particles with radius a_2. Only for particles with radius $a_1 < a_{1cr}(a_2)$ is this possible.

Let us consider a region in Figure 1.17 bounded below by curve $a_{1cr}(a_2)$ and by bisectors $a_1 = a_2$. Each point corresponds to a pair of particles with radii a_1 and a_2. For a pair of particles belonging to this region, secondary flocculation is impossible due to the sedimentation–hydrodynamic mechanism of stability. The position of the boundary of this region [i.e., of curve

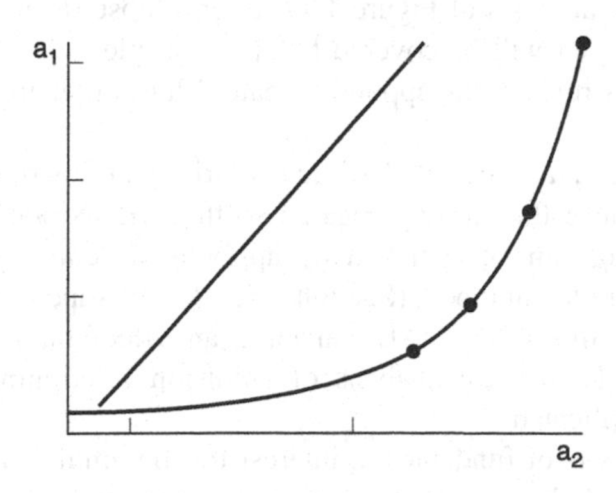

FIGURE 1.17 Flocculation in the secondary minimum is impossible for a pair of particles belonging to the region between bisectors $a_1 = a_2$ and curve $a_{1cr}(a_2)$.

$a_{1cr}(a_2)$ depends on values of $B]\Delta\rho$, electrolyte concentration, and Stern potential according to Equations 1.106 and 1.107.

1.5.3 Wet Fractionation Method and Sedimentation–Hydrodynamic Mechanism of Aggregate Stability

It is apparent that aggregation prevents powders from separation. The sedimentation–hydrodynamic aggregative stability mechanism (SHAS) prevents particles of rather large size from aggregation and thus creates necessary conditions for separation of narrow fractions. This explains the substantial success in the classification of wet powders [156]. The difference in sedimentation rate for particles of different size is used in wet classification. In the light of results obtained in the preceding section, pairs of particles belonging to region I (Figure 1.17) can be separated in principle using this method. We will divide the region under curve $a_{1cr}(a_2)$ into two zones. Zone 2 in which either a_1 or a_2 is reasonably small is characterized by a small depth of the remote potential well. Flocculation does not occur in zone 2. The boundary between zones 2 and 3 is fuzzy because the lifetime of a doublet due to secondary flocculation depends on the size of particles.

Pairs of particles corresponding to points in zone 2 are subjected to secondary aggregation. Liquid classification according to the method above is difficult. This difficulty is overcome by centrifuging. As can be seen from Equation 1.106, an ordinate of curve $a_{1cr}(a_2)$ decreases by a factor of 10 when we apply centrifugal acceleration of $100 \times \mathbf{g}$.

1.5.4 Peculiarities of Secondary Flocculation in Emulsion and Their Fractionation

According to Equation 1.106, a_{1cr} for the emulsion is $\sqrt{\Delta\rho_{TiO_2}/\Delta\rho_{em}}$ times higher than in the case of titanium carbide (i.e., approximately 7 μm at $a_1 \ll a_2$. At $\lambda = 0.9$, $a_{1c} \sim 30\,\mu\text{m}$ and $a_2 = 33\,\mu\text{m}$). Thus, due to the small $\Delta\rho$, emulsions are subjected to secondary flocculation to a substantially larger extent than microheterogeneous solid suspensions.

Nevertheless, the SHAS mechanism can presumably have a strong effect on secondary flocculation in emulsions and creaming and deserves systematic studies.

Considerations set forth in Sections 1.5.1–1.5.3 relate to the initial stage of flocculation. In going from doublets to multiparticle aggregates, the mechanism behind SHAS becomes more complicated.

In accordance with the analysis of Figure 1.17, after a lapse of time, the rear stagnant part of the surfaces of the big droplet will be covered by small droplets. It is apparent that applicability of Equation 1.106 needs a revision as applied to paired hydrodynamic interaction of such large particle aggregates.

Differing from powders, a substantial difficulty arises in fractioning micron-size drops as applied to emulsions because they are aggregated so that SHAS does not function in this case. It is likely that fractioning can be achieved by applying a centrifugal field. A change to the regime of fairly large Reynolds numbers (Section 1.4.5.3) and supercritical Stokes numbers (Section 1.4.4) is possible in this case. DAL can have an effect on interaction of drops in this situation (Section 1.4.6). Theoretical analysis of emulsion fractioning in a centrifugal field is thereby substantially complicated.

This range of questions is of fundamental interest for the emulsion stability problem, because in this way, it will most likely be possible to obtain narrow fractions of drops. This will allow us to apply the flocculation investigation methods to emulsions which have proved to be advantageous in studying flocculation kinetics of monodispersed systems (Section 1.3.4).

1.5.5 Gravitational Coagulation Kinetics in the Secondary Minimum

1.5.5.1 General

Flocculation of coarse emulsions in the secondary minimum proceeds at any electrokinetic potential and at fairly high electrolyte concentrations or even at low electrolyte concentrations when combined with thicker DL and small surface charge density (Section 1.5.1). In other words, secondary coagulation of coarse emulsions is more a rule than an exception. Because of the large drop size, this process is mainly a gravitational coagulation. For some intermediate drop size, Brownian coagulation might not be excluded; but gravity coagulation predominates for large drop sizes.

If we consider a general case of emulsion containing also small drops, the posed question has great importance. Secondary flocculation will be of minor importance for small drops because the depth of the secondary potential well is small. But the coarse part of drops will be involved in the secondary coagulation process, and after a period of time, the system will be represented by doublets rather than by singlets.

The formulated problem, in fact, defines concretely the problem of the initial stage of flocculation. Of course, this problem is presented in the general formulation and in general equations considered in articles and in the review by Melik and Fogler [15], but it is not treated as a separate problem. In a series of studies by Batchelor and Wen [143], the effect of surface forces on gravity coagulation was restricted to attractive forces. The same holds for works by Zhang and Davis [28] and Davis [145]. However, Spielman and Cukor [157] pay attention to the gravitational coagulation in the secondary minimum.

Derjaguin et al. [19], Rulyov [132,133], and Rulyov, Dukhin, and Semenov [155] raise the question about the importance of gravity coagulation in the secondary minimum. The necessary condition imposed on drop size is formulated [28]. The equation and the boundary condition are derived for finding the limiting trajectory but the equation is solved under the assumption of lack of repulsive forces, that is, in accordance with Refs. 28 and 143. Omitting the electrostatic component, Rulyov [133] was able to obtain an exact analytical solution of the equation. The inclusion of this term along with attractive forces excludes the possibility of an exact solution. However, an approximate method for analytical consideration of the problem can be proposed.

1.5.5.2 Similarity in Collision Efficiencies of Primary and Secondary Gravitational Coagulation

Secondary gravitational coagulation occurs in the force well of a larger drop and is correspondingly caused by the transport of the smaller droplet.

The larger the drops, the smaller the part of the well occupied by them. The drops of a finite dimension distribute symmetrically with respect to the maximum of the surface force h_{fm}. The maximum distance for the droplet of a finite dimension corresponds to the balance of the attaching surface force and detaching hydrodynamic force (see Equation 1.105).

The boundary condition for the grazing trajectory of the droplet is expressed by Equations 1.79 and 1.80, in which h_0 has to be substituted. The equation for the grazing trajectory in the case of the secondary flocculation differs from the treatment in Section 1.4.7 due to the additional repulsive electrostatic force. However, the electrostatic force can be neglected, as seen below.

Take into account that the electrostatic force is κh_m times smaller at the force maximum than the attraction force and that $\kappa h_{fm} \gg 1$ (Section 1.5.1). The distance to the surface increases along the grazing trajectory when the angle θ decreases. Thus, the maximum error in neglecting the electrostatic force corresponds to the point $\theta = \pi, h = h_0(a_1)$. Neglecting the electrostatic force corresponds to an increase in the molecular force or in the Hamaker constant by the multiplier $(1-\kappa h_{fm}^{-1})$. This approximation causes a negligible error in the collision efficiency because its dependence on the Hamaker constant is very weak. Taking into account Equation 1.98, one concludes that neglecting the electrostatic force is equivalent to omitting the multiplier $(1-\kappa h_{fm}^{-1})^{0.15}$. This estimation concerns the drop of the critical dimension a_{1cr} (Section 1.5.2) which occupies only one point in the well corresponding to force maximum. At $a_1 < a_{1cr}$, $h_0(a_1)$ exceeds h_{fm}. It means that the ratio of the electrostatic force to the attraction force for $a_1 < a_{1cr}$ is less than for a_{1cr}. Thus, for drops with radius less than a_{1cr}, the error decreases.

One concludes that Equation 1.94 (and, correspondingly, Figure 1.13) describes the collision efficiency of the gravitational coagulation in the remote well too. However, there is a large difference in the application of Equation 1.94 to the collision efficiency in the case of uncharged droplets (Section 1.4) and charged droplets (Section 1.5). This concerns the boundaries of the application of the equation. For the uncharged droplets, Equation 1.94 is valid for any $a_1 \ll a_2$. For the charged droplets, Equation 1.94 is valid for the case

$$a_1 < a_{1cr}(a_2) \tag{1.108}$$

1.6 GRAVITATIONAL COAGULATION IN THE PRIMARY MINIMUM

The pressing hydrodynamic force (Section 1.4.7) produces the pressure excess in the gap between sedimenting drops and causes the liquid to flow from the gap between them. The hydrodynamic pressing force can exceed the force barrier of the disjoining pressure, thereby allowing the possibility of coagulation in the primary minimum. This problem may be examined when considering the drop motion along their symmetry axis [19]. Under this condition, the hydrodynamic pressing force is a maximum; hence, one can obtain the necessary and sufficient condition for the primary gravitational coagulation of drops.

Along the axis of a drop pair, the tangential flow velocity is equal to zero. Thus, the duration of the deposition process can be indefinitely long. This means, in turn, that the viscous resistance of the interface film can be neglected in the balance of acting forces. It leads to an equation similar to Equation 1.105. The difference concerns the pressure hydrodynamic force instead of the detaching hydrodynamic force. Both forces are expressed by the same equation but of different sign.

Equation 1.105 indicates that for all values of h, the smaller droplet is subjected to a force directed toward the surface of bigger drop (front pole). Otherwise, the deposition and coagulation cannot occur. The same expression imposes limitations in the values of the parameters at which the disjoining pressure can be overcome. We represent it as the critical radius of the small drop for the gravitational primary coagulation. It occurs at

$$a_1 > a_{1cr}^{pc}(a_2) \tag{1.109}$$

where a_{1cr}^{pc} is the smaller root of Equation 1.107. Instead of only electrostatic forces, the repulsive hydration forces have to be considered too (Section 1.5.1.2).

1.7 CLASSIFICATION OF REGIMES OF GRAVITATIONAL COAGULATION

There are two regimes in the absence of repulsive forces. At condition 1.109,

$$a_1 < a_{1cr},$$

coagulation occurs. Due to SHAS (Section 1.4.7.4), there will be no repulsive forces in the opposite case.

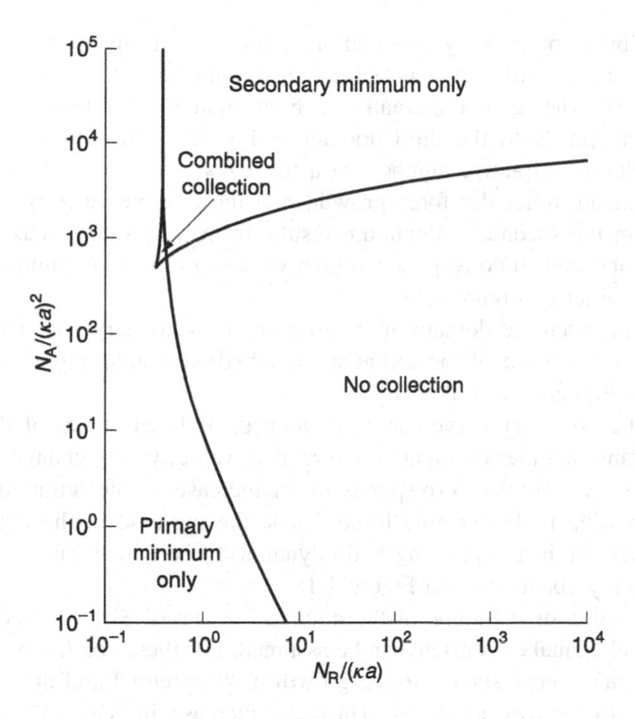

FIGURE 1.18 Stability diagram for a spherical collector showing the influence of electrostatic repulsion at constant potential. (After W.R. Russel, D.A. Saville, and W.R. Schowalter, *Colloidal Dispersions*, Cambridge University Press, Cambridge, 1989, Chap. 11.)

There are three regimes in the presence of repulsive forces: the primary coagulation, the secondary coagulation, and lack of coagulation. One can see in Figure 1.18 that at $a_1 > a_{1c}^{\det}(a_2)$ detachment occurs, at $a_1 < a_{1cr}^{pc}$ primary coagulation occurs, and the condition of the secondary coagulation is

$$a_{1cr}^{pc}(a_2) < a_1 < a_{1cr}^{\det}(a_2) \tag{1.110}$$

Again, Spielman and his co-workers [129,157] were first to classify the regimes of gravitational coagulation in systems of dimensionless criteria. In order to characterize the role of the attractive and the repulsive forces, they used the attraction Equation 1.99 and the repulsion numbers:

$$N_R = \frac{6\varepsilon\Psi^2 a}{A} = \frac{\kappa a}{16\beta}\left(\frac{\tilde{\Psi}}{th\tilde{\Psi}/4}\right)^2 \tag{1.111}$$

where β is given by Equation 1.99. The result is displayed in Figure 1.18 which is also presented in Ref. 110 (Figure 11.8 in that reference).

Naturally, the primary coagulation occurs at a sufficiently low Ψ potential when the electrostatic barrier is absent. It corresponds to low values of the repulsive number and to the left domain in Figure 1.18. The growth in Ψ and N_R leads to a growth in electrostatic barrier and excludes the possibility of primary coagulation. However, at any high value of the repulsive number, the secondary coagulation is possible, corresponding to the domain in the higher right corner of Figure 1.18. The second alternative at high repulsive numbers is the absence of coagulation, which corresponds to the third domain in Figure 1.18. Which alternative is realized depends on the value of attractive number. At a fixed N_R value (i.e., at fixed Ψ and a values), N_A increases the mean molecular force growth, and the relative velocity of droplet decreases. Drop coagulation in the secondary minimum results from a flow too weak to carry the droplet through the secondary well. It corresponds to high values of attractive numbers and to the domain in the upper right corner of Figure 1.18.

The boundary between the domain of stability and of secondary coagulation is given by the curve of the critical condition of the sedimentation–hydrodynamic aggregative stability [SHAS (Section 1.5.4) and Equation 1.106].

The slope of the boundary curve can be explained. At fixed values of the drop dimensions, the Hamaker constant, and the Ψ-potential, the relative velocity only changes along the boundary curve. The decrease in velocity corresponds to an increase in the attraction number and to a decrease in the pressing hydrodynamic force. Thus, the increase in the attraction number corresponds to the decrease in the pressing hydrodynamic force which causes the shrinkage of the domain of the primary coagulation in Figure 1.18.

For an interpretation of the slope of the boundary curve of SHAS, fixed droplet dimension, relative velocity, and Hamaker constant can be assumed. For these conditions, N_A is invariant and the growth in $N_R(\kappa a)^{-1}$ corresponds to the growth in Ψ-potential and the decrease in attraction force between the droplets in a doublet. Thus, the increase in $N_R(\kappa a)^{-1}$ is favorable for the sedimentation–hydrodynamic stability; it explains the shrinkage of the secondary coagulation domain in Figure 1.18 upon growing $N_R(\kappa a)^{-1}$ values.

The evaluation of the coagulation kinetics for a specific emulsion starts with the determination of the stability diagram, that is, the calculation of the boundary curves for the primary and secondary coagulation. The attraction number for emulsions can be specified by expressing the relative velocity by means of the Stokes equation and using the known value of the density difference $\Delta\rho$, that is,

$$N_A = \frac{A}{2\pi g \Delta\rho a^4}, \tag{1.112}$$

$$N_A(\kappa a)^2 = \frac{A\kappa^2}{2\pi g \Delta\rho a^2},$$
$$N'_R = N_R(\kappa a)^{-1} = \frac{3\varepsilon\Psi^2}{4\pi A\kappa} \tag{1.113}$$

In emulsions, the value of A changes in a narrow range and the information concerning $\Delta\rho, \kappa$, and $\Psi \cong \varsigma$ is usually available too. For a specific emulsion, $\Delta\rho$, κ, Ψ, and A are invariants. However, a can vary over a broad range.

The evaluation of N'_R enables one to discriminate between the gravitational stability and the gravitational coagulation. For sufficiently high ς-potentials and sufficiently low electrolyte concentrations, the gravitational stability or secondary coagulation will take place. These two regimes can be discriminated by the N'_R evaluation. As the electrolytic concentration increases, the secondary minimum approaches the surface, which is accompanied by an increase in the molecular attraction. Thus, experimental evaluation of the N_R and N_A criteria enables one to determine both boundary curves and identify the coagulation regime. Depending on the values of the parameters, the regimes of primary or secondary coagulation or gravitational stability can be identified. However, these simple regimes exist for emulsions with narrow drop size distributions. Due to the strong dependence of N'_R on real emulsions with a broad droplet distribution, mixed regimes will occur. For the largest droplets, SHAS becomes of increasing importance and droplets satisfying condition 1.110 participate in secondary coagulation.

The stability diagram (Figure 1.18) does not account for Brownian coagulation. In principle, Brownian coagulation manifests itself at any value of the repulsion number.

For an application of the kinetic theory of emulsion stability, the stability diagram is extremely important. Unfortunately, the exactness of the boundary curves plotted in Figure 1.18 is rather low. The boundary curve of primary coagulation reflects the interaction constant of both charge and constant potential. The curve in Figure 1.18 is plotted for an invariant potential. General criteria for the discrimination between the two cases are known (Section 1.2) and formulated on the basis of the Debora number. However, a quantitative determination is difficult due to the lack of information concerning the charge relaxation.

The exactness of the boundary curve of SHAS is limited due to the restriction in knowledge concerning the retardation of the van der Waals attraction (Sections 1.2 and 1.5.4).

The boundary curve in Figure 1.18 is calculated for the extreme case of $\lambda = 0$. The incorporation of the multiplier $\Lambda(\lambda)$ (Sections 1.4.9 and 1.5.2) into the equation for the balance of the hydrodynamic detaching force and attractive surface force equation (1.110) enables a generalization of the theory. It leads to a splitting of the boundary curve into a family of curves.

1.8 DELINEATION OF DIFFERENT PARTICLE LOSS MECHANISMS. RAPID COAGULATION

In emulsions, flocculation will be significant for smaller particles and creaming will be significant for larger particles. However, it is extremely important to have quantitative relationships that can predict the state of an emulsion or suspension; that is, whether the particles are creaming or flocculating, and if both are occurring, determine the predominant one. If they are flocculating, what type of flocculation dominates, if any? This information can be used to predict the behavior of the colloidal system from the solution of the convective–diffusion equation and population balance equations by including only those destabilizing factors which dominate the particle breakdown process. This type of analysis has the potential of greatly simplifying the process of predicting the stability of a given emulsion or suspension. Melik and Fogler [15] accomplished this for both uncharged and charged emulsions. Taking into account the results of the preceding section, a specification is necessary. In this section, the delineation concerns a weakly charged emulsion where the role of the electrostatic component of disjoining pressure can be neglected. As to emulsions stabilized by electrostatic interaction, the delineation procedure as a whole cannot be realistic due to the large discrepancy between theory of slow coagulation and experiment.

1.8.1 GRAVITY-INDUCED FLOCCULATION VERSUS BROWNIAN FLOCCULATION

This section is primarily based on the review by Melik and Fogler [15]. As a first approximation, the relative strength of gravity-induced flocculation as compared to Brownian flocculation is given by the gravity number

$$\mathrm{Gr} = \frac{(u_{02} - u_{01})(a_1 + a_2)}{2D_0} = \frac{2\pi g \Delta\rho a_2^4}{3kT}\lambda(1-\lambda)^2 \tag{1.114}$$

and represents the relative strength of gravitational to Brownian forces.

In Figure 1.19, the gravity number Gr is shown as a function of the particle size ratio λ for the case $\Delta\rho = 0.1$ g cm^{-3} at $T = 298$ K under normal gravity. As λ deviates from unity, the difference in particle size causes the differential creaming contribution to Gr to be enhanced at a rate faster than the increase of the Brownian motion of the smaller particle. A further decrease in λ only results in a small increase in differential creaming, whereas the diffusivity of the smaller particle starts becoming quite vigorous. The maximum value of Gr at $\lambda = (1/3)^{1/2}$ reflects the balance point between these two competing effects. That is, one can estimate the relative importance of gravity-induced flocculation and Brownian flocculation from a plot similar to Figure 1.19. For example, if the colloidal particles are about 1.0 µm in diameter with $\Delta\rho = 0.1$ g cm^{-3} and $T = 298$ K, Brownian flocculation will be approximately 100 times more

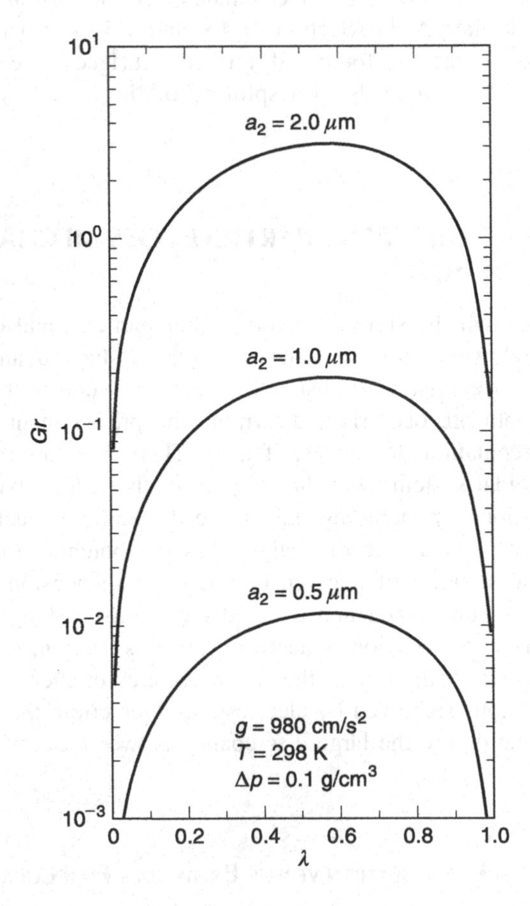

FIGURE 1.19 Effect of particle size ratio on the gravity number Gr for various type 2 particle sizes. (After D.H. Melik and H.S. Fogler, in *Encyclopedia of Emulsion Technology*, Vol 3, Marcel Dekker, Inc., New York, 1988, pp. 3–78.)

important than gravity-induced flocculation. On the other hand, for particles about 4.0 μm in diameter, gravity-induced flocculation will be about three times larger than Brownian flocculation. A plot of this type can provide useful qualitative information on the stability of a colloidal dispersion. This approach was used by Melik and Fogler [15] to delineate the various particle loss mechanisms in quiescent media.

However, an underlying assumption that is invoked when using the parameter Gr is that the stability factor for the loss mechanism of each particle is identical. The gravity number implies that the particle loss mechanisms due to differential creaming and Brownian motion can be quantified independently. Although there are regimes for which the assumption of linear independence is quite good, one should be careful because we can only account for the differences in the gravity-induced and Brownian flocculation stability factors. The ratio of the particle loss due to gravity-induced flocculation to the particle loss due to Brownian flocculation is given by

$$R_{GB} = \frac{J_{Gr}}{J_{Br}} = \frac{Gr}{2}\frac{W_{Br}}{W_{Gr}} \tag{1.115}$$

where J_{Br} is given by Equations 1.13 or 1.14. As shown by Melik and Fogler [15] for Gr $\ll$ 1.0, additivity is justified only in the absence of electrostatic repulsion. Consequently, we will restrict our discussion to conditions of rapid flocculation.

The effect of gravitational forces on the flocculation ratio R_{GB} is shown in Figure 1.20 for various strengths of interparticle attraction. As expected, the effect of gravity-induced flocculation

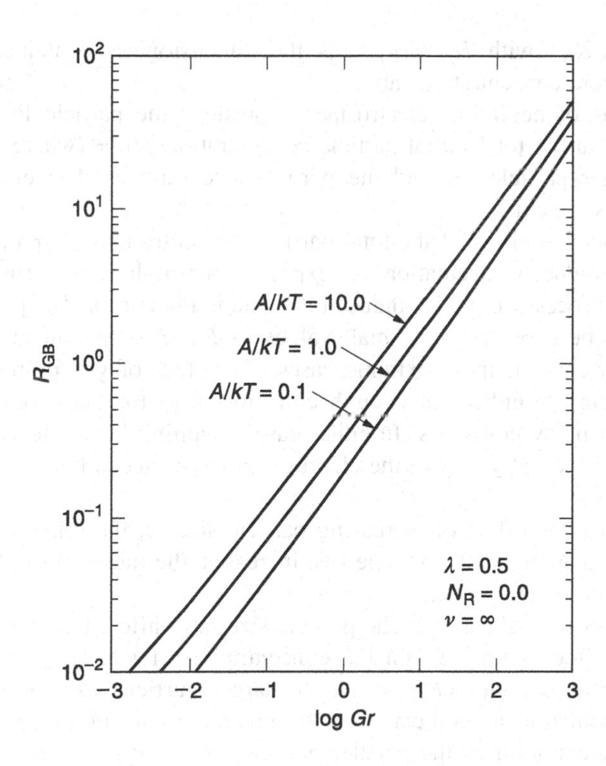

FIGURE 1.20 The importance of gravity-induced flocculation and Brownian flocculation. (After D.H. Melik and H.S. Fogler, in *Encyclopedia of Emulsion Technology*, Vol 3, Marcel Dekker, Inc., New York, 1988, pp. 3–78.)

becomes more pronounced as the gravitational force is increased. It is interesting to note that the gravity number Gr does not give a good estimate of the relative importance of gravity-induced flocculation as compared to Brownian flocculation. For example, if Gr = 10.0, one would probably assume that the rate of particle loss due to gravity-induced flocculation is approximately 10 times higher than the particle loss due to Brownian flocculation. However, as can be seen from Figure 1.20 for Gr 10.0, the particle loss ratio is only approximately 1.5 for a dimensionless Hamaker constant of $A/kT = 1.0$, typical of colloidal dispersions. This result clearly shows that hydrodynamic interactions reduce the rate of gravity-induced flocculation more than the rate of Brownian flocculation. This effect is absent when using the gravity number to delineate the mechanisms of gravity-induced and Brownian flocculation.

1.8.2 Gravity-Induced Flocculation versus Creaming

Whenever a colloidal system is destabilized due to gravity-induced flocculation, a breakdown due to particle creaming occurs simultaneously. The ratio of the total rate of particle loss due to creaming to the net rate of particle loss as a result of gravity-induced flocculation R_{CG} was determined by Melik and Fogler [15]. Their result is given by the following analytical expression:

$$R_{CG} = \frac{f(a_2, N_{02}, \lambda, R_N)}{\pi H E_{1,2}(1 - \lambda^2)(1 + \lambda)^2 R_N} \tag{1.116}$$

where $f(a_2, N_{02}, \lambda, R_N)$ with $H = h/a_2$ is the dimensionless container height and $R_N = N_{01}/N_{02}$ is the particle concentration ratio.

Under conditions of negligible electrostatic repulsion, the particle loss ratio R_{CG} is most sensitive to changes in the total initial particle concentration N_{TOT} (where $N_{TOT} = N_{01} + N_{02}$), the size of the larger particle a_2, and the particle size ratio λ. The effect of each of these parameters is discussed next.

Figure 1.21 shows the effect of the total particle concentration N_{TOT} on the relative rates of creaming to gravity-induced flocculation. As expected, the higher the particle concentration, the more significant the process of gravity-induced flocculation becomes. The presence of a minimum in Figure 1.20 is to be expected. For small values of R_N, $N_{02} \gg N_{01}$ and for large values of R_N, $N_{02} \ll N_{01}$. In each of these extreme cases, the effect of gravity-induced flocculation is reduced because there are either not enough collectors (a_1) for the given number of particles (a_2) or there are too many collectors. In either case, creaming begins to dominate the colloidal breakdown process. When $N_{02} \cong N_{01}$, the effect of gravity-induced flocculation is at a maximum in the breaking process.

Figure 1.22 shows the effect of increasing particle size on the relative rates of creaming to gravity-induced flocculation. As the particle size increases, the rate of flocculation increases faster than the creaming rate of the particles.

Figure 1.23 shows how changes in the particle size ratio affect the relative rates of creaming and gravity-induced flocculation. When the concentration of smaller particles is less than the concentration of larger particles ($R_N < 1$), the larger particle size ratio reduces the rate of gravity-induced flocculation as compared to the creaming rate of the particles. On the other hand, when the concentration of the smaller particles is slightly higher than that of the larger particles ($R_N \geq\sim 3.2$), the situation is reversed. Smaller particle size ratios favor gravity-induced flocculation over creaming more than the larger particle size ratios.

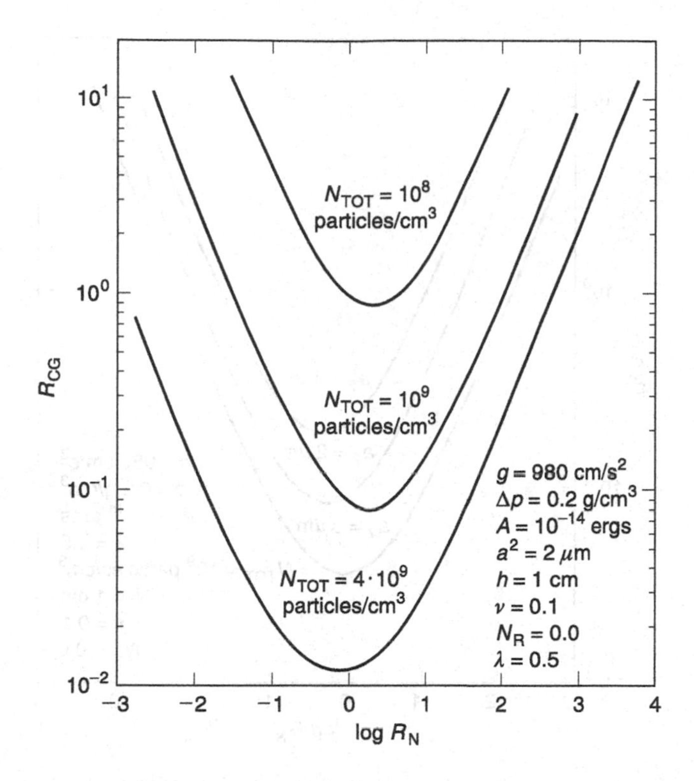

FIGURE 1.21 Effect of particle concentration ratio on the relative particle loss rates of creaming and gravity-induced flocculation for various total particle concentrations. (After D.H. Melik and H.S. Fogler, in *Encyclopedia of Emulsion Technology*, Vol 3, Marcel Dekker, Inc., New York, 1988, pp. 3–78.)

1.8.3 Domains of Dominant Particle Loss Mechanisms

Regimes of the various limiting cases of particle loss can be delineated quantitatively by manipulating the equations for the creaming rate and the Brownian and gravity-induced flocculation rates. For a given fluid, the creaming rate is strongly depending on particle size. For very small particles, the creaming velocity will be negligible as compared to the Brownian motion of the particles. It is generally accepted that the gravitational forces acting on a particle can be neglected for creaming velocities of 1.0 mm per day or less [15] and the case of negligible interactions (surface and hydrodynamic):

$$u_{0i} = \frac{2g\Delta\rho a_i^2}{9\mu_f} < 1.0 \ \text{mm/day} \tag{1.117}$$

or

$$a_c = \left(\frac{(1.0 \ \text{mm/day})9\mu_f}{2g\Delta\rho}\right)^{1/2} \tag{1.118}$$

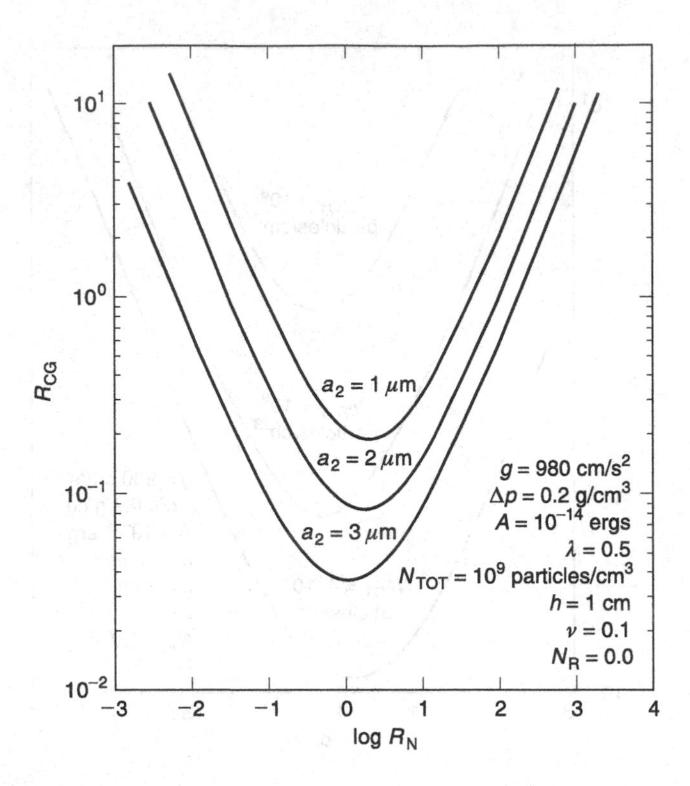

FIGURE 1.22 Effect of particle concentration ratio on the relative particle loss rates of creaming and gravity-induced flocculation for various type 2 particle sizes. (After D.H. Melik and H.S. Fogler, in *Encyclopedia of Emulsion Technology*, Vol 3, Marcel Dekker, Inc., New York, 1988, pp. 3–78.)

Particles with a radius smaller than a_c will disappear from the system primarily due to Brownian flocculation. For values larger than a_c particles can disappear by any of the three particle loss mechanisms.

Brownian flocculation will be negligible for large values of the flocculation ratio R_{GB}, whereas gravity-induced flocculation will be negligible for small values of R_{GB}. Consequently, Equation 1.115 can be used to define quantitatively the regions where only one mechanism will be predominant. Melik and Fogler [15] define Brownian flocculation as predominant if the flocculation ratio R_{GB} is less than 0.1, and gravity-induced flocculation as predominant if R_{GB} is higher than 10.0. As discussed in Section 1.8.1, this rationale does not account for any possible coupling between differential creaming and Brownian motion:

$$\frac{\mathrm{Gr}}{2}\frac{W_{\mathrm{Br}}}{W_{\mathrm{Gr}}} \leq 0.1 \text{ (Brownian flocculation)}, \tag{1.119}$$

$$\frac{\mathrm{Gr}}{2}\frac{W_{\mathrm{Br}}}{W_{\mathrm{Gr}}} \geq 10.0 \text{ (gravity-induced flocculation)} \tag{1.120}$$

The key parameters in Equations 1.118 to 1.120 are the particle radius a_2, the net gravitational force $g\Delta\rho$, and the particle size ratio λ. Consequently, these parameters can be used to map the various domains for the particle loss mechanisms.

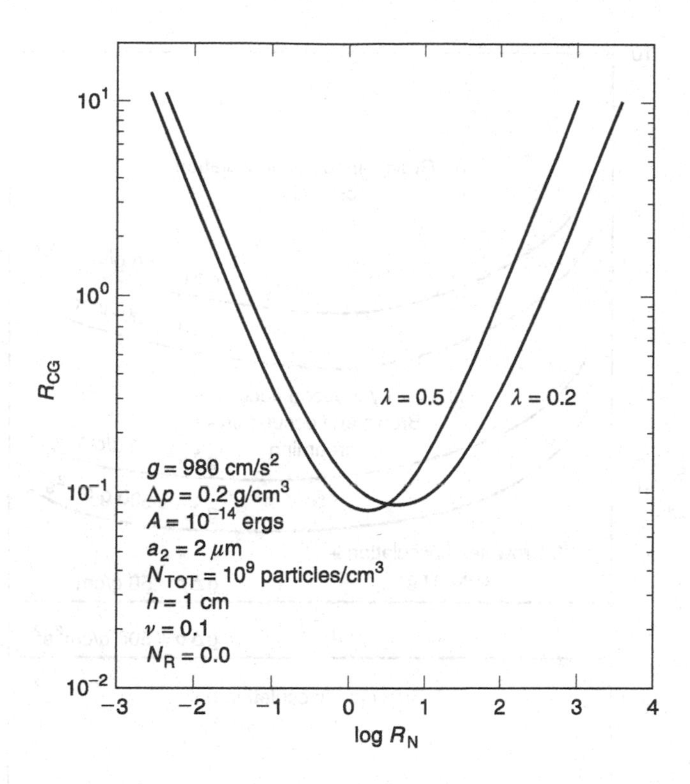

FIGURE 1.23 Effect of particle concentration ratio on the relative particle loss rates of creaming and gravity-induced flocculation for various particle size ratios. (After D.H. Melik and H.S. Fogler, in *Encyclopedia of Emulsion Technology*, Vol 3, Marcel Dekker, Inc., New York, 1988, pp. 3–78.)

In a fairly general manner, plots of a_2 versus λ for a constant value of $g\Delta\rho$ can be prepared by using Equations 1.118 to 1.120. An example of this analysis is shown in Figure 1.24 for the case of negligible electrostatic repulsion ($N_R = 0$) and for a Hamaker constant $A/kT = 1.0$. Equation 1.118 was used to obtain the upper boundary of region I, and Equations 1.119 and 1.120 to obtain the lower and upper boundaries of region III. A plot of this type can also be used for centrifugal creaming instead of gravitational creaming. For centrifugal creaming, g is simply replaced with $\omega^2 X$, where ω is the angular velocity of the centrifuge rotor and X is the distance from the axis of rotation.

Various regions in Figure 1.24 represent the possible dominant particle loss processes in quiescent media. For example, in region I, creaming and gravity-induced flocculation will be negligible and particles disappear primarily by Brownian flocculation. Brownian flocculation in region I will depend on the electrostatic properties of the particles and also the total particle concentration. If the colloidal system is very dilute, Brownian flocculation will also be negligible.

For colloidal systems in region II, gravity-induced flocculation will be negligible as compared to Brownian flocculation. In region III, creaming is bound to occur and the rate of Brownian and gravitational flocculation will depend on the total particle concentration. Similarly, in region IV, Brownian flocculation will be negligible as compared to gravity-induced flocculation. Creaming will also occur in this region and the rate of gravity-induced flocculation will depend on particle concentration, and even on the polydispersity of the colloidal system. Figure 1.24 is useful in

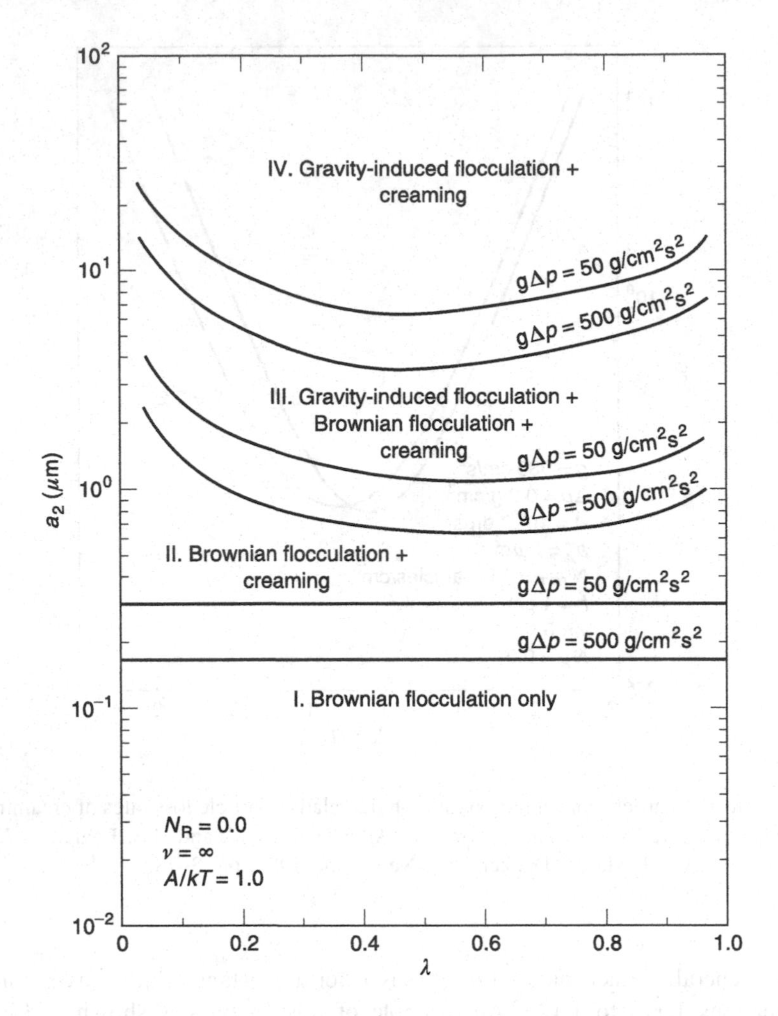

FIGURE 1.24 Regimes of dominant particle loss mechanisms in quiescent media. (After D.H. Melik and H.S. Fogler, in *Encyclopedia of Emulsion Technology*, Vol 3, Marcel Dekker, Inc., New York, 1988, pp. 3–78.)

determining the state of a colloidal system; that is, whether the colloidal particles are creaming or flocculating.

1.8.4 Experiment

A methodology of experimental investigation of the kinetic and aggregative instability of mini-emulsions is elaborated by Ostrovsky and Good [158]. The methodology is illustrated by Figures 1.25A and 1.25B. The invariant slope of section c of the experimental curve characterizes the rate of coalescence. Its value depends only on the composition of the system and is called the coalescence time parameter. More than 50 different compositions were studied.

By means of the linear extrapolation, as shown in Figure 1.25B, the authors introduced the notion of τ_{sed}. The authors neglect the flocculation and coalescence during stages b and c and

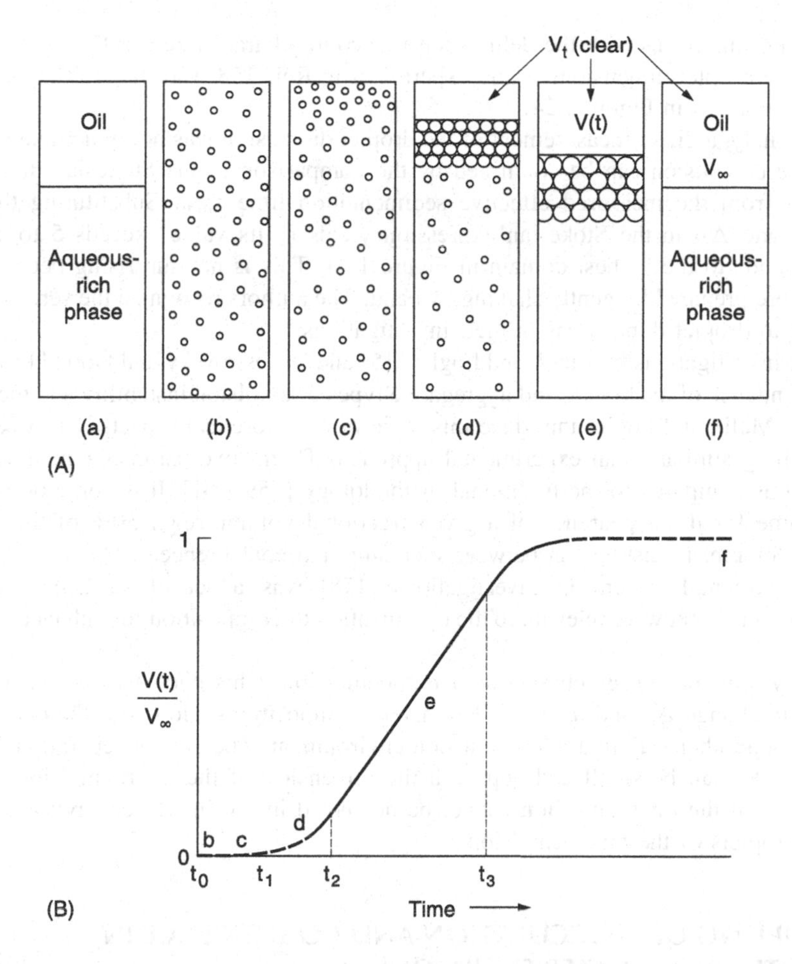

FIGURE 1.25 (A) Commonly observed sequence of events in separation processes. Schematic sequence of direct observations: (a) Before shaking: equilibrated volumes. (b) Just after stopping shaking: $t = 0$. Macroemulsion of oil phase in aqueous phase. (c) Droplets of oil have migrated toward top of aqueous phase, but no appreciable coalescence is detectable. $t_0 < t < t_1$. (d) Appreciable coalescence of oil droplets; V_t increasing. Note larger droplet size just below clear layer, and sedimentation from lower part of aqueous phase still in progress. $t_1 < t < t_2$. (e) Sedimentation effectively complete. Coalescence of droplets continues in layer at top of aqueous phase. $t_2 < t < t_3$. (f) Separation complete. $t \gg t_3$. (B) Schematic kinetic curve of separation, corresponding to conditions in (A). (Parts A and B after I.B. Ivanov, R.K. Jain, P. Somasundaran, and T.T. Traykov, in *Solution Behavior of Surfactants*, Vol 2 (K.L. Mittal, ed.), Plenum Press, New York, 1979, p. 817.)

at $t < \tau_{sed}$. They interpret the time $t < \tau_{sed}$ as being determined by the mean velocity of the translating droplet and identify averaged velocity with the velocity corresponding to averaged radius $\bar{r}$.

For the droplet velocity characterization, the radius, the density difference $\Delta\rho$, and the viscosity of the external phase μ were varied and measured in a wide range. For τ_{min} a rather good correlation between the measured and calculated values is established. It means that the flocculation influence on the droplet dimension and velocity was weak.

Comparing this result with the delineation procedure characterized in Figure 1.24, one concludes that the droplet dimensions in the experiment in Ref. 158 were large. They corresponded to the higher domain in Figure 1.24.

Unfortunately, a direct measurement of the droplet dimension was not undertaken in Ref. 158. However, the conclusion can be confirmed by the comparison of the Stokesian droplet velocity and velocity from the measured effective sedimentation time τ_{min}. Substituting the measured values of μ and $\Delta\rho$ in the Stokesian expression yields $\bar{r}$. Its value exceeds 5 to 10 μm; that is, it corresponds to the highest domain in Figure 1.24. This is not surprising because the mini-emulsions were prepared by gently shaking by hand. The authors recognize the serious restrictions with respect to droplet dimensions in their investigations.

Both the investigations by Melik and Fogler [15] and by Ostrovsky and Good [158] emphasize the interdependence of the kinetic and aggregative types of emulsion instability and the importance of this topic. Melik and Fogler introduce this topic in its theoretical aspect. Ostrovsky and Good make something similar in an experimental approach. Their "two-parameter methodology" is a step forward as compared to the traditional methodology [159–161]. It is common [159–161] to report the time for the appearance of a given fractional volume (e.g., 50% of the clear phase) without any attempt to distinguish between creaming and coalescence.

The experimental system in investigations [158] was a set of surfactant–brine–oil–co-surfactant mixtures that were relevant to the optimization of composition for enhanced oil recovery [162,163].

In this system, the lower phase was a microemulsion. This choice of external phase gave opportunity to change $\Delta\rho$ and μ in a wide range of compositions. However, the buoyant droplets can interact in another way in a microemulsion environment. The closest separation between two drop surfaces h_0 can be small and approach the dimension of the microemulsion droplet. The discrete nature of the microemulsion cannot be neglected in the interlayer environment between interacting droplets of the macroemulsion.

1.9 COUPLING OF FLOCCULATION AND COALESCENCE IN DILUTE OIL-IN-WATER EMULSIONS

1.9.1 General

1.9.1.1 Kinetic and Thermodynamic Stability in Macroemulsions and Mini-Emulsions

The majority of emulsion technology problems relate to the stabilization and destabilization of emulsions [29,164–169]. Despite the existence of many fundamental studies related to the stability of emulsions, the extreme variability and complexity of the systems involved in any specific application often pushes the oil industry to achieve technologically applicable results without developing a detailed understanding of the fundamental processes. Nevertheless, since in most cases technological success requires the design of emulsions with a very delicate equilibrium between stability and instability, a better understanding of the mechanisms of stabilization and destabilization might lead to significant break-throughs in technology.

Notwithstanding their thermodynamic instability, many emulsions are kinetically stable and do not change appreciably for a prolonged period. These systems exist in the metastable state [1,2,30,34,170–173]. The fundamentals of emulsion stability (destabilization) comprise emulsion surface chemistry and physicochemical kinetics.

In contrast to the large success in industrial applications of emulsion surface chemistry the potential of physicochemical kinetics as the basis for emulsion dynamics modeling is almost not used in emulsion technology. This situation has started to change during the last decade. Although the coupling of the subprocesses in emulsion dynamics modeling (EDM) continues to represent a large problem yet not solved, models are elaborated for (1) macroemulsions [172,174–180] and (2) mini-emulsions [181–188], for long and short lifetimes of thin emulsion films.

1. For large droplets (larger than 10 to 30 μm) in macroemulsions the rate of thinning of the emulsion film formed between two approaching droplets is rather low and, correspondingly, the entire lifetime of an emulsion need not be short, even without surfactant stabilization of the film. For this case the notion of *kinetic stability* is introduced [172, 174–177] to denote the resistance of the film against *rupture during thinning*. The droplet deformation and flattening cause this strong resistance, described by the Reynolds equation [189,190]. According to theory, the role of this deformation [165,191,192] decreases rapidly with decreasing droplet dimension.
2. For small droplets (smaller than 5 to 10 μm) in mini-emulsions droplet deformation can be neglected, because the Reynolds drainage rate increases as R_d^{-5} [172,193] (R_d, the Reynolds film radius) and because the smaller the droplets, the smaller is their deformation [165,191,192].

In distinction from macroemulsions, where the kinetic stability is the manifestation of droplet–droplet hydrodynamic interaction and droplet deformation, in mini-emulsion the kinetic stability is the manifestation of the interplay between surface forces and Brownian movement [181]. As the molecular forces of attraction decrease linearly with decreasing droplet dimension, namely approximately 10 times at the transition from macroemulsions to mini-emulsions, the potential minimum of droplet–droplet interaction (secondary minimum) decreases and for mini-emulsions thus depth can be evaluated as 1.1 to 1.5 kT [173,194]. At this low energy Brownian movement causes droplet doublet disaggregation after a short time (the doublet fragmentation time, τ_d). If this time is shorter than the lifetime of the thin film, rapid decrease in the total droplet concentration (t.d.c) is prevented (restricted by the coalescence time, τ_c), i.e., stability is achieved due to this kinetic mechanism [181].

1.9.1.2 Current State of Emulsion Stability Science

A large imbalance exists between knowledge concerning kinetic stability and thermodynamic stability. Main attention has been paid to *kinetic stability* for both macroemulsions [174–180] and mini-emulsions [181–188]. As a result, the droplet–droplet interaction and the collective processes in dilute emulsions are quantified [10,195] and important experimental investigations are made [185,186,196]. Some models are elaborated for the entire process of coalescence in concentrated emulsions as well [197,198]. Given thermodynamic stability, a thin interdroplet film can be metastable.

In contrast to the large achievements in investigations of kinetic stability modest attention has been paid to the fundamentals of the thermodynamic stability in emulsions, especially regarding the surfactant adsorption layer's influence on the coalescence time. There are several investigations devoted to the surface chemistry of adsorption related to emulsification and demulsification. However, the link between the chemical nature of an adsorption layer, its structure, and the coalescence time is not yet quantified.

A premise for such quantification is the theory of a foam bilayer lifetime [199]. The main notions of this theory is similar to the theory of Derjaguin et al. [200,201]. However, the theory [199] is specified for amphiphile foam films, it is elaborated in detail, and is proven by experiment with water soluble amphiphiles, such as sodium dodecylsulfate [202]. As the dependence of the rupture of the emulsion film on surfactant concentration is similar to that for a foam film, the modification of theory with respect to emulsions may be possible. Although this modification is desirable the specification of theory for a given surfactant will not be trivial, since the parameters in the equation for the lifetime [201] are unknown and their determination not easy. As the theory [199,203] is proposed for amphiphiles and since a wider class of chemical compounds can stabilize emulsions, the film rupture mechanism [200] is not universal regarding emulsions.

Thus, in contrast to the quantification of kinetic stability, the empirical approach continues to predominate regarding thermodynamic stability. Meanwhile, thermodynamic stability provides greater opportunity for long-term stabilization of emulsions, than does kinetic. This means that the experimental characterization of thermodynamic stability, i.e., the measurement of coalescence time, is of major importance.

1.9.1.3 The Specificity of Emulsion Characterization

Generalized emulsion characterization, i.e., measurement of droplet size distribution, electrokinetic potential, Hamaker constant, etc., is not always sufficient. Thermodynamic stability with respect to bilayer rupture cannot be quantified with such a characterization procedure alone. Consequently, measurement of the coalescence time τ_c is of major importance for an evaluation of emulsion stability; it is an important and specific parameter of emulsion characterization.

The current state of mini-emulsion characterization neglects the importance of τ_c measurement. A practice for τ_c measurement is practically absent with the exception of only a few papers, considered in this chapter. Meanwhile, many papers devoted to issues more or less related to emulsion stability do not discuss τ_c measurement. One reason for this scientifically and technologically unfavorable situation in which emulsions are incompletely characterized may originate from a lack of devices enabling τ_c measurement.

1.9.1.4 Scope of Section 9

This section is focused on kinetic stability in mini-emulsions with emphasis on the coupled destabilizing subprocesses. In general there are three coupled subprocesses that will influence the rate of destabilization and phase separation in emulsions. These are aggregation, coalescence, and floc fragmentation. Often, irreversible aggregation is called coagulation and the term flocculation is used for reversible aggregation [199,204]. Ostwald ripening [205,206] coupled [182] with aggregation and fragmentation is a separate topic which will be not considered here.

A simplified theory is available for the coupling of coalescence and flocculation in emulsions void of larger flocs. This theory is considered in Section 1.9.2 and will assist in the consideration of the more complicated theory of coupling of coalescence and coagulation (Section 1.9.3). The experimental investigations are described in parallel. Section 1.9.4 is devoted to the theory of doublet fragmentation time and its measurement, as this characterizes an emulsion regarding fragmentation and because its measurement is an important source of information about surface forces and the pair interaction potential. The discrimination between conditions for coupling of coalescence with coagulation or with flocculation is considered in Section 1.9.5. The quantification of kinetic stability creates new opportunities for long-term prediction of mini-emulsion stability, for stability optimization, and for characterization with the standardization of τ_c and τ_d measurements. This forms the base for emulsion dynamics modeling (Section 1.9.6).

1.9.2 Coupling of Coalescence and Flocculation

1.9.2.1 Singlet–Doublet Quasi-Equilibrium

Each process among the three processes under consideration is characterized by a characteristic time, namely τ_{Sm}, τ_d, and τ_c. The Smoluchowski time [3], τ_{Sm}, gives the average time between droplet collisions. If the time between two collisions is shorter than τ_d a doublet can transform into a triplet before it spontaneously disrupts. In the opposite case, i.e., at

$$\tau_{Sm} \gg \tau_d \tag{1.121}$$

the probability for a doublet to transform into a triplet is very low because the disruption of the doublet occurs much earlier than its collision with a singlet. The rate of multiplet formation is very low for

$$\mathrm{Rev} = \frac{\tau_d}{\tau_{Sm}} \ll 1 \tag{1.122}$$

where we introduce the notation Rev for small values of the ratio corresponding to the reversibility of aggregation and a singlet–doublet quasi-equilibrium.

The kinetic equation for reversible flocculation in a dilute monodisperse o/w emulsion when neglecting coalescence is [14,207,208]

$$\frac{dn_2}{dt} = \frac{n_1^2}{\tau_{Sm}} - \frac{n_2}{\tau_d} \tag{1.123}$$

where n_1 and n_2 are the dimensionless concentrations of doublets and singlets, $n_1 = N_1/N_{10}$, $n_2 = N_2/N_{10}$, N_1 and N_2 are the concentrations of singlets and doublets and N_{10} is the initial concentration, and

$$\tau_{Sm} = \left(\frac{4kT}{3\eta} N_{10}\right)^{-1} = \left(K_f N_{10}\right)^{-1} \tag{1.124}$$

where k is the Boltzmann constant, T is the absolute temperature, η is the viscosity of water. For aqueous dispersions at 25 °C, $K_f = \frac{4kT}{3\eta} = 6 \cdot 10^{-18}\ \mathrm{m^3\ s^{-1}}$. The singlet concentration decreases with time due to doublet formation, while the doublet concentration increases. As a result, the rates of aggregation and floc fragmentation will approach each other. Correspondingly, the change in the number of doublets $dn_2/dt = 0$. Thus, a dynamic singlet–doublet equilibrium (s.d.e.) is established:

$$n_{2eq} = \frac{\tau_d}{\tau_{Sm}} n_{1eq}^2 \tag{1.125}$$

At condition 1.122 it follows from Equation 1.125 that

$$n_{1eq} \cong 1, \ N_{1eq} = N_{10} \tag{1.126}$$

$$N_2 = \mathrm{Re}\ v \cdot N_{10} \text{ or } n_2 \ll 1 \tag{1.127}$$

Thus, at small values of Rev the s.d.e. is established with only small deviations in the singlet equilibrium concentration from the initial concentration (Equation 1.126). The doublet concentration is very low compared to the singlet concentration, and the multiplet concentration is very low compared to the doublet concentration. The last statement follows from a comparison of the production rates of doublets and triplets. The doublets appear due to singlet–singlet collisions, while the triplets appear due to singlet–doublet collisions. The latter rate is lower due to the low doublet concentration. The ratio of the number of singlet–doublet collisions to the number of singlet–singlet collisions is proportional to Rev.

1.9.2.2 Kinetic Equation for Coupling of Flocculation and Intradoublet Coalescence in Monodisperse Emulsions

Both the rate of doublet disaggregation and the rate of intradoublet coalescence are proportional to the momentary doublet concentration. This leads [181,187] to a generalization of Equation 1.123:

$$\frac{dn_2}{dt} = \frac{n_1^2}{\tau_{Sm}} - n_2\left(\frac{1}{\tau_d} + \frac{1}{\tau_c}\right) \tag{1.128}$$

There are two unknown functions in Equation 1.128, so an additional equation is needed. This equation describes the decrease in the droplet concentration caused by coalescence:

$$\frac{d}{dt}(n_1 + 2n_2) = -\frac{n_2}{\tau_c} \tag{1.129}$$

The initial conditions are

$$n_2|_{t=0} = 0 \tag{1.130}$$

$$\left.\frac{dn_2}{dt}\right|_{t=0} = \frac{n_{10}}{\tau_{Sm}} \tag{1.131}$$

Condition 1.131 follows from Equations 1.128 and 1.129. The solution of the set of Equations 1.128 and 1.129 with account for boundary conditions 1.130 and 1.131 is a superposition of two exponents [181,187]. In the case

$$\tau_c \gg \tau_d \tag{1.132}$$

the solution simplifies [181,187] to

$$n_2(t) = \frac{\tau_d}{\tau_{Sm}}\left[\exp\left(-\frac{2\tau_d t}{\tau_{Sm}\tau_c}\right) - \exp\left(-\frac{t}{\tau_d}\right)\right] \tag{1.133}$$

Equation 1.133, as compared to Equations 1.125 and 1.126, corresponds to the s.d.e. if the expression in the second brackets equals one. In the time interval

$$\tau_d < t < \tau \tag{1.134}$$

where

$$\tau = \tau_c \frac{\tau_{Sm}}{2\tau_d} \tag{1.135}$$

the first term in the second brackets approximately equals 1, while the second one decreases from 1 to a very small value. Thus, the s.d.e. is established during the time τ_d and preserves during the longer time interval (Equation 1.134).

For times longer than τ there is no reason to apply Equation 1.133 since the condition to linearize Equation 1.128 is no longer valid with the concentration decrease. At the beginning of the process the doublet concentration increases, while later coalescence predominates and the doublet concentration decreases. Thus, function 1.133 has a maximum [181,187].

1.9.2.3 Coalescence in a Singlet–Doublet System at Quasi-Equilibrium

After a time t_{max} a slow decrease in the doublet concentration takes place simultaneously with the more rapid processes of aggregation and disaggregation. Naturally, an exact singlet–doublet equilibrium is not valid due to the continuous decrease in the doublet concentration. However, the slower the coalescence, the smaller is the deviation from the momentary dynamic equilibrium with respect to the aggregation–disaggregation processes.

It is reasonable to neglect the deviation from the momentary doublet–singlet equilibrium with the condition

$$\frac{dn_2}{dt} \ll \frac{n_2}{\tau_d} \tag{1.136}$$

Indeed, for this condition the derivative in Equation 1.123 can be omitted, which corresponds to s.d.e. characterized by Equation 1.125.

It turns out [181,185–187] that the deviation from s.d.e. is negligible as the condition 1.136 is valid, i.e., for conditions 1.122 and 1.132. For these conditions the fragmentation of flocs influences the coalescence kinetics which can be represented as a three-stage process, as illustrated in Figure 1.26. During a rather short time τ_d the approach to s.d.e. takes place, i.e., a rather rapid increase in the doublet concentration (stage 1). During the next time interval $\tau_d < t < t_{max}$ the same process continues. However, the rate of doublet formation declines due to coalescence (stage 2). The exact equilibrium between the doublet formation and their disappearance due to coalescence takes place at the time t_{max} when the doublet concentration reaches its maximum value $n_2(t_{max})$. During the third stage, when $t > t_{max}$, the rate of doublet fragmentation is lower than the rate of formation, because of the coalescence within doublets. This causes a slow monotonous decrease in the concentration. Taking into account the s.d.e. (Equations 1.125 and 1.127) Equation 1.129 can be expressed as

$$\frac{dn_1}{dt} = -\frac{\tau_d}{\tau_{Sm}\tau_c} n_1^2 \tag{1.137}$$

The result of the integration of Equation 1.137 can be simplified to

$$n_1(t) = \frac{n_1(t_{max})}{1 + n_1(t_{max})\dfrac{t}{2\tau}} \cong \left(1 + \frac{t}{2\tau}\right)^{-1} \tag{1.138}$$

with a small deviation in $n_1(t_{max})$ from 1. Differing from the preceding stages when the decrease in the droplet concentration caused by coalescence is small, a large decrease is now possible during the third stage. Thus, this is the most important stage of the coalescence kinetics.

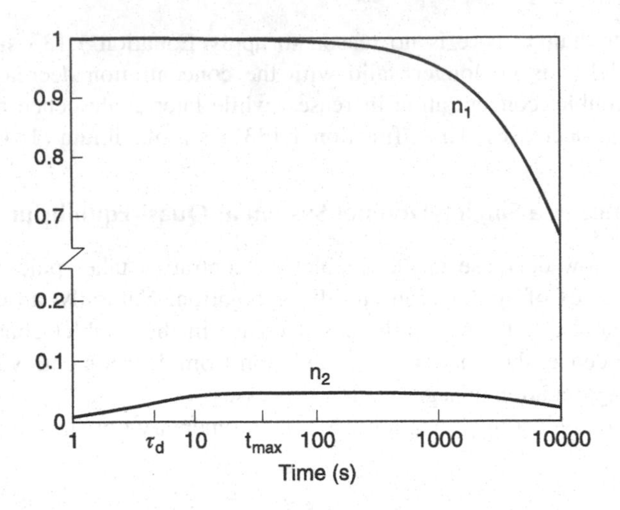

FIGURE 1.26 Three stages in the coupling of aggregation, fragmentation, and coalescence at the condition $\tau_d \ll \tau_{Sm} \ll \tau_c$. Initially, the doublet concentration n_2 is very low and the rates of doublet fragmentation and of coalescence are correspondingly low compared to the rate of aggregation (first stage, no coupling). Due to increasing n_2 the fragmentation rate increases and equals the aggregation rate at $t_{\max}$ (exact singlet–doublet equilibrium). The growth in n_2 stops at $t_{\max}$ (second stage, coupling of aggregation and fragmentation). Intradoublet coalescence causes a slight deviation from exact s.d.e. to arise at $t > t_{\max}$, and the singlet concentration n_1 and the doublet concentration decrease due to intradoublet coalescence (third stage, coupling of aggregation, fragmentation, and coalescence). n_1 and n_2 are dimensionless, $n_1 = N_1/N_{10}$, $n_2 = N_2/N_{10}$, N_{10} is the initial singlet concentration. (From S.S. Dukhin and J. Sjöblom, *J. Dispersion Sci. Technol.* 19: 311 (1998).)

1.9.2.4 Reduced Role of Fragmentation with Decreasing τ_c

With decreasing τ_c, condition 1.132 is violated and new qualitative features of the destabilization process not discussed in Refs. 185–187 arise. As the ratio τ_c/τ_d diminishes and

$$\tau_c < \tau_d \tag{1.139}$$

the s.d.e. is violated because a larger part of the doublets disappear due to coalescence. Correspondingly, the smaller the ratio τ_c/τ_d the smaller is the fragmentation rate in comparison with the aggregation rate, i.e., the larger the deviation from s.d.e. In the extreme case

$$\tau_c \ll \tau_d \tag{1.140}$$

the fragmentation role in s.d.e. can be neglected. This means that almost any act of aggregation is accompanied by coalescence after the short doublet lifetime. Neglecting this time in comparison with τ_{Sm} in agreement with condition 1.121 one concludes that any act of aggregation is accompanied by the disappearance of one singlet:

$$\frac{dn_1}{dt} = -\frac{n_1^2}{\tau_{Sm}} \tag{1.141}$$

This leads to a decrease in the singlet concentration described by an equation similar to the Smoluchowski equation for rapid coagulation:

$$n_1(t) = (1 + t/\tau_{Sm})^{-1} \tag{1.142}$$

The Smoluchowski equation describing the singlet time evolution does not coincide with Equation 1.142. The peculiarity of Equation 1.142 is that it describes the kinetics of coupled aggregation and coalescence with a negligible fragmentation rate. Due to rapid coalescence, doublet transformation into multiplets is almost impossible at condition 1.121.

The coupling of aggregation, fragmentation, and coalescence in the more general case described by condition 1.139 leads to equation

$$n_1(t) = \frac{n_1(t_{\max})}{1 + n_1(t_{\max})t/2\tau_g} \cong \left(1 + t/2\tau_g\right)^{-1} \qquad (t > \tau_d + \tau_c) \tag{1.143}$$

with a small deviation of $n_1(t_{\max})$ from 1 and

$$\tau_g = \frac{\tau_{Sm}(\tau_d + \tau_c)}{2\tau_d} \tag{1.144}$$

At conditions 1.121 and 1.132 $\tau_g \approx \tau$ and Equation 1.143 transforms into Equation 1.138. At conditions 1.121 and 1.140 $\tau_g \approx \tau_{Sm}$ and Equation 1.143 transforms into Equation 1.142. Equation 1.144 demonstrates the reduction of the role of fragmentation with decreasing τ_c. It is seen that at the transition from condition 1.139 to condition 1.140 τ_d cancels in Equation 1.141, i.e., the fragmentation role diminishes.

1.9.2.5 Experimental

1.9.2.5.1 Application of Video Enhanced Microscopy Combined with the Microslide Technique for Investigation of Singlet–Doublet Equilibrium and Intradoublet Coalescence [185–187]

Direct observation of doublets in the emulsion bulk is difficult because the doublets tend to move away from the focal plane. The microslide preparative technique can, however, be successfully applied, providing pseudo-bulk conditions. A microslide is a plane-parallel glass capillary of rectangular cross section. The bottom and top sides of the capillary are horizontal, and the gravity induced formation of a sediment or cream on one of the inner normal surfaces is rapidly completed due to the modest inner diameter of the slide. If both the volume fraction of droplets in an emulsion and the capillary height are small, the droplet coverage on the inside surface amounts to a few percent, and the analysis of results is rather simple. It can be seen through the microscope that the droplets which have sedimented onto the capillary surface participate in chaotic motion along the surface. This indicates that a thin layer of water separating the surface of the microslide from the droplets is preventing the main portion of droplets from adhering to the microslide surface, an action which would stop their Brownian motion.

During diffusion along the microslide ceiling the droplets collide. Some collisions lead to the formation of doublets. Direct visual observation enables evaluation of the doublet fragmentation time which varies in a broad range [183]. Another approach to doublet fragmentation time determination is based on the evaluation of the average concentration of singlets and doublets and using the theory outlined above.

The application of the microslide preparative technique combined with video microscopy is promising and has enabled the measurement of the coupling of reversible flocculation and coalescence [185]. However, some experimental difficulties were encountered. Droplets could sometimes be seen sticking to the glass surface of the microslide.

1.9.2.5.2 *Improving the Experimental Technique with the Use of Low Density Contrast Emulsions [186]*

The sticking of droplets indicates a droplet–wall attraction and the existence of a secondary potential pit as that for the droplet–droplet attraction in a doublet. The droplet concentration within the pit is proportional to the concentration on its boundary. The latter decreases with a decrease in the density contrast.

The electrostatic barrier between the potential pit and the wall retards the rate of sticking. The lower the droplet flux through this barrier, the lower is the potential pit occupancy by droplets. Thus, an essential decrease in the rate of sticking is possible with decreasing density contrast.

O/W emulsions were prepared [186] by mixing dichlorodecane (DCD, volume fraction 1%) into a 5×10^{-5} M sodium dodecylsulfate (SDS) solution with a Silverson homogenizer. The oil phase was a 70:1 mixture of DCD, which is characterized by an extremely low density contrast to water and decane.

The droplet distribution along and across the slide was uniform [186]. This indicates that there was no gravity induced rolling either. One slide among 4 was examined for 2 weeks without any sticking being observed. The absence of the rolling and sticking phenomena allowed acquisition of rather accurate data concerning the time dependence of the droplet size distribution.

1.9.2.5.3 *The Measurement of Coalescence Time and Doublet Fragmentation Time*

The doublet fragmentation time was measured by direct real-time observation of the doublets on the screen and by analysis of series of images acquired with 1- to 3-min time intervals [201]. The formation and disruption/coalescence of a doublet could thus be determined.

The general form of the concentration dependence agrees with the theory. At $C \sim 3 \times 10^{-3}$ M, both theory and experiment yield times of about 1 min; at $C = 9 \times 10^{-3}$ M, these times exceed 10 min. For calculation of the doublet fragmentation time the electrokinetic potential was measured [187,202].

In experiments with different droplet concentrations it was established that the higher the initial droplet concentration, the higher the doublet concentration. This corresponds to the notion of singlet–doublet equilibrium. However, if the initial droplet concentration exceeds 200 to 300 per observed section of the microslide, multiplets predominate. Both the initial droplet concentration and size affect the rate of decrease in the droplet concentration. The larger the droplets, the smaller the concentration sufficient for the measurement of the rate of decrease in the droplet concentration. This agrees with the theory of doublet fragmentation time which increases with droplet dimension. Correspondingly, the probability for coalescence increases. These first series of experiments [185,187] were accomplished using toluene-in-water emulsions without the addition of a surfactant and decane-in-water emulsions stabilized by SDS. The obtained data concerning the influence of the electrolyte concentration and surface charge density were in agreement with the existing notions about the mechanism of coalescence. With increasing SDS concentration, and correspondingly increasing surface potential, the rate of decrease in the droplet concentration is reduced.

Two methods were used for the measurement of the coalescence time [186,187]. Measurement of the time dependence for the concentrations of singlets and doublets and a comparison with Equation 1.129 enables an evaluation of the coalescence time. Further, information about the time dependence for singlets and the doublet fragmentation time may be used as well. These results in combination with Equations 1.138 and 1.135 determine the coalescence time. The good agreement between results obtained by these very different methods indicates that the exactness of the theory and experiments is not low.

In recent years several research groups have improved significantly the theoretical understanding of coalescence of droplet or bubbles. The new results [14,208–211] together with results of earlier investigations [212–216] have clarified the role of double-layer interaction in the elementary act of coalescence.

DLVO theory was applied [217,218] for the description of "spontaneous" and "forced" thinning of the liquid film separating the droplets. These experimental results and DLVO theory were used [217] for the interpretation of the reported visual study of coalescence of oil droplets 70 to 140 μm in diameter in water over a wide pH interval. A comparison based on DLVO theory and these experimental data led the authors to conclude [217] that "if the total interaction energy is close to zero or has a positive slope in the critical thickness range, i.e., between 30 and 50 nm, the oil drops should be expected to coalesce." In the second paper [218], where both ionic strength and pH effects were studied, coalescence was observed at constant pH values of 5.7 and 10.9, when the Debye thickness was less than 5 nm. The main trend in our experiments and in Refs. 217 and 218 are in accord, because it was difficult to establish the decrease in t.d.c. at NaCl concentrations lower than 5×10^{-3} M, i.e., DL thicknesses larger than 5 nm. An almost quantitative coincidence in the double layer influence on coalescence established in our work for micrometer-sized droplets and in Refs. 217 and 218 for almost 100 times larger droplets is important for general knowledge about coalescence.

1.9.2.6 Perspective for Generalization of the Theory for Coupling of Coalescence and Flocculation

The proposed theory for coupling of coalescence and flocculation at s.d.e. enables the proposal of some important applications (Section 1.9.6). At the same time generalized theory is necessary, since the role of multiplets increases after a long time or with a higher initial concentration. At least two approaches to this difficult task are seen.

According to our videomicroscopic observations there are large peculiarities in the structure and behavior of multiplets arising at conditions near to s.d.e. These peculiarities can be interpreted as the manifestation of quasi-equilibrium, comprising singlets, doublets, and multiplets. Similar to doublets, the lifetime of triplets, tetraplets etc. can be short due to fragmentation and coalescence. This can be valid for multiplets with an "open" structure, in distinction from another structure which can be called "closed." In open multiplets any droplet has no more than 1 to 2 contacts with other droplets, which corresponds to a linear chain-like structure. This causes easy fragmentation, especially for the extreme droplets within a chain. The "closed" aggregates have a more dense and isometric structure, in which droplets may have more than 2 contacts with neighboring droplets. As a result, fragmentation is more difficult and the frequency is lower.

The recent progress in theory of aggregation with fragmentation [219–223] for a suspension creates a premise for a theoretical extension towards emulsions. However, the necessity in accounting for coalescence makes this task a difficult one.

1.9.3 Coupling of Coalescence and Coagulation

1.9.3.1 General

For emulsion characterization the notation n_1 represents the number density of single droplets and n_i the number density of aggregates comprising i droplets ($i = 2, 3 \ldots$). The total number density of single droplets and all kinds of aggregates is given by

$$n = \sum_{i=1}^{\infty} n_i \tag{1.145}$$

This characterization corresponds with Smoluchowski theory [3]. To characterize coalescence, the total number of individual droplets moving freely, plus the number of droplets included in all kinds of aggregates, n_T

$$n_T = \sum_{i=1}^{\infty} i n_i \tag{1.146}$$

is introduced as well.

In distinction from the Smoluchowski theory for suspensions, which predicts the time dependence of the concentration of all kinds of aggregates, the time dependence for the total droplet number can be predicted at the current state of emulsion dynamics theory.

The quantification of coagulation within the theory of coupled coagulation and coalescence (CCC theory) is based on the Smoluchowski theory of perikinetic coagulation. Correspondingly, all restrictions inherent to Smoluchowski theory of Brownian coagulation preserve in the CCC theory. This means that creaming and gravitational coagulation are not accounted for. A variant of Smoluchowski theory specified with regard for gravitational coagulation is well known [4]. However, its application is very difficult because the rate constant of collisions induced by gravity depends on droplet dimension [173]. Due to the weak particle (aggregate) dimension dependence of the rate constants for Brownian collisions Smoluchowski theory is valid for polydisperse suspensions and remains valid as polydisperse aggregates arise. Unfortunately, this advantage of the Smoluchowski theory can almost disappear when combined with the coalescence theory, because the coalescence rate coefficients are sensitive to droplet dimension. Thus, droplet and aggregate polydispersity does not strongly decrease the exactness of the description of coagulation in the CCC theory, while the exactness of coalescence description can be severely reduced.

Although the coalescence influence on the Brownian coagulation rate coefficient can be neglected, its influence on the final equations of the Smoluchowski theory remains. It can be shown that Smoluchowski's equation for the total number of particles

$$n(t) = \left(1 + \frac{t}{\tau_{Sm}}\right)^{-1} \tag{1.147}$$

remains valid, while in parallel the equations for the singlet and aggregate concentrations cannot be used with account for coalescence. Regarding coupled coagulation and coalescence, the Smoluchowski equation for $n_1(t)$ is not exact because it does not take into account the singlet formation caused by coalescence within doublets.

The coalescence within an aggregate consisting of i droplet is accompanied by the aggregate transforming into an aggregate consisting of $(i-1)$ droplets. As coalescence changes the aggregate type only, the total quantity of aggregates and singlets does not change. This means that the Smoluchowski function $n(t)$ does not change during coalescence, since Smoluchowski defined the total quantity of particles as consisting of aggregates and singlets.

1.9.3.2 Average Models

Average models do not assign rate constants to each possibility for coalescence within the aggregates, but deal with certain averaged characteristics of the process. The models in Refs. 9 and 10 introduce the average number of drops in an aggregate m, because the number of films in an aggregate n_f and m are interconnected. For a linear aggregate

$$n_f = m - 1 \tag{1.148}$$

As the coalescence rate for one film is characterized by τ_c^{-1}, the decrease in the average droplet quantity in an aggregate is n_f times larger. This is taken into account in the model by van den Tempel for simultaneous droplet quantity increase due to aggregation and decrease due to coalescence. Van den Tempel formulates the equation which describes the time dependence for the average number of droplets in an aggregate as

$$\frac{dm}{dt} = K_f N_{10} - \tau_c^{-1}(m - 1) \tag{1.149}$$

where the first term is derived using Smoluchowski theory.

The total number of droplets n_T is the sum of single droplets $n_1(t)$ and the droplets within aggregates:

$$n_T(t) = n_1(t) + n_v(t)m(t) \tag{1.150}$$

where n_v is the aggregate number. The latter can be expressed as

$$n_v(t) = n(t) - n_1(t) \tag{1.151}$$

Both terms are expressed by Smoluchowski theory. The integration of Equation 1.149 and the substitution of the result into Equation 1.150 yield the time dependence $n_T(t)$ according to the van den Tempel model.

1.9.3.2.1 The Model by Borwankar et al.

In Ref. 10 the van den Tempel model is criticized and improved through the elimination of Equation 1.149. The authors point out that the "incoming" aggregates which cause the increase in m have themselves undergone coalescence. This is not taken into account in the first term of the r.h.s. of Equation 1.149. Instead of taking a balance on each aggregate (as van den Tempel did) Borwankar et al. took an overall balance on all particles in the emulsion. For linear aggregates, the total number of films in the emulsion is given by

$$n_f n_v = (m - 1)n_v \tag{1.152}$$

Thus, instead of Equation 1.149 the differential equation for n_T follows:

$$-\frac{dn_T}{dt} = \tau_c^{-1}(m-1)n_v \tag{1.153}$$

where m can be expressed through n_T using Equation 1.150. The advantage of this equation in comparison with Equation 1.149 is obvious. However, there is a common disadvantage of both theories, caused with the use of the Smoluchowski equation for $n_1(t)$. Coalescence does not change the total particle concentration $n(t)$, but changes $n_1(t)$ and correspondingly $n_V(t)$, according to Equation 1.151.

The application of Smoluchowski theory in the quantification of the coupling of coalescence and coagulation has to be restricted with the use of the total particle concentration $n(t)$ only. The average models of van den Tempel and Borwankar et al. do not meet this demand.

The theory by Danov et al. [195] does not contradict this demand, which makes it more correct than the preceding theories. Among the Smoluchowski results the function $n(t)$ only is present in the final equations of this theory. Although the exactness of averaged models is reduced due to the violation of the restriction in the use of Smoluchowski theory, results for some limiting cases are not erroneous.

1.9.3.2.2 The Limiting Cases of Fast and Slow Coalescence

Two limiting cases can be distinguished:

1. The rate of coalescence is much greater than that of flocculation (rapid coalescence):

$$\tau_c^{-1} \gg \tau_{Sm}^{-1} \tag{1.154}$$

2. The rate of flocculation is much greater than that of coalescence (slow coalescence):

$$\tau_c^{-1} << \tau_{Sm}^{-1} \tag{1.155}$$

According to general regularities of physicochemical kinetics the slowest process is rate controlling. If the coagulation step is rate-controlling, namely when condition 1.154 is valid, then the coalescence is rapid and the general equation of the theory in Ref. 10 is reduced to second order kinetics, i.e., to Smoluchowski's equation (1.147). Flocs composed of three, four, etc. droplets cannot be formed, because of rapid coalescence within the floc. In this case the structure of the flocs becomes irrelevant.

At first glance the coagulation rate has not manifested itself in the entire destabilization process in the case of slow coalescence (condition 1.155). At any given moment the decrease in the total droplet concentration is proportional to the momentary total droplet concentration (first order kinetics), which causes an exponential decrease with time:

$$n = \exp\left(-\frac{t}{\tau_c}\right) \tag{1.156}$$

However, this equation cannot be valid for an initial short period of time, because at the initial moment there are no aggregates and their quantity continues to be low during a short time. This means that the coagulation is limiting during an initial time at any slow coalescence rate. This example illustrates the necessity of a more exact approach than that which uses average models. This was done by Danov et al. [195].

1.9.3.3 DIGB Model for the Simultaneous Processes of Coagulation and Coalescence

This kinetic model, proposed by Danov, Ivanov, Gurkov, and Borwankar, is called the DIGB model for the sake of brevity. Danov et al. [195] generalized the Smoluchowski scheme (Figure 1.27(a)) to account for droplet coalescence within flocs. Any aggregate (floc) composed of k particles can partially coalesce to become an aggregate of i particles ($1 < i < k$), with the rate constant being $K_c^{k,i}$ (Figure 1.27(b)). This aggregate is further involved in the flocculation scheme, which makes the flocculation and coalescence processes interdependent. Therefore, the system exhibiting both flocculation and coalescence is described by a combination of the schemes 1 and 2.

$$\frac{dn_k}{dt} = \sum_{i=1}^{k-1} K_f^{i,k-i} n_i n_{k-i} - 2\sum_{i=1}^{\infty} K_f^{k,i} n_k n_i + \sum_{i=k+1}^{\infty} K_c^{i,k} n_i - \sum_{i=1}^{k-1} K_c^{k,i} n_k \tag{1.157}$$

Equation 1.157 is multiplied by k and summed up for all k, that yields the equation for n_T which is expressed through double sums. The change of the operation sequence in these sums leads to the important and convenient equation

$$\frac{dn_T}{dt} = \sum_{k=1}^{\infty} k \sum_{i=k+1}^{\infty} K_c^{i,k} n_i - \sum_{i=2}^{\infty} k \sum_{i=1}^{k-1} K_c^{k,i} n_k \tag{1.158}$$

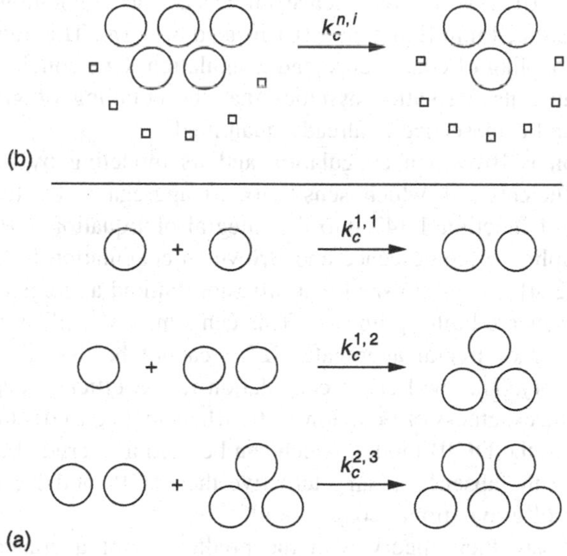

FIGURE 1.27 (a) Model of flocculation according to the Smoluchowski scheme. (b) Coalescence in an aggregate of k particles to become an aggregate of i particles, with a rate constant $K_c^{k,i}$, $1 < i < k$. (From K.D. Danov, I.B. Ivanov, T.D. Gurkov, and R.P. Borwankar, *J. Colloid Interface Sci.* 167: 8 (1994).)

Afterwards, a total rate coefficient referring to complete coalescence of the *i*th aggregate

$$K_{c,T}^{i} = \sum_{k=1}^{i-1} (i-k) K_c^{i,k} \quad i = 2, 3 \tag{1.159}$$

is introduced. For linearly built aggregates is derived

$$K_{c,T}^{i} = K_c^{2,1}(i-1) \tag{1.160}$$

With this expression for $K_{c,T}^{i}$, using also Equations 1.145 and 1.146, Equation 1.158 is transformed into

$$\frac{dn_T}{dt} = -K_c^{2,1}(n_T - n) \tag{1.161}$$

The integration result of this first order linear differential equation is well known and is represented in general form without specification of $n(t)$ (equation 1.18 in Ref. 195). An interesting peculiarity of this important derivation is the disappearance of terms, related to coagulation at the transition from the equation set 1.157 to the main equation (1.158). This corresponds to the fact that the total quantity of droplets does not change due to coagulation; it decreases due to coalescence only.

The coagulation regularity manifests itself in the $n(t)$ dependence, arising in Equation 1.161. It creates the illusion that Equation 1.161 can be specified for any $n(t)$ function corresponding to any subprocess affecting the droplet aggregate distribution. For example, the gravitational coagulation theory leads to a function $n_g(t)$ [4], but it does not create the opportunity to describe the gravitational coagulation coupling with coalescence by means of substituting $n_g(t)$ into the integral of Equation 1.161. As the coalescence influences the gravitational coagulation another function has to be substituted into Equation 1.161 instead of $n_g(t)$. This function has to be derived with account for the coupling of coalescence and coagulation. One concludes that Equation 1.161 cannot be used, because its derivation assumes that the coupling of gravitational coagulation (or another process) and coalescence is already quantified.

A happy exception is Brownian coagulation and its modeling by Smoluchowski with the coagulation rate coefficients, of which sensitivity to aggregate structure and coalescence is low. The substitution of function 1.147 into the integral of Equation 1.161 yields the equation, characterizing the coupling of coalescence and Brownian coagulation [195].

In fractal theory [224] it is established that diffusion limited aggregates and diffusion limited cluster–cluster aggregates are built up linearly. This can simplify application of the DIGB model. However, the diffusivity of fractal aggregates [225] cannot be described by simple equations and Smoluchowski theory. This will cause coagulation rate coefficient dependence on aggregate structure, decreasing the exactness of Equation 1.161 when applied to fractal aggregates. However, there is no alternative to the DIGB model, which can be used as a crude but useful approximation in this case as well. In the absence of an alternative the DIGB model can be recommended for evaluation in the case of gravitational coagulation.

Danov et al. compare their theory with the predictions of averaged models for identical conditions. It turns out that if coalescence is much faster than flocculation, the predictions of the different models coincide. Conversely, for slow coalescence the results of the averaged models deviate considerably from the exact solution. These two results of the comparison are in agreement with the qualitative considerations in Section 1.9.3.2.

Data for the relative change in the total number of droplets as a function of time are presented in Figure 1.28 (from Ref. 195). Figures a to c refer to $K_F N_{10} = 0.1\ \mathrm{s}^{-1}$ and the coalescence constant $K_c^{2,1}$ varies between $0.1\ \mathrm{s}^{-1}$ (a) and $0.001\ \mathrm{s}^{-1}$ (c). It is seen that the agreement between the Danov et al. and Borwankar et al. models is better the faster the coalescence, as was explained qualitatively above. The van den Tempel curves deviate considerably from the other two solutions.

For very long times, and irrespective of the values of the kinetic parameters, the model by Borwankar et al. [10] is close to the numerical solution. This is probably because the longer the time, the smaller is the concentration of single droplets. In this extreme case the error caused in the average models due to the influence of coalescence on the singlet concentration (not taken into account in the equation for $n(t)$) is negligible.

The shortcomings of the averaged models [9,10] and the advantages of the DIGB model are demonstrated in Ref. 195. However, the range of applicability of this model is restricted by many simplifications and the neglect of other subprocesses (see Section 1.9.3.1). An efficient analytical approach was made possible due to the neglect of the coalescence rate coefficient's dependence on the dimensions of both interacting droplets.

The model by Borwankar et al. was examined experimentally in Ref. 196. The emulsions were oil-in-water with soybean oil as the dispersed phase, volume fraction 30%, and number concentration 10^7 to $10^{10}\ \mathrm{cm}^{-3}$. The emulsions were gently stirred to prevent creaming during the aging study. A sample was placed on a glass slide, all aggregates were broken up, and the size of the individual droplets was measured. A rather good agreement with the theory was established. However, the fitting of the experimental data was accomplished using two model parameters, namely the coalescence and coagulation rate coefficients. For the last coefficient the optimal values (different for two emulsions) were obtained, strongly exceeding the Smoluchowski theory value (Section 1.9.2.1). An interpretation is that orthokinetic and perikinetic coagulation took place simultaneously due to stirring. Several experiments are known (discussed in Ref. 14) which demonstrate better agreement with the value for the coagulation constant predicted in Smoluchowski theory.

1.9.4 Doublet Fragmentation Time

1.9.4.1 Theory of Doublet Fragmentation Time

A doublet fragmentation was described by Chandrasekhar [226] by the diffusion of its droplets from the potential minimum, characterizing their attraction. The time scale for this process takes the form [110]

$$\tau_d = \frac{6\pi\eta a^3}{\kappa T}\exp(-U_{\min}/\kappa T) \tag{1.162}$$

where $U_{\min}$ is the depth of the potential minimum.

To derive the formula for the average lifetime of doublets Muller [227] considered the equilibrium in a system of doublets and singlets: that is, the number of doublets decomposing and forming are equal. Both processes are described by the standard diffusion flux J of particles in the force field of the particle that is regarded as central.

Each doublet is represented as an immovable particle with the second singlet "spread" around the central one over a spherical layer, which corresponds to the region of the potential well. The diffusion flux J of "escaping" particles is described by equations used in Fuchs' theory of slow coagulation. The first boundary condition corresponds to the assumption that the escaping particles

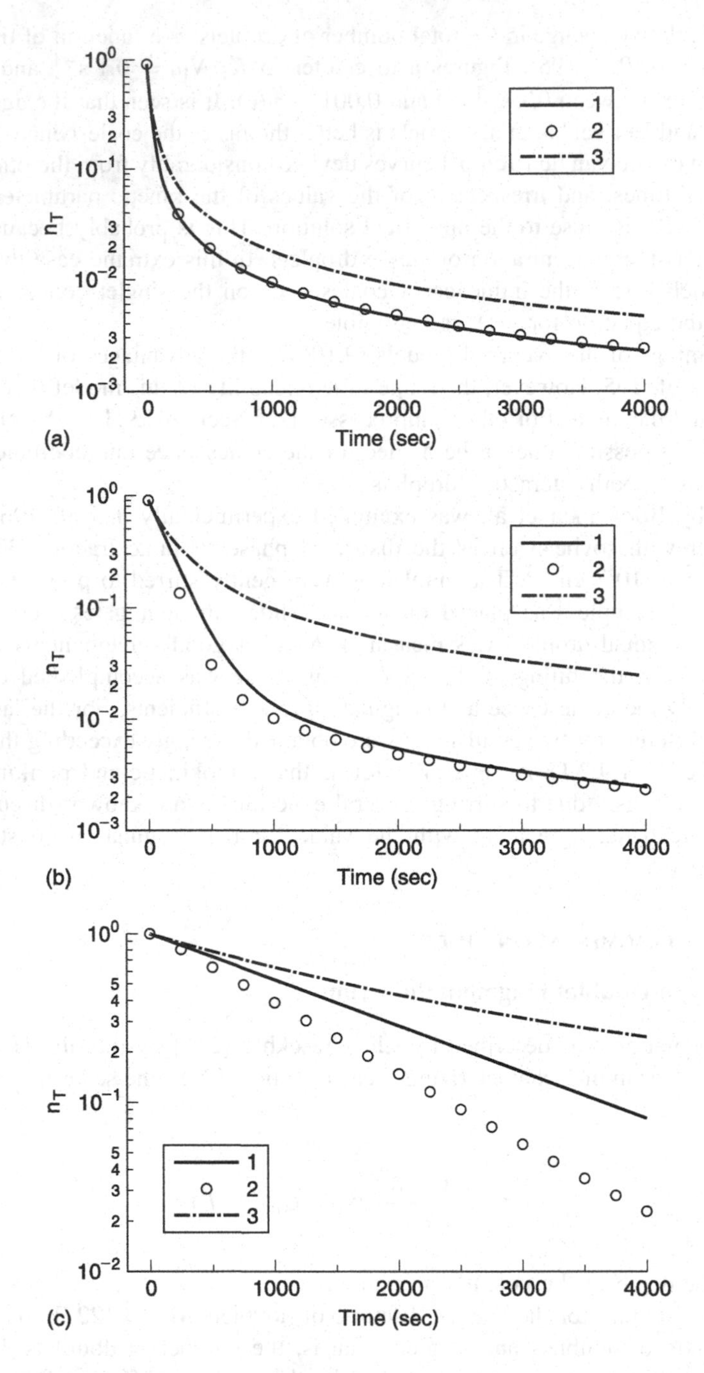

FIGURE 1.28 Relative change in the total number of droplets vs. time: initial number of primary particles $N_{10} = 1 \times 10^{10}$ cm^{-3}; curve 1, the numerical solution of the set Equation 1.157; curve 2, the model of Borwankar et al. [10] for diluted emulsions; curve 3, the model of van den Tempel [9]: (a) coalescence rate constant $K_c^{2,1} = 1 \times 10^{-1}$ s^{-1}; (b) $K_c^{2,1} = 1 \times 10^{-2}$ s^{-1}; (c) $K_c^{2,1} = 1 \times 10^{-3}$ s^{-1}. (From K.D. Danov, I.B. Ivanov, T.D. Gurkov, and R.P. Borwankar, *J. Colloid Interface Sci.* 167: 8 (1994).)

do not interact with other singlets. The second condition reflects the fact that the potential well contains exactly one particle.

At small separation between the droplets in a doublet the droplet diffusivity reduces because of the increasing hydrodynamic resistance during the droplet approach. A convenient interpolation formula was used [227] for the description of the influence of hydrodynamic interaction on the mutual diffusivity. The difference between the more exact Muller equation and Equation 1.162 is caused mainly on account of this hydrodynamic interaction.

1.9.4.2 Doublet Fragmentation Time of Uncharged Droplets

In this section we consider a doublet consisting of droplets with a nonionic adsorption layer. The closest separation between two droplet surfaces s exceeds the double thickness of the adsorption layer ($2h_a$). As a crude approximation h_0 can be identified with $2h_a$. In the case of small surfactant molecules $2h_a \approx 2$ nm.

In this case, the potential well has a sharp and deep minimum. This means that the vicinity of this minimum determines the value of the integral (Equation 1.162). For examination of this assumption Equation 1.162 was calculated numerically and according to the approximate equation [184]:

$$\int_{\alpha}^{\beta} \varphi(t) \exp[f(t)]dt = \left[\frac{2\pi}{f''(t_m)}\right]^{1/2} \varphi(t_m) \exp[f(t_m)] \tag{1.163}$$

where t_m corresponds to the potential well minimum.

The difference in results was small and enabled application of Equation 1.163 for calculation and substitution of the asymptotic expression [1,170]:

$$U(h) = -\frac{A}{12} \cdot \frac{a}{h} \tag{1.164}$$

which is valid at small distances to the surface. The result of calculations according to Equations 1.163 and 1.164 (the Hamaker constant $A = 1.3 \times 10^{-20}$ J) are shown in Figure 1.29. The chosen value of the Hamaker constant is consistent with those reported elsewhere [25,147]. In addition to the value of $A = 1.3 \times 10^{-20}$ J, we mention other values of the Hamaker constant which were employed elsewhere. For example, in food emulsions [25] the Hamaker constant lies within the range 3×10^{-21} to 10^{-20} J. The results of calculations for smaller Hamaker constants are also presented in Figure 1.29.

The influence of the adsorption layer thickness on doublet lifetime is shown in Figure 1.30 for one value of the Hamaker constant. There is high specificity in the thickness of a polymer adsorption layer. β-casein adsorbed onto polystyrene latex causes an increase in the radius of the particle of 10 to 15 nm [228]. A layer of β-lactoglobulin appears to be in the order of 1 to 2 nm thick, as compared to 10 nm for the caseins [229].

When adsorbed layers of hydrophilic nature are present the repulsive hydration forces must be taken into account. At low ionic strengths, the repulsion follows the expected exponential form for double-layer interaction:

$$U(h) = K_s e^{-(h/h_s)} \tag{1.165}$$

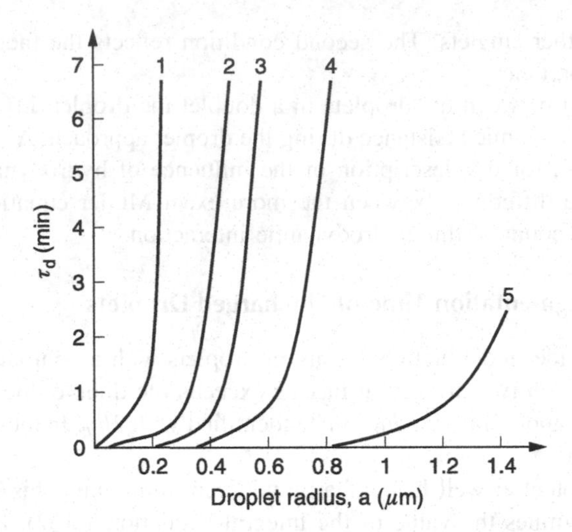

FIGURE 1.29 Dependence of doublet lifetime on droplet dimension at different values of the Hamaker constant A: 1: $A = A_1 = 1.33 \times 10^{-20}$ J; 2: $A = 0.5 \times A_1$; 3: $A = 0.35 \times A_1$; 4: $A = 0.25 \times A_1$; 5: $A = 0.1 \times A_1$. The shortest interdroplet distance 2 nm. (From S.S. Dukhin, J. Sjöblom, D.T. Wasan, and Ø. Sæther, *Colloids Surfaces*, 180: 223 (2001).)

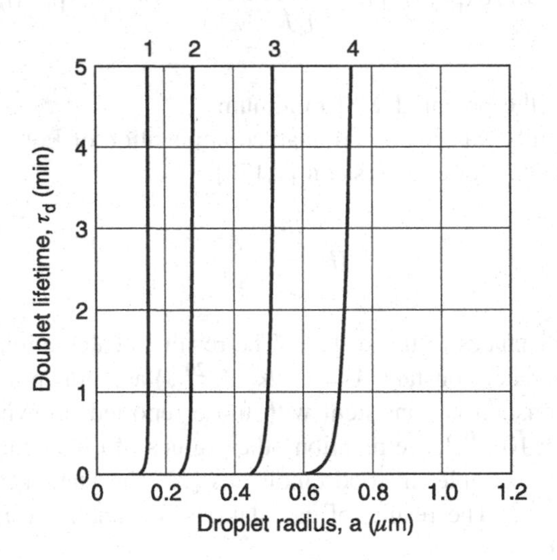

FIGURE 1.30 The adsorption layer thickness influence on the droplet lifetime of an uncharged droplet. Adsorption layer thickness: 1: $h_0 = 1$ nm; 2: $h_0 = 2$ nm; 3: $h_0 = 4$ nm; 4: $h_0 = 6$ nm. $A = 1.33 \times 10^{-20}$ J. (From S.S. Dukhin, J. Sjöblom, D.T. Wasan, and Ø. Sæther, *Colloids Surfaces*, 180: 223 (2001).)

In Ref. 31 the authors emphasize that the surface charge in food emulsions is low, electrolyte concentrations are high, and, hence, the DL is not responsible for emulsion stability. The stabilization can be caused by the hydration forces. However, the flocculation to the secondary minimum remains. Meanwhile, this conclusion must be specified with account for droplet dimension.

1.9.4.3 Lifetime of a Doublet of Charged Droplets and Coagulation/Flocculation

As seen in Figure 1 of Ref. 194 the coordinates of the secondary minimum corresponds to $\kappa h_{\min} =$ 5 to 12. Due to this rather large distance the frequency dependence of the Hamaker constant may be of importance, and the Hamaker function $A(h)$ characterizing molecular interaction should be introduced.

In Refs. 37 and 38 the distance independent interaction at zero frequency and interaction at non-zero frequency is considered separately:

$$A(h) = A_0 + [A(h) - A_0] \tag{1.166}$$

The results from 36 systems in Refs. 37 and 38 are in a rather good accordance with the calculations of other papers. According to Rabinovich and Churaev, the system polystyrene–water–polystyrene can be used to estimate the Hamaker function for oil–water systems. However, with increasing droplet separation the importance of A_0 is increasing on account of $[A(h) - A_0]$. The component A_0 is screened in electrolyte concentrations, because of dielectric dispersion [230–231]. At a distance of $h_{\min} \approx 3$ to 5 nm the authors [231] found that molecular interaction disappeared at zero frequency. Experimental evidence concerning this statement is discussed in Ref. 170. When evaluating the secondary minimum coagulation, A_0 can be omitted as illustrated in Ref. 232.

For the illustration of the influence electrolyte concentration, Stern potential, and particle dimension some calculations of doublet lifetime are made and their results are presented in Figure 1.31. The potential well depth increases and in parallel doublet lifetime increases with increasing particle dimension and electrolyte concentration and decreasing surface potential.

1.9.5 Coalescence Coupled with Either Coagulation or Flocculation in Dilute Emulsions

Limited attention is paid to the role of fragmentation in emulsion science. A comparison of the prediction of coalescence with and without accounting for fragmentation (Sections 1.9.2 and 1.9.3) enables evaluation of the fragmentation significance. This comparison will be done in Section 1.9.5.1 below.

The theories [181,195] have different areas of applicability (not specified in the papers) and are complementary. Naturally, this complicates the choice between these theories with account for concrete conditions of experiments. An approximate evaluation of the aforesaid areas of applicability is given in Section 1.9.5.2.

1.9.5.1 Fragmentation of Primary Flocs in Emulsions and the Subsequent Reduction of Coalescence

Floc fragmentation decreases the quantity of interdroplet films and correspondingly reduces the entire coalescence process. This reduction can be characterized by comparison of Equation 1.138 with theory [195], which neglects fragmentation. The longer the time, the greater the reduction which enables the use of the simpler theory [10] for comparison. The results for longer times coincide with the predictions of the more exact theory [195].

The results of theory [195] concerning slow coalescence are illustrated by curve 1 in Figure 3c in Ref. 195 which is redrawn as Figure 1.32(a). It can be seen that for a low value of the coalescence rate constant, the semilogarithmic plot is linear, indicating that the process follows a coalescence rate-controlled mechanism according to Equation 1.156. Differing from the simple

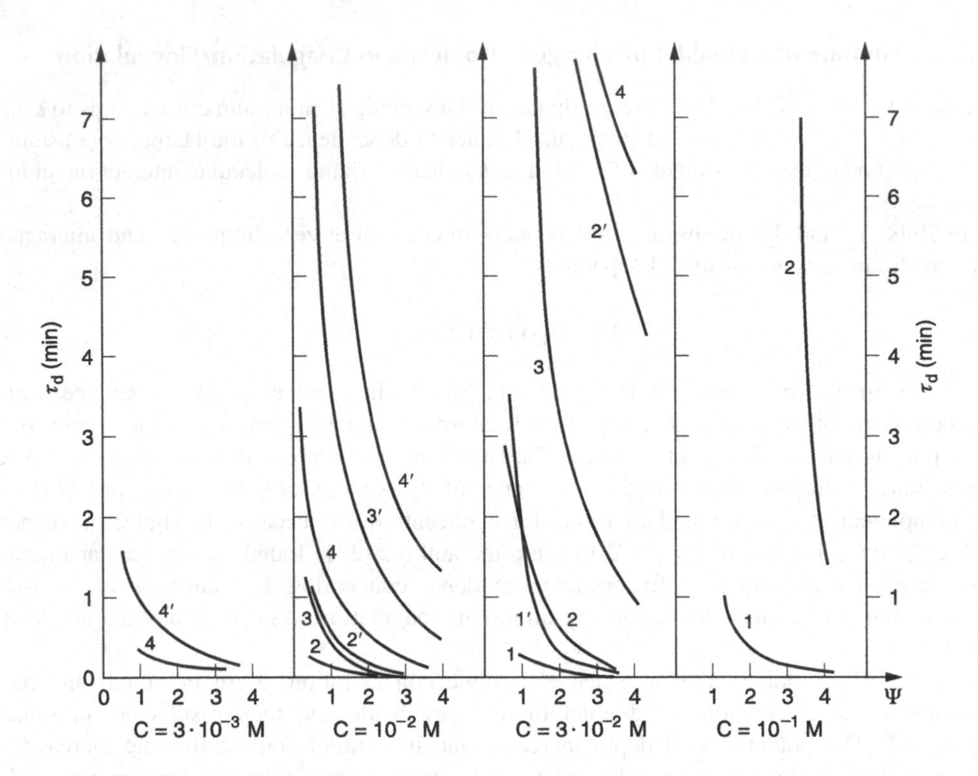

FIGURE 1.31 The dependence of doublet lifetime on the Stern potential for different electrolyte concentrations and droplet dimensions. Numbers near curves correspond to droplet radius. 1. Curves 1–4 without account for retardation of molecular forces of attraction, $\Psi = e\psi/kT$. 2. Curves 1′–4′ with account for retardation. (From S.S. Dukhin, J. Sjöblom, D.T. Wasan, and Ø. Sæther, *Colloids Surfaces*, 180: 223 (2001).)

exponential time dependence in Equation 1.156, second order kinetics dominate at rapid doublet fragmentation, even if coalescence is very slow. The physical reason becomes clear when considering how Equation 1.138 is derived. As seen from Equation 1.137 the rate of decline in the droplet concentration is proportional to the doublet concentration. The latter is proportional to the square of the singlet concentration at s.d.e., which causes second order kinetics. Thus, at slow coalescence the disaggregation drastically changes the kinetic law of the coalescence, i.e., from the exponential law to second order kinetics.

In the second stage coagulation becomes the rate controlling process because of the decrease in the collision rate accompanying the decrease in the droplet concentration. Thus, at sufficiently long times second order kinetics characterize both reversible and irreversible aggregation. Nevertheless, a large difference exists even when identical functions describe the time dependence, as the characteristic times are expressed through different equations for irreversible and reversible aggregation. In the first case it is the Smoluchowski time, in the second case it is the combination of three characteristic times, i.e., Equation 1.138.

Let us now try to quantitatively characterize the reduction in coalescence caused by doublet disintegration. For this purpose the calculations are performed according to Equation 1.126 at $\tau_{Sm} = 10$ s and $\tau_c = 10^3$ s (Figure 1.32(a)), 10^2 s (Figure 1.32(b)) and 10 s (Figure 1.32(c)). For all figures the same value of the ratio $2\tau_d/\tau_{Sm} = 0.1$ is accepted, satisfying condition 1.122. In all these figures the calculations according to Equation 1.138 are illustrated by curve 2.

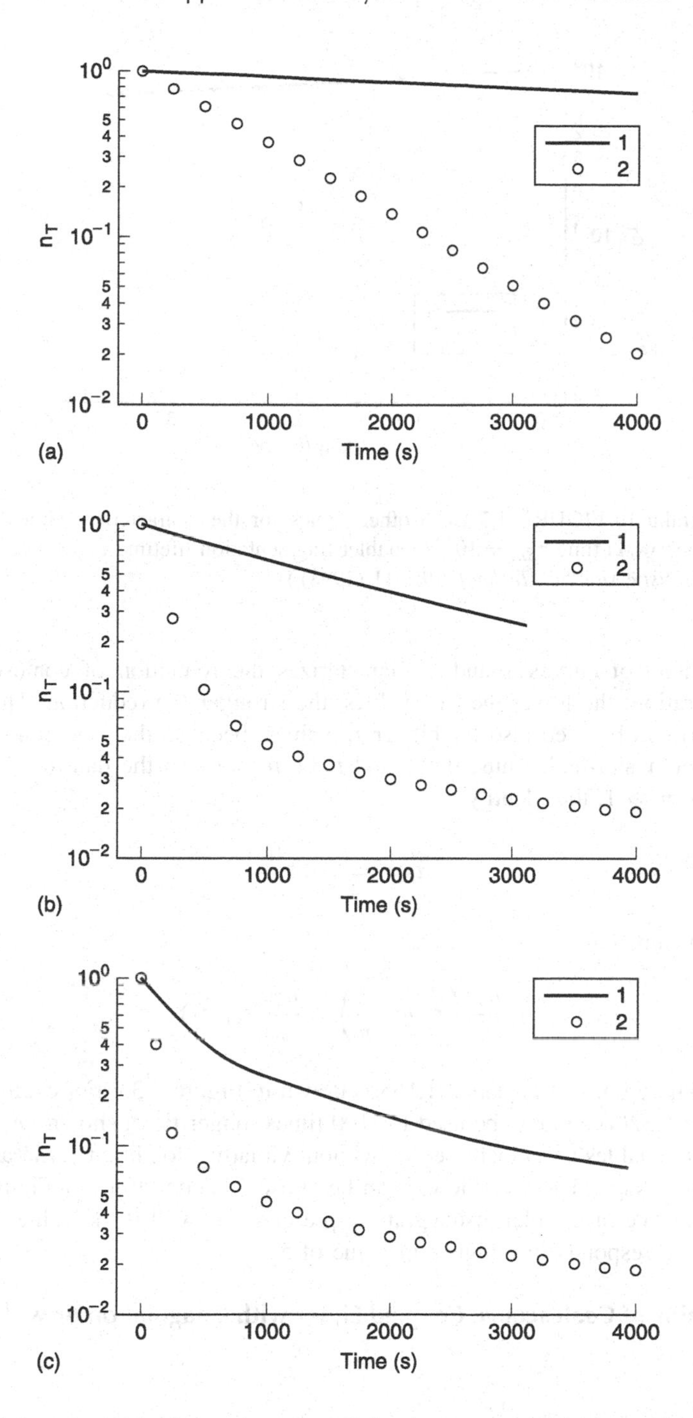

FIGURE 1.32 Relative change in the total number of droplets vs. time; initial number of droplets $N_{10} = 1 \times 10^{10}$ cm^{-3}; flocculation rate constant $K_f = 1 \times 10^{-11}$ cm^3 s^{-1}; curve 1: calculations according to Equation 1.138; curve 2: the model of Borwankar et al. [10] for dilute emulsions, coalescence rate constant (a) $K_c^{2,1} = 1 \times 10^{-1}$ s^{-1}, (b) $K_c^{2,1} = 1 \times 10^{-2}$ s^{-1}, (c) $K_c^{2,1} = 1 \times 10^{-3}$s^{-1}. Coalescence time $\tau_c = 10^3$ s (a), $\tau_c = 10^2$ s (b), $\tau_c = 10$ s (c). Smoluchowski time $\tau_{Sm} = 10$ s. Doublet lifetime $\tau_d = 0.5$ s. n_T is the dimensionless total droplet concentration, $n_T = N_T/N_{10}$. (From S.S. Dukhin and J. Sjöblom, *J. Dispersion Sci. Technol.* 19: 311 (1998).)

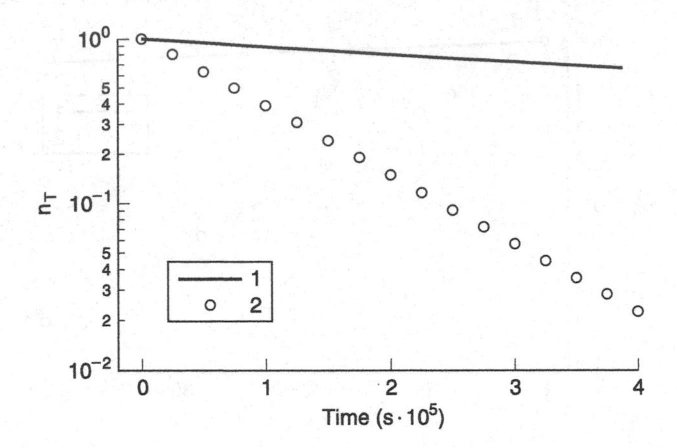

FIGURE 1.33 Similar to FIGURE 1.7, with other values for the characteristic times. Coalescence time $\tau_c = 10^5$ s. Smoluchowski time $\tau_{Sm} = 10^3$ s. Doublet fragmentation lifetime $\tau_d = 50$ s. (From S.S. Dukhin and J. Sjöblom, *J. Dispersion Sci. Technol.* 19: 311 (1998).)

The comparison of curves 1 and 2 characterizes the reduction of coalescence caused by doublet disintegration; the lower the Rev values, the stronger the reduction. The simple curve 1 in Figure 1.32(a) can be used also for higher τ_c values, because then the condition of Equation 1.156 is even better satisfied. Thus, if τ_{c1} and t_1 correspond to the data of Figure 1.32(a) and $\tau_{c2} = m\tau_{c1}$ with $m \gg 1$, the identity

$$\tau_{c2}t_2 = \tau_{c1}m\frac{t_1}{m} \tag{1.167}$$

is useful. This means that

$$\frac{n_T}{n_o}\left(\tau_{c1}m, \frac{t_1}{m}\right) = \frac{n_T}{n_o}(\tau_{c1}, t_1) \tag{1.168}$$

i.e., $t_2 = t_1/m$ where r.h.s. of Equation 1.168 is drawn in Figure 1.33. For example, Figure 1.8 is similar to Figure 1.32(a) and can be used for 100 times longer time, shown on the abscissa axis. The increase in τ_c enables us to increase τ_{Sm} without violating condition 1.155 and with Equation 1.156 valid. Thus, $\tau_{Sm} = 1000$ s or lower can be chosen as condition for Figure 1.33. Curve 2, characterizing the rate of doublet disintegration, preserves as well if the value of $2\tau_d/\tau_{Sm} = 0.1$ remains; now it corresponds to a higher τ_d value of 5 s.

1.9.5.2 Domains of Coalescence Coupled Either with Coagulation or with Flocculation

The condition

$$\text{Rev} \gg 1 \tag{1.169}$$

corresponds to coagulation. The theory for the intermediate case

$$\text{Rev} \sim 1 \tag{1.170}$$

when part of the droplets participate in flocculation and another coagulate is absent. To specify the conditions 1.122 and 1.169 the doublet lifetime must be expressed through surface force characteristics, namely through the surface electric potential and the Hamaker function, and droplet dimension, as was described in Section 1.9.4.

In the equation for the Smoluchowski time (1.124) the droplet numerical concentration N_{10} can easily be expressed through the droplet volume fraction φ and the average droplet radius a (we replace a polydisperse emulsion by an "equivalent" monodisperse emulsion). The resulting analysis respective to a and φ is easier than relating to N_{10} because the boundary of application of different regularities are usually formulated respective to a and φ. The Smoluchowski time is

$$\tau_{Sm} = \kappa_F^{-1} \varphi^{-1} \frac{4}{3} \pi a^3 \tag{1.171}$$

We exclude from consideration a special case of extremely dilute emulsions. Comparing Figure 1.31 and the results of calculations according to Equation 1.171 one concludes that condition 1.169 is mainly satisfied. It can be violated if simultaneously the droplet volume fraction and the droplet dimension are very small. This occurs if $\varphi < 10^{-2}$ and $a < (0.2 - 0.3)$ μm. Discussing this case we exclude from consideration the situation when $a \leq 0.1$ μm, corresponding to microemulsions and $\varphi \ll 10^{-2}$. With this exception one concludes that for uncharged droplets flocculation is almost impossible because condition 1.122 cannot be satisfied. A second conclusion is that at

$$a < (0.2 - 0.3)\ \mu\text{m} \tag{1.172}$$

theory 195 cannot be applied without some corrections made necessary by the partially reversible character of the aggregation. The main conclusion is that when

$$a > 0.2\ \mu\text{m and } \varphi > 10^{-2} \tag{1.173}$$

theory 195 does not need corrections respective to the reversibility of flocculation. However, this conclusion will change at the transition to a thicker adsorption layer. As described in Section 1.9.4, the thicker the adsorption layer, the shorter is the doublet fragmentation time.

The electrostatic repulsion decreases the depth of the potential well and correspondingly decreases the doublet lifetime. As a result, flocculation becomes possible for submicrometer droplets as well as for micrometer-sized droplets, if the electrolyte concentration is not too high, the surface potential is rather high, and the droplet volume fraction is not too high. This is seen from Figure 1.31.

The reversibility criterion depends on many parameters in the case of charged droplets. To discriminate and to quantify the conditions of coagulation and flocculation let us consider Rev values lower than 0.3 as low and values higher than 3 as high. In other words, coagulation takes place when Rev > 3, while at Rev < 0.3 there is flocculation; that is, the conditions

$$\text{Rev} = \frac{\tau_d(c_o, \varphi, a)}{\tau_{Sm}(a, \varphi)} > 3 \tag{1.174}$$

$$\text{Rev} < 0.3 \tag{1.175}$$

determine the boundaries for the domains of coagulation and flocculation. These domains are characterized by Figure 1.34 and correspond to fixed values of the droplet volume fraction.

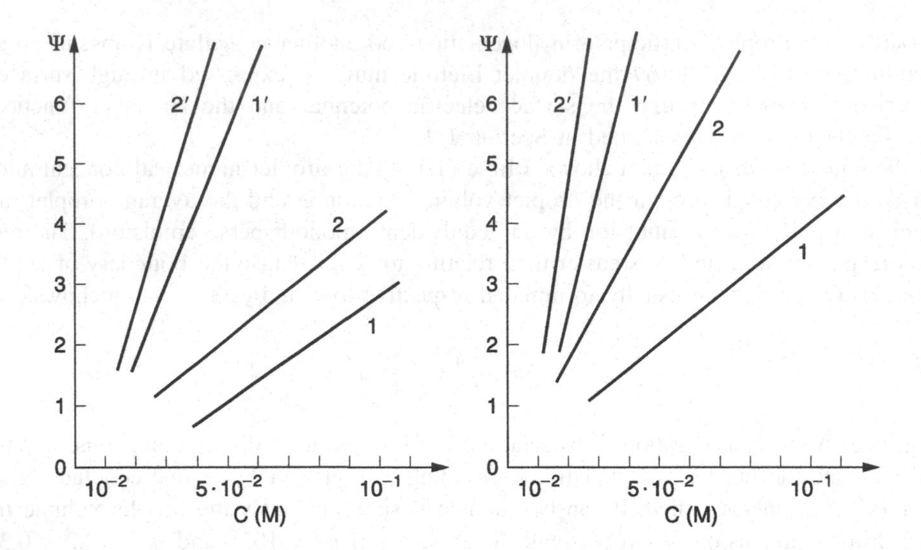

FIGURE 1.34 Domains of coagulation and flocculation. Curves 1 and 2 are calculated with the Rabinovich–Churaev Hamaker function; twice higher value is used for calculation of curves 1′ and 2′. The domain of flocculation is located above curve 1, while the domain of coagulation is located beneath curve 2. Volume fraction $\varphi = 0.01$ (*a*); $\varphi = 0.1$ (b). Particle dimension $2a = 4$ μm. (From S.S. Dukhin, J. Sjöblom, D.T. Wasan, and Ø. Sæther, *Colloids Surfaces* 180: 223 (2001).)

In addition, a definite and rather large droplet dimension $2a = 4$ μm is fixed. After fixation of the values of volume fraction and droplet dimension the domains are characterized in coordinates Ψ and C.

The domain of flocculation is located above and to the left of curve 2; the domain of coagulation is located beneath and to the right of curve 1. To characterize the sensitivity of the domain boundaries to the Hamaker function value, curves 1′ and 2′ are calculated using values two times higher than those of curves 1 and 2.

In distinction from uncharged droplets flocculation in the range of micrometer-sized droplets is possible. As seen in Figure 1.34, even rather large droplets (4 μm) aggregate reversibly if the electrolyte concentration is lower than $(1 - 5) \times 10^{-2}$ M and the Stern potential is higher than 25 mV. For smaller droplets the domain of flocculation will extend while the domain of coagulation will shrink. For submicrometer droplets flocculation takes place even at high electrolyte concentrations (0.1 M).

1.9.5.3 Hydration Forces Initiate Flocculation

Due to similar dependence on the distance h between hydration forces and electrostatic interaction the doublet lifetime decrease caused by hydration forces of repulsion can be calculated on account of this similarity. It is sufficient to use the substitution h_s for κ^{-1} and K_s for

$$16\varepsilon\left(\frac{kT}{e}\right)^2 \tanh\left(\frac{e\psi}{kT}\right)^2 \tag{1.176}$$

where k is the Boltzmann constant, T is absolute temperature, and e is the elementary charge. The doublet lifetime can be determined with the use of the results presented in Figure 1.31.

For the sake of brevity, the similar figure with K_s given on the ordinate axis and h_s on the abscissa axis is not shown. It turns out that the decrease in τ_d caused by hydration forces leads to flocculation of submicrometer droplets. As to micrometer-sized droplets, coagulation takes place with the exception for the case when both h_s and K_s are rather large.

1.9.6 Applications

The restrictions in Equations 1.121 and 1.135 corresponding to strong retardation of the rate of multiplet formation and slow intradoublet coalescence are not frequently satisfied. Nevertheless, these conditions are important because they correspond to the case of very stable emulsions. As the kinetics of the retarded destabilization of rather stable emulsions is of interest, attention has to be paid to provide these conditions and thus the problem of coupled coalescence and flocculation arises.

There are large qualitative distinctions in the destabilization processes for the coupling of coalescence and coagulation, and coalescence and flocculation. In the first case, rapid aggregation causes rapid creaming and further coalescence within aggregates. In the second case, the creaming is hampered due to the low concentration of multiplets and coalescence takes place both before and after creaming. Before creaming singlets predominate for a rather long period of gradual growth of droplet dimensions due to coalescence within doublets. The discrimination of conditions for coupling of coalescence with either flocculation or coagulation is accomplished in Ref. 184.

The creaming time is much shorter in the coagulation case and correspondingly the equation describing the coupling of coalescence and flocculation preserves its physical sense for a longer time than is the case for coagulation. One concludes that the theory of the coupling of coalescence and flocculation provides a new opportunity for long-term prediction of emulsion stability, although creaming restricts the application of this theory as well. Note that this restriction weakens in the emulsions of low density contrast and in w/o emulsions with a high viscosity continuum.

Long-term prediction is a two-step procedure. The first step is the determination of whether an emulsion exhibits coagulation or flocculation. It means that the characteristic time τ_d must be measured and compared with τ_{Sm}, the value of which is easily evaluated with account for measured concentration using Equation 1.124. A comparison of these times enables the choice between condition 1.121 and the opposite condition ($\tau_{Sm} \ll \tau_d$). The second step is the prediction of the evolution in time for the t.d.c. If condition 1.121 is valid Equation 1.138 has to be used for the prediction. τ in Equation 1.138 has to be specified in accordance with Equation 1.135. In the opposite case DIGB theory must be used.

1.9.6.1 Long-Term Prediction of Emulsion Stability

It is possible, in principle, to give a long-term prediction of emulsion stability based on the first indications of aggregation and coalescence. The next example clarifies the principal difficulty in a reliable long-term prediction, if a dynamic model of the emulsion is not available.

The first signs of aggregation and coalescence can always be characterized by a linear dependence, if the investigation time t is small in comparison with a characteristic time τ for the evolution of the total droplet concentration $n(t)$:

$$n(t) = n_0\left(1 - \frac{t}{\tau}\right) \qquad (1.177)$$

This short time asymptotic corresponds to many functions, for example to Equation 1.138 or to Equation 1.156. The first can arise in the case of coalescence coupled with coagulation [195], while the second can arise for coalescence coupled with flocculation [187]. The discrimination between irreversible and reversible aggregations is only one component of emulsion dynamics modeling (EDM) and it is seen that without this discrimination the difference in the prediction of the time necessary for a droplet concentration decrease, for example 1000 times, can be 7τ and 1000τ.

1.9.6.2 Perfection of Methods for Emulsion Stabilization (Destabilization) by Means of the Effect on Both Coalescence and Flocculation

Stability (instability) of an emulsion is caused by the coupling of coalescence and flocculation. Meanwhile, for emulsifiers (or deemulsifiers) the elaboration of their influence on the elementary act of coalescence only is mainly taken into account. The coupling of coalescence and flocculation is reflected in Equation 1.135 and one concludes that it follows the multiplicativity rule and not the additivity rule. This means that the total result of the application of a stabilizer (destabilizer) depends very much on both flocculation and fragmentation. The development of a more efficient technology for emulsion stabilization (destabilization) is possible by taking into account the joint effect on both the coalescence and the aggregation (disaggregation) processes.

1.9.6.2.1 Combining Surfactants and Polymers in Emulsion Stabilization

The coalescence rate depends mainly on the thin (black) film stability and correspondingly on the short-range forces. The flocculation depends on the long-range surface forces. Due to this large difference synergism in the dependence of these processes on the different factors can be absent.

The use of one surfactant only may not provide both the optimal fragmentation and optimal stability of an emulsion film. Probably the use of a binary surfactant mixture with one component which provides the film stability, and a second one which prevents the flocculation, may provide perfect emulsion stabilization. Naturally their coadsorption is necessary. For such an investigation a measurement method for both the doublet fragmentation time and the coalescence time is necessary.

1.9.6.2.2 Strong Influence of Low Concentrations of Ionic Surfactant on Doublet Fragmentation Time and Coalescence Time

Let us consider the situation when an emulsion is stabilized against coalescence by means of an adsorption layer of nonionic surfactant and is strongly coagulated because of the subcritical value of the Stern potential that is usual for inorganic electrolytes [202] at moderate pH. In a large floc any droplet has many neighbors, meaning a rather high number of interdroplet films per droplet. The coalescence rate is proportional to the total number of films and can be rather high. It can be strongly decreased by adding a low concentration of an ionic surfactant. This can be sufficient to provide a supercritical Stern potential value that will be accompanied by a drastic decrease in the doublet lifetime compared to that of weakly charged droplets.

At shorter doublet lifetimes flocculation can become reversible and it can stop at the stage of singlet–doublet equilibrium. It will provide a strong decrease in the coalescence rate because coalescence occurs within doublets only and their concentration can be very low.

Thus, a small addition of an ionic surfactant to higher concentration of a nonionic surfactant sufficient to provide an almost saturated adsorption layer can make the overall emulsion stabilization more efficient. The nonionic surfactant suppresses coalescence but cannot prevent flocculation, while the ionic surfactant retards the development of flocculation.

We can give an example when both coalescence and flocculation are affected by an ionic surfactant (SDS). In Ref. 233 it is established that coalescence is suppressed at SDS concentrations exceeding 6×10^{-5} M. Meanwhile, the C.C.C. is 2×10^{-2} M NaCl at 10^{-6} M SDS. Thus, SDS concentrations slightly above 10^{-6} M are sufficient to retard flocculation. In this example it is essential that the concentrations needed to retard flocculation are very low compared to those needed to prevent coalescence.

It is noteworthy that low concentrations of an ionic surfactant can increase emulsion stability due to the simultaneous manifestation of three mechanisms. First, the depth of the secondary potential minimum decreases due to the electrostatic repulsion that is accompanied by a τ_d decrease. Second, the transition from the secondary minimum through an electrostatic barrier and into the primary minimum extends the coalescence time. Third, the time of true coalescence, i.e., the time necessary for thin film rupture, increases due to electrostatic repulsion as well [185,217].

1.9.6.3 Standardization of the Measurement of τ_c and τ_d

Direct investigation of the coalescence subprocess in emulsions is difficult. Instead, the entire destabilization process is usually investigated. Meanwhile, the rate of the destabilization process depends on the rates of both flocculation and disaggregation and on the floc structure as well. All these characteristics vary in a broad range. At a given unknown value for the time of the elementary act of coalescence τ_c the different times can be measured for the integrated process and different evaluations of τ_c are possible.

The rate of coalescence in an aggregate essentially depends on the number of droplets within it and the packing type, i.e., on the number of films between the droplets. This complication is absent when considering the case of the s.d.e.

The possible advantage of τ_c measurement at s.d.e. is in avoiding the difficulty caused by polydispersity of droplets appearing during preceding coalescence within large flocs. At the s.d.e. the initial stage of the entire coalescence process can be investigated when the narrow size distribution of an emulsion preserves.

At s.d.e. the determination of the time dependence of the t.d.c. is sufficient for the investigation of coalescence. In Refs. 185 and 186 this was accomplished through direct visual observation. By using video enhanced microscopy and computerized image analysis the determination of the t.d.c. can be automated. Such automated determination of total droplet number in a dilute DCD-in-water emulsion at the s.d.e. can be recommended as a standard method for the characterization of the elementary act of coalescence.

In parallel, the second important characteristic, namely the doublet fragmentation time, is determined by the substitution of τ_c, τ_{Sm}, and measured τ_d into Equation 1.138.

1.9.6.4 Experimental–Theoretical Emulsion Dynamics Modeling

1.9.6.4.1 General

To predict the evolution of the droplet (floc) size distribution is the central problem in emulsion stability. It is possible, in principle, to predict the time dependence of the distribution of droplets (flocs) if information concerning the main subprocesses (flocculation, floc fragmentation, coalescence, creaming), constituting the whole phenomenon, is available. This prediction is based on consideration of the population balance equation (PBE).

The PBE concept was proposed by Smoluchowski. He specified this concept for suspensions and did not take into account the possibility of floc fragmentation. Even with this restriction he succeeded in the analytical solution neglecting gravitational coagulation and creaming, and obtained the analytical time dependence for a number of aggregates n_i comprising i particles ($i = 2, 3\ldots$).

In the most general case the equation for the evolution of the total droplet number takes into account the role of aggregation, fragmentation, creaming, and coalescence. There is no attempt to propose an algorithm even for a numerical solution to such a problem.

The usual approach in the modeling of an extremely complicated process is the consideration of some extreme cases with further synthesis of the obtained results. The next three main simplifications are inherent to the current state of emulsion dynamics modeling: the neglect of the influence of the gravitational field, i.e., neglect of creaming/sedimentation; in a first approximation it is possible to consider either coagulation or flocculation; finally, the neglect of the rate constant dependence on droplet dimension.

1.9.6.4.2 Combined Approach in Investigations of Dilute and Concentrated Emulsions

The modeling of collective processes in concentrated emulsions is extremely complicated. Recently, the efficiency of computer simulation in the systematic study of aggregates, gels, and creams has been demonstrated [204]. Monte Carlo and Brownian dynamics are particularly suited to the simulation of concentrated emulsions. However, information about droplet–droplet interaction is necessary. The reliability of this information is very important to provide reasonable results concerning concentrated emulsions. In other words, the assumption concerning pair additive potentials for droplet/droplet interaction and the thin emulsion film stability must be experimentally confirmed. The extraction of this information from experiments with concentrated emulsions is very difficult. On the other hand, measurement of the doublet fragmentation time in dilute emulsions is a convenient method to obtain information about pair additive potentials. Information about pair potentials and the elementary act of coalescence obtained in experiments with dilute emulsions preserves its significance for concentrated emulsions as well.

One concludes that modeling of concentrated emulsions becomes possible by combining experimental investigation of the simplest emulsion model system with computer simulation accounting for the characteristics of a concentrated emulsion (high droplet volume fraction, etc.).

1.9.6.4.3 Kernel Determination

Kernel determination is the main task which must be solved to transform the PBE into an efficient method for emulsion dynamics modeling. The levels of knowledge concerning kernels describing different subprocesses differ strongly. There exists a possibility for quantification of kernels related to aggregation and fragmentation [173,194,227]. On the other hand, the current state of knowledge is not sufficient for prediction of the thin film disruption time.

The deficit in knowledge about thin film stability makes purely theoretical modeling of emulsion dynamics impossible. As a result a complex semi-theoretical approach to EDM is necessary. The PBE is the main component of both the experimental and the theoretical stages of this approach. In the experimental stage the PBE simplified regarding s.d.e. provides the background for the determination of the coalescence kernels with the use of the experimental data [185,186].

For the determination of the coalescence kernels the more complicated reverse task must be solved, namely their determination based on the comparison of experimental data about the emulsion evolution in time with the PBE solution. In the absence of an analytical solution the reverse task is usually very difficult. The most efficient way to overcome this difficulty is the experimental realization with the use of the universally simplest conditions for emulsion time evolution, which can be described analytically.

1.9.6.4.4 Singlet–Doublet Quasi-Equilibrium

Singlet–doublet quasi-equilibrium with slow coalescence within doublets is the simplest emulsion state for which investigation can provide information about coalescence.

The simplest singlet–doublet emulsion can exist at singlet–doublet quasi-equilibrium and slow coalescence within doublets. Its simplicity results in a very simple kinetic law for the entire kinetics of coupled flocculation and coalescence, namely Equation 1.138. Thus, s.d.e. provides the most convenient conditions for investigations of the elementary act of coalescence and the doublet fragmentation time.

The main simplification in all existing models for emulsion dynamics [181,195] is the neglect of the coalescence time dependence on droplet dimensions. This simplification is not justified and decreases very much the value of the prediction, which can now be made with use of the PBE. For elimination of this unjustified simplification it is necessary to determine the coalescence time for emulsion films between droplets of different dimensions i and j, namely τ_{cij}, similar to the existing analytical expressions for the doublet fragmentation time, τ_{dij} [173]. The determination of a large set of τ_{cij} values by means of a comparison of experimental data obtained for an emulsion consisting of different multiplets and the PBE numerical solutions for it is impossible. On the other hand, this paramount experimental–theoretical task can be solved for a dilute emulsion at s.d.e. and slow intradoublet coalescence.

1.9.6.4.5 Substitution of the Coalescence Kernels

Substitution of the coalescence kernels makes the PBE equation definite and ready for the prediction of emulsion time evolution with the restriction of low density contrast and without account for gravitational coagulation and creaming.

With application of the scaling procedure for the representation of the kinetic rate constants for creaming and gravitational coagulation the PBE is solved analytically in Ref. 234. This scaling theory creates a perspective for the incorporation of creaming in the emulsion dynamics model in parallel with coalescence, aggregation, and fragmentation.

1.10 SUMMARY

The mechanisms of kinetic stability in macroemulsions and mini-emulsions are completely different. The strong droplet deformation and flattening in a macroemulsion cause the Reynolds mode of drainage which prolongs the life of the emulsion. This mechanism is not important for mini-emulsion droplet interaction, because either the deformation and flattening are weak (charged droplets) or the Reynolds drainage is rapid due to the small dimension of the interdroplet film (uncharged droplets). The kinetic stability of a mini-emulsion can be caused by floc fragmentation if the electrokinetic potential is not too low and the electrolyte concentration is not too high, corresponding to some electrostatic repulsion.

The potential strength of physicochemical kinetics with respect to emulsions is the population balance equation (PBE), enabling prediction of the time evolution of the droplet size distribution (d.s.d.) when the subprocesses (including droplet aggregation, aggregate fragmentation, droplet coalescence, and droplet (floc) creaming) are quantified. The subprocesses are characterized in the PBE by the kinetic coefficients. The coupling of the four subprocesses, the droplet polydispersity, and the immense variety of droplet aggregate configurations causes extreme difficulty in EDM. The three processes of aggregation, fragmentation, and creaming can be quantified. In contrast, only the experimental approach is now available for efficient accumulation of information concerning emulsion film stability and coalescence kernel quantification for EDM.

Correspondingly, EDM may be accomplished by combining experiment and theory: (1) the determination of coalescence and fragmentation kernels with the use of emulsion stability experiments at low density contrast (l.d.c.) and singlet–doublet equilibrium (s.d.e.), because this enables the omittance of creaming and gravitational terms in PBE, simplifying it and making solution of the reverse task possible; and (2) the prediction of the droplet size evolution with time by means of solution of the PBE, specified for the determined coalescence and fragmentation kernels. This mathematical model has to be based on the PBE supplemented by terms accounting for the role of creaming and gravitational coagulation in the aggregation kinetics.

Emulsion dynamics modeling with experiments using l.d.c. emulsions and s.d.e. may result in: (1) the quantification of emulsion film stability, namely the establishment of the coalescence time dependence on the physicochemical specificity of the adsorption layer of a surfactant (polymer), its structure, and the droplet dimensions. This quantification can form a basis for the optimization of emulsifier and demulsifier selection and synthesis for emulsion technology applications, instead of the current empirical level applied in this area; (2) the elaboration of a commercial device for coalescence time measurement, which in combination with EDM will represent a useful approach to the optimization of emulsion technology with respect to stabilization and destabilization.

Symbols

LATIN

a	Radius (m)
A	Hamaker constant
A_{ij}	Hamaker constant for interaction
b	Target distance
c_i	Bulk concentration of surfactant
C	Concentration (mol m^{-3})
D	Diffusion coefficient of a droplet (m^2 s^{-1})
D	Diffusion coefficient of an ion or a surfactant molecule
e	Elementary charge (C)
f	Friction coefficient (kg s^{-1})
F	Faraday constant (C mol^{-1})
F	Force (N)
g	Standard acceleration of free fall ($m\,s^{-2}$)
h	Shortest distance between colloidal particles, drops (m)
J	Flux ($m^2\,s^{-1}$)
k_f	Flocculation rate constant
K	Boltzmann constant ($J\,K^{-1}$)
n_i	Number concentration of particles (aggregates) of size i
N_A	Avogadro's number
p_{12}	Pair distribution function
Pe	Pecklet number
$\mathbf{r}(r)$	Distance (m)

R	Gas constant ($J\,K^{-1}\,mol^{-1}$)
R_{ij}	Collision radius
Re	Reynolds number
$R_{ij} = a_i + a_j$	Collision radius
t	Time (s)
t_F	Characteristic flocculation time
T	Temperature (K)
$u_0(\hat{\mu})$	Instantaneous creaming velocity (Eq. 1.42)
u_0	Stokes creaming rate (Eq. 1.117)
U	Energy (J)
U, v	Velocity ($m\,s^{-1}$)
U_{12}	Potential function between drops 1 and 2
V	Volume (m^3)

GREEK

γ	Interfacial tension ($N\,m^{-1}$)
Γ	Surface (excess) concentration ($mol\,m^{-2}$)
δ	Diffusion layer thickness (m)
ε	Relative dielectric permittivity (dielectric constant)
ε_0	Dielectric permittivity of vacuum
χ_r	Retardation coefficient (Eq. 1.45)
ζ	Electrokinetic potential
θ	Angle of rotation
κ	Reciprocal Debye length (m^{-1})
λ	Size ratio of two drops, $\lambda = a_1/a_2 < 1$
μ	Kinematic viscosity ($J\,mol^{-1}$)
$\hat{\mu} = \mu_1/\mu_2$	Viscosity ratio
$\Pi(h)$	Disjoining pressure ($N\,m^2$)
ρ	Density ($kg\,m^{-3}$)
σ	Surface charge ($C\,m^{-2}$)
τ	Characteristic time (s)
Ψ	Electric potential (V)

REFERENCES

1. B. V. Derjaguin, Theory of Stability of Colloids and Thin Films, Nauka, Moscow, 1986 [in Russian]; translation: Theory of Stability of Colloids and Thin Films, Plenum, New York, 1989.
2. J. Gregory, Crit. Rev. Environ. Control 19:185 (1989).
3. M. von Smoluchowski, Phys. Z. 17:557, 585 (1916).
4. M. von Smoluchowski, Phys. Chem. 92:129 (1917).
5. Th.F. Tadros and B. Vincent, in Encyclopedia of Emulsion Technology, Vol. 1 (P. Becher, ed.), Marcel Dekker, Inc., New York, 1983, p. 57.
6. E. E. Kumacheva, E. A. Amelina, A. V. Prtsev, and E. D. Shchukin, Kolloidn. Zh. 40:1214 (1989).
7. K. Larsen, in Emulsions – A Fundamental and Practical Approach (J. Sjöblom, ed.), NATO ASI Series Vol. 363, 1991, p. 41.
8. H. J. Junginger, Emulsions – A Fundamental and Practical Approach (J. Sjöblom, ed.), NATO ASI Series Vol. 363, 1991, p. 207.
9. M. van den Tempel, Rec. Trav. Chim. 72:419, 433 (1953).
10. R. P. Borwankar, L. A. Lobo, and D. T. Wasan, Colloids Surfaces 69:135 (1992).

11. Ph.T. Jacger, J. J. M. Janssen, F. G. Groeneweg, and W. G. M. Agterof, Colloids Surfaces 85:255 (1994).
12. B. V. Derjaguin, N. V. Churaev, and V. M. Muller, Surface Forces, Nauka, Moscow, 1985 [in Russian]; translation: Surface Forces, Plenum, New York, 1987.
13. R. J. Hunter, Foundations of Colloid Science, Vol. 1, Oxford Science Publication, Oxford, 1987.
14. H. Sonntag and K. Strenge, Coagulation Kinetics and Structure Formation, VEB Dentscher Verlag der Wissenschaften, Berlin, 1987.
15. D. H. Melik and H. S. Fogler, in Encyclopedia of Emulsion Technology, Vol. 3, Marcel Dekker, Inc., New York, 1988, pp. 3–78.
16. N. A. Fuchs, Mechanika aerosoley, Izd-vo AN SSSR, 1955: The Mechanics of Aerosols, Pergamon, London, 1964.
17. V. M. Voloshchuk and Yu. S. Sedunov, Coagulation Processes in Disperse Systems, Hydrometeoizdat, Leningrad, 1975 [in Russian].
18. S. S. Dukhin, N. N. Rulyov, and D. S. Dimitrov, Coagulation and Dynamics of Thin Films, Naukova Dumka, Kiev, 1986 [in Russian].
19. B. V. Derjaguin, S. S. Dukhin, and N. N. Rulyov, in Surface and Colloid Science, Vol. 13 (E. Matijevich, ed.), Wiley Interscience, New York, 1983, p. 71.
20. H. J. Schulze, Physicalisch-Chemische Elementarvorgange des Flotations Processes, VEB Verlag der Wissenschaft, Berlin, 1981.
21. J. Sjöblom (ed.), J. Disper. Sci. & Technol. 15 (1994).
22. W. Stumm and J. J. Morgan, Aquatic Chemistry, Wiley, New York, 1981.
23. E. Dickenson, in Emulsions – A Fundamental and Practical Approach (J. Sjöblom, ed.), NATO ASI Series Vol. 363, 1991, p. 25.
24. M. P. Aronson, in Emulsions – A Fundamental and Practical Approach (J. Sjöblom, ed.), NATO ASI Series Vol. 363, 1991, p. 75.
25. P. Walstra, in Gums and Stabilizers for the Food Industry, Vol. 4 (G. O. Phillips, P. A. Williams, and D. I. Wedlock, eds.), IRL Press, Oxford, 1988, p. 233.
26. S. R. Reddy, D. H. Melik, and H. S. Fogler, J. Colloid Interface Sci. 82:116 (1981).
27. D. H. Melik and H. S. Fogler, J. Colloid Interface Sci. 101:72 (1984).
28. X. Zhang and R. Davis, J. Fluid. Mech. 230:479 (1991).
29. S. E. Friberg, in Emulsions – A Fundamental and Practical Approach (J. Sjöblom, ed), NATO ASI Series Vol. 363, 1991, p. 1.
30. J. A. Kitcherer and P. R. Musselwhite, The Theory of Stability in Emulsions in Emulsion Science (P. Sherman, ed.), Academic Press, London, 1968.
31. B. Bergenståhl and P. M. Claesson, in Food Emulsions (K. Larsson and S. Friberg, eds.), Marcel Dekker, Inc., 1989, pp. 41–96.
32. Ya. I. Rabinovich and N. V. Churaev, Kolloidn. Zh. 52:309 (1990).
33. H. C. Hamaker, Physica. 4:1058 (1937).
34. B. A. Pailtorpe and W. B. Russel, J. Colloid Interface Sci. 89:56 (1982).
35. N. F. H. Ho and W. I. Higuichi, J. Pharm. Sci. 57:436 (1968).
36. J. Lyklema, Fundamentals of Interface and Colloid Science, Academic Press, London, 1993.
37. Ya. I. Rabinovich and N. V. Churaev, Kolloidn. Zh. 41:468 (1979).
38. Ya. I. Rabinovich and N. V. Churaev, Kolloidn. Zh. 46:69 (1984).
39. R. J. Hunter, Zeta Potential in Colloid Science, Academic Press, London, 1981.
40. S. S. Dukhin, Adv. Colloid Interface Sci. 44:1 (1993).
41. S. S. Dukhin, B. V. Derjaguin, and N. M. Semenikhin, Dokl. Akad. Nauk SSSR 192:357 (1970).
42. B. V. Derjaguin, Theory of Stability of Colloids and Thin Films, Plenum, New York, 1989, Chap. 5, Sec. 5.
43. S. S. Dukhin and V. N. Shilov, Dielectric Phenomena and the Double Layer in Disperse Systems and Polyelectrolytes, Wiley, Toronto, 1974.
44. J. Kijlstra, H. P. van Leuven, and J. Lyklema, J. Chem. Soc. Faraday Trans. 88:3441 (1992).
45. B. R. Midmore and R. I. Hunter, J. Colloid Interface Sci. 122:521 (1988).
46. J. N. Israelachvili and R. M. Pashley, J. Colloid Interface Sci. 98:500 (1984).

47. P. M. Claesson and H. K. Christenson, J. Phys. Chem. 92:1650 (1988).
48. H. K. Christenson, P. M. Claesson, J. Berg, and P. C. Herder, J. Phys. Chem. 93:1472 (1989).
49. R. M. Pashley and J. N. Israelachvili, J. Colloid Interface Sci. 97:446 (1984).
50. B. V. Derjaguin and Z. M. Zorin, Zh. Fiz. Khim. 29:1010 (1955).
51. G. Frens and J. Th. G. Overbeek, J. Colloid Interface Sci. 38:376 (1972).
52. T. W. Healy, A. Homola, and R. O. James, Faraday Discuss. Chem. Soc. 65:156 (1978).
53. P. Meakin, Phys. Rev. Lett. 51:1119 (1983).
54. M. Kilb, Phys. Rev. Lett. 53:1653 (1984).
55. T. Vicsek and F. Family, Phys. Rev. Lett. 52:1669 (1984).
56. P. Meakin, I. Vicsek, and F. Family, Phys. Rev. B31:564 (1985).
57. A. Einstein, Ann. Phys. 17:549 (1905); 19:371 (1906).
58. A. Einstein, The Theory of the Brownian Movement, Dover, New York, 1956.
59. N. A. Fuchs, Z. Phys. 89:739 (1934).
60. B. V. Derjaguin, Dokl. AN SSSR 109:967 (1956).
61. B. V. Derjaguin and N. A. Krotova, Physical Chemistry of Adhesion. Izd-vo AN SSSR. Moscow, 1949.
62. G. I. Taylor, Proc. Roy. Soc. London A 108:12 (1924).
63. L. Spielman, J. Colloid Interface Sci. 33:562 (1970).
64. E. P. Honig, G. J. Roberson, and P. H. Wiersena, J. Colloid Interface Sci. 36:97 (1971).
65. G. K. Batchelor, J. Fluid Mech. 119:379 (1982).
66. A. Z. Zinchenko, Prikl. Mat. Mech. 42:955 (1978).
67. A. Z. Zinchenko, Prikl. Mat. Mech. 44:30 (1980).
68. A. Z. Zinchenko, Prikl. Mat. Mech. 46:58 (1982).
69. G. Hetsroni and S. Haber, Int. J. Multiphase Flow 4:1 (1978).
70. Y. O. Fuentes, S. Kim, and D. J. Jeffrey, Phys. Fluids 31:2445 (1988).
71. Y. O. Fuentes, S. Kim, and D. J. Jeffrey, Phys. Fluids A1:61 (1989).
72. S. Haber, G. Hetsroni, and A. Solan, Int. J. Multiphase Flow 1:57 (1973).
73. E. Rushton and G. A. Davies, Appl. Sci. Res. 28:37 (1973).
74. R. H. Davis, J. A. Schonberg, and J. M. Rallison, Phys. Fluid A1:77 (1989).
75. X. Zhang and R. H. Davis, J. Fluid Mech. 230:479 (1991).
76. X. Zhang, Ph.D. thesis. University of Colorado, 1992.
77. H. Sonntag, in Coagulation and Flocculation (Dobias, ed.), Marcel Dekker, Inc., New York, 1993.
78. E. B. Vadas, H. L. Goldsmith, and S. G. Mason, J. Colloid Interface Sci. 43:630 (1973).
79. E. B. Vadas, R. G. Cox, H. L. Goldsmith, and S. G. Mason, J. Colloid Interface Sci. 57:308 (1976).
80. H. Siedentopf and R. Zsigmondy, Ann. Phys. (Leipzig) 10:1 (1903).
81. R. Zsigmondy, Z. Phys. Chem. 93:600 (1918).
82. A. Tuorilla, Kolloidchem. Beihefte 22:193 (1926).
83. B. V. Derjaguin and N. M. Kudravtseva, Kolloidn. Zh. 26:61 (1964).
84. A. Watillon, M. Romerowski, and F. van Grunderbeek, Bull. Soc. Chim. Belg. 68:450 (1959).
85. R. H. Ottewill and M. C. Rastogi, J. Chem. Soc. Trans. Faraday Soc. 56:866 (1960).
86. N. H. G. Penners and L. K. Koopal, Colloids Surfaces 28:67 (1987).
87. J. Cahill, P. G. Cummins, E. J. Staples, and L. Thompson, J. Colloid Interface Sci. 117:406 (1987).
88. H. Gedan, H. Lichtenfeld, H. Sonntag, and H.-J. Krug, Colloids Surfaces 11:199 (1984).
89. N. A. Fuchs, Uspechi Mechaniki Aerozoley, lzd-vo AN SSSR. Moscow, 1961 [in Russian].
90. L. A. Spielman and J. A. Fitzpatrick, J. Colloid Interface Sci. 42:607 (1973).
91. L. A. Spielman and J. A. Fitzpatrick, J. Colloid Interface Sci. 43:350 (1973).
92. J. Langmuir and K. Blodgett, Gen. Elec. Comp. Rep., July 1945, pp. 45–58.
93. W. Rybczynski, Bull. Cracovie (A) 40 (1911).
94. J. Hadamard, Comp. Rend. 152:1735 (1911).
95. P. F. Saffman and J. S. Turner, J. Fluid Mech. 1:16 (1956).
96. V. G. Levich, Physicochemical Hydrodynamics, Prentice-Hall, Englewood Cliffs, NJ, 1962.
97. A. N. Frumkin and V. G. Levich, Zh. Fiz. Chim. 21:1183 (1947).

98. S. S. Dukhin, in Modern Theory of Capillarity (A. I. Rusanov and F. Ch. Goodrich, eds.), Springer-Verlag, Berlin, 1981.
99. R. D. Vold and R. C. Groot, J. Colloid Sci. 19:384 (1964).
100. S. J. Rehfild, J. Phys. Chem. 66:1969 (1962).
101. E. R. Garret, J. Am. Pharm. Assoc. 51:35 (1962).
102. H. Brauer, Grandlagen der Ein-und Mehrphasenstronumgen, Verlag Sanerlander, Aarau, 1971.
103. R. G. Boothroyd, Flowing Gas–Solid Suspensions, Chapman & Hall, London, 1971, Sec. 2.2.
104. H. S. Muralidhara, R. B. Beard, and N. Senapti, Filt. Sep. 24:409 (1987).
105. K. L. Sutherland, J. Phys. Chem. 58:394 (1948).
106. B. V. Derjaguin and S. S. Dukhin, Trans. Inst. Mining Metall. 70:221, 231 (1960).
107. L. I. Levin, Research into the Physics of Coarsly Dispersed Aerosols, Izdvo AN SSSR, Moscow, 1961 [in Russian].
108. G. L. Natanson, Dokl. AN SSSR 116:109 (1957).
109. A. Fonda and H. Herne, in Aerosol Science (C. N. Davies, ed.), Academic Press, New York, 1969.
110. W. R. Russel, D. A. Saville, and W. R. Schowalter, Colloidal Dispersions, Cambridge University Press, Cambridge, 1989, Chap. 11.
111. S. S. Dukhin, Kolloidn. Zh. 44:431 (1982); 45:207 (1983).
112. S. S. Dukhin and B. V. Derjaguin, Kolloidn. Zh. 20:326 (1958).
113. M. B. Weber, J. Sep. Technol. 2:29 (1981).
114. A. Nguen Van and S. Kmet, Int. J. Miner. Process. 35:205 (1992).
115. R. C. Clift, J. R. Grace, and M. E. Weber, Bubbles, Drops and Particles, Academic Press, New York, 1978, p. 27.
116. R. H. Yoon and G. H. Luttrel, Miner. Process. Extractive Metal Rev. 5:101 (1989).
117. G. H. Luttrel and R. H. Yoon, J. Colloid Interface Sci. 159:129 (1992).
118. I. O. Protodiaconov and S. V. Uljanov, Hydrodynamics and Mass Transport in Disperse Systems, Liquid–Liquid, Nauka, Leningrad, 1986, pp. 28–48 [in Russian].
119. V. Ya. Rivkind and G. M. Riskin, Izv. AN SSSR MZhG N1:8 (1976).
120. M. B. Weber and D. Paddock, J. Colloid Interface Sci. 94:328 (1983).
121. J. H. Masliyah, Ph.D. dissertation, University of British Columbia, Vancouver, Canada, 1970.
122. S. W. Woo, Ph.D. dissertation, McMaster University, Hamilton, Canada, 1971.
123. N. N. Rulyov and E. S. Leshchov, Kolloidn. Zh. 42:1123 (1980).
124. A. E. Hamielec and A. I. Jonson, Can. J. Chem. Eng. 40:41 (1962).
125. J. F. Anfruns and J. A. Kitchener, in Flotation (M. C. Fuerstenau, ed.), A. M. Gaudin Memorial Volume, SME-AIME, New York, pp. 626–637.
126. E. S. R. Gopal, in Emulsion Science (P. Sherman, ed.), Academic Press, London, 1969, pp. 69–70.
127. S. S. Dukhin and M. V. Buikov, Zh. Fiz. Chim. 39:913 (1965).
128. L. A. Spielman, Annu. Rev. Fluid Mech. 9:297 (1977).
129. S. L. Goren and M. E. O'Neil, Chem. Eng. Sci. 26:325 (1971).
130. S. L. Goren, J. Fluid Mech. 41:613 (1971).
131. A. I. Goldman, R. G. Cox, and H. Brenner, Chem. Eng. Sci. 22:637 (1967).
132. N. N. Rulyov, Kolloidn. Zh. 40:898 (1978).
133. N. N. Rulyov, Kolloidn. Zh. 40:1202 (1978).
134. V. Shubin and P. Kekicheff, J. Colloid Interface Sci. 155:108 (1993).
135. G. A. Martinov and V. M. Muller, Dokl. AN USSR 207:1161 (1972).
136. D. Reay and G. A. Ratelif, Can. J. Chem. Eng. 51:178 (1973).
137. D. Reay and G. A. Ratelif, Can. J. Chem. Eng. 53:481 (1975).
138. G. L. Collins and G. J. Jameson, Chem. Eng. Sci. 31:985 (1976).
139. G. L. Collins and G. J. Jameson, Chem. Eng. Sci. 31:239 (1977).
140. K. Okada, Y. Akagi, M. Kogure, and N. Yoshioka, Can. J. Chem. Eng. 68:393 (1990).
141. K. Okada, Y. Akagi, M. Kogure, and N. Yoshioka, Can. J. Chem. Eng. 68:614 (1990).
142. Wen Jongsong and G. K. Batchelor, Scintia Sinica 28:172 (1985).
143. G. K. Batchelor and C. S. Wen, J. Fluid Mech. 124:495 (1982).
144. D. J. Jeffrey and Y. Onishi, J. Fluid Mech. 139:261 (1984).
145. R. H. Davis, J. Fluid Mech. 145:179 (1984).

146. L. D. Reed and F. A. Morrison, Int. J. Multiphase Flow 1:573 (1974).
147. Ya. I. Rabinovich and A. A. Baran, Colloids Surfaces 59:47 (1991).
148. V. A. Parsegian, N. Fuller, and R. P. Rand, Proc. Natl. Acad. Sci. USA, 2750 (1979).
149. R. P. Rand, Annu. Rev. Biophys. Bioeng. 10:277 (1981).
150. R. P. Rand, V. A. Parsegian, J. A. Henry, L. J. Lis, and M. McAlister, Can. J. Biochem. 58:959 (1980).
151. H. R. Kruyt, Colloid Science, Vol. 1, Elsevier, Amsterdam, 1952.
152. A. S. Dukhin, Kolloidn. Zh. 50:441 (1988).
153. A. V. Bochko, A. S. Dukhin, E. I. Moshkovski, and A. A. Baran, Kolloidn. Zh. 49:543 (1987).
154. M. Stimson and G. B. Jeffrey, Proc. Roy. Soc. London A110:1117 (1926).
155. N. N. Rulyov, S. S. Dukhin, and V. P. Semenov, Kolloidn. Zh. 41:263 (1979).
156. P. S. Kisliy, Yu. I. Nikitin, V. M. Melnik, and S. M. Uman, Poroshkovaja Metallurgija 6:92 (1982).
157. L. A. Spielman and P. M. Cukor, J. Colloid Interface Sci. 43:51 (1973).
158. M. V. Ostrovsky and R. J. Good, J. Disper. Sci Technol. 7:95 (1986).
159. F. S. Milops and D. T. Wasan, Colloids Surfaces 4:91 (1981).
160. M. Boenrel, A. Graciaa, R. S. Shechtel, and W. H. Waade, J. Colloid Interface Sci. 72:161 (1979).
161. J. Vinatieri, Paper SPE 6675, J. Petrol. Technol. (1977).
162. D. O. Shah and R. S. Shechter, Improved Oil Recovery by Surfactant and Polymer Flooding, Academic Press, New York, 1977.
163. H. K. van Bollen and Associates, Inc., Fundamentals of Enhanced Oil Recovery, Pen Well Publishing Co, Tulsa, OK, 1980.
164. P. Becher (ed.), Encyclopedia of Emulsion Technology, Vol. 1, Marcel Dekker, New York, 1983.
165. P. Becher (ed.), Encyclopedia of Emulsion Technology, Vol. 2, Marcel Dekker, New York, 1985.
166. R. A. Mohammed, A. I. Bailey, P. F. Luckham, and S. E. Taylor, Colloids Surfaces 80:223 (1993); 80:237 (1993); 83:261 (1994).
167. C. I. Chitewelu, V. Hornof, G. H. Neale, and A. E. George, Can. J. Chem. Eng. 72:534 (1994).
168. L. Bertero, A. Di. Lullo, A. Lentini, and L. Terzi, SPE 28543, Society of Petroleum Engineers (1994).
169. J. F. McCafferty and G. G. McClaflin, SPE 24850, Society of Petroleum Engineers (1992).
170. J. N. Israelachvili, Intermolecular and Surface Forces, 2nd ed., Academic Press, London, 1991.
171. J. Sjöblom (ed.), Emulsions and Emulsion Stability, Marcel Dekker, New York, 1996.
172. I. B. Ivanov and P. A. Kralchevsky, Colloids Surfaces 128:155 (1997).
173. S. S. Dukhin and J. Sjöblom, in Emulsions and Emulsion Stability (J. Sjöblom, ed.), Marcel Dekker, New York, 1996, p. 41.
174. P. A. Kralchevsky, K. D. Danov, and N. D. Denkov, in Handbook of Surface and Colloid Chemistry, CRC Press, London, 1996.
175. Chapter 11.4 in Ref. 174.
176. R. K. Prud'homme (ed.), Foams: Theory, Measurement and Applications, Marcel Dekker, New York, 1995.
177. P. A. Kralchevsky, K. D. Danov, and I. B. Ivanov, in Ref. 176.
178. S. A. K. Jeelany and S. Hartland, J. Colloid Interface Sci. 164:296 (1994).
179. P. D. I. Fletcher, in Drops and Bubbles in Interface Research (D. Mobius and R. Miller, eds.), Vol. 6, Elsevier, New York, 1998.
180. P. J. Breen, D. T. Wasan, Y. H. Kim, A. D. Nikolov, and C. S. Shetty, in Emulsions and Emulsion Stability (J. Sjöblom, ed.), Marcel Dekker, New York, 1996.
181. S. S. Dukhin and J. Sjöblom, J. Dispersion Sci. Technol. 19:311 (1998).
182. Ø. Holt, Ø. Sæther, J. Sjöblom, S. S. Dukhin, and N. A. Mishchuk, Colloids Surfaces 123–124:195 (1997).
183. Ø. Sæther, S. S. Dukhin, J. Sjöblom, and Ø. Holt, Colloid. J. 57:836 (1995).
184. S. S. Dukhin, J. Sjöblom, D. T. Wasan, and Ø. Sæther, Colloids Surfaces 180:223 (2001).
185. Ø. Sæther, J. Sjöblom, S. V. Verbich, N. A. Mishchuk, and S. S. Dukhin, Colloids Surfaces 142:189 (1998).
186. Ø. Sæther, J. Sjöblom, S. V. Verbich, and S. S. Dukhin, J. Dispersion Sci. Technol. 20:295 (1999).
187. Ø. Holt, Ø. Sæther, J. Sjöblom, S. S. Dukhin, and N. A. Mishchuk, Colloids Surfaces 141:269 (1998).

188. S. S. Dukhin, Ø. Sæther, and J. Sjöblom, Proc. Symposium on Emulsions, Foams and Thin Films, in honor of D. T. Wasan at Penn State, June 1998.
189. O. Reynolds, Phil. Trans. R. Soc. (London), A177:157 (1886).
190. A. Scheludko, Adv. Colloid. Interface Sci. 1:391 (1967).
191. K. D. Danov, D. N. Petsev, N. D. Denkov, and R. Borwankar, J. Chem. Phys. 99:7179 (1993).
192. D. N. Petsev, N. D. Denkov, and P. A. Kralchevsky, J. Colloid Interface Sci. 176:201 (1995).
193. I. B. Ivanov and D. S. Dimitrov, in Thin Liquid Films (I. B. Ivanov, ed.), Marcel Dekker, New York, 1998.
194. N. A. Mishchuk, J. Sjöblom, and S. S. Dukhin, Colloid J. 57:785 (1995).
195. K. D. Danov, I. B. Ivanov, T. D. Gurkov, and R. P. Borwankar, J. Colloid Interface Sci. 167:8 (1994).
196. L. A. Lobo, D. T. Wasan, and M. Ivanova, in Surfactants in Solution (K. L. Mittal and D. O. Shah, eds.), Plenum Press, New York, 1992, Vol. 11, p. 395.
197. L. Lobo, I. Ivanov, and D. Wasan, Materials Interfaces and Electrochemical Phenomena 39:322 (1993).
198. G. Narsimhan and E. Ruckenstein, in Ref. 176.
199. D. Exerova, D. Kashchiev, and D. Platikanov, Adv. Colloid Interface Sci. 40:201 (1992).
200. B. V. Derjaguin and Yu. V. Gutop, Kolloidn. Zh. 24:431 (1962).
201. A. V. Prokhorov and B. V. Derjaguin, J. Colloid Interface Sci. 125:111 (1988).
202. S. V. Verbich, S. S. Dukhin, A. Tarovsky, Ø. Holt, Ø. Sæther, and J. Sjöblom, Colloids Surfaces 23:209 (1997).
203. D. Kashchiev and D. Exerowa, J. Colloid Interface Sci. 203:146 (1998).
204. E. Dickinson and S. R. Euston, Adv. Colloid Interface Sci. 42:89 (1992).
205. P. Taylor, Adv. Colloid Interface Sci. 75:107 (1998).
206. H. W. Yarranton and Y. H. Masliyah, J. Colloid Interface Sci. 196:157 (1997).
207. G. A. Martynov and V. M. Muller, Kolloidn. Zh. 36:687 (1974).
208. V. M. Muller, Kolloidn. Zh. 40:85 (1978).
209. J. D. Chan, P. S. Hahn, and J. C. Slattery, AIChE J. 34:40 (1988).
210. S. A. K. Jeelani and S. Hartland, J. Colloid Interface Sci. 156:467 (1993).
211. R. W. Aul and W. I. Olbriht, J. Colloid Interface Sci. 115:478 (1991).
212. A. Scheludko and D. Exerova, Colloid J. 168:24 (1960).
213. A. Scheludko, Proc. K. Ned. Akad. Wet. Ser. B. 65:87 (1962).
214. D. Platikanov and E. Manev, in Proceedings 4th International Congress of Surface Active Substances, Plenum, New York, 1964, p. 1189.
215. K. A. Burrill and D. R. Woods, J. Colloid Interface Sci. 42:15 (1973).
216. K. A. Burrill and D. R. Woods, J. Colloid Interface Sci. 42:35 (1973).
217. S. R. Deshiikan and K. D. Papadopoulos, J. Colloid Interface Sci. 174:302 (1995).
218. S. R. Deshiikan and K. D. Papadopoulos, J. Colloid Interface Sci. 174:313 (1995).
219. M. E. Costas, M. Moreau, and L. Vicente, J. Phys. A: Math. Gen. 28:2981 (1995).
220. F. Le Berre, G. Chauveteau, and E. Pefferkorn, J. Colloid Interface Sci. 199:1 (1998).
221. B. J. McCoy and G. Madras, J. Colloid Interface Sci. 201:200 (1998).
222. J. Widmaier and E. Pefferkorn, J. Colloid Interface Sci. 203:402 (1998).
223. I. M. Elminyawi, S. Gangopadhyay, and C. M. Sorensen, J. Colloid Interface Sci. 144:315 (1991).
224. P. Meakin, Adv. Colloid Interface Sci. 28:249 (1988).
225. R. Nakamura, Y. Kitada, and T. Mukai, Planet Space Sci. 42:721 (1994).
226. S. Chandrasekhar, Rev. Mod. Phys. 15:1 (1943).
227. V. M. Muller, Colloid J. 58:598 (1996).
228. D. G. Dalgleish, Colloids Surfaces 46:14 (1990).
229. D. G. Dalgleish and Y. Fang, J. Colloid Interface Sci. 156:329 (1993).
230. D. J. Mitchell and P. Richmond, J. Colloid Interface Sci. 46:128 (1974).
231. V. N. Gorelkin and V.P. Smilga, Kolloidn. Zh. 34:685 (1972).
232. V. N. Gorelkin and V.P. Smilga, in Poverkhnostnye Sily v Tonkikh Plenkakh i Ustoichivost Kolloidov (B. V. Derjaguin, ed.), Nauka, Moskow, 1974, p. 206.
233. S. Usui, Y. Imamura, and E. Barough, J. Disp. Sci. Technol. 8:359 (1987).
234. S. B. Grant, C. Poor, and S. Relle, Colloids Surfaces 107:155 (1996).

2 Spontaneous Emulsification: Recent Developments with Emphasis on Self-Emulsification

Clarence A. Miller

CONTENTS

Abstract

The phenomenon of spontaneous emulsification is reviewed with emphasis on results obtained in recent years. Several recent studies have provided insight into the mechanism of self-emulsification of oils brought into contact with water with gentle stirring to form oil-in-water emulsions consisting of small droplets. In particular, diffusion and/or chemical reactions can cause changes in composition or environment and hence in spontaneous curvature of surfactant films between oil and water, which in turn cause inversion from an oil-continuous to a water-continuous microemulsion. Since the latter is not able to solubilize all of the oil present, local supersaturation and subsequent nucleation of oil droplets occurs. Under suitable conditions in some systems the lamellar liquid crystalline phase forms during the inversion process and coats the small droplets, thereby hindering coalescence and reducing the initial surfactant concentration required to obtain small droplets.

Self-emulsification can also occur without large changes in spontaneous curvature when the surfactant films are initially near the balanced state of zero spontaneous curvature. Formation of the lamellar phase during emulsification appears to be essential for obtaining small droplets in this case. Sometimes most, if not all, of the oil is converted to the lamellar phase before oil droplets nucleate. In other situations vesicles of an oil-rich lamellar phase have been observed to disintegrate during swelling by a process resembling an explosion, which is thought to disperse oil drops into the aqueous phase.

"Nanoemulsions" with drop sizes of order 100 nm or less have been produced by self-emulsification processes involving spontaneous emulsification accompanied by gentle stirring. When polymer is dissolved in the droplets, removal of the solvent yields small polymer nanoparticles, which are of interest in applications such as drug delivery.

The results of these studies show that knowledge of equilibrium phase behavior is important in choosing suitable systems and conditions for self-emulsification.

2.1 INTRODUCTION

When oil and water phases are mixed, high shear rates are typically required to generate emulsions consisting entirely of small drops with diameters of order 1 μm or less. If it is necessary or desirable to obtain such emulsions without using high shear rates, spontaneous emulsification is an option. Compositions of the initial oil and water phases are chosen in such a way that small droplets form spontaneously when the phases are brought into contact, i.e., no external energy of agitation is required.

Because it is an intriguing phenomenon, an extensive literature exists on spontaneous emulsification whether or not the drops formed are small. Various mechanisms have been suggested, some based on breakup of an interface between oil and water to produce drops, others based on formation of drops during phase transformation, e.g., by nucleation or spinodal composition in regions of local supersaturation produced by diffusion and/or chemical reaction. Davies and Rideal (1) reviewed the older literature on mechanisms of both types. Lopez-Montilla et al. (2) presented a recent comprehensive review of spontaneous emulsification. Here a brief discussion of mechanical mechanisms with emphasis on recent work is followed by a more extensive account of mechanisms involving phase transformation and local supersaturation. Particular attention is given to so-called self-emulsification of oil–surfactant mixtures leading to small droplets, indeed in some cases to nanoscale droplets or particles. This formation of nanoscale droplets and particles has attracted considerable attention in recent years owing to potential applications in delivery of poorly water soluble active ingredients in pharmaceutical, personal care, plant protection and other products (3).

2.2 MECHANICAL MECHANISMS

Prominent among proposed mechanical mechanisms of spontaneous emulsification is breakup produced by transient local negative values of interfacial tension. Davies and Rideal's review includes their own ideas on this mechanism (1). Gopal (4) presented an analysis of interfacial instability produced by negative tensions. More recently, Granek et al. (5) developed an improved instability model, and Theissen and Gompper (6) presented an analysis utilizing the lattice Boltzmann method. It is difficult to say whether transient negative tensions between oil and water are actually achieved in experiments because in oil–water–surfactant systems where ultralow tensions have been observed, intermediate microemulsion and/or liquid crystalline phases typically form at the interface when tensions reach very low positive values, i.e., before they become negative. Also since some mass transfer, e.g., of surfactant, invariably accompanies the reduction in tension, drop formation by nucleation in regions of local saturation produced by diffusion could be responsible for the emulsification.

Vigorous Marangoni flow leading to breakoff of drops from an interface has also been proposed as a mechanism of spontaneous emulsification. Such flow is produced by interfacial tension gradients arising from an instability that frequently develops when diffusion occurs near an interface (7). Here too, however, it is difficult to rule out emulsification produced by local supersaturation. When this latter mechanism is active, the simultaneous occurrence of Marangoni flow greatly increases the rate of emulsification by increasing rates of mass transport near the interface, which produces the supersaturation (1,8). Pfennig (9) employed molecular simulation techniques to investigate Marangoni flow. Some emulsification was predicted to develop near the interface for the particular system investigated. The relevant phase diagram suggests – at least to the present author – that the source of the drops was, in fact, local supersaturation. It is noteworthy that in surfactant systems mass transport leading to both Marangoni flow and local supersaturation may involve transport not only of individual molecules but also of micelles or other aggregates or within liquid crystalline or microemulsion phases.

Yeung et al. (10) observed instability leading to formation of water droplets when a larger drop of water some 20 μm in diameter held at the end of a micropipette and immersed for a few minutes in a toluene/heptane mixture containing 10% bitumen was suddenly deflated. The rather flexible interface nevertheless had some structure and developed multiple undulations projecting into the oil phase as its area decreased, eventually causing many water droplets having diameters of order 1 μm to detach and become dispersed in the oil phase. The authors suggested that these observations were related to "budding," which has been seen in giant unilamellar vesicles having phospholipid bilayer membranes (11). Such budding, which typically exhibits only one smaller vesicle forming and detaching at a time from the initial vesicle, has been explained theoretically as resulting from curvature elasticity effects which arise when a temperature change or some other change in environment imposes a high spontaneous curvature on the membrane (12). Perhaps also related to the emission of water droplets during the bitumen experiment is another theoretical analysis (13), which predicts that an interface having low interfacial tension and actual curvature much less than its spontaneous curvature will buckle under compression, generating long fingers which can break up into droplets. If spontaneous curvature in such a situation favored a water-in-oil arrangement, as seems likely in the bitumen system, the fingers would be of water projecting into the oil phase and would be expected to break up into water droplets.

Sequential budding events, each involving release of a single droplet or vesicle, were observed by Tungsubutra (14) using videomicroscopy when a drop of oleic acid or an oleic acid/triolein mixture some 50 μm in diameter was contacted with a 2 wt% solution of the pure nonionic surfactant $C_{14}E_6$ in a thin rectangular glass cell maintained at 35 °C (Figure 2.1). A thin layer of lamellar phase formed at the surface of the drop following contact of the phases. Diffusion of surfactant to the drop surface with subsequent adsorption could have generated the spontaneous curvature in the lamellar layer required by the budding mechanism developed for giant vesicles.

Finally, as is well known, the Rayleigh–Taylor instability can cause spontaneous emulsification due to instability of an oil–water interface which is nearly planar and horizontal when the denser phase is located above the lighter phase and the lateral extent of the interface is much greater than the capillary length $[\gamma/\Delta\rho g]^{1/2}$. For instance, a shallow oil pool on a horizontal solid surface can exhibit emulsification when contacted with an aqueous solution containing a surfactant which reduces interfacial tension to low values, thereby greatly reducing the capillary length. A striking example of such behavior was described recently by Hirasaki and Zhang (15). A solution containing sodium carbonate, sodium chloride, and a small amount of anionic surfactant was placed carefully on top of a thin layer of crude oil resting on a solid surface strongly wetted by the oil. The alkali converted the naphthenic acids in the oil to soap, which acted together with the added surfactant to reduce interfacial tension to very low values. At a suitable NaCl concentration all the oil was emulsified.

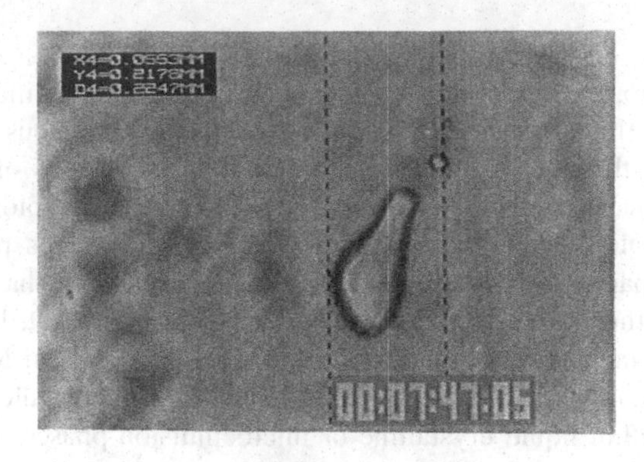

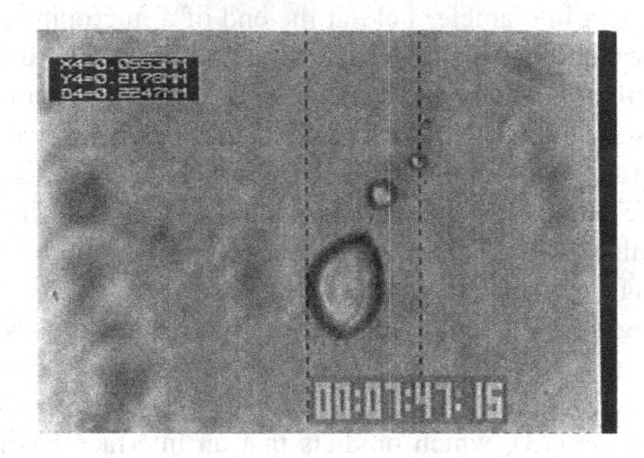

FIGURE 2.1 Budding for a drop initially containing 25% triolein and 75% oleic acid in a 2 wt% solution of $C_{14}E_6$ at 35 °C. Times after drop injection are indicated at bottom of frames (h:min:sec:frame), the lower frame being taken approximately 1/3 sec after the upper. Distance between dashed lines is approximately 50 μm.

2.3 LOCAL SUPERSATURATION: GENERAL CONSIDERATIONS

As indicated previously, the alternate to spontaneous emulsification caused by mechanical instabilities is formation of droplets in regions of local supersaturation produced by diffusion. That diffusion is responsible for emulsification in some systems has long been recognized, as discussed by Davies and Rideal (1), who described it as "diffusion and stranding." Ruschak and Miller (16) presented an analysis which clarified how and when regions of local supersaturation could develop. They started with the diffusion equations in a ternary system for semi-infinite immiscible phases brought into contact without mixing. When plausible assumptions are made, a similarity solution exists, and it is readily shown that the set of compositions in the system is independent of time and thus can be plotted directly on the ternary phase diagram, the so-called "diffusion path." For example, abcd2 in Figure 2.2 represents the diffusion path in an oil/water/alcohol system when an oil/alcohol mixture a is contacted with pure water 2. In some cases the diffusion path in one or both phases passes through the two-phase region of the phase diagram, e.g., cd in Figure 2.2, indicating that local supersaturation can develop, in this case in the aqueous phase, even though both initial compositions are in single-phase regions. They suggested that spontaneous emulsification would occur in these supersaturated regions and showed that this method successfully predicted both when emulsification would occur and in

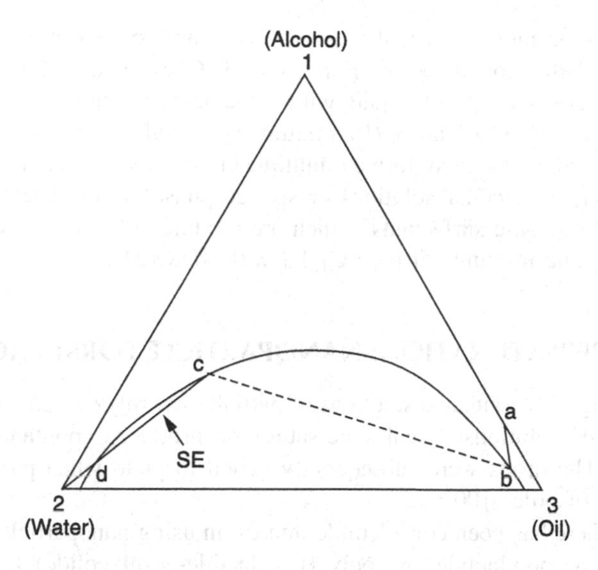

FIGURE 2.2 Diffusion path in a typical alcohol (1) – water (2) – oil (3) system indicating spontaneous emulsification (SE) in the aqueous phase.

which phase(s) for several oil/water/alcohol systems. As Figure 2.2 suggests, spontaneous emulsification is favored when the solute, here the alcohol, diffuses into the phase in which it is more soluble. The reader is referred to previous reviews for further discussion of spontaneous emulsification produced by this mechanism in oil/water/alcohol systems and in oil/water/surfactant systems with the surfactant initially located in the aqueous phase (8,17,18). The emulsions formed were more stable in the surfactant systems, as one would expect.

The review articles just cited describe situations where oil and the water–surfactant mixtures, some of which were dispersions of the lamellar liquid crystalline phase, were contacted without application of external mixing, and where the resulting emulsification was visible, either with the naked eye or using optical microscopy. However, stable "mini-emulsions" having much smaller droplets some 100 nm in diameter were formed when toluene or styrene was added with gentle stirring to water containing 10 mM SDS and 20 mM n-hexadecanol (19). The water/surfactant/alcohol mixture had previously been stirred vigorously at 65 °C, then cooled to 40 °C where an alcohol-rich birefringent phase had formed as a dispersion in an aqueous micellar solution. The added oil penetrated the alcohol-rich phase, presumably causing local supersaturation leading to nucleation of oil droplets containing dissolved hexadecanol. As this extraction of alcohol into oil droplets increased the SDS/alcohol ratio in the birefringent phase, the lamellar structure could eventually no longer be maintained, leaving an oil-in-water emulsion. Emulsification was truly spontaneous as shown by experiments with no stirring, which, however, produced larger drops and a turbid emulsion. Evidently, application of gentle stirring separated the growing oil droplets while they were still surrounded by the birefringent phase, thereby minimizing coalescence and maintaining droplet size near 100 nm. In later sections related phenomena also yielding small droplets and involving dynamics of appearance and/or disappearance of birefringent phases will be described for situations where the surfactant is initially dissolved in oil or an oil-containing phase.

It should perhaps be mentioned that it is not necessary to have both oil and water present for spontaneous emulsification to occur. For instance, Chen et al. (20) observed spontaneous formation of nonspherical drops of liquid within the lamellar liquid crystalline phase during dissolution in water at 35 °C of an 85/15 mixture by weight of the pure nonionic surfactant $C_{12}E_8$ and n-decanol. Spontaneously formed multiple emulsions of water and isotropic surfactant-containing liquid phases (micellar solutions or sponge phase) were observed during dissolution of some commercial nonionic surfactants, which are mixtures of many species; mixtures of pure nonionic surfactants; and mixtures of pure $C_{12}E_4$ with SDS (21).

2.4 LOCAL SUPERSATURATION: NANOPARTICLE FORMATION

Minehan and Messing (22) contacted solutions of partially hydrolyzed tetraethoxysilane in ethanol with aqueous NH_4OH solutions. Local supersaturation produced spontaneous emulsification in the aqueous phase. The drops were subsequently gelled to yield silica particles, which in some cases had diameters of order 100 nm.

In recent years there has been considerable interest in using nanoparticles (NPs) of biodegradable polymers such as polylactide and poly (D,L-lactide-co-glycolide) for drug delivery (e.g., 23–26). Among methods investigated for preparing such NPs are those in which a solution of the polymer and drug in an organic solvent is contacted with water containing polyvinyl alcohol (PVA) or a surfactant with gentle stirring. As water is a poor solvent for the polymer, it has been suggested that diffusion of solvent into the aqueous phase creates a region of local supersaturation where a polymer-rich phase nucleates as "protonanoparticles," which are stabilized by the PVA or surfactant (24). Later the remaining solvent is removed to yield the desired NPs, whose diameters are typically of order 100 nm although in some cases even smaller.

Some data on phase behavior such as cloud point curves have been published for a few of these systems (25). However, the present author is not aware of a system where sufficient quantitative information on phase behavior is available to clearly locate diffusion paths, although plausible phase diagrams where local supersaturation would be expected can be constructed. Moreover, some workers add the water rapidly, others slowly, and in at least one system NP size was not significantly affected by the rate of addition (26). In any case it seems clear that system composition is shifted by addition of water in such a way that new polymer-rich phase nucleates, i.e., the protonanoparticles are not the same phase as the original polymer solution in the solvent but instead are a phase with a much higher polymer content immersed in a water-rich continuous phase, which is a poor solvent for the polymer. That they remain small is the result of stabilization by PVA or surfactant immediately after nucleation, a process similar in concept but not in detail to the mini-emulsion formation process described above.

2.5 LOCAL SUPERSATURATION: SELF-EMULSIFICATION

The remainder of this paper will deal mainly with "self-emulsification" of oils, which is basically spontaneous emulsification of a phase containing oil, one or more surfactants, and frequently an active ingredient such as a pesticide or drug when it is brought into contact with a relatively large quantity of water, usually with gentle mixing. As in the mini-emulsion case, emulsification is spontaneous, but mixing speeds up the rate at which the initial phases are contacted and can significantly influence the final drop size distribution. Sometimes it is desirable to dissolve water in the oil phase to form a microemulsion before mixing with excess water. In any case the objective is generally to obtain a dispersion of "small" oil droplets in water, small meaning diameters of order 1 μm in some cases but 100 nm or even smaller in others. The latter are sometimes called mini-emulsions, as indicated above, or nanoemulsions.

Self-emulsification of oil is of interest in the use of various products such as *emulsifiable concentrates* of agricultural chemicals (27,28), soluble oils for machining, and bath oils and other personal care products. It has also been investigated for drug delivery (29–31). Self-emulsification of water in an oil phase is also possible although it seems not to be widely used.

The mechanism of self-emulsification has been poorly understood prior to work of the past several years, which is reviewed below. Previously, Lee and Tadros (32) found that the occurrence of low interfacial tensions could not fully explain the emulsification phenomena they observed. Self-emulsification was also sometimes attributed to formation of intermediate liquid crystalline phases near the surface of contact between oil and water phases (33,34), although the precise role of the liquid crystal was not made clear by these authors. Accordingly, selection of surfactants for self-emulsification and determination of the minimum surfactant concentration needed has been largely empirical.

2.6 SELF-EMULSIFICATION PRODUCED BY PREFERENTIAL DIFFUSION OF SOME SPECIES INTO WATER

One way to produce self-emulsification is to start with a surfactant or surfactant mixture in the initial oil phase that is rather lipophilic, favoring formation of oil-continuous microemulsions. The surfactant or mixture is chosen so that on contact with water diffusion and/or chemical reaction occur which make the surfactant more hydrophilic, favoring water-continuous microemulsions. As these have much less ability to solubilize oil, some oil nucleates due to local supersaturation and forms small drops. Provided that coalescence can be controlled, the desired emulsion with small oil drops is obtained.

Specific features of the inversion process just described depend on the equilibrium phase behavior and dynamics of phase transformation for the system being considered. For example, Rang and Miller (35) used videomicroscopy to study spontaneous emulsification in water at 30 °C of individual drops some 100 μm in diameter which contained n-hexadecane, n-octanol, and the pure linear alcohol ethoxylate $C_{12}E_6$. A thin hypodermic needle was used to inject an oil drop into water contained in a rectangular capillary cell approximately 400 μm thick (see Figure 2.3). Spontaneous emulsification yielding only small oil droplets was observed when the initial weight ratio of alcohol to hydrocarbon was greater than that (approximately 1/9) in the excess oil phase in equilibrium with a microemulsion and water at 30 °C, i.e., above the ratio existing at the phase inversion temperature (PIT), and when surfactant concentration exceeded 15 wt%.

The mechanism of emulsification is shown schematically in Figure 2.4. Initially water diffused rapidly into the oil phase, converting it to an oil-continuous microemulsion. Although little surfactant left the drop because its solubility was very low – near the CMC – for conditions above the PIT, the solubility of octanol was high enough for it to diffuse gradually into the aqueous phase following injection. As a result, the ratio of alcohol to surfactant in the films covering the microemulsion droplets decreased, making them more hydrophilic and leading to an increased capability of the microemulsion to solubilize water and a decreased capability to solubilize oil. Eventually, the microemulsion was no longer able to solubilize all the oil present, and oil droplets nucleated. Moreover, the microemulsion itself inverted to become water-continuous and miscible with water, so that the final state was oil droplets dispersed in an aqueous phase, the size distribution of the droplets depending largely on their rate of coalescence. When surfactant content of the initial drop was less than 15 wt%, coalescence was rapid and several large drops were

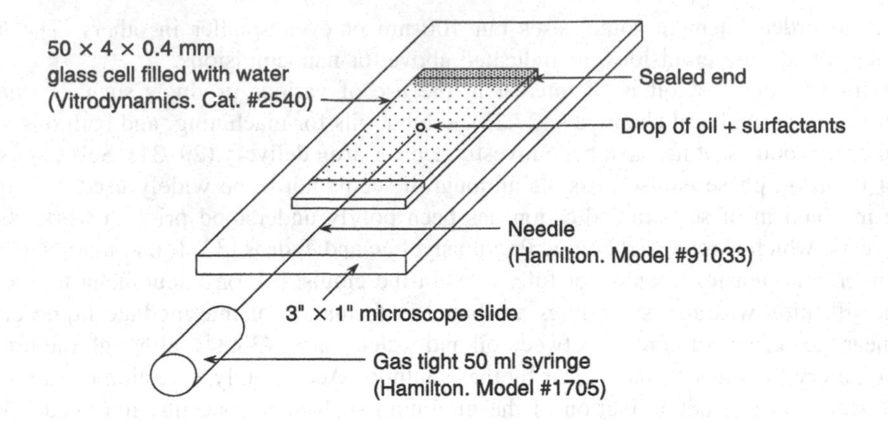

FIGURE 2.3 Schematic illustration of contacting experiment where a small drop of oil is injected into water.

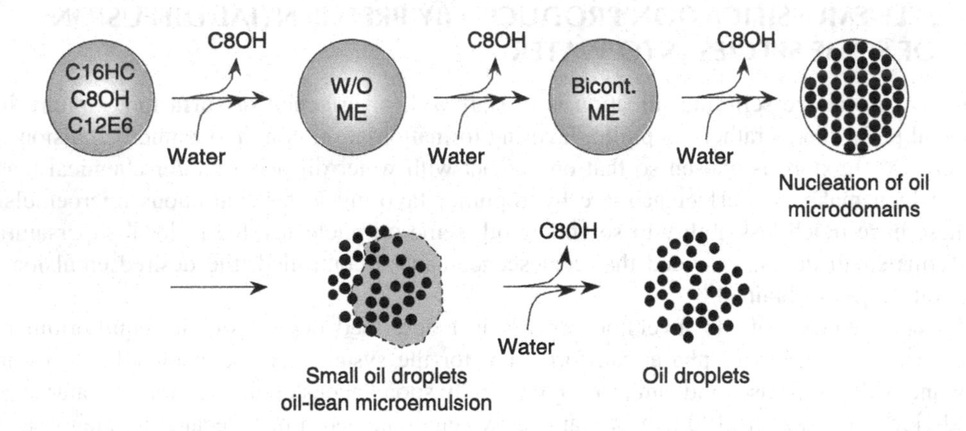

FIGURE 2.4 Schematic behavior showing spontaneous emulsification process for drops of hydrocarbon/alcohol/surfactant contacting water.

observed. However, for initial surfactant concentrations exceeding 15 wt%, coalescence was relatively slow, and the droplets remained small. A series of video frames showing the emulsification process in a system with 20 wt% surfactant is shown in Figure 2.5 (35). The higher surfactant concentrations yielded droplets with more hydrophilic surfactant films, which are well known to promote stability of oil-in-water emulsions.

No intermediate liquid crystalline phases were formed for the conditions of Figure 2.5. However, an intermediate lamellar phase *was* seen for a system where the injected drops were mixtures of n-decane, tetradecyldimethylamine oxide (C_{14}DMAO), and n-heptanol (36). Here too the initial ratio of alcohol to hydrocarbon had to be above that in the excess oil phase at the PIT to obtain satisfactory results. Emulsions with small oil droplets were observed for initial drops containing only 5 wt% surfactant in this system, apparently because the lamellar phase, which formed before oil nucleation began, reduced the rate of coalescence of the droplets. Nevertheless, both the liquid crystal and the microemulsion eventually became miscible with water as they lost alcohol by its diffusion into the aqueous phase, so that the final state was an oil-in-water emulsion.

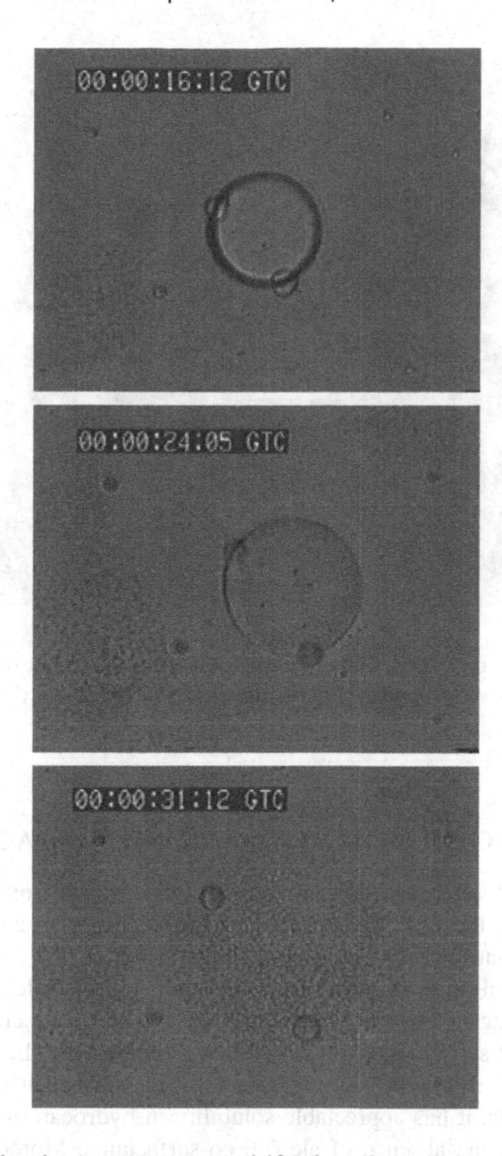

FIGURE 2.5 Video frames showing spontaneous emulsification of an oil drop with an initial composition of 20 wt% $C_{12}E_6$ and 80 wt% of a 90/10 mixture by weight of hexadecane and octanol. Behavior at approximately 16, 24, and 31 s after drop injection is shown. The initial drop is approximately 60 μm in diameter. The two small drops at its periphery did not influence the emulsification process.

It should be emphasized that *all* the oil initially present in both these systems was converted to microemulsion and/or liquid crystal. Hence the oil drops in the final emulsion were formed by nucleation during the inversion process. Extensive equilibrium phase behavior was obtained for both of these systems. That is, the mechanisms described above for the two surfactants were confirmed using a combination of videomicroscopy observations and phase behavior. For example, results of the combined techniques showed that no liquid crystal would form in the nonionic surfactant system for several compositions where satisfactory emulsification occurred.

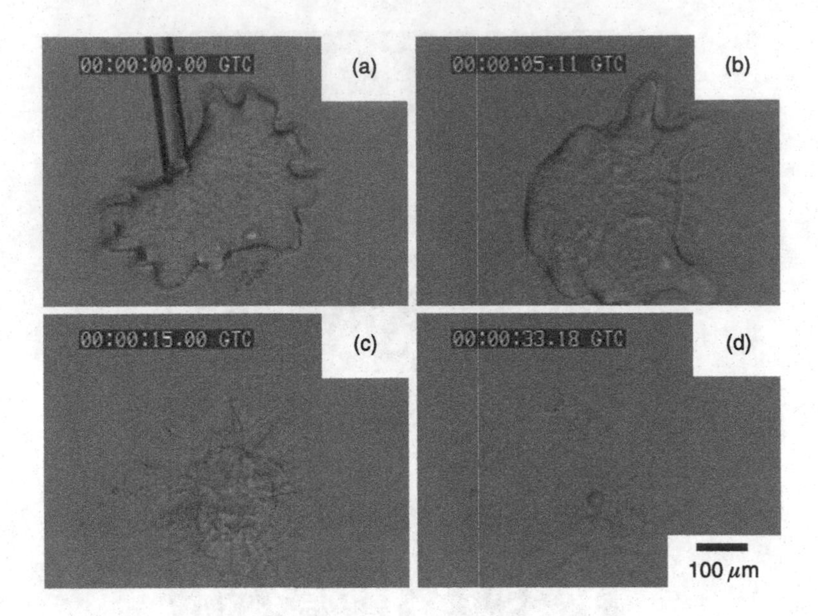

FIGURE 2.6 Video frames for a drop with proportions 60/30/10 by weight of n-octane/AOT/water injected into water at 30 °C. Times are as indicated in the upper part of the frame (h:min:sec:frame).

2.7 SELF-EMULSIFICATION WITH AN ANIONIC SURFACTANT

Ionic surfactants may also be used for self-emulsification. Aside from providing greater flexibility in surfactant choice, the electrical repulsion between drops in an emulsion formed with an ionic surfactant may contribute to emulsion stability. Greiner and Evans (37) observed spontaneous formation of an oil-in-water (o/w) emulsion with small droplets when a water-in-oil (w/o) microemulsion containing a potassium rosinate surfactant was contacted with water. Nishimi and Miller (38) investigated self-emulsification of alkanes containing the anionic surfactant Aerosol OT (AOT; sodium bis(2-ethylhexyl) sulfosuccinate). This surfactant was chosen because, unlike many anionic surfactants, it has appreciable solubility in hydrocarbons and forms oil-continuous microemulsions even in the absence of alcohol co-surfactants. Moreover, considerable information on its equilibrium phase behavior is available in the literature. Behavior of this well-defined ternary system is described in the following paragraphs although the basic cause of inversion leading to emulsification, i.e., reduction in ionic strength for an initially lipophilic system, was the same in both Refs. 37 and 38.

Figure 2.6 presents a series of video frames showing the behavior observed when a drop containing proportions 60/30/10 by weight of n-octane/AOT/water was injected into water at 30 °C. Upon injection the drop became cloudy and highly nonspherical. In addition, its surface appeared rough and exhibited numerous "bulges" (Figure 2.6(a,b)), which grew rapidly outward and then just as rapidly shrank, leaving behind in the aqueous phase small oil droplets (diameters of order 1 μm) and possibly some vesicles or particles of the lamellar phase. The droplets moved radially outward over the lower part of the cell's 400 μm thickness (Figure 2.6(b–d)). At the same time, oil which had not been emulsified and remained in the upper part of the cell experienced a radially inward flow to form a drop which was larger than the emitted droplets but much smaller than the initial drop (Figure 2.6(c,d)). This drop remained at the end of the experiment along with many small oil droplets but no visible particles of the lamellar phase.

Similar behavior was seen whenever initial water content in the drop was 10 wt% or lower and AOT concentration exceeded 25 wt%, but virtually no emulsification occurred for lower AOT concentrations. Emulsification also occurred when the 10 wt% water in the drop of Figure 2.6 was replaced by the same amount of NaCl solution. Indeed, when the solution contained 0.75 wt% NaCl, emulsification was observed even when AOT concentration was reduced to 20 wt%. However, when salt-free octane drops containing 30 wt% AOT and 10 wt% water were injected into a 0.5 wt% NaCl solution, emulsification was minimal.

These results can be understood by recognizing that surfactant films containing AOT have spontaneous curvature favoring a w/o configuration at high ionic strength and an o/w configuration at low ionic strength. In the experiment of Figure 2.6 ionic strength inside the drop decreased as water diffused in and AOT out, causing inversion via the lamellar liquid crystalline phase and nucleation of oil droplets. The decrease in ionic strength was even larger for drops which contained NaCl. But when salt-free drops were injected into sufficiently concentrated salt solutions, inversion and hence spontaneous emulsification did not occur.

As noted above, a small amount of the original oil was not converted to droplets. The reason is that the w/o microemulsion present initially or formed immediately after injection did not invert continuously to an o/w microemulsion as with the systems containing short-chain alcohols discussed above. Instead it was converted to a lamellar phase which itself was subsequently transformed to an o/w microemulsion. The lamellar phase was able to solubilize considerable oil but had a somewhat higher surfactant-to-oil ratio than the w/o microemulsion (see Ref. 38 for references on and discussion of phase behavior). As a result, nearly all of the surfactant was converted to the lamellar phase but not all of the oil.

The higher initial AOT concentrations which produced emulsification also produced lower interfacial tensions. Hence, when drops with high AOT concentrations rose to the upper surface of the cell after injection owing to their low density, gravity caused much more flattening than for drops containing less AOT, producing a generally radial outward flow during which diffusion caused inversion via the lamellar liquid crystalline phase and emulsification. Not all of the oil was converted to the lamellar phase, however, as indicated previously. The later retraction of this oil, which had not been emulsified (Figure 2.6c), was probably the result of an increase in interfacial tension caused by a decrease in AOT concentration in the oil. With the higher tension and smaller volume of this remaining oil, gravity no longer dominated interfacial effects and free energy was minimized by formation of a nearly spherical drop. The retraction in the upper part of the cell to form this drop produced the outward flow of the oil droplets in the lower part seen in Figure 2.6(c,d).

It is noteworthy that the basic inversion mechanism for ionic surfactants involving a decrease in ionic strength makes possible nearly complete emulsification in a ternary oil–water–surfactant system as in Figure 2.6. In contrast, complete or nearly complete emulsification was not observed for ternary n-hexadecane/water/$C_{12}E_6$ (or $C_{12}E_4$) systems. Because there is no analog to the ionic strength effect for nonionic surfactants, their phase behavior did not permit complete conversion of an oil–surfactant mixture to another phase which later became supersaturated in oil.

Dicharry et al. (39) obtained complete dispersion of oil as small droplets for a mixture of a cationic and a nonionic surfactant in their study of cutting fluid formation. They started with microemulsions – perhaps in some cases containing dispersed lamellar phase – having equal amounts of oil and water, 10% surfactant, and various amounts of dissolved monoethanolamine borate (MEAB). Although the water added contained some dissolved calcium and magnesium, ionic strength was reported to nevertheless decrease owing to dilution of the MEAB solution, so that inversion leading to emulsification occurred as in the AOT experiments described above. Some stirring was applied during water addition, and nearly monodisperse drops having diameters of order 100 nm were obtained. Contributing to the absence of larger oil drops was that, in contrast to the AOT experiment, no bulk oil phase was present initially and thus all the drops in the final emulsion must have nucleated as small droplets during inversion.

2.8 SELF-EMULSIFICATION PRODUCED BY CHEMICAL REACTION OF SURFACTANTS

As demonstrated above, inversion of a w/o to an o/w microemulsion can lead to spontaneous emulsification of oil. The inversion occurs because the surfactant film experiences changes in composition and/or environment which effect a reversal in spontaneous curvature. In the systems discussed above diffusion brought about these changes. For nonionic surfactants a reduction in temperature can produce similar effects, the phase inversion temperature (PIT) method of emulsification (40–42). Of interest in this section, however, are chemical reactions which convert lipophilic to hydrophilic surfactants and thereby cause inversion. For example, reaction might produce ionization of an uncharged amphiphilic compound or convert a double-chain to a single-chain surfactant.

Perhaps the simplest reactions which cause self-emulsification are those where ionization of fatty acids, amines, or other amphiphilic compounds dissolved in an oil phase occurs when the oil contacts an aqueous phase having a suitable pH. It has long been known that contacting oil containing a dissolved fatty acid with an alkaline solution produces spontaneous emulsification (1,43), which is often accompanied by interfacial turbulence generated by interfacial tension gradients (44). Emulsification has been attributed to this turbulence, to the occurrence of transient negative interfacial tensions, or both. Raney (45) observed spontaneous emulsification using videomicroscopy when n-decane/oleic acid/2-pentanol mixtures were contacted by comparable volumes of aqueous solutions containing different amounts of NaOH and NaCl. Information on equilibrium phase behavior was available for this system (46). For a composition where the equilibrium state was an oil-in-water microemulsion in equilibrium with excess oil, spontaneous emulsification was observed in the aqueous phase at some distance from the surface of contact and thus clearly due to local supersaturation. Similarly, spontaneous emulsification produced by local supersaturation was seen in the oil phase for a composition where the equilibrium state was a water-in-oil microemulsion in equilibrium with excess brine. For a composition where a bicontinuous microemulsion was present with both excess oil and excess brine at equilibrium, interfacial turbulence featuring vigorous roll cells was seen on initial contact between oil and aqueous phases along with emulsification of drops of an intermediate microemulsion phase in the oil. Some lamellar liquid crystal was formed as well, and soon continuous layers of both microemulsion and liquid crystal developed near the initial surface of contact, while the roll cells disappeared. The initial interfacial turbulence likely contributed to the emulsification, but the phase dispersed was not one of those initially present but the microemulsion phase, which formed at the initial interface due to mass transfer.

Nishimi and Miller (47) investigated self-emulsification produced by other types of reactions which convert lipophilic to hydrophilic surfactants. In one case a drop containing 5 wt% $C_{12}E_8$, 5 wt% monoolein, and 90 wt% n-dodecane was injected into Tris buffer solution (pH = 10.4) containing 0.05 wt% phenylboronic acid at 40 °C. The reaction converted the rather lipophilic monoglyceride into a much more hydrophilic ionic surfactant, thereby producing local supersaturation and extensive, though not complete, emulsification (Scheme 2.1).

$CH_3(CH_2)_7CH{=}CH(CH_2)_8C(=O)O$–glyceryl–OH (OH) + $\bar{B}(OH)_3$–phenyl

$\longrightarrow CH_3(CH_2)_7CH{=}CH(CH_2)_8C(=O)O$–(cyclic boronate ester O–B(OH)–O, phenyl)

Separate experiments showed that the PIT was approximately 35 °C for the same proportions of nonionic surfactant, monoglyceride, and hydrocarbon in the absence of phenylboronic acid. Hence the system was slightly lipophilic in its initial state but rapidly inverted to hydrophilic conditions as a result of the reaction. No emulsification was seen when a similar drop was injected into the buffer solution without added phenylboronic acid.

Self-emulsification yielding nearly all small droplets was observed when drops containing 10 wt% $C_{12}E_4$, 10 wt% of the calcium salt of a commercial dodecylbenzene sulfonic acid, and 80 wt% n-octane were injected into water containing 0.5 wt% of the sodium salt of EDTA at 50 °C (Figure 2.7). As is well known, this EDTA salt complexes calcium ions, simultaneously releasing sodium ions. The PIT was found to be approximately 45 °C for the same oil phase and water containing no EDTA. Thus, at 50 °C the system was again slightly lipophilic initially but inverted when EDTA complexed calcium ions, thereby converting double-chain calcium sulfonates to more hydrophilic single chain sodium sulfonates. No emulsification was seen in the absence of EDTA.

Finally, self-emulsification was observed after several minutes when a drop of ethyl oleate, a nontoxic oil, containing 5 wt% of the zwitterionic phospholipid DOPC (1,2-dioleylglycero-3-phosphorylcholine) was injected into an aqueous phase maintained at pH 9 by a borate buffer and containing 5000 units per gram of the enzyme phospholipase A_2. This enzyme catalyzed splitting of the double-chain phospholipid into two single-chain and hence more hydrophilic surfactants, the corresponding lysolecithin and oleic acid. In addition, much of the acid was ionized since the aqueous phase was alkaline.

2.9 SELF-EMULSIFICATION FOR COMPOSITIONS NEAR THE PHASE INVERSION TEMPERATURE

In the systems of the preceding three sections self-emulsification of oil occurred because diffusion and/or chemical reaction caused spontaneous curvature of the surfactant films at oil–water interfaces in a microemulsion to change from that favoring a w/o microstructure to that favoring o/w. During this process the associated decrease in the ability to solubilize oil produced local supersaturation leading to nucleation of small oil droplets.

However, self-emulsification of oil has also been observed in some systems where the above explanation seems inadequate. That is, the transformation from an oil-rich isotropic phase to an oil-in-water emulsion appears to be due primarily to an increase in the water-to-oil ratio (WOR) rather than to a change in spontaneous curvature. The existence of two types of inversion

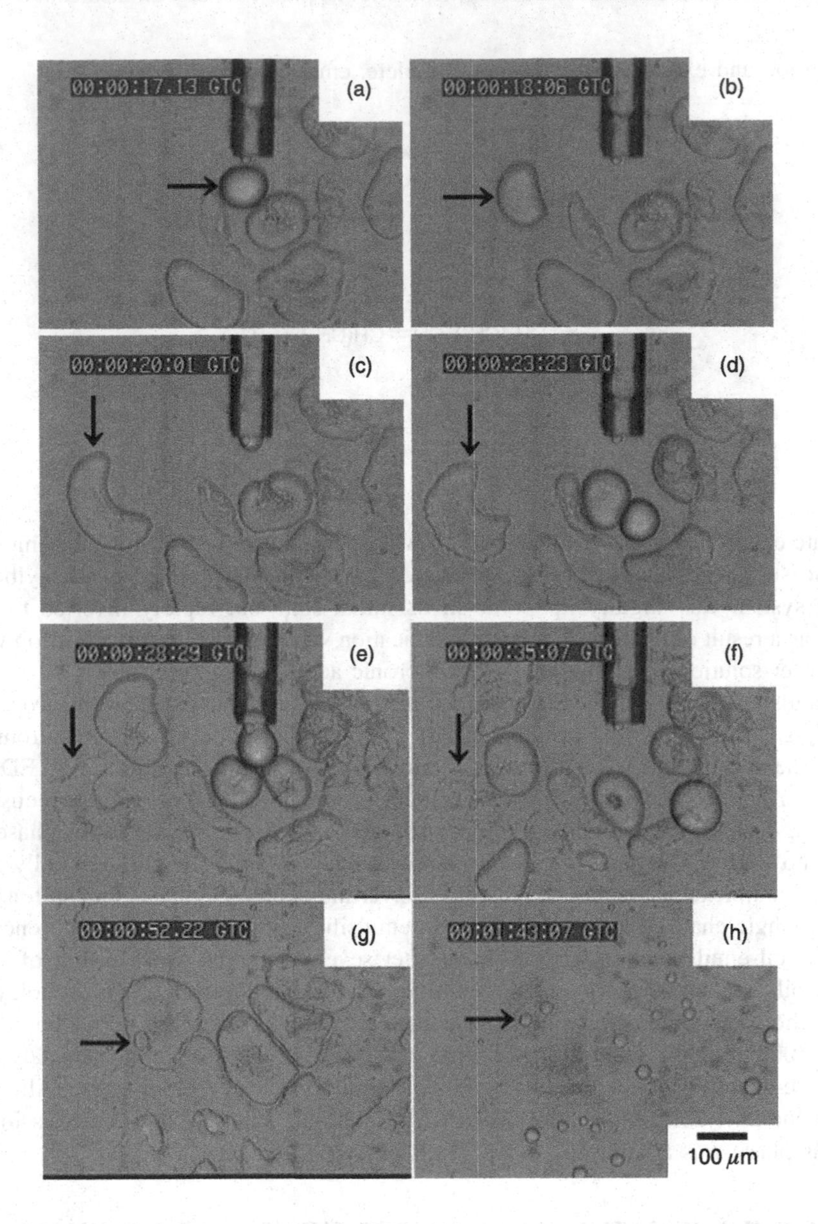

FIGURE 2.7 Video frames showing spontaneous formation and exit from the micropipette of drops consisting of 10 wt% $C_{12}E_4$, 10 wt% Ninate 401, and 80 wt% n-octane in contact with water containing 0.5 wt% of the sodium salt of EDTA at 50 °C. The arrows show behavior of a single drop as nearly all of it is emulsified. Times are indicated as (h:min:sec:frame).

processes produced respectively by changes in spontaneous curvature and WOR is well known for conventional emulsions (48). However, in such cases inversion caused by variation in WOR has been studied mainly for systems where the surfactant is either rather hydrophilic or rather lipophilic and not near the balanced condition where spontaneous curvature is zero. The self-emulsification phenomena described next occurred for systems near this balanced condition.

Rang and Miller (49) investigated emulsification systematically for the system n-hexadecane/oleyl alcohol/$C_{12}E_6$/water, where the alcohol is nearly insoluble in water. Complete emulsification to form small oil droplets was seen only when initial surfactant concentration was at least 20 wt% and when the hexadecane/oleyl alcohol ratio in the injected drop was between 95/5 and 90/10 by weight. Phase behavior studies suggested that the higher surfactant concentrations were necessary in order for the drop to be completely converted to the lamellar phase as it took up water following injection. Approximately 10 wt% water was required. As yet more water entered the drop, a microemulsion phase formed as well. One or both of these phases eventually became supersaturated in oil, and spontaneous emulsification occurred. Complete conversion to the lamellar phase assured that all oil droplets would be formed by nucleation, i.e., that no larger drops of the original oil phase remained. It is noteworthy that, according to the phase behavior experiments, the minimum surfactant concentration in this oil-rich lamellar phase occurred near the balanced condition or PIT where spontaneous curvature was small, as would be expected.

Figure 2.8 illustrates the effect of hexadecane/oleyl alcohol ratio on the size of droplets formed with 25 wt% surfactant present. The condition that this ratio be within the range cited

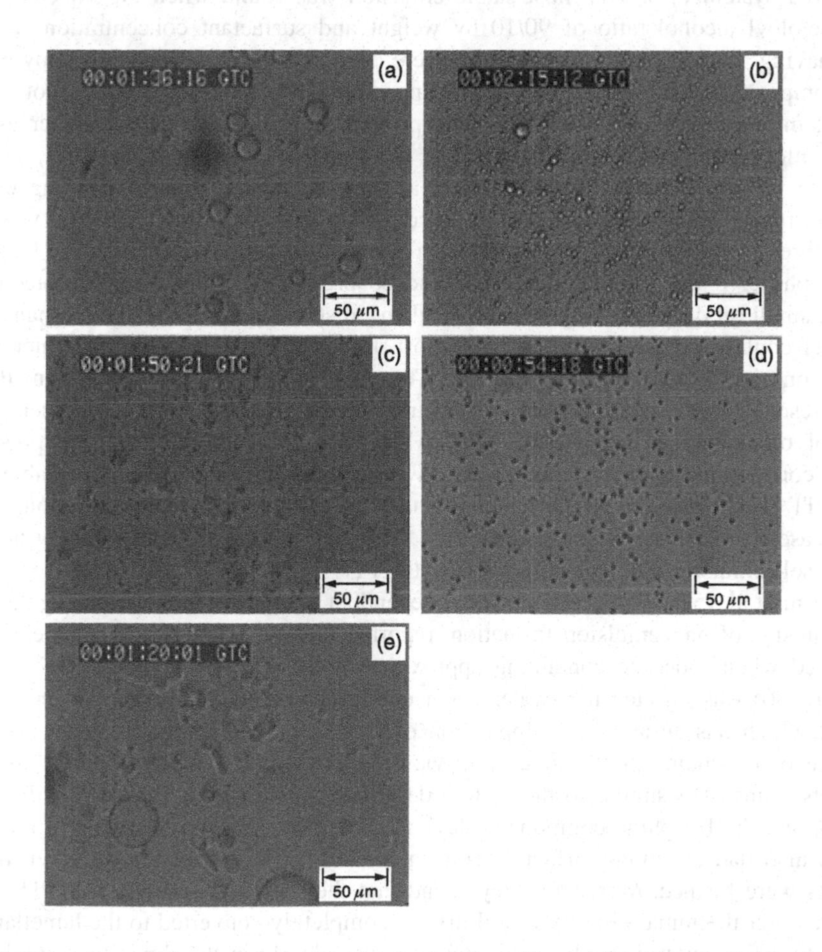

FIGURE 2.8 Video frames showing drop size distribution for self-emulsification of oil drops initially containing 25 wt% $C_{12}E_6$ and having various hexadecane/oleyl alcohol ratios by weight: (a) 97/3, (b) 95/5, (c) 93/7, (d) 90/10, and (e) 87/13.

above is basically a requirement that it be near that corresponding to the PIT. Because for these conditions both oleyl alcohol and $C_{12}E_6$ have low solubility in water, little diffusion into the aqueous phase occurs with the result that there is little change in spontaneous curvature. Hence, unlike the situation in a similar system with octanol discussed above, inversion to an oil-in-water microemulsion does not occur for initial states considerably more lipophilic than the PIT, as shown in Figure 2.8e where particles of the lamellar phase are present along with some oil drops in the final state. Pautot et al. (50) recently reported interesting observations of the mechanism by which particles of the lamellar phase are formed and subsequently ejected into water when oil containing a lipophilic surfactant (phospholipid) contacts water.

Emulsification also occurred more quickly for the system containing oleyl alcohol than for that with octanol. In fact, it was not possible to observe details of the emulsification process. The reason is that the initial state was near the PIT and intermediate phases formed immediately on injection. In contrast, no emulsification or intermediate phase formation occurred for the octanol system until sufficient alcohol had diffused from the injected drop into the surrounding water.

The stability of the emulsions so formed was studied using turbidity measurements in a gently stirred system (49). The most stable emulsion was found when the injected drop had a hexadecane/oleyl alcohol ratio of 90/10 by weight and surfactant concentration was 20 wt%. Droplets having diameters of a few microns were observed in the videomicroscopy experiments for this composition when there was no stirring. Of course, smaller droplets not visible with the optical microscope might also have been present. However, droplet diameter as measured by light scattering for the emulsion formed in the stability experiment was only 60 ± 20 nm. Possibly the mixing dispersed small droplets as they nucleated, thereby limiting coalescence. Equilibrium phase behavior indicated that the composition of the dilute emulsion was in or very close to a three-phase region consisting of oil, a dilute surfactant solution, and the lamellar liquid crystalline phase. If the lamellar phase was indeed present, it could form a protective coating around the small oil droplets and provide the enhanced stability observed. This enhanced stability near the PIT contrasts sharply with results of most studies of emulsion stability, which have found that emulsions are least stable near the PIT. The difference is that in this system the lamellar phase is present even at relatively low surfactant concentrations near the PIT when comparable amounts of oil and water are present (49), whereas only microemulsions are present at low surfactant concentrations in systems which have been studied and exhibit unstable emulsions near the PIT. The existence of the lamellar phase at low surfactant concentrations seems also to be the case for the systems discussed in the following paragraphs and is likely necessary for successful self-emulsification for systems near the PIT.

Forgiarini et al. (51), who conducted an independent investigation emphasizing the conditions and mechanisms of nanoemulsion formation, reported that oil droplets with diameters of 50 nm were formed when n-decane containing approximately 20 wt% of a commercial nonionic surfactant (Brij 30) was injected into water under conditions of gentle stirring. As this commercial surfactant, which has an average composition of $C_{12}E_4$, contains various species with a range of ethylene oxide chain lengths, it can be viewed as a mixture of hydrophilic and lipophilic components. Thus, it is similar to the system described above where the hydrophilic component was $C_{12}E_6$ and the lipophilic component oleyl alcohol. Other similarities exist in that Forgiarini et al. (51) also had an initial surfactant concentration of 20–25% in the oil when the smallest oil droplets were formed. Moreover, they found that their system was near the PIT and that on addition of water the initial mixture would first be completely converted to the lamellar phase and ultimately to a final emulsion whose composition was clearly in the three-phase region containing oil, a surfactant solution, and the lamellar liquid crystal. They noted in this and subsequent work that the liquid crystal stabilized the emulsion with respect to coalescence. Nevertheless, drop size increased with time owing to Ostwald ripening (52). The same group formed stable nanoemulsions by cooling (PIT method), apparently for conditions where no liquid crystal was present (42). Evidently nucleation occurred at a temperature low enough that surfactant films on the drops were sufficient to minimize coalescence.

Mohlin et al. (53) studied self-emulsification at 60 °C of alkyl ketene dimer containing a mixture of a nonionic surfactant and calcium dodecylbenzene sulfonate. Here too compositions of the mixtures yielding nanoemulsions were near the PIT, the system was stirred during the emulsification process, and stability was enhanced because the lamellar phase was present and apparently coated the oil droplets in the final state. It is not clear whether significant dissociation of the calcium sulfonate occurred. If so, the ionic surfactant formed would modify spontaneous curvature to favor oil-in-water emulsions.

Shahidzadeh et al. (54) presented intriguing phase contrast microscopy results on self-emulsification of n-hexadecane drops containing mixtures of $C_{12}E_5$ and n-dodecanol. They observed that the lamellar phase formed as vesicles within the oil phase upon contact with water and that these vesicles "exploded" when they reached the oil–water interface, apparently dispersing oil drops with diameters of order 1 μm into the aqueous phase. They considered that when such behavior occurred, complete transformation of the original oil phase to the lamellar phase was not required to obtain only small oil drops in the final emulsion although phase behavior showing the minimum surfactant content of the lamellar phase was not provided. In any case the authors were able to produce emulsification with only 5–10% surfactant in the oil, in contrast to the work described in the preceding paragraphs where typical concentrations were of order 20%. It may be that, since both surfactant and alcohol had straight hydrocarbon chains of the same length, their lamellar phase was more coherent than those in the other systems and hence dispersed more oil when it exploded. It is noteworthy that the same group observed similar explosions in experiments in which dispersions of the lamellar phase of AOT in water were contacted with oil (55).

Lopez-Montilla et al. (2) also observed phenomena resembling explosions when a drop of n-hexadecane containing Brij 30 was contacted with a drop of water. Some explosions were seen before the lamellar phase became visible. However, as the observations were made at relatively low magnification, it seems possible that some liquid crystal formed at the interface on initial contact and may have been involved in the explosive phenomena.

All of the systems reported to yield nanoemulsions were subjected to modest stirring during the emulsification process. It seems unlikely that shear stresses generated by such stirring was sufficient to break up oil into nanodroplets (although homogenizers which *can* generate nanodroplets exist). It seems more likely the stirring helped disperse nanodroplets formed by nucleation before substantial coarsening could occur due to coalescence or possibly Ostwald ripening, the rate of which can be enhanced for droplets in contact. Further investigation of the effect of stirring is desirable. Nevertheless, the above results suggest that nanoemulsions do form spontaneously but may exist only transiently, especially in the absence of stirring.

2.10 SUMMARY

The results summarized above have helped clarify the mechanism of self-emulsification of oils yielding small droplets. One method is to start with oil containing a surfactant or surfactant mixture which forms interfaces whose spontaneous curvature favors water-in-oil microemulsions. When suitable mixtures of this type contact water, diffusion and/or chemical reaction causes inversion to conditions where spontaneous curvature favors oil-in-water microemulsions. During inversion the microemulsion becomes unable to solubilize all the oil present with the result that oil droplets nucleate. Self-emulsification is possible without formation of the lamellar liquid crystal during inversion, but the lamellar phase can coat the small droplets and prevent coalescence from generating large drops. Higher surfactant concentrations may also be able to minimize coalescence but with the disadvantage of added expense.

Self-emulsification of surfactant/oil mixtures or microemulsions near the balanced condition or PIT seems to occur in some systems with little change in spontaneous curvature but with a large increase in water-to-oil ratio (WOR). Conventional emulsions are known to invert with changing WOR, but such behavior has not been studied extensively near the PIT. In the self-emulsification studies described above, formation of the lamellar liquid crystalline during addition of water seems to be essential, both because oil incorporated into the lamellar phase must renucleate as small droplets if an oil phase is to be present in the final state and because striking phenomena involving ejection of oil have been observed when the anisotropic lamellar phase is swollen by uptake of water.

Emulsions having nanosize droplets can form spontaneously, but gentle stirring seems to be required during the process to prevent coarsening, even when the droplets are coated by the lamellar phase.

REFERENCES

1. Davies, J.T.; Rideal, E.K. *Interfacial Phenomena*, 2nd ed., Academic Press, New York, 1963, 360–366.
2. Lopez-Montilla, J.C.; Herrera-Morales, P.E.; Pandey, S.; Shah, D.O. Spontaneous emulsification: mechanisms, physicochemical aspects, modeling, and applications. J. Disp. Sci. Technol., **2002**, *23*(1–3), 219–268.
3. Horn, D.; Rieger, J. Organic nanoparticles in the aqueous phase – theory, experiment and use. Angew. Chem. Int. Ed., **2001**, *40*, 4330–4361.
4. Gopal, E.S.R. Hydrodynamic aspects of the formation of emulsions. In *Rheology of Emulsions*, Sherman, P. Ed.; Macmillan, New York, 1962, 15–25.
5. Granek, R.; Ball, R.C.; Cates, M.E. Dynamics of spontaneous emulsification. J. Phys. II (France), **1993**, *3*, 829–849.
6. Theissen, O.; Gompper, G. Lattice-Boltzmann study of spontaneous emulsification. Eur. Phys. J. B, **1999**, *11*, 91–100.
7. Sternling, C.V.; Scriven, L.E. Interfacial turbulence: hydrodynamic instability and the Marangoni effect. AIChEJ, **1959**, *5*, 514–523.
8. Miller, C.A. Spontaneous emulsification produced by diffusion – a review. Colloids Surf., **1988**, *29*, 89–102.
9. Pfennig, A. Mass transfer across an interface induces formation of micro droplets in lattice systems. Chem. Eng. Sci., **2000**, *55*, 5333–5339.
10. Yeung, A.; Dabros, T.; Czarnecki, J.; Masliyah, J. On the interfacial properties of micrometer-sized water droplets in crude oil. Proc. R. Soc. Lond. A, **1999**, *455*, 3709–3723.
11. Lipowsky, R. The conformation of membranes. Nature, **1991**, *349*, 475–481.
12. Seifert, U.; Lipowsky, R. Shapes of fluid vesicles. In *Handbook of Nonmedical Applications of Liposomes*, Lasic, D.D., Barenholz, Y. Eds.; CRC Press, 1996, 43–68.
13. Hu, J.G.; Granek, R. Buckling of amphiphilic monolayers induced by head–tail asymmetry. J. Phys. II (France), **1996**, *6*, 999–1022.
14. Tungsubutra, T. Solubilization–emulsification processes in nonionic surfactant–water–liquid triglyceride systems. PhD Thesis, Rice University, 1994, 121–125.
15. Hirasaki, G.; Zhang, D.L. Surface chemistry of oil recovery from fractured, oil–wet carbonate formations. SPEJ, **2004**, *9*, 151–162.
16. Ruschak, K.; Miller, C.A. Spontaneous emulsification in ternary systems with mass transfer. Ind. Eng. Chem. Fundam., **1972**, *11*, 534–540.

17. Miller, C.A.; Rang, M.J. Spontaneous emulsification in oil–water–surfactant systems. In *Emulsions, Foams, and Thin Films*, Mittal, K.L., Kumar, P. Eds.; Marcel Dekker, New York, 2000, 105–120.
18. Miller, C.A. Solubilization and intermediate phase formation in oil–water surfactant systems. Tenside Surf. Det., **1996**, *33*, 191–196.
19. El-Aasser, M.S.; Lack, C.D.; Vandefhoff, J.W.; Fowkes, F.M. The miniemulsification process – different form of spontaneous emulsification. Colloids Surf., **1988**, *29*, 103–118.
20. Chen, B.H.; Miller, C.A.; Garrett, P.R. Dissolution of nonionic surfactant mixtures. Colloids Surf. A, **2001**, *183–185*, 192–202.
21. Chen, B.H. Rates of solubilization of triolein and triolein/fatty acid mixtures and dissolution processes of neat surfactants. PhD Thesis, Rice University, 1998.
22. Minehan, W.T.; Messing, G.L. Synthesis of spherical silica particles by spontaneous emulsification. Colloids Surf., **1992**, *63*, 181–187.
23. Niwa, T.; Takeuchi, H.; Hino, T.; Kunou, N.; Kawashima, Y. Preparations of biodegradable nanospheres of water-soluble and insoluble drugs with lactide/glycolide copolymer by a novel spontaneous emulsification solvent diffusion method, and the drug release behavior. J. Controlled Release, **1993**, 89–98.
24. Quintanar-Guerrero, D.; Allemann, E.; Doelker, E.; Fessi, H. A mechanistic study of the formation of polymer nanoparticles by the emulsification–diffusion technique. Colloid Polym. Sci., **1997**, *275*, 640–647.
25. Murakami, H.; Kobayashi, M.; Takeuchi, H.; Kawashima, Y. Further application of a modified spontaneous emulsification solvent diffusion method to various types of PLGA and PLA polymers for preparation of nanoparticles. Powder Technol., **2000**, *107*, 137–143.
26. Kwon, H.Y.; Lee, J.Y.; Choi, S.W.; Jang, Y.; Kim, J.H. Preparation of PLGA nanoparticles containing estrogen by emulsification–diffusion method. Colloids Surf. A, **2001**, *182*, 123–130.
27. Becher, D.Z. Applications in agriculture. In *Encyclopedia of Emulsion Technology*, vol. 2, Becher, P. (Ed.); Marcel Dekker, New York, 1985, 239–320.
28. Tadros, T.F. *Surfactants in Agrochemicals,* Marcel Dekker, New York, 1995, Ch. 4.
29. Li, Y.; Groves, M. The influence of oleic acid on the formation of a spontaneous emulsion. S.T.P. Pharma Sci., **2000**, *10*, 341–344.
30. Pouton, C.W. Lipid formulations for oral administration of drugs: non-emulsifying, self-emulsifying and "self-microemulsifying" drug delivery systems. Eur. J. Pharma. Sci., **2000**, *11*, Suppl. 2, S93–S98.
31. Constantinides, P.P. Lipid microemulsions for improving drug dissolution and oral absorption – physical and biopharmaceutical aspects. Pharm. Res., **1995**, *12*, 1561–1572.
32. Lee, G.W.J.; Tadros, Th.F. Formation and stability of emulsions produced by dilution of emulsifiable concentrates. Parts 1–3. Colloids Surf., **1982**, *5*, 105–115, 117–127, 129–135.
33. Groves, M.J. Spontaneous emulsification. Chem. Ind., **1978** (June 17), 417–419.
34. Wakerly, M.G.; Pouton, C.W.; Meakin, B.J.; Morton, F.S. Self-emulsification of vegetable oil–nonionic surfactant mixtures. In *Phenomena in Mixed Surfactant Systems*, Scamehorn, J.F., Ed., ACS Symp. Ser. No. 311, American Chemical Society, Washington, D.C., 1986, 242–255.
35. Rang, M.J.; Miller, C.A. Spontaneous emulsification of oil drops containing surfactants and medium-chain alcohols. Progr. Colloid Polym. Sci., **1998**, *109*, 101–117.
36. Rang, M.J.; Miller, C.A.; Hoffmann, H.H.; Thunig, C. Behavior of hydrocarbon/alcohol drops injected into dilute solutions of an amine oxide surfactant. Ind. Eng. Chem. Res., **1996**, *35*, 3233–3240.
37. Greiner, R.W.; Evans, D.F. Spontaneous formation of a water–continuous emulsion from a w/o microemulsion. Langmuir, **1990**, *6*, 1793–1796.
38. Nishimi, T.; Miller, C.A. Spontaneous emulsification of oil in Aerosol OT/water/hydrocarbon systems. Langmuir, **2000**, *16*, 9233–9241.
39. Dicharry, C.; Bataller, H.; Lamaallam, S.; Lachaise, J.; Graciaa, A. Microemulsions as cutting fluid concentrates: structure and dispersion into hard water. J. Disp. Sci. Technol., **2003**, *24*, 237–248.
40. Sagitani, H. Phase-inversion and D-phase emulsification. In *Organized Solutions*, Friberg, S.E., Lindman, B., Eds.; Marcel Dekker, New York, 1992, 259–271.

41. Förster, T.; von Rybinski, W.; Wadle, A. Influence of microemulsion phases on the preparation of fine-disperse emulsions. Adv. Colloid Interface Sci., **1995**, *58*, 119–149.
42. Izquierdo, P.; Esquena, J.; Tadros, T.F.; Dederen, C.; Garcia, M.J.; Azemar, N.; Solans, C. Formation and stability of nanoemulsions prepared using the phase inversion temperature method. Langmuir, **2002**, *18*, 26–30.
43. von Stackelberg, M.; Klockner, E.; Mohrhauer, P. Spontane Emulgierung infolge negativer Grenzflächenspannung. Kolloid-Z., **1949**, *115*, 53–66.
44. Rudin, J.; Wasan, D.T. Interfacial turbulence and spontaneous emulsification in alkali–acidic oil systems. Chem. Eng. Sci., **1993**, *48*, 2225–2238.
45. Raney, K.H. Studies of nonequilibrium behavior in surfactant systems using videomicroscopy and diffusion path analysis. PhD Thesis, Rice University, 1985, 71–92.
46. Qutubuddin, S.; Miller, C.A.; Fort, Jr., T. Phase behavior of pH-dependent microemulsions. J. Colloid Interface Sci., **1984**, *101*, 46–58.
47. Nishimi, T.; Miller, C.A. Spontaneous emulsification produced by chemical reaction. J. Colloid Interface Sci., **2001**, *237*, 259–266.
48. Salager, J.L.; Forgiarini, A.; Marquez, L.; Peña, A.; Pizzino, A.; Rodriguez, M.P.; Rondon-Gonzalez, M. Using emulsion inversion in industrial processes. Adv. Colloid Interface Sci., **2004**, *108–109*, 259–272.
49. Rang, M.J.; Miller, C.A. Spontaneous emulsification of oils containing hydrocarbon, nonionic surfactant, and oleyl alcohol. J. Colloid Interface Sci., **1999**, *209*, 179–192.
50. Pautot, S.; Frisken, B.J.; Cheng, J.X.; Xie, X.S.; Weitz, D.A. Spontaneous formation of lipid structures at oil/water/lipid interfaces. Langmuir, **2003**, *19*, 10281–10287.
51. Forgiarini, A.; Esquena, J.; Gonzalez, C.; Solans, C. Formation of nano-emulsions by low-energy emulsification methods at constant temperature. Langmuir, **2001**, *17*, 2076–2083.
52. Pons, R.; Carrera, I.; Caelles, J.; Rouch, J.; Panizza, P. Formation and properties of miniemulsions formed by microemulsion dilution. Adv. Colloid Interface Sci., **2003**, *106*, 129–146.
53. Mohlin, K.; Holmberg, K.; Esquena, J.; Solans, C. Study of low energy emulsification of alkyl ketene dimer (AKD) related to the phase behavior of the system. Colloids Surf. A, **2003**, *218*, 189–200.
54. Shahidzadeh, N.; Bonn, D; Meunier, J.; Nabavi, M.; Airiau, M.; Morvan, M. Dynamics of spontaneous emulsification for fabrication of oil in water emulsions. Langmuir, **2000**, *16*, 9703–9708.
55. Shahidzadeh, N.; Bonn, D; Aguerre-Chariol, O.; Meunier, J. Spontaneous emulsification: relation to microemulsion phase behavior. Colloids Surf. A, **1999**, *147*, 375–380.

3 Symmetric Thin Liquid Films with Fluid Interfaces

Dimo Platikanov and Dotchi Exerowa

CONTENTS

3.1 INTRODUCTION

Thin liquid films arise spontaneously in all colloid and disperse systems with a liquid disperse media. They are formed always when two particles of the disperse phase (solid particles, liquid drops, or gas bubbles) come close to each other. The liquid film is a symmetric film when both approaching particles originate from the same disperse phase with the same composition of substances, i.e., the symmetric liquid film divides two equal micro-phases. Such thin liquid films are the emulsion films, the foam films, and the films between the particles of a sol. In more complicated cases, when two particles with different composition of substances approach each other, an asymmetric liquid film is formed. These are all cases of heterocoagulation. The most important asymmetric films are the wetting films – thin liquid films which divide a solid and a gas phase.

The *emulsion films* are symmetric liquid films between two liquid phases, i.e., L′LL′ films, L′ being a liquid immiscible with the liquid L. One can distinguish two types of emulsion films: *OWO films* and *WOW films*, which correspond to the two emulsion types. Accordingly the *foam films* are GLG films. Both emulsion and foam films are thin liquid films with fluid interfaces. Only these types of films are considered in the present chapter. Since the foam films are more extensively studied, most of the examples which illustrate different aspects of the thin liquid films with fluid interfaces are foam films. The similarity between emulsion and foam films allows information obtained about foam films to be used for emulsion films and vice versa. This similarity is specially discussed in Section 3.7.

Obviously the properties and the behavior of the thin liquid films determine the stability or instability of the corresponding disperse system. That is why the thin liquid film is one of the most important objects of the colloid and interface science. Furthermore the liquid films play a central role in many technological processes. All this caused the vast expansion of the fundamental and applied research of thin liquid films, especially in the second half of the twentieth century. However, the first observation on the properties and behavior of thin liquid films were reported in the eighteenth century.

The prominent scientists Hook and Newton described the process of thinning of foam films and the formation of black spots and black films in a thicker film [1]. This is the reason that the current IUPAC nomenclature refer to the thinnest emulsion or foam films as “Newton black films.” In the middle of the nineteenth century the Belgian scientist Plateau performed extensive investigations of the properties of the thin liquid films [2]. He established the important fact that the thin liquid film is always connected either with a bulk liquid phase or with a solid wall through a liquid body with concave surfaces. We call now this liquid body “Plateau border.” Marangoni studied [3] the role of the expansion or compression of the surfactant adsorption layers for the stability of the films, known now as the “Marangoni effect.” The great Gibbs gave significant ideas concerning the liquid film’s stability [4], among them the so-called “Gibbs elasticity.”

During the twentieth century, studies on thin liquid films expanded at accelerated rates. In the first decades Perrin studied extensively the different types of black foam films as well as the interesting phenomenon of stratification [5]. In 1936 Derjaguin introduced [6] the quantity *disjoining pressure* – the most important thermodynamic parameter which characterizes the thin liquid film. The great success of the DLVO theory for the stability of the lyophobic colloids [7,8] put forward the decisive role of the thin liquid films for the stability of all types of disperse systems with liquid disperse media. Significant contributions for the quantitative description of thin liquid films have been carried out by the scientific schools of Derjaguin (Russia), Overbeek (Netherlands), Mysels (U.S.A.), Scheludko (Bulgaria), and many others.

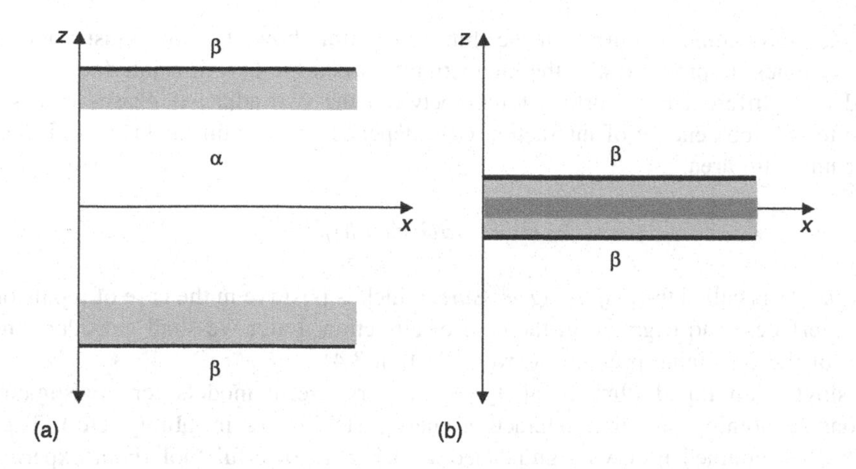

FIGURE 3.1 (a) Scheme of a "thick" liquid film and (b) scheme of a thin liquid film formed from the liquid phase α between two equal fluid phases β.

What exactly should we call a *thin liquid film*? Let us consider a symmetric liquid film with parallel plane interfaces (Figure 3.1). The Cartesian coordinate system is oriented so that the x and y axes lie in the film plane, while the z-axis is normal to the film interfaces. Obviously, the film dimension along z is much smaller than the dimensions along x and y which determine the film area. However, the question arises: what is much smaller?

The liquid phase from which the film is formed is denoted in Figure 3.1 by α and the adjacent phase (another liquid or gas) by β. The α/β interface is actually not a mathematical plane but an interfacial layer with final thickness. Most of the physical quantities inside this interfacial layer vary along the z-axis, due to the mutual influence of the phases α and β. When the distance between both interfaces is rather large (Figure 3.1(a)), the two interfacial layers are far away from each other. There is liquid between them with the same properties as in the bulk liquid phase α. For instance such a L′LL′ system should be considered as a system of three bulk liquid phases, although one of the liquid phases' shape is a film. Such a liquid film is a *thick film* from the viewpoint of the thermodynamics; nevertheless its thickness could be few micrometers only.

In the case when the film thickness is small enough so that both interfacial layers overlap (Figure 3.1(b)), there is no liquid inside the film with the properties of the bulk liquid phase α. Just like in a single interfacial layer most of the physical quantities vary along the z-axis over the entire liquid film from the bulk of the one phase β up to the other phase β. As we shall see later, this is the physical ground for the rise of the disjoining pressure in the film as well as for the change of the surface tension of the film interfaces. Some properties of the film become dependent on its thickness. Such a film is referred to as a thin liquid film.

It should be noted that this is a thermodynamic definition of the thin liquid film and it reflects the peculiar properties of the thin film and their thickness dependence. However, when kinetic properties (e.g., film drainage, rheology, dynamic elasticity, etc.) are considered, often the liquid film is called "thin" simply because of its very small thickness, not taking into account that it could be a thermodynamically "thick" film. For instance, Gibbs has considered the foam film elasticity assuming the same surface tension of the film surfaces as that of the bulk liquid.

The thermodynamic definition of the thin liquid film shows that the most important factor, which determines its properties, is the interaction between the two film interfaces, i.e., the interactions due to different type surface forces between the two adjacent phases across the liquid film. The total Gibbs energy of interaction $G(h)$ depends on the film thickness h. It determines a force per unit film area:

$$\Pi = -[\partial G(h)/\partial h]_{p,T} \tag{3.1}$$

The pressure Π is called the *disjoining pressure*, which is positive in the case of repulsion between the film interfaces, and negative in the case of attraction. Later we shall consider a mechanical definition of the disjoining pressure as well (Section 3.4).

The single thin liquid films of all types are very useful models for investigation of pair interactions (disjoining pressure, interaction energy, stability or instability, etc.). Thus the thin liquid film has attained its own significance as being a powerful tool (both experimental and theoretical) in colloid and interface science. On the other hand, as already noticed, the properties of thin liquid films are decisive for the properties of the corresponding disperse system, e.g., an emulsion. Certainly, the emulsion system is not simply a sum of many liquid films. It is much more complicated. Nevertheless, knowledge of film's properties significantly advances the elucidation of a problem when investigating a real emulsion.

3.2 EXPERIMENTAL METHODS FOR RESEARCH OF EMULSION FILMS

The peculiar properties of thin liquid films cannot be studied using the well known physicochemical methods for investigation of bulk phases and even of single interfaces. Original and unique methods have been developed for liquid film research. The increase of our knowledge about thin liquid films is due, to a great extent, to the effective experimental techniques created for their investigation. This chapter focusses on symmetric thin liquid films with fluid interfaces, i.e., emulsion films and foam films. Accordingly, the experimental methods for mainly emulsion films are considered in this section.

3.2.1 THE PROBLEM: "FILM THICKNESS"

Film thickness is the most important quantity to be experimentally determined. It not only characterizes the thin liquid film but also most film properties depend on the thickness. Obviously, the thickness h of a thin liquid film is defined as the distance between the two film interfaces. The quantity h would be exactly defined if the film interfaces were mathematical faces. However, this is not the case for symmetric films with fluid interfaces. The definition of their film thickness is very difficult. The transition from the liquid film to the adjacent liquid (or gas) phase is not abrupt. There is an interfacial layer with finite thickness even in a one-component liquid film. However, the emulsion and foam films contain more components; at least a surfactant which forms adsorption layers at the film interfaces. The complicated structure of the film interfacial layers requires special definition of the mathematical face that could be accepted as a film boundary.

One possibility is the mechanical definition: the film boundaries are two plane-parallel mathematical faces, each of which is subject to a uniform, isotropic tension. These faces are the *surfaces of tension*. The *mechanical film thickness* is defined as the distance h between the two surfaces of tension [9].

The thermodynamic description of an interface is based on the so-called *Gibbs dividing plane*. The thin liquid film thermodynamics involves two Gibbs dividing planes (see Section 3.4).

The position of these planes is defined by a special condition, which can be chosen by us. It is most appropriate for thin liquid films to accept zero interfacial excess of the main component in the liquid film, i.e., of the liquid which is solvent in the bulk phase from which the film is formed. Under this condition the position of both Gibbs dividing planes of the film is defined and the *thermodynamic film thickness* is the distance between these two planes [10,11]. The thermodynamic film thickness should be used when data for the quantities characterizing the film are treated with thermodynamic relations.

The big problem is that neither the mechanical nor the thermodynamic film thickness exactly coincides with the thickness which is experimentally determined using different physical methods. Most of these methods are optical methods and the measurement of the film thickness is based on the abrupt change of properties such as the refractive index and electron density at the film boundary. In the case of electric methods for film thickness measurement, again an abrupt change at the film boundary of specific conductivity, dielectric permitivity, for example, is required. The thin liquid film definition presented in Section 3.1 shows that most of the physical quantities vary along the normal to the film over the entire liquid film and some film properties are thickness dependent. Moreover there is no abrupt transition at the interface film/adjacent phase. That is why the real physical thickness is not directly measured through the experimental methods used.

Let us consider the most widely used optical method – the interferometric method, based on measurement of the intensity of visible light reflected from an emulsion film. Classical optics provides relations that link the thickness of the film with its optical characteristics [12]. If emulsion films are observed in white light, it can be seen that during thinning their coloration changes periodically. Initially, the process runs rather rapidly and gradually slows down. Such a course of the interference can be registered as a curve photocurrent/time in which the extrema correspond to the interference maxima and minima, i.e., film thicknesses are divisible to $\lambda/4n$ (where λ is the wavelength of light and n is the refractive index of the liquid in the film). Thus, knowing the order k of interference, it is easy to determine the film thickness at these points. Film thickness (between a maximum and a minimum) is calculated from the ratio between the intensities measured of the reflected monochromatic light I, corresponding to a certain thickness, and I_{max}, corresponding to the interference maximum, according to the formula [13]

$$h_w = \frac{\lambda}{2\pi n}\left(k\pi \pm \arcsin\sqrt{\frac{I/I_{max}}{1+\left[4N/(1-N)^2\right](1-I/I_{max})}}\right) \tag{3.2}$$

with $N = (n-n')^2/(n+n')^2$, n' being the refractive index of the liquid in the adjacent phases; h_w is called equivalent film thickness, i.e., the thickness of an emulsion film with uniform refractive index n of the bulk solution from which the film is formed. For foam films Equation 3.2 transforms [14] into

$$h_w = \frac{\lambda}{2\pi n}\left(k\pi \pm \arcsin\sqrt{\frac{I/I_{max}}{1+\left[(n^2-1)/2n\right]^2(1-I/I_{max})}}\right) \tag{3.3}$$

The accuracy of thickness measurements with the interferometric technique is ± 0.2 nm.

It is clear that h_w is not equal to the real physical film thickness. Although h_w does not coincide with the thermodynamic thickness h, h_w is close to it, if the last is defined through the condition of zero interfacial excess of the solvent in the film. The difference between h_w and h could be neglected for thicknesses larger than 30 nm. However, for thinner liquid films it is necessary to account also for the film structure. The three-layer film model ("sandwich model")

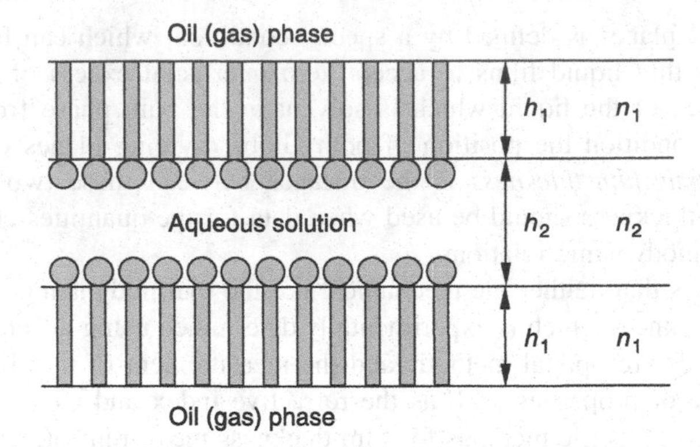

FIGURE 3.2 A three-layer model ("sandwich model") of an OWO emulsion film or a foam film: an aqueous layer with thickness h_2 and refractive index n_2 between two hydrocarbon layers with thickness h_1 and refractive index n_1.

with an aqueous core of thickness h_2 and refractive index n_2 and two homogenous layers of hydrocarbon chains of the adsorbed surfactant of thickness h_1 each and refractive index n_1 is often used for foam films (Figure 3.2). The thickness of the aqueous core h_2 is most often determined, according to the formula [15]

$$h_2 = h_w - 2h_1 \frac{n_1^2 - 1}{n_2^2 - 1} \tag{3.4}$$

The same film model and Equation 3.4 can be adapted to aqueous emulsion films as well.

Since each model involves some assumptions, the calculation of h_2 always renders certain inaccuracy. The most important problem in the three-layer model concerns the position of the plane that divides the hydrophobic and hydrophilic parts of the adsorbed surfactant molecule. In some cases it seems reasonable to have this plane passing through the middle of the hydrophilic head of the molecule, in others the head does not enter into the aqueous core. A more detailed model is the five-layer model in which the aqueous layer (Figure 3.2) is divided into three layers: two layers contain the hydrophilic parts of the adsorbed molecules and the inner core contains aqueous solution only. The calculations of the film thickness based on different film models do not solve the general thickness problem. However, they provide possibilities for reasonable interpretation of the experimental results.

3.2.2 Microscopic Emulsion or Foam Films

Small circular emulsion or foam films, the radius of which is within the range of 10 to 500 μm, are considered to be microscopic films. The experimental technique for their study has been most successful and is constantly being improved. It allows thermodynamic quantities to be measured, kinetic behavior to be followed, the formation of black films to be recorded, and the formation of metastable states to be realized, for example. An advantage is the possibility to work at very low surfactant concentrations.

The measuring cell of Scheludko and Exerowa [16] has proven to be a suitable and reliable tool for the formation of microscopic horizontal emulsion or foam films. It is shown in Figure 3.3,

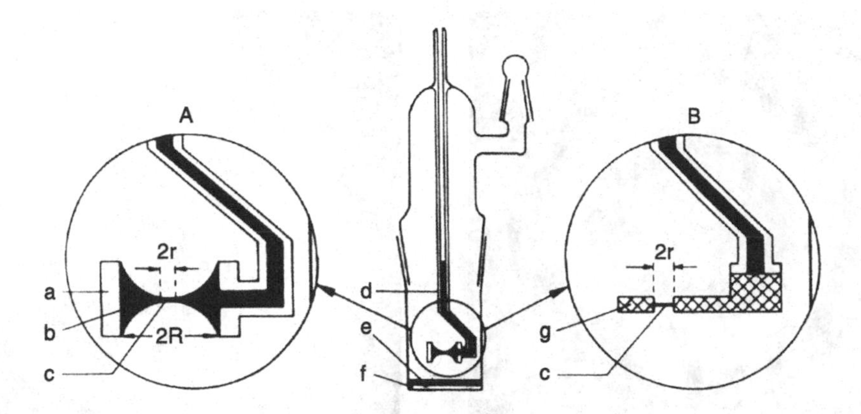

FIGURE 3.3 Scheme of the measuring cell for the study of microscopic emulsion or foam films. A: in a glass tube; B: in a porous plate; **a** is the glass tube film holder; **b** is a biconcave drop; **c** is the microscopic emulsion or foam film; **d** is a glass capillary; **e** is surfactant solution (in case of foam film only); **f** is optically flat glass; **g** is the porous plate; *r* is the film radius; *R* is the film holder radius. (After Refs. 16, 18, and 38.)

variants A and B. The microscopic film c is formed in the middle of a biconcave drop b, situated in a glass tube a of radius *R*, by withdrawing liquid from it (variant A) and in the hole of porous plate g (variant B). Typical photographs of different states of an emulsion microscopic film (non-equilibrium film with dimpling, relatively thick film, black spots, black film) taken under a microscope are presented in Figure 3.4. The suitable tube radius *R* in variant A is 0.2 to 0.6 cm and the film radius ranges from 100 to 500 μm. In variant B the hole radius can be considerably smaller, for instance 120 μm, and, respectively, the film radius can be 10 μm. The film holder with the film is situated in the closed space of the cell, filled with the second liquid (or saturated with the solution vapor in the case of a foam film). The periphery of the film is in contact with the first liquid phase, of the solution from which the film is formed. The film holders, the tube or the porous plate, are welded to capillary d. In the case of aqueous emulsion or foam films, the inner part of the tube a, carrying the biconcave drop, is finely "furrowed" with vertical lines, closely situated to one another, which improve wetting [17]. In the case of "oil" films the film holder should be made hydrophobic. A constant capillary pressure acts on the film formed in part A in Figure 3.3. It is determined by the radius of curvature of the meniscus.

Porous plates of various pore radii can be used in part B [18], Figure 3.3. If the meniscus penetrates into the pores, their radius determines the radius of curvature, i.e., the small pore size allows to increase the capillary pressure until the gas phase can enter in them (in the case of foam films). The capillary pressure can be increased to more than 10^5 Pa, depending on the pore size and the surface tension of the solution.

Another measuring cell for microscopic films has been specially designed [19] for emulsion films (Figure 3.5). Vessel 1 contains the first liquid – the solution from which the emulsion film is formed. The tubes 2 situated co-axially contain the second liquid of the adjacent phases. This liquid forms two menisci at the orifices of both tubes. Using the micrometric pistons 3 the two menisci can be very precisely put close to each other until the microscopic emulsion film is obtained between them. The film is observed and its thickness can be interferometrically measured through the plane-parallel glass window 4 by the microscope 5.

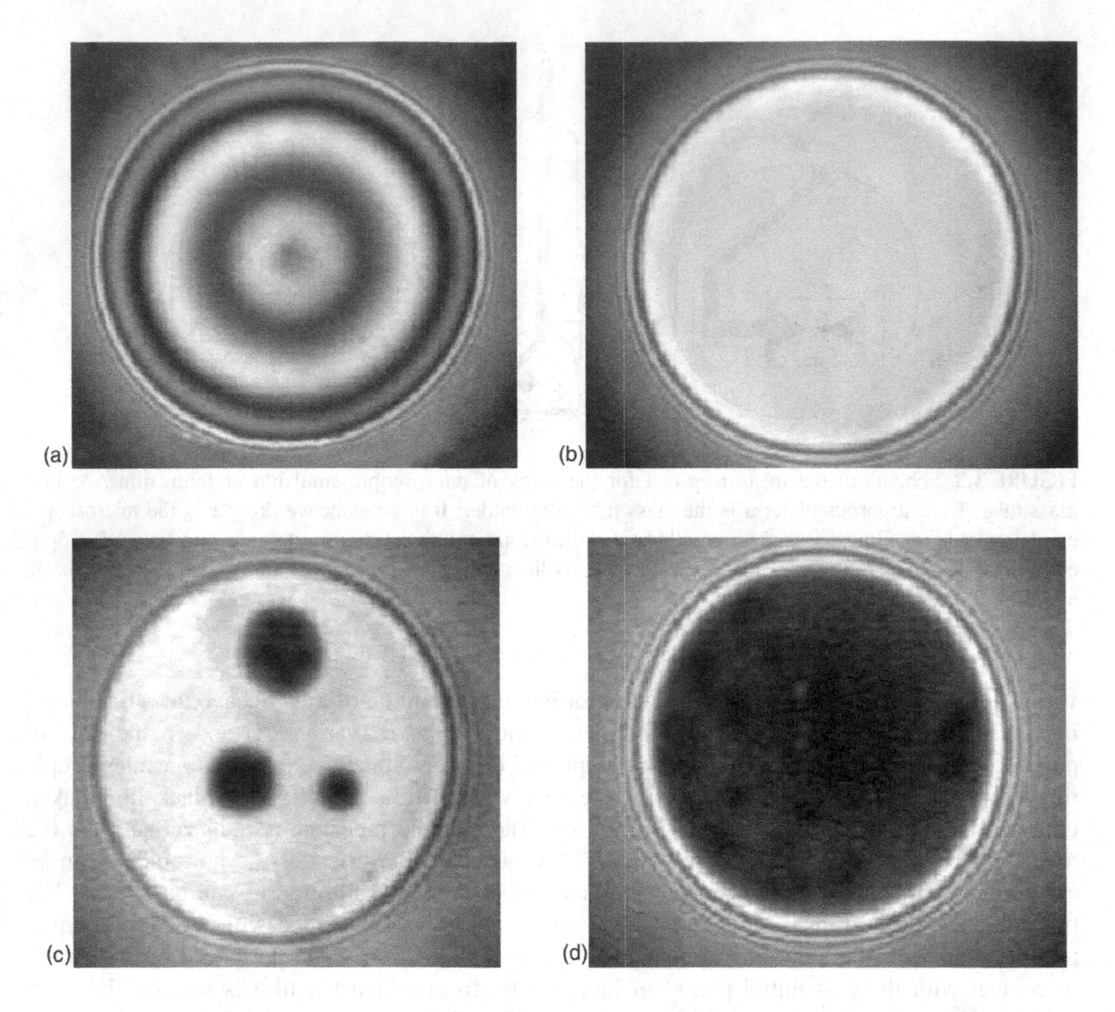

FIGURE 3.4 Typical photographs of a microscopic thin emulsion film taken under a microscope: (a) non-equilibrium film with dimpling; (b) relatively thick equilibrium film; (c) non-equilibrium film with black spots in it; (d) equilibrium black film.

3.2.3 Black WOW Emulsion Films

Black "oil" emulsion films between aqueous phases (often denoted as *bilayer lipid membranes* or BLM) are usually obtained in the cell [20] schematically presented in Figure 3.6. The vessel which contains the aqueous solution 1 is divided in two compartments by the teflon wall 3. There is a small hole with sharp edges in this wall. A drop of the "oil" – a surfactant solution in a nonpolar solvent – is put in the hole using a fine pipette 4. A small vertical emulsion film from this nonpolar liquid is formed in the hole. It drains quickly due to gravity until a *Newton black emulsion film* is formed in the hole. In case the surfactant used is a phospholipid the Newton black emulsion film is called bilayer lipid membrane. It is observed and investigated using the light source 6 and the microscope 7. For research of some electrical properties of such films the two electrodes 5 are used.

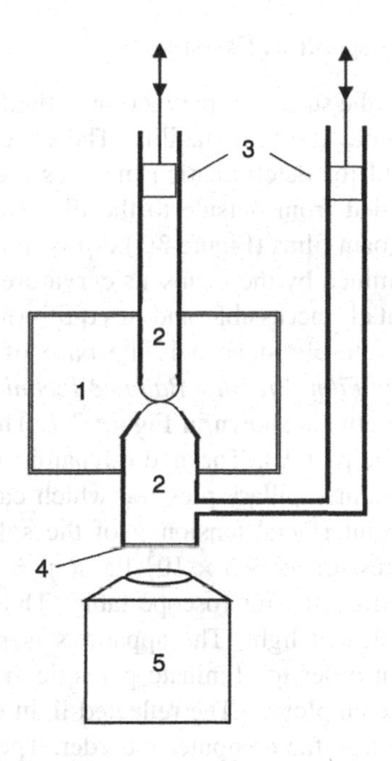

FIGURE 3.5 Principle scheme of a measuring cell for the study of microscopic emulsion films. 1 is a glass vessel with the liquid (phase α) from which the emulsion film is formed; 2 are glass tubes with the other liquid (phase β) of the adjacent phases; 3 are fine micrometric pistons; 4 is a plane-parallel glass plate; 5 is a microscope.

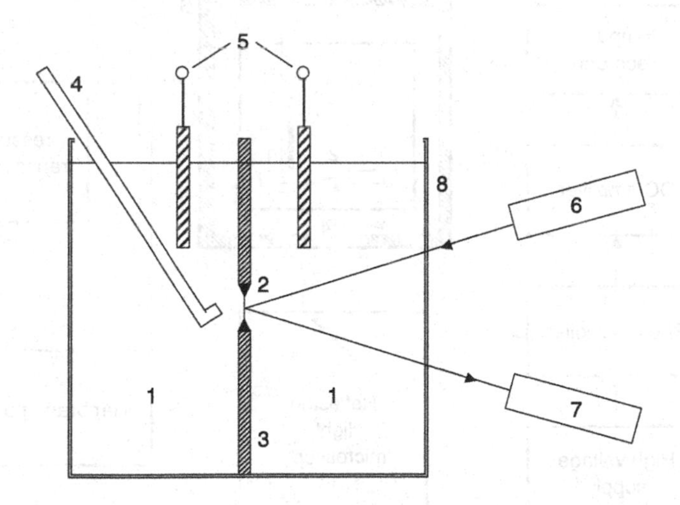

FIGURE 3.6 Principle scheme of a measuring cell for the study of black WOW emulsion films including bilayer lipid membranes (BLM). 1 are two compartments of the vessel 8 which contain aqueous solution – the liquid of the adjacent phases β; 2 is the WOW black emulsion film formed in a small hole in the Teflon wall 3; 4 is a fine pipette filled with the nonpolar liquid solution (phase α); 5 are electrodes; 6 is a light source; 7 is a microscope.

3.2.4 Measurement of the Disjoining Pressure

The disjoining pressure Π due to the surface forces acting in the film is the most important thermodynamic quantity which determines the film stability. The experimental techniques described in subsection 3.2.2 are usually used for determination of Π as well. At equilibrium Π is equal to the pressure difference Δp applied from outside to the film (see Section 3.4). For microscopic horizontal circular emulsion or foam films (Figure 3.3) Δp is the capillary pressure of the concave meniscus around the film, determined by the meniscus curvature and the interfacial tension of the bulk solution. Δp is experimentally accessible and at equilibrium $\Pi = \Delta p$ can be determined. Let us consider this case more closely since it is the base of the successful and widely used experimental method *Thin Liquid Film–Pressure Balance Technique* [21].

A block scheme of the apparatus is shown in Figure 3.7. The films are formed in the porous plate measuring cell (Figure 3.3, part B). The hydrodynamic resistance in the porous plate is sufficiently small and the maximum capillary pressure which can be applied to the film is determined by the pore size and the interfacial tension γ of the solution. When the maximum pore size is 0.5 μm, the capillary pressure is $\sim 3 \times 10^5$ Pa at $\gamma = 70$ mN/m. The cell is placed in a thermostating device, mounted on the microscope table. Thus the film can be monitored and measured photometrically in reflected light. The apparatus is placed in a special anti-vibration table in a thermostated room. In order to eliminate parasitic stray light, special diaphragms for the falling and reflected light are employed. The reflected light enters the photomultiplier and its signal is amplified and registered by the computer recorder. The regulation of the capillary pressure is achieved by a special membrane pump which allows a gradual and reversible change in the pressure p in the closed cell. The pump and the manometer are placed close to the measuring cell and are connected to it by thick wall tubing, thus ensuring a good thermostating.

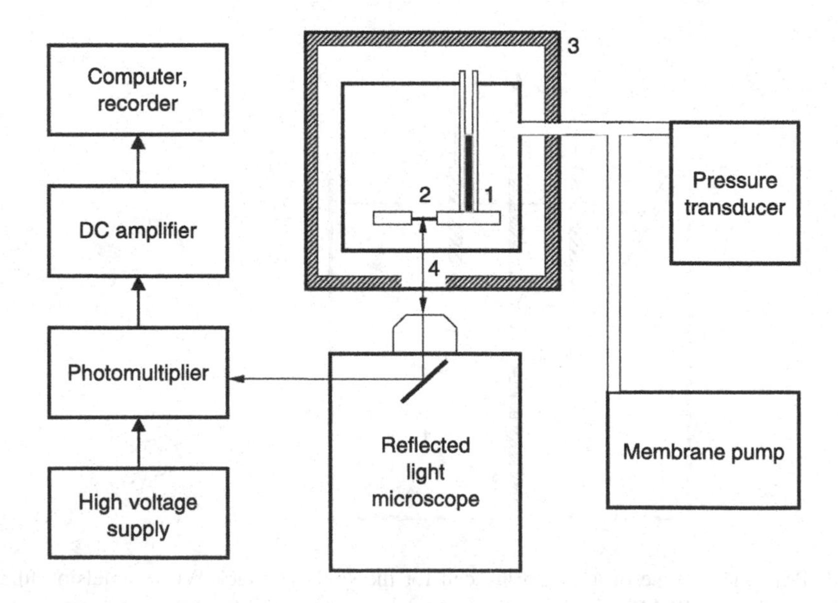

FIGURE 3.7 Block scheme of the thin liquid film–pressure balance technique: 1 is the measuring porous plate cell (cf. Figure 3.3(B)); 2 is the microscopic emulsion or foam film; 3 is a thermostating device; 4 is a plane-parallel glass window.

The thin liquid film–pressure balance technique has been used by a number of researchers who introduced several technical improvements. For example, values of Π less than 100 Pa prove much more difficult to measure, so there should be entire conformity with the equation giving the balance of pressures acting in the film [22] and the geometry of the measuring cell:

$$\Pi = p - p_r + \frac{2\gamma}{r} - \Delta\rho g h_c \tag{3.5}$$

where p_r is the external reference pressure; r is the radius of the capillary tube; $\Delta\rho$ is the density difference between the adjacent liquid (or gas) and the surfactant solution from which the film is formed; h_c is the height of the solution in the capillary tube above the film.

3.2.5 Measurement of the Contact Angles

The microscopic emulsion or foam film's experimental techniques are used for determination of two other important thermodynamic characteristics of the films: the *contact angle* α_o appearing at the contact of the film with the bulk solution from which it is formed and the *film tension* γ^f related to it. Two methods for the measurement of α_o have been developed: a topographic method and a film expansion method [23,24]. The topographic method is most suitable for small contact angles and is used in the study of black films of thickness 6 to 8 nm. This technique is based on the measurement of the radii of the interference Newton rings when the film is observed in a reflected monochromatic light (Figure 3.8). Knowing the film thickness, the contact angle is calculated according to:

$$\mathrm{tg}^2\alpha_o = B^2 - 4A\,(C - h/2) \tag{3.6}$$

where $A = \dfrac{l}{2x_1x_2}\dfrac{2x_1 - x_2}{x_2 - x_1}$; $B = \dfrac{l}{2x_1x_2}\dfrac{x_2^2 - 2x_1}{x_2 - x_1}$; $C = \dfrac{l}{2}$; $l = \dfrac{\lambda}{4n}$

x_1 is the distance between the first and the second Newton rings; x_2 is the distance between the first and the third Newton rings; λ is the light wavelength; n is the refractive index of the film liquid; h is the equilibrium thickness of the film. The applicability of the topographic method can be extended by using an approximated solution of the Laplace equation. Here it is necessary to measure the radius r_k of the k-th Newton ring of the meniscus surrounding the film, and the thickness at which this ring emerges.

The expansion method allows accurate determination of larger contact angles and is suitable for studying Newton black films of thickness about 5 nm. It is based on the ratio between the parameters of the thicker film (r_1 and α_1) and the black film, which results from the thicker film at constant volume of the meniscus. An equilibrium black film of radius r_2 and contact angle $\alpha_2 = \alpha_o$ is formed from the initial non-equilibrium thicker film having parameters r_1 and $\alpha_1 = 0$ during its "expansion" at $V = \text{const}$. The value of α_o can be determined from the experimentally measured values of r_1, r_2, α_1, and R (the tube radius) according to the formula

$$V = \frac{R}{3}E - RuE\,(R + 2/3u) + RuF\,(R + 1/3u) - \frac{r}{3}\sqrt{(r^2 - u^2)\,(R^2 - r^2)} \tag{3.7}$$

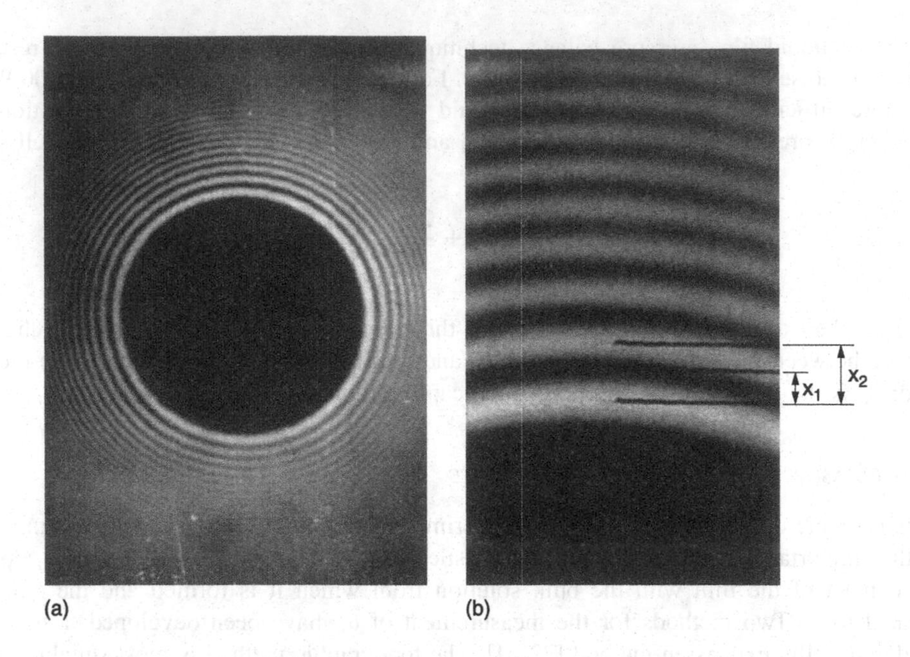

FIGURE 3.8 (a) Microscopic common black film, magnification ~300 times, (b) a segment of its meniscus, additionally enlarged. CBF from 0.277 mM Na dodecylsulphate + 0.2 M NaCl aqueous solution. x_1 and x_2 are the distances between the interference rings. (From T. Kolarov, A. Scheludko, and D. Exerova, *Annuaire Univ. Sofia, Fac. Chim.* 62: 75 (1967–68).)

where $E(\varphi, k^2)$ and $F(\varphi, k^2)$ are elliptic integrals of second and first order and

$$\varphi = \arcsin\sqrt{\frac{R^2 - r^2}{R^2 - u^2}}, \quad k = \frac{R^2 - u^2}{R^2} \quad \text{at} \quad u = \frac{R\sin\alpha_o - r}{R - r\sin\alpha_o}$$

The contact angles can be determined also by the technique of *floating lens* [25], which is widely used in the study of emulsion black films from hydrocarbons in aqueous medium.

There is a simple relation (see Section 3.4) between α_o and γ^f at equilibrium:

$$\gamma^f = 2\gamma\cos\alpha_o \tag{3.8}$$

Hence the film tension is often determined from experimental data about α_o and γ for a given emulsion or foam film.

Certainly the great variety of experimental techniques can not be described in this short section. Besides many variants of the methods reviewed above, there are also other original methods for studying specific aspects of the behavior and properties of thin liquid films.

3.3 THINNING AND RUPTURE OF THIN LIQUID FILMS; BLACK SPOT FORMATION

When a liquid film is formed from a bulk liquid phase α in a surrounding liquid (or gas) β, it is initially a thick liquid film. It becomes a thin liquid film as a result of the *process of thinning* under the driving force $\Delta p = p^\beta - p^\alpha$. This pressure difference can be: the capillary pressure of the meniscus, a hydrostatic pressure difference, a Δp created by experimental device, etc. An equilibrium thin liquid film is obtained only if Δp is counterbalanced by a positive disjoining pressure Π. However, when Π remains lower than Δp or it is even negative, no equilibrium can be established. The process of thinning leads to either *film rupture* or to jump-like formation of a much thinner *black film* stabilized by enough high positive Π.

The processes film thinning, film rupture, and jump-like formation of black film are the most important non-equilibrium behavior of the thin liquid films. In this section the non-equilibrium properties are considered only for microscopic, circular, horizontal, emulsion, or foam films, surrounded by double-concave meniscus. Such films are obtained in the cylindrical holder of the Scheludko–Exerowa cell (Figure 3.3) and obviously they have a cylindrical symmetry. Microscopic films offer certain advantages with respect to treatment and investigation under strictly defined conditions.

3.3.1 Kinetics of Thinning of Plane-Parallel Liquid Films

An important factor determining the kinetics of thinning of thin liquid films is their anisodiametricity, i.e., their radii are much larger than their thickness, $r \gg h$. Two quantities are introduced for quantitative description of the kinetics of thinning: the lifetime τ of thinning and the rate v of film thinning, defined according to the relations

$$\tau = \int_{h_0}^{h_{cr}} \frac{dh}{v}; \quad v = -\frac{dh}{dt} \tag{3.9}$$

where h_0 is the initial thickness; h_{cr} is the critical thickness at which the film ruptures; t is time.

Let us first consider the kinetics of thinning of plane-parallel, circular films with tangentially immobile interfaces. This is the case of the high degree of tangential blocking of the fluid film's interfaces by a surfactant adsorption layer. Under such conditions the hydrodynamics of thin liquid films is very well described by the Reynolds lubrication theory [26] leading to the well known Stephan–Reynolds relation [27] for the rate v_{Re} of thinning/thickening of a liquid layer of thickness h and viscosity η between two solid circular plates of radius r under a pressure drop Δp

$$-\frac{dh}{dt} = \frac{2h^3 \Delta p}{3\eta r^2} \equiv v_{Re} \tag{3.10}$$

Equation 3.10 was first applied for the kinetics of thinning of circular, horizontal, microscopic liquid films by Scheludko [28]. In his interpretation the pressure drop $\Delta p = \Delta p_c - \Pi$ represents the difference between the capillary pressure Δp_c of the meniscus and the disjoining pressure Π in the film. Thus,

$$\frac{dh^{-2}}{dt} = a\Delta p, \qquad a = \frac{4}{3\eta r^2} \tag{3.11}$$

Since the interfaces of emulsion or foam films considered here are far from rigid and in some cases can deviate considerably from the requirements of the lubrication theory, the applicability of Equation 3.11 to the process of thinning imposes the requirements: (1) viscosity η should not depend on film thickness; (2) interfaces should be tangentially immobile (zero surface velocity); (3) film interfaces should remain plane-parallel (should not deform); (4) capillary pressure Δp_c of the meniscus should not be affected by film thinning.

There are neither theoretical nor experimental data indicating that there is a dependence of bulk viscosity on film thickness h within the range of 0.01 to 1 μm [29]. One specific feature that distinguishes emulsion films from foam films and films between solid interfaces is the coupling of fluid motion in the adjacent phases. This interdependence of the flows is closely related to the mobility of the liquid/liquid interfaces. The latter is determined by two major factors: (1) the relative viscosity of the droplets η_d and of the film η; and (2) the presence of surfactants.

At close separations between the liquid droplets, the impact of the viscosity factor [30] $\bar{\eta} = \eta_d/\eta$ is modified due to the hindered outflow in the thin liquid film. In flow characteristics, it appears always in a combination with the geometry of the emulsion film (h, the film thickness, and r, the radius of the film) in the form of the hydrodynamic factor $f_h = \bar{\eta}(h/r)$. The rate of film thinning might then be estimated by the following scaling relation [31]:

$$\frac{v}{v_{Re}} \approx \frac{1 + f_h}{(1 + f_h)\left(\frac{h}{r}\right)^2 + f_h} \tag{3.12}$$

The specifications for the cases of emulsion and foam films are

$$\frac{v}{v_{Re}} \Rightarrow 1 + \frac{1}{f_h} \tag{3.12$'$}$$

for emulsion films, and

$$\frac{v}{v_{Re}} \Rightarrow \left(\frac{r}{h}\right)^2 \tag{3.12$''$}$$

for foam films. The above equations show that in pure systems the viscosity ratio cannot sustain the pressure drop Δp and the films are thinning always with a velocity higher than v_{Re}.

The other very important factor for the behavior of the emulsion systems is the presence of surfactants. The most popular observation is the so-called Bancroft rule: "The phase in which the stabilizing agent is more soluble will be the continuous phase." The kinetic aspect of this rule is related to the effect of the surfactant mass transfer on the hydrodynamics of emulsion films. This point of view stems from the physicochemical hydrodynamics [30] and was applied for the specific interactions in emulsion films [32].

The key idea is that at small separations the coupling of the surfactant mass transfer and the fluid motion is modified by the formation of thin emulsion films between the approaching droplets. In certain cases, this results in an enhanced suppression of the tangential mobility of the film interfaces, which is affected considerably by the presence of a surfactant through the Marangoni effect [3]. When the liquid drains from a film stabilized by a surfactant, a gradient of surface tension is created at its interfaces which counterbalances the viscous tensions. This gradient, equivalent to the surfactant adsorption gradient, causes the surfactant mass transfer: diffusion flow from the bulk to the film surface, and surface flow in the direction of the adsorption gradient. The tangential velocity of film interfaces is determined by the surface flow and it is always different from zero, i.e., condition "zero surface velocity" is never fulfilled [33].

This effect might be accounted for as a specification of the interfacial mobility scaling by introducing the so-called Marangoni factor f_c and the most general form of the estimate for the drainage rate of the emulsion film acquires [34] the form of

$$\frac{v}{v_{\text{Re}}} \approx \frac{1+f_h+f_c}{(1+f_h+f_c)\left(\dfrac{h}{r}\right)^2+f_h+f_c} \approx \frac{1+f_h+f_c}{f_h+f_c} \tag{3.13}$$

For the majority of the cases of correlation between the flow motion and the surfactant fluxes, both effects, f_h and f_c, act in an autonomous additive manner.

The extensive studies on the coupling of film hydrodynamics and the mass transfer of the surfactant during the drainage process have shown that the only case when there is a surfactant influence on the velocity of thinning is when the surfactant is soluble predominantly in the film phase and f_c acquires the form of

$$f_c = \frac{\dfrac{\Gamma}{\eta D}\left(\dfrac{\partial \gamma}{\partial c}\right)}{1+\dfrac{D_s}{Dh}\left(\dfrac{\partial \Gamma}{\partial c}\right)} \tag{3.14}$$

where D, D_s are, respectively, the bulk and surface diffusion coefficients of the surfactant; γ is the surface tension. The surfactant might block effectively the fluid interfaces even in the case of foam films. The deviation from Equations 3.10 and 3.13 resulting from Marangoni effect has been experimentally observed and theoretically predicted [35] in the expression

$$\frac{v}{v_{\text{Re}}} = 1-\left[1-\frac{2D_s}{Dh}\left(\frac{\partial \Gamma}{\partial c}\right)\right]\frac{3D\eta}{\Gamma\left(\partial\gamma/\partial c\right)} \tag{3.15}$$

The quantity $Ma = |\Gamma(\partial\gamma/\partial c)|/D\eta$ is called the Marangoni number and plays an important role in all transport processes at phase interfaces. Other factors (e.g., surface viscosity) affecting surface mobility of films and, hence, kinetics of their thinning are also analyzed [36]. The general criterion for the validity of the Reynolds equation (lubrication theory equations) for the description of emulsion film hydrodynamics [37] states:

$$\frac{f_h+f_c}{1+f_h+f_c} > \left(\frac{h}{r}\right)^2 \tag{3.16}$$

Equation 3.15 describes correctly the film's hydrodynamics only when factor Ma has large values – equivalent to small deviation from zero surface velocity. The general analysis of the role of surface mobility on the rate of thinning indicates that at high surface mobility ($Ma < 1$) the hydrodynamics of film thinning changes strongly, thus leading to a deviation from the lubrication approximation, Equation 3.10, i.e., the rate of thinning v strongly increases compared to that of films with immobile surfaces v_{Re} [34]. In the limiting case of completely free surfaces (films without a surfactant) the inertia effect prevails to the viscous one (see also Equation 3.12″).

3.3.2 Deviations from the Plane-Parallel Film During its Thinning

The Reynolds relation (Equation 3.10) requires liquid drainage from the film to follow strictly the axial symmetry between parallel walls. Rigid surfaces ensure such drainage through their non-deformability, while non-equilibrium emulsion or foam films are in fact never plane-parallel. This is determined by the balance between hydrodynamic and capillary pressure. Experimental studies have shown that only microscopic films of radii less than 100 μm retain their quasi-parallel surfaces during thinning, which makes them particularly suitable for model studies. Films of larger radii exhibit significant deviations from the plane-parallel shape which affect both the kinetics of thinning and their stability [38].

Not very large films ($r \geq 100$ μm) keep the axial symmetry of thinning but lose their plane-parallel shape. In the center there forms a typical thickening, known as a *dimple* (thicker lens-like formation), whose periphery is a thinner *barrier ring* (Figure 3.4(a)). The dimple forms spontaneously as a result of the hydrodynamic resistance to thinning in the periphery of the circular liquid film [39,40]. Experimental investigations proved that the rate of thinning is practically equal in both the dimple's center and barrier ring, i.e., the difference in thickness of the thickest and the thinnest domains does not decrease up to the critical thickness of rupture. This leads to an increase in the non-uniformity by thickness. On the other hand the non-uniformity by thickness increases with the increase in film size (film radius) as well. These results can be very useful in the interpretation of the experimental data about the dependence of the rate of film thinning and film lifetime on film radius. The experiments indicate [41] a weaker dependence of τ on r (Figure 3.9) than theoretically predicted by Equations 3.11 and 3.15.

A very interesting, in principle different, phenomenon has been observed at the thinning of microscopic, circular O/W/O emulsion films: spontaneous cyclic formation of a dimple occurs in an emulsion film from aqueous solution of the nonionic surfactant Tween 20 between oil droplets [42]. This phenomenon was described as a *diffusion dimple formation* in contrast to the dimple due to the hydrodynamic resistance to thinning in liquid films. The dimple shifted from the center to the periphery and periodically regenerated. During dimple growth the thickness of the circular, almost plane-parallel portion of the film between the dimple and the meniscus remains approximately constant (no change in the reflected light intensity). Photos of the different periods of a dimple growth are shown in Figure 3.10 and the process is schematically presented in Figure 3.11. An explanation has been suggested related to Marangoni effect of continuous

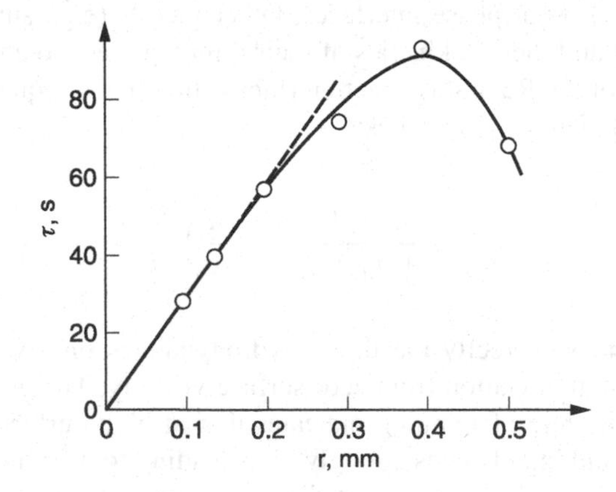

FIGURE 3.9 Film lifetime τ vs. film radius r; foam films from aqueous isovalerianic acid + KCl solution. (From D. Exerova and T. Kolarov, *Annuaire Univ. Sofia, Fac. Chim.* 59: 207 (1964–65).)

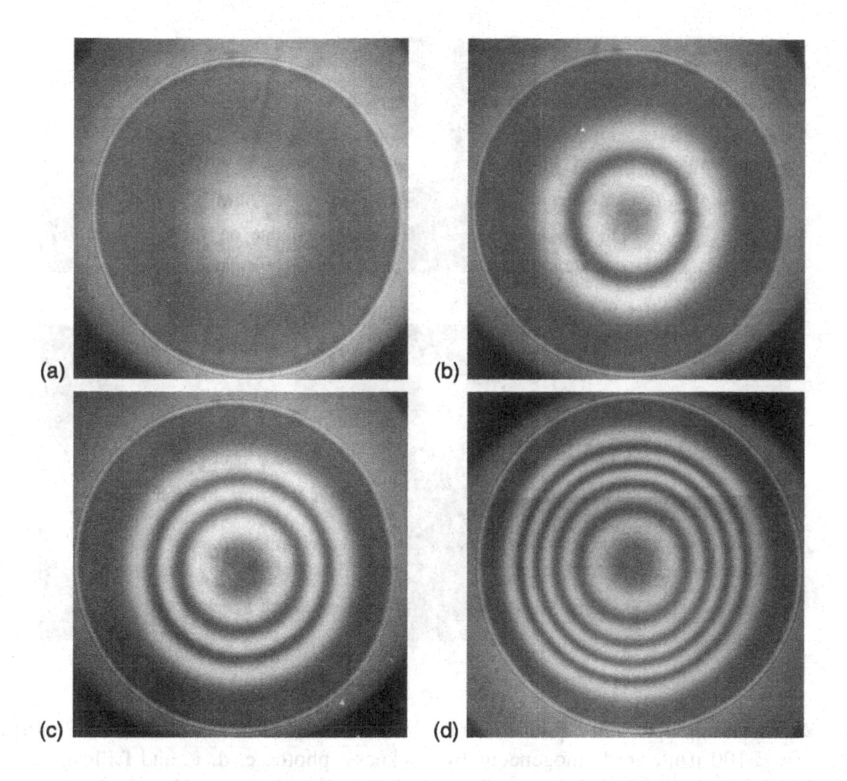

FIGURE 3.10 Four consecutive pictures, (a), (b), (c), and (d), of a spontaneous growth of the dimpling in an aqueous emulsion film with diameter 330 μm. (From O. Velev, T. Gurkov, and R. Borwankar, *J. Colloid Interface Sci.* 159: 497 (1993).)

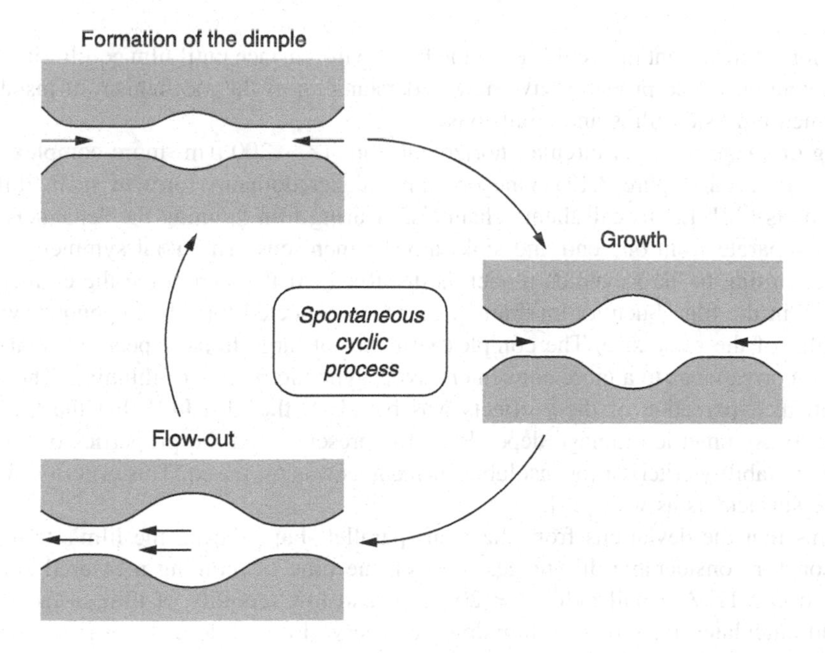

FIGURE 3.11 A cross-section of an emulsion film perpendicular to its interfaces: scheme of the main stages of the spontaneous cyclic process "diffusion dimple formation"; the film and dimple dimensions are not to scale. (From O. Velev, T. Gurkov, and R. Borwankar, *J. Colloid Interface Sci.* 159: 497 (1993).)

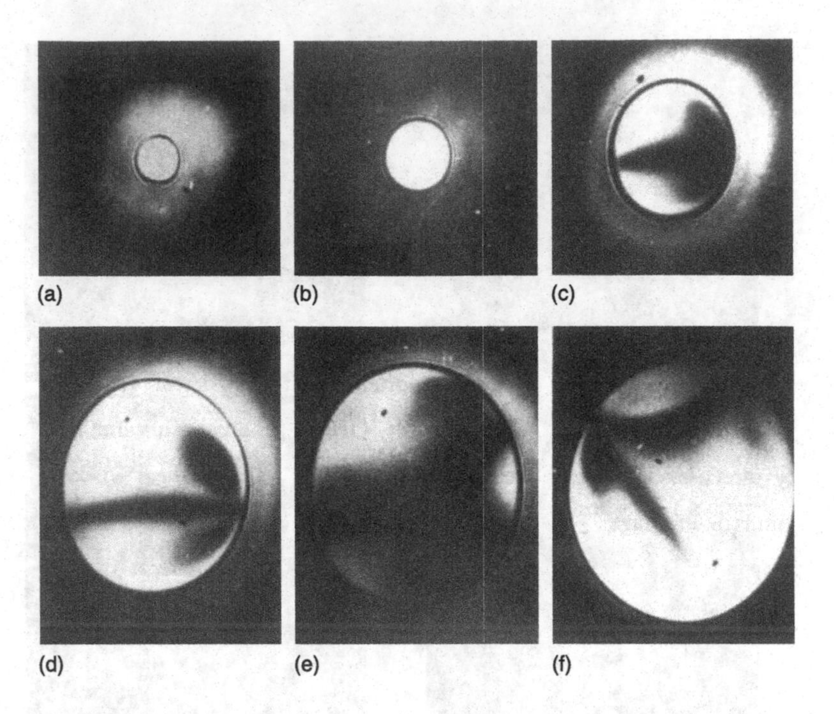

FIGURE 3.12 Six microscopic thin liquid films ($h \sim 100$ nm) with different diameters. Photos **a** and **b**, the smaller films ($r < 100$ μm), are homogeneous by thickness; photos **c**, **d**, **e**, and **f** illustrate the formation of "channels" in larger films ($r > 100$ μm). (From A. Scheludko, *Annuaire Univ. Sofia, Fac. Chim.* 59: 263 (1964–65).)

redistribution of surfactant molecules from the bulk to the surface until film equilibrium is reached. This phenomenon would probably give new understanding of the mechanism of instability in the newly formed emulsion films and emulsions.

During drainage of larger circular, horizontal films ($r > 200$ μm) more complex phenomena have been observed (Figure 3.12). One or more thicker domains form in such films, dividing them into parts [38]. Let us call them "channels." During film thinning the "channels" move and sometimes separate from one end and sink into the meniscus. The axial symmetry of drainage, assumed according to the Reynolds model, is disturbed. At the same time there emerge "centers of thinning" in the film. Such a film drains faster than expected for a homogenous symmetrically draining film of the same size. The complex structure of large films appears spontaneously and, therefore, it corresponds to a more convenient hydrodynamic regime of thinning. The first attempt for theoretical explanation of these effects was based on the idea [43] that the transition from symmetric to asymmetric drainage depends on the presence and the properties of the stabilizing surfactant. A stability criterion for insoluble surfactants was proposed. This criterion was modified for soluble surfactants as well [44].

It seems that the deviations from the plane-parallel shape during the film's thinning are the main reason for considerable differences between the time of thinning measured and calculated from Equation 3.11. At small radii ($r < 200$ μm) and low mobility of film surfaces the experimental and calculated rates of film thinning were very close while at large radii ($r > 200$ μm) the experimentally determined values were ten or more times higher than the calculated ones.

This extremely large difference in the rates and time of thinning cannot be explained only with the tangential mobility of film interfaces (estimated to be 5 to 80%). Moreover, this mobility does not depend on film radius and does not affect the function rate of thinning/radius.

The theoretical model assuming non-deforming film surfaces requires a linear dependence of the rate of thinning on the inverse square of film radius. In all experiments performed the $v(r)$ relation was weaker. These experimental results clearly indicate that Reynolds relation can be applied only to sufficiently small films ($r < 100$ μm), i.e., to films of uniform thickness. Obviously, the main reasons for strong deviations from the theoretical relation should be attributed to the deformations of the film interfaces. With further increase in radius the thinning films lose their axial symmetry and disintegrate into individual subdomains. A theory of the dynamics of large, circular, horizontal films with developed subdomains in them derives an equation [45] about thinning of films nonhomogenous by thickness:

$$v = \frac{1}{6\eta}\sqrt[5]{\frac{h^{12}\Delta p^{8}}{4\gamma^{3}r^{4}}} \tag{3.17}$$

This equation strongly differs from Reynolds equations (3.10 and 3.11) and has been checked with respect to rate of thinning as a function of film radius (v vs. $r^{-4/5}$). The experimental results are in good agreement with the theoretical prediction. Equation 3.17 holds for large films with strongly expressed nonhomogeneity by thickness. When the film radius decreases the film becomes plane-parallel and its rate of thinning asymptotically satisfies the Reynolds relation. The theory predicts also that the transition between the two regimes of drainage should occur at radius $r^* = 4\sqrt{\gamma h/\Delta p}$. A typical value is $r^* = 50$ μm which agrees well with the experimental observations [46].

3.3.3 Rupture of Thin Liquid Films

One substantial property of the emulsion films is their coalescence stability. The stability of the thin liquid film plays a decisive role. The theory of film rupture was first formulated for the case of foam films. With a slight modification, it is also applicable for emulsion films [47]. The studies show that the true stability of the emulsion films is related to their thermodynamic properties (surface tension, disjoining pressure), exactly as for foam films. The intrinsic coupling of hydrodynamics and mass transfer of surfactants concern the so-called kinetic stability, related to the time evolution of the draining emulsion film. It may be expressed by the interfacial mobility of thin liquid film alone (see Ref. 47, for example). The interfacial mobility of the droplets is important, insofar as it ensures sufficient slowdown of the emulsion film drainage. For higher interface mobility, the drainage velocity is also increased. This increase is also related to the onset of well-defined nonhomogeneities of the film thickness, and, therefore, results in a more rapid coalescence. As a rule, higher interface mobility is observed when surfactants are soluble inside the droplets, or almost equally soluble in the contiguous phases. The basic result is that while for some characteristics (such as lifetime of films) the interfacial mobility is of utmost importance, the thickness of rupture is actually independent on mobility factors.

The study of processes leading to rupture of thin liquid films is useful for understanding the reasons for their stability. The simplest explanation of film rupture involves reaching a thermodynamically unstable state. A typical example of thermodynamically unstable systems are symmetric thin liquid films in which the van der Waals contribution to disjoining pressure obeys Hamaker's relation, Equation 3.74. Such are films from some aqueous surfactant solutions containing sufficient amount of an electrolyte to suppress the electrostatic component of disjoining pressure as well as films from nonaqueous solutions (see also Section 3.5).

During thinning the thermodynamically unstable films keep their shape in a large thickness range until a rather small thickness is approached, at which the film ruptures. This thickness is called *critical thickness of rupture* h_{cr}. Therefore, the thermodynamic instability is a necessary but not sufficient condition for film instability. There are other factors determining instability which at thicknesses smaller than the critical one cease to act. Two are the possible processes involved in film instability – film thinning while retaining film shape, and film rupture. Which of these is realized when thermodynamic instability is reached requires analysis of the various mechanisms of film rupture.

Contemporary understanding of liquid film rupture is based on the concept of existence of fluctuational waves on liquid surfaces [48]. According to this approach the film is ruptured by unstable waves, i.e., waves the amplitudes of which increase with time. The rupture occurs at the moment when the amplitude Δh or its root mean square value $\sqrt{\overline{(\Delta h)^2}}$ of a certain unstable wave grows up to the order of the film thickness

$$\sqrt{\overline{(\Delta h)^2}} \approx h_{cr} \quad (3.18)$$

The basis of this model has been developed by Scheludko [33,49] who showed that the condition of increase in the amplitude of a wave is equivalent to the condition of increase in local pressure

$$\delta\left(\Delta p_c + \Pi\right) > 0 \quad (3.19)$$

where

$$\delta \Delta p_c = -\gamma k^2 \Delta h \quad (3.20)$$

is a capillary pressure corresponding to wavelength $\lambda = 2\pi/k$ and amplitude Δh; $\delta\Pi = (\mathrm{d}\Pi/\mathrm{d}h)\Delta h$ is the respective perturbation of disjoining pressure Π. Hence, Equation 3.19 yields

$$k^2 < [\mathrm{d}\Pi/\mathrm{d}h]/\gamma = k_{cr}^2 \quad (3.21)$$

The upper limit k_{cr} of the unstable spectrum range is known as the *Scheludko wave number*.

In thick films ($h > 0.5$ μm) only capillary forces act against the surface deformations, i.e., $\delta\Delta p_c \gg \delta\Pi$ and fluctuation waves are practically stable for the whole wavelength spectrum determined by Equation 3.18. Moreover, the steady state amplitudes of the capillary waves determined from the equipartial law $\sqrt{\overline{(\Delta h)^2}} \approx \sqrt{kT/\gamma}$ at typical conditions ($\gamma \sim 50$ mN m^{-1}; $kT = 4 \times 10^{-21}$ J) have values of the order of $\sqrt{\overline{(\Delta h)^2}} \sim 0.1$ nm, i.e., thick films are not only stable but remain practically unaffected by thermal fluctuations.

In the process of film thinning the interactions due to surface forces in the film become stronger and the attractive components corresponding to the negative disjoining pressure have a destabilizing effect (deepening of amplitude). When only van der Waals forces act in the film, Equations 3.21 and 3.74 give [49] for k_{cr}

$$k_{cr} = \sqrt{\frac{3K_{VW}}{\gamma h^4}} \tag{3.22}$$

Once formed the unstable waves grow until one of them (the fastest) conforms with Equation 3.18 and then the film ruptures. During this time the film thins additionally, depending on the conditions under which it is produced. This kinetic part of the theory of the critical thickness of rupture has been formulated and partially solved by Vrij [50].

An important feature of the kinetics of film rupture is the random character of the process. Here the question is about the correct description of the effect of fluctuations on the evolution of single waves. The further development of the theory gives an expression [46] which seems suitable from an experimental point of view:

$$h_{cr} = \frac{(kT)^{1/10} K_{VW}^{2/5}}{(\nu\eta)^{1/5} \gamma^{3/10}} \tag{3.23}$$

The experimental data are in good agreement with Equation 3.23. The K_{VW} value recalculated from the experimental data about h_{cr} is very close to the theoretically calculated according to the Lifshitz theory $K_{VW} = 10^{-21}$ J. There are also some corrections introduced in the theory of film rupture later.

Another approach [51] to the rupture of thin liquid films is based on stochastic modeling of this critical transition. Auto-correlation functions for steady state and for thinning liquid films are obtained. A method for calculation of the lifetime τ and h_{cr} of films is introduced. It accounts for the effect of the spatial correlation of waves. The existence of sub-domains leads to decrease in τ and increase in h_{cr}, i.e., increase in the probability for film rupture. Coupling of surface wave dynamics and the rate of drainage v leading to stabilization of thinning films is also accounted for [52].

The mechanism of film rupture proposed by Scheludko and Vrij has stimulated experimental work for determination of h_{cr}. Most successful are those employing the method of microscopic circular thin liquid films with radii $r \sim 100$ μm (Section 3.2). Since rupture is a process with a clearly pronounced random character, reliable measurements of h_{cr} are possible only with microscopic films in which nonfluctuation disturbances are eliminated. The probability character of rupture is illustrated [41] by the curves in Figure 3.13. As can be seen the most probable critical thickness h_{cr} of a film's rupture increases with the increasing film radius, its value being about 30 nm at $r = 0.1$ mm.

3.3.4 Jump-Like Formation of Black Spots in an Emulsion or Foam Film

As shown in the previous section, during thinning, thin liquid films reach a certain critical thickness h_{cr} at which they lose their stability. There are two possibilities: either the film ruptures or a local thinning in the film occurs jump-like. Since at very small thicknesses an emulsion or foam film looks black in reflected light, the jump-like local thinning appearance is called

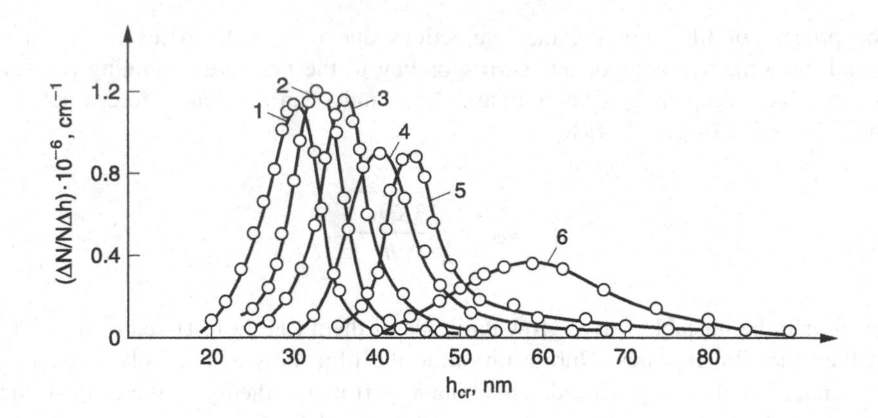

FIGURE 3.13 Distribution curves of the critical thickness of rupture h_{cr}: ΔN is the number of ruptured films with thickness between h and $h + \Delta h$; N is the total number of critical thicknesses measured. Microscopic foam films from isovalerianic acid + KCl aqueous solution with different film radius r. Curve 1: $r = 98$ μm; curve 2: $r = 138$ μm; curve 3: $r = 197$ μm; curve 4: $r = 295$ μm; curve 5: $r = 394$ μm; curve 6: $r = 492$ μm. (From D. Exerova and T. Kolarov, *Annuaire Univ. Sofia, Fac. Chim.* 59: 207 (1964–65).)

formation of *black spots*. The black spot is a very small round area of *black film* which is much thinner than the surrounding area of the unstable thin liquid film. Figure 3.14 shows black spots at different stages of their development.

It appears that Scheludko's theory of rupture of thin liquid films, as well as the new concepts further introduced on that basis (see the previous subsection), are applicable not only to the process of rupture by local thinning but also to the formation of black spots. Hence, black spots can serve to detect the mechanism of local flexion in the film which allows a rough estimation of the fluctuation in wavelength ($\lambda/2$ is about 1 μm). Such a general treatment of instability, including the formation of black spots, can be employed as an additional tool to verify the theory of rupture.

Another important result should also be mentioned: the rupture of unstable films and formation of black spots occur at the same critical thickness: h_{cr} is about 30 nm for films from aqueous surfactant solutions. Emulsion and foam films look gray at this thickness.

What happens in a thinning liquid film when it loses its stability – film rupture at h_{cr} or black spot formation – depends on the surfactant type and concentration in solution. On that basis a new parameter has been introduced [53] for quantitative characterization of the surfactants – emulsion or foam stabilizers. This is the minimal bulk surfactant concentration at which black spots begin to form in the microscopic, unstable, thin liquid film. It is called the *concentration of black spots formation* c_{bl}. This concentration c_{bl} is a very important characteristic of the emulsifiers and foaming agents. Its determination is done by observing emulsion or foam films under a microscope in reflected light. The films are obtained from surfactant solutions with different, gradually increasing, concentration while all other parameters are maintained constant. The lowest concentration at which black spots first appear is c_{bl}.

The concentrations c_{bl} at which various kinds of surfactants ensure formation of stable towards rupture films can be very different depending on the nature of the surfactant molecule. Within a homologous series there is a regular change in c_{bl} – the c_{bl} values decrease with increasing chain length of the molecule. On the other hand c_{bl} is a function of temperature, electrolyte content,

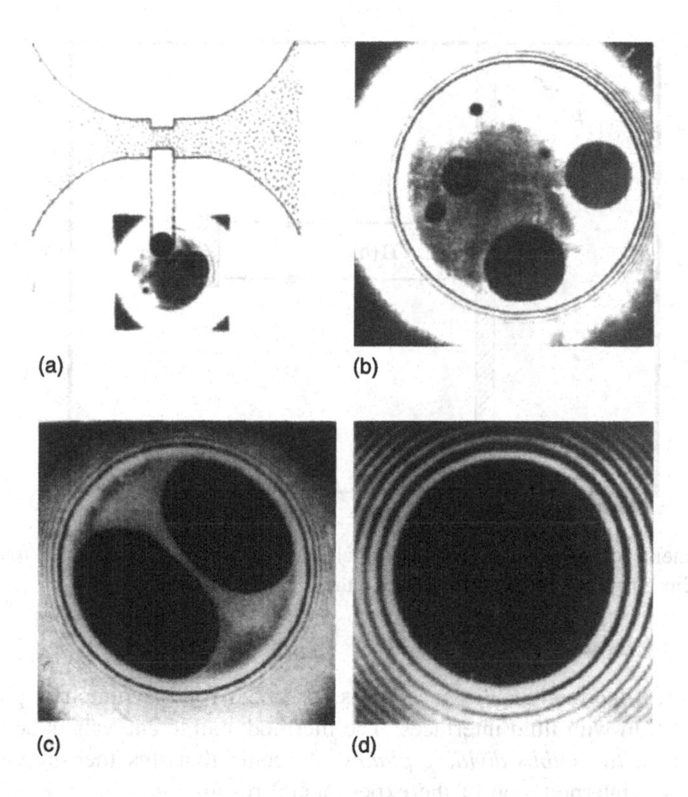

FIGURE 3.14 Formation of black spots and black film in a thicker unstable microscopic liquid film. (a) A schema of the thin black spot in the thicker film; (b) and (c) the growth and coalescence of black spots; (d) the end of the process: the entire film is black. (From D. Exerowa, D. Kashchiev, and D. Platikanov, *Adv. Colloid Polymer Sci.* 40: 201 (1992).)

pH, and presence of other surfactants in the solution. For emulsion films the nature of the second liquid phase also influences c_{bl}. That is why c_{bl} should be determined at standard values of all these parameters.

After the jump-like formation of the black spots, they expand, merge, and at the end the whole film area is occupied by the black film (Figure 3.14). The kinetics of expansion of the black spots in the gray film is also considered. The characteristic concentration c_{bl}, which is a specific for each surfactant parameter, is also important for the emulsion, respectively foam, stability.

3.4 THERMODYNAMICS OF THIN EMULSION AND FOAM FILMS

By definition a thin liquid film is a liquid phase with very small thickness, so that there is no liquid inside the film with the bulk properties of the liquid phase α from which the film has been formed (Figure 3.1). Most of the physical properties vary along the z-axis normal to the film. Hence the thin liquid film is a typical *small* thermodynamic phase. The thermodynamic description of such a system is based, according to Gibbs, on a simplified model. The differences between the values of physical quantities of the model and of the real system are introduced as excess quantities.

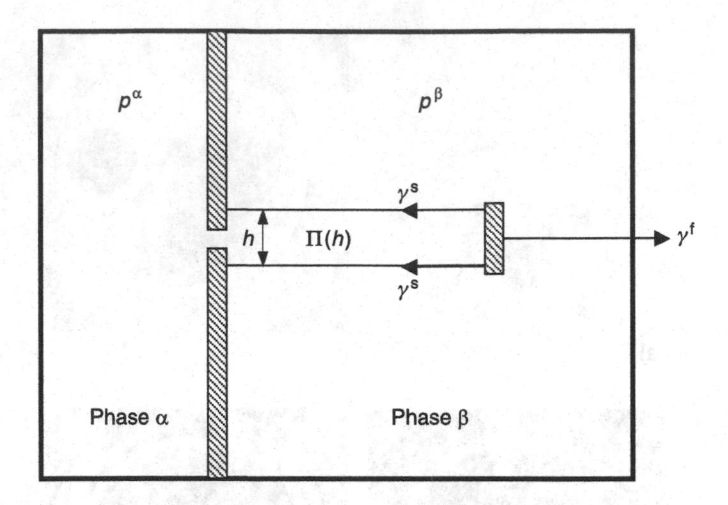

FIGURE 3.15 Scheme of a symmetric plane-parallel thin liquid film (emulsion or foam film) in a frame formed from the liquid phase α and surrounded by the fluid phase β.

In this section a thermodynamic analysis is presented for a symmetric, plane-parallel, horizontal thin liquid film with fluid interfaces. The thermodynamic analysis is developed [54] using only the *model with two Gibbs dividing planes*. It seems that this thermodynamic approach is most convenient for interpretation of the experimental results.

3.4.1 The Simplified Film Model

The entire thermodynamic system (Figure 3.15) consists of: (1) the liquid phase α, in general a multi-component solution; (2) the phase β, either liquid, immiscible with the liquid α (in the case of an emulsion film) or gas (in the case of a foam film); (3) the small phase f, a thin liquid film formed from and connected with the liquid phase α. The thin liquid film is drawn out in a solid frame; the solid material of the vessel walls and the frame is not soluble in phases α and β. The following simplifications are accepted in the film model:

- The Plateau borders, respectively the menisci, at the film contact with the solid walls are neglected. The transition film/meniscus/bulk liquid will be considered later.
- Both interfaces between the film and phase β are replaced by two parallel, horizontal Gibbs dividing planes. The whole space outside the dividing planes is filled by fluid with the bulk properties of phase β.
- The space between the dividing planes is filled by liquid with the bulk properties of the reference phase α.
- The distance h between the dividing planes is defined as thermodynamic thickness h of the thin liquid film.

3.4.2 Mechanical Equilibrium of a Thin Emulsion or Foam Film

Initially a thick liquid film is drawn by the frame from the liquid phase α. It can wear thin only if $p^\alpha < p^\beta$; the pressure difference $\Delta p = p^\beta - p^\alpha$ is the driving force for the film thinning.

Why an equilibrium thin liquid film can be obtained at the end of the process of thinning if Δp = constant?

The pressure p in an isotropic bulk fluid phase (like phases α and β) is the same everywhere – the Pascal's law. The following components of the pressure tensor are identical:

$$p_{xx} = p_{yy} = p_{zz} = p \tag{3.24}$$

This is not the case of a real interfacial layer as well as a real thin liquid film (Figure 3.1). There only the component p_{zz} of the pressure tensor normal to the film (or single interface) remains constant, while the components parallel to the film (or single interface) are functions of z:

$$p_{zz} = p_{\mathrm{n}} = \text{constant} \tag{3.25}$$

$$p_{xx} = p_{yy} = p_{\mathrm{t}}(z) \tag{3.26}$$

As a result the *interfacial tension* γ arises at each interface between two bulk phases, its mechanical definition being the Bakker's formula:

$$\gamma = \int_{-\infty}^{+\infty} [p_{\mathrm{n}} - p_{\mathrm{t}}(z)]\,\mathrm{d}z = \int_{-\infty}^{+\infty} \left[p^{\beta} - p_{\mathrm{t}}(z)\right]\mathrm{d}z \tag{3.27}$$

The thin liquid film, just as an interface, is anisotropic along the z-axis (Figure 3.1) and by analogy the *film tension* γ^{f} (Figure 3.15) can be defined by the same expression:

$$\gamma^{\mathrm{f}} = \int_{-\infty}^{+\infty} \left[p^{\beta} - p_{\mathrm{t}}(z)\right]\mathrm{d}z \tag{3.28}$$

Here the integration is carried out from the bulk of phase β over the entire film up to the bulk of phase β on the other film side. The film tension is (like the interfacial tension) a force per unit length acting tangential to a surface which is called *surface of tension* of the thin liquid film. In the case of a symmetric, plane-parallel film the surface of tension coincides with the mid-plane of the film. The property film tension of the thin liquid film requires that an external force per unit length γ^{f} has to pull the frame with the film at equilibrium (Figure 3.15).

It is also possible to introduce on the base of the simplified film model (described in the previous subsection) a *surface (interfacial) tension of the film* γ^{s}. Its mechanical definition [11] is given by the equation

$$\gamma^{\mathrm{s}} = \int_{0}^{h/2} [p^{\alpha} - p_{\mathrm{t}}(z)]\,\mathrm{d}z + \int_{h/2}^{\infty} \left[p^{\beta} - p_{\mathrm{t}}(z)\right]\mathrm{d}z \tag{3.29}$$

In Equation 3.29 h is the thermodynamic film thickness, i.e., the distance between the two Gibbs dividing planes. Hence γ^{s} depends on the location of the dividing planes, in contrast to the surface (interfacial) tension γ between two bulk phases, which is independent of the location of the single dividing plane. Both γ^{s} and γ^{f} determine the mechanical equilibrium of the film in *tangential* direction.

We shall consider now the mechanical equilibrium of the film in *normal* direction. In the simplified film model the space between the two dividing planes is filled by liquid with pressure p^{α} of the reference phase α (Figure 3.15). The film is thinning because of the pressure difference $\Delta p = p^{\beta} - p^{\alpha}$. Obviously at equilibrium Δp must be balanced by any additional pressure. This pressure, due to the action of surface forces in the film (or pair interactions between the adjacent phases β through the film), has been introduced as *disjoining pressure* Π. Its mechanical definition [55] is given by

$$\Pi = p_{\mathrm{n}} - p^{\alpha} \tag{3.30}$$

i.e., the disjoining pressure is the difference between the normal component of the film's pressure tensor (which is constant) and the pressure in the reference phase α. Taking into account that

$$\Delta p = p^{\beta} - p^{\alpha}, \quad p^{\beta} = p_{\mathrm{n}} \tag{3.31}$$

it follows for symmetric, plane-parallel, horizontal thin liquid films

$$\Pi = \Delta p \tag{3.32}$$

Equation 3.32 is the condition for mechanical equilibrium of the film in normal direction. Positive Π means repulsion between the two film interfaces, i.e., both interfaces are "disjoined" by the interactions due to surface forces. The resultant of all interactions per unit film area is actually Π. It balances the applied external pressure difference Δp. For a plane-parallel, horizontal, real thin liquid film, which contacts trough a Plateau border the frame wall, Δp is the *capillary pressure* at the curved surface of the corresponding meniscus.

Equation 3.28 can be transformed into

$$\begin{aligned}\gamma^{\mathrm{f}} &= 2\int_0^{\infty}\left[p^{\beta} - p_{\mathrm{t}}(z)\right]\mathrm{d}z = 2\int_0^{h/2}\left[p^{\beta} - p_{\mathrm{t}}(z)\right]\mathrm{d}z + 2\int_{h/2}^{\infty}\left[p^{\beta} - p_{\mathrm{t}}(z)\right]\mathrm{d}z \\ &= 2\int_0^{h/2}(p^{\beta} - p^{\alpha})\mathrm{d}z + 2\int_0^{h/2}\left[p^{\alpha} - p_{\mathrm{t}}(z)\right]\mathrm{d}z + 2\int_{h/2}^{\infty}\left[p^{\beta} - p_{\mathrm{t}}(z)\right]\mathrm{d}z\end{aligned} \tag{3.33}$$

The combination of Equation 3.33 with Equations 3.29, 3.31, and 3.32 gives [56]

$$\gamma^{\mathrm{f}} = 2\gamma^{\mathrm{s}} + \Pi h \tag{3.34}$$

This is a very important relation between film tension γ^{f}, surface tension of the film γ^{s}, disjoining pressure Π, and film thickness h – the most important thermodynamic characteristics of a thin liquid film.

3.4.3 Fundamental Thermodynamic Equations of a Thin Liquid Film

The fundamental thermodynamic equation for the internal energy U^f of the thin liquid film (Figure 3.15) follows directly from the First and Second Laws:

$$dU^{\mathrm{f}} = TdS^{\mathrm{f}} - p^{\beta}dV^{\mathrm{f}} + \gamma^{\mathrm{f}}dA^{\mathrm{f}} + \sum_i \mu_i dn_i^{\mathrm{f}} \tag{3.35}$$

Here and further all superscripts denote the phase to which the respective quantity belongs. The first term to the right reflects the heat reversibly transferred to the film from outside, S^f being the entropy of the film. The second term is the volume work done on the film by its surroundings, V^f being the film volume; note that this work is done by the pressure p^β which is an external pressure with respect to the film. The work done for reversible increase of the film area A^f is given by the third term to the right. The last term gives the "chemical" work done when the moles n_i^f of each component in the film are increased. The temperature T and the chemical potential μ_i of each component i are considered constant in all three phases f, α, and β (Figure 3.15). Using the Legendre transformations we can obtain the fundamental thermodynamic equations for the Helmholtz free energy F^f and for the Gibbs energy G^f of the thin liquid film:

$$dF^{\mathrm{f}} = -S^{\mathrm{f}}dT - p^{\beta}dV^{\mathrm{f}} + \gamma^{\mathrm{f}}dA^{\mathrm{f}} + \sum_i \mu_i dn_i^{\mathrm{f}} \tag{3.36}$$

$$dG^{\mathrm{f}} = -S^{\mathrm{f}}dT + V^{\mathrm{f}}dp^{\beta} + \gamma^{\mathrm{f}}dA^{\mathrm{f}} + \sum_i \mu_i dn_i^{\mathrm{f}} \tag{3.36'}$$

The integration of Equation 3.35 according to Euler's theorem at constant T, p^β, γ^f, and μ_i gives the total internal energy U^f of the thin liquid film:

$$U^{\mathrm{f}} = TS^{\mathrm{f}} - p^{\beta}V^{\mathrm{f}} + \gamma^{\mathrm{f}}A^{\mathrm{f}} + \sum_i \mu_i n_i^{\mathrm{f}} \tag{3.37}$$

The differentiation of Equation 3.37 gives

$$\begin{aligned} dU^{\mathrm{f}} &= TdS^{\mathrm{f}} + S^{\mathrm{f}}dT - p^{\beta}dV^{\mathrm{f}} - V^{\mathrm{f}}dp^{\beta} + \gamma^{\mathrm{f}}dA^{\mathrm{f}} + A^{\mathrm{f}}d\gamma^{\mathrm{f}} \\ &\quad + \sum_i \mu_i dn_i^{\mathrm{f}} + \sum_i n_i^{\mathrm{f}} d\mu_i \end{aligned} \tag{3.38}$$

The combination of Equations 3.38 and 3.35 leads to the very important Gibbs–Duhem relation of the thin liquid film:

$$S^{\mathrm{f}}dT - V^{\mathrm{f}}dp^{\beta} + A^{\mathrm{f}}d\gamma^{\mathrm{f}} + \sum_i n_i^{\mathrm{f}} d\mu_i = 0 \tag{3.39}$$

The Gibbs–Duhem relation expressed per unit film area reads:

$$\begin{gathered} d\gamma^{\mathrm{f}} = -S_A^{\mathrm{f}}dT + hdp^{\beta} - \sum_i \Gamma_i^{\mathrm{f}} d\mu_i \\ \frac{V^{\mathrm{f}}}{A^{\mathrm{f}}} = h, \quad \frac{S^{\mathrm{f}}}{A^{\mathrm{f}}} = S_A^{\mathrm{f}}, \quad \frac{n_i^{\mathrm{f}}}{A^{\mathrm{f}}} = \Gamma_i^{\mathrm{f}} \end{gathered} \tag{3.40}$$

All fundamental equations as well as Gibbs–Duhem relations considered above are the most general thermodynamic equations of the thin liquid film. They do not depend on the film model chosen since the extensive film properties have not been specified yet. These equations are valid for symmetric, horizontal, plane-parallel films with fluid interfaces under the condition that the Plateau borders, respectively, the menisci, at the film contact with phase α, or the solid walls are neglected. However, the equations can be used also in the more complicated system which includes Plateau borders, menisci, etc. if the thermodynamic state of all parts of the system, except the film itself, is kept constant.

3.4.4 Thermodynamic Approach with Two Gibbs Dividing Planes

According to this model the thin liquid film consists of two parallel mathematical planes (Gibbs dividing planes), the space between them being filled by liquid with the bulk properties of the reference phase α; the space outside the dividing planes is filled by fluid with the bulk properties of phase β (Figure 3.15). The differences between the extensive thermodynamic properties of the model and of the real film are introduced as surface excess quantities. So the surface excess internal energy U_A^{sf} of the film per unit area of each film surface is presented as

$$U_A^{\mathrm{sf}} = \int_0^{h/2} \left[U_v(z) - U_v^{\alpha}\right] \mathrm{d}z + \int_{h/2}^{\infty} \left[U_v(z) - U_v^{\beta}\right] \mathrm{d}z \tag{3.41}$$

In Equation 3.41 U_v^{α} and U_v^{β} are the volume densities of the internal energy in the homogeneous bulk phases α and β respectively; $U_v(z)$ is the volume density of the internal energy in the inhomogeneous real film. The excess surface concentration Γ_i^{sf} of each component i in the film, per unit area of each film surface, is given by

$$\Gamma_i^{\mathrm{sf}} = \int_0^{h/2} \left[n_{\mathrm{vi}}(z) - n_{\mathrm{vi}}^{\alpha}\right] \mathrm{d}z + \int_{h/2}^{\infty} \left[n_{\mathrm{vi}}(z) - n_{\mathrm{vi}}^{\beta}\right] \mathrm{d}z \tag{3.42}$$

n_{vi}^{α} and n_{vi}^{β} being the volume densities of each component i in the homogeneous bulk phases α and β respectively and $n_{\mathrm{vi}}(z)$ the volume density of each component i in the inhomogeneous real film.

The surface excess quantities as defined by Equations 3.41 and 3.42, as well as other surface excess extensive properties of the film, depend on the location of the Gibbs dividing planes. Just as in the case of a single interface different conditions which determine the dividing plane location can be accepted. The best possibility is when the same condition determines the location both of a single dividing plane between the phases α and β and of the film dividing planes. So the surface excess properties of the film and of the single interface can directly be compared with each other. Most often the condition $\Gamma_1^{\mathrm{sf}} = 0$ is used, i.e., the surface excess of component 1 in the film is zero, component 1 being the liquid which is solvent in phase α.

According to the simplified model the total thin liquid film consists of two Gibbs dividing planes and a volume part filled by liquid with the bulk properties of the reference phase α. Hence the total extensive properties of the film can be obtained as a sum of the doubled surface excess quantity plus the corresponding extensive quantity of the volume part of the film. So the internal energy U^{f} of the total film is presented as

$$U^{\mathrm{f}} = 2U^{\mathrm{sf}} + U^{\alpha\mathrm{f}} = A^{\mathrm{f}}\left(2U_A^{\mathrm{sf}} + hU_{\mathrm{v}}^{\alpha}\right) \tag{3.43}$$

where U^{sf} is the surface excess internal energy at one dividing plane and $U^{\alpha\mathrm{f}}$, the internal energy of the volume part of the total film. The total amount n_i^{f} of each component i in the film is expressed by

$$n_i^{\mathrm{f}} = 2n_i^{\mathrm{sf}} + n_i^{\alpha\mathrm{f}} = A^{\mathrm{f}}\left(2\Gamma_i^{\mathrm{sf}} + hn_{\mathrm{v}i}^{\alpha}\right) \tag{3.44}$$

n_i^{sf} being the surface excess amount of each component i in the film and $n_i^{\alpha\mathrm{f}}$ the amount of component i in the volume part of the total film.

If we accept that the surface excess of component 1 in the film is zero, as a condition which determines the location of the Gibbs dividing planes, we obtain from Equation 3.44 an equation which determines the thermodynamic film thickness h:

$$n_1^{\mathrm{f}} = A^{\mathrm{f}}hn_{\mathrm{v}1}^{\alpha}, \quad \Gamma_1^{\mathrm{sf}} = 0 \tag{3.45}$$

$$h = \frac{n_1^{\mathrm{f}}}{A^{\mathrm{f}}n_{\mathrm{v}1}^{\alpha}} \tag{3.46}$$

According to Equation 3.46 the thermodynamic film thickness h is equal to the thickness of a layer of the solution in phase α, which contains the same amount of component 1 as the film. The real emulsion (or foam) films are usually stabilized by dense surfactant adsorption layers at both interfaces. That means the thermodynamic film thickness is closer to the thickness of the inner liquid layer of the film, rather than to the whole film thickness. However, the surfactant molecules can be very different and the film structure as well. Hence the comparison of the thermodynamic and the real film thickness should be done for each system separately. In general the physical film thickness is usually larger than the thermodynamic thickness determined under condition 3.45.

Equation 3.43 after differentiation can be transformed into

$$2\mathrm{d}U^{\mathrm{sf}} = \mathrm{d}U^{\mathrm{f}} - \mathrm{d}U^{\alpha\mathrm{f}} \tag{3.47}$$

The fundamental thermodynamic equation for the internal energy $U^{\alpha\mathrm{f}}$ of the volume part of the thin liquid film reads

$$\mathrm{d}U^{\alpha\mathrm{f}} = T\mathrm{d}S^{\alpha\mathrm{f}} - p^{\alpha}\mathrm{d}V^{\mathrm{f}} + \sum_i \mu_i \mathrm{d}n_i^{\alpha\mathrm{f}} \tag{3.48}$$

The combination of Equation 3.47 with 3.35 and 3.48, taking into account Equations 3.31, 3.32, and 3.44, and the relation $S^{\mathrm{f}} = 2S^{\mathrm{sf}} + S^{\alpha\mathrm{f}}$, gives

$$2\mathrm{d}U^{\mathrm{sf}} = 2T\mathrm{d}S^{\mathrm{sf}} - \Pi\mathrm{d}V^{\mathrm{f}} + \gamma^{\mathrm{f}}\mathrm{d}A^{\mathrm{f}} + 2\sum \mu_i \mathrm{d}n_i^{\mathrm{sf}} \tag{3.49}$$

This is the fundamental thermodynamic equation for the surface excess of the internal energy of the film. Following the usual procedure, the corresponding important Gibbs–Duhem relation of the Gibbs dividing planes of the film can be obtained

$$2S^{sf}dT - V^{f}d\Pi + A^{f}d\gamma^{f} + 2\sum n_i^{sf}d\mu_i = 0 \tag{3.50}$$

which, assigned to unit film area, reads

$$d\gamma^{f} = -2S_A^{sf}dT + hd\Pi - 2\sum_i \Gamma_i^{sf}d\mu_i \tag{3.51}$$

The substitution of Equation 3.34 into 3.49 leads to another form of the fundamental thermodynamic equation for the surface excess of the internal energy of the film:

$$2dU^{sf} = 2TdS^{sf} - \Pi A^{f}dh + 2\gamma^{s}dA^{f} + 2\sum \mu_i dn_i^{sf} \tag{3.52}$$

Respectively, the Gibbs–Duhem relation of the dividing planes of the film takes another form as well:

$$2S^{sf}dT + \Pi A^{f}dh + 2A^{f}d\gamma^{s} + 2\sum n_i^{sf}d\mu_i = 0 \tag{3.53}$$

or assigned to unit film area:

$$2d\gamma^{s} = -2S_A^{sf}dT - \Pi dh - 2\sum \Gamma_i^{sf}d\mu_i \tag{3.54}$$

On the basis of both Gibbs–Duhem equations (3.54 and 3.51) very useful relations between the main thermodynamic characteristics (h, Π, γ, γ^f, γ^s) of the thin liquid film can be obtained. From Equation 3.54 it follows that

$$2\left(\frac{\partial \gamma^{s}}{\partial h}\right)_{T,\mu_i} = -\Pi \tag{3.55}$$

which after integration at constant temperature and all chemical potentials (for thick film $h \to \infty$, $\Pi = 0$, $\gamma^s = \gamma$) gives

$$2(\gamma^{s} - \gamma) = -\int_{\infty}^{h} \Pi dh = \Delta F(h)$$

or

$$2\gamma^{s} = 2\gamma + \Delta F(h) \tag{3.56}$$

The quantity $\Delta F(h)$ is usually called *interaction free (Helmholtz) energy of the film.* Obviously this is the isothermal, reversible work per unit film area done against the disjoining pressure when thinning of a thick film down to a small equilibrium thickness h.

Equation 3.55 shows that when disjoining pressure Π appears due to interactions in the thin liquid film, the surface tension of the film γ^s depends on the film thickness and it differs from the surface tension γ of the bulk liquid phase. This result was not accepted by some authors who claim that one liquid surface (in our case film/transition zone/bulk meniscus) should have everywhere a constant surface tension at equilibrium; otherwise a flow in the liquid surface layer would arise due to the surface tension gradient. It has been shown [57] that this is an apparent contradiction and both approaches – the one assuming variable surface tension and the other based on constant surface tension – are completely equivalent. The apparent differences are due to the simplified thermodynamic approach chosen.

Other similar relations can be obtained from Equation 3.51:

$$\left(\frac{\partial \gamma^{\mathrm{f}}}{\partial \Pi}\right)_{T,\mu_i} = h \tag{3.57}$$

which after integration at constant temperature and all chemical potentials gives

$$\gamma^{\mathrm{f}} = 2\gamma - \int_{\infty}^{h} \Pi \mathrm{d}h + \Pi h$$

or

$$\gamma^{\mathrm{f}} = 2\gamma + \Pi h + \Delta F(h) \tag{3.58}$$

Equation 3.58 can be used to determine the interaction free energy of the film $\Delta F(h)$ from experimental data for γ and γ^{f} (the last two quantities can be directly measured). The product Πh is usually very small since h is extremely small for the so-called *black films* – the case when 2γ and γ^{f} noticeably differ. Then Πh is neglected and $\Delta F(h) \approx \gamma^{\mathrm{f}} - 2\gamma$.

From the fundamental thermodynamic equations (3.52 and 3.49) for the surface excess of the internal energy of the film, using the Legendre transformations we can obtain the corresponding fundamental thermodynamic equations for the surface excess of the Helmholtz free energy of the film:

$$2\mathrm{d}F^{\mathrm{sf}} = -2S^{\mathrm{sf}}\mathrm{d}T - \Pi \mathrm{d}V^{\mathrm{f}} + \gamma^{\mathrm{f}}\mathrm{d}A^{\mathrm{f}} + 2\sum \mu_i \mathrm{d}n_i^{\mathrm{sf}} \tag{3.59}$$

$$2\mathrm{d}F^{\mathrm{sf}} = -2S^{\mathrm{sf}}\mathrm{d}T - \Pi A^{\mathrm{f}}\mathrm{d}h + 2\gamma^{\mathrm{s}}\mathrm{d}A^{\mathrm{f}} + 2\sum \mu_i \mathrm{d}n_i^{\mathrm{sf}} \tag{3.60}$$

A thermodynamic definition of the film tension γ^{f} follows from Equation 3.59:

$$\gamma^{\mathrm{f}} = 2\left(\frac{\partial F^{\mathrm{sf}}}{\partial A^{\mathrm{f}}}\right)_{T,V^{\mathrm{f}},n_i^{\mathrm{sf}}} \tag{3.61}$$

By analogy thermodynamic definitions of the surface tension of the film γ^{s} and of the disjoining pressure Π can be derived from Equation 3.60:

$$\gamma^{\mathrm{s}} = \left(\frac{\partial F^{\mathrm{sf}}}{\partial A^{\mathrm{f}}}\right)_{T,h,n_i^{\mathrm{sf}}} \tag{3.62}$$

$$\Pi = -\frac{2}{A^{\mathrm{f}}}\left(\frac{\partial F^{\mathrm{sf}}}{\partial h}\right)_{T,A^{\mathrm{f}},n_i^{\mathrm{sf}}} \tag{3.63}$$

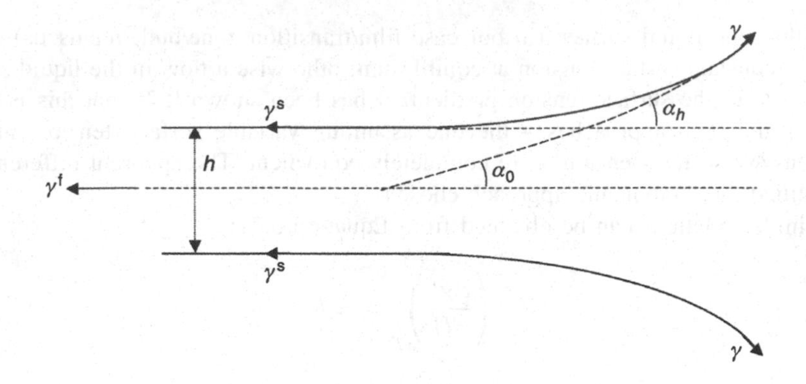

FIGURE 3.16 Scheme of part of a plane-parallel symmetric thin liquid film (left side) connected through a transition zone with a meniscus of the liquid phase α.

Equations 3.61 to 3.63 show that all three quantities characterizing the film, γ^f, γ^s, and Π, are defined in terms of the surface excess of the Helmholtz free energy, a main feature of the thermodynamic approach with two Gibbs dividing planes.

3.4.5 Contact Between a Thin Emulsion or Foam Film and the Adjacent Bulk Liquid

A thin liquid film is always connected with a bulk liquid phase or with a solid wall through a *Plateau border* – a liquid body with concave interfaces like a meniscus. The Plateau border contains bulk liquid α from which the film has been formed. The contact film/Plateau border is schematically presented in Figure 3.16. On the left side there is a part of the horizontal, plane-parallel, symmetric, thin liquid film (emulsion or foam film) and on the right side, a part of the Plateau border. A *transition zone film/bulk* always exists between them [58]. At equilibrium the disjoining pressure Π acts in the film and the surface tension of the film γ^s differs from the interfacial (surface) tension γ of the bulk liquid phase. Both Π and γ^s are due to the surface force interactions in the thin liquid film. These interactions operate in the transition zone film/bulk as well, but they decrease with increasing thickness of the transition zone towards the meniscus. This means that both Π and $(\gamma - \gamma^s)$ gradually decrease along the transition zone in direction from the film to the Plateau border, becoming zero in the bulk liquid.

In a real system the curved meniscus interface gradually becomes a flat film interface and γ gradually turns into γ^s. The macroscopic thermodynamic description neglects the presence of a transition zone and assumes an abrupt change of γ (bulk liquid) into γ^s (film). This simplified thermodynamic approach requires introduction of two other thermodynamic quantities: the *contact angle film/bulk* and the *line tension* of the contact line. Thus the simplified macroscopic thermodynamic description becomes equivalent to the real system.

In the Plateau border the interfacial (surface) tension of the bulk liquid is γ and the shape of its concave interface is determined by the Laplace equation. If we consider γ constant also in the transition zone and extrapolate the meniscus interface according to Laplace equation towards the film (the dotted curve in Figure 3.16), it intersects the mid-plane of the film (the horizontal dotted straight line). The line of intersection is the *contact line* of the film and the angle between both dotted lines is the *contact angle* α_o [59,60].

If we consider a cylindrical meniscus, i.e., a straight contact line, the balance of forces in tangential direction, taking into account Equation 3.58, is given by

$$\gamma^{\mathrm{f}} = 2\gamma \cos \alpha_0 = 2\gamma + \Pi h + \Delta F(h) \tag{3.64}$$

In the case of a curved contact line (e.g., a round film in a spherical meniscus) the balance of forces reads [61]

$$2\gamma \cos \alpha_0 = \gamma^{\mathrm{f}} + \frac{\kappa}{r^{\mathrm{f}}} \tag{3.65}$$

where κ is the line tension and r^{f} is the radius of curvature of the contact line. The term κ/r^{f} is actually a two-dimensional capillary pressure, a surface pressure difference on both sides of the curved contact line. However, this term is usually very small because of the very small κ value; it should be taken into account at extremely small r^{f} only.

Another definition of the contact angle film/bulk has been introduced as well: the angle α_h between the extrapolated according to Laplace equation meniscus interface towards the film (the dotted curve in Figure 3.16) and the extrapolated surface of the film (the upper horizontal dotted straight line). Obviously the contact angles α_h and α_o are different but this difference is usually very small. In this case the balance of forces in tangential direction (straight contact line), taking into account Equation 3.56, is given by

$$\gamma^{\mathrm{s}} = \gamma \cos \alpha_h, 2\gamma \cos \alpha_h = 2\gamma + \Delta F(h) \tag{3.66}$$

The contact angles α_h and α_o can be measured experimentally. From these experimental values with a measured γ value one can calculate values of the interaction free (Helmholtz) energy $\Delta F(h)$ of the film:

$$\Delta F(h) = -2\gamma(1 - \cos \alpha_h) \tag{3.67}$$

$$\Delta F(h) = -2\gamma(1 - \cos \alpha_0) - \Pi h \tag{3.68}$$

As already noted in the previous section, the term Πh is usually very small. In such a case if we accept that $\Pi h \approx 0$, it follows that $\alpha_h \approx \alpha_o$.

3.5 SURFACE FORCES IN THIN EMULSION AND FOAM FILMS

According to the thermodynamics of the thin liquid films, the most important factor which determines their properties is the interaction between the two film interfaces, i.e., the pair interactions between the two adjacent phases across the liquid film. The thermodynamic quantity *disjoining pressure* Π is a result of these interactions, due to different types of *surface forces* acting in the thin liquid films. The dependence of the disjoining pressure on the film thickness, the so-called *disjoining pressure isotherm*, is closely related with the more general problem of the stability of a disperse system. Hence knowledge about the surface forces in the thin liquid films has developed in connection with the theories of colloid stability, the most important being the theory of Derjaguin, Landau, Verwey, and Overbeek – the DLVO theory [7,8].

Major efforts are directed to understanding the nature and origin of surface forces acting in thin liquid films, including bilayer films. On the other hand explanations are sought for the cases in which the classical DLVO theory contradicts experimental findings. Two categories of surface forces are usually distinguished: DLVO and non-DLVO surface forces. The *van der Waals molecular interactions* or *dispersion molecular forces* as well as the *electrostatic* or *double layer forces* are called DLVO forces (both long-range forces), the balance of which lays the foundation for the DLVO theory. The non-DLVO forces are of different nature [62,63]. Most important for the symmetric thin liquid films with fluid interfaces (emulsion and foam films) are the *steric surface forces*, long range for macromolecules, short range for small molecules. Theoretical as well as experimental considerations are put forth for the existence of *structural* (*solvation, hydration, or hydrophobic*) *surface forces* in some cases; they originate from modifications in the liquid structure adjacent to the film interfaces. In specific film cases other surface forces could be introduced as well.

The precise direct measurement of surface forces is a subject of current interest, since it provides sufficiently reliable distinction of the forces, along with the elucidation of their mutual influence, their dependence on the distance between the interacting surfaces (the film thickness) in systems of different composition, temperature, etc. All this enables a more critical application of the theories of the known surface forces. On the other hand, the direct measurement of surface forces stimulates theoretical analyses. Surface force measurements in emulsion and foam films, in particular in microscopic films, stabilized with amphiphile molecules (surfactants, phospholipids, polymers), enable the study of these forces at large film thickness, long-range surface forces as well as interaction forces at the formation of a bilayer film. The long-range/short-range interaction transitions, including the reversal transition, are also studied in detail.

3.5.1 DLVO Surface Forces: Theoretical

The dependence of the disjoining pressure Π on film thickness h, the $\Pi(h)$ isotherm, is at the root of the DLVO theory of stability of lyophobic colloids. According to this theory the disjoining pressure in thin liquid films is considered as a sum of an electrostatic component Π_{el} and a van der Waals component Π_{VW}:

$$\Pi = \Pi_{\text{el}} + \Pi_{\text{VW}} \tag{3.69}$$

A complete analysis of the theory can be found in several monographs (e.g., Refs. 8, 62, and 63). Here, only a few expressions, which are used for interpretation of the experimental results, are presented.

3.5.1.1 Electrostatic Disjoining Pressure

The electrostatic or double layer disjoining pressure is given by the equation:

$$\Pi_{\text{el}} = 2cRT(\cosh zy^m - 1) \tag{3.70}$$

with $m = h/2$ and $y^m = F\psi^m/RT$. Here c is the concentration of a symmetric electrolyte of valence z, F is Faraday's constant; ψ^m is the potential at a distance $h/2$ (at the central plane of the symmetric film) which is connected to the potential ψ^0 of the double electric layer at the film/adjacent phase interface through the relation:

$$-\frac{\kappa h}{2} = \int_{y=y^0}^{y=y^m} \frac{zdy}{\sqrt{2\,[\cos h(zy) - \cos h(zy^m)]}} \tag{3.71}$$

with

$$y = \frac{F\psi}{RT}, \quad y^0 = \frac{F\psi^0}{RT}$$

and

$$\kappa = \sqrt{F^2 z^2 c / \varepsilon_0 \varepsilon RT} \tag{3.72}$$

where ε is the relative dielectric permittivity and $\varepsilon_0 = 8.854 \times 10^{-12}\ \mathrm{C^2V^{-1}m^{-1}}$ is the dielectric permittivity of free space (vacuum).

An approximated expression of the electrostatic component of disjoining pressure can be derived from Equation 3.70 at small values of ψ^m

$$\Pi_{\mathrm{el}} = \frac{64cRT}{z^2}\left[\tanh(zy^0/4)\right]^2 e^{-\kappa h} \tag{3.73}$$

Equation 3.73 is quite adequate for prompt calculations of Π_{el}. For a wide range of ψ^0 and c values, it is necessary to use the more general equations (3.70 and 3.71).

3.5.1.2 van der Waals Disjoining Pressure

Two theories, macroscopic and microscopic, are involved in the calculation of the van der Waals component Π_{VW} of disjoining pressure in thin liquid films. According to the *microscopic theory*, the total interaction force in a flat gap between two semi-infinite phases decreases with distance much slower than the interaction force between two individual molecules. The following expression for the van der Waals component of disjoining pressure in a symmetric film bordering gas or condensed phases is obtained:

$$\Pi_{\mathrm{VW}} = -K_{\mathrm{VW}}/h^n \tag{3.74}$$

where K_{VW} is the van der Waals–Hamaker constant (introduced by A. Scheludko); $K_{\mathrm{VW}}(h) = A(h)/6\pi$, $A(h)$ being the Hamaker constant for symmetric films, a weak function of film thickness. In general $n = 3$; but if a correction for the electromagnetic retardation of dispersion forces is taken into account [64], $n = 4$.

A general formula for calculation of the dispersion molecular interactions in any type of condensed phases has been derived in the *macroscopic theory*. The attraction between bodies results from the existence of a fluctuational electromagnetic field of the substance. If this field is known for a thin liquid film, it is possible to determine the disjoining pressure in it. The more strict macroscopic theory avoids the approximations assumed in the microscopic theory, i.e., additivity of forces; integration; extrapolation of interactions of individual molecules in the gas to interactions in the condensed phase. The macroscopic theory is successfully used for calculation of the dispersion molecular interactions, respectively the $\Pi_{\mathrm{VW}}(h)$. Several approximations have been derived since the exact equations are too complicated. A good approximation for symmetric flat films is

$$\Pi_{\mathrm{VW}} = -\frac{\hbar}{8\pi^2 h^3}\int_0^\infty \left(\frac{x^2}{2} + x + 1\right) e^{-x} \frac{(\varepsilon_{\mathrm{f}} - \varepsilon)^2}{(\varepsilon_{\mathrm{f}} + \varepsilon)^2} d\omega \tag{3.75}$$

with $x = 2\omega h \varepsilon_f^{1/2}$, ε_f, and ε being the relative dielectric permittivities (functions of the frequency ω) of the film and the adjacent phases respectively; $\hbar$ is Planck's constant divided by 2π. For small film thickness a more simple approximation can be derived:

$$\Pi_{VW} = -\frac{\hbar}{8\pi^2 h^3} \int_0^{\infty} \frac{(\varepsilon_f - \varepsilon)^2}{(\varepsilon_f + \varepsilon)^2} d\omega \tag{3.76}$$

Another approximation for a foam film with relatively large thickness reads

$$\Pi_{VW} = \frac{\hbar c}{h^4} \frac{\pi^2}{240\sqrt{\varepsilon_0}} \left(\frac{1-\varepsilon_0}{1+\varepsilon_0}\right)^2 \varphi\left(\frac{1}{\varepsilon_0}\right) \tag{3.77}$$

where ε_0 is the static value of the dielectric permittivity of the film; c is the speed of light. Here Π_{VW} is inversely proportional to h^4, similar to the Hamaker's formula which accounts for the electromagnetic retardation of dispersion forces [64].

For practical use of the macroscopic theory equations with empirical constants are also appropriate for calculation of Π_{VW}, for instance [65]

$$\Pi_{VW} = h^{-3}\left[b + (a + ch)/(1 + dh + eh^2)\right] \tag{3.78}$$

where a, b, c, d, and e are empirical constants.

3.5.2 DLVO Surface Forces: Experimental

Quantitative experimental studies of the DVLO surface forces have been performed with microscopic emulsion and foam films. As shown above, the disjoining pressure is given as a sum of Π_{el} and Π_{VW}, i.e., they are considered independent. In the symmetric emulsion or foam film Π_{el} is always positive while Π_{VW} is always negative. In an equilibrium film, the state of which is determined by the action of the positive disjoining pressure, Π can be calculated from the balance of the forces experimentally obtained. When Π is negative, it is possible to use the dynamic method [33] according to which $\Pi(h)$ is calculated using the hydrodynamic equation for the rate of film thinning (see Equation 3.11). During film thinning at $h > \approx 120$ nm the constant capillary pressure only determines the rate of thinning and $1/h^2$ vs time t is a straight line according to Equation 3.11. Below thickness of ≈ 120 nm the disjoining pressure appears and it accelerates or decelerates the liquid outflow. From the experimental $1/h^2(t)$ dependence the pressure drop $\Delta p = \Delta p_c - \Pi$ is calculated and with a known Δp_c value the experimental Π values are determined for different film thickness. This method enables the study of the $\Pi(h)$ isotherm within a large thickness range. It also provides a possibility to verify the different theories of disjoining pressure, including the cases in which Π is always negative and no equilibrium films are formed.

3.5.2.1 Experiments with Foam Films

The first quantitative proof of the DLVO theory has been conducted [66] on a model system – microscopic foam film using the microinterferometric technique (Figures 3.3 and 3.7). Independent studies have been performed of the $\Pi_{el}(h)$ and $\Pi_{VW}(h)$ as well as of their joint action at various electrolyte concentrations c. At very low c equilibrium films of large thickness formed

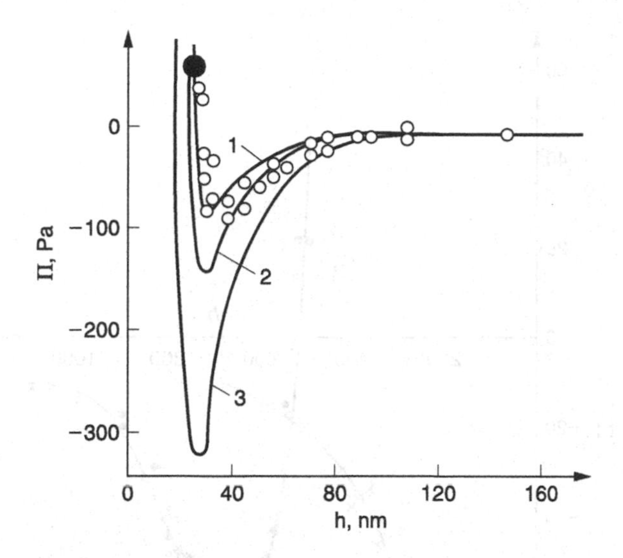

FIGURE 3.17 Disjoining pressure Π vs. film thickness h isotherm at 25 °C. Thin liquid film from 5×10^{-4} wt.% saponin + 0.01 M KCl aqueous solution. Solid lines: calculated according to Equations 3.73 and 3.74 with $\psi^0 = 90$ mV and $K_{VW} = 3.3 \times 10^{-21}$ J (curve 1), $K_{VW} = 5.1 \times 10^{-21}$ J (curve 2), $K_{VW} = 8.5 \times 10^{-21}$ J (curve 3); the point marked with • corresponds to the equilibrium film. (From A. Scheludko and D. Exerowa, *Kolloid-Z*, 165: 148 (1959); and 168: 24 (1960).)

in which the electrostatic interaction was prevailing and their behavior could be described completely with this interaction. At such film thickness Π_{VW} was still very low so that the equilibrium film state was reached at the balance of the electrostatic disjoining pressure and the capillary pressure, $\Pi_{el} = \Delta p_c$. Non-equilibrium thinning films were formed at high electrolyte concentration with prevailing action of the negative Π_{VW}. At intermediate electrolyte concentrations (10^{-2} to 10^{-3} mol dm^{-3}) Π_{el} and Π_{VW} are competitive. At large film thickness accelerated drainage was initially observed after which the process was delayed until equilibrium was established. This delay was due to the rise of Π_{el} with decreasing thickness. Figure 3.17 plots the $\Pi(h)$ isotherm obtained by the dynamic method for saponin aqueous films. The experimental point marked with full circle corresponds to an equilibrium film at $\Pi = \Delta p_c = 73$ Pa. In spite of the considerable data scattering an expressed minimum can be clearly seen, similar to the dependence calculated from Equations 3.69, 3.73, and 3.74 with $\psi^0 = 90$ mV and $n = 3$. The value of the van der Waals–Hamaker constant has been determined to be $K_{VW} = 4 \times 10^{-21}$ J, although one must keep note of the dependence of K_{VW} on the film thickness, predicted by Lifshitz' theory and calculated by some authors using approximate methods.

The DLVO surface forces have been studied [67] with equilibrium macroscopic vertical foam films from sodium octylsulfate aqueous solution containing KCl. The film thickness comprised a large range from 80 to 8 nm and electrolyte concentrations from 10^{-3} to 1 mol dm^{-3}. A smooth fall of the equilibrium thickness is clearly seen when the counter-ion concentration is increased. Studying films of small thickness, the authors considered for the first time film structure, i.e., the presence of an aqueous core and adsorption layers (Figure 3.2). This is important for the treatment of not only the results from optical thickness measurements but of $\Pi_{el}(h)$ and $\Pi_{VW}(h)$ as well. On the basis of these and other experimental data one can conclude that deviations from the DLVO theory appear at foam film thickness less than 20 nm.

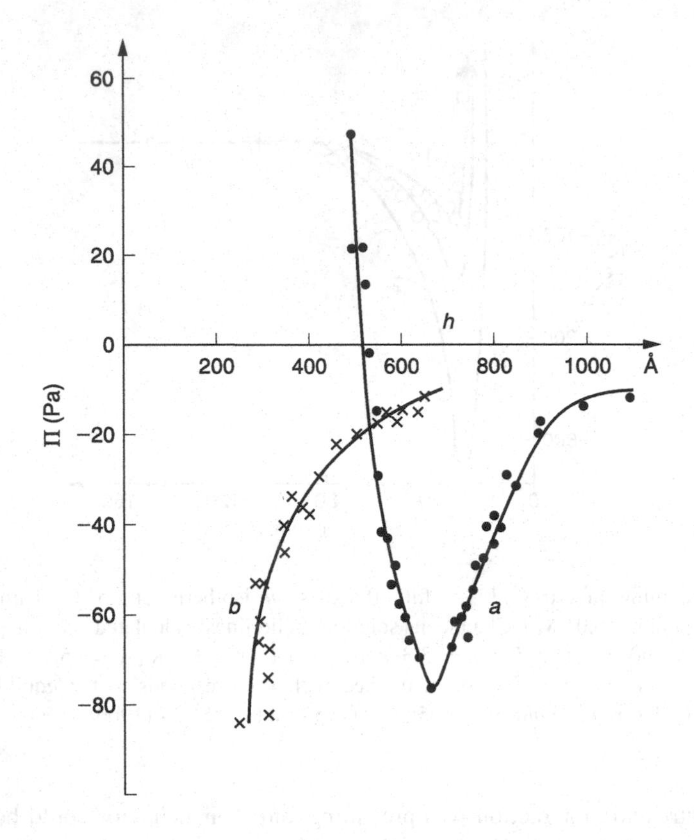

FIGURE 3.18 Disjoining pressure Π vs. film thickness h isotherm. Microscopic OWO emulsion films from NP20 aqueous solution between cyclohexane phases. Electrolyte concentration: 0.005 M KCl (curve *a*) and 2 M KCl (curve *b*). (From H. Sonntag, J. Netzel, and H. Klare, *Kolloid-Z Z Polym.* 211: 121 (1966).)

3.5.2.2 Experiments with Emulsion Films

Microscopic emulsion films have been studied [19] in the special measuring cell (Figure 3.5) using the microinterferometric technique. Both OWO and WOW films have been investigated. The dynamic method for assessment of the $\Pi(h)$ isotherms, based on Equation 3.11, has been used. Such experimental $\Pi(h)$ isotherms are presented in Figure 3.18 for emulsion films from aqueous solutions of NP20 (nonylphenylpolyethylenoxide20) + KCl in two different concentrations; the oil phase is cyclohexane. The right-hand curve *a* (with a minimum) has been obtained for an emulsion film from NP20 solution with 0.005 M KCl. At this low electrolyte concentration a significant Π_{el} arises with decreasing thickness after the minimum and the entire disjoining pressure $\Pi = \Pi_{\text{el}} + \Pi_{\text{VW}}$ becomes positive. At film thickness of 51.5 nm (Figure 3.18) $\Delta p = \Pi$ and an equilibrium film is established. The equilibrium film thickness h depends on the electrolyte concentration c, namely h decreases with increasing c.

Completely different $\Pi(h)$ dependence (Figure 3.18, left-hand curve *b*) has been obtained [19] for an emulsion film from the same NP20 solution, but with high electrolyte concentration $c = 2$ M KCl. The Π_{el} is now very low and can be neglected. The measured Π is practically Π_{VW} only; it is negative at all thicknesses and no equilibrium film can be obtained. The emulsion film wears thin until its rupture.

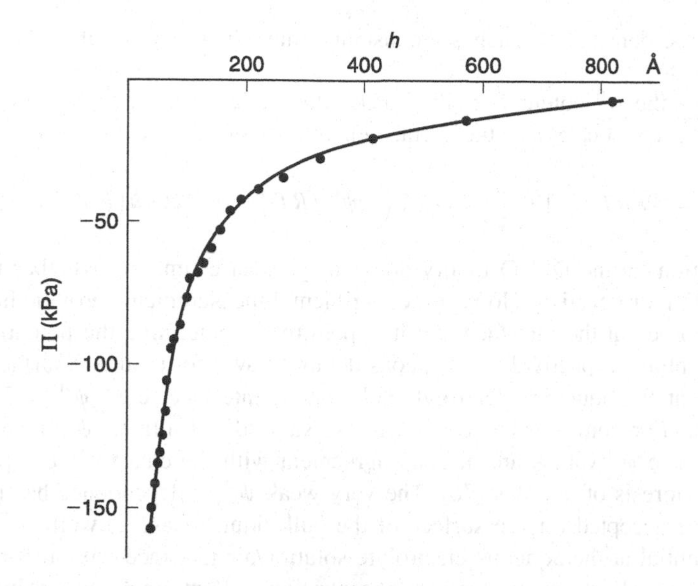

FIGURE 3.19 Disjoining pressure Π vs. film thickness h isotherm. Microscopic WOW emulsion films from 0.2 g dm^{-3} Span80 in octane solution between aqueous phases with electrolyte concentration 0.05 M KCl. (From H. Sonntag, J. Netzel, and H. Klare, *Kolloid-Z Z Polym.* 211: 121 (1966).)

The behavior of the WOW emulsion films from solutions of a nonpolar liquid ("oil") is similar. There are no ions dissolved in such a liquid, no double layers, and hence no Π_{el}. Figure 3.19 presents a $\Pi(h)$ isotherm obtained [19] for an emulsion film from Span80 solution in octane. The adjacent liquid phases are from 0.05 M KCl aqueous solution. Again the measured Π is Π_{VW} only, negative at all thicknesses, and the emulsion films always rupture since no equilibrium film can be obtained. All these results are in agreement with the DLVO theory.

3.5.2.3 Potential of the Diffuse Double Layer at the Film's Interface

Under certain conditions aqueous electrolyte solutions form equilibrium thin foam or OWO emulsion films. Their equilibrium thickness is determined by the positive electrostatic disjoining pressure Π_{el} which depends on the potential ψ^0 of the diffuse double layer at the liquid film/gas (or oil) interface. The relation between film thickness h and electrolyte concentration c obeys the DLVO theory of electrostatic disjoining pressure (Equations 3.70 and 3.71). The potential ψ^0 at the film's interface can be calculated from the measured equilibrium film thickness h and known c value. This is the "method of the equilibrium thin film" for determining the ψ^0-potential [68] and for studying the electric properties at the interface. The method provides a valuable possibility since an equilibrium potential and, respectively, surface charge density σ^o can be evaluated and all complications occurring at kinetic measurements are avoided.

Detailed study with the "method of equilibrium foam film" of the $h(c)$ and $h(\mathrm{pH})$ dependence in the absence of a surfactant [68], as well as $h(c_s)$ at very low surfactant concentrations c_s, gave $\psi^0 \approx 30$ mV at the interface aqueous electrolyte solution/air. The ψ^0-potential at low electrolyte concentrations c increases with increasing c_s from its value for an aqueous electrolyte solution without surfactant to a constant value, typical for each surfactant kind [17]. The increase of ψ^0 at low concentrations c_s is connected with the surfactant's adsorption at the solution/air interface.

The surface charge density σ^o reaches a constant value when or just before the adsorption layer becomes saturated.

The values of the ψ^0-potential of the diffuse double layer at the surfactant aqueous solution interface can be used to calculate the surface charge density σ^o, according to

$$\sigma^o = [2\varepsilon RTc/\pi]^{1/2}\left[1/2\cos h\left(F\psi^0/RT\right) - 1/2\cos h\left(F\psi^m/RT\right)\right]^{1/2} \tag{3.79}$$

The calculation by the DLVO theory does not give an estimation whether the values of ψ^0 and σ^o are positive or negative. However, experimental measurements provide information which are the ions adsorbed at the interface and it is possible to determine the potential sign.

The ψ^0-potential (respectively σ^o) is considerably lower for nonionic surfactants but slightly higher than ψ^0 at the aqueous electrolyte solution/air interface, e.g., $\psi^0 = 39$ mV for decyl-methylsulfoxide. For ionic surfactants such as Na dodecylsulfate, ψ^0 is essentially higher ($\psi^0 = 82$ mV). The ψ^0 values are in good agreement with the data for the ζ-potential obtained through electrophoresis of bubbles [70]. The very weak $\psi^0(h)$ dependence has indicated that the ψ^0 values can be accepted for the surface of the bulk liquid phase as well.

The ψ^0-potential at the aqueous electrolyte solution/air interface without surfactant has been measured very precisely at low electrolyte concentration and extremely pure solutions and vessels. The ψ^0(pH) dependence for aqueous solutions at constant ionic strength (HCl + KCl) shows that at pH > 5.5, the potential becomes constant and equal to about 30 mV. At pH < 5.5 the potential sharply decreases and becomes zero at pH $\approx$ 4.5, i.e., an isoelectric state at the solution surface is reached [68]. It is clear that the isoelectric state is controlled by the change in pH. This is very interesting, for it means that the charge at the surface of the aqueous solutions is mainly due to the adsorption of H^+ and OH^- ions. The estimation of the adsorption potential of these ions in the Stern layer showed that the adsorption potential of the OH^- ions is higher. It follows that the ψ^0-potential at a solution/air interface appears as a result of adsorption of OH^- ions [68].

3.5.3 Steric Surface Forces

The dispersion systems emulsion and foam contain practically always different surfactants as stabilizing agents. Hence for thin liquid films with fluid interfaces (emulsion or foam films) the most important *non-DLVO surface forces* are the *steric interactions*: short range for small molecules (to be discussed in the next section) and long range for macromolecules, which are considered here.

Microscopic foam films have been used to study the steric interaction between two liquid/gas interfaces [71]. Two ABA triblock copolymers have been employed: P85 ($PEO_{27}PPO_{39}PEO_{27}$, $M = 4600$ Da) and F108 ($PEO_{122}PPO_{56}PEO_{122}$, $M = 14000$ Da). Blocks A are hydrophilic polyethylene oxide (PEO) chains, while block B is a hydrophobic polypropylene oxide (PPO) chain. The film thickness has been measured by the micro-interferometric method. Figure 3.20 presents the dependence of the equivalent film thickness h_w on the electrolyte concentration c for aqueous solutions of the two copolymers. The h_w values initially decrease with increasing c and then level off at a constant value. The plateau starts at a *critical electrolyte concentration* $c_{cr} = 3 \times 10^{-3}$ mol dm^{-3} NaCl for F108 and $c_{cr} = 3 \times 10^{-2}$ mol dm^{-3} NaCl for P85, similar to the low molecular mass surfactants. The bulk properties of aqueous solutions of PEO–PPO–PEO copolymers are essentially similar to those of nonionic surfactants. It is very likely that the interaction behavior is determined by the hydrophilic PEO chains protruding into the solution.

The left branch of the $h_w(c)$ dependence (Figure 3.20) is dominated by electrostatic repulsion. The copolymers are nonionic and charge creation may be attributed to preferential adsorption of

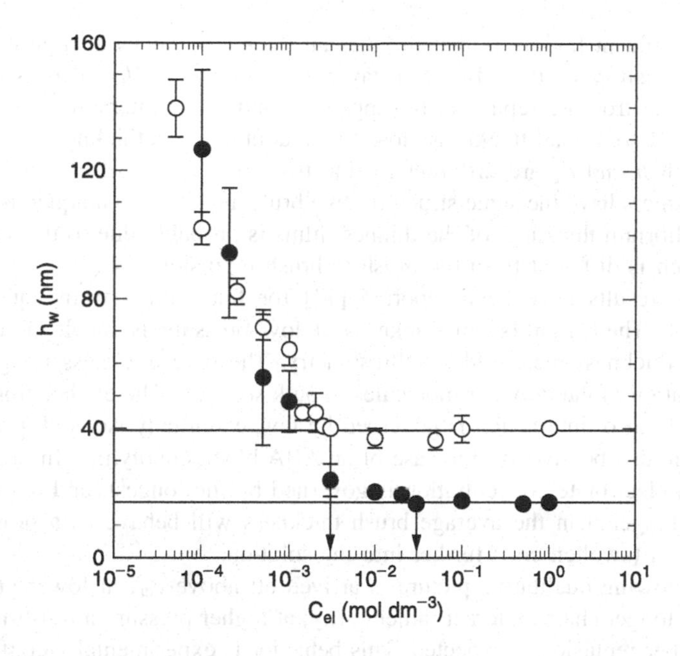

FIGURE 3.20 Equivalent film thickness h_W vs. NaCl concentration c_{el} at 23 °C. (●): 7 × 10^{-5} M P85 aqueous solutions; (○): 7 × 10^{-7} M F108 aqueous solutions. The arrows show the corresponding c_{cr}. (From D. Exerowa, R. Sedev, R. Ivanova, T. Kolarov, and ThF. Tadros, *Colloids Surfaces* 123; 277 (1997).)

OH^- ions at the air/water interface. The data for both copolymers located below their c_{cr} roughly follow the same trend (Figure 3.20). On the contrary, the plateau values for the two copolymers are very different: 39 ± 3 nm and 17 ± 1 nm, respectively. Since the higher copolymer gives thicker films a surface force component of steric origin may be evoked. However, the thickness h_W is an effective parameter which is too crude. As a reasonable compromise between physical relevance and tractability, the three-layer model is adopted (Figure 3.2). The equivalent film thickness h_W is larger than the total film thickness h, since the refractive index of the adsorption layer is higher than that of water ($n_1 > n_2$). Allegedly, the electrolyte concentration affects only the aqueous core thickness h_2 which can be calculated using Equation 3.4.

The thickness h_1 of the adsorption layer is determined by the conformation of the macromolecules at the interface. The radius of gyration R_F of a flexible neutral chain in good solvent is given [72] by the Flory relation:

$$R_F = aN^{3/5} \tag{3.80}$$

where a (~0.2 nm) is an effective monomer size and the number $N = N_{PO} + 2N_{EO}$. If a macromolecule adsorbs at the interface as a separate coil, it should occupy an area of the order of the projected area of the molecule in the bulk solution, i.e., R_F^2. The figures obtained (~2 nm^2) are much lower and, therefore, the PEO chains are crowding the solution/air interface and stretched, i.e., they form a brush [73]. Its thickness h_1 can be calculated from the simple brush model:

$$h_1 = aN_{EO}\left(a^2/A_o\right)^{1/3} \tag{3.81}$$

A_o being the area per molecule.

The most significant finding is that the plateau thickness values at high electrolyte concentration are larger than twice the adsorption layer thickness: $h > 2h_1$. This is rather unexpected since above c_{cr}, electrostatic repulsion is suppressed and steric interaction alone is expected to stabilize the film. If so, a total thickness close to the double brush thickness, i.e., $h \approx 2h_1$, would be expected. Both h and h_1 are different for the two copolymers. However, the thinnest films from both copolymers have the same structure: two brush layers and an aqueous core of thickness $\sim 3R_F$. The equilibrium thickness of the thinnest films is probably due to the same type of steric stabilization which is different from the brush-to-brush repulsion.

Quite similar results have been reported [74] for foam films from aqueous solutions of polyvinyl alcohol. Their equilibrium thickness at low pressure is much larger than twice the adsorption layer thickness measured by ellipsometry. The core thickness is again several times the radius of gyration of the polymer molecules in bulk solution. The explanation given in Ref. 74 is that the upper limit of interaction is governed by few, essentially isolated, polymer tails. Similar arguments can also be given in the case of an ABA block copolymer. In other words the film thickness at high electrolyte concentration is governed by the longest, and not the average, PEO chains. A chain longer than the average brush thickness will behave as a brush only up to h_1, but as a mushroom (swollen coil) further into the solution.

Thus, the following qualitative picture is arrived at: above c_{cr}, at lower pressure the mushroom tails of the longer chains interact rather softly; at higher pressure a brush-to-brush contact is realized and steeper repulsion is expected. This behavior is experimentally confirmed. At lower c electrostatic repulsion dominates and decreases until steric repulsion (which is independent of c) becomes operative at c_{cr}. The transition from electrostatic to steric repulsion occurs at c_{cr} given by

$$\frac{1}{\kappa_{cr}} \equiv \left(\frac{\varepsilon\ kT}{8\pi\ e^2 N_A c_{cr}}\right)^{1/2} = R_F \tag{3.82}$$

where ε is the dielectric constant of the medium; e is the elementary charge; N_A is Avogadro's number; $1/\kappa_{cr}$ is the Debye length at c_{cr}.

The most detailed information about the steric surface forces can be obtained [75] from the disjoining pressure vs. thickness isotherm $\Pi(h)$ for films from polymer aqueous solutions with $c > c_{cr}$. The two cases discussed above can be clearly distinguished: (1) at low pressures, $\Pi < 10^2$ Pa, the "soft" steric repulsion determines quite large film thickness of about 40 to 50 nm; (2) at high pressures, $\Pi > 10^2$ Pa, a strong steric repulsion due to the brush-to-brush contact determines considerably smaller thickness. Presumably, the PEO brushes extending from the two surfaces come into contact and repel each other. Under these conditions the de Gennes scaling theory [73] for interaction between two surfaces carrying polymer brushes applies. Accordingly the steric disjoining pressure is given by

$$\Pi_{st} \cong \frac{kT}{D^3}\left(H^{-9/4} - H^{3/4}\right) \tag{3.83}$$

where $H = h/2h_1$ is the dimensionless film thickness (h_1 is the brush layer thickness at infinite separation); D is distance between two grafted sites. The first term is osmotic pressure arising from the increased polymer concentration in the two compressed layers. The second one is an elastic restoring force (polymer molecules always tend to coil and hence the negative sign of this term). De Gennes' theory gives a satisfactory description of the steric surface forces at pressures and film thickness where brush-to-brush contact is realized.

3.5.4 Oscillatory Disjoining Pressure

A consecutive stepwise film thinning has been observed at thin liquid films with fluid interfaces, formed from solutions with surfactant concentration much higher than CMC. This is the phenomenon known as *stratification*. During this process the initially formed films thin to smaller and smaller thickness, reaching black films in most of the cases, sometimes even down to bilayers. The stratification phenomenon in foam films has been described at the beginning of the twentieth century [5,76]. Later the stepwise thinning was studied by many authors [77–81] and it became evident that this phenomenon is universal and, in addition to being observed in emulsion and foam films, it has also been seen in asymmetric films of the air/water/oil type, films from latex suspensions, liquid crystalline films, etc.

The stratification phenomenon is connected with the layered ordering of molecules or micelles inside the film. During drainage these ordered layers of molecules or micelles flow out towards the film's periphery (the meniscus that surrounds the film), i.e., the film thinning occurs stepwise, layer after layer. It has been pointed out that this phenomenon is caused by oscillation in the Gibbs free energy of the film when its thickness is changed. Two theories have emerged to account for these Gibbs energy oscillations: bilayers of amphiphile molecules in the film structure or cubic lattice of ordered micelles. The "combined" structuring of films is also possible.

Very precise measurements of $\Pi(h)$ isotherms have been performed [22] with foam films from aqueous high concentrated solutions of Na dodecylsulfate, employing both the pressure balance technique and the dynamic method. The disjoining pressure isotherms were established down to pressures of 10 Pa with specially constructed film holders and careful pressure isolation and control. The typical $h(t)$ curve shows a film thinning with a step of 10 nm. The stepwise thinning has been explained by the oscillatory form of the $\Pi(h)$ isotherm (Figure 3.21). At constant capillary pressure an oscillating driving force develops that produces the observed stepwise thinning.

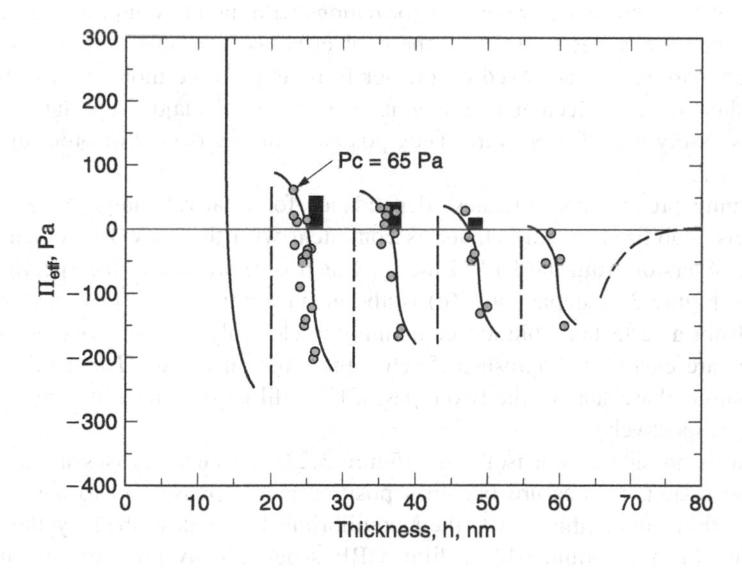

FIGURE 3.21 Oscillatory disjoining pressure Π vs. film thickness h isotherm at 24 °C. Thin liquid films from 0.1 M Na dodecylsulfate aged aqueous solution. (○): dynamic method; (■): equilibrium films. (From V. Bergeron and C.J. Radke, *Langmuir* 8: 3020 (1992).)

The authors introduced the term *oscillatory disjoining pressure* [22]. As seen from Figure 3.21, the results obtained by the dynamic and the equilibrium methods conform well. The former gives also negative values of disjoining pressure which is an advantage of this method. It is worth noting that the length scale of the oscillations was large, about 10 nm, and even reached about 50 nm. A theoretical analysis was based on the equilibrium "oscillatory structural component" of disjoining pressure. It is clear that the oscillatory disjoining pressure is due to structures of surfactant molecules, including micelles, in the thin liquid film. However, it is commonly accepted that the structural component of disjoining pressure is determined by the structuring of solvent molecules. Hence the question if the oscillatory disjoining pressure could be considered as a kind of structural disjoining pressure is open.

3.6 BLACK EMULSION AND FOAM FILMS

3.6.1 Two Equilibrium Phase States of Black Films

During their thinning, thin liquid films lose their stability at a certain critical thickness h_{cr}. Depending on the type of surfactant and its concentration a jump-like formation of *black spots* in the emulsion or foam films can occur. The black spots expand, merge, and at the end the entire film area is occupied by *black film* (Figure 3.14). These films can reach extremely small thicknesses. Observed under a microscope they reflect very little light and appear black when their thickness is below 20 nm. Therefore, they could be called nanofilms as well. The IUPAC nomenclature [82] distinguishes two equilibrium phase states of black films: *common black film* (CBF) and *Newton black film* (NBF). There is a pronounced transition between them, i.e., CBF can transform into NBF or the reverse [83].

The CBF, just as the common thin liquid films, can be described by the three layer film model or "sandwich model," a liquid core between two adsorption layers of surfactant molecules (Figure 3.2). The NBF, however, are bilayer formations without a free liquid core between the two layers of surfactant molecules [83]. Thus, the contact between droplets, respectively bubbles, in the emulsion or foam can be achieved by bilayer from amphiphile molecules. In the behavior of the latter the short-range molecular interactions prove to be of major importance. The definition "liquid film" is hardly valid for bilayers. They possess a higher degree of ordering similar to that of liquid crystals.

The disjoining pressure vs. thickness dependence for relatively large h values of common thin liquid films, stabilized by surfactants, is consistent with the DLVO theory. However, black films exhibit a diversion from the DLVO theory which is expressed in the specific course of the $\Pi(h)$ isotherm. Figure 3.22 depicts a $\Pi(h)$ isotherm (in arbitrary scale) of an aqueous emulsion or foam film from a surfactant solution containing an electrolyte. The two states of black films, CBF and NBF, are clearly distinguished. Such a presentation of the $\Pi(h)$ isotherm can explain the thermodynamic phase state of the two types of black films stabilized by long- and short-range surface forces, respectively.

In the right-hand side of the isotherm (Figure 3.22) the curve passes a shallow minimum, after which the disjoining pressure becomes positive and increases up to a maximum. In this range common thin liquid films exist, their equilibrium being described by the DLVO theory. If $h < h_{cr}$, the film is a common black film, CBF, schematically presented in the figure. Such a film forms via black spots and at the equilibrium film thickness h_1 the disjoining pressure equals the external (capillary) pressure, $\Pi = \Delta p$. The equilibrium of CBF is also described by the DLVO theory, although discrepancies between the theoretical and experimentally obtained

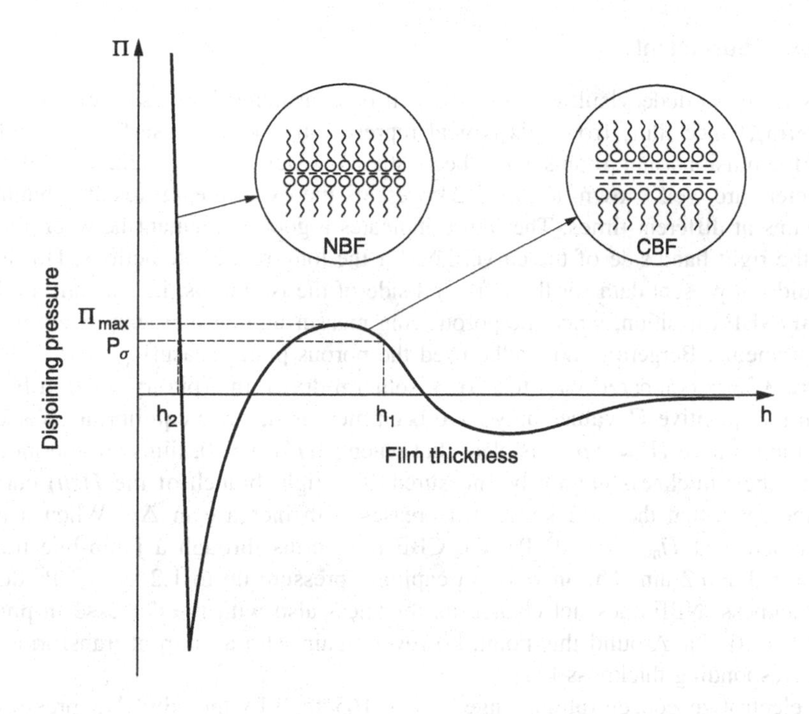

FIGURE 3.22 Schematic presentation of a disjoining pressure Π vs. film thickness h isotherm of a symmetric thin liquid film in arbitrary scale.

$\Pi(h)$ isotherms at film thickness below 20 nm are established [83,84]. The pressure difference $\Pi_{max} - \Delta p$ is the barrier which hinders the transition to a film of smaller thickness.

According to the DLVO theory, after Π_{max} the disjoining pressure should decrease infinitely in direction of thickness decrease. However, the experimental results [24,85] show the existence of a second minimum in the $\Pi(h)$ isotherm after which the disjoining pressure sharply ascends. Another equilibrium is established on the rising left hand side of the isotherm, again under the condition $\Pi = \Delta p$. This black film with a smaller equilibrium thickness h_2 is actually the bilayer Newton black film (NBF), schematically presented in the figure as well. The left branch of the $\Pi(h)$ curve, like the preceding minimum, is not described by the DLVO theory. Obviously at this extremely small film thickness short range non-DLVO surface forces, for instance steric interactions, determine this part of the $\Pi(h)$ isotherm.

3.6.2 Disjoining Pressure in Black Films

The direct experimental measurement of the disjoining pressure Π and the film thickness h (the $\Pi(h)$ isotherm) is of major importance for understanding the nature of surface forces which determine the equilibrium state of both types of black film as well as to establish the CBF/NBF transition. It turned out that the Thin Liquid Film–Pressure Balance Technique (Figure 3.7), employing the porous plate measuring cell of Exerowa–Scheludko (Figure 3.3, B), provides the equilibrium values in a wide range of pressures and thicknesses. This technique has been successfully applied by many authors for plotting $\Pi(h)$ isotherms of black films from various surfactant solutions. The different kinds of surfactants (emulsion or foam stabilizers) show some peculiarities in the $\Pi(h)$ isotherms.

3.6.2.1 Ionic Surfactants

Black films from Na dodecylsulfate (NaDoS) will be considered here as an example for plotting $\Pi(h)$ isotherms, since films from this typical representative of ionic surfactants and very good stabilizer of emulsions and foams have been studied extensively [21,22,86] and most of the film parameters are well known. Figure 3.23 presents the experimental results obtained by three different teams at different times. The figure indicates a good agreement between the data of all authors in the right-hand side of the curve, i.e., for the long-range interactions. However, Mysels and Jones did not present data for the left-hand side of the isotherms, i.e., within the NBF region and the CBF/NBF transition, since the porous ring measuring cell they employed does not allow such measurements; Bergeron and Radke used the porous plate measuring cell.

If Figure 3.23 is compared with the $\Pi(h)$ isotherm drawn in arbitrary scale in Figure 3.22, it can be seen that positive Π values only have been measured, since equilibrium black films form only when the positive $\Pi = \Delta p$. It is clear that when $(\mathrm{d}\Pi/\mathrm{d}h) > 0$, films are thermodynamically unstable and their thickness cannot be measured. The right branch of the $\Pi(h)$ curve refers to CBF and indicates that their thickness h decreases with increase in Δp. When a thickness of 7.1 nm is reached at $\Pi_{\max} \approx 10^5$ Pa, the CBF transforms through a jump-like transition into NBF of $h = 4.3 \pm 0.2$ nm. The increase in capillary pressure up to 1.2×10^5 Pa does not alter the NBF thickness. NBF does not change its thickness also with the decrease in pressure down to $\Delta p = 0.1 \times 10^5$ Pa. Around this point, however, again with a jump, it transforms into a CBF with the corresponding thickness [21].

In the electrolyte concentration range $c = 0.165$ to 0.18 mol dm^{-3} a pressure region of fluctuating NBF/CBF transition has been found [21]. At higher electrolyte concentrations when a certain pressure value is reached, a jump-like transition CBF to NBF occurs. The film thickness of such a transition does not depend on c which means that there is a fluctuation zone where the energy needed to overcome the barrier $\Pi_{\max}$ in the $\Pi(h)$ isotherm (Figure 3.22) is of the order of kT. Films that satisfy this energy requirement are called metastable films. Upon increasing c the pressure, at which the transition occurs, reduces as a result from lowering of the barrier $\Pi_{\max}$.

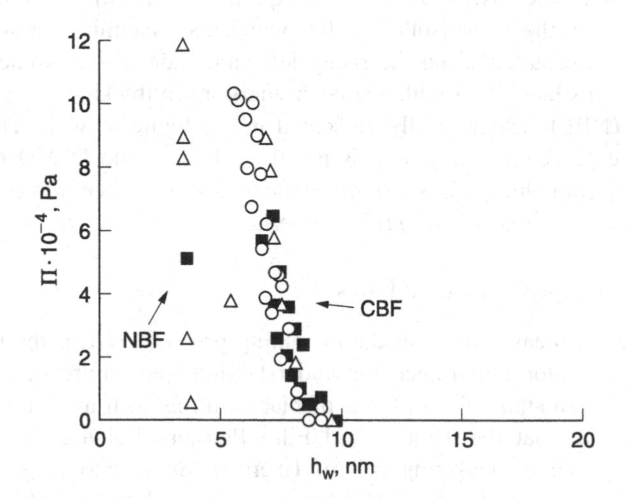

FIGURE 3.23 Disjoining pressure Π vs. film thickness h_w isotherms at 23 to 24 °C. Thin liquid films from 1 mM Na dodecylsulfate + 0.18 M NaCl aqueous solution. (Δ): data from Ref. 21; (■): data from Ref. 22; (○): data from Ref. 86.

At $c > 0.2$ mol dm^{-3} there is no NBF to CBF transition occurring, which means that the left minimum in the $\Pi(h)$ isotherm lies deeply into the region of negative Π values. Thus, an electrolyte concentration range from 0.165 to 2.0 mol dm^{-3} confines the isotherm region in which both CBF/NBF and NBF/CBF transitions can take place. The NBF thickness measured (within the experimental accuracy) does not depend on capillary pressure and its average value is 4.3 nm. This value will be further discussed again. Similar $\Pi(h)$ isotherms have been depicted also for other anionic surfactants [87].

3.6.2.2 Nonionic Surfactants

Two type $\Pi(h)$ isotherms have been established [88] for black films from aqueous solutions of nonionic surfactants. The first one can be illustrated by solutions of $C_{10}(EO)_4$ (decyl 4-oxyethyl), the surfactant and electrolyte concentrations being chosen so that equilibrium films within a large thickness range are obtained, including the CBF/NBF transition region. The change in film thickness has been achieved by gradually and isothermally increasing the capillary pressure. The initial thickness of the $C_{10}(EO)_4$ film decreases down to 11 nm and black spots appear, gradually invading the whole film and turning it black. Its thickness falls to 6 nm and does not change upon further increase in capillary pressure. This result indicates that the transition to NBF occurs via a jump-like transition overcoming the $(\Pi_{max} - \Delta p)$ barrier, i.e., the $\Pi(h)$ isotherm has the same shape as that in Figure 3.23.

The other type $\Pi(h)$ isotherm can be illustrated by black films of NP20 (20-oxyethyl nonylphenol). The experimentally obtained curve follows another course. With the increase in Π the film thickness decreases gradually without any jump-like transition and an equilibrium thickness of 9 nm is reached. It remains constant with further increasing of the pressure. Since the ψ^0-potential is almost equal for both surfactants, there should be an additional repulsion force in the NP20 films. Knowing that the thickness of these films does not depend on c but is a function of Π such an assumption seems reasonable. That is why it is interesting to see whether the experimental isotherms conform with those derived from the DLVO theory.

The experimental $\Pi(h)$ curve for films from $C_{10}(EO)_4$ lies between the two theoretical curves obtained at $\sigma^o = \text{const}$ and $\psi^0 = \text{const}$. Therefore, it can be supposed that in this case the DLVO theory describes well the electrostatic and van der Waals interactions in the common black films. The experimental $\Pi(h)$ isotherms of NP20 films are in agreement with the theory only in the right-hand side of the isotherm, at relatively large thickness. The curve follows a smooth course which probably reflects the existence of gradually increasing force additional to Π_{el} repulsive force. As far as this force appears in films from surfactants with long oxyethylene chains, it can be considered as steric interaction.

The $\Pi(h)$ isotherms of black films from NP20 solutions measured at pH = 5.7, 6.1, and 4.0 have clearly shown the influence of OH^- ions [88] and the isotherms are in a very good agreement with the h(pH) dependence determined at $\Delta p = \text{const}$. Detailed quantitative analysis (using the nonlinear Poisson–Boltzmann equation) of the disjoining pressure experimental values obtained with black films from nonionic sugar-based surfactant [89] at various c_s, pH, and ionic strength have also shown that the OH^- ions create the surface charge at the film/air interface.

3.6.2.3 Phospholipids

Formation of thin aqueous films from phospholipids is a difficult task since many of them are insoluble in water. Sonicated dispersions of insoluble phospholipids is an option [90], and direct measurement of interaction forces in films stabilized by neutral phospholipids was first done with films from suspensions of unilamellar DMPC vesicles [91]. The $\Pi(h)$ isotherm indicates a barrier transition to NBF and Newton black spots appear in the CBF. The NBF thickness of 7.6 nm remains constant with further increase in pressure. This isotherm is similar to the one obtained for nonionic surfactants such as $C_{10}(EO)_4$. The right-hand side of the isotherm can be interpreted in the same way with DLVO surface forces.

Interesting results have been obtained for black films stabilized by soluble zwitterionic phospholipids [92]: lysophosphatidylcholine (lyso PC) and lysophosphatidyl ethanolamine (lyso PE) in the presence of Na^+ and Ca^{2+}. Besides the $\Pi(h)$ isotherm, another type of isotherm proves to be very informative, namely the dependence of film thickness h on electrolyte concentration c, at c_s = const, Δp = const, and T = const. This $h(c)$ dependence allows the action of electrostatic disjoining pressure to be distinguished clearly and the electrolyte concentration at which the CBF/NBF transition occurs to be found.

The effect of Ca^{2+} on the equilibrium thickness of films from lyso PC aqueous solution [93] is shown in Figure 3.24. Unlike the results obtained with the monovalent Na^+ ions, black films of constant thickness $h = 7.6$ nm have always been formed at low Ca^{2+} concentrations. At 0.001 mol dm^{-3} $CaCl_2$, however, a dramatic increase in film thickness to about 35 nm has been observed, i.e., a transition to relatively thick films occurred. Further increase in c leads to a monotonous decrease in film thickness until c reached 0.02 mol dm^{-3} $CaCl_2$ where gradually expanding black spots appeared, until the whole film became black. The thickness of these black films (obviously CBF) continues to decrease with increase in c up to 0.2 mol dm^{-3} $CaCl_2$ when again black films of the same thickness as that in the low concentration range are obtained. This is indicated in Figure 3.24 by a small thickness jump. Further increase in c up to 0.5 mol dm^{-3} has no effect on the thickness of these black films. The $\Pi(h)$ isotherm at 0.002 mol dm^{-3} $CaCl_2$ is also similar to that for $C_{10}(EO)_4$: relatively thick films are formed at low pressures and their

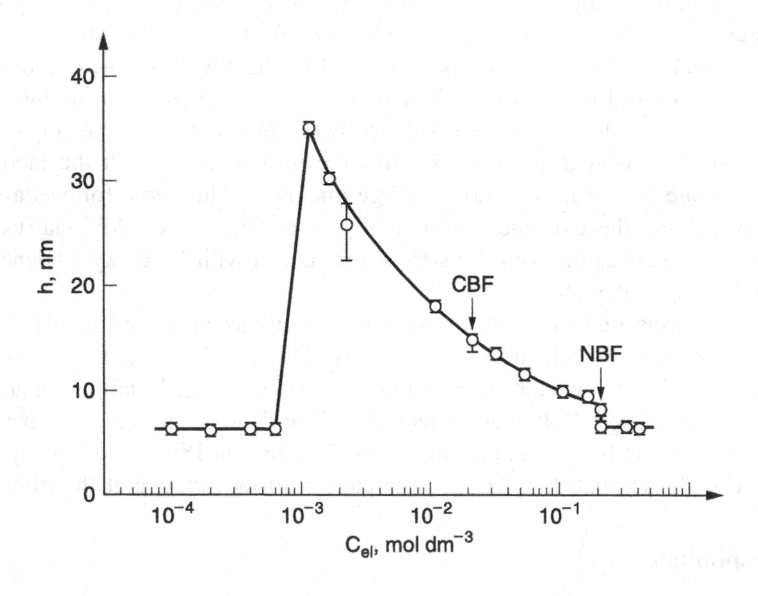

FIGURE 3.24 Equivalent film thickness h vs. $CaCl_2$ concentration c_{el} at 30 °C. Microscopic foam films from lyso-phosphatidylcholine aqueous solution, pH = 5.5, Δp = 35 Pa. (From R. Cohen, D. Exerowa, T. Kolarov, T. Yamanaka, and V.M. Muller, *Colloids Surfaces* 65: 201 (1992).)

thickness decreases with increase in Π. After a jump-like CBF/NBF transition the NBF thickness does not change with further increase in pressure.

Measurements involving different techniques have shown that monovalent ions practically do not bind to zwitterionic phospholipids. In contrast, the observed thickness transition in Figure 3.24 clearly demonstrates that the divalent Ca^{2+} ions have a specific effect on the properties of the black films studied. Most probably this is due to the specific binding of Ca^{2+} ions to the adsorbed lyso PC [94]. The binding of the positive Ca^{2+} ions in the low concentration range can lead to low ψ^0-potentials, consequently, to a decrease in the electrostatic repulsion, and to formation of NBF. The transition from NBF to thicker films can be related to further Ca^{2+} ion binding at higher c that induces higher ψ^0 values. The corresponding Π_{el} is evidently sufficient to allow for the formation of thicker films. Further increase in c leads to film thickness diminishing as a result of the competitive action of further Ca^{2+} binding (higher ψ^0-potential) and higher ionic strength of the solution (lower Π_{el}).

3.6.3 Properties of CBF and NBF and the Transition between Them

The two types of black film, common black film (CBF) and Newton black film (NBF), seem to be similar: obtained from the same surfactant solution, both are very thin ($h < 20$ nm). However, they show rather different properties due to their different structure and they are two different equilibrium phase states of a black emulsion or foam film [83]. This difference determines the CBF/NBF transition as well. A number of thermodynamic parameters determine if CBF or NBF is in equilibrium: temperature T, electrolyte concentration c, surfactant concentration c_s, pH, and capillary pressure Δp. Besides the values of these parameters, the kind of the surfactant (including amphiphile polymers) is also decisive for the existence of the black films. The parameter values, at which a CBF or an NBF is stable, are in definite (sometimes narrow) intervals and only the combination of all parameters determines a given type of black film.

The following example illustrates this. The good emulsion or foam stabilizer NaDoS (Na dodecylsulfate) provides formation of stable black films only if its concentration $c_s > c_{bl}$ (c_{bl} is the concentration of black spot formation; see Section 3.3.4). However, the c and $\Delta p = \Pi$ values are also decisive. They determine if the films either rupture or a CBF/NBF transition occurs at a given film thickness h. In the NaCl concentration range 10^{-4} to 0.15 mol dm^{-3}, the films rupture in a certain pressure interval which becomes narrower with the rise in c.

The temperature is another very important parameter: systematic studies [95,96] of black foam films from NaDoS aqueous solutions have shown that at given c, c_s, and Δp the change of temperature causes CBF/NBF transition; at high temperatures the equilibrium black films are CBF and at low temperatures, NBF. In Figure 3.25 the line splits up the diagram between the area of CBF stability (left and upper part) and the area of NBF stability (right and lower part).

The influence of temperature is well demonstrated by measurements [97] of the longitudinal specific electrical conductivity κ_f of black foam films from NaDoS solution. The temperature dependence of κ_f of CBF in Arrhenius coordinates has the same slope as the temperature dependence of the specific electrical conductivity κ of the bulk solution. The slope of the line for NBF is much steeper. The cross-point of both lines at 31.6 °C indicates the temperature of the CBF/NBF transition at the given concentrations of the initial solution. The slope of these lines in Arrhenius coordinates is proportional to the activation energy of the ions' movement. The almost equal slopes of the lines for CBF and bulk solution indicate that the liquid core of CBF contains practically the same solution as in the bulk. On the contrary the 2.5 times larger slope of the line for NBF proves that the NBF does not contain any liquid solution.

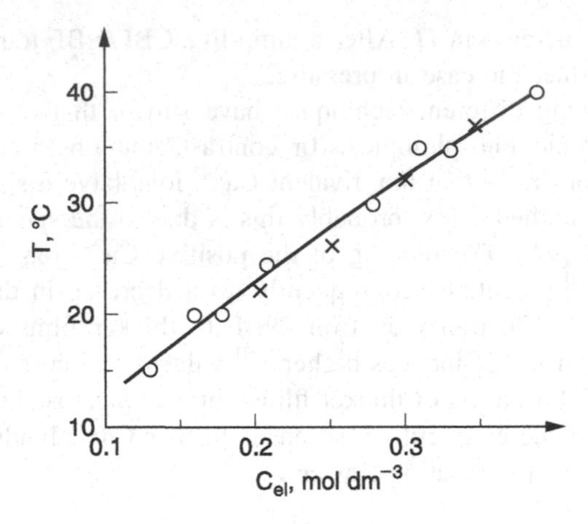

FIGURE 3.25 Temperature T of the CBF/NBF transition vs. electrolyte concentration c_{el}. Black films from Na dodecylsulfate aqueous solutions. (×): data from Ref. 95; (○): data from Ref. 96.

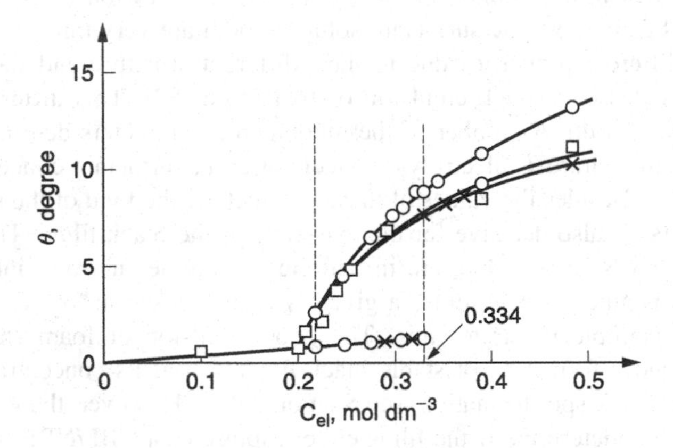

FIGURE 3.26 Contact angle θ vs. NaCl concentration c_{el} at 22 °C. Black films from Na dodecylsulfate aqueous solutions. (×): data from Ref. 24; (□): data from Ref. 85; (○): data from Ref. 98.

Another very informative fact is that the transport numbers of the counter-ions in NBF from ionic surfactants are close to 1, i.e., the electric current is due to the surfactant counter-ions only; the film contains practically no other ions although high electrolyte concentration is present in the bulk solution. The transport numbers of ions in CBF are almost the same as in the bulk.

Thermodynamic quantities which show rather different values for CBF and NBF are the contact angle α_o and film tension γ^f related through Equation 3.64. Systematic measurements of α_o have been performed with black foam films from NaDoS aqueous solutions [24,85,98]. Figure 3.26 presents the results obtained by three different teams. A very sharp jump in the $\alpha_o(c)$ curve is observed at the CBF/NBF transition. The critical electrolyte concentration c_{cr} at which the transition occurs from one type of black film to the other can be determined from this jump. For instance, for microscopic black films from NaDoS + NaCl solutions, $c_{cr} = 0.334$ mol dm^{-3}. At lower c values, CBF with $\alpha_o < 1°$ are stable while at higher c values, NBF with α_o of the order of 10° are stable. The data obtained for macroscopic flat films in a frame indicate a bit lower values, i.e., $c_{cr} = 0.2$ mol dm^{-3}.

Hence, a range of NaCl concentration (0.2 to 0.334 mol dm^{-3}) is distinguished in which the $\alpha_o(c)$ curves follow a different course. This is due to the metastable state of CBF; in this concentration range the microscopic CBF transform into NBF after a certain time. In the metastable region, indicated by the dashed lines in Figure 3.26, both NBF and CBF are observed; respectively, large or small values of the contact angles. The α_o values obtained with microscopic NBF are close to those obtained with macroscopic NBF. When there is no external influence, NBF is only formed at electrolyte concentrations at which the maximum in the $\Pi(h)$ isotherm is overcome: $\Pi_{max} \leq \Delta p$. However, if the film area is large or the film is subjected to an external disturbance, NBF can be obtained at lower c when $\Pi_{max} > \Delta p$.

The film tension γ^f being related through Equation 3.64 with the contact angle α_o show similar behavior. Its value for CBF ($\alpha_o < 1°$) is close to 2γ of the bulk solution, while for NBF (α_o of the order of 10°) it is essentially different. Accordingly, the values of the interaction Helmholtz energy $\Delta F(h)$ of the film are rather different for CBF and NBF.

A completely different behavior exhibit the two types of black films from NaDoS solutions in another aspect. Contrary to CBF, the NBF do not change their thickness with Δp and c alterations. However, their properties depend on the composition of the bulk surfactant solution, e.g., α_o depends on c. The thickness of NBF determined from $h(c)$ dependence is approximately equal to the doubled thickness of the adsorption layer as assumed by Perrin. This is confirmed by NBF obtained from other surfactants. It has been proved, however, by infrared spectroscopy [99] and electrical conductivity measurements [97], that there is water in the NBF. It is most probable that the adsorption layers contain a certain quantity of water but they are not separated by an aqueous core. This is confirmed also by ellipsometric measurements [100].

The data for the film thickness of NBF from NaDoS solutions obtained by several authors with different techniques are rather scattered between 33 and 45 Å. Precise X-ray reflectivity measurements with CBF and NBF films from NaDoS + NaCl aqueous solutions provide more details about their structure [101]. The thickness of the respective layers which detail the film structure is: hydrocarbon chains layer 10.8 Å, polar groups layer 3.75 Å, water layer 3.75 Å, total film thickness 32.9 Å. Although these data indicate an aqueous core, this water is most probably not a liquid but hydration water and the NBF can be considered as bilayers.

3.6.4 Stability and Rupture of Bilayer Black Films

The Newton black foam films, the Newton black emulsion films, and the bilayer lipid membranes (BLM) are bilayers of amphiphile molecules. Their stability in respect to rupture and their permeability can be considered from a unified point of view [102,103]. Figure 3.27 represents schematically the generally accepted molecular model of such bilayers. The description of the fluctuation formation of microscopically small holes responsible for the bilayer stability and permeability can be based on both thermodynamic and molecular models.

The driving force or supersaturation $\Delta\mu$ for bilayer rupture is a thermodynamic quantity, since it is defined by

$$\Delta\mu = \mu_b - \mu_s \tag{3.84}$$

where μ_b and μ_s are, respectively, the chemical potentials of the amphiphile molecules in the bilayer and in the solution. For sufficiently dilute solutions μ_s is known to depend logarithmically on the concentration c_s of the monomer amphiphile molecules in the solution. Equation 3.84 can

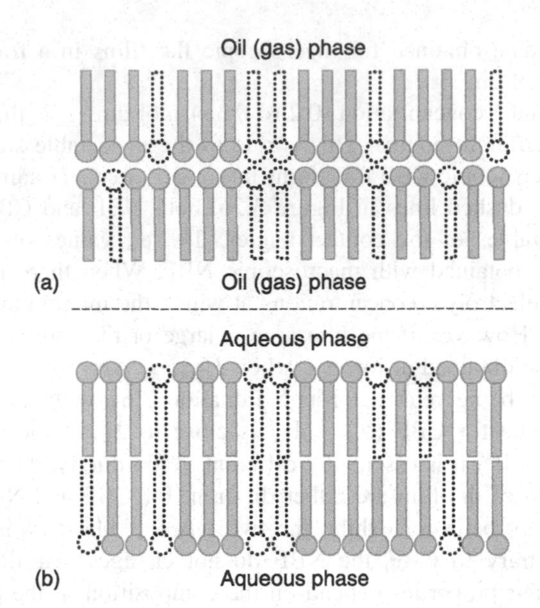

FIGURE 3.27 (a) Molecular model of OWO emulsion or foam bilayer film; (b) molecular model of WOW emulsion bilayer film or bilayer lipid membrane (BLM). The filled symbols denote amphiphile molecules; the dotted symbols schematize vacancies and holes.

be approximated by

$$\Delta\mu(c_s) = kT \ln\left(c_e/c_s\right) \tag{3.85}$$

where c_e is a certain concentration of monomer amphiphile molecules in the solution, called bilayer equilibrium concentration.

Thermodynamics can also be used for determining of the work w_i for fluctuation formation of an i-sized hole, $i = 1,2,3,\ldots$ being the number of amphiphile molecules that would fill the hole. Since the hole appears as a result of the passage of i molecules from the bilayer into the solution, the work associated with this process is $-i\Delta\mu$. On the other hand, work equal to the hole total peripheral free energy $P_i > 0$ is done in creating the hole periphery. P_i is proportional to the length of the hole periphery, which for circular holes of bilayer depth is $(2\pi A_0 i)^{1/2}$. Hence, w_i is given by

$$w_i = -i\Delta\mu + \tau_L\,(2\pi A_0 i)^{1/2} \tag{3.86}$$

where τ_L (J m^{-1}) is the hole specific edge free energy which, in principle, can be a function of i; A_0 is the area per molecule.

When $\Delta\mu > 0$ (for example, when $c_s < c_e$), the competition between the two energy terms in Equation 3.86 causes w_i to pass through a maximum of value. The hole of size i^* is the so-called nucleus hole, and w^* is the nucleation work. While the subnucleus holes tend to decay (w_i decreases with decreasing $i < i^*$), the supernucleus holes can grow spontaneously (w_i decreases with increasing $i > i^*$). For this reason, the bilayer can rupture only after the fluctuation appearance of at least one nucleus hole per unit time and, accordingly, w^* is the energy barrier for bilayer rupture. If $\Delta\mu = 0$, i.e., if $c_s = c_e$, the bilayer is truly stable (and not metastable) in respect to rupture by hole nucleation. It must be emphasized that the bilayer also retains this true (or infinite) stability for $\Delta\mu < 0$, i.e., for $c_s > c_e$, since both terms in Equation 3.86 are then positive and w_i can only increase with increasing i. The bilayer cannot rupture despite the presence of a certain population of holes in it.

Let us turn to results obtained on the basis of the molecular model of amphiphile bilayer illustrated in Figure 3.27. The basic idea in the theoretical description is to regard the bilayer as consisting of two monolayers of amphiphile molecules mutually adsorbed on each other. Each of the monolayers can be filled with a maximum of $N_m = 1/A_0 \sim 10^{18}$ molecules m^{-2}, but the thermal motion of the molecules reduces their density below N_m. This means that vacancies of amphiphile molecules (i.e., molecule-free sites) exist in the bilayer, their density being n_1 (m^{-2}). At high values of n_1 the vacancies cluster together to form holes. If these holes are sufficient in number and size, they can make the bilayer permeable to molecular species. When $\Delta\mu > 0$, i.e., $c_s < c_e$, nucleus hole can come into being and, by irreversible overgrowth, cause the rupture of the bilayer. In other words, rupture occurs as a result of a two-dimensional first-order phase transition of the "gas" of amphiphile vacancies in the bilayer into a "condensed phase" of such vacancies which is equivalent to a ruptured bilayer.

The lattice model of the bilayer (Figure 3.27) can also be used to express c_e in terms of intermolecular bond energies, the result being

$$c_e = c_0 \exp(-Q/2kT) \tag{3.87}$$

c_0 is a reference concentration and $Q > 0$ given by

$$Q = z\varepsilon + z_0\varepsilon_0 \tag{3.88}$$

is the binding energy of an amphiphile molecule in the bilayer; ε and ε_0 (positive for attraction, negative for repulsion) are, respectively, the energies of the lateral and normal bonds (due to short range surface forces) between the nearest neighbor amphiphile molecules in the bilayer; z and z_0 are the lateral and normal coordination numbers of these molecules (e.g., $z = 6$, $z_0 = 1$ or 3 for hexagonal packing). For steady-state nucleation the nucleation theory derives the explicit $\tau(c_s)$ dependence of the bilayer mean lifetime τ on the bulk concentration c_s

$$\tau(c_s) = \mathrm{A} \exp[\mathrm{B}/\ln(c_e/c_s)] \tag{3.89}$$

with

$$\mathrm{A} = 1/Z\omega N_0 A$$

and

$$\mathrm{B} = \pi A_0 \tau_L^2 / 2(kT)^2 \tag{3.90}$$

where Z is the so-called Zeldovich factor (a number from about 0.01 to 1); A is the bilayer area; ω (s^{-1}) is a frequency of a vacancy joining the nucleus hole; N_0 (m^{-2}) is the density of available lattice sites on which a vacancy can be formed.

In some cases τ can be so short that experimental observation of the bilayer after its formation is possible only with a certain probability W depending on the resolution time t_r of the particular equipment used. In direct visual observation of bilayer rupture, for instance, $t_r \approx 0.5$ s, which is the reaction time of the eye. Since observation of the bilayer is possible only if the bilayer has ruptured during the time $t > t_r$, the nucleation theory yields W as a function of $c_s \leq c_e$:

$$W(c_s) = \exp\{-t_r/\mathbf{A}\exp[\mathbf{B}/\ln(c_e/c_s)]\} \tag{3.91}$$

Equations 3.89 and 3.91 show that both τ and W increase sharply with the bulk surfactant concentration c_s in a relatively narrow range. For this reason, a critical concentration c_c for bilayer rupturing in less than τ_c seconds can be defined by the condition $\tau(c_c) = \tau_c$. From Equation 3.89 we obtain

$$c_c = c_e \exp[-\mathbf{A}/\ln(\tau_c/\mathbf{B})] \tag{3.92}$$

The theoretical $\tau(c_s)$ dependence Equation 3.89 can be easily checked since both τ and c_s are measurable quantities. The bulk concentration c_s of the amphiphile molecules is determined when preparing the surfactant solution and the mean lifetime τ is measured as the time elapsing from the moment of formation of a bilayer with a given radius until the moment of its rupture. Due to the fluctuation character of the film rupture, the film lifetime τ is a random parameter. Experimentally, τ has been determined by averaging from a great number of measurements.

A comparative investigation [104] of the rupture of microscopic emulsion and foam bilayers obtained from the same aqueous solutions of $Do(EO)_{22}$ (dodecyl 22-oxyethyl, a nonionic surfactant) at electrolyte concentration $c > c_{cr}$ has been carried out. The emulsion bilayer was formed between two oil phases of nonane. Figure 3.28 shows the $\tau(c_s)$ dependence for foam (open circles) and emulsion (black circles) bilayer films. In both cases τ depends strongly on the

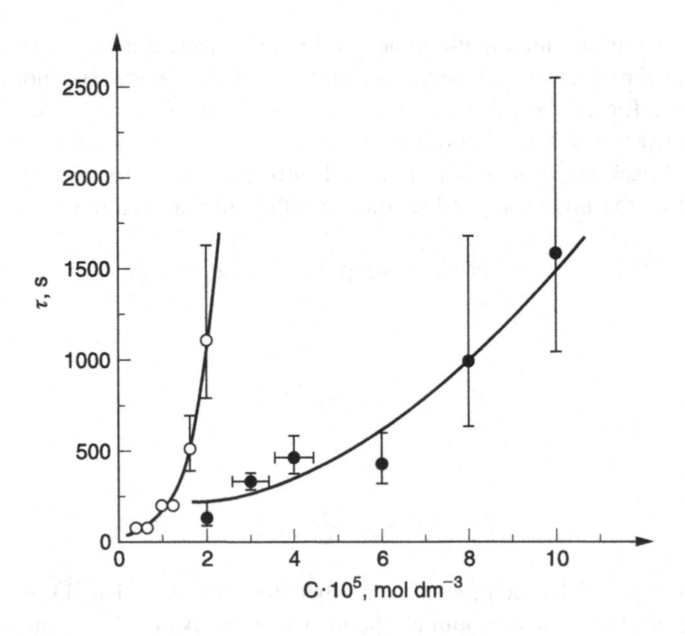

FIGURE 3.28 Mean lifetime τ of Newton black films (bilayer films) vs. surfactant concentration c. Microscopic ($r = 100$ μm) emulsion (●) and foam (○) bilayers from dodecyl 22-oxyethyl + 0.3 M KCl aqueous solutions. (From H.J. Müller, B. Balinov, and D. Exerowa, *Colloid Polymer Sci.* 266: 921 (1988).)

surfactant concentration. The solid curve for the foam bilayer represents the best-fit dependence of Equation 3.89 with A = 0.145 s, B = 25, and $c_e = 3.3 \times 10^{-5}$ mol dm^{-3}. It is seen that the stability of the foam bilayers is greater than that of the emulsion bilayers and that c_e is much higher for the emulsion bilayers.

The same experiment can be also used to check Equation 3.91, W being determined as the ratio between the number of bilayer films observed and number of all films studied. Studying the $W(c_s)$ dependence is possible and particularly convenient at low c_s values when the bilayer mean lifetime τ is comparable with t_r. A characteristic feature of W, according to Equation 3.91, is its sensitivity to changes of c_s in a very narrow range only. The $W(c_s)$ dependence obtained fits well to Equation 3.91 and its shape is different for different types of surfactants [105].

It follows from the theory that $c_c = c_e$ in the case of a missing metastable region when only thermodynamically stable foam bilayers are formed. The high stability of DMPC foam bilayers gives reason to assume that $c_c = c_e$, thus permitting to calculate the binding energy Q from the experimental dependence of c_c on temperature. The values of Q obtained from the best fit of Equation 3.87 to the experimental data are $(1.93 \pm 0.04) \times 10^{-19}$ J at $T < 23\,°C$ and $(8.03 \pm 0.19) \times 10^{-20}$ J at $T > 23\,°C$.

The good agreement between theory and experiment allows the determination of several quantities of the theory: w^*, I^*, τ_L, c_c, c_e, etc. This offers an interesting possibility for the evaluation of some molecular characteristics of the bilayers. For example, from the hole edge energy τ_L the bond energy ε between two nearest-neighbor surfactant molecules in the bilayer plane may be estimated. This important parameter for the bilayer stability and permeability has been determined for various surfactants.

3.7 SIMILARITY BETWEEN EMULSION FILMS AND FOAM FILMS

The comparison of the results for foam films with those for emulsion films, especially the OWO films, is reasonable since in both cases the thin aqueous film is in contact with two hydrophobic phases. It is anticipated that the effects related to adsorption and orientation of surfactant molecules at the film/hydrophobic phase interface are very similar. Hence, some regularities established for aqueous foam films can be applied to OWO emulsion films and vice versa.

Analogous to foam films, stable (or metastable) emulsion films are formed only in the presence of surfactants (emulsifiers) at concentrations higher than the critical concentration of formation of black films c_{bl}. For each emulsifier the latter is a function of temperature, electrolyte content, pH, and presence of other surfactants in the water or organic phase. Unlike the foaming agents c_{bl} of emulsifiers depends also on the nature of the nonpolar ("oil") phase.

The comparison of the concentrations c_{bl} corresponding to formation of black spots in emulsion and foam films, obtained from solutions of one and the same surfactant (which is both emulsifier and foaming agent), indicates that c_{bl} for foam films is considerably lower than c_{bl} for emulsion films. This means that stable foam films (usually black films) can be obtained at lower surfactant concentrations than stable emulsion films even from the nonpolar organic phase. With increasing polarity of the molecules of the organic phase c_{bl} for emulsion aqueous films increases which is analogous to the increase in c_{bl} for hydrocarbon emulsion films. The concentration c_{bl} for emulsion films is close to the emulsifier concentration at which it is possible to disperse a small quantity of the organic phase in a certain volume of the aqueous surfactant solution under definite conditions resulting in formation of stable emulsions.

The process of expansion of a black emulsion film is quite similar to that one in a foam film: at low electrolyte concentrations the black spots in both types of films expand slowly; at high electrolyte concentrations the process is very fast (within a second or less) and ends up with the formation of a black film with large contact angle film/bulk liquid phase (meniscus). In the process of transformation of the black spots into a black film, the emulsion film is very sensitive to any external effects (vibrations, temperature variations, etc.) in contrast to the black foam film.

As in the case of foam films the thickness of aqueous emulsion films depends on the electrolyte concentration c. Depending on c two types of black films can be formed: CBF and NBF. In contrast to foam films, the OWO emulsion films become NBF at much higher electrolyte concentrations. This peculiarity can be explained with the weaker molecular attraction in emulsion films and the resulting increase in the barrier of the $\Pi(h)$ isotherm. The Hamaker constant A for emulsion films is by a decimal order of magnitude lower than that of aqueous foam films and is within the range between 10^{-21} and 10^{-20} J depending on the surfactant kind and the nature of the organic phase. It should be noted that in the absence of a surfactant, the values of the Hamaker constant of aqueous and nonaqueous emulsion films are equal.

At large surfactant concentrations emulsion films as well as foam films exhibit a layer-by-layer thinning (stratification) and metastable black films are formed. Compared to foam films of analogous composition, the respective emulsion films were thicker, due to the weaker molecular attraction, and the stratification occurred at lower surfactant concentrations. Such a behavior has been reported also for hydrocarbon films (WOW emulsion films).

The dependence of the lifetime of NBF on surfactant concentration of emulsion films stabilized with 22-oxythylated dodecyl alcohol is in conformity with the theory of bilayer stability (see Section 3.6.4). The values of the equilibrium concentrations c_e calculated for emulsion films are higher than those for foam films. It is worth noting that $c_e <$ CMC of foam films from certain surfactants, while for emulsion films $c_e >$ CMC. That is why it is impossible to obtain thermodynamically stable emulsion NBF. This result is of particular importance for the estimation of stability of O/W emulsions. There is a correlation between the stability of single OWO emulsion films, single drops under oil/water interfaces, and real O/W emulsions.

It should be noted as well that the difference between aqueous emulsion films and foam films involves also the dependences of the various parameters of these films (potential of the diffuse double electric layer, surfactant adsorption, surface viscosity, etc.) on the nature of the organic phase, the distribution of the emulsifier between water and organic phase, and the relatively low, as compared to the water/air interface, interfacial tension.

REFERENCES

1. I Newton. Optics. London: Smith & Watford, 1704.
2. I Plateau. Statique Expérimentale et Théorétique des Liquides Soumis aux Seulles Forces Moléculaires. Paris: Dauthier-Villars, 1873.
3. C Marangoni. Nuovo Cimento, Ser 2: 239, 1872.
4. JW Gibbs. Collected Works. London: Longmans Green, 1928.
5. J Perrin. Ann Phys Paris 10: 160, 1918.
6. BV Derjaguin, EV Obukhov, Acta Physicochim USSR 5: 1, 1936.
7. BV Derjaguin, LD Landau. Acta Physicochim USSR 14: 633, 1941.
8. EJW Verwey, JThG Overbeek. Theory of Stability of Lyophobic Colloids. Amsterdam: Elsevier, 1948.
9. JC Eriksson, BV Toshev. Colloids Surfaces 5: 241, 1982.
10. A Scheludko. Annuaire Univ Sofia, Fac Chimie 62: 47, 1970.
11. BV Toshev, IB Ivanov. Colloid Polymer Sci 253: 558, 593, 1975.
12. CJ Vašiček. Optics of Thin Films. Amsterdam: North Holland Publish Co, 1960.
13. A Scheludko, D Platikanov. Kolloid-Z 175: 150, 1961.
14. A Scheludko. Proc Koninkl Nederl Akad Wetenschap B65: 76, 1962.
15. EM Duyvis. PhD Thesis. University of Utrecht, 1962.

16. A Scheludko, D Exerowa. Comm Dept Chem, Bulg Acad Sci 7: 123, 1959.
17. D Exerowa, M Zacharieva, R Cohen, D Platikanov. Colloid Polymer Sci 257: 1089, 1979.
18. D Exerowa, A Scheludko. Compt Rend Acad Bulg Sci 24: 47, 1971.
19. H Sonntag, J Netzel, H Klare. Kolloid-Z Z Polym 211: 121, 1966.
20. PM Kruglaykov, YuG Rovin. Fizikokhimiya chernykh uglevodorodnykh plenok (in Russian). Moscow: Nauka, 1978.
21. D Exerowa, T Kolarov, Khr Khristov. Colloids Surfaces 22: 171, 1987.
22. V Bergeron, CJ Radke. Langmuir 8: 3020, 1992.
23. A Scheludko, B Radoev, T Kolarov. Trans Faraday Soc 64: 2213, 1968.
24. T Kolarov, A Scheludko, D Exerowa. Trans Faraday Soc 64: 2864, 1968.
25. DA Haydon, JT Taylor. Nature 217: 739, 1968.
26. O Reynolds. Phil Trans Royal Soc A177: 157, 1886.
27. JS Stephan. Math Natur Akad Wiss 69: 713, 1884.
28. A Scheludko. Kolloid-Z 155: 39, 1957.
29. J Lyklema, PC Scholten, KJ Mysels. J Phys Chem 69: 116, 1965.
30. V Levich. Physicochemical Hydrodynamics. Engelwood Cliffs: Prentice Hall, 1962.
31. E Mileva, B Radoev. Colloid Polymer Sci 263: 587, 1985; 264: 823, 1986.
32. J Lee, T Hodgson. Chem Eng Sci 23: 1375, 1968.
33. A Scheludko. Adv Colloid Interface Sci 1: 391, 1967.
34. E Mileva, B Radoev. Colloid Polymer Sci 264: 965, 1986.
35. B Radoev, E Manev, I Ivanov. Kolloid-Z 234: 1037, 1969.
36. A Barber, S Hartland. Canad J Chem Eng 54: 279, 1976.
37. E Mileva, B Radoev. Colloids Surfaces A 74: 259, 1993.
38. A Scheludko. Annuaire Univ Sofia, Fac Chim 59: 263, 1964/65.
39. S Frankel, K Mysels. J Phys Chem 66: 190, 1962.
40. D Platikanov. J Phys Chem 68: 3619, 1964.
41. D Exerova, T Kolarov. Annuaire Univ Sofia, Fac Chim 59: 207, 1964/65.
42. O Velev, T Gurkov, R Borwankar. J Colloid Interface Sci 159: 497, 1993.
43. J Joye, G Hirasaki, C Miller. Langmuir 10: 3174, 1994.
44. E Mileva, D Exerowa, in Proceedings of Second World Congress on Emulsion, Bordeaux 2: 2-2-235/1-10, 1997.
45. E. Manev, R Tsekov, B Radoev. J Dispers Sci Technol 18: 769, 1997.
46. B Radoev, A Scheludko, E Manev. J Colloid Interface Sci 95: 254, 1983.
47. E Mileva, B Radoev. Chapter 6 in D Petzev, ed. Emulsions: Structure, Stability and Interactions. Amsterdam: Elsevier–Academic Press, 2004.
48. L Mandelstam. Ann Physik 41: 609, 1913.
49. A Scheludko. Proc Koninkl Nederl Akad Wetenschap B65: 87, 1962.
50. A Vrij. Faraday Discussion Chem Soc 42: 23, 1966.
51. R Tsekov, B Radoev. Adv Colloid Interface Sci 38: 353, 1992.
52. A Sharma, E Ruckenstein. Langmuir 3: 760, 1987.
53. D Exerowa, A Scheludko, in JThG Overbeek, ed. Chemistry, Physics and Application of Surface Active Substances. London: Gordon & Breach, 2: 1097, 1964.
54. JA de Feijter. Chap.1 in I Ivanov, ed. Thin Liquid Films. New York: Marcel Dekker, 1988.
55. BV Deryagin, NV Churaev. Kolloidny Zhur 38: 402, 1976.
56. AI Rusanov. Kolloidny Zhur 28: 583, 1966.
57. BV Toshev, D Platikanov. Adv Colloid Interface Sci 40: 157, 1992.
58. T Kolarov, Z Zorin, D Platikanov. Colloids Surfaces 51: 37, 1990.
59. HM Princen, SG Mason. J Colloid Sci 20: 156, 1965.
60. BV Derjaguin, GA Martynov, YuV Gutop. Kolloidny Zhur 27: 298, 1965.
61. VS Veselovskij, VN Pertsov. Zhur Fiz Khim 8: 245, 1936.
62. BV Derjaguin, NV Churaev, VM Muller. Surface Forces. New York: Consult Bureau, 1987.
63. JN Israelachvili. Intermolecular and Surface Forces, New York: Academic Press, 1991.
64. HBG Casimir, D Polder. Phys Rev 73: 360, 1948.

65. WA Donners. PhD Thesis. University of Utrecht, 1976.
66. A Scheludko, D Exerowa. Kolloid-Z 165: 148, 1959; 168: 24, 1960.
67. J Lyklema, KJ Mysels. J Am Chem Soc 87: 2539, 1965.
68. D Exerowa. Kolloid-Z 232: 703, 1969.
69. D Exerowa, D Kashchiev, D Platikanov. Adv Colloid Polymer Sci 40: 201, 1992.
70. RW Huddelston, AL Smith, in AJ Akers, ed. Foams. New York: Academic Press, 1976.
71. D Exerowa, R Sedev, R Ivanova, T Kolarov, ThF Tadros. Colloids Surfaces 123: 277, 1997.
72. P Flory. Principles of Polymer Chemistry. New York: Cornell University Press, 1956.
73. PG de Gennes. Macromolecules 13: 1069, 1980.
74. J Lyklema, T van Vliet. Faraday Discuss Chem Soc 65: 25, 1978.
75. R Sedev, D Exerowa. Adv Colloid Interface Sci 83: 111, 1999.
76. ES Johnonnott. Phil Mag 11: 746, 1906.
77. HG Bruil, J Lyklema. Nature Phys Sci 233: 19, 1971.
78. A Nikolov, D Wasan. Langmuir 8: 2985, 1992.
79. JW Keuskamp, J Lyklema, in KL Mittal, ed. Adsorption at Interfaces. Washington: ACS Symposium Series 8: 191, 1975.
80. AD Nikolov, PA Kralchevsky, IB Ivanov, DT Wasan. J Colloid Interface Sci 138: 13, 1989.
81. V Bergeron, CJ Radke. Colloid Polymer Sci 273: 165, 1995.
82. L Ter Minassian-Saraga, ed. Thin Films Including Layers: Terminology in Relation to Their Preparation and Characterization. Pure & Appl Chem 66: 1667, 1994.
83. D Exerowa, PM Kruglyakov. Foam and Foam Films. Amsterdam: Elsevier, 1998.
84. J de Feijter, A Vrij. J Colloid Interface Sci 70: 456, 1979.
85. F Huisman, KJ Mysels. J Phys Chem 73: 489, 1969.
86. KJ Mysels, M Jones. Discuss Faraday Soc 42: 42, 1966.
87. IJ Black, RM Herrington. J Chem Soc, Faraday Trans 91: 4251, 1995.
88. T Kolarov, R Cohen, D Exerowa. Colloids Surfaces A 42: 49, 1989; 129–130: 257, 1997.
89. V Bergeron, A Waltermo, P Claesson. Langmuir 12: 1336, 1996.
90. T Yamanaka, M Hayashi, R Matuura. J Colloid Interface Sci 88: 458, 1982.
91. R Cohen, R Koynova, B Tenchov, D Exerowa. Eur Biophys J 20: 203, 1991.
92. R Cohen, D Exerowa, T Yamanaka. Langmuir 12: 5419, 1996.
93. R Cohen, D Exerowa, T Kolarov, T Yamanaka, VM Muller. Colloids Surfaces 65: 201, 1992.
94. T Yamanaka, T Tano, O Kamegaya, D Exerowa, R Cohen. Langmuir 10: 1871, 1994.
95. M Jones, K Mysels, P Scholten. Trans Faraday Soc 62: 1336, 1966.
96. D Platikanov, M Nedyalkov. Annuaire Univ Sofia, Fac Chim 64: 353, 1969/70.
97. D Platikanov, N Rangelova, in BV Deryagin, ed. Research in Surface Forces. New York: Consultants Bureau 4: 246, 1972.
98. D Exerowa, Khr Khristov, M Zacharieva, in Poverkhnostnye sily v tonkikh plenok (in Russian). Moscow: Nauka, 1979.
99. J Corkill, J Goodman, C Orgden, J Tate. Proc Roy Soc A 273: 84, 1963.
100. D den Engelsen, G Frens. J Chem Soc Faraday Trans I 70: 237, 1974.
101. O Belorgey, JJ Benattar. Phys Rev Lett 66: 313, 1991.
102. D Exerowa, D Kashchiev. J Colloid Interface Sci 77: 501, 1980.
103. D Exerowa, D Kashchiev. Contemp Phys 27: 429, 1986.
104. HJ Müller, B Balinov, D Exerowa. Colloid Polymer Sci 266: 921, 1988.
105. D Exerowa, B Balinov, A Nikolova, D Kashchiev. J Colloid Interface Sci 95: 289, 1983.
106. T Kolarov, A. Scheludko, D Exerowa. Annuaire Univ Sofia, Fac Chim 62: 75, 1967/68.

4 Emulsion Phase Inversion Phenomena

Jean-Louis Salager

CONTENTS

4.1 INTRODUCTION

Emulsions are by definition dispersions of a liquid phase in another, which exhibit a certain stability, in most cases thanks to the presence of an adsorbed surfactant at interface. Hence, at least two immiscible liquid phases should be present to make an emulsion. As a consequence, the phase behavior of the system made by the surfactant (S) and the two immiscible liquids, referred to as oil (O) and water (W) in a very general way, is of paramount importance [1].

As a general trend, it can be said that oil and water alone produce a two-phase system which leads to an unstable dispersion upon stirring. As surfactant is added to the oil/water binary, it tends to make the liquids more compatible, it reduces the interfacial tension and facilitates the emulsification, and often promotes a stabilization mechanism at a concentration level down to 0.3 wt% or even less. As far as the phase behavior is concerned, the added surfactant also tends to reduce the oil–water immiscibility gap, just as it would do with an increase in temperature, but this time through a so-called solubilization process associated with the presence of micelles and other self-association structures.

However, a large amount of surfactant, say 10 to 20%, is generally required to attain a critical point at which a single phase SOW system is obtained. As a consequence, it can be said that the surfactant concentration range over which two-phase systems are likely to occur and to produce reasonably stable emulsions is from less than 1 wt% up to 5 or 10 wt% and even more in some applications. This would be the range over which the phase behavior has to be scrutinized.

The emulsion type, either oil-in-water (O/W) or water-in-oil (W/O) or more complex morphologies, and therefore the type swap (so-called phase inversion), depends on many different variables, which can be classified in three categories: (1) field variables associated with the equilibrium state of the system, e.g., its phase behavior, also called **formulation** variables because they include the nature of the components as well as temperature and pressure; (2) **composition** variables which deal with quantities or proportions, for instance surfactant concentration and water/oil ratio; and (3) emulsification **protocol** characteristics which indicate the way the emulsion is made or modified, or how the formulation or composition are changed as a function of time or space.

The attainment of an emulsion with given properties is a very difficult practical problem which often requires, firstly, a step by step disaggregation of the current information on the phenomena involved. This is the approach presented in the following section sequence: phase behavior of SOW systems at equilibrium, methods to reduce the number of independent variables and to pack them into a generalized formulation concept, emulsion types, standard inversion, and dynamic processes.

4.2 PHASE BEHAVIOR

Phase behavior is an equilibrium thermodynamic property which depends on the formulation variables, i.e., the nature of the different components, plus temperature and pressure. Since surfactants are amphiphilic substances with an affinity toward both water and oil, at high surfactant concentration the system usually exhibits a single phase behavior, but by dilution either by water, oil, or both, it splits in a two-phase behavior system, which is the proper kind to generate an emulsion upon stirring. Hence the boundary between the single and multiphase regions is of primary importance in practice, because it sets the limits to make emulsions.

4.2.1 Two-Phase Behavior

In what follows, the two immiscible phases are referred to as water (W) for the polar one, generally an aqueous solution, and oil (O) for the less polar or apolar one, which can be any non-water-soluble organic liquid from an n-alkane to a natural triglyceride or a polydimethyl siloxane polymer. In most cases, one of these phases contains the majority of the surfactant, as

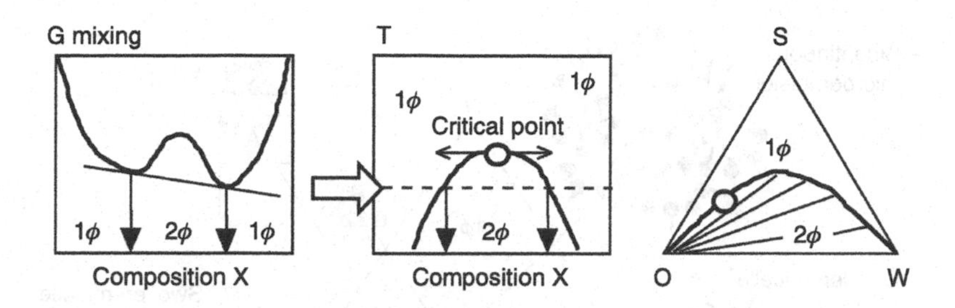

FIGURE 4.1 Phase behavior of binary and ternary systems.

a molecular solution or as a microstructure such as a micellar solution, a microemulsion, or a liquid crystal.

In the simplest case of mismatched mixture, i.e., a binary, the free energy function G (composition X) exhibits at constant temperature and pressure two minima and the splitting between the corresponding phases reduces the overall free energy of the system according to the Maxwell convention (Figure 4.1, left graph).

The situation is often similar in a SOW ternary, at least at low surfactant concentration, e.g., the system splits into two phases, one of them closer to pure W and the other closer to pure O. The phase diagram of a binary versus temperature (Figure 4.1, center) has essentially the same aspect that the SOW ternary diagram (Figure 4.1, right), which means that the surfactant plays the same role as temperature in making the components compatible. As the amount of surfactant increases, the compatibility gap diminishes and the conjugate phase compositions get closer, until they merge at a so-called critical point, beyond which the system is monophasic. In a ternary SOW diagram the critical point is often located on one side because the tie lines are slanted, as in Figure 4.1's triangular plot indicating that the surfactant preferentially partition in one of the conjugate phases.

There are essentially two cases of two-phase behavior with SOW systems, so-called Winsor I and Winsor II (abbreviated W I and W II in Figure 4.3), [2] which correspond respectively to a surfactant-rich water phase in equilibrium with a surfactant-depleted oil phase or vice versa. In W I (respectively II) case the critical point (white circle in Figure 4.1, right graph) is located close to the W (respectively O) vertex.

Two-phase systems would lead to different dispersion morphologies, the simplest ones being the so-called O/W and W/O types in which one of the phases is dispersed as drops in the other, which is continuous. More intricate morphologies like the so-called multiple emulsions, i.e., emulsion in emulsion, could theoretically become very complex although in practice there are essentially two cases, i.e., o/W/O and w/O/W, in which the first lower case letter indicates the phase dispersed as droplets in drops, the next letter indicates the drop phase, and the last one the continuous phase. Multiple emulsions could be somehow considered as a mixture of both types of emulsions, and as such they might be found as intermediate morphology in inversion processes, as will be discussed later.

4.2.2 Three-Phase Behavior

Three-phase behavior (Figure 4.2) is allowed by the phase rule for a true SOW ternary system, and any higher order multicomponent system, but it is very unlikely to happen by chance. However, it has to be considered in some detail because it is important as an intermediate between the two classical biphasic cases.

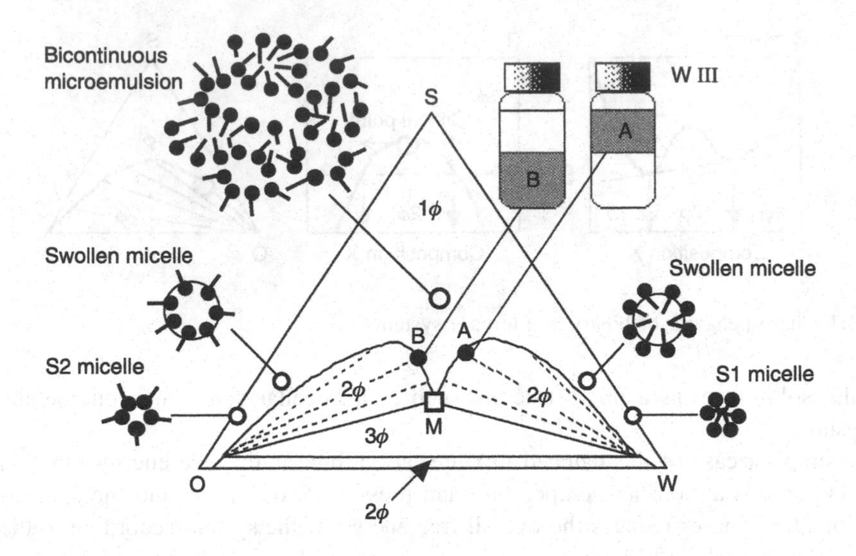

FIGURE 4.2 Phase behavior in Winsor type III diagram.

When the amphiphile **equally likes** O and W, there would be a two-phase splitting with equal surfactant activities X_S^o and X_S^w (or equal concentrations as a first approximation) in both the water and oil phases, so that at equilibrium:

$$\mu^o = \mu^{*o} + \text{RT}\ \ln[X_S^o] = \mu^w = \mu^{*w} + \text{RT}\ \ln[X_S^w] \tag{4.1}$$

and

$$\Delta\mu^*_{O\to W} = \mu^{*w} - \mu^{*o} = \text{RT}\ \ln[X_S^o/X_S^w] = 0 \tag{4.2}$$

Such a situation, with equal surfactant concentrations in oil and water, which is quite common with low molecular weight amphiphiles like alcohols, should result in a two-phase behavior region with horizontal tie-lines in a ternary diagram. However, it is not absolutely correct to say that the surfactant equally likes oil and water since its "tail" does not like to be in water and its "head" group does not like to be in oil either. Consequently, it is more accurate to state that this is a case in which the surfactant **equally dislikes** W and O, and thus might look for a third alternative. In such a case, which is common with surfactants, the splitting into three phases (MOW) [2,3] can be the best way to reduce the overall free energy, with the third phase being a microemulsion (M), a somehow strangely organized medium of so-called percolated micelles, bicontinuous or sponge structure as described in detail elsewhere [4–10].

The corresponding SOW ternary diagram is the so-called Winsor type III, illustrated in Figure 4.2, which contains a three-phase splitting triangle with vertices at essentially pure O and pure W, and at a microemulsion M which contains most of the surfactant and similar amounts of oil and water. The three-phase zone is surrounded by three two-phase regions, the lower one being a very thin strip close to the OW side of the diagram.

In this case it is worth labeling the microemulsion as M to avoid a terminology inconsistency in the two-phase lobes located above the three-phase triangle. In the right lobe the system splits into two phases: one which is essentially pure W, and the other one (A) which is close to M and thus a microemulsion, and may be called the "oil" phase since it is in equilibrum with W. Similarly, in the left two-phase lobe the splitting is between essentially pure oil O and a microemulsion (B) which may be called the "water" phase since it is in equilibrium with O. The paradox is that a look at the phase diagram confirms that "oil" A contains more water than "water" B. Hence, it is better to refer to this type of phase as a microemulsion and to label it as M, or Om and Wm as some authors do.

The bicontinuous structure of M means that its conductivity can be intermediate between those of water and oil, and this would make it very difficult to determine the dispersed system morphology type from conductivity measurements [11]. However, it does not matter very much, because in practice, three-phase W III systems result in extremely unstable emulsions, which do not persist more than a few minutes.

On the other hand, the extremely low interfacial tension associated with W III phase behavior probably generates a dispersed phase which is not made of drops but rather of threads which elongate in the direction of flow as soon as the system is submitted to shear [12]. This is essentially what was observed 30 years ago in enhanced oil recovery (EOR) methods by surfactant flooding [13].

4.2.3 Variables that Influence the Phase Behavior

4.2.3.1 Variable Types

Extensive studies associated with the EOR research drive in the 1970s showed that the phase behavior of SOW systems depends on many different parameters [3]. The typical equation for chemical potential of a surfactant ($\mu = \mu^* + \mathrm{RT} \ln \mathrm{X}$) indicates that it depends on intensive variables or field variables, so-called **formulation** variables, which are independent of quantities, i.e., the nature of the different components (at least SOW), as well as temperature and pressure, which all influence the standard chemical potential μ^* of a surfactant in any phase.

There are also the extensive **composition** variables, like concentrations or proportions, which in the case of a ternary (three components, thus two independent composition variables) are generally taken as the surfactant concentration (represented here by X) and the water/oil ratio WOR.

4.2.3.2 Generalized Formulation

In real systems that additionally contain electrolytes and alcohols, the number of degrees of freedom is appalling. However, the extensive research work carried out to rationalize the observed findings has shown that the formulation effect can be gathered in a single parameter, either theoretical such as Winsor's R ratio of interaction energies [2,3] or experimental such as Shinoda's phase inversion temperature (PIT) [14], whose value command the phase behavior. This generalized formulation has been numerically expressed as the surfactant affinity difference (SAD) [9,10], which takes into account the numerical contribution of all formulation variables, including temperature and eventually pressure [15,16].

These are defined from thermodynamic equilibrium of the surfactant in the oil and water phases, using concentrations instead of activities, provided that the concentrations are low enough so that there is no micelle or other self-assembly structure [17].

$$\mu^{\mathrm{o}} = \mu^{*\mathrm{o}} + \mathrm{RT}\ \ln[\mathrm{C}_{\mathrm{S}}^{\mathrm{O}}/\mathrm{C}^{*\mathrm{o}}] = \mu^{\mathrm{w}} = \mu^{*\mathrm{w}} + \mathrm{RT}\ \ln[\mathrm{C}_{\mathrm{S}}^{\mathrm{w}}/\mathrm{C}^{*\mathrm{w}}] \tag{4.3}$$

$$\mathrm{SAD} = \Delta\mu^*_{\mathrm{O}\to\mathrm{W}} = \mu^{*\mathrm{w}} - \mu^{*\mathrm{o}} = \mathrm{RT}\ \ln[\mathrm{K}_{\mathrm{ow}}/\mathrm{K}^*_{\mathrm{ow}}] = \mathrm{RT}\ \ln \mathrm{K}_{\mathrm{ow}} - \mathrm{RT}\ \ln \mathrm{K}^*_{\mathrm{ow}} \tag{4.4}$$

where the μ^* indicates the standard chemical potential of the surfactant in some reference state (superscript *) at which the dimensionless concentration C/C^* is unity, with C^* to be defined later. K_{ow} is the partitioning coefficient of the surfactant between oil and water, and K^*_{ow} its value in the reference state. If the affinity is defined as the negative of the chemical potential of the surfactant in the reference state, the surfactant affinity difference (SAD) is the variation of chemical potential when a molecule of surfactant is transferred from oil to water, and it can be estimated from the measurement of the partitioning coefficient according to Equation 4.4. For the sake of simplicity the reference state is taken at the Winsor III three-phase behavior, so-called **optimum formulation** because it corresponds to the best formulation to displace petroleum in enhanced oil recovery techniques, and the surfactant concentrations in excess O and W phases, which are at or below the CMC, are taken as reference C^{*o} and C^{*w}, and $K^*_{ow} = C^{*o}/C^{*w}$. For ionic surfactants K^*_{ow} is essentially unity at optimum formulation [18] so that $\text{SAD} = RT \ln K_{ow}$.

However, this is not the general case and K^* may be different from unity, in particular if the surfactant CMC is quite different in water and oil, as for ethoxylated alkylphenol for which K^*_{ow} has been found to be about 30 [19–21].

Equation 4.4 has been recently expressed as a dimensionless parameter so-called hydrophilic–lipophilic deviation (HLD), some kind of "system HLB," which measures the departure from optimum formulation [22,23]:

$$\text{HLD} = \text{SAD}/RT = (1/RT)\Delta\mu^*_{O\to W} = \ln K_{ow} - \ln K^*_{ow} \tag{4.5}$$

It was found that HLD may be written as a linear form of the different formulation variables, with essentially the same expression as the empirical equations proposed 25 years ago to attain the optimum formulation to produce an ultra-low tension for enhanced oil recovery. For ionic surfactant systems [18,24]:

$$\text{HLD} = \ln S - K\ \text{ACN} + f(A) - \sigma + a_T\ (T\text{-}25) \tag{4.6}$$

and for ethoxylated nonionic systems [25]:

$$\text{HLD} = \alpha - \text{EON} - k\ \text{ACN} + b\ S + \phi(A) - c_T\ (T\text{-}25) \tag{4.7}$$

These linear expressions have been found to provide quite a good match [26], with the exception of the temperature term for ethoxylated nonionic systems, for which the effect is more complex than the linear variation proportional to c_T [22]. More detailed information on these correlations and their extension to many systems may be found in the literature [9–10,27–30].

HLD = 0 is equivalent to the case of unit Winsor's R ratio or of a temperature equal to Shinoda's PIT. In practice the formulation studies are carried out by changing a single formulation variable at the time, which is often the salinity of the aqueous phase for ionic systems, and the degree of ethoxylation (EON) or the temperature (T) for ethoxylated nonionics.

When the scanned variable in an ionic surfactant system is for instance the salinity of the aqueous phase (see Figure 4.3) at all other variables constant, there is a certain value of the salinity, so-called optimum salinity S^*, which correspond to W III phase behavior:

$$\text{HLD} = \ln S^* - K\ \text{ACN} + f(A) - \sigma + a_T\ (T\text{-}25) = 0 \tag{4.8}$$

At salinity lower (respectively higher) than S^*, $\text{HLD} < 0$ (respectively > 0) and a W I (respectively II) phase behavior is observed. At HLD = 0 a W III phase behavior is found.

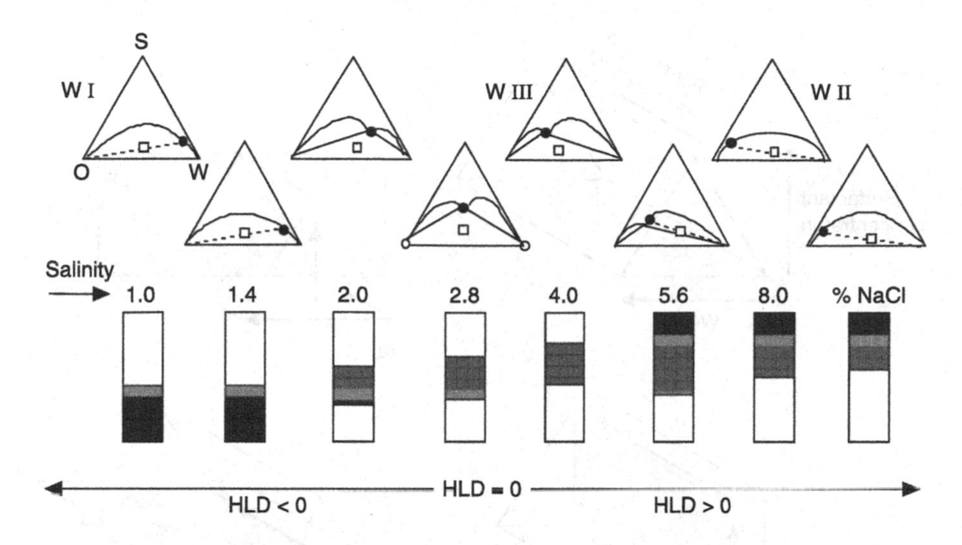

FIGURE 4.3 Phase behavior of a SOW system along a formulation (salinity) scan.

If the scanned variable is the temperature with an ethoxylated nonionic system, then W III phase behavior is attained at a so-called optimum temperature T*:

$$\text{HLD} = \alpha - \text{EON} - \text{k ACN} + \text{b S} + \phi(\text{A}) - c_T\ (\text{T}^*\text{-}25) = 0 \tag{4.9}$$

At temperature lower (respectively higher) than T*, HLD < 0 (respectively > 0) and a Winsor type I (respectively II) phase behavior is observed, i.e., T* is the PIT or the HLB-temperature of the system.

The advantage of HLD over any specific variable is that it gathers the numerical contributions of all formulation variables, and allows the calculation of compensation effects, a trade-off between variables, and the combined effects due to the concomitant variations of two or more variables.

HLD is thus a single generalized formulation variable, whose value is associated with the phase behavior of the SOW system at equilibrium, and whose meaning is related to a free energy of transfer of a surfactant molecule from one phase to the other, thus having the same physicochemical meaning whatever the specific values of the variables.

In practice the use of HLD represents quite a reduction in the degrees of freedom, and for the most simple SOW ternary there are now **only** three independent variables to describe the system at equilibrium:

- HLD generalized formulation (nature of components plus T and P)
- Surfactant concentration
- Water/oil composition, e.g., WOR

4.2.4 2D/3D Representations of Phase Behavior

The handling of three independent variables implies a three-dimensional (3D) representation of the phase behavior and emulsion properties. This is not easy to deal with even in the simplest

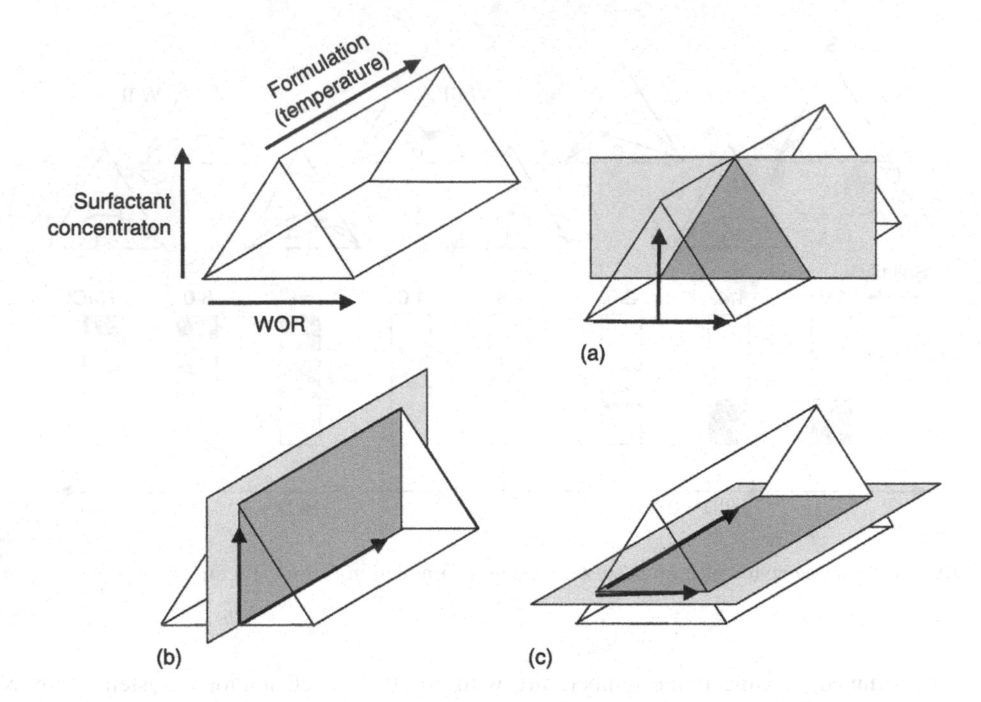

FIGURE 4.4 Different 2D cuts in the 3D formulation-WOR-surfactant concentration space.

case, and in real cases it can be unmanageable. Hence it is generally more convenient to use a two-dimensional (2D) cut of the 3D space.

There are three kinds of 2D cuts, illustrated in Figure 4.4:

- At constant formulation it is a triangular phase diagram (a) with two independent composition variables, e.g., surfactant concentration and WOR.
- At constant WOR it is a surfactant concentration-formulation (or temperature) so-called gamma and fish diagram (b) for reasons that will become clear later.
- At constant surfactant concentration, it is the formulation (or T)-WOR diagram (c), so-called F-WOR or HLD-WOR.

All these diagrams have been used in practice to represent the phase behavior and emulsion properties, and each of them has some advantages and some drawbacks, and the selection of the best 2D representation depends on what matters more in each case.

Surfactant concentration is the variable which determines whether there is a single or a multiple phase behavior case. Since emulsions are multiple phase systems, surfactant concentration should be low enough for the system to split in two phases. Moreover, the surfactant concentration has in practice a higher limit in accordance to cost, toxicity, or environmental constraints. It has also a lower limit which is associated with a minimum performance. As a consequence, the surfactant concentration is often fixed in practice and little variation is allowed. Accordingly, the most significant of the two composition variables is often the WO proportion expressed as WOR or water or oil fraction.

The generalized formulation dictates the phase behavior, which can be seen as a preference of the surfactant for O or W. Nevertheless, the WO composition might also dictate a preference

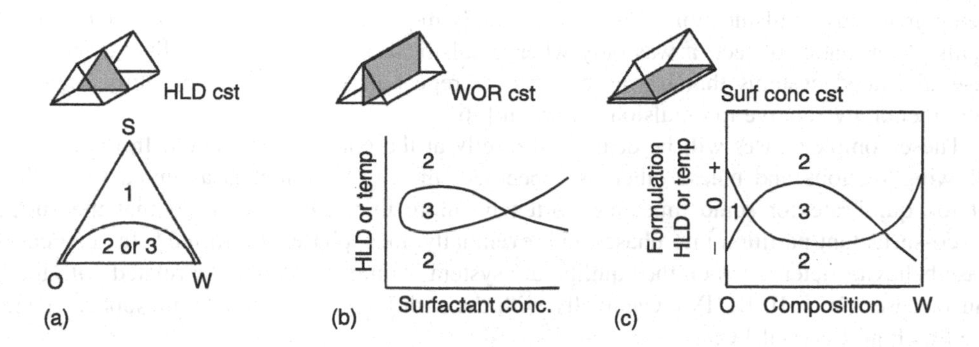

FIGURE 4.5 Phase behavior in the different 2D representations.

toward the phase which is present in higher proportion; indeed, if the WO proportion is 1% W and 99% O it might be difficult for all the surfactant to migrate to the W phase in spite of a strong affinity. Hence the WOR is the second most important variable, and the F-WOR diagram at constant surfactant concentration, i.e., (c) in Figure 4.4, is often the most useful in practice, as will be shown in the following.

Figure 4.5 shows the phase behavior maps in all types of 2D diagrams. The triangular diagram (a) allows the effect of surfactant concentration and WOR to be dealt with, but not the effect of formulation, nor temperature. As seen previously, depending on the value of the generalized formulation HLD, there are three basic types of diagrams, so-called Winsor types I, II, or III, but in practice real phase diagrams are much more complex, particularly if liquid crystals are likely to exist [31].

The center diagram (b) allows the combined effect of formulation (including temperature) and surfactant concentration to be dealt with [32]. The effect of the WOR is not easy to handle in this type of diagram because it results in an apparent change in formulation, e.g., a twisting of the fish, and there are very few detailed reports in the literature [33]. On the contrary this diagram is quite useful to study the temperature effect, in particular on the phase inversion, which depends on the crossing of the single phase region (1) and the eventual presence of liquid crystal phases in it [34,35].

Finally, the right diagram (c), sometimes called X-cut, allows the handling of problems in which the formulation (including temperature) and WOR are of paramount importance, which is the case of the different instances of emulsion inversion. Therefore, it will be the diagram chosen for dealing with the topic of the present chapter.

4.3 STANDARD INVERSION BOUNDARY

4.3.1 What is the Standard Inversion?

An emulsion is a system which is out of equilibrium from the point of view of surface area and its free energy can be reduced by shrinking the surface area, i.e., by the coalescence of the drops. If the two phases exhibit a difference in chemical potential due for instance to a difference in some concentration, a non-equilibrium situation of physicochemical nature arises, whose effects often drive a mass transfer between phases and are much more difficult to ascertain. Such transfer, particularly of surfactant or other amphiphilic molecules, can produce very complex phenomena like spontaneous emulsification, which considerably impairs the interpretation of the experimental results. As a matter of fact, it was only when emulsification was systematically carried out from pre-equilibrated systems that it was possible to repeat experiments and to fully organize the phenomenology relative to emulsion inversion [36].

These complex cases will be dealt with briefly at the end of this chapter. In the present and following sections and unless otherwise specified, the SOW system is assumed to be the first left to equilibrate for some time, e.g., often no more than 24 to 48 h, so that the surfactant and co-surfactant partition into phases and, eventually, mesophases are formed. In such cases the phase behavior determined on the equilibrated system is unique and may be related with the basic Winsor cases (W I, II, III, IV), eventually with the additional presence of a mesophase generally noted as liquid crystal LC.

If the pre-equilibration is achieved, it is generally accepted that the formulation is able to favor one or other curvature, hence one type of morphology, when the emulsification energy is supplied. This trend has been presented in different ways such as the empirical Bancroft's rule [37,38], the intuitive Langmuir's wedge theory [39], the mechanical dual tension model [40], Davies' competing emulsions coalescence [41], or more recent and sophisticated approaches by Ivanov [42] or Kabalnov [43].

The second most important parameter to influence emulsion type has been found to be the WOR, and its effect, sometimes referred to as Ostwald's rule [44,45], has to be combined with the formulation rule.

In experimental practice, an equilibrated SOW system is emulsified according to a standard typical procedure, e.g., a turbulent stirring during 30 sec or 10 min in a mixing device, and its electrolytical conductivity is measured. Since the aqueous phase contains some electrolyte in most practical cases, the emulsion would conduct electricity only if the aqueous phase is continuous. From the conductivity value, the morphologies O/W or W/O could be established. Multiple emulsion and even more complex morphologies may be deduced from conductivity measurements, though not always in a simple way [11].

The standard inversion line is by definition the frontier which separates the W-external from the O-external emulsions. Since multiple emulsions may have an intermediate conductivity as well as bicontinuous microemulsions and LC, a conductivity cut-off has to be specified, sometimes with some difficulty in marginal situations.

4.3.2 Bidimensional Mapping of Emulsion Type

The standard inversion line can be drawn in the two-phase regions of any of the three types of diagrams shown in Figure 4.5, as indicated for typical cases in Figure 4.6. Some authors have used the triangular diagram whereas others, particularly those working with temperature as a formulation variable, have reported the fish type diagram [46–51].

However, the most convenient diagram is the formulation (or temperature) versus WOR diagram (c), because it deals with the two most significant variables as far as emulsion type is concerned. As will be seen later there is an additional and quite powerful reason, which is that the effect of all other variables is very simple and easy to systematize on this kind of map.

This diagram was introduced for studying nonionic systems in relation with the PIT concept [52] and then was systematically used with any of the formulation variables in the past twenty years [34,36,53–60].

To sum up the basic features of this F-WOR diagram, as found in the literature, it can be said that the formulation effect dominates unless the WO composition is too unequal, e.g., there is more than 70% of water or oil. The concept of disproportionate OW amounts might apply not

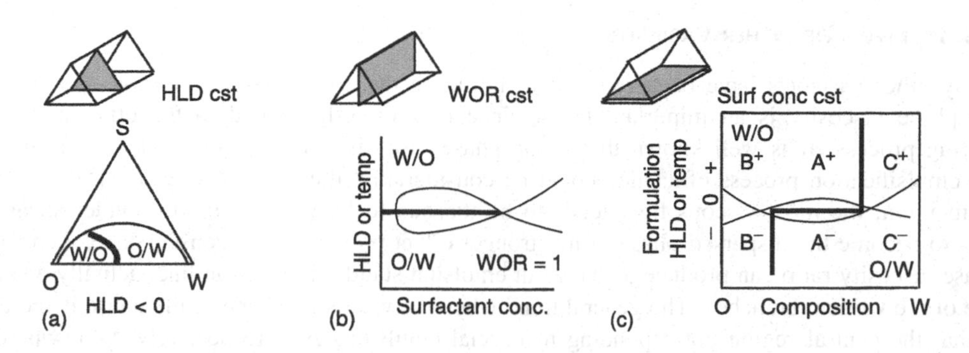

FIGURE 4.6 Emulsion type mapping in the different 2D representations.

only to the system as a whole but also to the local situation and hence could depend on the way the emulsification is carried out, the shape of the vessel or stirrer, the position of the impeller etc. The excess of one phase over the other is essentially the same intuitive argument as Ostwald's rule related to the maximum packing (74% for most dense packing of equal spheres).

Generally speaking it can be said that the phase in smaller amount is more likely to be dispersed in the other than the reverse. It means that at high W (respectively O) content O/W (respectively W/O) emulsions are produced. In each diagram the region with high W and high O content, of more than 70%, exhibits a WOR-controlled morphology, whereas the formulation, i.e., the bending tendency, determines the type when the WOR is not extreme, e.g., from 30 to 70% of one phase.

Formulation-WOR maps have essentially the same aspects, whatever the formulation variable, e.g., salinity of the aqueous phase, surfactant hydrophilicity, temperature, pH, oil type etc. As seen in Figure 4.6(c) the standard inversion frontier (bold) exhibits an almost horizontal branch, which follows the optimum formulation (HLD = 0) line in the central region of the map (from typically 30 to 70% of O or W), and two almost vertical branches located at about 30% W in the HLD > 0 region and at about 30% O in the HLD < 0 region. This results in a step-like shape of the inversion frontier illustrated in Figure 4.6(c) in which six regions are labeled: A, B, C to indicate the WOR, with + or − sign according to the HLD value.

The regions where Bancroft's rule applies, i.e., where the emulsion type is in accordance with the formulation (A+, B+ and A−, C−), are called **normal**, whereas the two regions in which the WOR imposes the morphology (C+ and B−) are called **abnormal** [36]. Abnormal regions are often associated with the presence of multiple morphologies, i.e. w/O/W in C+ and o/W/O in B−, where the lower case letter indicates the droplet phase dispersed in drops. Abnormal emulsions thus consist of a combination of both morphologies: the most internal emulsion (droplets in drops) follows Bancroft's rule and is thus normal, whereas the most external one is abnormal. It seems that this dual emulsion system is a way to settle the conflict between formulation and WOR effects.

In some cases, related to the fractionation of surfactant mixtures, the optimum formulation line (at the center of the three-phase behavior region) is slanted, and the "horizontal" branch of the inversion line essentially follows it and becomes slanted too [54,59]. In some other cases the vertical lines are not exactly vertical. However, if the formulation is taken as the deviation to three-phase behavior, i.e., the actual HLD value at each WOR, then the plot takes a stepwise shape, and this is the general phenomenology for essentially all known systems, even with very viscous resins [61–63] or very nonviscous "oils" such as supercritical carbon dioxide [64,65].

4.3.3 Effect of Other Variables

Many other variables have been reported to influence the emulsification process. For instance, the phase viscosity is an important factor since it is directly related to the efficiency of the stirring process. It is well known that if the phase viscosity ratio is very different from unity, the emulsification process efficiency could be considerably altered [66]. Because it can transfer momentum, the most viscous fluid tends to be the one to be put into motion under shear, and thus to become the dispersed phase, if no stronger effect prevails. As a consequence, a change in phase viscosity ratio can produce a change in emulsion standard inversion line, actually a shift in one of the vertical branches. The general trend when the viscosity of one of the phase is increased is that the central region corresponding to normal emulsions A+ (respectively A−), where the increased viscosity phase is the external phase, i.e., O (respectively W) in A+ (respectively A−), tends to shrink at the expense of the adjacent abnormal emulsion C+ (respectively B−) region [36,67–69].

Other variables have been shown to influence the position of the inversion line. For instance in many applications the surfactant concentration may vary in some range, and may be increased, in particular to produce a finer or more stable emulsion [70,71].

Hence, more surfactant should logically be associated with more command of the formulation variable, i.e., a wider A region range in a formulation-WOR map. This effect can be analyzed as well in a triangular diagram, in which it is obvious that an increase in surfactant concentration reduces the gap between the phases, until a limit, i.e., the critical point, is reached and above which there is a single phase and no emulsion. Such critical points (plait point in the triangular diagram in Figure 4.1, right graph) are on one side and at a low surfactant concentration in W I and W II types. In formulation-WOR diagrams the surfactant concentration changes the overlapping of the W I and W II regions which produces the W III region (see Figures 4.5 and 4.6). At high surfactant concentration the near optimum formulation region is associated with a single phase W IV behavior, a situation which may be used to produce phase inversion such as the PIT method, but implies a high surfactant cost.

Another parameter able to influence the standard inversion line is the stirring energy. The few available studies [72–75] show that an increased stirring energy tends to reduce the domain where formulation dominates, i.e., the A region. In the extreme extrapolated infinite stirring, the mixing is likely to dominate, because the formulation effect should fade as in absence of surfactant, and hence the phase in lesser amount will be dispersed in the phase of higher proportion. This is why the inversion line tends to the center (WOR = 1) when stirring is increased in the triangular and F-WOR diagram. This may be quite important in practice, as for instance in the case of heavy crude oil fuel emulsions which contain 70% or more internal oil phase stabilized by a hydrophilic surfactant [76,77].

The representative point of these normal O/W emulsions are located in the A− region of the map close to the vertical branch of the inversion line. If such emulsions are pumped with a high shear device, e.g., a centrifugal pump, they are found to invert and this is due to the shift of the vertical branch toward the center of the map so that the representative point of the emulsion is now in the B− region.

The initial position of the impeller with respect to the original oil–water interface might result in a shift of the vertical branches of the inversion frontier which is easily related with a local WOR concept discussed elsewhere [78].

The effect of these extra variables (phase viscosity, stirring energy, surfactant concentration) can be rendered in any of the three types of 2D diagrams. First a diagram is experimentally built

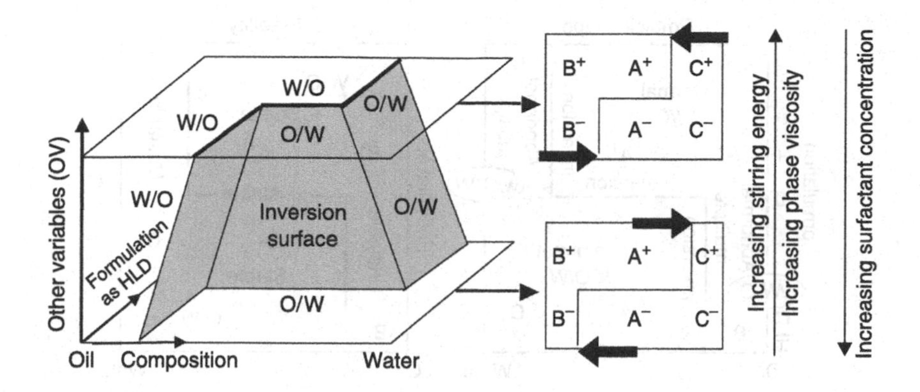

FIGURE 4.7 3D mapping of emulsion type.

at a constant value of the third variable, then it is built again at another constant value, and so forth to produce a stacking of diagrams.

This generally produces a 3D diagram in which the extra variable is represented in the third dimension. This 3D diagram is particularly simple in the F-WOR plot in which the standard inversion line becomes a surface whose shape is illustrated in Figure 4.7.

It is worth remarking that a characteristic of the generalized formulation-WOR 2D map is that the central branch of the standard inversion frontier is unchanged by the variation of any of the other variables (indicated OV in Figure 4.7), and is always located at HLD = 0. As a consequence, the effect of any other variable is essentially to shift the vertical branches of the standard inversion line one way or the other, i.e., to increase or decrease the normal emulsion A range (right diagrams in Figure 4.7). This is quite advantageous because the qualitative effect of all other variables can be rendered by moving upward or downward in Figure 4.7's left graph, whose third axis (OV) accounts for the change, individual or combined, of other variables.

Hence, it can be said that all cases can be qualitatively treated in a 3D map in which the inversion surface exhibits the typical shape indicated in Figure 4.7, left graph, with the three axes representing (1) a generalized formulation variable such as HLD, (2) a WO composition variable, and (3) an "other effects" generalized variable OV. Although it is still an approximation, it is quite an improvement because it allows interpretations and elaborate predictions to be carried out, in particular about the relative position of the system with respect to the inversion. From what is currently known this simplification does not happen with the other diagram types, and this is why it is convenient to discuss dynamic inversion issues on the 2D and 3D diagrams involving HLD-WOR-OV.

4.3.4 Importance of 2D (HLD-WOR) and 3D (HLD-WOR-OV) Maps

The previous F-WOR 2D/3D maps indicating the location of the standard inversion frontier not only delimit the regions where one type of emulsion or the other is attained, but also allow the mapping of other emulsion properties such as stability, viscosity, and drop size. The typical property patterns associated with the different regions are beyond the scope of the present chapter and are thus only schematically indicated in Figure 4.8. More details can be found elsewhere [54,79,80].

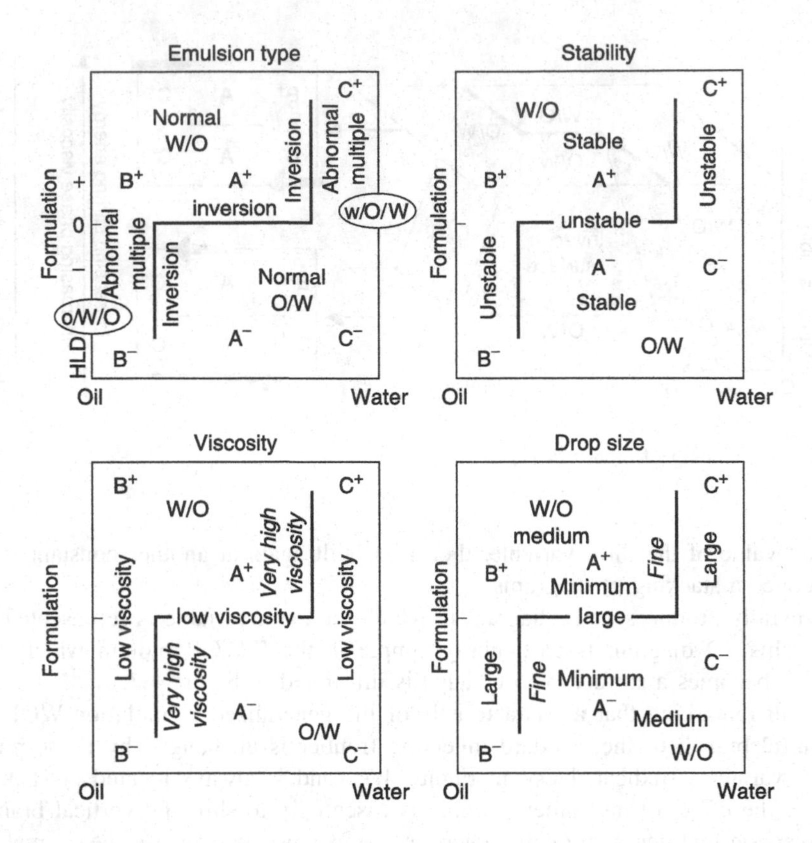

FIGURE 4.8 Emulsion property maps in a 2D formulation-WOR diagram. (Reprinted with permission from J.L. Salager. In: A. Chattopadhay and K.L. Mittal, eds. *Surfactants in Solution*. New York: Marcel Dekker, 1996, pp. 261–295. Copyright (1996) Dekker.)

As far as the following discussion on emulsion inversion is concerned it is worth remarking here a few features exhibited by Figure 4.8 maps, which are determinants to understanding the phenomenology of dynamics. First of all, it should be noted that emulsions are very unstable near optimum formulation [43,81–84].

It was found that in all cases they coalesce very quickly close to HLD $= 0$, and their persistence may be shorter than the timescale of any manipulation or property measurement. Since optimum formulation is also associated with a very low interfacial tension which favors drop breaking, there are two opposite trends: rapid breakage and rapid coalescence which favor quick changes. It has been shown recently [85,86] that the combination of opposite tendencies tends to produce a minimum drop size on both sides of optimum formulation, i.e., at HLD slightly positive and HLD slightly negative, as indicated in Figure 4.8's drop size map.

On the other hand, it is worth noting that the high internal phase emulsions formed in each A region close to the vertical branch exhibit a high relative viscosity and in many cases pseudoplastic or viscoelastic rheology, which are favorable conditions for the formation of fine and monodispersed emulsions [87–89].

The different maps shown in Figure 4.7 confirm that the inversion frontier is of paramount importance because it delimits the regions where specific properties are attained. Hence, an eventual shift or alteration of the inversion frontier would be accompanied by a modification of the property map, and thus can be used as a tool to custom-made emulsions. Up to now, it has been seen that "other" variables such as stirring energy, surfactant concentration, or phase viscosity could alter the position of the standard inversion frontier. In what follows, a new way to set the inversion frontier is discussed. It is associated with the displacement of the representative point of the emulsion in the 2D and 3D maps until the emulsion morphology changes, in a so-called dynamic inversion.

4.4 DYNAMIC INVERSION

4.4.1 Experimental Description of Dynamic Inversion

The standard inversion frontier indicated in Figures 4.6 and 4.7 is the limit between the regions where the HLD-WOR-OV conditions result upon stirring the pre-equilibrated SOW system, in a W/O or O/W emulsion in A+/B+ or A−/C− regions, and eventually in a multiple morphology in C+/B− zones.

In a dynamic experiment one or more of the variables (HLD, WOR, or OV) is altered as time elapses, e.g., temperature is changed or oil is added, so that the representative point of the emulsion moves in the map. At some time the representative point of the emulsion might pass from one side to the other of the standard inversion frontier, and the question is to know whether the emulsion morphology changes, and eventually how. The answer is not simple and depends on many factors. In some cases the morphology change takes place instantly, as soon as the representative point trespasses the frontier, whereas in other cases, the original morphology is retained for some time, eventually for a long time, and it is said that the inversion is delayed, as if the emulsion were keeping the "memory" of its type.

Although dynamic inversion phenomena can be followed in the three types of maps, it has been mostly discussed in the HLD-WOR 2D map, in which the coordinates correspond to the two most important variables, as far as the phenomenon is concerned. In other maps (formulation–surfactant concentration or surfactant concentration-WOR) only one of the variables is able to trigger the dynamic inversion, while the other one (surfactant concentration) produces the formation of the emulsion through another mechanism similar to a nucleation process which will be discussed in the PIT method section.

Figure 4.9 indicates several paths of change in the HLD-WOR 2D map, most of them crossing the standard inversion frontier indicated as a bold dashed line. All these paths correspond to a change in variable value, such as the formulation (e.g., the salinity of the aqueous phase or the

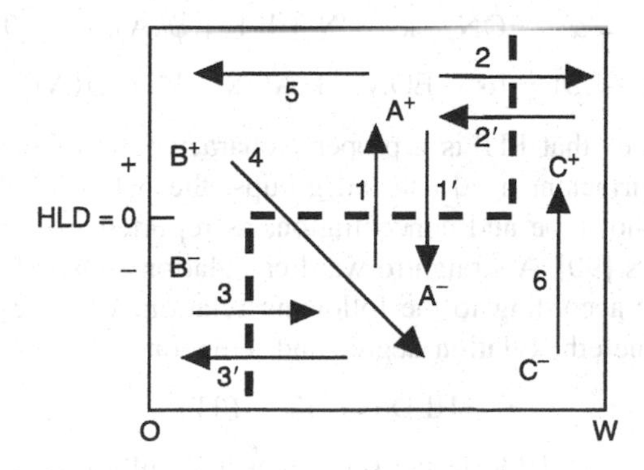

FIGURE 4.9 Paths of change in a formulation-WOR 2D map.

temperature), or the water/oil composition (e.g., by addition of oil or water), while the system is kept under stirring according to a specific procedure, whose characteristics are nevertheless determinant as far as the dynamic inversion is concerned.

For instance, path 1 could correspond to an increase in salinity of the aqueous phase of an anionic system, or an increase in temperature of a polyethoxylated nonionic system, at (essentially) constant WOR and surfactant concentration. On the other hand, path 2 could be representative of the addition of water to a W/O emulsion, whereas path 4 implies two concomitant changes, e.g., adding cold water to a nonionic W/O emulsion.

Paths like 5 and 6 do not cross the standard inversion line, and are not expected to result in an inversion, provided that all other variables are kept constant. However, the emulsion properties might be altered quite a lot to accommodate to the changes indicated in Figure 4.8 maps. For instance, along path 5 the W/O emulsion is "diluted" and its relative viscosity is expected to decrease, and along path 6 a O/W emulsion is likely to become a multiple w/O/W emulsion in which the O/W external emulsion would coalesce quickly.

The 1–1′ (respectively 2–2′ or 3–3′) crossing of the horizontal (respectively vertical) branch of the inversion line would produce, if it occurs, a so-called **transitional** (respectively **catastrophic**) dynamic inversion, whereas a crossing like path 4 would be called a combined process. The labels come from the modeling of the inversion process by a catastrophe theory model, as discussed later.

4.4.2 Transitional Inversion

Transitional inversion corresponds to a change in HLD at all other variables held constant, including WOR and surfactant concentration.

Historically, the first transitional inversion was reported by Shinoda and collaborators [14] in the 1960s when they introduced the concept of phase inversion temperature (PIT), as an improvement over the cloud point temperature and the HLB, and an intent to numerically describe a formulation concept. The phase inversion temperature was initially the temperature at which an emulsion containing a polyethoxylated nonionic surfactant switches from O/W to W/O. Afterwards, it was assimilated into the temperature at which the affinity of the surfactant changes from hydrophilic to lipophilic, an essentially coincident concept which was later relabeled HLB-temperature. The fact is that in most cases of temperature scans, the emulsion inversion takes place at optimum formulation, which is in this case an optimum temperature, and has been labeled T^* in Equation 4.10 when HLD = 0, and is called PIT in Equation 4.11, which is actually the same expression.

$$\text{HLD} = \alpha - \text{EON} - \text{k ACN} + \text{b S} + \phi(\text{A}) - c_T\ (\text{T*25}) \tag{4.10}$$

$$(\text{PIT} - 25) = [\alpha - \text{EON} - \text{k ACN} + \text{bS} + \phi(\text{A})]/c_T \tag{4.11}$$

Equation 4.11 indicates that PIT is a property characteristic of the SOW system. It is seen that it depends on the surfactant head and tail groups, the nature of the oil, the salinity of the aqueous phase, and alcohol type and concentration, as reported from experimental evidence by Shinoda and collaborators [90]. A straightforward calculation shows that the deviation from PIT is an expression of HLD according to the following relation, where c_T is not really a constant, but a weak function of the ethoxylation degree and temperature [22]:

$$\text{HLD} = c_T(\text{T} - \text{PIT}) \tag{4.12}$$

Temperature is a handy variable in the sense that it is fully reversible, and if evaporation is controlled, it can be changed at will at all other variables held constant. However, temperature affects essentially all interactions and it is sometimes difficult to interpret the results in a simple way. Moreover, temperature exhibits a strong formulation effect for polyethoxylated nonionic surfactants, but much less for ionic or polyol-based nonionic ones like glycosides, for which other scanning variables are more helpful.

HLD may be changed by varying any other variable in Equation 4.10, whenever it is practically feasible to do so at constant WOR and surfactant concentration. Since a change in any of the other variables, but temperature, implies the addition or substitution of some substance, care has to be taken not to alter significantly the WOR and surfactant concentration, as well as other variables susceptible to alteration by a dilution process. In practice, high concentration solutions of surfactant or electrolytes are used not to significantly change the phase volumes. Alternatively, the added phases contain surfactant, oil, and water in constant amounts while the nature of some of them is changed, or the total system volume is kept constant by a suction device. It is worth noting that a slight change, say 20 to 30%, in WOR (around unity) or surfactant concentration (around 1 to 2 wt%) is generally not significant, and can be tolerated.

When the HLD is continuously changed along a path such as 1 or 1′ in Figure 4.9 (near WOR = 1) at constant surfactant concentration, and at a slow rate of change, the experience shows that the dynamic inversion takes place essentially at optimum formulation, i.e., at HLD = 0, whatever the variable used to produce the formulation variation. This is probably due to the fact that emulsions are very unstable near HLD = 0, so that any morphology would be destroyed quickly, while the morphology corresponding to the actual condition would be generated at once by the stirring, as if some constant renewal process were taking place. Consequently, a slight variation of formulation (including temperature) able to shift the sign of HLD would result in the inversion as indicated by the variation of emulsion conductivity and viscosity shown in Figure 4.10. The large variation of conductivity indicating inversion takes place at or close to HLD = 0, in the region associated with W III or W IV phase behavior [91]. In some cases an intermediate conductivity in this region could be interpreted as the formation of a dispersed system in which a bicontinuous microemulsion is the continuous phase.

The changes in emulsion viscosity at optimum formulation are associated with the change in morphology, and eventually in the emulsion internal phase ratio. The two viscosity maxima which appear in Figure 4.10 slightly above and slightly below PIT are related with the minimum drop size locations previously mentioned in Figure 4.8.

In some cases, the kinetics of morphology renewal and inversion might be slower than the formulation change, and a multiple emulsion resulting from some residual memory effect could occur just after trespassing HLD = 0. If the direction of change is appropriate, it is easily detected by a conductivity lower than expected for an O/W morphology and a viscosity higher than expected for the given WOR, as shown in Figure 4.10 from 26 to 23 °C [91].

These trends may be considered as quite general for a transitional inversion, whatever the scanned variation variable, with very few exceptions which will be discussed later.

This inversion resulting from a temperature scan has been used as an emulsification technique so-called PIT emulsification method [90,92,93]. This term is, however, misleading because the essence of the method is not the emulsion inversion in itself but the production of the finest drop emulsion at one of the two "best" locations on one side or the other of HLD = 0 (see "minimum" in Figure 4.8 drop size map) and the preservation of this fine drop size by some thermal or physicochemical quench. Such a quench essentially consists of a quick change in temperature or formulation to move the representative point of the emulsion far away from the instability zone, i.e., some distance inside the A+ or A− region, according to path 1 in Figure 4.11. The starting point might also be in the W III region as indicated in path 2 in Figure 4.11.

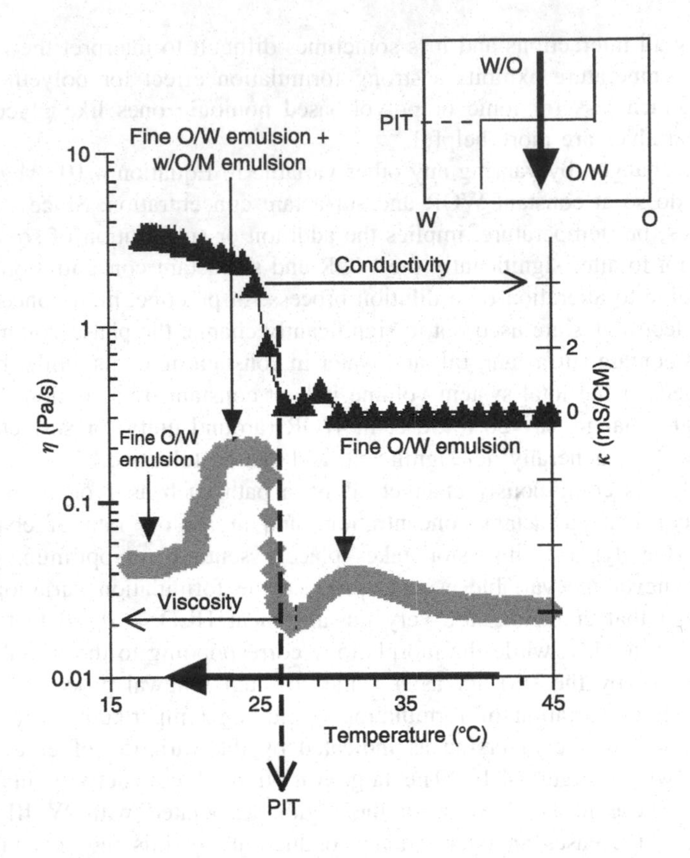

FIGURE 4.10 Changes in conductivity and viscosity along a transitional inversion. (Reprinted with permission from J. Allouche, E. Tyrode, V. Sadtler, L. Choplin, and J.L. Salager. *Langmuir* 20: 2134 (2004). Copyright (2004) American Chemical Society.)

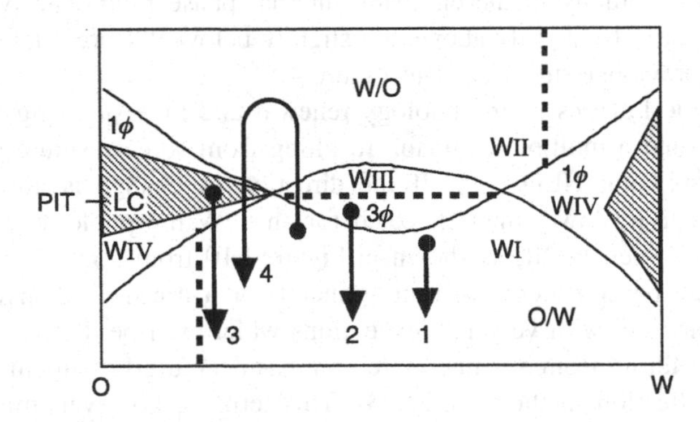

FIGURE 4.11 Path for PIT method of emulsification to produce an O/W morphology.

Another kind of quenching may be accomplished by the formation of a lamellar liquid crystal layer to encapsulate the droplets [94]. In practice, this happens not as an inversion but as a change from a W IV single phase system to a W I or W II one, with some residual structured layer at the drop interface. A single phase behavior at optimum formulation is associated with a high surfactant concentration in the system or in one of the phases, and it is often easier to attain at WOR different from unity, in one of the W IV "triangles" located at HLD = 0 on the O and W sides (shaded in Figure 4.11). Figure 4.11 illustrates the several paths (1, 2, and 3) in which fine emulsions are made according to the PIT method, but without actually exhibiting any inversion. In path 3 the LC structure is likely to encapsulate the droplets formed when the representative point enters the W I region. Path 4, which crosses the inversion zone twice, is often cited [95], but this back and forth displacement is not really needed since the fine emulsion is produced just after the second crossing through the liquid crystal zone. It is worth remarking that the kinetics of changing temperature is of utmost importance [96].

4.4.3 Catastrophic Inversion

When the WOR is changed by addition of a phase under constant stirring, this results either in dilution or in an increase in internal phase content, the latter one being the only case of interest, with two subcases: a change from normal morphology (A+ or A−) to abnormal one (C+ or B−) or vice versa.

In both subcases the increase in internal phase content results in an increase in the number of drops which then tend to get closer and closer to each other until they finally touch with one another.

It is worth noting here that the hexagonal close packing of identical rigid spheres, i.e., the most dense 3D way to stack spheres, accounts for 74% of the space volume. If the spheres are not identical, the packing could reach a higher percentage of the volume, but if the packing is random, as expected for an emulsion, the drops might start being in contact at a lower volume fraction, e.g., 60 to 65%.

If the emulsion is abnormal, the thin film between approaching drops is likely to break at once upon touching and the drops would coalesce quickly to become the continuous phase of the inverted emulsion. This would occur as soon as the drops are close enough for the contact to be highly probable, which could happen as early as 60% of internal phase content. In the previous figures the vertical branches of the standard inversion frontier were located at about 30 and 70%, which are typical values. Hence, the path from abnormal to normal morphology would cross the standard inversion frontier at 30%, whereas the previous discussion suggests that the critical packing for drop contact could be much higher, at about 60%. Consequently the contact occurrence and resulting inversion is likely to be delayed from 30 to 60% in the abnormal to normal morphology case.

Contrary to the previous case, the interdrop thin films in normal emulsions are expected to resist the contact, and the drops are not likely to coalesce instantly upon touching. As a consequence, normal emulsions can attain an internal phase content much higher than the hexagonal packing by deformation of the drops, sometimes so substantial that they result in polyhedral structures called biliquid foams. Therefore, the inversion from normal to abnormal morphology is likely to take place later than the typical 70% of internal phase content which corresponds to the location of the vertical branch of the standard inversion line.

The previous discussion based on fairly reasonable arguments points out that a dynamic inversion process due to an increase in internal phase content is expected to be delayed with respect to the position of the standard inversion line in all both subcases, i.e., in both directions of change. Experiments carried out in a variety of conditions of addition of internal phase show that, indeed, the dynamic inversion is always delayed, i.e., it takes place often well after the standard inversion frontier is crossed. The arrows in Figure 4.12 show the direction of change and the bold line passing at the tip of the black arrows indicates the point where the dynamic

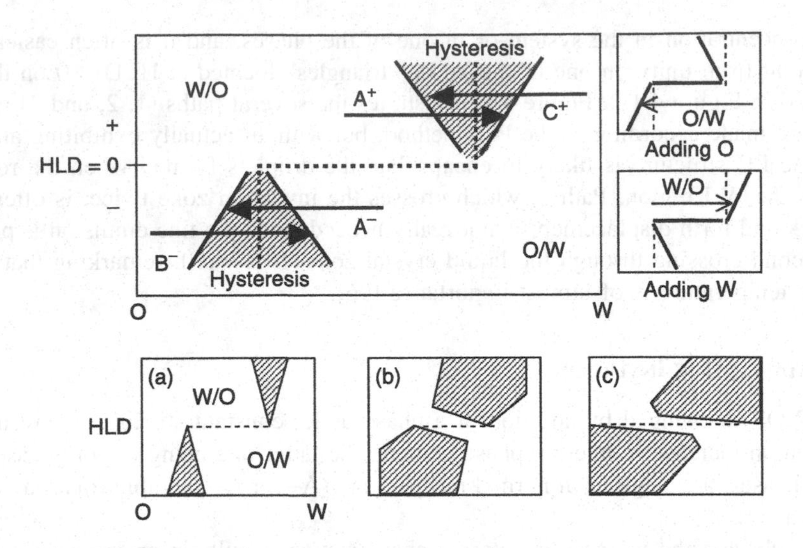

FIGURE 4.12 Catastrophic inversion 2D map (top) and typical cases of hysteresis regions (bottom).

inversion takes place. The region between the two dynamic inversion lines corresponds to one type of emulsion or the other depending on the direction of change. It is called an **ambivalent** or **hysteresis** zone, in which any of the two morphologies can occur, and thus belongs to the normal or abnormal region depending on the case.

This means that the extension of the regions, particularly the A− and A+ normal ones, can be considerably modified by this so-called **memory** phenomenon, as well as the resulting emulsion properties associated with the regions [22].

The shape and extension of hysteresis regions might vary from case to case, and seems particularly dependent on the protocol by which WOR is changed, as discussed later. Figure 4.12 (lower graphs) illustrates different typical hysteresis patterns drawn after experimental data. Very close to HLD = 0 (dashed line in Figure 4.12, lower graphs) the emulsions produced are very unstable and the experimental data from different sources exhibit discrepancies. However, it is probable that the delay vanishes in this region, where any morphology change would take place as in a transitional inversion process, that is reversibly [97].

Since the handling of the experimental procedure could result in hysteresis regions with very different shape and extension, controlling this feature is of primary importance as far as the emulsion ultimate properties are concerned.

The issue can become more complex if the initial abnormal emulsion is, as often the case, a multiple emulsion. In effect the droplets which are inserted in the drops tend to swell them, hence to apparently increase the dispersed phase content, and to reduce the drop–drop distance. Consequently, the interdrop contact in the abnormal outer emulsion is more probable and the inversion will take place earlier, particularly if many droplets become inserted in drops.

As far as the droplets-in-drops inner emulsion is concerned, it is a normal emulsion according to formulation and the droplets are not likely to coalesce upon contact. If the conditions are appropriate this inner emulsion would resist coalescence and hence more and more droplets would be inserted in drops, which end up so swollen with droplets that they could reach the 50 to 60% critical packing condition to coalesce. In such cases inversion could take place after a small amount, say 10%, of the added phase is incorporated, i.e., even before the standard inversion frontier is crossed at about 30%, with a puzzling apparent anticipation instead of delay with respect to the standard inversion frontier.

Things could turn out even more complex because multiple emulsions frequently evolve when they are submitted to continued stirring. In effect, the competition between the inner and outer emulsions is the result of the opposite mechanism: drop breakup and coalescence. Since the inner emulsion is more stable than the outer one, the former tends to increase at the expense of the latter. As a consequence, and just because of the stirring process, and without adding any phase, more and more droplets are incorporated in the drops as time elapses [98]. Hence the volume of the drops tends to increase and at some point the packing reaches the critical value and the inversion takes place.

Catastrophic inversion has been reported in triangular diagrams for very few systems, most of them with n-butyl monoglycol ether, which is not really a surfactant, though its behavior seems to be a fair model [46,47]. On the other hand, fish diagrams have been used to interpret transitional inversion by a change in temperature with ethoxylated nonionics, particularly n-alkyl polyglycol ethers [99]. In both kinds of diagrams the surfactant concentration axis's only important use is to locate the tri-critical point, and no effect of WOR can be followed. Hence, these later diagrams are essentially equivalent to 1D plots as far as the information is concerned.

4.4.4 Basic Modeling and Interpretation

From the experimental evidence discussed in the previous paragraph, it is seen that the characteristics of the transitional and catastrophic inversions are quite different.

4.4.4.1 Mechanism of Transitional Inversion

Transitional inversion is associated with a swap in phase behavior at equilibrium, and if the time scale of change is slow enough, mass transfer takes place during the process and equilibrium is approached at all times. Experimental evidence indicates that close to optimum formulation, surfactant diffusional mass transfer between phases is particularly quick [100,101] and on the other hand a complete transfer is not required for the emulsion morphology to change [102]. Moreover, the presence of alcohol cosurfactant seems to speed up the kinetics effects [103] and the low tension prevailing close to optimum formulation insure the transient formation of small drops, i.e., high interfacial area of transfer, even if stirring is not very energetic. Finally the very quick coalescence rate provides mixing which favors convective transfer.

At optimum formulation the phase behavior is either a single phase microemulsion (W IV) or a three-phase one (W III), whereas it is diphasic (W I or W II) at HLD slightly different from zero. Hence, the crossing of the HLD = 0 region involves a change in the number of phases, whose analysis provides an interpretation of the mechanism of transitional inversion.

Figure 4.13 illustrates the case of a temperature scan with a polyethoxylated nonionic system at WOR = 1 and a surfactant concentration which is about half what is required to attain the single phase behavior at optimum formulation. When the temperature is increased, the phase behavior changes from W I (T < PIT) to W III (T = PIT) and W II (T > PIT) as indicated in the upper scan. If the temperature change is slow enough, the mass transfer, which proceeds quickly in the optimum formulation region, essentially maintains the phases in a pseudo-equilibrium state, and from left to right the microemulsion phase takes up oil and releases water. If the system is stirred, as shown in the lower scan, the dispersed phase is the one containing most of the surfactant, i.e., the microemulsion. At low temperature the emulsion is O/W, and as temperature increases, the external phase microemulsion (gray phase) takes up oil and the oil drop size decreases.

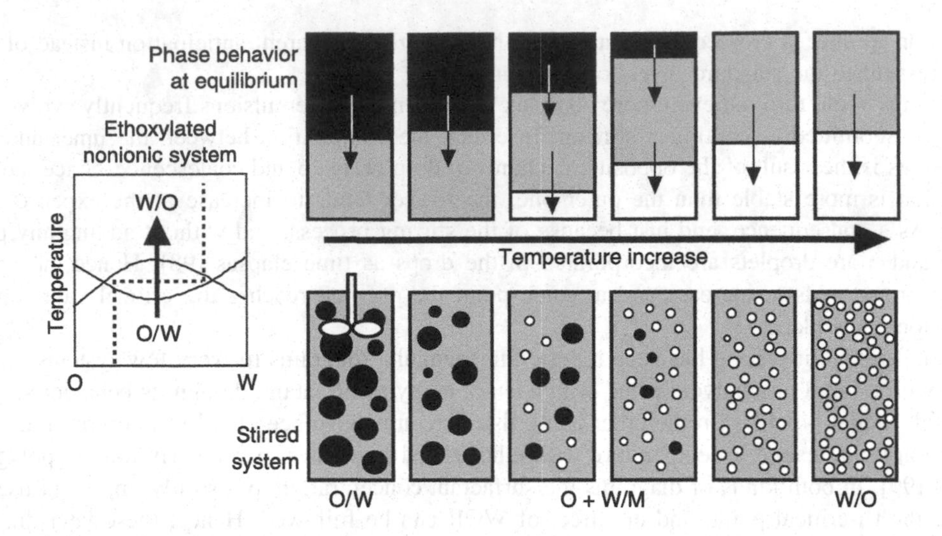

FIGURE 4.13 Phase behavior at equilibrium and emulsion evolution in a transitional inversion process.

At some point close to PIT the microemulsion is in equilibrium with two excess phases, which means that the microemulsion is exuding tiny water droplets by some nucleation process and, consequently, the emulsified system contains oil drops and water droplets. As the temperature increases further, the microemulsion swallows more and more oil from the drops and expels more and more water as droplets, until a temperature higher than PIT is reached at which the microemulsion has incorporated all the oil and the emulsified system is now of the W/O type as tiny water droplets in oil. This mechanism corroborates that the transitional inversion is not really an inversion but a continuous change of the content of the external phase, from water to oil in the Figure 4.13 case. This mechanism is obviously reversible and can take place exactly in the same way in the other direction, provided that the assumptions of quasi-equilibrium are met.

The water droplets are particularly small because they are formed by a nucleation process produced by the splitting of a single phase. However, close to optimum formulation the droplets are likely to coalesce very quickly if something is not done to prevent it. As previously mentioned, there are essentially two practical tricks to keep the fine emulsion. The first one consists of applying a thermal [104] or physicochemical quench, i.e., to quickly change the value of HLD away from zero as soon as the droplets are formed, in the Figure 4.13 case to increase HLD by at least one unit, e.g., by increasing the temperature 20 °C or so, or by adding a small amount of lipophilic surfactant or cosurfactant. The second clever procedure is to use a system in which the phase behavior at optimum formulation includes or consists of a lamellar liquid crystal able to encapsulate the droplets and prevent their coalescence. This second alternative requires a high surfactant concentration, e.g., at least 10 to 15%. At such a concentration or maybe a little higher, the phase behavior at optimum formulation is likely to be monophasic (W IV), eventually with a LC zone in it (path 1 in Figure 4.14). In this case the emulsion comes from a nucleation process which is expected to produce tiny and relatively monodispersed droplets if the quench process is efficient in inhibiting their coalescence.

In contrast, at low surfactant concentration the phase behavior at optimum formulation is W III and no liquid crystal is usually found along the followed path (path 2 in Figure 4.14). The volume of the microemulsion in such a three-phase system is probably small, e.g., 10% or less,

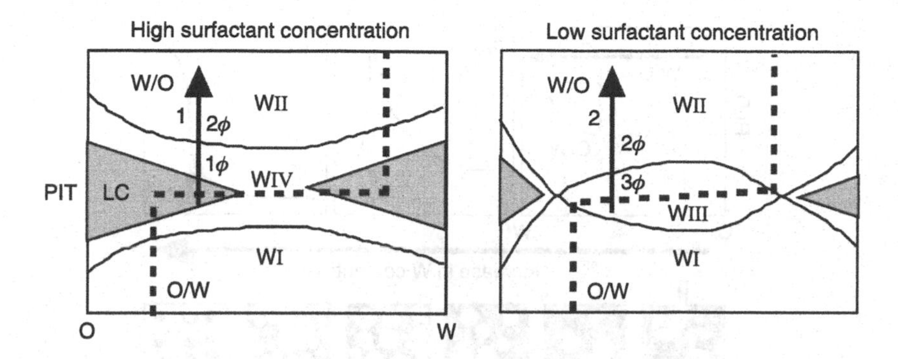

FIGURE 4.14 Transitional inversion at high and low surfactant concentration.

hence the stirring of the system is likely to produce an emulsion whose morphology may be quite complex and sometimes results in some intermediate conductivity changes which are often difficult to interpret. Since the tension is low, any stirring, even at low energy, might be enough to disperse water-in-oil or vice versa. This is another drop formation process, this time due to stirring, and most often associated with the formation of larger drops, because of the inherent instability prevailing in this region. Consequently, at low surfactant concentration the inverted emulsion often exhibits a bimodal drop size distribution, the smaller droplets coming from the nucleation process and the larger ones from the stirring [71,96].

Although this bimodality may be corrected by controlling the stirring and quench processes, an increased surfactant concentration is often unavoidable.

4.4.4.2 Modeling of Catastrophic Inversion

Catastrophic inversion is triggered by the addition of (too much) internal phase to an emulsion submitted to stirring. Main features involve the possibility of two states (O/W and W/O), sudden jump, irreversibility, hysteresis, divergence, all characteristic of the so-called cusp catastrophe [105]. An even better match can be provided by a higher order catastrophe model, so-called butterfly [106], which accounts for the HLD, WOR, and surfactant concentration effects on both the phase behavior at equilibrium (with a Maxwell convention of multistate overall minimum) [107] and emulsion inversion (with a maximum delay convention of local minimum seeking). A full account of these models may be found elsewhere [106], and it is enough to say here that catastrophe theory supplies an elegant nonconventional qualitative model to interpret a complex physical process. The fact that the same potential allows an interpretation of both the phase behavior and the emulsion inversion corroborates that some link exists between them, in spite of the fact that thermodynamics and kinetics are usually independent [108,109].

Trials to reach a quantification of the experimental data with a catastrophe theory model have achieved limited success, particularly through the use of coordinate change [46,110], but they failed to provide a deeper insight on some aspects [111].

Kinetic models on emulsion inversion were proposed a long time ago [41]. They are based on the competitive coalescence rate of the two types of emulsions, the most stable one leading to the prevailing morphology. These models can match the influence of both by the formulation and WOR and lead to a more rational description from the physical point of view [75]. However, they

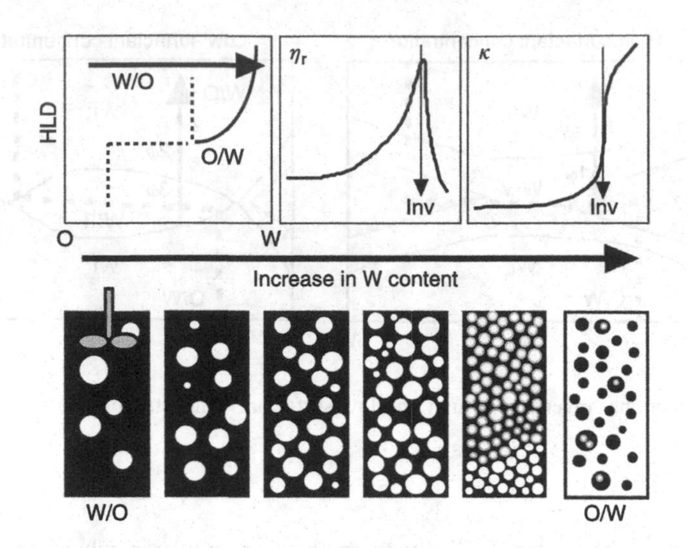

FIGURE 4.15 Catastrophic inversion from normal to abnormal morphology (A+ → C+). Variation of emulsion relative to viscosity, conductivity, and drop distribution.

have failed to provide a good description of some emulsion inversion basic features such as irreversibility and hysteresis.

Consequently, it may be said that the description of emulsion inversion still relies on experimental data, with a large variety of cases depending mostly on the experimental protocol.

Figure 4.15 illustrates the case of catastrophic inversion from normal to abnormal morphology, produced by the addition of water to a W/O emulsion in conditions in which HLD > 0, at constant surfactant concentration. As water is added under continuous stirring it is dispersed into drops, which become more numerous and closer to each other. When the amount of water attains some critical packing, say about 65 to 70%, the drops start touching, but coalescence is not expected to take place upon contact because the location in the map (A+ region) corresponds to stable emulsions (see Figure 4.8). Hence, the internal phase amount continues to increase and at some W content, say 75 to 85% depending on the drop size distribution, drops are likely to become distorted. At some higher W content, which depends on the applied shear and on the performance of the surfactant, the interdrop films would break up and drops would coalesce to produce a continuous W phase, resulting in the inversion of the emulsion into a O/W abnormal morphology of low internal phase content, sometimes with a few residual droplets in drops.

In some cases the resistance of the interdrop film is high enough to hamper their rupture in the conditions of shear, and the normal emulsion does not invert (see Figure 4.12, lower right case). Actually the inversion likelihood is essentially set by the experimental protocol, and the fact that high internal phase content emulsions become viscoelastic and cannot be handled like a fluid any more.

It is worth noting that this extreme situation is not uncommon, as is the case of mayonnaise or cold creams. Moreover a viscoelastic emulsion may be a quite useful medium to increase the slow mixing emulsification performance in a variety of cases ranging from cosmetics and foods to heavy crude oil fuels [76].

Figure 4.16 illustrates the case of catastrophic inversion from abnormal to normal morphology, produced by the addition of water to an oil phase in conditions in which HLD < 0, at constant

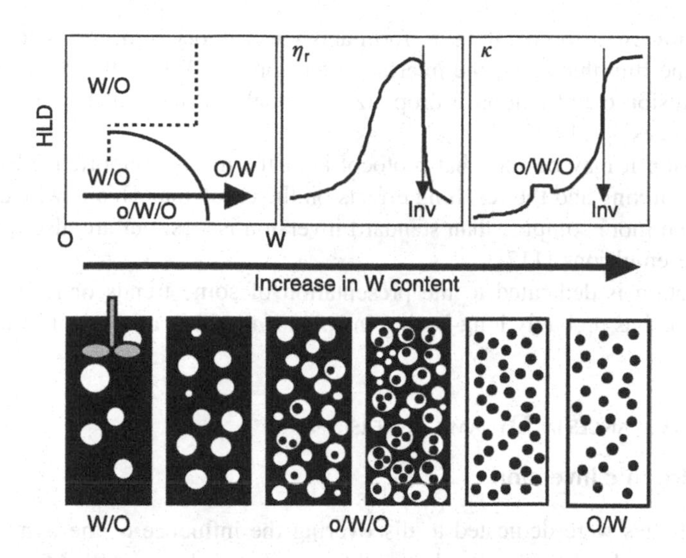

FIGURE 4.16 Catastrophic inversion from abnormal to normal morphology (B− → A−). Variation of emulsion relative to viscosity, conductivity, and drop distribution.

surfactant concentration. When water is added upon stirring of the oil phase, a W/O dispersion results since the water content is very low at first in this B− region. Nevertheless, it is worth noting that the formulation favors an O/W morphology, which means that the W/O type is abnormal and that W drops are prone to coalesce upon contact. Because the stirring device applies some kind of shear on the W drops, these can be elongated or flattened, and the formulation might locally favor the curvature leading to the O/W morphology. This process generally results in the formation of "o" droplets inside the W drops, i.e., the attainment of an o/W/O multiple emulsion, by a mechanism so-called inclusion [98,112–114]. If many "o" oil droplets are included in the W drops internal coalescence can take place [115], or the mechanism opposite to inclusion, so-called escape process, might happen too [116].

It is worth noting that the "o" droplets inside the W drops are not expected to coalesce upon contact since they are the normal type according to the formulation. Hence, the inclusion mechanism is expected to proceed, and eventually result in a high internal phase content o/W inner emulsion. This "stuffing" of W drops with "o" droplets produces their swelling and the apparent (W+o) drop volume increases (much) faster than predicted from the water content, since the volume of "o" droplets might be much larger than the W drops one. As a consequence, the W drops get closer and their contact might take place at a very low water content, e.g., 15% instead of the typical 50 to 60% critical value of the internal phase content at which an abnormal emulsion drops start to coalesce to trigger the inversion.

Since the critical W drop volume can be attained with more or less swollen W drops, the crucial and determining issue is to know what to do to favor or not the swelling of the W drops during the process, i.e., how to favor the formation of a multiple emulsion, a topic which will be discussed in the next section.

When the W drops come into contact and coalesce to form the continuous phase, the small oil droplets are released, and often result in an O/W nanoemulsion, which is quite stable according to the location of the HLD-WOR map. On the other hand the oil films of the continuous O phase in the pre-inversion emulsion break up to form another category of drops, whose size depends on factors such as the film thickness, the interfacial tension, the WOR, the shear etc. Consequently, the inverted emulsion could exhibit a drop size bimodal distribution that accounts for this dual emulsification process [71].

As a conclusion it may be said that protocol is of the utmost importance because it can alter the conditions of mixing and the resulting effects on the emulsion. This is why dynamic inversion concepts are much more complex than standard inversion issues, but are also much more flexible for custom-made emulsions [117].

The next section is dedicated to the presentation of some trends or recent advances which have been identified as probably long-lasting in spite of the little amount of information available to support them.

4.4.5 Recent Advances in Dynamic Inversion

4.4.5.1 Catastrophic Inversion

Several recent studies were dedicated to discovering the influence of the dynamic (catastrophic) inversion protocol under continuous addition of one of the phases [69]. Most experiments were followed by conductivity monitoring in the case of inversion from abnormal O/W or w/O/W morphology (C+) to normal W/O type (A+), but the results seem to be valid as well for the inversion from B− to A− which was followed by rheological monitoring [12,118] since conductivity does not contribute much information.

As the rate of addition of internal phase increases, the dynamic inversion is delayed according to the conductivity variation shown in Figure 4.17 [119], in which the addition of oil to water (A+ ← C+) results in a displacement from right to left, until the inversion from W-external to O-external takes place at 35% O for a low (20 ml/min) addition rate and at 63% O for a high (100 ml/min) addition rate. Figure 4.17 also shows that in the case of a high addition rate, the conductivity follows Bruggeman's law, which means that the abnormal emulsion is a simple O/W morphology. In contrast, in the case of a slow addition rate, the conductivity of the abnormal

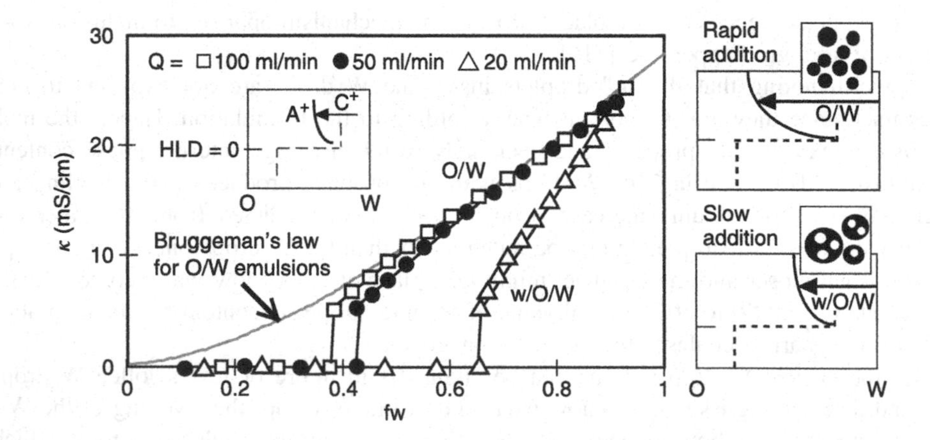

FIGURE 4.17 Effect of addition rate of internal phase on catastrophic inversion from abnormal to normal morphology (A+ → C+). (Reprinted with permission from N. Zambrano, E. Tyrode, I. Mira, L. Marquez, M.P. Rodriguez, and J.L. Salager. *Ind. Eng. Chem. Res.* 42: 50 (2003). Copyright (2003) American Chemical Society.)

emulsion is much lower than expected from Bruggeman's law, indicating that the emulsion is a w/O/W multiple morphology.

The interesting result in Figure 4.17 is that the conductivity at inversion is the same in all cases, i.e., the one which corresponds to 35% of external water, whether the drop phase consists either of pure oil or a w/O emulsion. This 65% of internal (drop) phase volume seems to be the critical packing at which the abnormal to normal inversion is triggered in this case [12,98,120,121]. It is worth remarking that the inversion takes place at a earlier time (after 2.3 min versus 5 min) if the addition rate is quicker (100 ml/min versus 20 ml/min) and essentially no multiple emulsion is produced. However, the inversion times are not in the same proportion as the addition rates, and the added amount of internal phase at inversion is much lower when a multiple emulsion is generated; consequently the resulting inverted emulsion would exhibit a higher internal phase amount and a higher viscosity.

This is particularly important for the emulsification of resins (epoxy, alkyd, silicone viscous polymers) in an O/W emulsion [57,122]. In this case water is added to the viscous oil at HLD < 0 as shown in Figure 4.18. The conductivity measurement is of no use to detect pre-inversion phenomena and only shows the occurrence of multiple or simple emulsion after inversion. Before inversion the emulsion morphology has to be monitored by other means such as its viscosity [118]. Fast addition produces a delayed inversion from W/O to O/W morphology with large drops, whereas a slow addition results in the formation of an intermediate o/W/O emulsion which inverts to a high internal phase content O/W emulsion with fine drops. Drop size was found to depend on the addition rate, because of the different mechanism illustrated in Figure 4.18, as well as stirring energy [57].

In practice the slow addition can be substituted by the addition of a small quantity of internal phase (say 10%) to start making a multiple emulsion followed by the uninterrupted stirring of the emulsion after the addition is stopped [118]. The continuous stirring shear tends to produce both types of dispersion but the favored morphology is the one which corresponds to Bancroft's rule,

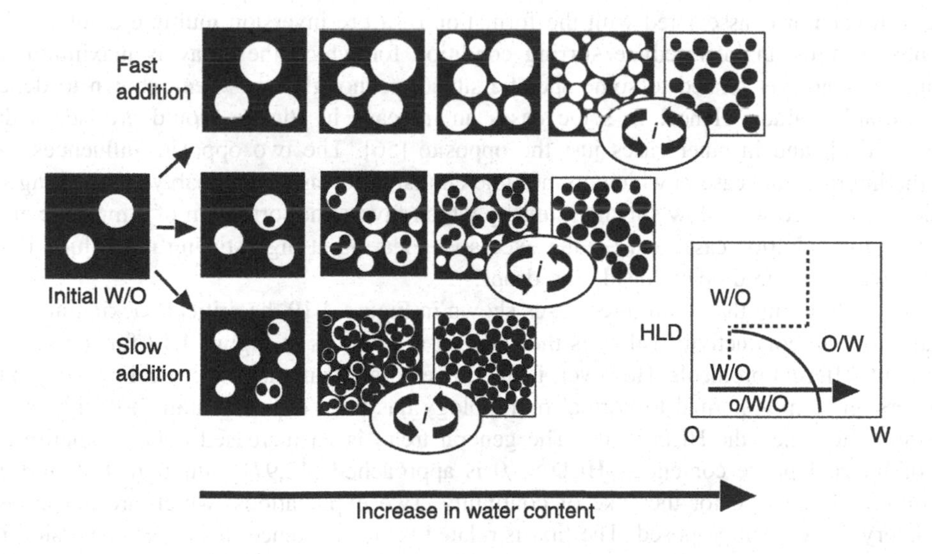

FIGURE 4.18 Influence of addition rate on the catastrophic inversion from abnormal to normal morphology (B− → A−). (Reprinted with permission from E. Tyrode, J. Allouche, L. Choplin, and J.L. Salager. *Ind. Eng. Chem. Res.* 44: 67 (2005). Copyright (2005) American Chemical Society.)

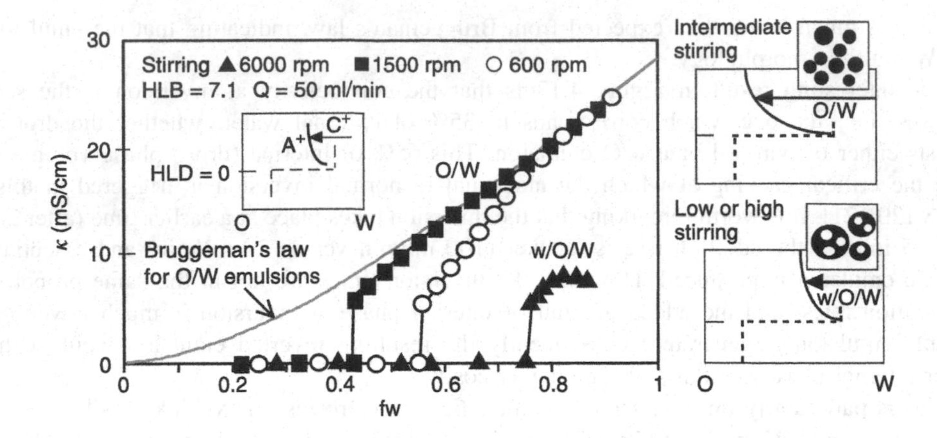

FIGURE 4.19 Effect of stirring energy on catastrophic inversion from abnormal to normal morphology (A+ → C+). (Reprinted with permission from I. Mira, N. Zambrano, E. Tyrode, L. Marquez, A.A. Peña, A. Pizzino, and J.L. Salager. *Ind. Eng. Chem. Res.* 42: 57 (2003). Copyright (2003) American Chemical Society.)

i.e., the inner emulsion. As a consequence, the multiple emulsion evolution under continuous stirring is always toward the inclusion of more droplets into drops until the critical packing is attained. This means that slower addition is equivalent to longer stirring and more inclusion [12,118,123]. It also means that an abnormal emulsion is likely to invert, soon or later, if it is stirred long enough in an appropriate way.

The influence of the stirring intensity is more complex because it deals with two opposite effects [73]. Figure 4.19 indicates that when the stirring intensity increases from slow mixing (600 rpm) to high Couette shear (1500 rpm) the inversion is first delayed, then at a high turbulent regime (6000 rpm) it takes place sooner. As indicated in Figure 4.17 the lower addition of internal phase at inversion is associated with the formation of a pre-inversion multiple emulsion.

There is thus an intermediate stirring condition for which the delay is maximum, but its attainment is not yet related to some specific situation, though it has been shown to depend on other variable values. Hence, in some cases an increase in stirring would produce a delayed inversion [72], and in other cases just the opposite [56]. The two opposite influences, starting from the intermediate case in which no multiple emulsion forms, are probably (1) the long stirring duration associated with slow stirring intensity which favors the formation of a multiple emulsion as in the slow addition case, and (2) the excessive shearing at high stirring rate which is able to break down even the droplets inside the drops.

It is worth noting that in all three cases shown in Figure 4.19 the critical packing at inversion, indicated by the conductivity value, is the same (5 mS/cm) as in Figure 4.17 for the same SOW system but different protocols. However, it should be pointed out that the critical packing to trigger the inversion from abnormal to normal morphology has been experimentally found to vary with the formulation, i.e., the HLD value. The general trend is an increased delay, sometimes up to 90% of internal phase content as HLD = 0 is approached [12,97], although it should end up vanishing at HLD = 0 for the sake of continuity. Two explanations, which are not completely satisfactory, have been proposed. The first is related to the existence of fibrous dispersion instead of spherical drop emulsion, because of the low interfacial tension and permanent renewal of the dispersion by a breakup and coalescence of fast equilibrium [12]. The second is the existence of only a transitional inversion mechanism close to optimum formulation [94]. It should be noted that in practice it is very difficult to carry out foolproof experiments on emulsions close to HLD = 0 because of their extreme instability [113], hence this actual mechanism might not be settled soon.

Another issue which has been addressed is the scale of the WOR variation in a stirred system, i.e., the micromixing, and the introduction of the local WOR instead of the global one. In a system containing equal amounts of oil and water, the position and fluid mechanics action of the stirrer has been found to influence the emulsion type, particularly if the convection pattern produces a thrust of one of the phases into the other, thus resulting in a local WOR value quite different from the overall one. Hence the apparent (global WOR) position of the catastrophic branches of the inversion line is shifted right or left depending on the difference between the global and local WOR. This shift can be harnessed to produce a multiple emulsion with a dual stirring process [78].

This notion of local WOR is probably linked to the very onset of emulsion inversion, since the position of the catastrophic inversion in both directions has been found to be dependent on the way the internal phase is added. Reported inversion studies used a variety of addition protocol from continuous flow to lump-wise aliquot addition. A recent study [124] has pointed out that incipient inversion might require, particularly in the normal to abnormal direction, a local excess of added phase of some extension and persistence. This can be attained either by adding a large aliquot or using a fast continuous addition rate at high stirring or a smaller aliquot or a slower continuous addition rate at lower stirring, so that the heterogeneity does not dissipate so quickly.

Nevertheless, it is yet to be demonstrated that a spatial and temporal WOR heterogeneity could be the key starting point of catastrophic inversion.

As a general rule of thumb, it can be said that in the case of inversion from abnormal to normal morphology, anything that promotes the development or persistence of a spatial or temporal WOR heterogeneity tends to (1) inhibit the formation of multiple emulsion, (2) tends to delay the inversion in the sense that a higher amount of added internal phase is required, (3) but tends to shorten the time scale at which inversion is likely to take place. Consequently, a slow addition would result in the formation of multiple emulsion before the inversion, which would take place after only a small amount is added. In contrast, in the normal to abnormal direction a slow addition is not expected to result in a multiple emulsion but to produce a simple normal morphology, but inversion is delayed, sometimes to an extremely high internal phase content, and in some extreme cases it never takes place (Figure 4.12, lower right).

4.4.5.2 Transitional Inversion

As was discussed in the previous sections, transitional inversion takes place when the generalized formulation is close to optimum, i.e., when the simultaneous low interfacial tension and high instability result in a pseudo-equilibrium between drop breakdown and coalescence. However, a high rate of formulation change or a reduced equilibration rate might defeat these assumptions [125].

PIT experiments carried out at different rates of temperature change were recently reported [126] to show a delay of transitional inversion in both cooling and heating directions (see Figure 4.20, left). The rate of change was also reported to alter the delay, which was noteworthy, say more than 20 °C, in some cases. The system contained a mixture of surfactants, among them high molecular weight ethoxylated sorbitan trioleate, which is not likely to diffuse rapidly. The addition of sec-butanol cosurfactant was found to eliminate the delay (Figure 4.20, right) whatever the rate of change, a result which is consistent with its role of speeding up equilibration processes, as shown in other kinds of experiments [100,103].

The conductivity record along a transitional inversion path often exhibits irregularities, particularly excessive conductivity values which are not due to random errors of measurement.

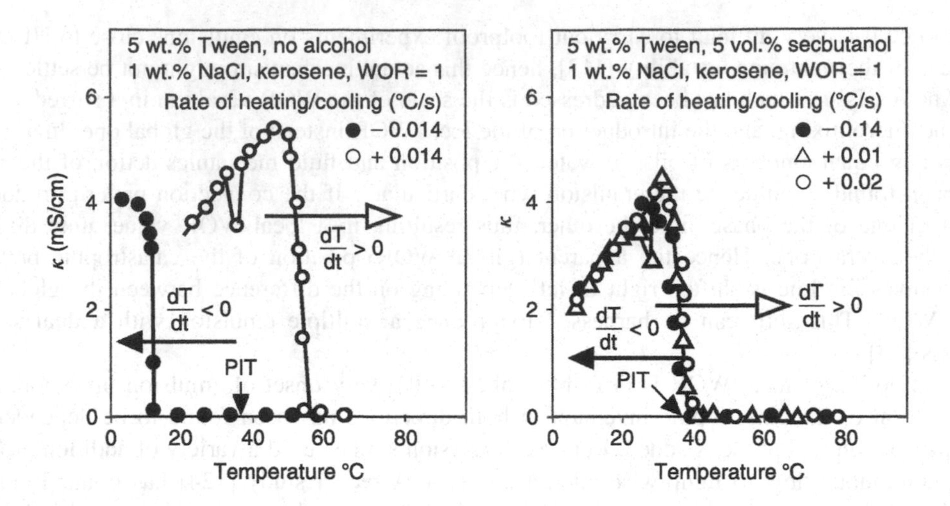

FIGURE 4.20 Transitional inversion with and without delay (Reprinted with permission from L. Marquez, A. Graciaa, J. Lachaise, J.L. Salager, and N. Zambrano. *Polym. Int.* 52: 590 (2003). Copyright (2003) John Wiley and Sons.)

Such conductivity bumps, as seen in Figure 4.10 at 29 °C, takes place at surfactant concentration above 10%, and are likely to be associated with the presence of lamellar liquid crystals which confer a significant, though still low, conductivity [51,127–129]. Because of the fast dynamic balance between breakup and coalescence, any alteration of one of these opposed effects might result in a large alteration of the drop size. This is the case of the presence of a liquid crystal at the drop interface at the moment the drop is generated either by shear or by nucleation. Because of the proximity of HLD = 0, the liquid crystal likely to exist is of the zero curvature lamellar type, which takes time to be formed or to dissolve. Hence the liquid crystal should be pre-existent to the formation of the drop. If the drops are formed by stirring, the LC might be in either phase, but since the formulation (Bancroft rule) favors the surfactant-rich phase as the external phase, the LC should be in it. When the drops come from a nucleation due to a change in formulation, the W IV region containing LC is somehow crossed during the process. Obviously, the time scale of the LC region crossing is of utmost importance for the formation and layering of LC at the drop interface [94]. Slight variations in the starting point and in the formulation or temperature protocol of change result in a considerable change in drop size [35,96] because of their effect on the kinetics of formation of the protective layer around the drops. On the other hand the stirring seems not to have any effect on drop size [57].

4.4.5.3 Inversion Produced by Combined Formulation-WOR Variation

Transition and catastrophic mechanisms are not always easy to separate in real cases. In some instances both formulation or temperature and WOR are changed simultaneously in order to cross the inversion line in some slanted way (path 4 in Figure 4.9). This is very easy to realize in practice, e.g., by adding a phase with a composition or temperature which alters the formulation. Depending on the exact position of the crossing of the inversion frontier, it is the transitional or catastrophic branch which is mainly crossed, with the intermediate case of crossing just at the corner where transitional and catastrophic regimes overlap. Experimental evidence [130] indicates

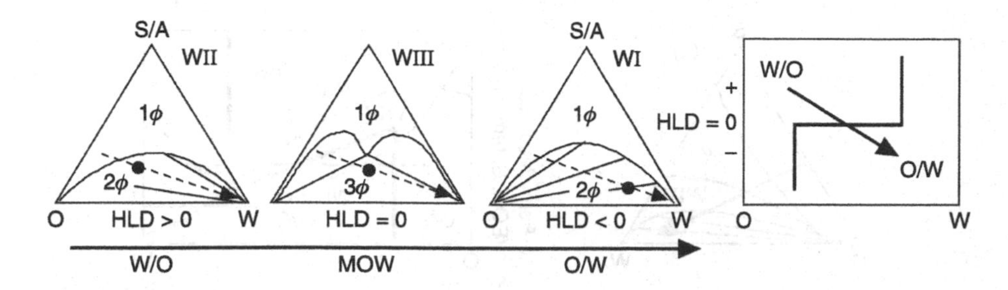

FIGURE 4.21 Transitional inversion produced by adding water in the case in which it results in a formulation change.

that very slight variations in such corner paths produce huge variations in the resulting drop size after inversion.

In most cases the mixed concomitant variations of formulation and WOR, and sometimes surfactant/cosurfactant concentration, is not voluntary, but comes from the protocol. For instance, the addition of one phase can result in different effects which are likely to change the formulation. The first one is a dilution effect, e.g., adding water to an electrolyte or alcohol solution, which would result in a formulation effect through the S and f(A) term in HLD expression. In practice it means that the composition change results in a simultaneous formulation change, or in other words a diagram change as illustrated in Figure 4.21 for the case of dilution with pure water of a system containing an aqueous brine and some slightly lipophilic alcohol. As water is added, the path of the representative point in the ternary diagram heads to the W vertex along the dashed arrow. However, the salinity decrease and the resulting dilution of alcohol both tends to decrease the HLD value, which results in change of ternary diagram type WII → V III or IV → WI transition accompanied by the W/O → O/W emulsion inversion.

In a HLD-WOR diagram, the change in both the water content and formulation would result in a slanted path indicated in Figure 4.21 (right map), with the same transitional inversion result, but with a much more straightforward interpretation.

In most cases the phase behavior SOW ternary diagram is more complex than any of the classical Winsor types, and exhibits liquid crystal region (Figure 4.22(a)) [131–138] or/and complex variation due to the selective partitioning of different amphiphiles, e.g., surfactant mixtures [139] or surfactant and alcohol [140], as in Figure 4.22(b), in which case the increase of surfactant + alcohol tends to increase HLD [141].

Such a selective partitioning [139] results in the slanting of the three-phase band and thus of the optimum formulation line in a F-WOR diagram (Figure 4.22(c)) [52–54,59,142]. In the case illustrated, in which the formulation variable could be, for instance, the surfactant EON, the salinity of the aqueous phase or the temperature, a dilution with water along the arrow generates a path that crosses the inversion line three times: first a catastrophic inversion from abnormal to normal morphology, then a transitional one from normal to normal, and finally another catastrophic inversion, but this time from normal to abnormal. Provided that the crossing of the "vertical" branches could be delayed, additional twisting of the frontier could result, with an apparently whimsical morphology sequence if the proper diagram is not available to interpret the data [143].

It is worth noting that if the formulation is expressed as HLD (at any value of WOR) the optimum formulation becomes the horizontal line at HLD = 0, but now it is the path which is

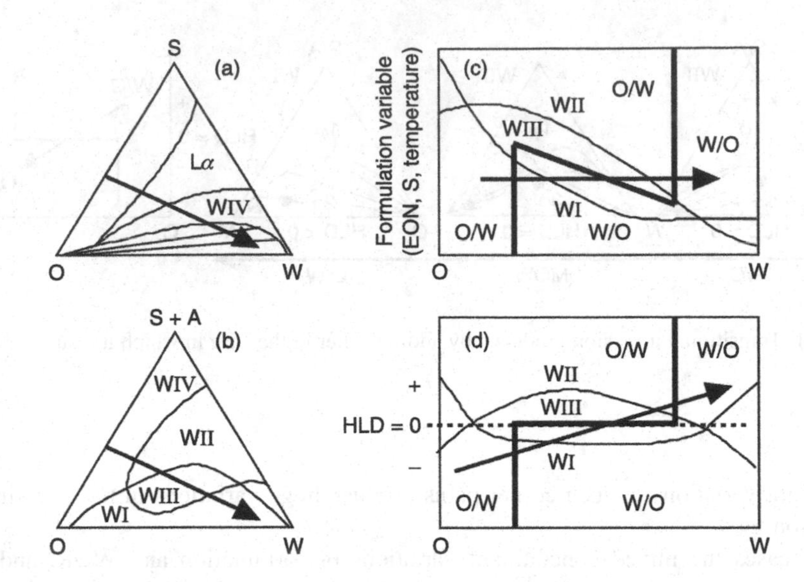

FIGURE 4.22 Complex inversion processes produced by adding water.

slanted, because the change in WOR changes the formulation, i.e., the HLD value, as illustrated in Figure 4.22(d). As a consequence the crossing features are essentially the same.

In real complex phase behavior cases such as in Figure 4.22(a) and (b), the path of dilution by water (arrow) could cross liquid crystal [133,144] or microemulsion regions or both and could result in emulsification by phase inversion or more complex mechanisms such as spontaneous emulsification [145–147], if the rate of change is rapid enough for non-equilibrium transfer processes to take place. Several such processes have been reported and are used in practice [142,148,149].

4.4.5.4 Inversion Produced by Other Variables in 3D Representation

Protocol variables such as internal phase addition rate or stirring energy were found to alter the location of the dynamic inversion produced by a WOR change, which results in the shifting of the catastrophic branches of the inversion line in a 2D F-WOR map.

This shift may probably be represented as well in a 3D map if the change is now along the direction of the third variable, i.e., vertically in Figure 4.23 plot. However, it seems that the only case to have been reported of such an inversion is the effect of increased stirring on Orimulsion®, a fuel substitute marketed by Venezuela National Oil Company. This product is an emulsion of heavy crude oil-in-water emulsion containing 70% of extremely viscous hydrocarbon dispersed as 10 to 20 μm drops, which has to be transported by pipeline for over 300 km from the emulsification plant located in the oil field to the tanker embarking terminal. Pumping by high shear devices such as centrifugal pumps has been reported to produce an increase in drop size and to end up in emulsion inversion [150]. Since the emulsion is located in the A– region very close to the "vertical" branch of the inversion line, an increase in stirring reported to shrink the A– region and the hysteresis zone as well [72] would pass the F-WOR representative point of the emulsion on the other side of the frontier, i.e., in the B– region. According to Figure 4.23 notation, the O/W emulsion at point D (at low stirring) inverts into a W/O emulsion at point E

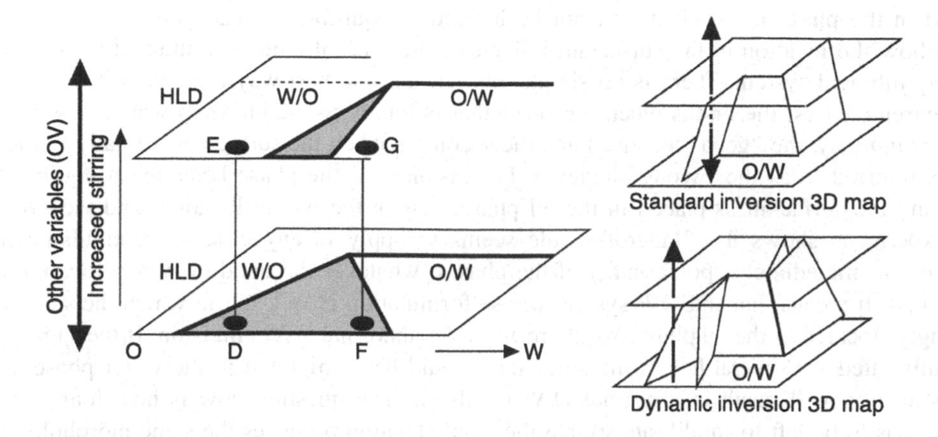

FIGURE 4.23 An increase in stirring energy may or may not produce emulsion inversion. (Reprinted with permission from J.L. Salager, L. Marquez, A. Peña, M. Rondon, F. Silva, and E. Tyrode. *Ind. Eng. Chem. Res.* 39: 2665 (2000). Copyright (2000) American Chemical Society.)

at high stirring. The hysteresis of the change has not been reported in detail but it is known from experience that the reverse inversion from E to D does not happen in practice, which is a confirmation of the existence of a memory effect, as illustrated in Figure 4.23, lower right graph. Consequently, such high internal phase content emulsions have to be pumped with a low shear device at a low Reynolds number.

Figure 4.23 allows a discussion of the apparently capricious effect of any third variable, such as an increase in stirring energy along an upward arrow. D and F are located in the hysteresis region at low stirring and may represent either a W/O or an O/W emulsion depending on the path used in the first place to reach this location. If the original normal O/W emulsion representative point is at F and the stirring energy is increased to G, there is no change in morphology. With a higher internal phase ratio of normal O/W emulsion at D, the same increase in stirring energy triggers the inversion to point E where an abnormal W/O emulsion is produced. If an abnormal W/O emulsion at D is submitted to increased stirring to reach E, it does not change morphology. In contrast, an abnormal W/O emulsion located at F would invert by increasing stirring. The moral of the tale is that an increase in stirring can produce any of the four possible situations depending on the circumstances. As a matter of fact, such variety of alternatives would be quite difficult to explain without the 3D representation of Figure 4.23 at hand.

An increase in surfactant concentration has been found to exhibit exactly the opposite effect in the 2D F-WOR map, i.e., it increases the extension of the A region and it extends the hysteresis zone [70,112]. However, it is not known if the addition of surfactant could trigger a dynamic inversion as an increase in stirring does.

4.4.5.5 Emulsification and Inversion of Non-Equilibrated Systems

In most of the previous cases, at least those dealt with in the three types of 2D diagrams, the SOW system is at equilibrium prior to the onset of the emulsification process. In such cases and provided that the WOR is not very different from unity, Bancroft's rule indicates that the external phase of the emulsion is the one which contains most of the surfactant at equilibrium.

In industrial practice, the SOW system is not necessarily at equilibrium and a phase contact might be needed for the surfactant to migrate from one phase to the other if it is originally placed in the phase in which it will not be located at equilibrium. The practical question is to know how the location of the surfactant influences the type of emulsion attained upon stirring a non-equilibrated system. There is no simple answer, and the best way to discuss it is to separate two extreme cases: the first is when the surfactant is totally in the phase in which it would be at equilibrium (say, the "good" location) and the second is when the surfactant is totally in the other phase, referred to as the "wrong" location. For instance, if the phase behavior at equilibrium is W I, and the surfactant is placed in the oil phase, it is in the wrong location, and vice versa.

Experience shows that Bancroft's rule seems to apply in any case when emulsification is carried out immediately upon contact of the phases, whatever the good or wrong location of the surfactant. It means that an SOW system whose formulation is W I, but in which the surfactant is (wrongly) located in the oil phase, would result in an abnormal W/O emulsion. If the same system is equilibrated for several hours, the surfactant would have migrated in the water phase and the emulsification will result in a normal O/W emulsion. The question now is how long the SOW system has to be left to equilibrate so that the emulsification produces the same morphology as in a completely equilibrated system, i.e., the normal emulsion. The minimum time of contact for this normal emulsion to be produced has been called the apparent equilibration time (t_{APE}). Initially, it was thought that t_{APE} was essentially linked to the time required for most of the surfactant to transfer to the other phase in the contacting conditions, e.g., several hours or days for a SOW at rest. The experimental evidence shows that it was not so and that in some cases only a very small amount of the surfactant had been transferred when the normal emulsion is produced from a quite non-equilibrated system [102].

In Figure 4.24 the apparent equilibration time (t_{APE}) is plotted as a function of the formulation (salinity) and it is found that the departure from optimum formulation (i.e., HLD = 0) is directly associated with the apparent equilibration. Figure 4.24 shows that t_{APE}, which represents the contact time for a change in morphology to take place, is zero for systems close to HLD = 0, which means that at near optimum formulation the emulsification takes place as if the SOW system

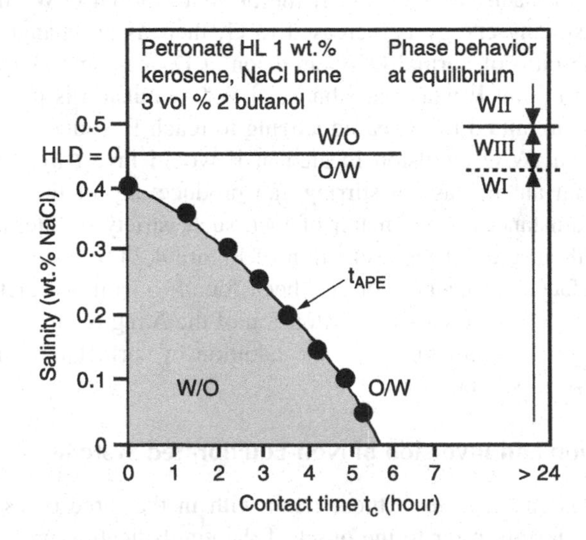

FIGURE 4.24 Apparent equilibration time (t_{APE}) as the minimum contact time to attain the normal O/W morphology, when the surfactant initial location is in the O phase, as a function of formulation (salinity). (Reprinted with permission from J.L. Salager, N. Moreno, R.E. Antón, and S. Marfisi. *Langmuir* 18: 607 (2002). Copyright (2002) American Chemical Society.)

were at equilibrium, even if only a very small amount of the surfactant has had the time to transfer to the other phase. On the contrary, in systems far from HLD = 0, e.g., at S = 0.1 wt% NaCl, more than 5 h of phase pre-equilibration time is required to produce a normal emulsion. Recent research has shown that the presence of low MW co-surfactant such as sec-butanol considerably reduces the pre-equilibration time, although it is still a function of the HLD [103].

This apparent pre-equilibration time is the only concept which has been proposed to handle the out-of-equilibrium cases, and further research is needed on these phenomena because many practical cases of great interest, such as crude oil dehydration or resin emulsification with ionomers, are intrinsically out-of-equilibrium processes.

4.4.6 Emulsion Inversion in Practice

In spite of being an instability phenomenon that is intrinsically difficult to harness, emulsion inversion has been used as the basis of different processes, particularly to make fine emulsions [117,143]. Cosmetic nanoemulsions of the O/W type are usually made starting with an oil phase and adding a water phase containing an hydrophilic surfactant. The original emulsion is of the W/O type in B− region, and it slowly evolves into an o/W/O emulsion (path 1 in Figure 4.25) and finally an O/W normal nanoemulsion (path 1′). In most instances, a small amount of water (10 to 20%) is slowly added first (path 2) and then slow motion stirring is continued for several hours without adding water (location 3) and an o/W/O emulsion forms, so that the water content at inversion is often less than the value expected from the standard inversion (dashed line). The process is long but does not consume much energy.

A similar inversion process (2–3) is used for making waterborne epoxy resin [151,152], polydimethyl siloxane [153], or alkyd emulsions [122,154]. This time the oil phase is extremely viscous and practically impossible to disperse as a nanoemulsion by brute force. The HLD < 0 formulation ensures that in the multiple o/W/O emulsion the stable one is the inner one, and the high viscosity external oil phase is particularly efficient in transferring the slow shear from the mobile device. The multiple emulsion becomes extremely viscous, actually viscoelastic, as more "o" droplets are included in drops, particularly if the W amount is low [57].

When inversion takes place, a very high internal phase content O/W fine emulsion is produced, which is still quite viscous in spite of being water external. Further slow mixing of such emulsion results in the formation of a monodispersed nanoemulsion according to a recently elucidated mechanism [89]. The experience shows that the surfactant concentration plays an important role in particular if a LC can be formed in the W phase [152]. Since the water phase might be in a very small amount, a 10% surfactant in water might represent only a 1% surfactant in the system, indeed a reasonable concentration.

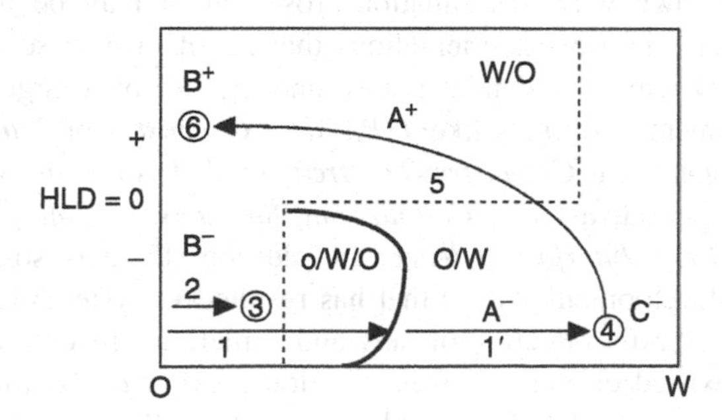

FIGURE 4.25 Paths in the F-WOR map illustrating complex inversion processes in industrial applications.

In some instances the oil phase is a polymer that contains ionomer or reactive molecules, which are able to activate as a surfactant in the presence of an alkaline water phase [154]. It is also the case of acidic crude oil in contact with alkaline brines. In most cases the *in situ* formation of the surfactant molecule at interface is accompanied by some spontaneous emulsification. Nevertheless, it can be considered that the process is essentially a B− → A− inversion in which the hydrophilicity of the surfactant at interface depends on the alkalinity of the water phase, i.e., on the percentage of potentially hydrophilic groups which are activated.

Emulsion inversion can take place as a part of the end use of an emulsion, as in so-called cutting or laminating oils. The commercial products are in most cases sold as oil phase concentrates which self-emulsify at the contact of water to produce a diluted fine O/W emulsion along path $1{+}1'$. This whitish emulsion is sprayed, in order to cool and lubricate, where there is a metal–metal high speed or high pressure contact. Hence, the original O/W emulsion is generally at HLD < 0 in the C− region (location 4). Because it contains an ethoxylated nonionic system which is temperature sensitive, the contact with the hot metal tends to increase HLD and to produce the evaporation of most of the water, so that the location of the representative point of the emulsion moves rapidly along path 5 to end up almost instantly at location 6 in the B+ region with a W/O morphology. The now continuous oil phase is able to wet the solid surface and to provide lubrication at the friction point, whereas the high latent heat of the remaining water can still contribute to the heat-removal process.

4.5 APPENDIX: HOW TO RETRIEVE INFORMATION

The technical literature on emulsion phase inversion is not very extensive, but it is not easy for the non-specialist reader to extract from it a good grasp of the state of the art, and there are several reasons for this. Firstly, there is a logical but long and tortuous sequence, which must be followed as a guideline. It should begin with a good understanding of the phase behavior of SOW systems, and not only the simple cases close to Winsor diagrams. Secondly, it is necessary to understand how phase behavior is related to formulation in different maps, then how emulsion type is related to formulation, composition, and other variables. Finally, knowledge of how to handle more complex dynamic processes with memory features is required.

A second problem is that the few research groups which have dealt with these issues have developed their own methodology independently of each other. Consequently, the reader is confronted with different ways of approaching and presenting the same phenomena, with several notations and terminologies, dealing with completely different applications and often quite different systems, so that the integration and generalization of the available information is not a straightforward matter.

It is not actually known why this situation arose, but it may be related to very different applications and different sponsoring orientations that result in diverse systems, different goals and interests particularly with confidential issues, and the use of a large assortment of divulgation media from fundamental journals like *J. Physical Chemistry* or *Langmuir*, to applied ones like *Industrial and Engineering Chemistry Research*, or *J. Dispersion Science and Technology*, or other wide ranging periodicals like *Colloids and Surfaces*, *J. Colloid and Interface Science*, or *Advances in Colloid and Interface Science*. In addition, there is some kind of competition between research and development groups that has resulted in systematic self-citations and little cross-references, and in the introduction of new and sometimes incompatible terminology when a proper one already existed elsewhere. This isolation prevented a unified approach to develop spontaneously, and some very old and outdated concepts like HLB are still used when generalized new ones have been available for decades.

This section gives an overview of the development of the current knowledge and mentions the names of the main researchers who contributed to the field. Current computer search by author name is probably the best way to quickly get a fair listing of relevant publications, and is certainly much better than using a keyword search which is likely to confront different difficulties such as variations in terminology and excessive use of some terms in other fields of research. Another way is to start with the few review papers which were written with a pedagogical approach [60,69,106,117,155–157].

Phase behavior of surfactant–oil–water systems was pioneered by P. Winsor 50 years ago. In the 1970s the enhanced oil recovery (EOR) research drive financed a great deal of research and development (R&D) toward the attainment of ultra-low tension, as a condition to displace crude oil trapped in the pores of an oil reservoir. Optimum formulation for EOR was found to coincide with the attainment of three-phase behavior of SOW systems, in which case Winsor's R ratio of interactions of the surfactant for the oil and water phase is unity. Extensive research effort focused on anionic surfactant systems in university and industrial R&D groups mostly in the U.S.A. and Europe, and can be retrieved with the following names: R. S. Schechter, W. H. Wade, D. O. Shah, D. T. Wasan, L. E. Scriven, C. A. Miller, R. L. Reed, R. N. Healy, G. Hirasaki, J. E. Vinatieri, S. Friberg, H. T. Davis, D. Langevin, M. Bourrel, P. Doe, and A. M. Bellocq. Complementary studies on polyethoxylated nonionics indicated that the phase inversion temperature (PIT) concept introduced in the mid-1960s by K. Shinoda and collaborators (H. Arai, H. Saito, and H. Kunieda) was a particular case of the multivariate correlations for optimum formulation presented in the late 1970s by R. S. Schechter, W. H. Wade, and collaborators (J. L. Salager, M. Bourrel, and M. Hayes), and transformed into a generalized formulation concept in the next decade (surfactant affinity difference (SAD)) by J. L. Salager, or as a diagram correlation by J. C. Ravey, or other similar equations by H. Kunieda, T. Förster, and A. Wadle. A source of difficulty was, with the exceptions of Winsor SOW ternary which was used by most people, that different groups used dissimilar representations: formulation–surfactant concentration or "gamma" plot (M. Bourrel and C. Chambu), and its equivalent (with 90° rotation) "fish" diagram, mostly used by people working with ethoxylated commercial or pure nonionics and a variation of temperature (K. Shinoda, H. Kunieda, M. Kalhweit, and R. Strey), bidimensional optimum formulation plots (R. S. Schechter, W. H. Wade, J. L. Salager, and M. Bourrel), formulation-WOR map (J. L. Salager) or its equivalent temperature-WOR map (K. Shinoda).

Some groups worked with commercial surfactants of assorted types with typical applications in mind (A. M. Bellocq), often with alcohol to avoid the formation of mesophases; others voluntarily studied the behavior of surfactant mixtures (A. Graciaa, J. Lachaise, J. L. Salager, and R. E. Anton), while others focused on isomerically pure surfactant series such as the CiEj alcohol ethoxylate series to work on well defined as-simple-as-possible systems (P. Ekwall, M. Kalhweit, R. Strey, and H. Kunieda); others focused their studies on a single amphiphile like dihexyl sulfosuccinate (D. A. Sabbatini and E. Acosta) or butyl monoglycol ether (D. H. Smith and K. H. Lim). Oil phases were mostly alkanes and alkylbenzenes, and sometimes polar oil like triglycerides (M. Miñana), fatty esters (T. Förster), or chlorinated hydrocarbons (D. A. Sabatini and E. Acosta).

As far as emulsion inversion is concerned, its treatment is somehow influenced by the phase behavior approach, and the type of system and variables. A variety of ionic and nonionic systems with oils ranging from pure alkanes, to petroleum cuts, edible, and other polar oils (more or less viscous), were studied in formulation-WOR maps by J. L. Salager, R. E. Anton, and M. Miñana, with a variety of formulation variables: surfactant HLB in series or mixture, salinity of aqueous phase for ionic systems, ethoxylation degree for nonionic, alcohol effect, temperature, even pH for fatty acid or amine, and referred to generalized formulation SAD and generalized property maps.

A system containing a short amphiphile (butyl monoglycol ether C4E1), i.e., a solvent with a surfactant-like behavior, was extensively studied by D. H. Smith and K. H. Lim, mostly with ternary and fish diagram representations in which the effect of WOR is not easy to see. Another group (B. W. Brooks, H. N. Richmond, and S. Sajjadi) dealt with polymer applications (polyisobutene, polyurethane ionomer) with a variety of surfactants, also in a formulation/temperature-WOR map.

Transitional inversion has most often been studied by scanning temperature, according to the PIT method proposed by K. Shinoda, and refined by people (T. Förster, W. von Rybinski, C. Solans, P. Izquierdo, J. Esquena) who were involved in the preparation of nanoemulsions.

A special type of transitional process that is often associated with inversion is a half-transition from WIV to WI or WII, consisting essentially in the dilution of a single phase microemulsion or Lα LC into a nanoemulsion. This is particularly used to make nanoemulsions without stirring energy, and the original PIT method can be extended to the case in which the formulation transition is produced by dilution and associated change in formulation, e.g., dilution by water that decreases the salinity or the alcohol or surfactant concentration (C. Solans, J. Esquena, and A. Forgiarini).

Catastrophic inversion data has been reported mostly by three groups for many different systems and conditions (J. L. Salager, B. W. Brooks, and D. H. Smith). The modeling of catastrophic inversion first proposed by E. Dickinson (1981–82) was then presented in detail but still as a qualitative interpretation by J. L. Salager (1986–88), and fitted with actual data by D. H. Smith (1990). Then, G. E. J. Vaessen and H. N. Stein showed that the catastrophe theory model was not completely satisfactory and proposed a competitive coalescence kinetic model along J. T. Davies' (1957) classical approach, but failed to provide a better approach from the practical point of view. Hence, the catastrophe theory model, which applies both to describe the phase behavior with the Maxwell convention and to interpret the dynamic emulsion inversion with the perfect delay convention, is probably the description that makes more sense to account for the thermodynamics underlayer behind emulsion inversion. Some purely physical model based on mixing considerations (A. W. Pacek and A. W. Nienow) allowed an explanation for some features but failed to explain and predict the effect of the surfactant and physicochemical formulation, which cannot be neglected in any realistic approach.

As far as experimental measurements are concerned, D. H. Smith and K. H. Lim have provided the only detailed analysis related to the value of conductivity of an emulsion according to Maxwell relationships in different cases of two- and three-phase emulsions. Unfortunately, they use a different terminology from other main groups which does not help the reader in following their complex discussion. On the other hand, most other groups use a high/low value of the conductivity to define the type of emulsion, and take for granted a cut-off at a fraction of mS/cm, although this is still a controversial issue in some cases. Rheological measurements have been carried out by L. Choplin, J. Allouche, and E. Tyrode. Drop sizes have been studied by B. W. Brooks and S. Sajjadi, by conventional sampling techniques and recently by an in-situ method (F. B. Alban and S. Sajjadi).

Reported applications of emulsification by phase inversion concern waterborne epoxy (Z. Z. Yang), poly-dimethyl siloxanes (J. Rouviere and G. Guérin), and polyisobutene and ionomer systems (B. W. Brooks).

ACKNOWLEDGMENTS

The author thanks CDCHT-ULA (Grant 1-834-05-08-AA) and CONICIT-FONACIT (Grants AP-1997-3719, 51-2001-1156, F-2000-1629) for the financial support received in the past decade which has enabled him to carry out the research program on emulsion inversion at University of the Andes and which has led to some of the results presented here.

REFERENCES

1. R Laughlin. The Aqueous Phase Behavior of Surfactants. New York: Academic Press, 1994.
2. P Winsor. Solvent Properties of Amphiphilic Compounds. London: Butterworth, 1954.
3. M Bourrel, RS Schechter. Microemulsions and Related Systems. New York: Marcel Dekker, 1988.
4. AM Cazabat, D Chatenay, D Langevin, J Meunier. Faraday Disc Chem Soc 76:291, 1982.
5. S Scriven. Nature 263:123, 1976.
6. D Anderson, H Wennerström, U Olsson. J Phys Chem 93:4243, 1989.
7. LM Prince. Microemulsions – Theory and Practice. New York: Academic Press, 1977.
8. J Texter. Colloids Surf A 167:115, 2000.
9. JL Salager. In: G Broze, ed. Handbook of Detergents – Part A: Properties. New York: Marcel Dekker, 1999, pp 253–302.
10. JL Salager, RE Anton. In: P Kumar, KL Mittal, eds. Handbook of Microemulsion Science and Technology. New York: Marcel Dekker, 1999, pp 247–280.
11. JM Lee, HJ Shin, KH Lim, J Colloid Interface Sci 257:344, 2003.
12. E Tyrode, I Mira, N Zambrano, L Márquez, M Rondón-Gonzalez, JL Salager. Ind Eng Chem Res 42:4311, 2003.
13. DO Shah, RS Schechter. Improved Oil Recovery by Surfactant and Polymer Flooding. New York: Academic Press, 1977.
14. K Shinoda, H Arai. J Phys Chem 68:3485, 1964.
15. D Fotland, A Skauge. J Dispersion Sci Technol 7:563, 1986.
16. A Skauge, D Fotland. SPE Reservor Eng 5:601, 1990.
17. WH Wade, J Morgan, J Jacobson, JL Salager, RS Schechter. Soc Pet Eng J 18:242, 1978.
18. JL Salager, E Vasquez, J Morgan, RS Schechter, WH Wade. Soc Pet Eng J 19:107, 1979.
19. N Marquez, RE Anton, A Graciaa, J Lachaise, JL Salager. Colloids Surf A 100:225, 1995.
20. N Marquez, RE Anton, A Graciaa, J Lachaise, JL Salager. Colloids Surf A 131:45, 1998.
21. N Marquez, A Graciaa, J Lachaise, JL Salager. Langmuir 18:6021, 2002.
22. JL Salager, N Marquez, A Graciaa, J Lachaise. Langmuir 16:5534, 2000.
23. JL Salager, RE Anton, JM Anderez, JM Aubry. In: Techniques de l'Ingénieur, Paris: Techniques de l'Ingénieur, Vol. J2, Chap. 157, 2001.
24. RE Anton, N Garces, A Yajure. J Dispersion Sci Technol 18:539, 1997.
25. M Bourrel, JL Salager, RS Schechter, WH Wade. J Colloid Interface Sci 75:451, 1980.
26. C Pierlot, J Poprawski, M Catte, JL Salager, JM Aubry. Polym Int 52:614, 2003.
27. JL Salager, M Bourrel, RS Schechter, WH Wade. Soc Pet Eng J 19:271, 1979.
28. RE Anton, JL Salager. J Colloid Interface Sci 140:75, 1990.
29. RE Anton, A Graciaa, J Lachaise, JL Salager. J Dispersion Sci Technol 13:565, 1992.
30. RE Anton, D Gomez, A Graciaa, J Lachaise, JL Salager. J Dispersion Sci Technol 14:401, 1993.
31. M Buzier, JC Ravey. J Colloid Interface Sci 91:20, 1983.
32. M Kahlweit, R Strey, P Firman, D Haase, J Jen, R Shomäcker. Langmuir 4:499, 1988.
33. S Burauer, T Sachert, T Sottmann, R Strey. Phys Chem Chem Phys 1:4299, 1999.
34. T Förster, F Schambil, H Tesmann. Int. J Cosmetic Sci 12:217, 1990.
35. P Izquierdo, J Esquena, TF Tadros, C Dederen, MJ Garcia, N Azemar, C Solans. Langmuir 18:26, 2002.
36. JL Salager, M Miñana-Perez, M Perez-Sanchez, M Ramirez-Gouveia, CI Rojas. J Dispersion Sci Technol 4:313, 1983.
37. WD Bancroft. J Phys Chem 17:501, 1913.
38. WD Bancroft. J Phys Chem 19:275, 1915.
39. I Langmuir. J Am Chem Soc 39:1848, 1917.
40. JT Davies, EK Rideal. Interfacial Phenomena. New York: Academic Press, 1963.

41. JT Davies. In: Proceedings 2nd Int Congress Surface Activity. London: Butterworths, 1957, pp 426–438.
42. IB Ivanov, PA Kralchevsky. Colloids Surf A 128:155, 1997.
43. A Kabalnov, H Wennerström. Langmuir 12:276, 1996.
44. W Ostwald. Kolloid Z 6:103, 1910.
45. W Ostwald. Kolloid Z 7:64, 1910.
46. DH Smith. Langmuir 6:1071, 1990.
47. DH Smith, KH Lim. J Phys Chem 94:3746, 1990.
48. KH Lim, DH Smith. J Dispersion Sci Technol 11:529, 1990.
49. DH Smith, YHS Wang. J Phys Chem 98:7214, 1994.
50. DH Smith, R Sampath, D Dadybujor. J Phys Chem 100:17558, 1996.
51. KH Lim, JM Lee, DH Smith. Langmuir 18:6003, 2002.
52. K Shinoda, H Saito. J Colloid Interface Sci 26:70, 1968.
53. RE Anton, P Castillo, JL Salager. J Dispersion Sci Technol 7:319, 1986.
54. M Miñana-Perez, P Jarry, M Ramirez-Gouveia, JL Salager. J Dispersion Sci Technol 7:331, 1986.
55. P Jarry, M Miñana-Perez, JL Salager. In: KL Mittal, P Bothorel, eds. Surfactants in Solutions, vol. 6. New York: Plenum Press, 1987, pp 1689–1696.
56. BW Brooks, HN Richmond. Colloids Surf A 58:131, 1991.
57. BW Brooks, HN Richmond. Chem Eng Sci 49: 1065, 1994.
58. D Miller, T Henning, W Grünbein. Colloids Surf A 183:681, 2001.
59. M Zerfa, S Sajjadi, BW Brooks. Colloids Surf A 155:323, 1999.
60. HT Davis. Colloids Surf A 91:9, 1994.
61. ZZ Yang, Y Xu, S Wang, H Yu, W. Cai. Chin J Polym Sci 15:92, 1997.
62. ZZ Yang, DL Zhao, Y Xu, M Xu. Chin J Polym Sci 15:373, 1997.
63. ZZ Yang, D Zhao, Chin J Polymer Sci 18:33, 2000.
64. PA Psathas, SRP Da Rocha, CT Lee, KP Johnston. Ind Eng Chem Res 39:2655, 2000.
65. KP Johnston, D Cho, SRP Da Rocha, PA Psathas, W Ryoo. Langmuir 17:7191, 2001.
66. HP Grace. Chem Eng Com 14:225, 1982.
67. JL Salager, G Lopez-Catellanos, M Miñana-Perez. J Dispersion Sci Technol 11:397, 1990.
68. WB Brooks, H Richmond. Chem Eng Sci 49:1843, 1994.
69. JL Salager, L Marquez, A Peña, M Rondon, F Silva, E Tyrode. Ind Eng Chem Res 39:2665, 2000.
70. F Silva, A Peña, M Miñana-Perez, JL Salager. Colloids Surf A 132:221, 1998.
71. S Sajjadi, F Jahanzad, BW Brooks. Ind Eng Chem Res 41:6033, 2002.
72. A Peña, JL Salager. Colloids Surf A 181:319, 2001.
73. I Mira, N Zambrano, E Tyrode, L Marquez, AA Peña, A Pizzino, JL Salager. Ind Eng Chem Res 42:57, 2003.
74. WB Brooks, H Richmond. Chem Eng Sci 49:1053, 1994.
75. GEJ Vaessen, M Visschers, HN Stein. Langmuir 12:875, 1996.
76. JL Salager, MI Briceño, CL Bracho. In: J Sjoblöm, ed. Encyclopedic Handbook of Emulsion Technology. New York: Marcel Dekker, 2001, pp 455–495.
77. GA Nuñez, MI Briceño, C Mata, H Rivas. J Rheology 40:405, 1996.
78. SE Salager, EC Tyrode, MT Celis, JL Salager. Ind Eng Chem Res 40:4808, 2001.
79. JL Salager. In: F Nielloud, G Marti-Mestres, eds. Pharmaceutical Emulsions and Suspensions. New York: Marcel Dekker, 2000, pp 73–125.
80. JL Salager. In: A Chattopadhay, KL Mittal, eds. Surfactants in Solution. New York: Marcel Dekker, 1996, pp 261–295.
81. M Bourrel, A Graciaa, RS Schechter, WH Wade. J Colloid Interface Sci 72:161, 1979.
82. JL Salager, L Quintero, E Ramos, JM Anderez. J Colloid Interface Sci 77:287, 1980.
83. RE Antón, JL Salager. J. Colloid Interface Sci 111:54, 1986.
84. R Hazzlett, RS Schechter. Colloids Surf 29:53, 1988.
85. JL Salager, M Perez, Y Garcia. Colloid Polymer Sci 274:81, 1996.
86. M Pérez, N Zambrano, M Ramírez, E Tyrode, JL Salager. J. Dispersion Sci Technol 23:55, 2002.
87. G Goubault, K Pays, D Olea, P Gorria, J Bibette, V Schmitt, F Leal-Calderon. Langmuir 17:5184, 2001.

88. C Mabille, V Schmitt, P Gorria, F Leal-Calderon, V Faye, B Deminiere, J Bibette. Langmuir 16, 422, 2000.
89. TG Mason, J Bibette. Langmuir 13:4600, 1997.
90. K Shinoda. Proceedings 5th Int Congress Surface Activity. Barcelona, Spain, 1969, Vol. 2, pp 275–283.
91. J Allouche, E Tyrode, V Sadtler, L Choplin, JL Salager. Langmuir 20:2134, 2004.
92. T Förster. In: MM Rieger, LD Rhein, eds. Surfactants in Cosmetics. New York: Marcel Dekker, 1997, pp 105–125.
93. K Shinoda, H Saito. J Colloid Interface Sci 30:258, 1969.
94. T Engels, T Förster, W von Rybinski. Colloids Surf A 99:141, 1995.
95. T Förster, W von Rybinski, A Wadle. Adv Colloid Interface Sci 58: 119, 1995.
96. M Miñana-Perez, C Gutron, C Zundel, JM Anderez, JL Salager. J Dispersion Sci Technol 20:893, 1999.
97. S Sajjadi, F Jahanzad, M Yianneskis. Colloids Surf A 240:149, 2004.
98. S Sajjadi, M Zerfa, WB Brooks. Chem Eng Sci 57:663, 2002.
99. M Kahlweit, R Strey, P Firman. J Phys Chem 90:671, 1986.
100. L Fillous, A Cardenas, J Rouviere, JL Salager. J. Surfactants Detergents 2:303, 1999.
101. A Cardenas, L Fillous, J Rouviere, JL Salager. Ciencia 9:70, 2001.
102. JL Salager, N Moreno, RE Antón, S. Marfisi. Langmuir 18:607, 2002.
103. G Alvarez, R Antón, S Marfisi, L Marquez, JL Salager. Langmuir 20:5179, 2004.
104. L Taisne, B Cabane. Langmuir 14:4744, 1998.
105. E Dickinson. J Colloid Interface Sci 84:284, 1981.
106. JL Salager. In: P Becher, ed. Encyclopedia of Emulsion Technology. New York: Marcel Dekker, 1988, Vol 3, pp 79–134.
107. JL Salager. J Colloid Interface Sci 105:21, 1985.
108. E. Dickinson. J. Colloid Interface Sci 87:416, 1982.
109. E Ruckenstein. Adv Colloid Interface Sci 79:59, 1999.
110. KH Lim, DH Smith. J Colloid Interface Sci 142:278, 1991.
111. GEJ Vaessen, HN Stein. J Colloid Interface Sci 176:378, 1995.
112. S Sajjadi, M Zerfa, BW Brooks. Langmuir 16:10015, 2000.
113. JM Lee, KH Lim, DH Smith. Langmuir 18:7334, 2002.
114. S Sajjadi, F Jahanzad, M Yianneskis, BW Brooks. Ind Eng Chem Res 42:3571, 2003.
115. CH Villa, LB Lawson, Y Li, KD Papadopoulos. Langmuir 19:244, 2003.
116. JK Klahn, JJM Janssen, GEJ Vaessen, R de Swart, WGM Agterof. Colloids Surf A 210:167, 2002.
117. JL Salager, A Forgiarini, L Marquez, A Peña, A Pizzino, MP Rodriguez, M Rondon-Gonzalez. Adv Colloid Interface Sci 108–109:259, 2004.
118. E Tyrode, J Allouche, L Choplin, JL Salager. Ind Eng Chem Res 44:67, 2005.
119. N Zambrano, E Tyrode, I Mira, L Marquez, MP Rodriguez, JL Salager. Ind Eng Chem Res 42:50, 2003.
120. AW Pacek, AW Nienow, IPT Moore. Chem Eng Sci 49:3485, 1994.
121. AW Nienow. Adv Colloid Interface Sci 108–109:95, 2004.
122. DJ Watson, MR Mackley. Colloids Surf A 196:121, 2002.
123. A Nandi, DV Khakhar, A Mehra. Langmuir 17:2647, 2001.
124. F Bouchama, GA van Aken, AJE Autin, GJM Koper. Colloids Surf A 231:11, 2003.
125. S Sajjadi, M Zerfa, BW Brooks. Colloids Surf A 218:241, 2003.
126. L Marquez, A Graciaa, J Lachaise, JL Salager, N Zambrano. Polym Int 52:590, 2003.
127. H Kunieda, Y Fukuhi, H Uchiyama, C Solans. Langmuir 12: 2136, 1996.
128. L Huang, A Lips, CC Co. Langmuir 20:3559, 2004.
129. P Izquierdo, J Esquena, TF Tadros, JC Dederen, J Feng, MJ Garcia-Celma, N Azemar, C Solans. Langmuir 20:6594, 2004.
130. L Marquez, A Graciaa, J Lachaise, JL Salager. Proceedings 3rd World Congress on Emulsion, Lyon, France, Sept 23–27, 2002.
131. S Friberg, L Mandell, M Larsson. J Colloid Interface Sci 29:155, 1969.

132. K Mandani, S Friberg. Prog Colloid Polym Sci 65:164, 1978.
133. H Kunieda, K Shinoda. J Dispersion Sci Technol 3:233, 1982.
134. H Sagitani. J Dispersion Sci Technol 9:115, 1988.
135. J Rouviere, JL Razakarison, J Marignan, B Brun. J Colloid Interface Sci 133:293, 1989.
136. T Suzuki, H Takei, S Yamazaki. J Colloid Interface Sci 129:491, 1989.
137. A Forgiarini, J Esquena, C Gonzalez, C Solans. Prog Colloid Polym Sci 118:184, 2001.
138. A Forgiarini, J Esquena, C Gonzalez, C Solans. Langmuir 17:2076, 2001.
139. A Graciaa, J Lachaise, JG Sayous, P Grenier, S Yiv, RS Schechter, WH Wade. J Colloid Interface Sci 93:474, 1983.
140. AM Bellocq, D Bourbon, B Lemanceau, G Fourche. J Colloid Interface Sci 89:427, 1982.
141. AM Bellocq, J Biais, B Clin, A Gelot, P Lalanne, B Lemanceau. J Colloid Interface Sci 74:311, 1980.
142. T Nishimi, CA Miller. Langmuir 16:9233, 2000.
143. JL Salager, A Forgiarini, JC Lopez, S Marfisi, G Alvarez. Proceedings 6th World Surfactant Congress, CESIO, Berlin, Germany, June 20–23, 2004. See also S. Marfisi, MP Rodriguez, G Alvarez, MT Celis, A Forgiarini, J Lachaise, JL Salager. Langmuir 21:6712, 2005.
144. A Forgiarini, J Esquena, C Gonzalez, C Solans. J Dispersion Sci Technol 23:209, 2002.
145. K Ozawa, C Solans, H Kunieda. J Colloid Interface Sci 188:275, 1997.
146. N Shahidzahed, D Bonn, O Aguerre-Chariol, J Meunier. Colloids Surf A 147:375, 1999.
147. R Pons, I Carrera, J Caelles, J Rouch, P Panizza. Adv Colloid Interface Sci 106:129, 2003.
148. MJ Rang, CA Miller. J Colloid Interface Sci 2–9:179, 1999.
149. N Shahidzahed, D Bonn, J Meunier, M Nabavi, M Airiau, M Morvan. Langmuir 16:9703, 2000.
150. GA Nuñez, MI Briceño, C Mata, H Rivas. J Rheology 40:405, 1996.
151. ZZ Yang, YZ Xu, DL Zhao, M Xu. Colloid Polym Sci 278:1164, 2000.
152. ZZ Yang, YZ Xu, DL Zhao, M Xu. Colloid Polym Sci 278:1103, 2000.
153. G Guérin. Proceedings 2nd World Congress on Emulsion, Bordeaux, France, Sept. 1997.
154. LK Saw, BW Brooks, KJ Carpenter, DV Keight. J Colloid Interface Sci 257:163, 2003.
155. E Dickinson. Annual Reports C. London: Royal Society of Chemistry, 1986, pp 31–58.
156. GEJ Vaessen. PhD Dissertation. Eindhoven University, Netherlands, 1996.
157. JL Salager, L Marquez, I Mira, A Peña, E Tyrode, N Zambrano. In: KL Mittal, DO Shah, eds. Adsorption and Aggregation of Surfactants in Solution. New York: Marcel Dekker, pp 501–523.

5 Phase Inversion Studies as a Tool for Optimization and Characterization of Surfactant Mixtures in Specific Oil/Water Systems

Ingegärd Johansson and Ilja Voets

CONTENTS

5.1 INTRODUCTION

In the development work of new emulsifiers or new mixtures of emulsifiers the need for versatile characterization tools is urgent. In line with this many high throughput techniques are introduced especially within the industries. The tool described in this chapter quickly provides information and understanding of the nature of the investigated surface activity of mixtures of emulsifiers from the industrial world. Another advantage is the direct information obtained about the behavior of the system when the hydrophilic/lipophilic balance is changed.

In many applications the simultaneous presence of a polar and a nonpolar solvent is desirable. When oil and water are mixed under stirring, an emulsion is formed. The resulting emulsion often needs to be stable. This stability is achieved by the addition of one or several emulsifiers. The problem of selecting the best system for this purpose is ubiquitous. Solutions have been sought in describing the structure of the emulsifier in terms of hydrophilic–lipophilic balance (HLB) (1), or critical packing parameter (CPP) (2). These attempts have not included the influence of other parameters, such as oil, temperature, and salt, which play important roles for emulsion stability. Shinoda (3) works with temperature effects and uses phase inversion temperature (PIT) to describe the function of emulsifiers, mainly of ethoxylated nonionic surfactants. Lately, a systematic approach to use the PIT has been described (4) under the abbreviation CAPICO (calculation of phase inversion in concentrated emulsions).

Typical for an emulsion, stabilized by nonionic emulsifiers, which is heated or cooled beyond the phase inversion temperature, is its change in morphology: the emulsion inverts by going from an oil-in-water (O/W) to a water-in-oil (W/O) emulsion or vice versa. This means that the spontaneous curvature, H_O, of the interfacial film changes sign and passes a flat form where $H_O = 0$. This point is the balanced point or optimal point where, at thermodynamic equilibrium, a bicontinuous microemulsion is formed and the interfacial tension is at its minimum.

Davies writes: "The spontaneous mean curvature, H_O, determines whether the interfacial film wants to curve onto its oil or its water side or prefers to be flat" (1). That this tendency plays an important role in deciding if an O/W or a W/O emulsion will be formed, even though the curvature of an emulsion droplet is much less than that of a micelle, has been re-emphasized by Kabalnov and Wennerström (5) in the late 1990s.

It has also been shown by Pérez et al. (6) that the most stable emulsions are to be found at a specific distance from the balanced situation where $H_O = 0$ or, in his terminology, SAD = 0 (surfactant affinity difference). Thus it is important to find the balanced point of the actual emulsion or formulation that is being optimized.

The aim of the work reported here is to find a practical method to determine the balanced or optimal point of a system which can be employed with all types of emulsifiers, including those that are not temperature or salt sensitive. An automatic titration procedure is used by which a co-surfactant is added to a surfactant–oil–water (SOW) blend under stirring at constant temperature (see Figure 5.1). The morphology changes of the emulsion are monitored by means of conductometry. A steep conductivity decrease is observed when the emulsion inverts from O/W to W/O and vice versa when the morphology changes from W/O to O/W. Three important aspects are:

1. By using reference compounds like isomerically pure ethoxylates and the same co-surfactant and oil, the behavior of any other surfactants can be compared to systems with known HLBs or whatever traditional description concepts that are familiar to the user.
2. If instead the oil phase is changed and the same surfactant pair is used, the titration will provide information about the character of the oil, whether its behavior is more hydrophilic or more hydrophobic than the reference oil.
3. In many applications the surfactant blend resulting in the lowest interfacial tension is the most efficient. Since the inversion point is the point of minimum interfacial tension, the titration method is also a means of optimizing formulations in specific areas like cleaning or soil remediation.

Three practical examples will be given to illustrate the three aspects mentioned above.

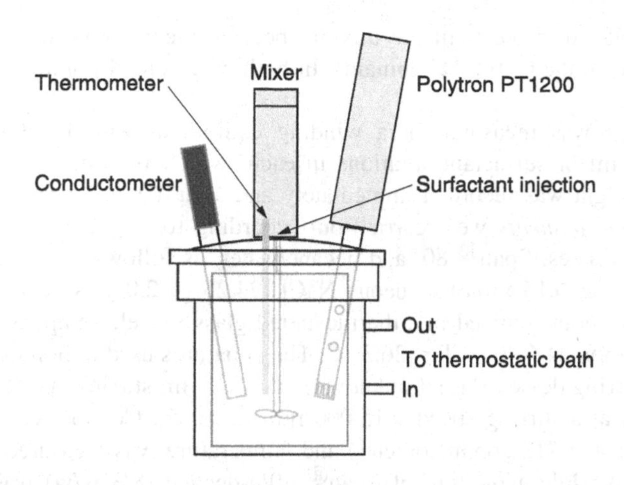

FIGURE 5.1 Experimental set-up for the co-surfactant titration followed with conductometry. Commercially available as the Multi Parameter Scanning (MPS) instrument from Scanalys (www.scanalys.com), with possibilities to measure turbidity, pH, conductometry, and viscosity simultaneously.

5.2 EXPERIMENTAL

Laboratory products from Akzo Nobel Surface Chemistry AB were:

- Branched C10 alcohol of Guerbet type with 3, 5, 8, and 10 ethylene oxide units, both narrow and broad range distribution, referred to as BrC10E⟨m⟩ narr or broad
- Straight C10 alcohol with 3, 5, 8, and 10 ethylene oxide units, both narrow and broad range distribution, referred to as 1D E⟨m⟩ narr or broad
- Glucosides of the same two alcohols with an average degree of polymerization of 1.6

Commercial products from Akzo Nobel Surface Chemistry AB were:

- Berol ox 91-6 and 91-8 which are made from C9–11 alcohol with 6 or 8 ethylene oxide units, broad range
- Berol 992, alkoxylated short chain fatty alcohol with cloud point 80 °C in 10% NaCl
- Berol 822, calcium dodecylbenzene sulfonate, 60% in butanol
- AG 6210, a C8/C10 glucoside with an average degree of polymerization of 1.6
- Berol 532, C11 with two ethylene oxide units

Other products were: isomerically pure C10E5, C10E7, and C10E8 (Sigma Aldrich and Fluka), Span® 80 and sorbitan monooleate (Fluka), n-decane (>95%) (Merck), Danish winter diesel (Statoil), Solvesso 150D (ExxonMobil), and methyl laurate (Merck). Water was distilled and de-ionized.

To determine the *critical micelle concentration*, surface tension was measured on a KSV unit with a Sigma 70 program using the Du Noüy ring method.

The *contact angle* was measured on Parafilm® with a FTÅ200 instrument in a climate controlled room (T = 21 ± 1 °C, 45 ± 10% rH).

To determine *cloud points*, mixtures of 1.0 wt% surfactant in distilled water, aqueous NaCl (2.0 g/l; 0.034 M), or 11% BDG were prepared. The mixtures that were clear at room temperature were heated in a water bath until the mixture became turbid. The mixtures that were turbid at room temperature were placed in ice until the mixture became clear (or no difference in appearance could be determined in the 0–100 °C domain). In both cases cloud points were determined upon cooling.

The *foam height* was measured in a winding equipment with fixed measuring cylinders (500 ml) with 200 ml of surfactant solutions in each, which is turned around 40 times during 1 min. The foam height was recorded immediately and after 1 min.

Conductivity measurements were carried out according to a typical example taken from the study of decyl ethoxylates, Span® 80, and decane/water, as follows.

n-Decane (24.25 g; 0.15 mole), aqueous NaCl (24.25 g; 2.0 g NaCl/L; 0.034 M), and surfactant (1.50 g) were homogenized in a thermostatted glass vessel. Temperatures at the inversion point were in the region 24.6 °C $\leq$ T $\leq$ 26.3 °C. The apparatus used to homogenize the mixture in the vessel was a stirring device (brand unknown) at maximum stirring speed and an Ultra Turrax (Polytron PT 1200) at a stirring speed >15,000 rpm (level 5). Conductivity was measured with a Pt electrode (Metrohm 712 conductometer) and temperature was measured with a thermometer (Physitemp BAT-10) while a mixture of Span® 80/n-decane (8/3 w/w) was titrated in using a dosimat (Metrohm 665) at a dosing rate (dr) of 0.15 ml/min. Figure 5.1 shows the experimental setup of the automatic titration technique.

Some measurements were performed on Scanalys® equipment. It is essentially the same experimental setup as above (www.scanalys.com), but one single stirrer was used at 225 or 300 rpm. Sometimes a complementary magnetic stirrer was introduced. The co-surfactant was titrated in with a dosing rate of 4.44 μl/sec.

5.3 RESULTS AND DISCUSSION

5.3.1 Characterization of Decyl Ethoxylates and Decyl Glucosides

A set of ethoxylates from straight chain and Guerbet branched decanol was made. Different catalysts were used to obtain broad or narrow distribution. Typical distributions are shown in Figure 5.2.

5.3.1.1 Emulsion Inversion

Since the total concentration of surfactant is increased during the titration, the titration path is somewhat diagonal as is shown in Figure 5.3. The inversion will take place when passing over the three phase region in the so-called "Kahlweit fish" (7). Both surfactant concentration and character is varied. The region where three phases, an excess oil phase, a surfactant phase, and an excess water phase, occur often has the form of a fish body. Polydispersity of the surfactant may result in a skewed fish body, for example bent upwards or downwards (8). To successfully apply the technique described here it is practical to know what the relative position in the phase diagram is of the initial composition compared to the fish body. Otherwise there is the risk of missing the inversion point, either starting too far below the required ratio of hydrophobic co-surfactant to total surfactant (α) (= below the fish body) or with too low total surfactant concentration (γ) (= before the fish body) to bring about the inversion.

In principle, the nature of the surfactant(s) can be varied by temperature (nonionics), salt (ionics), or by addition of a co-surfactant (all surfactants). We apply the latter, most universal, technique to invert an emulsion from O/W to W/O (addition of a relatively hydrophobic co-surfactant) or from W/O to O/W (addition of a more hydrophilic co-surfactant).

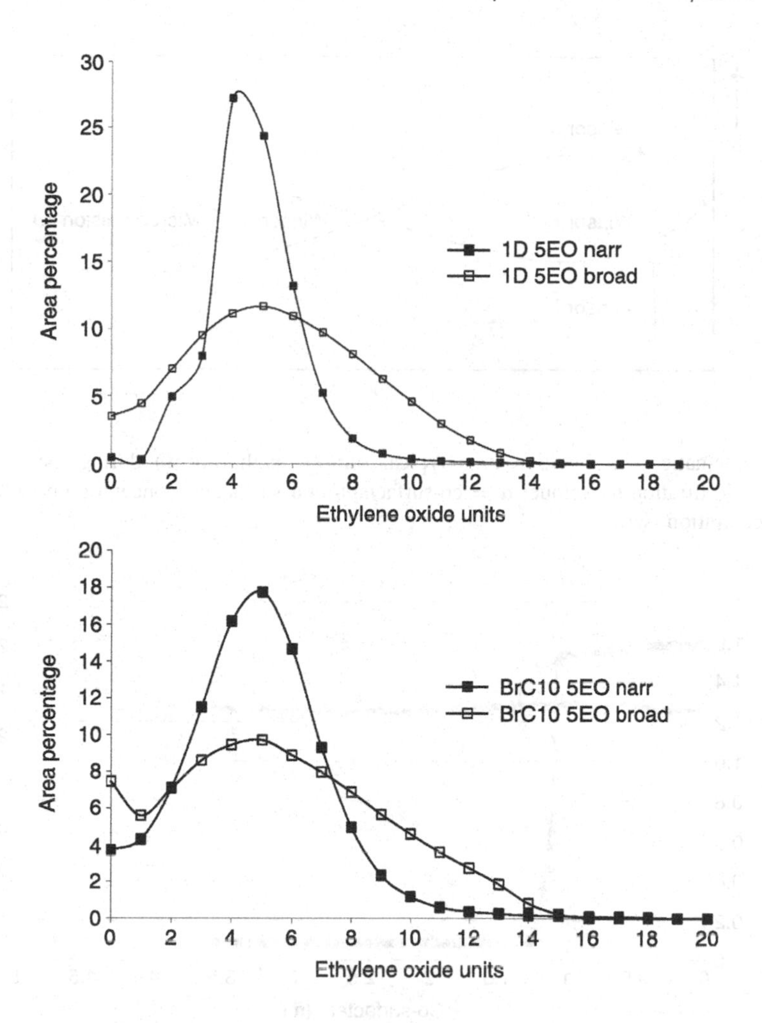

FIGURE 5.2 (a) Ethylene oxide distributions of decyl alcohol ethoxylates, broad and narrow range. (b) Ethylene oxide distributions of branched decyl alcohol ethoxylates, broad and narrow range.

Most of the nonionics studied here are hydrophilic enough to give a water continuous emulsion when stirred. Sorbitan monooleate, Span® 80, was chosen as co-surfactant (HLB = 4.3, Ref. 3) and decane as the oil phase. The inversion was followed from water continuous (O/W) to oil continuous (W/O). A typical well-behaved run is shown in Figure 5.4 for a 1D E3 broad range.

Many of the titrations showed fine structure (see Figure 5.5), indicating areas with liquid crystalline phases as one of several phases in the phase diagram. For a detailed understanding equilibrium studies have to be performed. However, as the systems become better known the interpretation of these fine structures become more and more direct.

To be able to work with the results of the titration an evaluation technique was established. An example of this is given in Figure 5.5. Linear fits to the first and last part of the steep conductivity drop in a plot of conductivity as a function of the amount of added co-surfactant are constructed (Figure 5.5). IP_{upper} is determined as the breakpoint of a horizontal line through the maximum in the conductivity curve and a linear fit to the first portion of the steep drop in

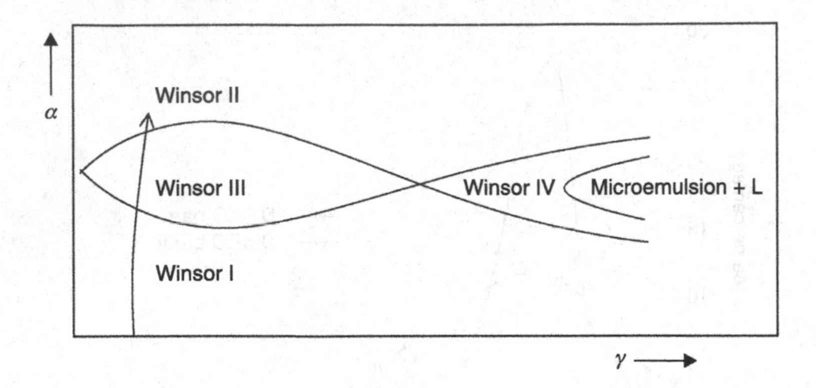

FIGURE 5.3 Schematic representation of a typical titration path (arrow) during the conductivity scan with the automatic titration technique; α = co-surfactant/total surfactant concentration (w/w%), γ = total surfactant concentration (w/w%).

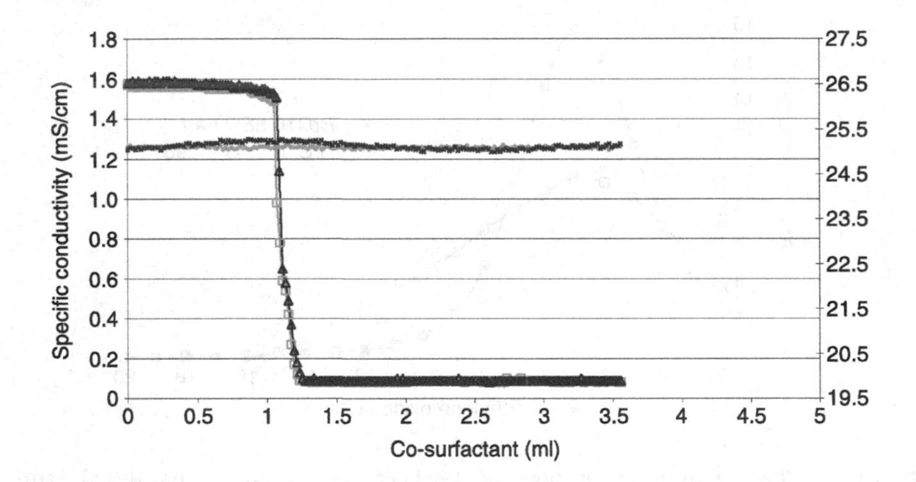

FIGURE 5.4 Conductivity scan for 1D E3 broad range with Span-80/n-decane/NaCl/water (T = 25.0 °C, dr = 4.44 μl/s, ss = 300 rpm); temperature is shown on the right axis, (□) and (Δ) correspond to two different measurements on the same system.

conductivity. IP_{lower} is the amount of added co-surfactant in milliliters at the breakpoint of a horizontal line through the minimum in conductivity and a linear fit to the last portion of the steep drop in conductivity. The average of these two values can then be used as the "inversion point."

In Figure 5.6 all the inversion points for the four series of ethoxylated surfactants, the isomerically pure ones and the alkyl glucosides, are gathered in terms of wt% of Span® 80 of the total surfactant blend (α) versus the degree of polymerization.

If less co-surfactant is needed to invert the emulsion, the surfactant (mixture) itself may be described as more hydrophobic. In this sense one could say that the surfactants based on a branched hydrophobe in most cases are more hydrophobic; see for instance Br C10 glucoside and Br C10E5 narrow or Br C10E8 narrow as compared to the corresponding straight C10 based ones.

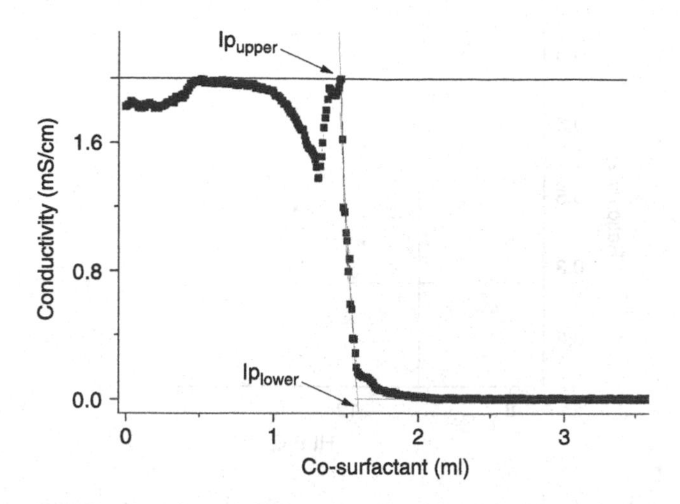

FIGURE 5.5 Linear fits to the first and last part of the steep conductivity drop in a plot of conductivity as a function of the amount of added co-surfactant with 1-decyl glucoside/n-decane/water/NaCl/Span® 80 (T = 25 °C, dr = 0.15 ml/min, ss > 15,000 rpm) to be used for establishing upper and lower inversion points.

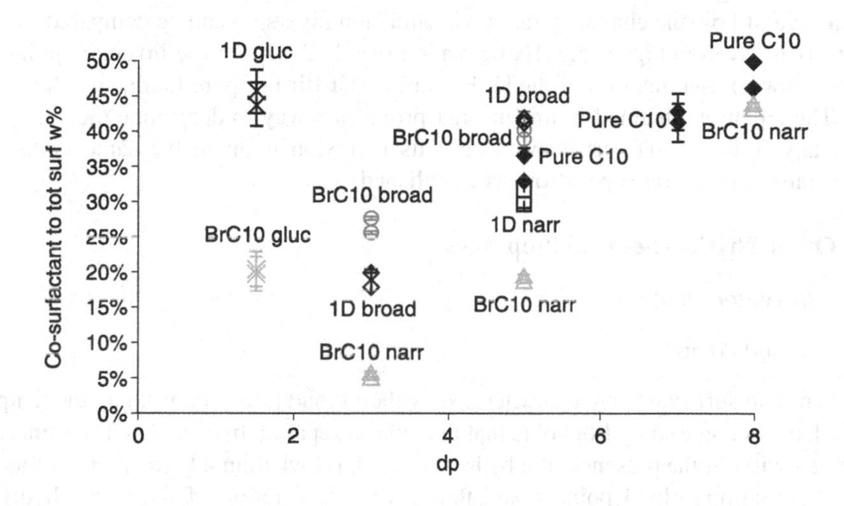

FIGURE 5.6 Inversion points (IP_{upper} and IP_{lower}) as a function of the degree of polymerization of technical grade alcohol ethoxylates, technical grade alkyl glucosides, and isomerically pure alcohol ethoxylates with 5, 7, or 8 EO units. Error bars indicate the deviation of two individual measurements on the system from the average value.

When only 3 moles of ethylene oxide are added the remaining, unreacted, alcohol in the mixture will disturb the overall picture since that alcohol will dissolve in the oil phase and not contribute to the interfacial layer of emulsifiers. Ethoxylation of a branched alcohol yields more unreacted alcohol compared to the reaction with a straight chain one due to steric hindrance. Thus the actual active content of emulsifier will be significantly decreased and the blend will behave in a more hydrophilic manner than is expected from the amount of ethylene oxide added.

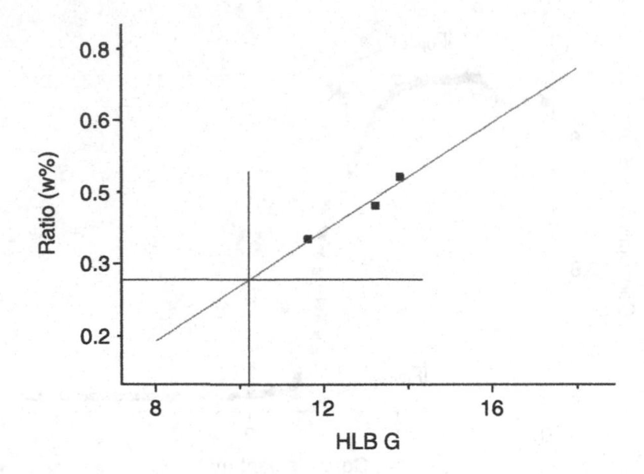

FIGURE 5.7 Extrapolation of a linear fit on the inversion points (average of upper and lower) as a function of the HLB number (Griffin) of pure $C_{10}E_m$ alcohol ethoxylates.

Another general conclusion is that the broad range surfactants seem to be more hydrophilic than the narrow range.

An example of how the characterization via emulsion inversion can be compared to commonly used concepts is given in Figure 5.7. Extrapolation of a linear fit on the inversion points (average of upper and lower) as a function of the HLB number (Griffin) of pure $C_{10}E_m$ alcohol ethoxylates is shown. The equation obtained in this manner provides a way to determine the equivalent HLB number of any (mixture of) surfactant(s) once its inversion point in the same system (i.e., the same temperature, pressure, type of oil) is established.

5.3.1.2 Other Physicochemical Properties

5.3.1.2.1 In Water Solution

5.3.1.2.1.1 Cloud Points

Often the nonionic surfactants are characterized by their cloud points (C.p.), i.e., the temperature at which they have become so hydrophobic that they phase separate from water. These measurements are performed without the presence of a hydrophobic phase wherein a high amount of alcohol may dissolve. The resulting cloud point would thus give a description of the actual hydrophobicity of the technical nonionic surfactant. The cloud points for the set of nonionics studied here is shown in Figure 5.8. The measurement is performed in a mixture of butyl diethylene glycol and water to get a C.p. between 0 and 100 °C for all the products. The general conclusion also here is that the branched hydrophobes give more hydrophobic products than the straight chain ones. Differences between distributions are leveled out in this medium.

The glucosides usually do not show any cloud points except in very special cases and thus can not be characterized in this way. Their solubility increases slightly with temperature.

5.3.1.2.1.2 Critical Micelle Concentration

In Figure 5.9 the dependence of the critical micelle concentrations (CMCs) on the chemical structures is shown. As expected (9), the CMC increases for the same type of surfactant with

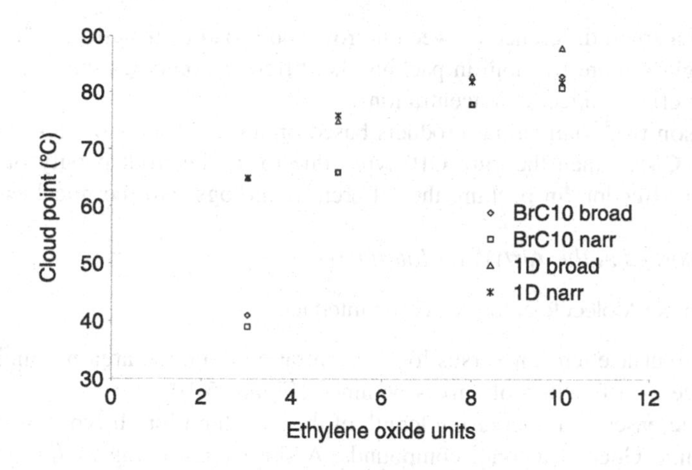

FIGURE 5.8 Cloud point (CP) as a function of the average number of ethylene oxide units ⟨m⟩ of 1.0 w% solutions of alcohol ethoxylates with ⟨m⟩ ≥ 5 in 11% butyl diethylene glycol.

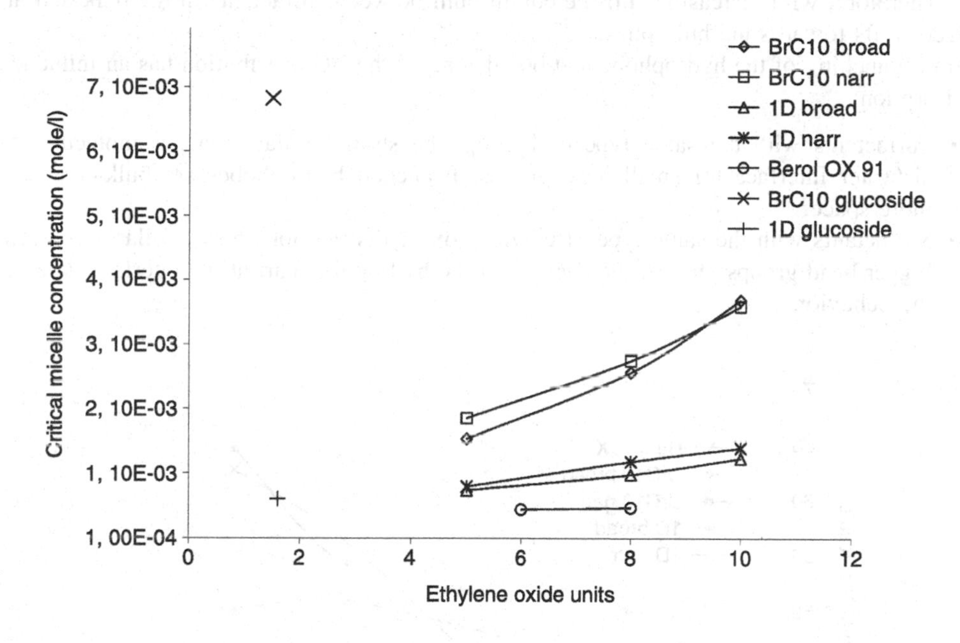

FIGURE 5.9 Critical micelle concentrations of technical grade alcohol ethoxylates and alkyl polyglucosides as a function of degree of polymerization of the head group (T = 21 ± 1 °C).

increasing ⟨m⟩ indicative of the increasing hydrophilicity of the surfactant monomer. Moreover, the steric repulsion between the surfactant head groups that form the micelles increases with increasing ⟨m⟩ (average length of the EO chain).

Generally the branched products have higher CMCs than the straight chain ones. The first explanation that comes to mind is the packing parameter (CPP). The packing parameter of surfactants with straight chain hydrophobes is more favorable for the formation of (spherical) micelles than is the case for branched hydrophobes.

There is also a small difference between narrow and broad distributions. The broad range surfactants have slightly more freedom in packing the different homologs into the micelles resulting in slightly lower critical micelle concentrations.

For comparison two commercial products based on a C9–11 alcohol blend are shown, which both have lower CMCs than the pure C10 types due to the heterodispersity of the hydrophobe, again giving more freedom in packing the different homologs into the micelles.

5.3.1.2.2 Properties at the Air/Water Interface

5.3.1.2.2.1 Area per Molecule at the Air/Water Interface

By replotting the surface tension versus log concentration data, the area per molecule (a_o) at the air/water interface as a function of ⟨m⟩ is obtained (Figure 5.10).

Clearly, a_o increases with increasing length of the EO chain for all types of surfactants except for the broad range Guerbet alcohol compounds. A similar reasoning as for the critical micelle concentration is valid here. Due to the increase in hydration of the head group with increasing length of the EO chain, there is less free energy to be gained upon transfer of the surfactant from the bulk phase to the air–water interface. Moreover, larger head groups simply need more space. Therefore, with increasing ⟨m⟩ the equilibrium between surfactants in the bulk and at the interface shifts towards the bulk phase.

Both branching of the hydrophobe and broadening of the EO distribution has an influence on the adsorption:

- Surfactants with the same type of hydrophobe show similar area per molecule at the air/water interface for small head groups. Branched hydrophobes are bulkier and need more space.
- Surfactants with the same type of ethylene oxide distribution show similar a_o values for bigger head groups, that is, the packing of the hydrophilic part at the interface determines the behavior.

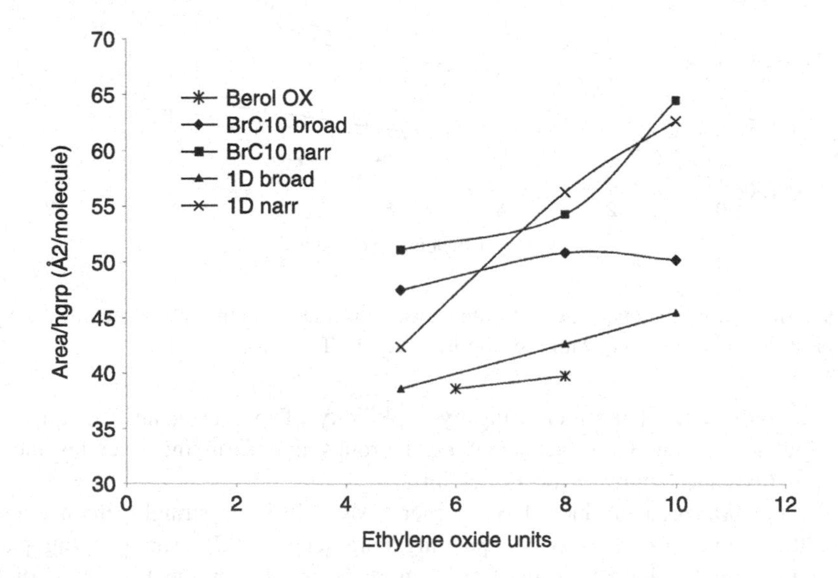

FIGURE 5.10 Area per molecule at the air/water interface as a function of the average number of ethylene oxide units.

Thus the influence of the hydrophobe dominates for smaller head groups and the heterodispersity of the EO chain is the dominant factor for larger head groups. This reflects the increase in the size of the head group relative to the hydrophobe with increasing amounts of EO. It is obvious that broad range surfactants have smaller areas per molecule than narrow range surfactants. Most probably this is a result of the packing advantages of a broader EO distribution.

5.3.1.2.2.2 Surface Tension at the Critical Micelle Concentration

More information to be gained from the CMC curves is the surface tension at or above CMC (see Figure 5.11). Surface tension increases with increasing amounts of EO, except for the broad range Guerbet alcohol based compounds. For the same type of hydrophobe in the surfactant, the smaller the head group, the more surfactant adsorbs at the interface and the lower the surface tension. Note, however, that the special behavior for the branched C10 broad range compounds is also found in the area/molecule graph.

The most striking feature is that the Guerbet alcohol based surfactants show lower surface tension values in spite of the relatively low adsorption when compared to the straight chain surfactants. Apparently, the branched surfactants form a more hydrophobic surfactant layer at the air–water interface.

It can be speculated that the reason for this might be as follows. The branch may be positioned planar to the interface, while the rest of the surfactant tail is positioned lateral to the interface. Therefore, the branch reduces the contact between water and air resulting in lower surface tensions for branched surfactants.

5.3.1.2.2.3 Foam

The foam was measured for the different C10E5s and the corresponding glucosides at 20 and 50 °C (Figure 5.12). The branched alcohol based surfactants all show much less stable foam, probably due to less elasticity of the foam lamella created from the less well packed branched surfactants.

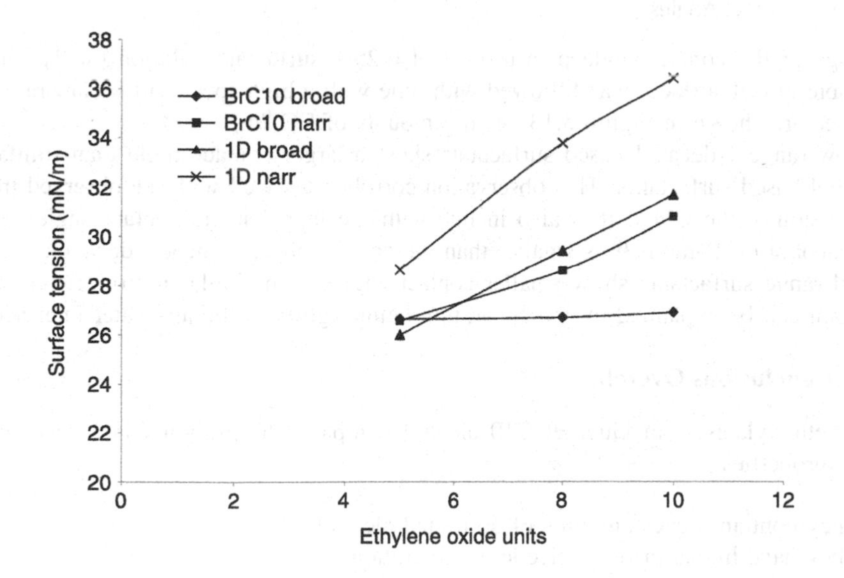

FIGURE 5.11 Surface tension at the critical micelle concentration as a function of the average number of ethylene oxide units.

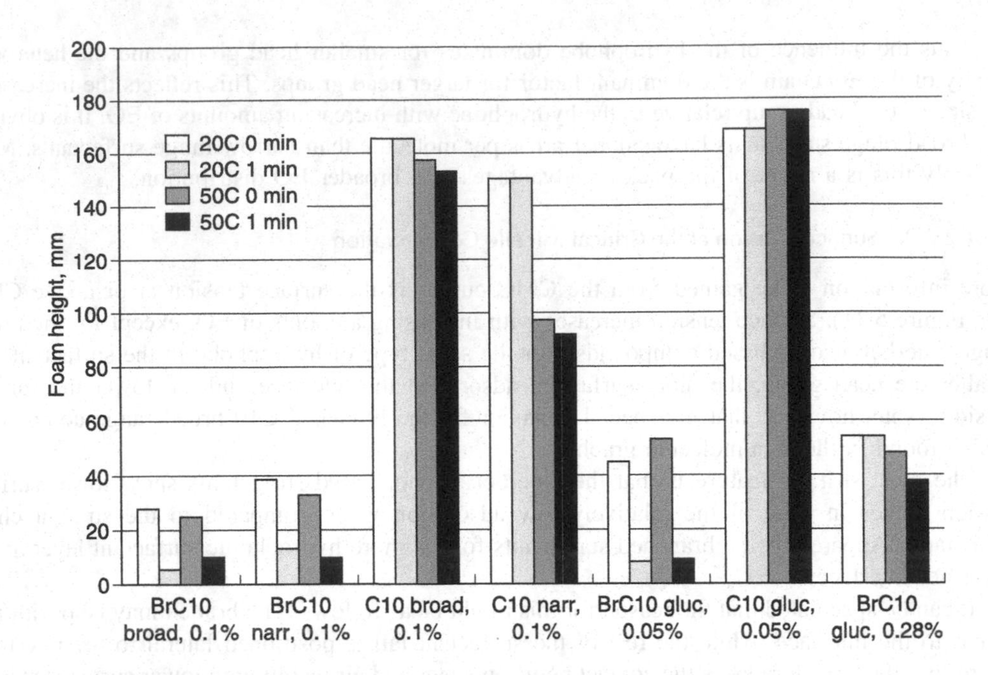

FIGURE 5.12 Foam height, winding equipment, for different C10 alcohols with 5 EO compared to alkyl glucosides from the same alcohols, immediately and after 1 min, at 20 °C and 50 °C.

Temperature effects are seen in one case, C10 narrow range, where the higher temperature gives lower foam depending on the solution having reached its cloud point at 50 °C.

5.3.1.2.3 At Hydrophobic Surfaces

5.3.1.2.3.1 Contact Angles

The change of the contact angle of a droplet of 0.25% surfactant solution on Parafilm® (as a hydrophobic model surface) was followed with time with a high speed video camera. The values after 60 sec are shown in Figure 5.13 versus amounts of EO ⟨m⟩.

Narrow range 1-decanol based surfactants show a larger contact angle than similar Guerbet C10 alcohol based surfactants. This observation corroborates well with the observed trend in the surface tension at the CMC. It is also in line with the fact that the contact angle of neat C10 Guerbet alcohol on Parafilm® is smaller than the contact angle of neat 1-decanol.

Broad range surfactants show smaller contact angles than similar narrow range surfactants, which again can be explained by the adsorption being tighter at the air–water interface.

5.3.1.3 Conclusions Overall

Technical ethoxylates from Guerbet C10 alcohol compared to straight chain decanol have the following properties:

- They contain higher amounts of unreacted alcohol
- They have higher critical micelle concentrations
- They have larger areas per molecule at the air–water interface

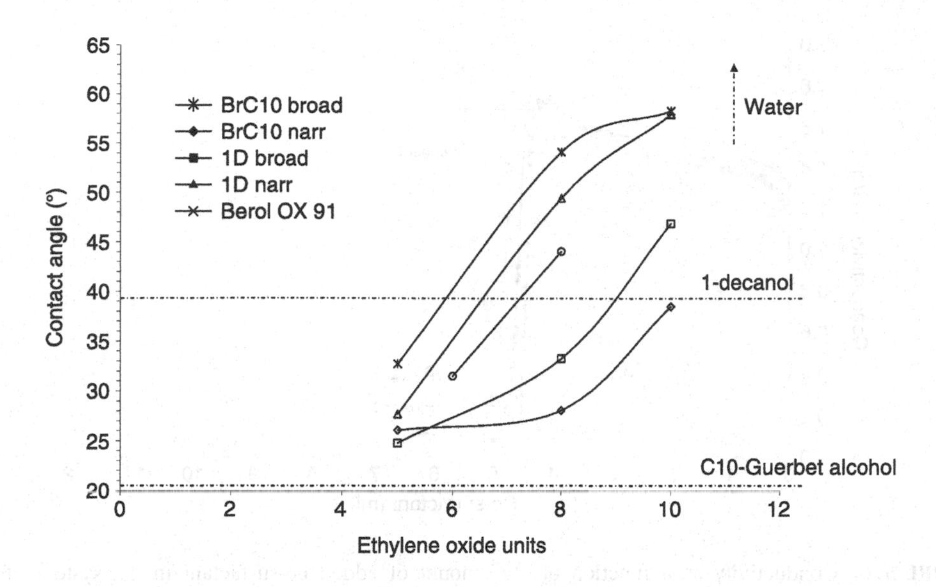

FIGURE 5.13 Contact angle (θ) on Parafilm® (60 s after drop formation) as a function of the average number of ethylene oxide units ($\langle m \rangle$) of 0.25 w% solutions of alcohol ethoxylates with $\langle m \rangle \geq 5$ (T = 21 ± 1 °C, 45 ± 10% rH).

However, they also have:

- Lower surface tension
- Lower cloud points
- Less foam
- Lower contact angle against hydrophobic surfaces
- Lower inversion points in emulsion systems

With the simple automatic titration technique described above the ratio of hydrophobic co-surfactant to total surfactant (α) necessary to invert an emulsion from O/W to W/O can be obtained and related to other characterization methods like HLB. However, it refers to a specific system of oil, temperature, stirring rate etc. In this respect it is more similar to the PIT launched by Shinoda (3) or SAD emanating from Wadle (4).

The deviating behavior of the branched structures giving rise to an expanded area per molecule at the interface and yet a lower surface tension and smaller contact angle is not revealed by titration. To understand structure–property relationships at the molecular level complementary measurements of physicochemical properties need to be done. Titration shows the behavior of the actual technical mixture and is thus also useful for optimizing blends for specific purposes.

5.3.2 Comparison between Solvesso 150 D and Methyl Laurate

With the same titration procedure as described above an investigation was made to fine tune the balance between two emulsifiers of different HLBs when the oil to be emulsified was changed from an aromatic solvent, Solvesso 150 D, to methyl laurate (model oils for an agrochemical formulation) due to environmental reasons.

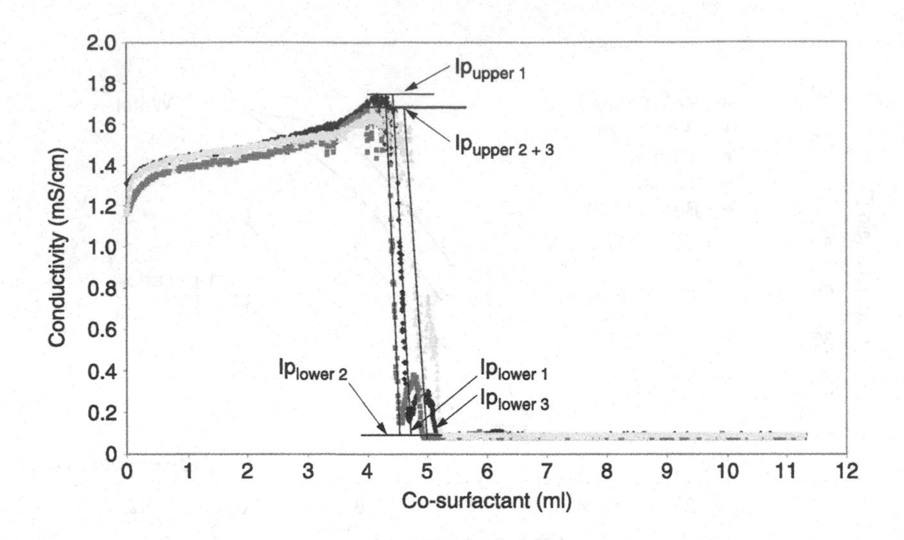

FIGURE 5.14 Conductivity as a function of the amount of added co-surfactant in the system of Berol 992/Solvesso 150 ND/water/NaCl/Berol 822 (Titration speed 4.44 μl/s).

As emulsifiers an alkoxylated short chain fatty alcohol (Berol 992) as the hydrophilic part and dodecyl benzene calcium sulfonate (Berol 822) as the hydrophobic part were chosen. The Scanalys® MPS-I system was used throughout the whole investigation. The measurements were made with a starting concentration of 6% Berol 992, to be closer to the emulsion system used in practice. The stirring speed was 300 rpm. Through the construction of the stirrer information about the rheology of the solution can be obtained. An additional magnetic stirrer at the bottom of the vessel was used.

With this emulsifier/oil system more fine structure was observed in the conductivity curves. The inversion remained clearly visible and the upper and lower inversion points could be deduced in the same way as before.

A typical run is shown in Figure 5.14 where Berol 822 is added slowly under stirring to the 50/50 Solvesso 150 D/water emulsion (25 °C). A very clear inversion is seen and the upper and lower inversion points can be determined. The average is taken as the inversion point. Three repetitive runs give smaller deviations but the fine structure and the inversion are reproducible within an uncertainty limit ($\pm$1.5% for the percent co-surfactant of the total surfactant). A series of tests with increasing amounts of methyl laurate, from 0 to 100% of the oil phase, was made. The results are shown in Figure 5.15.

As can be seen in the graph, more of the hydrophobic co-surfactant is needed when methyl laurate is used as the oil phase. Methyl laurate could then be described as more "hydrophobic" than the aromatic solvent, i.e., having a higher equivalent hydrocarbon number (EACN) (10). Approximate values for EACN for the two oils used here are around 3–4 for the aromatic solvent and for methyl laurate around 7 (by analogy with ethyl oleate) (11), which is consistent with the results in this investigation. It is well known that aromatic oils have a tendency to penetrate into the interfacial layer (12) thus changing its curvature. One could anticipate that this would happen here, thus decreasing the demand for a hydrophobic co-surfactant to achieve the balanced state for the aromatic oil.

It has been shown (13) that when mixtures of oils with very different polarity, like hexadecane (ACN = 16) and ethyl oleate (EACN = 6–7), are used there is an interfacial segregation of the oils

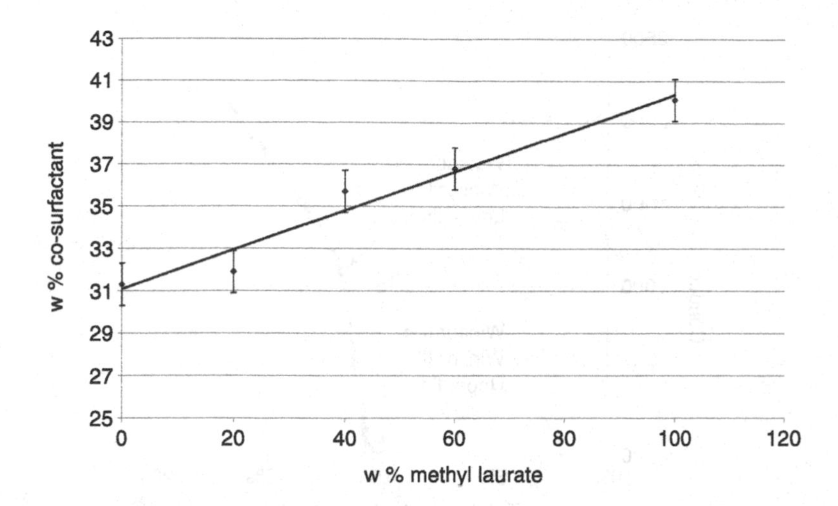

FIGURE 5.15 Inversion point dependence on weight percent of methyl laurate in the oil mixture for the emulsifier system Berol 992 and Berol 822.

causing an increased concentration of the most polar oil close to the interfacial layer. The results in this case do not indicate any such segregation but rather show that the stepwise increased ratio also causes a stepwise increase of hydrophobicity. Successive measurements with mixtures of solvents and titrations with known emulsifier pairs can obviously be used to classify and describe the balance of a practical emulsion system.

As has been pointed out in the introduction, the surfactant ratio at the inversion point is the point of minimal interfacial tension, i.e., where the formation of droplets is easiest, but it is also the most unstable emulsion situation (14). To find the stable area, tests have to be made at some distance from the inversion point. These studies are of course easier and more practical once the location of the inversion point is known.

5.3.3 Optimization of a Surfactant Combination for Soil Remediation

An application where the minimal interfacial tension is the optimum situation is soil remediation (15). The last example is taken from development work for that application.

To clean up contaminated soil different techniques can be applied. One of the most efficient is surfactant flushing, where surfactants are used to facilitate the transport of water insoluble contaminants like fuels, solvents, or halogenated hydrocarbons out of the ground. It is very important to find the lowest interfacial tension, i.e., to be within the microemulsion region (Winsor III) to obtain the best results. When balancing the surfactant system the oil phase can either be the dirt as such or a model oily dirt with similar properties.

In the example given here Danish winter diesel is the oil phase and the emulsifiers are a hydrophilic alkyl glucoside based on a 50/50 C8/C10 alcohol mixture (AG 6210), and a hydrophobic nonionic, C11E2 (Berol 532). When investigated in the system described in Section 5.3.1 the hydrophilicity of AG 6210 was very similar to pure straight chain C10 glucoside.

In this case the procedure was reversed as compared to the examples in Sections 5.3.1 and 5.3.2, starting with the hydrophobic surfactant, C11E2, in an oil/water 50/50 mixture and titrating in the hydrophilic surfactant until conductivity started to increase. The fine structure in the graph showing conductivity versus surfactant concentration was compared to the equilibrium behavior

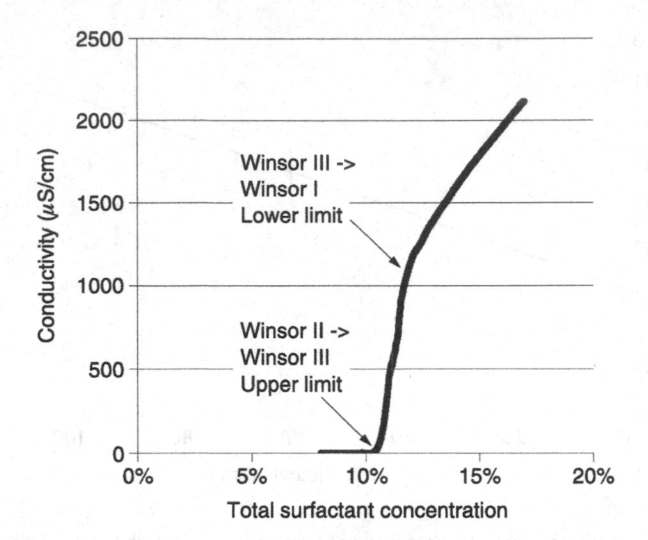

FIGURE 5.16 Breakpoints in the conductivity scan for Danish diesel/water 50/50, with C11E2 as starting emulsifier and C8–10 glucosides added as cosurfactant against total amount of surfactant.

of corresponding samples. In this case two bends in the curve were seen and found to correspond to the upper and lower limits of the Kahlweit fish body in the phase diagram. The analysis uses the well known terms from microemulsion studies (16), Winsor I, II, and III, to describe whether the surfactant phase is in the water phase, in the oil phase, or in the three-phase region.

Figure 5.16 shows a typical run with upper and lower limits marked and what they correspond to in an equilibrium system. A series of titrations with different starting concentrations was made and the resulting bends in the curve were related to the phase diagram as is seen in Figure 5.17.

To confirm that we had found the optimum situation with minimal interfacial tension the interfacial tension was measured between oil and water phase with 8% total surfactant and the part of the hydrophobic surfactant being varied from 0 to 0.9 (Figure 5.18). The minimum was found at 0.7 which corresponds well with the titration results and the phase diagram shown in Figure 5.17.

Later investigations on sand columns on a laboratory scale as well as pilot scale trials have confirmed the efficiency of the mixture investigated here as surfactants for flushing in soil remediation.

5.4 CONCLUSIONS

In this chapter various methods of using emulsion inversion as a technique for characterizing and fine tuning surfactant formulations have been discussed. The phenomenon, as such, with the inversion point being related to the minimum in interfacial tension and interfacial curvature equal to zero, is well known and applied in many areas (1,17,18). To know where your formulation is situated in the phase diagram, and especially the location of the optimum formulation with SAD = 0, is crucial for efficient and qualified product development. The tool described here opens a way for automatic HLB scanning and thus easy access to this important information. Examples have been given on how to characterize and compare surfactants of varying structures when they are used in similar surroundings. Effects of varying the surroundings, such as the

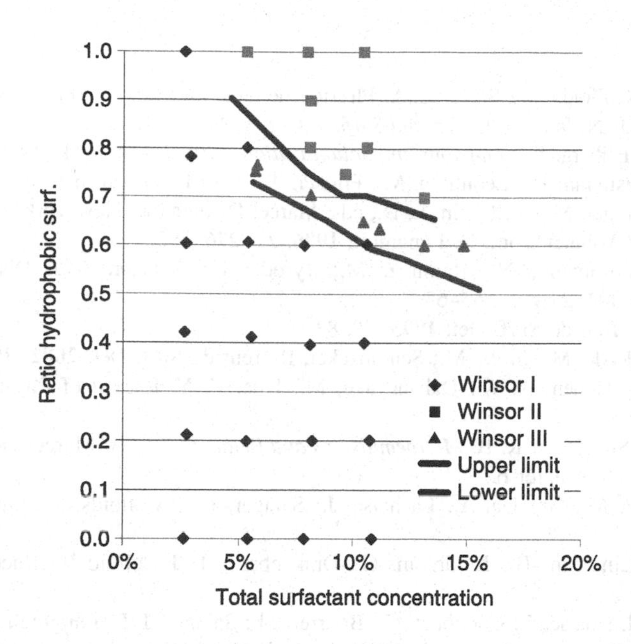

FIGURE 5.17 Conductivity scan together with equilibrium results for Danish diesel/water 50/50, with C11E2 and C8–10 glucoside as emulsifiers.

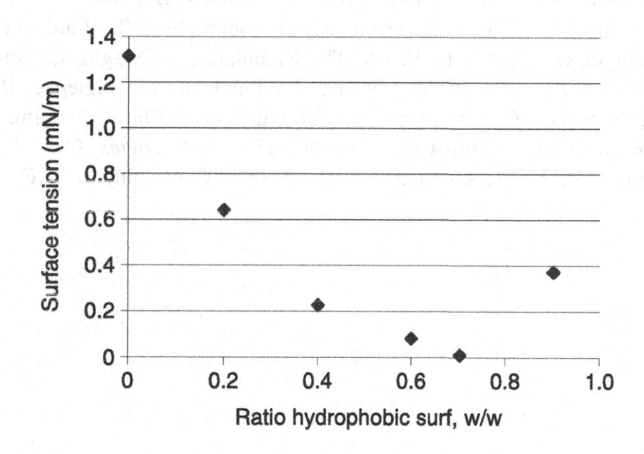

FIGURE 5.18 Interfacial tension dependence on emulsifier mixture for Danish diesel and water.

character of the oil, have been shown as well as optimization of surfactant mixtures for use in specific applications like soil remediation.

More work on other possible ways of performing the titration needs to be done to facilitate the interpretation of the inversion results in terms of phase behavior. Extensive comparisons with equilibrium samples will give access to explanations for fine structures in the titration curves and thus increase the feasibility of this technique. All in all we hope to see an increase in application of the principles underlying the results in this study as well as an increased usage of the equipment in development of new products and formulations for novel as well as established application areas.

REFERENCES

1. Davis, H. T. Colloids and Surfaces, A: Physicochemical and Engineering Aspects, **1994**, *91*, 9–24.
2. Israelachvili, J. N. *Intermolecular and Surface Forces*, Academic Press Ltd, London, 1992.
3. Shinoda, K.; Friberg, S. *Emulsions and Solubilization*, Wiley, New York, 1986.
4. Wadle, A.; Tesmann, H.; Leonhard, M.; Förster, T. Phase inversion in emulsions, in *Surfactants in Cosmetics*, Rieger, M. M.; Rhein, L. D., eds, Marcel Dekker Inc, New York, Basel, 1997, 207–224.
5. Kabalnov, A.; Wennerström, H. Langmuir, **1996**, *12*, 276–292.
6. Pérez, M.; Zambrano, N.; Ramirez, M.; Tyrode, E.; Salager, J. J. Dispersion Science and Technology, **2002**, *23(1–3)*, 55–63.
7. Kahlweit, M. Tenside Surf. Det. **1993**, *30*, 83.
8. Sottman, T.; Lade, M.; Stolz, M.; Schomäcker, R. Tenside Surf. Det. **2002**, *39(1)*, 20–28.
9. Rosen, M. J.; Cohen, A. W.; Dahanayake, M.; Hua, X. Y. Journal of Physical Chemistry, **1982**, *86(4)*, 541–545.
10. Bourrel, M.; Schechter, R. B. *Microemulsions and Related Systems*, Marcel Dekker, Inc, New York and Basel, 1998, chapter 6.
11. Minana-Perez, M.; Graciaa, A.; Lachaise, J.; Salager, J. L. Colloids and Surfaces, A, **1995**, *100*, 217–224.
12. Ulmius, J.; Lindman, B.; Lindblom, G.; Drakenberg, T. J Colloid Interface Science, **1978**, *65*, 88–98.
13. Graciaa, A.; Lachaise, J.; Cucuphat, C.; Bourrel, M.; Salager, J. L. Langmuir, **1993**, *9*, 1473–1478.
14. Kabalnov, A.; Weers, W. Langmuir, **1996**, *12(8)*, 1931–1935.
15. Sabatini, D. A.; Knox, R. C.; Harwell, J. H.; Wu, B. Journal of Contaminant Hydrology, **2000**, *45*, 99–121.
16. Winsor, P. A. *Solvent Properties of Amphiphilic Compounds*, Butterworth, London, 1954.
17. Kahlweit, M.; Strey, R.; Haase, D.; Kunieda, H.; Schmeling, T.; Faulhaber, B.; Borkovec, M.; Eicke, H. F.; Busse, G.; Eggers, F.; Funck, Th.; Richmann, H.; Magid, L.; Söderman, O.; Stilbs, P.; Winkler, J.; Dittrich, A.; Jahn, W. Journal of Colloid and Interface Science, **1987**, *118*(*2*), 436–453.
18. Johansson, I.; Boen Ho, O. Emulsions or microemulsions? Phase diagrams and their importance for optimal formulations, in *Mesoscale Phenomena in Fluid Systems*, Case, F.; Alexandridis, P. eds. ACS Symposium Series 861; American Chemical Society, Washington D.C., 2003, 394–412.

6 Highly Concentrated Emulsions as Templates for Solid Foams

Jordi Esquena and Conxita Solans

CONTENTS

6.1 INTRODUCTION

Highly concentrated emulsions are an interesting class of emulsions characterized by an internal phase volume fraction exceeding 0.74, the critical value of the most compact arrangement of uniform, undistorted spherical droplets [1,2]. Consequently, their structure consists of deformed (polyhedral) and/or polydisperse droplets separated by a thin film of continuous phase, a structure resembling gas–liquid foams, as shown in Figure 6.1. They are high-internal-phase (HIPE) emulsions which are also referred to in the literature as gel emulsions [3–6], hydrocarbon gels [7], biliquid foams [8], etc. In this text they will be referred to as highly concentrated emulsions.

The internal or dispersed phase of highly concentrated emulsions can be either polar or nonpolar and, as ordinary emulsions, they are classified in two categories: water-in-oil (W/O) and oil-in-water (O/W). However, they can be classified according to other criteria such as the microstructure of the continuous phase or the interaction forces between droplets. Phase behavior studies have shown [3,9–11] that highly concentrated emulsions separate into two phases: one phase is a submicellar surfactant solution in water (or a surfactant solution in oil) and the other phase can be either a microemulsion [3,9] or a cubic liquid crystalline phase [10,11]. Studies using

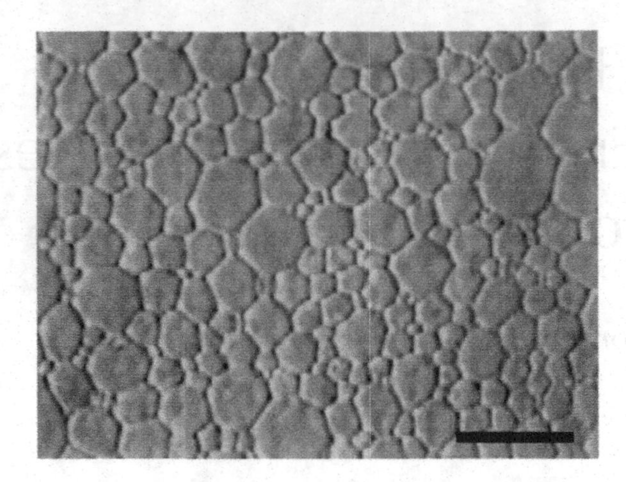

FIGURE 6.1 Example of a highly concentrated emulsion observed by optical microscopy. The scale bar indicates 20 μm.

different techniques, nuclear magnetic resonance (NMR) [9], electron spin resonance (ESR) [12], small angle X-ray scattering (SAXS) [11,13], etc., have confirmed that the continuous phase of these emulsions possess indeed such microstructures. Accordingly, they can be classified as microemulsion-based and cubic-phase-based [14]. Considering the interaction forces between droplets, they can also be classified as adhesive and nonadhesive [15]. Adhesive emulsions, in the presence of attractive forces between droplets, do not relax and maintain a high degree of packing. Nonadhesive emulsions, brought into contact with its continuous phase, relax in a way that the emulsion dilates to the state of spherical (nondeformed droplets). In this chapter, attention is focused to microemulsion-based W/O and O/W highly concentrated emulsions.

The conventional method of preparation for highly concentrated emulsions consists of dissolving a suitable emulsifier in the component that will constitute the continuous phase followed by stepwise addition of the component which will constitute the dispersed phase, with continuous moderate stirring. However, they can be also prepared by weighing all the components at the final composition, followed by shaking or stirring the sample [3–5]. This method of preparation has been termed the multiple emulsification method because at a certain step of the emulsification process, the mixture consists of a multiple W/O/W emulsion [5]. It should be noted that emulsification by the conventional and multiple emulsification methods results in rather large (over a micrometer in diameter) and polydisperse emulsions, as that depicted in Figure 6.1. An interesting method of preparation, which takes advantage of the phase transitions produced during the emulsification process, the so-called spontaneous formation method [16–18], produces emulsions with smaller droplet size and narrower size distributions than those obtained by the other methods.

The spontaneous formation method is based on the phase inversion temperature (PIT) emulsification method [19]. The PIT is the temperature at which polyoxyethylene nonionic surfactants change their preferential solubility from water (oil) to oil (water) and inversion from O/W to W/O emulsions or vice versa is produced [19]. Highly concentrated W/O emulsions form at temperatures above the PIT of the corresponding system while highly concentrated O/W emulsions form below this temperature [3]. The PIT is also designed for hydrophilic–lipophilic balance (HLB) temperature because the hydrophilic–lipophilic properties of the surfactant are balanced. At this temperature, maximum solubilization of oil and water is produced with the minimum amount of surfactant and the interfacial tensions in the system attain ultra-low values (e.g., 10^{-3} mN m^{-1}) [20]. Therefore, emulsification is favored: emulsions with small droplet size can be obtained with low energy input. However, emulsion stability is very low at this balanced temperature. The PIT emulsification method consists in preparing the emulsions at the HLB temperature (taking advantage of the low interfacial tension values achieved) followed by a rapid cooling or heating of the samples to produce O/W or W/O emulsions, respectively.

Formation of highly concentrated emulsions without the need for mechanical stirring can be achieved, by quickly cooling (or heating) a water-in-oil (or oil-in-water) microemulsion from a temperature higher (or lower) than the HLB temperature of the system to a temperature below (or above) it [5,16,18]. In practice, it may be difficult to obtain a single (microemulsion) phase in a given system at the starting emulsification temperature. Emulsification can be also carried out by the same method starting from a two- or multiple-phase system (i.e., emulsion). However, the temperature change process has to be accompanied by mechanical stirring in order to obtain uniform and small emulsion droplets [21].

The stability of highly concentrated emulsions is greatly affected, as in conventional emulsions, by the nature of the components, the volume fraction of the dispersed phase, the oil/surfactant weight ratio, presence of additives, and temperature [22]. The time taken for phase separation, which may vary from minutes to years, can be significantly retarded by appropriate selection of the different composition variables and temperature. Accordingly, highly concentrated emulsions with stabilities suitable for their use as reaction media can be obtained. In systems with polyoxyethylene type nonionic surfactants and aliphatic hydrocarbons, a relationship between the stability of highly concentrated emulsions and the HLB temperature of the corresponding system was observed [3]. The stability of O/W (or W/O) highly concentrated emulsions is maximum at about 25 to 30 °C lower (or higher) than the HLB temperature of the corresponding system. This fact was explained by the changes in structure of the continuous microemulsion phase with temperature on the basis of phase behavior, NMR, and conductivity studies [5,9]. At the HLB temperature (minimum stability), the structure of the microemulsion is bicontinuous, while when temperature is increased or decreased, the structure changes to globular W/O or O/W type, respectively. Additives, e.g., electrolytes, may also influence emulsion stability. It was found that salts with large salting-out effect are more effective in stabilizing these emulsions because they decrease more pronouncedly the HLB temperature of the system [12,22]. The hydrophilic part of the surfactant dehydrates in the presence of electrolytes (producing a decrease in the HLB temperature), the surfactant–surfactant interactions increase and, therefore, the interfacial film is more rigid.

One characteristic property of highly concentrated emulsions is their high viscosity as compared to that of the constituent phases. They are non-Newtonian fluids characterized by a yield stress below which they show a solid-like behavior [2,23,24]. Determination of the rheological properties by means of dynamic (oscillatory) measurements showed that they have a viscoelastic response that can be fitted to a Maxwell liquid element: an elastic modulus produced by interfacial area increase and a viscous modulus produced by the loss caused by slippage of droplets against droplets. The values of relaxation time were found to be proportional to the continuous phase viscosity and inversely proportional to the continuous phase volume fraction [25].

The characteristic properties of highly concentrated emulsions are of particular interest for theoretical studies and for applications. Their use for practical applications is well known since long ago. They are widely used as formulations in, for example, cosmetics, foods, and pharmaceuticals. One of the most promising applications is their use as reaction media. In this context, they have received a great deal of attention for the preparation of low-density organic and inorganic materials (solid foams, aerogels) [8,21,26–28] and in chemical and enzyme-catalyzed reactions [29–32] as alternative to conventional solvent media.

6.2 HIGHLY CONCENTRATED EMULSIONS AS TEMPLATES FOR MACROPOROUS FOAMS

Solid macroporous foams, which consist of interconnected sponge-like macropores, can be obtained by polymerization in the continuous phase of highly concentrated emulsions, followed by the removal of the dispersed phase components [33–39]. These materials can possess very high pore volume and very low bulk density. Both organic and inorganic monomers can be used to obtain macroporous solid foams.

6.2.1 Preparation of Organic Macroporous Foams in Highly Concentrated Emulsions

The use of highly concentrated emulsions to prepare solid foams with densities smaller than 0.1 $g\,ml^{-1}$ was first described in a patent from Unilever in 1982 [33]. The foams were obtained in a concentrated W/O emulsion, stabilized by sorbitan fatty esters, with organic monomers such as styrene and the cross-linker divinylbenzene in the continuous phase.

This process, which was later studied by Williams [34,35], has been applied to other highly concentrated emulsions to obtain a wide variety of different macroporous materials using, as monomers, styrene [26,27], styrene/divinylbenzene [26,27,36], divinylbenzene/ethylvinylbenzene [40], acrylamide, alkyl methacrylates, etc. [26,27,37]. These macroporous solid foams have found successful applications as supports for catalysts, immobilization of enzymes, selective membranes, templates for the preparation of other materials, etc. [26,27,36,37,41]. The properties (average cell size and cell size distribution, surface area, macroscopic density, etc.) of the materials obtained using this technology were improved with respect to those obtained by other methods, but their structures were still difficult to control. An example, corresponding to a polystyrene foam, obtained from a highly concentrated W/O emulsion, is shown in Figure 6.2. The polyhedral macropores are interconnected through narrower necks, because the films that separate adjacent droplets in concentrated emulsions are very thin, and polymerization takes place mainly in the emulsion plateau borders.

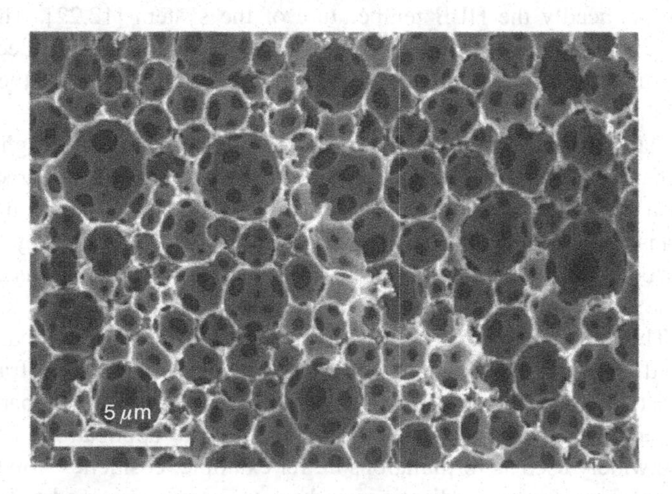

FIGURE 6.2 Micrograph of a polystyrene foam, obtained by polymerizing (60 °C, 48 h) in the continuous phase of a water-in-styrene emulsion, which contains 90 wt% aqueous solution, stabilized by 2 wt% $C_{16}(EO)_6$. The scale bar indicates 5 μm.

Williams studied the preparation of styrene–divinylbenzene cross-linked solid foams, by polymerization in water-in-monomer highly concentrated emulsions [34]. He demonstrated that the increase in both the volume fraction of the dispersed phase and the surfactant concentration produces a thinning of the films with separate adjacent droplets, increasing the necks which connect the cells that constitute the solid foams [34]. Williams and Sherington studied the formation of the connecting necks between the cells [35,36]. The monomers are adsorbed preferentially on the parts of the surfactant monolayer located in the interstices between adjacent drops [35]. Cameron et al. followed the polymerization process by cryo-SEM [36]. Their results suggested that the formation of necks between the adjacent cells is due to the contraction of the thin monomeric films during the conversion of monomer to polymer. Ruckenstein et al. reported the preparation of composite polymers by simultaneous polymerization of both hydrophilic and lipophilic monomers in highly concentrated emulsions, stabilized either by ionic or nonionic surfactants [37,38]. Some reviews have appeared on this subject [26,27].

The density of the foams can be as low as 0.02 g/ml and the specific surface area can be as high as 350 m^2/g, by addition of an inert oil (porogen) [41]. The cell size in the foams is typically in the range of 1 to 20 μm. Foam cell sizes are highly dependent on coalescence during polymerization and addition of electrolyte has been found to greatly decrease the cell size [42].

In all these reports [26,27,33–42], the cells were large, with diameters generally greater than 10 μm, and highly nonhomogeneous due to the high polydispersity of the precursor emulsion. These emulsions were prepared by the classical method of dissolving a suitable emulsifier in the component that will constitute the continuous phase, followed by the slow addition of the dispersed phase components, with stirring over a relatively long period of time. Solans et al. reported that highly concentrated emulsions with smaller and more homogeneous droplet size were obtained by making use of phase behavior [16–18]. Microemulsions were prepared at the PIT, and the temperature was rapidly changed. The PIT or hydrophile–lipophile temperature (THLB) is the temperature at which the hydrophilic and lipophilic properties of the surfactant are balanced in the system [19].

Figure 6.3(a) shows the aspect, as seen by optical microscopy, of a water-in-styrene highly concentrated emulsion that contains 90 wt% of dispersed phase, which has been obtained by a method based on the phase inversion principle, in which emulsification was achieved increasing the temperature, across the PIT, from a diluted O/W emulsion to a concentrated W/O emulsion. The droplets of this emulsion are relatively large, between 3 and 20 μm, similar to highly concentrated emulsions prepared by conventional methods [1–5], because the two-phase system is unstable at the PIT. The droplet size is much smaller, ≈ 1 μm, if stronger agitation is applied to the sample during the increase in temperature, as shown by Figure 6.3(b). The respective polystyrene macroporous foams are shown in Figures 6.3(c) and (d). It should be pointed out that the foam in Figure 6.3(d) consists of smaller and less polydispersed cells than foams from emulsions prepared by conventional methods, which are generally bigger than 10 μm [33–40]. Therefore, this emulsification method that is based on the PIT principle allows the preparation of homogeneous solid foams, with smaller cell size [21].

Studies by Esquena et al. on mechanical properties of macroporous organic monoliths, by means of compression tests, are described in Figure 6.4. Strength and toughness were determined from the stress/strain curve [21]. Both parameters were normalized by dividing by the monolith bulk densities, to allow better comparisons. The results of mechanical properties (Figure 6.4) clearly show that samples prepared by the PIT method lead to monoliths with higher mechanical properties than those prepared by conventional method, which consists in step-wise addition of

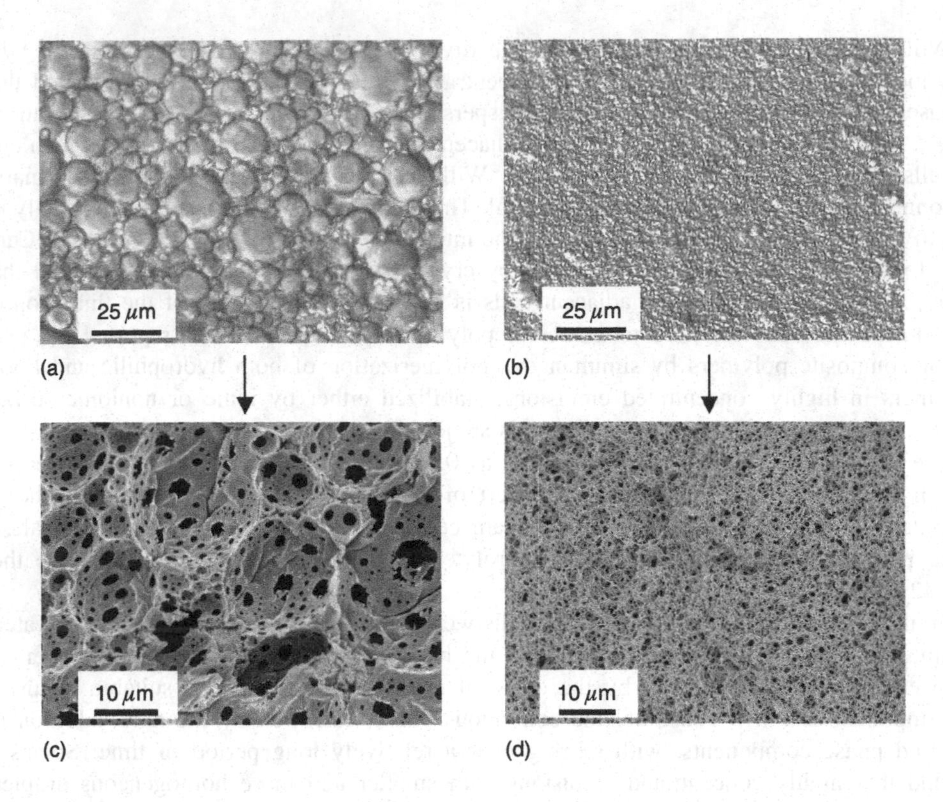

FIGURE 6.3 Preparation of macroporous polystyrene foams in highly concentrated W/O emulsions of the system $H_2O/K_2S_2O_8/C_{16}(EO)_6/C_{12}(EO)_8$/Synperonic L-64/styrene/divinylbenzene/tetradecane containing 90 wt% of dispersed phase. (a) W/O emulsion obtained by PIT method, after heating the sample from 0 °C to 70 °C, while stirring very gently by hand. (b) W/O emulsion obtained by PIT method, from 0 °C to 70 °C, and simultaneously applying strong agitation by hand. (c) Scanning electron micrographs of sample (a) after polymerization and drying. (d) Micrograph of sample (b) after polymerization and drying. (Reproduced from Esquena, J., Sankar, G.S.R.R., and Solans, C. *Langmuir* 19, 2983–2988, 2003. With permission of the American Chemical Society.)

the dispersed phase to the continuous phase, under continuous vibromixer agitation. Strength was approximately three times higher and toughness was approximately 50% higher.

A common issue, which can be solved according to different methods, is the purification process. Generally, organic solvents are removed by evaporation, but small quantities can remain trapped in the final foam. Macroporous organic foams cannot be purified by calcination at high temperature, which is the most general method to purify inorganic foams. Cooper et al. have reported an interesting approach, which produces solvent-free macroporous organic foams, by using supercritical CO_2-in-water highly concentrated emulsions as templates [43]. This method is described, schematically, in Figure 6.5. High pressure reactions (100 bar, 20 °C) were carried out in a stainless steel reactor. After polymerizations, CO_2 was removed by reducing the pressure. Finally, residual water was removed from the macroporous polymers by drying in air followed by drying under vacuum.

Macropore templating by supercritical CO_2 droplets has the advantage that it does not require volatile organic solvent, either in the synthesis or in the purification steps [43]. Unlike other

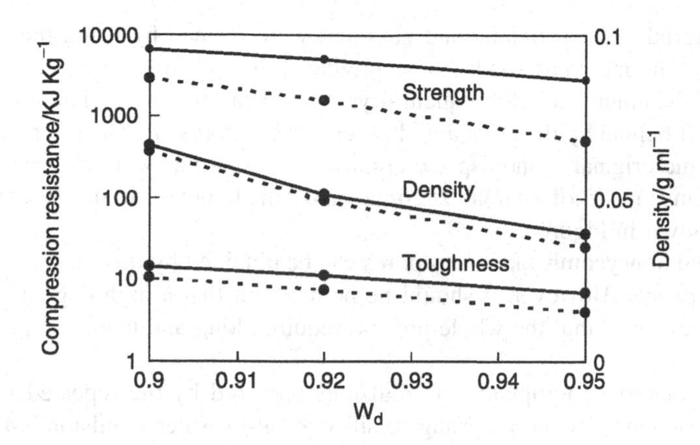

FIGURE 6.4 Density and normalized mechanical properties as a function of the weight fraction of the dispersed phase, for the two preparation methods. Solid line: PIT method, Dashed line: conventional method. (Reproduced from Esquena, J., Sankar, G.S.S.R., and Solans, C. *Langmuir* 19, 2983–2988, 2003. With permission of the American Chemical Society.)

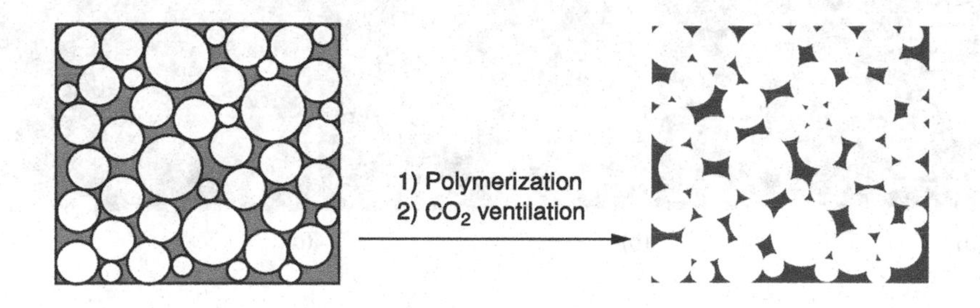

FIGURE 6.5 Preparation of macroporous materials by supercritical CO_2-in-water emulsion templating. (Redrawn from Butler, R., Davies, C.M., and Cooper, A.I. *Adv. Materials* 13, 1459–1463, 2001. With permission of Wiley-VCH Verlag GmbH.)

solvents, which may be difficult to remove completely, supercritical CO_2 is nontoxic, non-flammable, and reverts to the gaseous state upon de-pressurization. However, this method has the inconvenience that expensive equipment is required, and fluoro-containing surfactants are needed because conventional surfactants do not adsorb on supercritical CO_2 droplets.

6.2.2 Preparation of Inorganic Macroporous Foams in Highly Concentrated Emulsions

The previous section has described different methods, which involve the use of highly concentrated emulsions as templates, to obtain macroporous organic foams. Macroporous inorganic materials can also be templated in highly concentrated emulsions. Imhof and Pine described a method that makes use of monodisperse emulsions [44], which were obtained according to the Bibette repeated fractionation procedure [45]. Droplet volume fractions can exceed the close-packing limit of 74%. Starting with highly concentrated O/W emulsions of monodisperse droplets, macroporous materials (silica, titania, and zirconia) were formed by using the emulsion droplets as templates [44]. The inorganic oxides were produced through sol–gel processes, at the aqueous external phase of the emulsions. Subsequent drying and heat treatment yields solid materials with spherical pores left behind by the emulsion droplets. These pores are highly ordered, reflecting the self-assembly of the original monodisperse emulsion droplets into a nearly crystalline array [45]. Macroporous foams, made of titania, zirconia, and silica, obtained in isooctane-in-formamide emulsions, are shown in Figure 6.6.

Highly ordered honeycomb-like macropores can be obtained by this method, depending on the emulsion polydispersity. However, it should be pointed out that a high degree of monodispersity is difficult to achieve, and that the whole process required long and tedious repeated fractionation steps.

Macroporous materials, templated in emulsions prepared by the repeated fractionation technique, were also obtained by Yi and Yang in silicone oil-in-water emulsions [46]. The emulsion droplets were flocculated and tended to migrate to certain regions, which eventually led to the formation of highly concentrated emulsions. The aspect of macroporous silica, obtained in one such concentrated region, is shown in Figure 6.7.

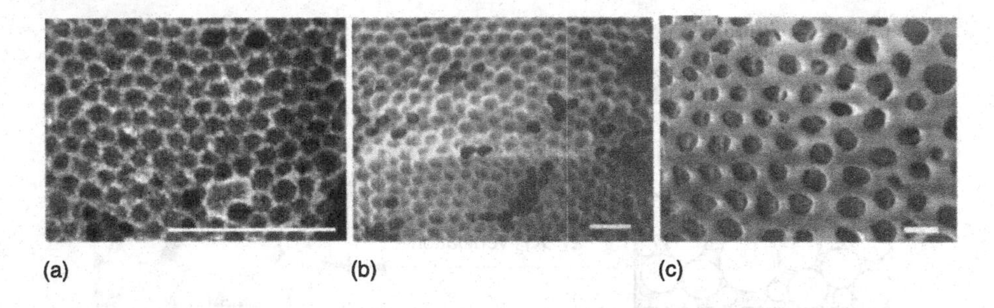

FIGURE 6.6 Scanning electron micrographs of macroporous inorganic oxides, obtained after calcinations. Scale bars, 1 μm. (a) Titania, (b) Zirconia, (c) Silica (89% porosity). (Reproduced from Imhof, A. and Pine, D.J. *Nature* 389, 948–951, 1997. With permission of the Nature Publishing Group.)

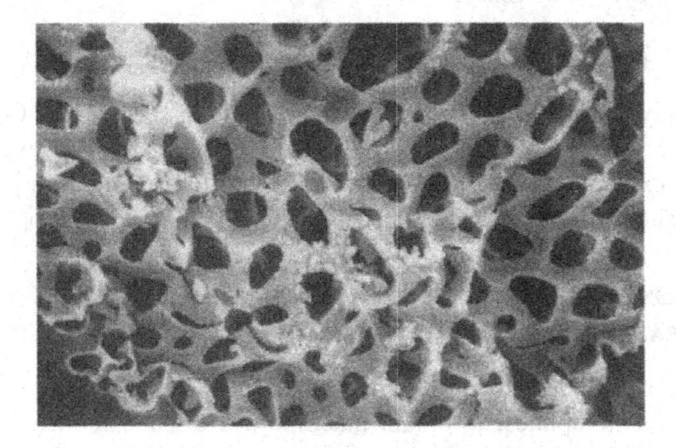

FIGURE 6.7 Image of macroporous silica, obtained by scanning electron microscopy, of a region that consisted of highly concentrated emulsion droplets. (Reproduced from Yi, G.R. and Yang, S.M. *Chem. Mater.* 11, 2322–2325, 1999. With permission of the American Chemical Society.)

These results show that macroporous foams of inorganic oxides can be obtained by using highly concentrated emulsions as templates, by carrying out the reactions in the external phase. It is worthwhile noting that many other materials, different from inorganic oxides, could be obtained by templating in highly concentrated emulsions, despite the fact that most of the studies deal with inorganic oxides, specially with silica.

6.3 HIGHLY CONCENTRATED EMULSIONS AS TEMPLATES FOR HIERARCHICALLY TEXTURED FOAMS

Materials with extremely high specific surface areas, such as zeolites or micro/mesoporous materials, are important in applications such as in catalysis and molecular separations [47]. However, these materials usually have rather poor pore volumes ($<2\ cm^3\ g^{-1}$), which restrict their performance in certain applications in which large molecules (enzymes, polymers, etc.) are involved.

The inclusion of macropores can provide enhanced mass transport into and out of the porous structure. This becomes very important for processes that have low diffusion rates. Recently, materials with dual meso- and macroporous properties have attracted much attention because they combine the advantages of high specific surface with the accessible diffusion pathways associated with macroporous structures. The large active surface of mesopore walls can be reached easily, with good diffusion coefficients, from the macropores. Because of these dual properties, these meso/macropororous materials are very important in many industrial applications, which include catalytic surfaces and supports [48], adsorbents, chromatographic materials, filters [49,50], light-weight structural materials [51], and thermal, acoustic [52], and electrical insulators [53]. Some novel applications include catalytic supports in fuel cells or energy storage in double-layer capacitors [54].

The capability to control the pore size and morphology at different length scales is of utmost importance. In addition, it is also very important to obtain such materials by simple and cost-effective methods. Several methods to prepare hierarchically textured materials have been described. Many of these use different kinds of emulsions as templates, because they allow the control of large macropores. Highly concentrated emulsions have the advantage that materials with a very high pore volume and with well-connected macropores can be obtained. The methods to obtain hierarchically textured foams, from highly concentrated emulsions, are reviewed in the next section.

6.3.1 Preparation of Hierarchically Textured Foams in Highly Concentrated Emulsions Stabilized by Surfactant Molecules

6.3.1.1 Two-Step Methods

Esquena et al. have used highly concentrated emulsions as templates for the preparation of materials, with dual meso/macroporous structures, by a two-step templating process [28]. The first step consists in the preparation of organic macroporous foams by polymerizing in the continuous phase of highly concentrated O/W emulsions [21], as described in a previous section. These solid foams can be used as scaffolds for the preparation of the meso/macroporous inorganic oxides. The second step consists in imbibing with ethanol sol–gel solutions that contain inorganic precursors and block copolymer surfactants [28]. The cooperative self-assembly of the surfactant molecules, together with the inorganic oxide precursor species, leads to the formation of ordered mesostructures. The resulting organic–inorganic composite materials were dried and then calcined in air at high temperature in order to remove all organic components and to obtain the meso/macroporous

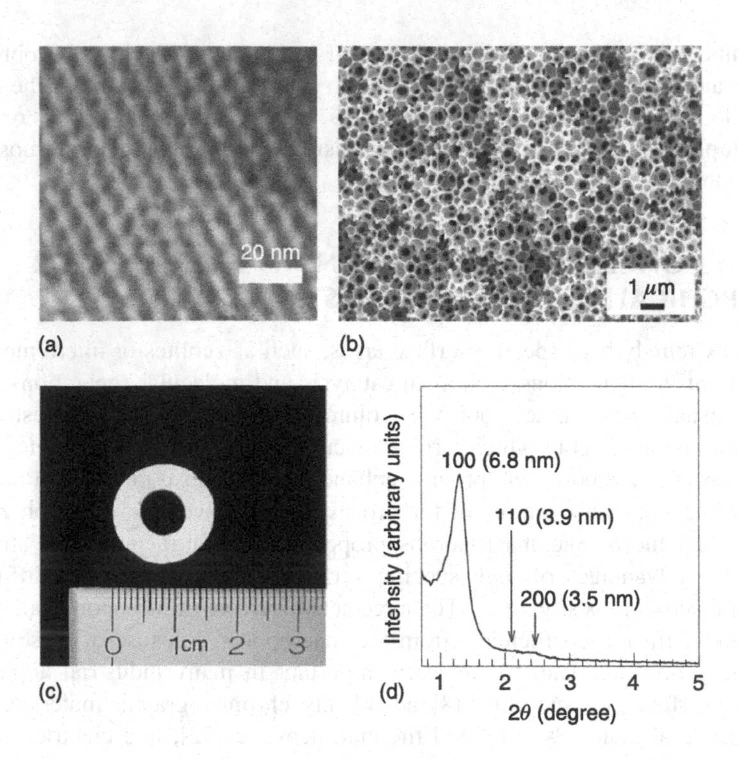

FIGURE 6.8 (a) Image, obtained by transmission electron microscopy, of a calcined meso/macroporous silica monolith. (b) Image, obtained by scanning electron microscopy, of a calcined meso/macroporous silica monolith. (c) An annulus-shaped silica monolith, after calcination. (d) X-ray diffraction spectra, at small angle, of the same sample shown in (a). (Reproduced from Maekawa, H., Esquena, J., Bishop, S., Solans, C., Chmelka, B.F. *Adv. Materials* 15, 591–596, 2003. With permission of Wiley-VCH Verlag GmbH.)

inorganic materials. Therefore, these materials are obtained by a double templating process: Highly concentrated emulsions for the macropores and supramolecular self-assembly for the mesopores [28]. The bimodal distribution of pores was charaterized by highly ordered mesopores, in the size regime 5 to 10 nm (Figure 6.8(a)), and by interconnected macropores between 0.1 and 5 μm (Figure 6.8(b)).

The meso/macroporous materials obtained by this method are moldable, since the macroscopic external shape depends on the size and morphology of the mold used to carry out the first polymerization process. For instance, an annulus-shaped meso/macroporous monolith is displayed in Figure 6.8(c). The inorganic oxide materials were also characterized by nitrogen sorption and by X-ray diffraction using small-angle techniques (Figure 6.8(d)). The results were consistent with those obtained by electron microscopy techniques.

The two meso/macro length scales can be adjusted independently because the surfactant molecules, which form the mesostructure, are introduced after the first step of polymerization of a highly concentrated emulsion. Therefore, the two templating systems do not interfere with each other, and consequently the meso- and macro-structures can be optimized separately [28]. Inorganic oxides, such as silica, titania, and zirconia, were obtained by this two-step process. High pore volumes (17 $cm^3\ g^{-1}$) could be obtained. The specific surface areas, determined by the Brumauer-Emmett-Teller (BET) method applied to the nitrogen adsorption isotherm, were

FIGURE 6.9 Photographs of a meso/macroporous silica monolith (a) and the templated meso/macroporous carbon monolith (b). The diameter of the coin is 2.3 cm. (Reproduced from Alvarez, S., Esquena, J., Solans, C., and Fuertes, A.B. *Engineering Materials* 6, 897–899, 2004. With permission of Wiley-VCH Verlag GmbH.)

between 250 and 750 $m^2 g^{-1}$. The higher surface areas were obtained at higher inorganic precursor concentrations and correlate with reduced macropore volumes.

The meso/macroporous inorganic oxide monoliths can be used as templates for obtaining porous carbon monoliths, as shown by Alvarez et al. [54]. Silica monoliths were impregnated with furfuryl alcohol, which was polymerized. Then, the materials were carbonized by heating at 800 °C in the absence of oxygen, obtaining an exact meso/macroporous carbon replica of the initial monolith. Finally, the silica was dissolved in HF. Examples of the initial silica monoliths and the final carbon meso/macroporous replicas are shown respectively in Figure 6.9(a) and (b).

Nitrogen sorption and electron microscopy results indicated that the meso/macroporous carbon is a perfect replica of the meso/macroporous silica monoliths [54]. The carbon monoliths have a macroporosity approximately equal to 80%, consisting of a fully interconnected macroporous network composed of hierarchically organized cells. The mesopores possess a narrow range of sizes, centered around 5 nm, which allow a high BET surface area (1500 $m^2 g^{-1}$).

Another type of hierarchically textured foam, which has been obtained in highly concentrated emulsions in a two-step process, was described by Zang et al. [47]. An oil-in-water highly

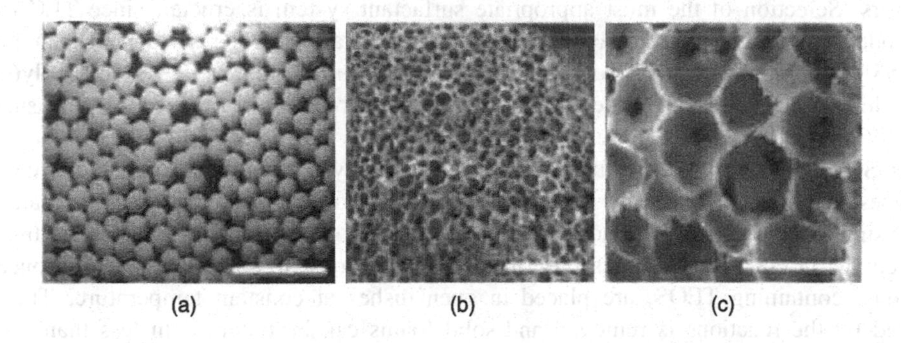

FIGURE 6.10 Images of silica beads at three different magnifications. (a) Optical microscopy, scale bar = 5 mm. (b) Scanning electron microscopy, scale bar = 200 μm. (c) Scanning electron microscopy, scale bar = 50 μm. (Reproduced from Zhang, H., Hardy, G.C., Rosseinsky, M.J., and Cooper, A.I. *Adv. Materials* 15, 78–81, 2003. With permission of Wiley-VCH Verlag GmbH.)

concentrated emulsion was formed by slowly adding mineral oil, while vigorously stirring, to a mixed solution containing acrylamide, a cross-linker, and a silica precursor. Next, the O/W emulsion was then injected as individual droplets into a hot oil sedimentation medium. Organic polymer/silica composite beads were obtained by polymerization of the mixture. This method allows the semi-continuous preparation of organic/inorganic composite beads, with uniform macroscopic particle sizes, high pore volumes, and interconnected emulsion-templated macropore structures [47]. Finally, the polymer was removed by calcination. Images of the calcined silica beads are shown in Figure 6.10. Silica porous beads, with an average bead diameter of 1.3 mm and high pore volume ($\approx$6 $cm^3\ g^{-1}$), have been obtained [47]. The structure of the highly concentrated emulsions was retained in the beads and the materials had a large specific surface area (420 $m^2\ g^{-1}$).

6.3.1.2 Single-Step Methods

Materials with meso/macroporous hierarchical porosity can be obtained in a simple single-step method, by directly carrying out polymerization reactions in the external phase of highly concentrated emulsions, as described by different authors [55,56]. This method makes use of the common structure of highly concentrated emulsions, which consist of emulsion droplets that template the formation of macropores and micellar aggregates, located in the external phase of the emulsions, that template the formation of mesopores [55,56]. This method can allow a high degree of control over both sizes and morphologies. All organic components can be removed, by calcinations in the presence of air, to prepare inorganic porous materials. Silicas with high specific surface areas, such as 800 $m^2\ g^{-1}$, associated with bulk density as low as 0.08 $g\,cm^{-3}$, have been obtained by Carn et al. [55]. These values are comparable to those obtained for silica aerogels obtained by reactions in the vapor phase.

It is worthwhile noting that powder materials are obtained rather than monoliths, at volume fractions of the dispersed phase above 0.8 [55,56]. No large monoliths have yet been obtained by this method, using highly concentrated emulsions. This feature has been attributed to the fact that the wall thickness decreases and does not have the expected mechanical strength, inducing the collapse of the inorganic scaffolds.

Overall, meso/macroporous inorganic materials consisting of well interconnected macropores and monodispersed mesopores can be obtained. Commercially available alkoxysilanes, such as tetraethoxysilane (commonly denoted tetraethyl orthosilicate, TEOS), can be used as inorganic precursors. Selection of the most appropriate surfactant system is crucial, since TEOS hydrolysis produces ethanol, which may decrease emulsion stability. The surfactants can be either ionic [55] or nonionic [56]. Tetradecyltrimethyl ammonium bromide (TTAB) and poly(ethylene oxide)-block-poly(propylene oxide)-block-poly(ethylene oxide) triblock copolymer surfactants (EO)n-(PO)m-(EO)n have been described [55,56].

TEOS reactions in the continuous phase of the highly concentrated emulsions may require long times, as much as 1-week periods [55]. A strategy to obtain meso/macroporous materials, by a faster single-step method, is to accelerate the condensation reactions by evaporating the ethanol produced by TEOS hydrolysis, as described by Esquena et al. [56]. The highly concentrated emulsions, containing TEOS, are placed in open dishes at constant temperature. The ethanol produced by the reactions is removed and solid foams can be obtained in less than 24 h. The highly concentrated emulsions remain stable, which allows the control of the macropore size.

Examples of the highly concentrated O/W emulsions and the meso/macroporous silica materials are shown in Figure 6.11(a) and (b), respectively. The emulsion droplets template the formation of the macropores, and the size and polyhedral morphology of the droplets remains preserved.

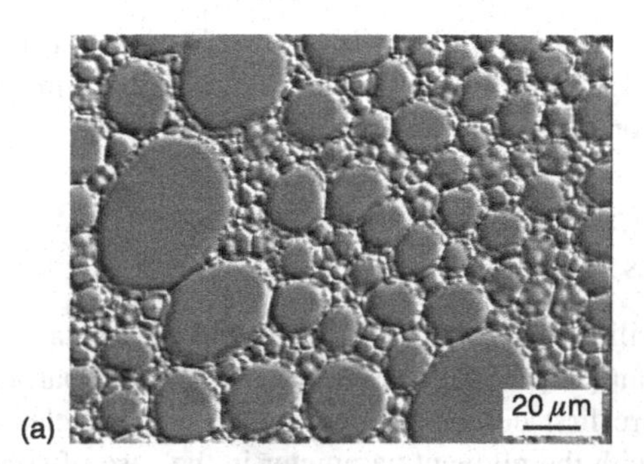

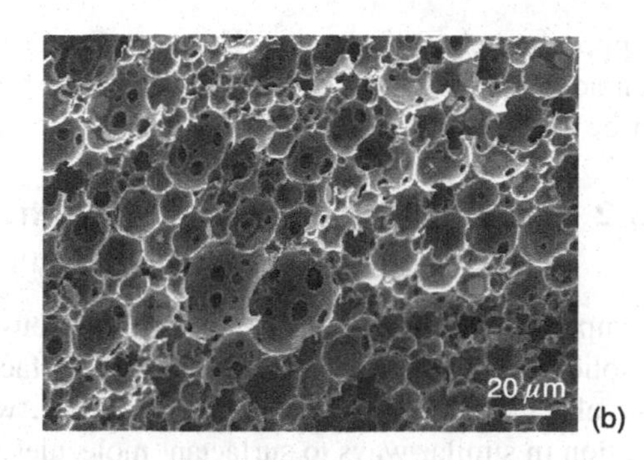

FIGURE 6.11 (a) Optical microscopy image of highly concentrated emulsion, consisting of HCl 0.2 M/$(EO)_{20}$-$(PO)_{70}$-$(EO)_{20}$/decane (8/2/90 mass ratios). (b) Scanning electron microscopy image of meso/macroporous silica obtained in the system HCl 0.2 M/$(EO)_{20}$-$(PO)_{70}$-$(EO)_{20}$/decane/TEOS (6/2/85/7 mass ratios). Condensation was favored by placing the sample in an open dish at 45 °C.

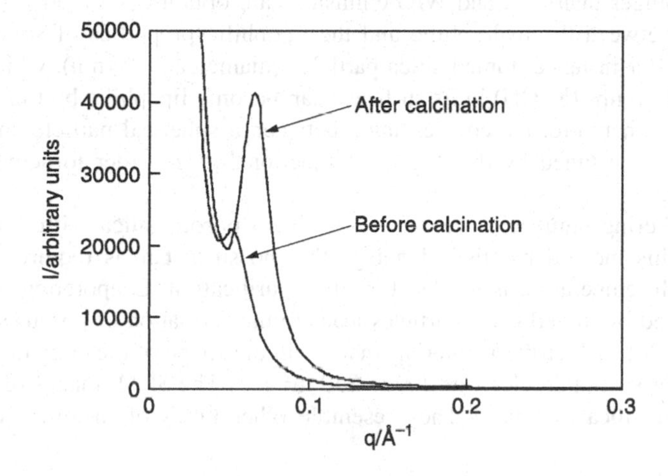

FIGURE 6.12 Small angle X-ray scattering (SAXS) spectra of a silica sample obtained in the system HCl 0.2 M/P123/decane/TEOS (6/2/85/7 mass ratios), before and after calcinations at 500 °C in the presence of air. The calcined sample shows a relatively intense peak, which is not observed in the noncalcined sample.

The spectra obtained by small-angle X-ray scattering is shown in Figure 6.12. The as-synthesized noncalcined sample produces a low-intensity peak, and no high order peaks are observed in the spectra. Therefore, the mesostructure probably consists of disordered pores, which does not lead to X-ray diffraction patterns. The peak is probably due to the pore–pore average correlation distance (d-spacing = 121.7 Å). The peak intensity significantly increased after calcination and its position was shifted to a higher q value (d-spacing = 95.8 Å). Again, no high order peaks are observed.

The reduction in d-spacing produced by calcination indicates a certain degree of shrinkage induced by the removal of the organic core or the mesopores. The increase in peak intensity may indicate a less polydispersed pore–pore correlation distance. The specific surface area of the calcined sample, determined by applying the BET method to the nitrogen adsorption isotherm, is approximately 300 $m^2\,g^{-1}$. This specific surface area is relatively high, considering that it has been achieved with only 2 wt% surfactant [56]. Larger specific surface areas, around 800 $m^2\,g^{-1}$, can be obtained, but higher surfactant concentrations (≈9 wt%) are required [55].

6.3.2 Preparation of Hierarchically Textured Foams in Highly Concentrated Emulsions Stabilized by Solid Particles

A simple and effective way of preparing porous silica materials arises from emulsions stabilized by solid particles alone, in the absence of surfactant [57]. Emulsions stabilized by solid particles, generally called Pickering emulsions [58,59], were first described long ago. Solid particles can function in similar ways to surfactant molecules, with the relevant parameter in the case of spherical particles being the contact angle of the particle with the liquid–liquid interface. Solid particles may possess high energy of attachment to interfaces, which produces irreversible adsorption. As a result, emulsion droplets coated with solid particles can be remarkably stable to coalescence.

Pickering emulsions can be oil-in-water or water-in-oil. Hydrophilic solid particles produce contact angles smaller than 90°, and the preferred emulsions are O/W. For lipophilic particles the contact angle is bigger than 90°, and W/O emulsions are obtained.

The balance between the hydrophilic and the lipophilic properties of solid particles can be easily controlled. For instance, fumed silica particles (diameter ≈ 10 nm), which are hydrophilic because of silanol groups (Si-OH) in its surfaces, can become lipophilic by functionalization with alkyl silanes [60]. Therefore, the contact angle between a spherical particle and a given liquid–liquid interface can be tuned by the degree of functionality, in order to achieve the maximum emulsion stability.

Recently, Pickering emulsions were used to obtain porous silica with controlled pore size by Binks [57]. This method has the advantage that no surfactant is required and therefore no calcination at high temperature is needed for silica purification. Evaporation in air of Pickering emulsions stabilized by fumed silica particles leads to the formation of porous silica, as shown in Figure 6.13. Monolith (tablet-like) materials that retain the shape of the container can be obtained. The surface area of silica particles was 200 to 250 $m^2\,g^{-1}$. The SEM images of Figure 6.13 show the macropores of silica materials. They resemble other kinds of materials obtained by other

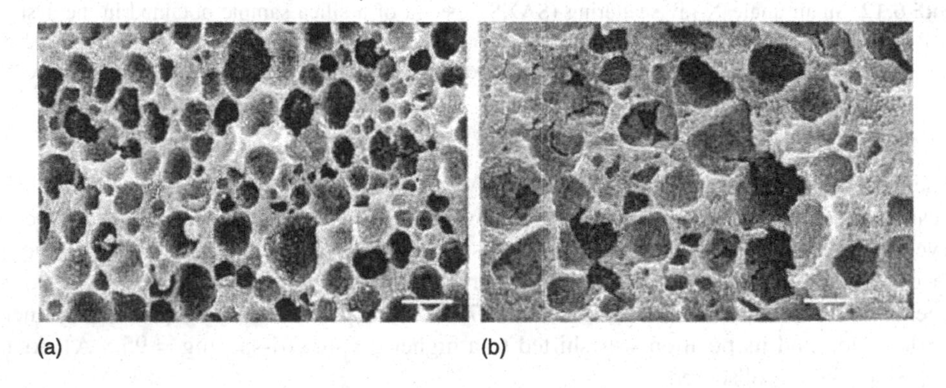

FIGURE 6.13 SEM images of porous silica following evaporation of both oil and water from silica particle-stabilized emulsions. Scale bar = 10 μm. (a) 10 vol% decane-in-water with 4 wt% silica (silanol content = 76%). (b) 10 vol% water-in-hexane with 4 wt% silica (silanol content = 50%). (Reproduced from Binks, B.P. *Adv. Materials* 14, 1824–1827, 2002. With permission of Wiley-VCH Verlag GmbH.)

means in highly concentrated emulsions. The fact that the most volatile emulsion component is the continuous phase induces the formation of highly concentrated emulsions during the solvent evaporation. Oil evaporation must be slow enough to allow the rearrangement and bonding of the silica particles before water is lost completely [57].

6.4 SUMMARY

Highly concentrated emulsions, which possess a volume fraction of the dispersed phase higher than 0.74 [1,2], are very appropriate systems for the preparation of low-density macroporous materials, also designated aerogels, which have very large pore volume and very low bulk density. These materials can be obtained by polymerization in the continuous phase of highly concentrated emulsions followed by the removal of the dispersed phase components [21,33–44]. The macroporous solid foams produced by this method consist of interconnected sponge-like macropores and possess low bulk densities. The droplet size distribution of the highly concentrated emulsion is a crucial factor in determining the mechanical properties of macroporous monoliths. Strength and toughness of organic foams can be improved by selecting the most appropriate system components and emulsification methods [21]. Highly ordered honeycomb-like macroporous materials can be obtained by making use of monodisperse emulsions prepared by fractionation techniques [44]. Both organic and inorganic monomers can be used to obtain macroporous solid foams, and the results show that a very wide variety of low-density macroporous materials can be obtained.

These macroporous materials have received widespread attention because of their successful applications as supports for catalysts, immobilization of enzymes, selective membranes, templates for the preparation of other materials, etc. Mesoporous materials that possess a high surface area and a uniform porosity are also very important because of their potential applications as catalytic surfaces and supports, adsorbents, energy storage devices, chromatographic materials, filters, etc. [48–54]. In order to obtain technological useful materials, the porous structure must have a dual meso/macroporous network. The presence of mesopores ensures large surface areas for reaction and/or adsorption, whereas the macropores guarantee surface accessibility, permitting the rapid transport of chemical species.

Highly concentrated emulsions are also very appropriate systems for the preparation of hierarchically textured foams [28]. These are materials that possess structural organization at different size ranges. Several methods for the preparation of materials with dual meso/macroporous properties have been described [28,47,54–57]. Mesopores are templated by supramolecular aggregates and macropores by the emulsion droplets. Materials that possess high specific surface area and high macropore volume have been obtained. The preparation methods can be classified as two-step and single-step methods.

Two-step methods permit an independent control of the mesopore and the macropore sizes. The first step consists of the preparation of organic macroporous foams by polymerizing in the continuous phase of highly concentrated O/W emulsions [21]. These solid foams are used as scaffolds for the preparation of the meso/macroporous inorganic materials. The second step consists of imbibing with ethanol sol–gel solutions that contain inorganic precursors and surfactants [28].

Hierarchically porous beads can be obtained by a two-step method. Emulsion-templated polymer/silica composite beads have been obtained by polymerization of the continuous phase of highly concentrated emulsions [47]. The emulsions are then injected as individual droplets into a hot oil sedimentation medium, allowing the semi-continuous preparation of organic/inorganic composite beads, with uniform macroscopic particle sizes, high pore volumes, and interconnected emulsion-templated macropore structures.

In the single-step methods polymerization reactions are carried out directly in the continuous phase of highly concentrated emulsions [55,56]. Some of these methods make use of the common structure of highly concentrated emulsions, which consist of emulsion droplets that template the formation of macropores and micellar aggregates, located in the external phase of the emulsions, that template the formation of mesopores. Another single-step method consists of the use of highly concentrated emulsions stabilized by solid particles (Pickering emulsions) [57]. No surfactant is required and therefore no calcination at high temperature is needed for silica purification.

ACKNOWLEDGMENTS

The authors gratefully acknowledge the financial support by the Spanish Ministry of Science and Education (Grant PPQ2002-04514-C03-03) and "Generalitat de Catalunya, DURSI" (Grant 2001SGR-00357).

REFERENCES

1. K.J. Lissant. J. Colloid. Interface Sci., 22:462, 1966.
2. H.M. Princen. J. Colloid Interface Sci., 71:55, 1979.
3. H. Kunieda, C. Solans, N. Shida and J.L. Parra. Colloids Surf. A, 24:225, 1987.
4. C. Solans, J.G. Domínguez, J.L. Parra, J. Heuser and S.E. Friberg. Colloid Polym. Sci., 266:570, 1988.
5. C. Solans, R. Pons and H. Kunieda. In: B.P. Binks (ed.), Modern Aspects of Emulsion Science. Cambridge: Royal Society of Chemistry, 1998, pp 367–394.
6. J.C. Ravey, M.J. Stébé and S. Sauvage. Colloids Surf. A, 91:237, 1994.
7. G. Ebert, G. Platz and H. Rehage. Berichte Der Bunsengesellschaft, 92:1158, 1988.
8. O. Sonneville-Aubrun, V. Bergeron, T. Gulik-Krzywicki, B. Jönsson, H. Wennerström, P. Lindner and B. Cabane. Langmuir, 16:1566, 2000.
9. C. Solans, R. Pons, S. Zhu, H.T. Davies, D.F. Evans, K. Nakamura and H. Kunieda. Langmuir, 9:1479–1482, 1993.
10. H. Kunieda, Md. H. Uddin, M. Horii, H. Furukawa and A. Harashima, J. Phys. Chem. B, 105: 5419–5426, 2001.
11. Md. H. Uddin, H. Kunieda and C. Solans. In: K. Esumi and M. Ueno (eds.), Structure–Performance relationships in Surfactants. New York: Marcel Dekker, 2003, pp 599–626.
12. V. Rajagopalan, C. Solans and H. Kunieda, Colloid Polym. Sci., 272:1166, 1994.
13. R. Pons, J.C. Ravey, S. Sauvage, M.J. Stébé, P. Erra and C. Solans, Colloids Surf. A, 76:171, 1993.
14. C. Solans, J. Esquena, N. Azemar, C. Rodríguez and H. Kunieda. In: D.N. Petsev (ed.), Emulsions: Structure, Stability and Interactions. Amsterdam: Elsevier, 2004, pp 367–394.
15. V.G. Babak and M.J. Stébé. J. Dispersion Sci & Tech., 23:1, 2002.
16. R. Pons, I. Carrera, P. Erra, H. Kunieda and C. Solans. Colloids Surf. A, 91:259, 1994.
17. H. Kunieda, Y. Fukui, H. Uchiyama and C. Solans. Langmuir, 12:2136, 1996.
18. K. Ozawa, C. Solans and H. Kunieda. J. Colloid Interface Sci., 188:275–281, 1997.
19. K. Shinoda and H. Saito. J. Colloid Interface Sci., 26:70, 1968.
20. H. Kunieda and K. Shinoda, J. Colloid Interface Sci., 75:601, 1980.
21. J. Esquena, G.S.R.R. Sankar and C. Solans. Langmuir, 19:2983–2988, 2003.
22. H. Kunieda, N. Yano and C. Solans. Colloids Surf. A, 36:313, 1989.
23. H.M. Princen. J. Colloid Interface Sci., 91:160, 1983.
24. H.M. Princen and A.D. Kiss, J. Colloid Interface Sci., 112:427, 1986.
25. R. Pons, P. Erra, C. Solans, J.C. Ravey and M.J. Stébé. J. Phys. Chem., 97:12320, 1993.
26. E. Ruckenstein. Adv. Polym. Sci., 127:3–58, 1997.
27. N.R. Cameron and D.C. Sherington. Adv. Polym. Sci., 126:163–214, 1996.
28. H. Maekawa, J. Esquena, S. Bishop, C. Solans and B.F. Chmelka. Adv. Materials, 15:591–596, 2003.

29. A. Pinazo, M.R. Infante, P. Izquierdo and C. Solans, J. Chem. Soc. Perkin Trans., 2:1535–1539, 2000.
30. C. Solans, A. Pinazo, G. Calderó and M.R. Infante. Colloids Surf. A, 176:101–108, 2001.
31. P. Clapés, L. Espelt, M.A. Navarro and C. Solans. J. Chem. Soc. Perkin Trans., 2:1394–1399, 2001.
32. L. Espelt, P. Clapés, J. Esquena, A. Manich and C. Solans. Langmuir, 19:1337–1346, 2003.
33. D. Barby and Z. Haq. European Patent 0060138 (Unilever), 1982.
34. J.M. Williams and D.A. Wrobleski. Langmuir, 4:656–662, 1988.
35. J.M. Williams. Langmuir, 4:44–49, 1988.
36. N.R. Cameron, D.C. Sherington, L. Albiston and D.P. Gregory. Colloid Polym. Sci., 274:592–595, 1996.
37. E. Ruckenstein and J.S. Park. Polymer, 33:405–417, 1992.
38. E. Ruckenstein and J.S. Park. J. Polym. Sci. (C), Polym. Lett., 26:529–536, 1988.
39. C. Solans, J. Esquena and N. Azemar. Curr. Opinion Colloid Interface Sci., 8:156–163, 2003.
40. A. Mercier, H. Deleuze and O. Mondain-Monval. Actualité Chimique, 5:10–15, 2000.
41. P. Hainey, I.M. Huxham, B. Rowatt, D.C. Sherington and L. Tetley. Macromolecules, 24:117–121, 1991.
42. J.M. Williams, A.J. Gray and M.H. Wilkerson. Langmuir, 6:437–444, 1990.
43. R. Butler, C.M. Davies and A.I. Cooper. Adv. Materials, 13:1459–1463, 2001.
44. A. Imhof and D.J. Pine. Nature, 389:948–951, 1997.
45. J. Bibette. J. Colloid Interface Sci., 147:474–478, 1991.
46. G.R. Yi, S.M. Yang. Chem. Mater., 11:2322–2325, 1999.
47. H. Zhang, G.C. Hardy, M.J. Rosseinsky and A.I. Cooper. Adv. Materials, 15:78–81, 2003.
48. M.P. Harold, C. Lee, A.J. Burggraaf, K. Keizer, V.T. Zaspalis and R.S.A. Delange. MRS Bull., 19:34–39, 1994.
49. R.R. Bhave. Inorganic Membranes Synthesis, Characteristics and Applications. New York: Van Nostrand Reinhold, 1991.
50. N. Ishizuka, H. Minakuchi, K. Nakanishi, N. Soga and N. Tanaka. J. Chromatography A, 797: 133–137, 1998.
51. M.X. Wu, T. Fujiu and G.L. Messing. J. Non-Cryst. Solids, 121:407–412, 1990.
52. E. Litovsky, M. Shapiro and A. Shavit. J. Am. Ceram. Soc., 79:1366–1376, 1996.
53. P. Singer. Semicond. Int., 19:88–96, 1996.
54. S. Alvarez, J. Esquena, C. Solans and A.B. Fuertes. Adv. Engineering Materials, 6:897–899, 2004.
55. F. Carn, A. Colin, M.F. Achard, H. Deleuze, E. Sellier, M. Birot and R. Backov. J. Mater. Chem., 14:1370–1376, 2004.
56. J. Esquena, H. Kunieda and C. Solans. XVIII European Colloid and Interface Society Conference, Almería, Spain 2004.
57. B.P. Binks. Adv. Materials, 14:1824–1827, 2002.
58. W. Ramsden. Proc. R. Soc. London, 72:156, 1903.
59. S.U. Pickering. J. Chem. Soc., 91:2001, 1907.
60. B.P. Binks and S.O. Lumsdon. Langmuir, 16:8622–8631, 2000.

7 Organic Reactions in Emulsions and Microemulsions

Krister Holmberg

CONTENTS

7.1 INTRODUCTION

Organic synthesis is normally performed in a homogeneous liquid reaction medium, consisting either of one solvent or of a mixture of solvents. When the reaction involves reactants with widely different polarity, such as a lipophilic organic molecule and an inorganic salt – which is common, for instance in oxidation reactions and in nucleophilic substitution reactions – a two-phase system is often employed. The common practice is to accelerate such a two-phase reaction by addition of a phase transfer catalyst, which may be either a quaternary ammonium salt or a crown ether (1).

Emulsions are oil–water two-phase systems with a large interfacial area stabilized by surfactant. The large interfacial area can be taken advantage of and emulsions have been successfully used as media for organic synthesis. The use of emulsions as vehicles for organic reactions is reviewed in this chapter.

A special case of the use of emulsions as reaction medium is emulsion polymerization, i.e., preparation of lattices. In emulsion polymerization micrometer-sized drops containing the monomer are dispersed in water using a relatively large amount of surfactant in order to obtain a high concentration of surfactant micelles in the bulk aqueous phase. A water-soluble initiator is added and free-radical polymerization is initiated in the aqueous phase. (The monomer has some solubility in water.) The growing oligomers become gradually less water soluble. Since the interfacial area of the many small micelles is orders of magnitude larger than that of the relatively few drops of monomer, the oligomers go into the micelles where continued polymerization occurs. The micelles grow at the expense of the monomer drops and the reaction proceeds until all monomers are consumed. The resulting product, the latex, typically consists of a concentrated dispersion of surfactant-stabilized particles in the size range 70 to 500 nm. Emulsion polymerization is an established process, thoroughly described in the literature (2,3). It is not dealt with further in this review. Free-radical polymerization may also be performed within the drops of a pre-formed emulsion of a monomer, such as styrene, in water (4,5). Both the free radical initiator and the transfer agent should be oil-soluble. Under these conditions, each particle behaves as an isolated bulk reactor. Because of compartmentalization of the radicals the conversion is fast, much faster than obtained in bulk. The process gives polymer particles of approximately the same size as the size of the starting monomer drops. The molecular weight distribution appears to be similar to that of bulk polymerization.

Microemulsions, i.e., oil–water–surfactant mixtures that unlike emulsions are thermodynamically stable systems, are used as media for a variety of chemical reactions. Microemulsions are macroscopic one-phase systems but microscopically they consist of oil and water domains, separated by a monolayer of surfactant. Depending on the oil-to-water ratio and on the choice of surfactant the interface can be curved towards oil or towards water or the curvature may be small, which is the case for so-called bicontinuous microemulsions. The aqueous droplets of water-in-oil microemulsions have been explored as minireactors for the preparation of nanoparticles of metals and metal salts and particles of the same size as the starting microemulsion droplets can often be obtained (6–8). Polymerization in microemulsions is an efficient way to prepare nanolattices and also to make polymers of very high molecular weight. Both discontinuous and bicontinuous microemulsions have been used for the purpose (9). Microemulsions are also of interest as media for enzymatic reactions. Much work has been done with lipase-catalyzed reactions and water-in-oil microemulsions have been found suitable for ester synthesis and hydrolysis, as well as for transesterification (10,11).

Microemulsions, being microheterogeneous mixtures of oil, water, and surfactant, are excellent solvents for both polar and nonpolar compounds. The capability of microemulsions to solubilize a broad range of substances in a one-phase formulation has been found useful in preparative organic synthesis. Microemulsions are one way, out of several, to overcome the reactant incompatibility problems that are frequently encountered in preparative organic chemistry. The capability of microemulsions to compartmentalize and concentrate reactants can lead to unusual reactivity in organic synthesis. Attempts have also been made to use the oil–water interface of microemulsions as a template to induce regiospecificity of organic reactions. The use of microemulsions as vehicles for organic reactions is reviewed in this chapter and special emphasis is put on the more recent literature.

7.2 REACTION IN EMULSIONS

The term "interfacial synthesis" relates to an organic synthesis performed at the oil–water interface. By using a surface-active catalyst that operates at the interface, interfacial synthesis permits a convenient separation of the product and the catalyst with the view to recover and reuse the latter. The process is somewhat between homogeneous and heterogeneous catalysis; homogeneous in the sense that the catalyst is a soluble molecule that can adopt its conformation in solution, heterogeneous in that the reaction takes place at the boundary between two phases. Performing interfacial synthesis in a system with a large oil–water interface, such as an emulsion, is a way to increase the reaction rate.

Lim and co-workers have performed a range of interfacial syntheses using hydrocarbons or chlorinated hydrocarbons as oil phase and sodium dodecyl sulfate as emulsifier. The reported syntheses include free-radical polymerization (12–14), oxidations (15), and carbonylation reactions (16). Oxidative polymerization of 2,6-dimethylphenol to yield poly(2,6-dimethyl-1,4-phenylene oxide) (Figure 7.1) is a particularly interesting example of the use of the procedure. The reaction has been performed both in toluene–water (13) and in chloroform–water (14) emulsions. Copper is used as the active catalyst and a range of ligands was evaluated to maximize the performance

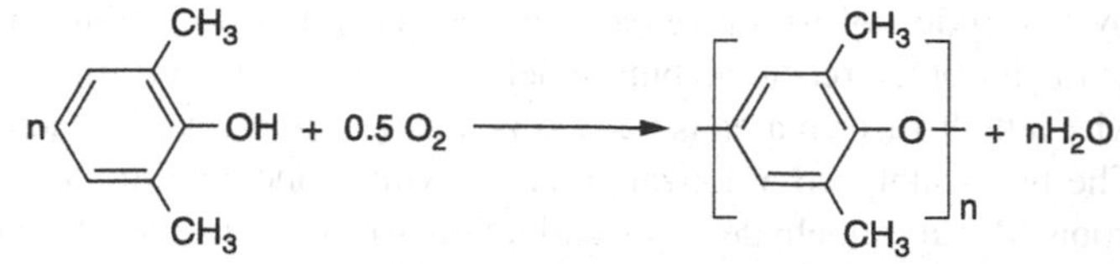

FIGURE 7.1 Oxidative polymerization of 2,6-dimethylphenol.

$$CH_3{-}CH_2{-}CH_2{-}CHO + HCHO \longrightarrow CH_3{-}CH_2{-}CH(CH_2OH){-}CHO$$

$$CH_3{-}CH_2{-}CH(CH_2OH){-}CHO + HCHO \longrightarrow CH_3{-}CH_2{-}C(CH_2OH)_2{-}CHO$$

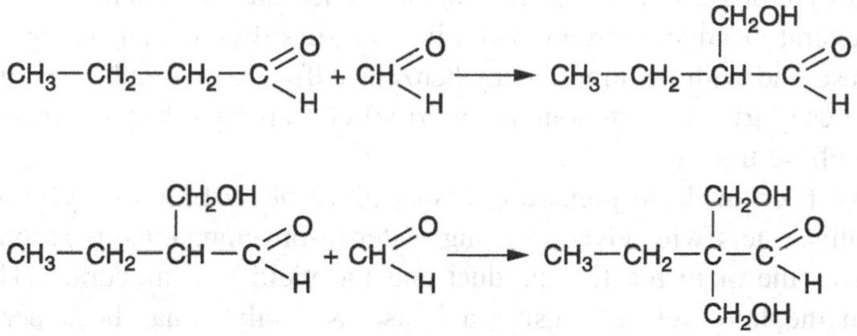

FIGURE 7.2 Crossed aldolization of butyraldehyde and formaldehyde.

of the catalyst at the interface. For the toluene–water system triethyl phosphite gave a proper reaction rate. Much higher reactivity was, however, found in the chloroform–water system when ethylenediamines were used as ligands. Ethylenediamines and substituted ethylenediamines can act as bidentate ligands, capable of forming binuclear copper complexes that can mimic the active site of enzymes that catalyze coupling polymerization of phenols in plants. The highest reactivity, measured as oxygen uptake rate, was obtained with the asymmetrically substituted diamine *N,N*-dibutylethylenediamine. A polymer with a molecular weight of 48,000 was obtained in 95% yield.

It was found that for a series of substituted ethylenediamines the rate of polymerization of 2,6-dimethylphenol correlated well with the surface activity of the copper–ligand complex (14). It is probable that increased surface activity, i.e., increased tendency to reside at the interface, will reduce the diffusion pathway of the reactants in the biphasic system. The surface activity of the complex may also affect the copper (I) $\rightarrow$ (II) redox cycle.

There are other examples of organic reactions based on a catalytic step that occur at the interface between an aqueous and an oil phase but where that interface has not been enlarged and stabilized by surfactant. The reaction is performed under stirring but without emulsifier the emulsion will be very crude, i.e., the interface will not be very large. A good example of the procedure is reaction between butyraldehyde and formaldehyde, using a solid anion exchange resin, in the form of a fine powder, as catalyst (17). The reaction, a so-called crossed aldolization, takes place in two subsequent steps in which the two α-hydrogens of butyraldehyde are replaced by hydroxymethyl groups (see Figure 7.2). The product formed is an important chemical intermediate. Reduction of the aldehyde group leads to trimethylolpropane, an important building block for alkyds and other polymers.

Butyraldehyde has low solubility in water while formaldehyde is readily water soluble. A liquid–liquid interface will form and the weakly basic catalyst containing tertiary amino groups goes to the boundary. The reaction medium is in effect a three-phase system. The reaction is believed to start by adsorption of formaldehyde on the catalyst surface; thus, a high surface area and the proximity to the other reactant, butyraldehyde, are parameters that favor the reaction. (Another reason why a high surface area is needed is that a byproduct, ethylacrolein, also adsorbs on the catalyst.) The base-catalyzed aldolization starts with a nucleophilic attack of the base catalyst at the α-carbon of butyraldehyde. The carbanion formed then reacts with formaldehyde. Thus, the reaction is a truly interfacial synthesis. It would be interesting to examine the effect of an increase of the liquid–liquid interfacial area, which could be obtained by addition of a suitable emulsifier.

Organic reactions in emulsions, involving one water-soluble and one oil-soluble reactant, can also be catalyzed by a phase transfer agent. This approach has been applied to the preparation of azo dyes (18). Several azo dyes were prepared in water-in-oil emulsions using chlorinated hydrocarbon as oil phase and with sodium dodecylbenzenesulfonate as emulsifier. Either a quaternary ammonium salt (benzyltrimethylammonium chloride) or a crown ether (dibenzo-18-crown-6) was employed as the phase transfer catalyst.

Attempts have been made to prepare esters of glycerol by base-catalyzed transesterification of fatty acid methyl esters with glycerol using water-in-oil emulsions as reaction medium (19). Monoglyceride was the main reaction product and the yield was moderate. The same reaction, but starting from the fatty acid and using a lipase as catalyst, has been performed in water-in-oil microemulsions (20,21). The product mixture varied with the hydrocarbon used as oil component. When a small hydrocarbon was used, monoglyceride was the main reaction product and only traces of triglyceride were obtained. The reason that triglycerides are not formed in these systems is that the diglyceride formed in the oil–water interfacial zone partitions into the continuous hydrocarbon domain. It is then inaccessible to the enzyme. The yield of triglyceride is increased by using longer hydrocarbons as the oil component. These have poorer solvency for mono- and diglycerides. When the hydrocarbon chain length exceeds that of the glyceride acyl groups, triglyceride was the main reaction product (21). The reaction has also been performed in a monolayer, i.e., at the air–water interface (22,23), as well as in foams (24). The yield of triglyceride is high in both monolayers and foams because the hydrophobic diglyceride intermediate will have no other choice than to stay at the interface. The situation at the oil–water and the air–water interfaces is illustrated in Figure 7.3.

A nucleophilic substitution reaction, preparation of alkyl sulfonate from decyl halide and sodium sulfite, has been performed in water-in-oil emulsions based on the cationic surfactant $R_2(Me)_2N^+ X^-$, where R is a long chain alkyl and X is Br or Cl (25,26). For a given amount of decyl halide and a given ratio of halide to sulfite a plot of reaction rate vs. surfactant concentration showed an optimum, as can be seen from Figure 7.4. The results are explained as follows. A certain surfactant concentration is needed to produce a good emulsion. However, increasing the amount of surfactant means increasing the ratio of surfactant counterion (in this case bromide) to reacting ion (sulfite). Since bromide is a large, polarizable anion, it associates with the interface and expels the smaller and less polarizable sulfite ion. Thus, an "unnecessary" large amount of surfactant has a detrimental effect on the reaction rate. Figure 7.5 shows the reaction profiles for different decyl halides. As can be seen, the reaction is fast for decyl iodide, somewhat slower for decyl bromide, and much more sluggish for decyl chloride. This is the expected order of reactivity because iodide is a good leaving group, bromide a slightly less good, and chloride is a poor leaving group. The fact that the order of reactivity is the same as obtained in single-phase systems was taken as an indication that the reaction in the emulsion is kinetically controlled (26).

An indirect way of using emulsions for organic synthesis is the use of foams derived from high internal phase emulsions (HIPE), so-called polyHIPE (27). Such foams are produced from a water-in-oil emulsion with a water-to-oil ratio above 95% and with the oil component being a

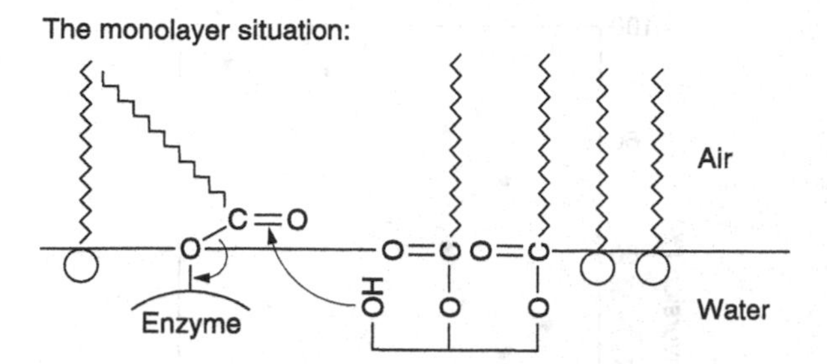

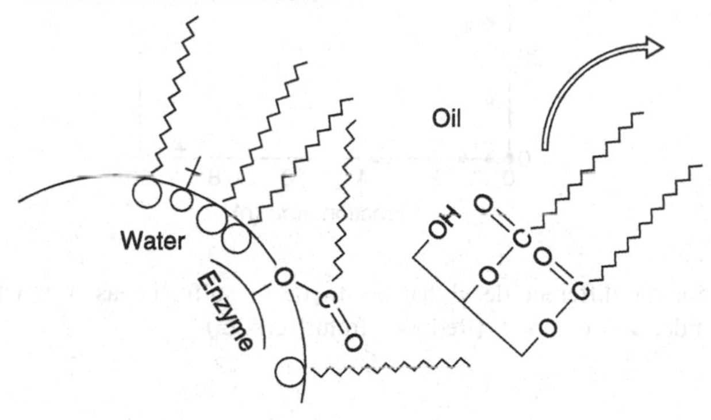

FIGURE 7.3 Arrangements of fatty acids and fatty acid glycerol esters at the air–water and the oil–water interfaces.

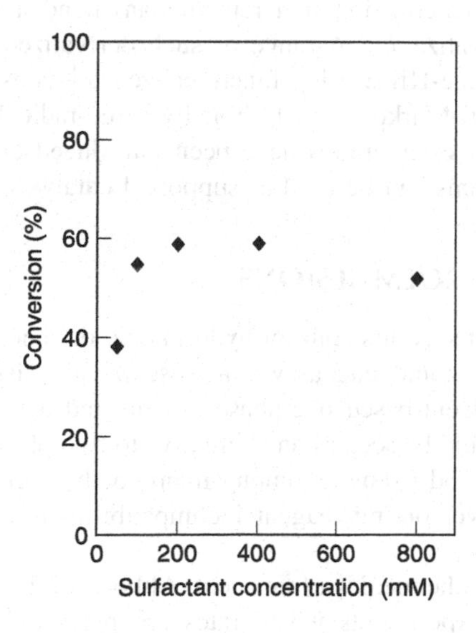

FIGURE 7.4 Conversion of decyl bromide to decyl sulfonate after 9 h at room temperature as a function of surfactant concentration. (Redrawn from Ref. 26.)

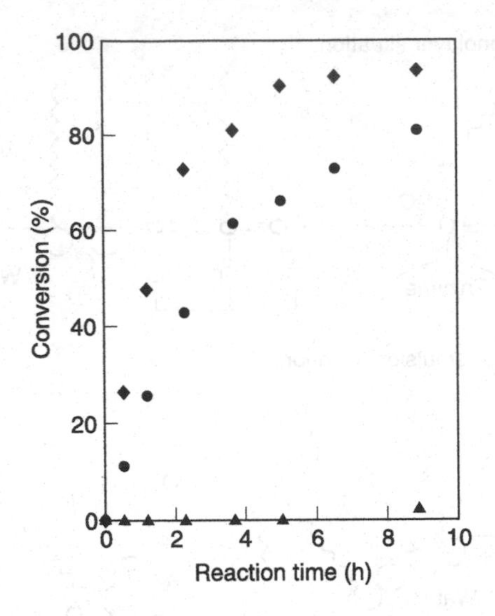

FIGURE 7.5 Conversion of different decyl halides to decyl sulfonate as a function of reaction time. ♦ = iodide, ● = bromide, ▲ = chloride. (Redrawn from Ref. 26.)

polymerizable entity, such as a combination of styrene and divinylbenzene. An oil-soluble emulsifier is used and the free-radical initiator should be water-soluble. An open structure is formed and the water can easily be removed after completed polymerization, resulting in a light foam with the foam lamella consisting of polymer and surfactant. The cells obtained from polymerized divinylbenzene are typically in the range of 10 to 20 μm in diameter, which corresponds rather well with the size of the water drops of the starting water-in-oil emulsion.

Even after completed polymerization there remain many pendant carbon–carbon double bonds. These can be used for derivatization. A range of such derivatized foams has been made from polystyrene/polydivinylbenzene-HIPE using functionalized thiols as derivatization reagents (28). Thiols add to olefins in an anti-Markovnikov fashion by a free-radical mechanism. Amine, alcohol, ester, thiol, halide, acid, and ester groups have been introduced on the foam by this approach. The functionalized HIPE foams can be used as supported catalysts for organic synthesis.

7.3 REACTION IN MICROEMULSIONS

Microemulsions are excellent solvents both for hydrophobic organic compounds and for inorganic salts. Being macroscopically homogeneous yet microscopically dispersed, they can be regarded as something between a solvent-based one-phase system and a true two-phase system. In this context, microemulsions should be seen as an alternative to two-phase systems with phase transfer agents added (29). A very good example, among many, of how effective a properly formulated microemulsion can be in overcoming reagent incompatibility is the detoxification of mustard, $ClCH_2CH_2SCH_2CH_2Cl$.

Mustard is a well-known chemical warfare agent. Although it is susceptible to rapid hydrolytic deactivation in laboratory experiments where rates are measured at low substrate concentrations, its deactivation in practice is not easy. Due to its extremely low solubility in water, it remains for months on a water surface. Addition of strong caustic does not markedly increase the

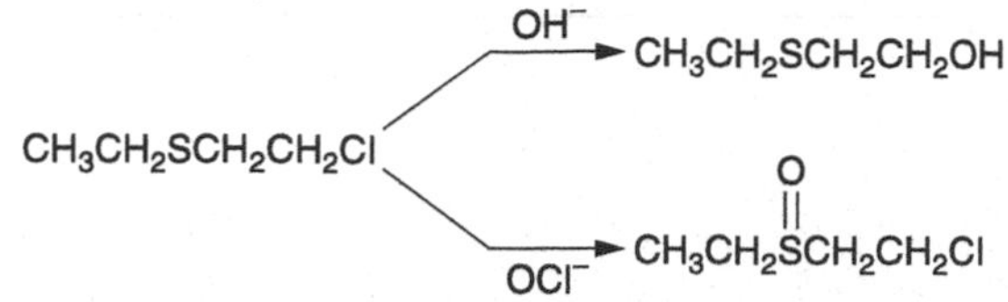

$$CH_3CH_2SCH_2CH_2Cl \xrightarrow{OH^-} CH_3CH_2SCH_2CH_2OH$$

$$CH_3CH_2SCH_2CH_2Cl \xrightarrow{OCl^-} CH_3CH_2S(=O)CH_2CH_2Cl$$

FIGURE 7.6 Transformation of 2-chloroethylethyl sulfide (half-mustard) into 2-hydroxyethylethyl sulfide (by alkali) or into 2-chloroethylethyl sulfoxide (by hypochlorite).

rate of reaction. Menger and Elrington have explored microemulsions as media for both hydrolysis and oxidation of "half-mustard," $CH_3CH_2SCH_2CH_2Cl$, a mustard model (Figure 7.6) (30). Oxidation with hypochlorite turned out to be extremely rapid in both o/w and w/o microemulsions. In formulations based on anionic, nonionic, or cationic surfactant oxidation of the half-mustard sulfide to sulfoxide was complete in less than 15 s. The same reaction takes 20 min when a two-phase system, together with a phase agent, is employed.

A microemulsion can also accelerate the reaction by other means than by supplying a large interfacial area. If the surfactant headgroup carries a charge, as is the case with anionic and cationic surfactants, reagents of opposite charge will be confined to the interior of the droplets of water-in-oil microemulsions. Such a compartmentalization and concentration of the reagents may lead to a rate enhancement, as has been demonstrated in several cases, e.g., in oxidation of iodide by persulfate (31,32). The surfactant monolayer may also accelerate the reaction by attracting reagents of opposite charge situated in the water domain, thus increasing its concentration in the interfacial zone, where the reaction occurs. This type of rate increase has been referred to as microemulsion catalysis in analogy with micellar catalysis for systems without an oil component (33).

Yet another use of microemulsions in organic synthesis is to take advantage of the oil–water interface as a template for one or more of the reagents. The presence of an oil–water interface may induce orientation of reactants in microemulsion systems, which in turn may affect the regioselectivity of organic reactions. All three aspects of the use of microemulsions as medium for organic synthesis, i.e., overcoming reagent incompatibility, providing specific rate enhancement (microemulsion catalysis), and inducing regiospecificity, will be covered in this chapter, which has focused on recent developments in the field. There are several reviews of organic reactions in microemulsions that cover the earlier literature on the topic (7,29,33–38).

7.3.1 Overcoming Reagent Incompatibility

Several studies have shown that microemulsions are suitable reaction media for organic reactions involving reactants of widely different polarity. A properly formulated microemulsion can dissolve both reactants and the large oil–water interfacial area enables contact between the two otherwise incompatible species. For some reactions the reactivity in a microemulsion and in a two-phase system with added phase transfer agent is similar (39) but in other instances the reaction is considerably faster in a microemulsion (30,40). The superiority of a microemulsion was particularly pronounced in the synthesis of a surface-active product, decyl sulfonate (39). Decyl bromide and sodium sulfite were used as starting materials and two different microemulsions were tested, one water-in-oil and one bicontinuous. Either a crown ether or a quaternary ammonium salt (Q salt) was employed as phase transfer agent in the two-phase system. As can be seen from Figure 7.7, the reaction proceeded at a reasonable rate in the microemulsions but ceased at an early state in

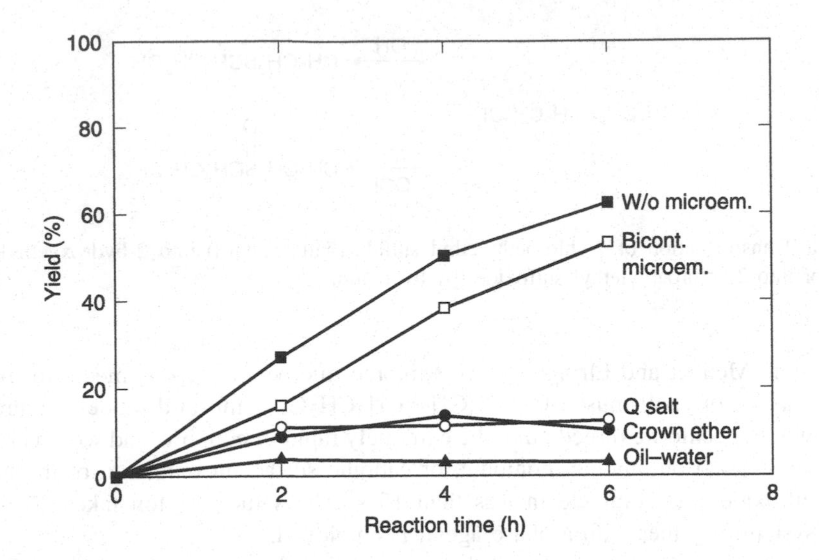

FIGURE 7.7 Reaction profiles for decyl sulfonate synthesis in various reaction media.

the two-phase systems. The poor reactivity in the latter systems has been interpreted as being due to the phase transfer agent partitioning entirely into the hydrocarbon phase due to ion pair formation with the lipophilic anion formed. Too strong association with a lipophilic anion will prevent the Q salt or crown ether to go back into the aqueous phase to pick up another reacting anion and transfer it into the organic phase where the reaction occurs (37,39). This is likely to be a general phenomenon and microemulsions may, thus, constitute a particularly interesting alternative to phase transfer catalysis in cases where the product is a lipophilic anion.

Microemulsions and two-phase systems with added phase transfer agent are both useful means of overcoming reagent incompatibility, but on entirely different accounts. In phase transfer catalysis, the nucleophilic reagent is carried into the organic phase where it becomes poorly solvated and highly reactive. In the microemulsion approach there is no transfer of reagent from one environment to another; the success of the method relies on the very large oil–water interface at which the reaction occurs. In an attempt to combine the two methods and take advantage of both the high reactivity of a poorly solvated anion in phase transfer catalysis and the very large oil–water interface of a microemulsion, Häger and Holmberg carried out a ring-opening reaction of a lipophilic epoxide in a microemulsion in the presence of a conventional Q salt, tetrabutylammonium hydrogen sulfate (40). Reactions were also performed in a two-phase system with and without added Q salt. It was found that the rate of the Q salt-catalyzed system is increased further when the reaction is carried out in a microemulsion instead of an oil–water two-phase system.

The above reaction was performed in a microemulsion based on a chlorinated hydrocarbon and a combination of two alkylglucoside surfactants was used for formulating the microemulsion. An attempt was also made to accelerate the same reaction performed in a microemulsion based on water, nonionic surfactant, and hydrocarbon oil (41). The reaction was performed in a Winsor III system and the same Q salt, tetrabutylammonium hydrogen sulfate, was added to the formulation. In this case the addition of the phase transfer agent gave only a marginal increase in reaction rate. Similar results had previously been reported for an alkylation reaction performed in different types of micellar media (42). The addition of a Q salt did not affect a system based on cationic surfactant, a marginal increase in rate for a system based on nonionic surfactant, and a substantial effect when an anionic surfactant was used. The last system, also with Q salt added, gave lower yield than the first two, however, most likely due to electrostatic repulsion of the negatively charged nucleophile by the anionic micelles.

In a more recent work the microemulsion approach and phase transfer catalysis was combined in a nucleophilic substitution, reaction between *t*-butylbenzyl bromide and potassium iodide (43). The microemulsion was based on a chlorinated hydrocarbon and a sugar surfactant, octyl glucoside, which means that there were no counter-ions competing with the nucleophile at the oil–water interface. A quaternary ammonium salt, tetrabutylammonium hydrogen sulfate, was used as phase transfer catalyst, both in an amount equimolar to the iodide and in a catalytic amount. The results are shown in Figure 7.8. As can be seen, the combined approach gives the highest reaction rate in both cases. Use of a phase transfer catalyst in a microemulsion-based reaction can be practically useful.

Reactions in microemulsions have been compared with reactions in liquid crystalline phases. Such comparisons are interesting because they may give information about what is most important for the high reactivity of microemulsion systems: the large oil–water interface or the high dynamics of the system. Both liquid crystalline phases and microemulsions have a high interfacial area but liquid crystals lack the high dynamics of microemulsions. Formation of 2-furfurylthiol from cysteine/furfural and cysteine/ribose mixtures proceeded much more rapidly in both water-in-oil microemulsions (44) and bicontinuous cubic phases (44,45) than in aqueous systems. The reaction was faster in the cubic phase than in the microemulsion, a fact that was mainly attributed to the larger interfacial area in the former system (45).

In another comparison between a liquid crystalline phase and a microemulsion as reaction medium an amphiphilic polymer, a poly(ethylene glycol) with hydrophobic chains at both ends, was synthesized in an oil-in-water microemulsion and in a lamellar liquid crystal. Both systems were based on penta(ethylene glycol)monododecyl ether (C12E5) as surfactant and decane as oil component. Using the same composition and changing only the temperature one could go from a microemulsion (at 22 °C) to a lamellar liquid crystalline phase (at 39 °C). Thus, in this case the interfacial area, which is largely determined by the surfactant content, was about the same in the two systems. The reaction rates in the two systems were approximately the same and very much higher than the rate of the same reaction performed in a two-phase system without neither surfactant, nor phase transfer agent (46).

Figure 7.9 shows yet another example of a comparison between a microemulsion and a liquid crystalline phase as reaction medium (47). The reaction studied, substitution of bromide by iodide, has been discussed above (see Figure 7.8). As can be seen from the figure, the rate in the lamellar liquid crystalline phase is somewhat lower than that in the microemulsion. An interesting observation that can be made from the figure is that the rate in a Winsor I system is very similar to that in the one-phase microemulsion. A Winsor I system is a two-phase system consisting of an oil-in-water microemulsion in equilibrium with excess oil. It was also shown that it was not necessary to stir the Winsor system. The rate with and without stirring was the same. The observation that a Winsor system is just as effective a reaction medium as a one-phase microemulsion has been observed before, for another nucleophilic substitution reaction (48). The fact that a Winsor I system can be used instead of a one-phase microemulsion is practically important. Formulation of a one-phase microemulsion is often a problem, particularly when one wants a high loading of reactants into the oil and water domains, and one may end up with various types of two- or three-phase systems. Evidently, such systems may be just as useful as reaction media, as long as one of the phases is a microemulsion. The excess phase (or phases) can

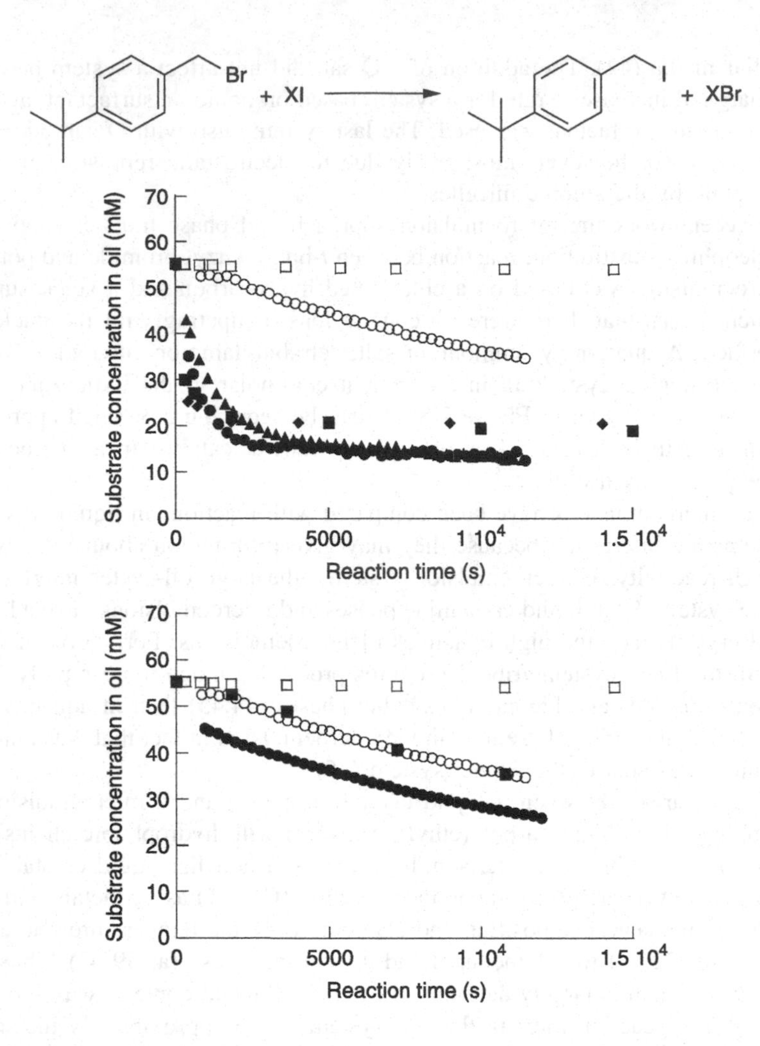

FIGURE 7.8 Reaction profiles for the reaction between 4-*t*-butylbenzyl bromide and potassium iodide in microemulsion based on dichloromethane, octyl glucoside, and deuterated water using either an equimolar (top) or a catalytic (bottom) amount of tetrabutylammonium hydrogen sulfate (TBAHS) or tetrabutylammonium bromide (TBAB) as phase transfer agent. The reaction was monitored by NMR observing the disappearance of the $-CH_2Br$ signal and appearance of the CH_2I signal. Top: in D_2O-C_8G_1-CH_2Cl_2: (○) without Q salt, (●) with TBAHS, (▲) with TBAB; and in D_2O-CH_2Cl_2: (□) without Q salt, (■) with TBAHS, (♦) with TBAB. Bottom: in D_2O-C_8G_1-CH_2Cl_2: (○) without Q salt, (●) with TBAHS; and in D_2O-CH_2Cl_2: (□) without Q salt, (■) with TBAHS.

be regarded as reservoirs for the reactant (or reactants) while the reaction occurs at the oil–water interface of the microemulsion phase.

One of the early examples of the usefulness of microemulsions to overcome the problem of reagent incompatibility was the work by Menger and Elrington on detoxification of mustard; see Figure 7.6 (30). Menger and Rourk have recently revisited the reaction, now exploring the use of cryochemistry for the purpose (49). Microemulsions that resist freezing and phase separation at −18 °C were developed using water–propylene glycol as aqueous component and fatty alcohol

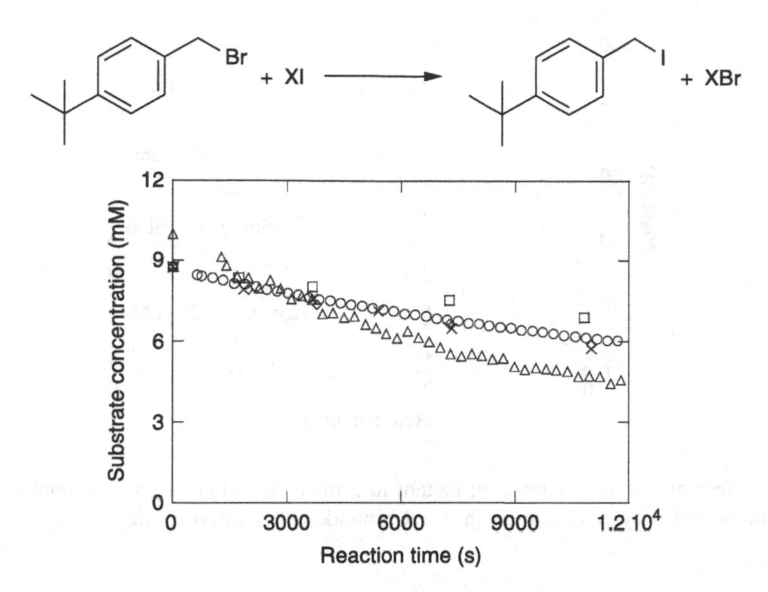

FIGURE 7.9 Reaction profiles for the reaction between 4-*t*-butylbenzyl bromide and potassium iodide performed in lamellar phase (□), oil-in-water microemulsion (○), micellar system (Δ), and Winsor I system (oil-in-water microemulsion in equilibrium with excess oil) (×), all based on alcohol ethoxylate surfactants.

ethoxylates with long polyoxyethylene chains as surfactants. Different types of chemical warfare agents were destroyed by reaction at −18 °C, either by hydrolysis or by oxidation.

An attempt was made to correlate the rate of hydrolysis of acetylsalicylic acid with the structure of the microemulsion used as reaction medium (50). Water was added to the formulation and it was found that the reaction rate changed abruptly when the system passed from water-in-oil to bicontinuous and again when the bicontinuous microemulsion was transformed into an oil-in-water microemulsion. For a microemulsion based on a nonionic surfactant the bicontinuous microemulsion gave the highest reaction rate, for a microemulsion based on an anionic surfactant the oil-in-water system was the most reactive, and for a cationics-based microemulsion the water-in-oil system gave the highest reaction rate. Similar results were obtained in another study of the same reaction (51). In previous studies of reactivity in different types of microemulsions based on nonionic surfactants, no large differences in reaction rate were observed, however (52).

Microemulsions have recently been used as medium for hydroformulation reactions (53). Water-soluble rhodium complexes have been found to be highly active in microemulsions based on alcohol ethoxylates. A series of internal olefins have been hydroformylated at temperatures of around 120 °C and a pressure of 100 bar. 7-Tetradecene as model alkene was hydroformulated to give 2-hexylnonal at high reaction rate and with a high degree of regioselectivity. It was also shown that the catalyst could be recycled in the work-up process.

7.3.2 Effect of Surfactant Charge on Reactivity

There are many examples of how the charge of the surfactant headgroup can affect the rate of reaction between a lipophilic organic molecule and a charged, polar nucleophile. The general rule is that cationic surfactants accelerate the reaction involving anionic nucleophiles and vice versa, an effect referred to as microemulsion catalysis. However, it has been found that for cationic

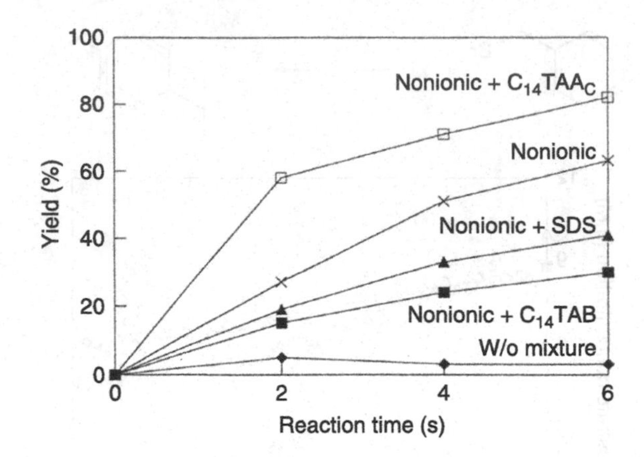

FIGURE 7.10 Effect of addition of ionic surfactant to a microemulsion based on a nonionic surfactant on the rate of formation of decyl sulfonate from decyl bromide and sodium sulfite.

surfactants the choice of counter-ion is decisive of the rate. The choice of counter-ion for anionic surfactants seems to be of less importance.

Oh et al. have shown that the addition of a small amount of the anionic surfactant sodium dodecyl sulfate (SDS) to a nonionics-based microemulsion increased the rate of decyl sulfonate formation from decyl bromide and sodium sulfite (54,55). Addition of minor amounts of the cationic surfactant tetradecyltrimethylammonium salt gave either a rate increase or a rate decrease depending on the surfactant counter-ion. A poorly polarizable counter-ion, such as acetate, accelerated the reaction. A large polarizable counter-ion, such as bromide, gave a slight decrease in reaction rate. The reaction profiles for the different systems are shown in Figure 7.10.

More recently, the effect of surfactant charge on the reaction rate was investigated for the ring opening of 1,2-epoxyoctane with sodium hydrogen sulfite (41). The reaction, which was performed in a Winsor III microemulsion, i.e., a microemulsion in equilibrium with both oil and water, was fast when a nonionic surfactant was used as the sole surfactant and considerably more sluggish when a small amount of SDS was added to the formulation. Another reaction in which the effect of added ionic surfactant was studied related to oxidation of azo dyes by hydrogen peroxide catalyzed by manganese porphyrins. The reaction was fast in a microemulsion based on nonionic surfactant. Addition of a small amount of cationic surfactant speeded up the reaction further while addition of an anionic surfactant led to retardation (56). These results imply that an anionic species is involved in the rate-determining step of the reaction.

An interesting example of a specific ion effect in microemulsions is the strong increase in reactivity found for large, polarizable anions such as iodide and bromide. The tendency for such ions to interact with, and accumulate at, the interface can be taken advantage of for preparative purposes. The increased concentration of such ions in the interfacial zone, where the reaction takes place, will lead to an increase in reaction rate. Expressed differently, the reactivity of iodide and other highly polarizable ions will be very high in such systems. The microemulsions need not be based on cationic surfactants that would drive the anions to the interface by electrostatic attraction. Also microemulsions based on nonionic surfactants display the effect because large, polarizable anions interact with the hydrophobic surface by dispersion forces (57,58). The mechanism of attraction (and the rational for the increased concentration of reacting species) is, thus, different from that of micellar catalysis, which is basically an ion exchange process.

An illustrative example of the effect of a large polarizable ion at the oil–water interface is the above-mentioned reaction between 4-*t*-butylbenzyl bromide and potassium iodide to give 4-*t*-butylbenzyliodide. The reaction was performed in a microemulsion based on the nonionic surfactant penta(ethylene glycol)monododecyl ether ($C_{12}E_5$) and the temperature was varied from 23 to 29 °C, which is the total temperature range of the isotropic oil-in-water region of this system (59). It was found that the reaction rate decreased considerably when the temperature was increased from 23 to 29 °C. ^{125}I NMR showed a marked temperature effect on line broadening; the lower the temperature the broader the signals (within the temperature range 23 to 29 °C). This indicates that the iodide ion interacts more strongly with the interface at lower temperature, which is likely to be the reason for the inverse temperature–reactivity relationship. Thus, ion binding to the microemulsion interface can have a pronounced effect on reactivity also in microemulsions based on uncharged surfactants. In another investigation of a nucleophilic substitution reaction in microemulsion, synthesis of 1-phenoxyoctane from 1-bromooctane and sodium phenoxide, no accumulation of the nucleophile at the interface could be observed based on the kinetics data (48,60). This is in line with the view that only the large polarizable anions, such as iodide, become attracted to the interface due to dispersion force interactions (57).

Thus, the use of microemulsions as media for organic substitution reactions can lead to a difference in reaction pattern due to differences in relative nucleophilicity as compared to reactions in homogeneous media. Bunton, Savelli, and others have shown this to be the case for reactions in micellar media (61–65). For instance, the type of reaction medium – homogeneous solution or micellar solution based on cationic surfactant – has been found to be decisive of the reaction of sulfonate esters in the presence of equimolar amounts of bromide and hydroxyl ions. Without surfactant present attack by hydroxyl ions dominates. In the micellar solutions, on the other hand, bromide is the dominating reacting species. Evidently, the larger and more polarizable bromide ion becomes more attracted than the hydroxyl ion to the micellar interface (61).

7.3.3 Selectivity in Microemulsion-Based Reactions

The large oil–water interface of microemulsions can be used as a template for organic reactions. Organic molecules with one more polar and one less polar end will accumulate at the oil–water interface of microemulsions. They will orient at the interface so that the polar part of the molecule extends into the water domain and the nonpolar part extends into the hydrocarbon domain. This tendency for orientation at the interface can be taken advantage of to induce regiospecificity of an organic reaction. A water-soluble reagent will react from the "water side," i.e., attack the polar part of the amphiphilic molecule, and a reagent soluble in hydrocarbon will react at the other end of the amphiphilic molecule. The principle, although applied to a micellar system, not a microemulsion, is nicely illustrated by an early study of the ability of aqueous micelles to control the regioselectivity of a Diels–Alder reaction in which both reactants were surface active (66). The diene, as well as the dienophile, were lipophilic molecules with a trimethylammonium headgroup. This Diels–Alder system should display no regiochemical preference in the absence of orientational effects since the substituents were close to being both electronically and sterically equivalent with respect to the diene and dienophile reaction centers. When the reactions were run in an organic solvent, i.e., in absence of micellar orientational effects, the two regioisomers were obtained in equal amounts. When, on the other hand, the reactions were carried out in an aqueous buffer in which the reactants form mixed micelles, a regioisomer ratio of 3 was obtained.

Another example, this time in a proper microemulsion, is stereoselective nitration of phenol. It has been claimed that the reaction in an oil-in-water microemulsion gives a preference for ortho nitration whereas conventional nitration in a homogeneous reaction system gives predominantly

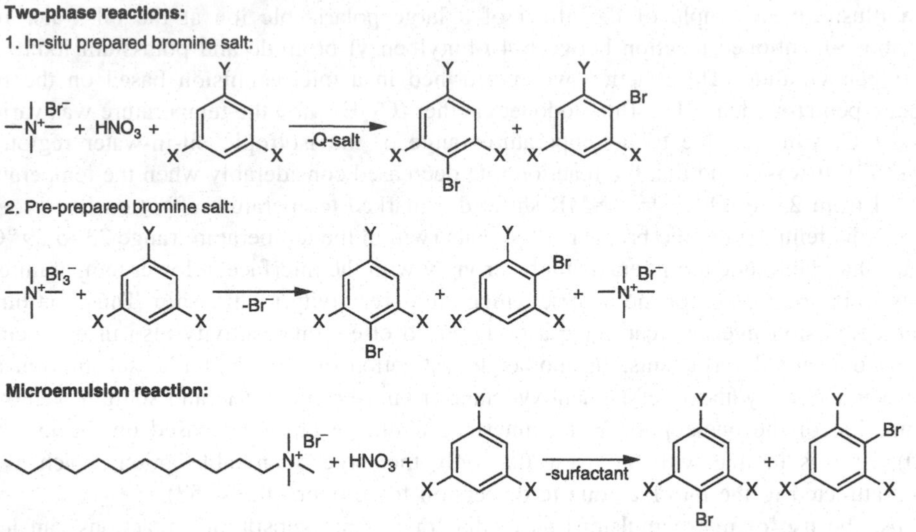

FIGURE 7.11 Reaction approaches used for bromination; x = H or CH_3; y = CH_3, OH, or OCH_3.

para nitration (67). The results have been questioned, however, since it has not been possible to repeat the experiments (68).

A related reaction, bromination of two phenols and two anisols, has been carried out in a cationic surfactant-based microemulsion using the surfactant counter-ion, i.e., bromide, as reagent (68). The bromide ion was oxidized to elemental bromine by dilute nitric acid, which in turn reacted with the aromatic compound. The results have been compared with two-phase procedures using either an in situ-prepared or a pre-prepared complex between bromine and a quaternary ammonium salt as oxidizing reagent and also with conventional bromination using elemental bromine. The different routes are shown in Figure 7.11. Reaction in the microemulsion gave a more selective para-bromination than the other procedures. In addition, the microemulsion-based reaction proceeded smoothly at room temperature.

Water-in-oil microemulsions have recently been employed to control the regioselectivity of the photocycloaddition of 9-substituted anthracenes (69). Reaction in microemulsion gave very different orientation from that in homogeneous solution. This difference is believed to be due to orientation of the anthracenes at the oil–water interface prior to the irradiation.

Several studies have dealt with the Diels–Alder reaction carried out in micellar media, but not in microemulsions. An early example is given in Ref. 66. Jaeger has continued the work on combining surface-active dienes with surface-active dienophiles and obtained a very high preference for the adduct with the quaternary ammonium groups situated on the same side, obviously due to an orientation in the micelles prior to the cycloaddition (70). The reaction is shown in Figure 7.12.

The adduct with the quaternary ammonium groups situated on the same side was formed predominantly. The regioselectivity is most likely due to the orientation of the diene and dienophile at the micelle–water interface, as illustrated in Figure 7.13.

It has recently been found that microemulsions can be used as a tool to differentiate between the first and the second step of a substitution of symmetrical bifunctional reactants, in this case α,ω-dibromoalkanes (71). In a homogeneous system, where both the lipophilic reactant, the

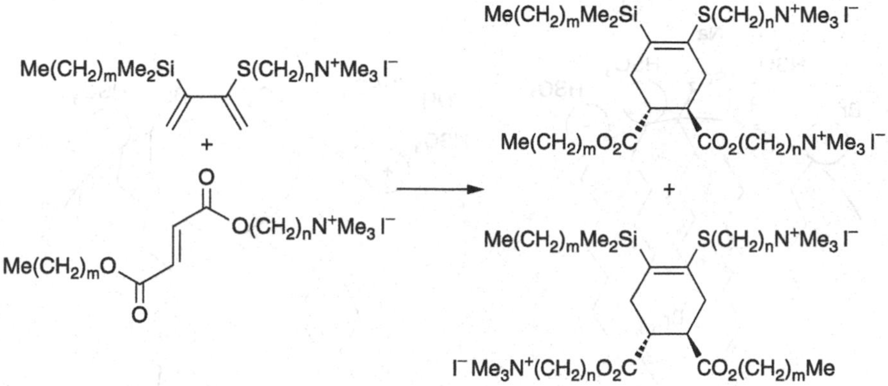

FIGURE 7.12 A Diels–Alder reaction between a surface-active diene and a surface-active dienophile.

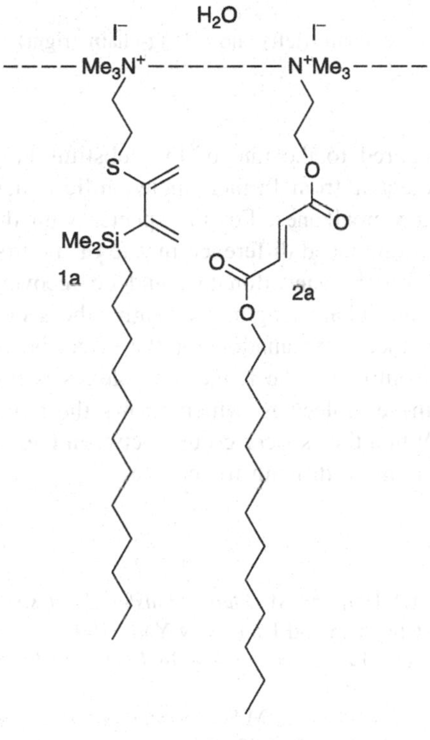

FIGURE 7.13 Preferred orientation of the reactants of Figure 7.12 in a micelle composed of the two surface-active cationic species.

α,ω-dibromoalkane, and the hydrophilic reactant, sodium sulfite, are dissolved the two substitution steps will occur at the same rate. The situation proved to be different in a microemulsion. The intermediate mono-substituted species, a bromoalkanesulfonate, has one polar and one nonpolar end; hence, it orients at the interface such that the sulfonate end points into the water domain, leaving the bromo end in the nonpolar environment. Provided that the alignment of

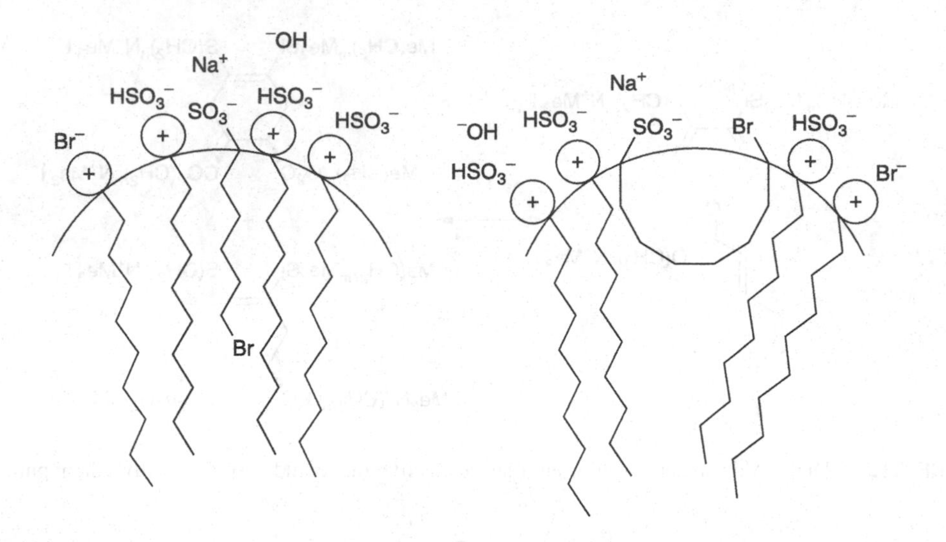

FIGURE 7.14 Alignment of a short-chain (left) and a long-chain (right) bromoalkanesulfonate at the oil–water interface of a microemulsion.

the intermediate is fast compared to the rate of the substitution reaction, such an orientation of the intermediate may protect it from further nucleophilic attack. This turned out to be the case for some of the α,ω-dibromoalkanes. For the species with the shortest alkane chain, 1,4-dibromobutane, there was a pronounced difference in rate of the first and the second substitution step and the intermediate, bromobutanesulfonate could be recovered in high yield. The selectivity decreased with the alkane chain length. Evidently, the second bromide is less protected for the longer derivatives. A likely explanation for the effect is illustrated in Figure 7.14. The more limited regiochemical control for the longer derivatives is probably due to a considerable conformational freedom of these molecules, which allows the remaining bromomethyl group to be exposed at the interface. When the "spacer group" between the sulfonate and the bromomethyl group is short, such folding is more difficult to achieve.

REFERENCES

1. Starks, C.M.; Liotta, C.L.; Halpern, M. *Phase Transfer Catalysis: Fundamentals, Applications and Industrial Perspectives*, Chapman and Hall: New York, 1994.
2. Gelbert, C.H.; Grady, M.C. Latex Technology. In *Encyclopedia of Chemical Technology*, 4th ed.; Vol. 15, 51–68.
3. Daniels, E.; Sudol, E.D.; El-Aasser, M.S. *Polymer Latexes: Preparation, Characterization and Applications*. ACS Symp. Series 492, ACS: Washington, 1992.
4. Lansalot, M.; Farcet, C.; Charleux, B.; Vairon, J.-P.; Pirri, R. Controlled free-radical miniemulsion polymerization of styrene using degenerative transfer. Macromol. **1999**, 32, 7354–7360.
5. Butté, A.; Storti, G.; Morbidelli, M. Miniemulsion living free radical polymerization of styrene. Macromol. **2000**, 33, 3485–3487.
6. Pileni, M.-P. Reverse micelles as microreactors. J. Phys. Chem. **1993**, 97, 6961–6973.
7. Sjöblom, J.; Lindberg, R.; Friberg, S.E. Microemulsions – phase equilibria characterization, structures, applications and chemical reactions. Adv. Colloid Interface Sci. **1996**, 95, 125–287.
8. Holmberg, K. Surfactant-templated nanomaterials synthesis. J. Colloid Interface Sci. **2004**, 274, 355–364.

9. Candau, F. Polymerization in microemulsions. In *Handbook of Microemulsion Science and Technology*; Kumar, P., Mittal, K.L., Eds.; Marcel Dekker: New York, 1999; 679–712.
10. Holmberg, K. Microemulsions in Biotechnology. In *Industrial Applications of Microemulsions*; Solans, C., Kunieda, H., Eds.; Surfactant Science Series 66, Marcel Dekker: New York, 1997; 69–95.
11. Holmberg, K. Enzymatic reactions in microemulsions. In *Handbook of Microemulsion Science and Technology*; Kumar, P., Mittal, K.L., Eds.; Marcel Dekker: New York, 1999; 713–742.
12. Lim, P.K.; Zhong, Y. Chemically-benign synthesis at organic–water interface. In *Designing Chemical Syntheses and Processing for the Environment*; Anastas, P., Williamson, T., Eds.; ACS Symp. Series 626, 1996, 168.
13. Zhong, Y.; Abrams, C.F.; Lim, P.K.; Brown, P.A. Biphasic synthesis of poly(2,6-dimethyl-1, 4-phenylene oxide) using a surface-active coupling catalyst. Ind. Eng. Chem. Res. **1995**, 34, 1529–1535.
14. Chung, Y.M.; Ahn, W.S.; Lim, P.K. Organic/aqueous interfacial synthesis of poly(2,6-dimethyl-1,4-phenylene oxide) using surface-active copper complex catalysts. Appl. Catalysis A **2000**, 192, 165–174.
15. Chung, Y.M.; Ahn, W.S.; Lim, P.K. Organic–water interfacial synthesis of an α-tetralone using nickel-tetraethylenepentamine complex catalyst. J. Catal. **1998**, 173, 210–218.
16. Zhong, Y.; Godfrey V.M.; Lim, P.K.; Brown, P.A. Biphasic synthesis of phenylacetic and phenylenediacetic acids by interfacial carbonylation of benzyl chloride and dichloro-p-xylene. Chem. Eng. Sci. **1966**, 51, 757–767.
17. Serra-Holm, V.; Salmi, T.; Kangas, M.; Mäki-Arvela, P.; Lindfors, L.P. Kinetics and phase equilibria of competing aldolization and elimination in a three-phase reactor. Catalysis Today **2001**, 66, 419–426.
18. Chirakadze, G.G.; Elizbarashvili, E.N.; Baidoshvili, P.O. Synthesis of new azo dyes under phase-transfer catalysis. Russ. J. Org. Chem. **2000**, 36, 1642–1645.
19. Kaufman, V.R.; Garti, N. Organic reactions in emulsions – preparation of glycerol and polyglycerol esters of fatty acids by transesterification reaction. J. Am. Oil Chem. Soc. **1982**, 59, 471–474.
20. Singh, C.P.; Shah, D.O.; Holmberg, K. Synthesis of mono- and diglycerides in water-in-oil microemulsions. J. Am. Oil Chem. Soc. **1994**, 71, 583–587.
21. Oh, S.-G.; Holmberg, K.; Ninham, B.W. Effect of hydrocarbon chain length on yield of lipase catalyzed triglyceride synthesis in microemulsion. J. Colloid Interface Sci. **1996**, 181, 341–343.
22. Singh, C.P.; Shah, D.O. Lipase-catalyzed esterification in monolayers and microemulsions. Colloids Surfaces A **1993**, 77, 219–224.
23. Singh, C.P.; Skagerlind, P.; Holmberg, K.; Shah, D.O. A comparison between lipase catalyzed esterification of oleic acid with glycerol in monolayer and microemulsion systems. J. Am. Chem. Soc. **1994**, 71, 1405–1409.
24. Oh, S.-G.; Singh, C.P.; Shah, D.O. Esterification of stearic acid with glycerol by lipase in foam. Langmuir **1992**, 8, 2846–2848.
25. Husein, M.M.; Weber, M.E.; Vera, J.H. Nucleophilic substitution sulfonation in microemulsions and emulsions. Langmuir **2000**, 16, 9159–9167.
26. Husein, M.M.; Vera, J.H.; Weber, M.E. Nucleophilic substitution sulfonation in emulsions: effect of the surfactant counterion and different decyl halide reactants. Colloids Surfaces A **2001**, 191, 241–252.
27. Cameron, N.R.; Sherrington, D.C. High internal phase emulsions (HIPEs) structure, properties and use in polymer preparation. Adv. Polym. Sci. **1996**, 126, 163.
28. Deleuze, H.; Maillard, B.; Mondain-Monval, O. Development of a new ultraporous polymer as support in organic synthesis. Bioorg. Medicinal Chem. Letters **2002**, 12, 1877–1880.
29. Schwuger, M.; Stickdorn, K.; Schomäcker, R. Microemulsions in technical processes. Chem. Revs. **1995**, 95, 849–864.
30. Menger, F.M.; Elrington A.R. Organic reactivity in microemulsion systems. J. Am. Chem. Soc. **1991**, 113, 9621–9624.
31. Izquierdo, C.; Casado, J.; Rodriguez, A.; Moya, M.L. Microemulsions as medium in chemical kinetics. J. Chem. Kinetics **1992**, 24, 19–30.

32. Luisi, P.L.; Giomini, M.; Pileni, M.-P.; Robinson, B.H. Reverse micelles as hosts for proteins and small molecules. Biochim. Biophys. Acta **1988**, 947, 209–246.
33. Holmberg, K. Organic and bioorganic reactions in microemulsions. Adv. Colloid Interface Sci. **1994**, 51, 137–174.
34. Fendler, J.H. *Membrane Mimetic Chemistry*. Wiley: New York, 1982.
35. Mackay, R.A. Chemical reactions in microemulsions. Adv. Colloid Interface Sci. **1981**, 15, 131–156.
36. Schomäcker, R. Mikroemulsionen als Medium für chemische Reaktionen. Nachr. Chem. Tech. Lab. **1992**, 40, 1344–1351.
37. Holmberg, K.; Häger, M. Organic synthesis in microemulsion; an alternative or a complement to phase transfer catalysis. In *Adsorption and Aggregation of Surfactants in Solution*; Mittal, K.L., Shah, D.O., Eds.; Marcel Dekker: New York, 2002; 327–342.
38. Häger, M.; Currie, F.; Holmberg, K. Organic Reactions in Microemulsions. In *Colloid Chemistry*; Antonietti, M., Ed. Topics in Current Chemistry, Springer-Verlag: Heidelberg, 2003; 53–74.
39. Gutfelt, S.; Kizling, J.; Holmberg, K. Microemulsions as reaction medium for surfactant synthesis. Colloids Surfaces A **1997**, 128, 265–271.
40. Häger, M.; Holmberg, K. A substitution reaction in an oil-in-water microemulsion catalyzed by a phase transfer agent. Tetrahedron Lett. **2000**, 41, 1245–1248.
41. Andersson, K.; Kizling, J.; Holmberg, K.; Byström, S. A ring-opening reaction performed in microemulsions. Colloids Surfaces A **1998**, 144, 259–266.
42. Siswanto, C.; Battal, T.; Schuss, O.E.; Rathman, J.F. Synthesis of alkylphenyl ethers in aqueous solutions by micellar phase-transfer catalysis. 1. Single-phase systems. Langmuir **1997**, 13, 6047–6052.
43. Häger, M.; Holmberg, K. Phase transfer agents as catalysts for a nucleophilic substitution reaction in microemulsion. Chem. Europ. J. **2004**, 10, 5460–5466.
44. Vauthey, S.; Milo, C.; Frossard, P.; Garti, N.; Leser, M.E.; Watzke, H.J. Structured fluids as microreactors for flavor formation by the Maillard reaction. J. Agric. Food Chem. **2000**, 48, 4808–4816.
45. Yagahmur, A.; Aserin, A.; Garti, N. Furfural-cystein model reaction in food grade nonionic oil/water microemulsions for selective flavor formation. J. Agric. Food Chem. **2002**, 50, 2878–2883.
46. Häger, M.; Holmberg, K.; Olsson, U. Synthesis of an amphiphilic polymer performed in an oil-in-water microemulsion and in a lamellar liquid crystalline phase. Colloids Surfaces A **2001**, 189, 9–19.
47. Häger, M.; Olsson, U.; Holmberg, K. A nucleophilic substitution reaction performed in different types of self-assembly structures. Langmuir, **2004**, 20, 6107–6115.
48. Bode, G.; Lade, M.; Schomäcker, R. The kinetics of an interfacial reaction in microemulsions with excess phase. Chem. Eng. Technol. **2000**, 23, 405–409.
49. Menger, F.M.; Rourk, M.J. Deactivation of mustard and nerve agent models via low-temperature microemulsions. Langmuir **1999**, 15, 309–313.
50. Qian, J.; Guo, R.; Zou, A. Effect of microemulsion structures on the hydrolysis of acetylsalicylic acid. J. Dispersion Sci. Technol. **2001**, 22, 541–549.
51. Hao, J. Effect of structures of microemulsions on chemical reactions. Colloid Polym. Sci. **2000**, 278, 150–154.
52. Oh, S.-G.; Kizling, J.; Holmberg, K. Microemulsions as reaction medium for synthesis of sodium decyl sulfonate. I: Role of microemulsion composition. Colloids Surfaces A **1995**, 97, 169–179.
53. Haumann, M.; Yildiz, H.; Koch, H.; Schomäcker, R. Hydroformylation of 7-tetradecene using Rh-TPPTS in a microemulsion. Appl. Catalysis A **2002**, 236, 173–178.
54. Oh, S.-G.; Kizling, J.; Holmberg, K. Microemulsions as reaction medium for synthesis of sodium decyl sulfonate. II: Role of ionic surfactants. Colloids Surfaces A **1995**, 104, 217–222.
55. Holmberg, K.; Oh, S.-G.; Kizling, J. Microemulsions as reaction medium for a substitution reaction. Progr. Colloid Polym. Sci. **1996**, 100, 281–285.
56. Häger, M.; Holmberg, K.; Rocha Gonsalves, A.M.; Serra, A. Oxidation of azo dyes in oil-in-water microemulsions catalyzed by metalloporphyrins in presence of lipophilic acids. Colloids Surfaces A **2001**, 183–185, 247–257.

57. Ninham, B.W.; Yaminsky, V. Ion binding and ion specificity: The Hofmeister effect and Onsager and Lifshitz theories. Langmuir **1997**, 13, 2097–2108.
58. Kabalnov, A.; Olsson, U.; Wennerström, H. Salt effects on nonionic microemulsions are driven by adsorption/depletion at the surfactant monolayer. J. Phys. Chem. **1995**, 99, 6220–6230.
59. Häger, M.; Currie, F.; Holmberg, K. A nucleophilic substitution reaction in microemulsions based on either an alcohol ethoxylate or a sugar surfactant. Colloids Surfaces A **2004**, 250, 163–170.
60. Tjandra, D.; Lade, M.; Wagner, O.; Schomäcker, R. The kinetics of an interfacial reaction in a microemulsion. Chem. Eng. Technol. **1998**, 21, 666–670.
61. Brinchi, L.; Di Profio, P.; Germani, R.; Savelli, G.; Bunton, C.A. Chemoselectivity in S_N2-E2 reactions induced by aqueous association colloids. Colloids Surfaces A **1998**, 132, 303–314.
62. Brinchi, L.; Di Profio, P.; Germani, R.; Marte, L.; Savelli, G.; Bunton, C.A. Micellar S_N2 reaction of methyl naphthalene-2-sulfonate and its 6-sulfonate derivative: Effect of the negative charge. J. Colloid Interface Sci. **2001**, 243, 469–475.
63. Ruan, K.; Zhao, Z.; Ma, J. Effect of bromide salts on cationic micellar catalysis. Colloid Polym. Sci. **2001**, 279, 813–818.
64. Foroudian, H.J.; Gillitt, N.D.; Bunton, C.A. Effect of nonionic micelles on dephosphorylation and aromatic nucleophilic substitution. J. Colloid Interface Sci. **2002**, 250, 230–237.
65. Romsted, L.S.; Bunton, C.A.; Yao, J. Micellar catalysis, a useful misnomer. Curr. Opin. Colloid Interface Sci. **1997**, 2, 622–628.
66. Jaeger, D.A.; Wang, J. Micellar effects on the regiochemistry of a Diels–Alder reaction. Tetrahedron Lett. **1992**, 33, 6415–6418.
67. Currie, F.; Holmberg, K.; Westman, G. Regioselective nitration of phenols and anisols in microemulsion. Colloids Surfaces A **2001**, 182, 321–327.
68. Currie, F.; Westman, G.; Holmberg, K. Bromination in microemulsion. Colloids Surfaces A. **2003**, 215, 51–54.
69. Wu, D.-Y.; Zhang, L.-P.; Wu, L.-Z.; Wang, B.; Tung, C.-H. Water-in-oil microemulsions as microreactors to control the regioselectivity in the photocyclization of 9-substituted anthracenes. Tetrahedron Lett. **2002**, 43, 1281–1283.
70. Jaeger, D.A.; Su, D.; Zafar, A.; Piknova, B.; Hall, S.B. Regioselectivity control in Diels–Alder reactions of surfactant 1,3-dienes with surfactant dienophiles. J. Am. Chem. Soc. **2000**, 122, 2749–2757.
71. Currie, F.; Holmberg, K.; Romsted, L.S. Regioselectivity of a substitution reaction in a micellar system. (To be submitted.)

8 NMR Characterization of Emulsions

Alejandro A. Peña and George J. Hirasaki

CONTENTS

8.1 INTRODUCTION

Emulsions are among man-made commodities such as food goods [1,2], cosmetics [3,4], paints [5], agricultural sprays [6,7], asphalt preparations [8], and pharmaceuticals [4,7]. Emulsions are also found as undesired byproducts of industrial processes, including crude oil production [9,10] and liquid–liquid extraction operations [11]. Information on the microstructure and composition of an emulsion is relevant whether its making is sought or unwanted, because such information can be related to important properties of the dispersion such as its viscosity [12–14] and its stability against phase separation [15,16], and also to quality control standards. For example, the color [17] and texture (creaminess) [18] of food emulsions are known to depend on the distribution of droplet sizes; crude oil streams with emulsified water content in excess of a given threshold are not suited for processing in refining equipment [19].

Experimental techniques such as microphotography and video-enhanced microscopy, gravitational/centrifugal sedimentation, Coulter counting, differential scanning calorimetry, turbidimetry, dynamic and static light scattering, low-intensity ultrasound, and nuclear magnetic resonance (NMR) have been used to characterize emulsions [20–22]. NMR-based techniques are becoming increasingly popular due to practical advantages that they offer over other above-mentioned techniques. For example, the same emulsion sample can be tested via NMR as many times as desired. Dilution, cooling/heating, centrifugation, or confinement in a narrow gap are not necessary. Measurements are not influenced by the optical or dielectric properties of the system. Therefore, clear and opaque emulsions, and dispersions in which the continuous phase is non-conducting, can be tested. A typical NMR test is fast ($\sim$ 1 to 10 min), and it requires a small sample ($\geq\sim$ 0.5 g). Furthermore, the composition of the emulsion can be resolved from the NMR data.

Important progresses have been made in the characterization of emulsions via NMR since the publication of the first edition of this book in 1996. Efforts in recent years have been mainly focused on (1) expanding the theory and techniques used to interpret self-diffusion data obtained from the classic pulsed field gradient (PGSE) experiments for the determination of the microstructure and stability of emulsions, to overcome limitations that were inherent to classic views; (2) use alternative NMR pulsed sequences to allow data acquisition in shorter timescales, both for quiescent and flowing emulsions; and (3) utilize NMR-based techniques in applications involving the usage of emulsions, and for which such methods had not been previously considered. This chapter provides a critical compendium of such progress. The discussion is developed while presenting a brief introduction of NMR principles and techniques that are relevant to emulsion characterization.

8.2 BASIC NMR PRINCIPLES

NMR microscopy on emulsions is based on the following few physical principles [23]: Some nuclei, such as protons (^{1}H), exhibit a magnetic moment μ. When a steady uniform magnetic field $\mathbf{B_0}$ is applied on these nuclei, μ precesses around the direction of $\mathbf{B_0}$ at the Larmor frequency $\omega_0 = \gamma B_0$, where γ is a constant. Nuclei with precessing μ are termed *spins*. The ensemble of spins exhibits net magnetization $\mathbf{M}$ in the direction of $\mathbf{B_0}$. If a *radio frequency (rf) pulse* of a second magnetic field $\mathbf{B_1}$ orthogonal to $\mathbf{B_0}$ is applied, the net magnetization is rotated to an extent (typically 90° or 180°) that depends on the duration of the pulse. When the rf pulse ceases, $\mathbf{M}$ will *relax* toward and eventually reach some equilibrium state. Relaxation of $\mathbf{M}$ can be measured from the spins present in the emulsion, either in the direction of $\mathbf{B_0}$ (longitudinal magnetization), or transverse to it (transverse magnetization), whence characteristic relaxation rates can be quantified in terms of specific sets of longitudinal (T_1) and transverse (T_2) relaxation times, respectively.

In addition, the precession of spins at the same frequency is referred to as coherent or in-phase. If the steady magnetic field is not uniform as above the frequency depends on the position of the nuclei [$\omega(\mathbf{r}) = \gamma \mathrm{B}(\mathbf{r})$]. Two spins at positions $\mathbf{r_a}$ and $\mathbf{r_b}$, such that $\mathbf{B}(\mathbf{r_a}) \neq \mathbf{B}(\mathbf{r_b})$, precess incoherently out of phase. Magnetic field gradients are commonly applied to create a non-uniform steady magnetic field and adjust coherence.

Structural information of the emulsion, such as droplet sizes, relative concentration of the immiscible phases present, and morphology (water-in-oil, W/O vs. oil-in-water, O/W), can be inferred from the time-resolved signal as discussed later. Furthermore, Fourier transformation of the signal renders the NMR spectrum of the sample, whence individual chemical components can be identified and their concentrations determined [24–26]. Low-frequency NMR spectrometers operating at 2 to 20 MHz suffice for time-resolved signal applications and are most commonly used in industrial applications. High-frequency (typically 400 to 800 MHz) NMR spectrometers are required to obtain NMR spectra with adequate resolution for detailed spectroscopic applications.

Whereas most NMR experiments on emulsions rely on measurements on ^{1}H nuclei due to their natural abundance, spectroscopic studies on emulsions targeting ^{19}F [27,28], ^{31}P [29–34], ^{13}C [34], and ^{129}Xe [35,36] nuclei have also been reported, as described in recent reviews on this subject [37,38]. For the sake of brevity, the descriptions of basic NMR pulsed sequences given in Sections 8.3 and 8.4 are supplemented with data from ^{1}H studies. The reader should keep in mind, however, that such sequences can be applied, in principle, to fluids containing other nuclei such as those listed above.

8.3 RELAXATION MEASUREMENTS ON EMULSIONS VIA CPMG EXPERIMENTS

8.3.1 The CPMG Pulsed Sequence

The CPMG pulsed sequence bears the initials of its developers, Carr and Purcell [39], who first introduced it, and Meiboom and Gill [40], who further refined it to make it suitable for routine experiments. It consists of a rf 90° pulse, followed by N rf 180° pulses that induce successive phase recoveries and generate a train of N spin-echoes (Figure 8.1). As time proceeds, relaxation of the magnetization occurs and the amplitude of the spin-echo that is generated after each 180° re-phasing decays. In this experiment, the transverse component of the magnetization vector $M_{xy}(2n\tau)$ is measured, and the resulting relaxation curve is fitted to a discrete multi-exponential function of the form:

$$\frac{M_{xy}(2n\tau)}{M_{xy}(0)} = \sum_{i=1}^{m} f_i \exp\left(-\frac{2n\tau}{T_{2,i}}\right); \quad 0 \leq n \leq N; \quad m < N; \quad \sum_{i=1}^{m} f_i = 1 \tag{8.1}$$

$M_{xy}(0)$ is the amplitude of signal that corresponds to the initial transverse magnetization and f_i is the fraction of nuclei with characteristic relaxation time $T_{2,i}$. The fitting procedure consists of calculating f_i values for a pre-established set of $T_{2,i}$, whence the T_2 distribution is obtained. Fitting data to a multi-exponential sum is an ill-posed problem, i.e., multiple sets of f_i values can render a satisfactory fit [41]. For this reason, a regularization method must be implemented to calculate the most representative T_2 distribution [42].

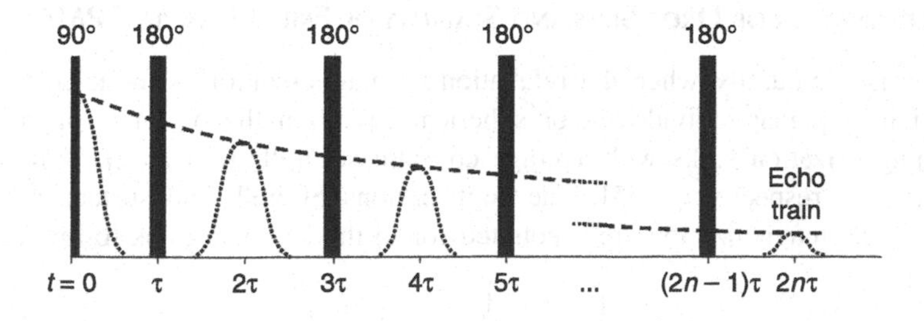

FIGURE 8.1 Sequence of events in a CPMG experiment.

8.3.2 Determination of the Water/Oil Composition in Emulsions via CPMG

In a CPMG test, the amplitude of the signal that is obtained for a given sample is proportional to the number of spins present in such sample. This principle allows relating the water/oil ratio of common O/W and W/O emulsions to the T_2 distribution obtained from ^{1}H CPMG experiments through the expression:

$$\frac{\phi_W}{\phi_O} = \frac{\sum(f_i)_W/HI_W}{\sum(f_i)_O/HI_O} \tag{8.2}$$

whence

$$\phi_W = \frac{\left[\sum(f_i)_W/HI_W\right]}{\left[\sum(f_i)_W/HI_W\right]+\left[\sum(f_i)_O/HI_O\right]}; \quad \phi_O = 1-\phi_W \tag{8.3}$$

ϕ_k is the volume fraction of phase k present in the emulsion. *HI* is the hydrogen index, which is defined as the number of protons in a sample divided by the number of protons present in the same volume of water [43]. The calculation of *HI* for pure fluids using this definition is straightforward. Empirical correlations and diagrams to estimate *HI* are available for "water" phases such as brines and "oil" phases such as light and heavy crude oils [44–46]. In general, $HI \sim 1.0$ for aqueous solutions and $HI \sim 0.9$ to 1.0 for most crude oils except for aromatic oils, which exhibit *HI* between 0.6 and 0.8 due to the low H/C ratio of aromatic compounds.

Figure 8.2 summarizes results from ^{1}H CPMG experiments performed with a 2-MHz spectrometer on thirty water-in-crude oil emulsions prepared at different water/oil compositions and containing the same crude oil ($HI = 0.928$) [47]. Figure 8.2(a) shows the T_2 distributions that were obtained for six of these samples, normalized with respect to the oil signal. It is seen in Figure 8.2(a) that the signal from the water phase increases with the water content, as might be expected. Figure 8.2(b) compares the actual water content of these and other mixtures with the water content that is calculated using Equation 8.3. The excellent agreement that was found between actual and measured water contents illustrates the usefulness of the technique to resolve for the water/oil composition of emulsions.

Low-field NMR-CPMG has been regarded as superior to all other available techniques for the routine determination of water content in heavy oil, bitumen, and oilfield emulsions [48,49]. This application of the CPMG method has been discussed by LaTorraca et al. [43] and Hills et al. [50]. Allsopp et al. [48] have developed and successfully tested *in situ* a low-field spectrometer suitable for usage in the field. The method was accurate to ± 5% and measuring times were typically 4 min or less. This application is a natural extension of the usage of NMR relaxation measurements for the determination of porosity in minerals and rocks [51–54].

8.3.3 Determination of Drop Sizes and Stability of Emulsions via CPMG

Equation 8.1 arises naturally when the relaxation of magnetization is modeled for an isotropic fluid confined in a planar, cylindrical, or spherical cavity in the presence of volume-like and surface-like magnetization sinks with average constant strength $1/T_{2,bulk}$ (*bulk relaxivity*) and ρ (*surface relaxivity*), respectively [55]. The contributions of bulk and surface relaxivity to the decay of transverse magnetization are accounted for in the $T_{2,i}$ values as follows:

$$\frac{1}{T_{2,i}} = \frac{1}{T_{2,bulk}} + \rho\left(\frac{S}{V}\right)_i \tag{8.4}$$

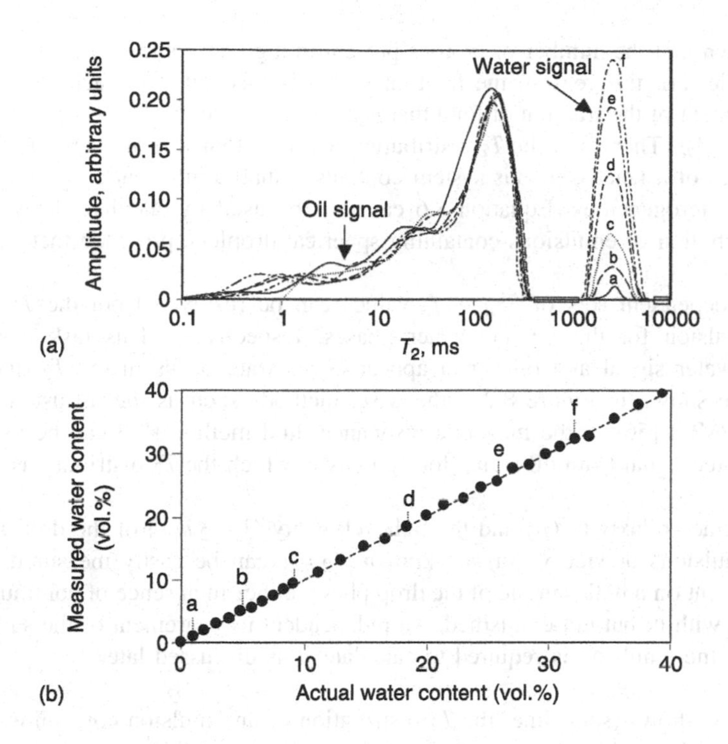

FIGURE 8.2 Determination of the water/oil composition of emulsions via low-field NMR CPMG experiments. (a) The signal from the water phase increases with water content. (b) Comparison of actual water content with that measured by using Equation 8.3.

$(S/V)_i$ is the surface-to-volume ratio of the cavity i. For a sphere of radius a_i, $(S/V)_i = 3/a_i$. Hence,

$$\frac{1}{T_{2,i}} = \frac{1}{T_{2,bulk}} + \rho\frac{3}{a_i} \tag{8.5}$$

and

$$a_i = 3\rho\left(\frac{1}{T_{2,i}} - \frac{1}{T_{2,bulk}}\right)^{-1} \tag{8.6}$$

Equations 8.4 to 8.6 are valid in the fast diffusion limit, in which the characteristic timescale for diffusion of the molecules confined in the drops (t_D) is much smaller than the characteristic timescale for surface relaxation (t_ρ):

$$\frac{t_D}{t_\rho} = \frac{a_i^2/D}{a_i/\rho} \ll 1, \text{ whence } \frac{\rho a_i}{D} \ll 1 \tag{8.7}$$

D is the self-diffusion coefficient of the drop phase. In practice, the relaxation of magnetization of fluids confined in spherical cavities occurs in the fast diffusion mode whenever $\rho a/D <\sim 0.25$.

It was shown that the number of protons present in a given volume of sample determines the signal amplitude. For this reason, the fraction f_i that is associated to each $T_{2,i}$ value renders a direct measurement of the fraction of fluid that is confined in cavities of corresponding surface-to-volume ratio $(S/V)_i$. Therefore, the T_2 distribution that is obtained from isotropic fluids contained in the interstices of a heterogeneous system contains valuable information on the distribution of sizes of such heterogeneities. Equation 8.6 can thus be used to calculate the volume-weighted drop size distribution of emulsions containing spherical droplets, provided that:

1. Two independent sets of $T_{2,i} - f_i$ values can be resolved from the T_2 distribution of the emulsion for the oil and water phases, respectively. This task is straightforward if the water signal and oil signal appear as separate peaks in the T_2 distribution, such as those shown in Figure 8.2. Otherwise, methods such as the diffusion-editing CPMG (DE-CPMG) [56] or the magnetic resonance fluid method [57] can be used to discriminate water signal from oil signal for systems in which the T_2 distributions of these phases overlap.
2. The surface relaxivity (ρ) and the bulk relaxivity ($1/T_{2,bulk}$) of the drop phase (water in *w/o* emulsions or vice versa) are known. $T_{2,bulk}$ can be easily measured from a CPMG experiment on a bulk sample of the drop phase, either in absence of continuous phase or in contact with it, but not emulsified. An independent measurement of the surface-to-volume ratio of the emulsion is required to calculate ρ as discussed later.

Figure 8.3(a) shows (solid line) the T_2 distribution of an emulsion containing 30 vol% water in crude oil. Details on the preparation of samples and test conditions are given in Ref. 47. The T_2 distribution of the water and oil phases were also measured from individual samples of each phase (dotted lines). The most significant feature of Figure 8.3(a) is the fact that the signal of the water phase present in the emulsion shifted towards lower relaxation times. Such shift is caused by relaxation of the magnetization at the water–oil interfaces that are created once the water phase was dispersed as droplets within the oil phase. Figure 8.3(b) shows a comparison of NMR-CPMG measurements with microphotography measurement for the same emulsion. The NMR data were calculated by replacing the T_2 data for the water phase of the emulsion shown in Figure 8.3(a) in Equation 8.6 using a surface relaxivity of 0.88 μm/s, which was determined concomitantly with data acquired using the NMR-PGSE method as described in Section 8.4. For the particular tests shown here, it is seen that the CPMG method adequately describes both the numerical values for drop sizes and also the distribution of such sizes. An important feature of the NMR-CPMG method shown in Figure 8.3 is that it allows the resolution for drop size distributions with arbitrary shape, i.e., it does not require assuming a priori a mathematical expression such as the log–normal distribution that is commonly used to interpret drop size distribution measurements using the NMR-PGSE technique [26,58–65].

It has been proposed [47] that the minimum and maximum drop sizes that can be determined via CPMG are, respectively:

$$d_{MIN} \cong 6\rho T_{2,\min} \tag{8.8}$$

and

$$d_{MAX} \cong \min\{D/2\rho, 2e^{-1} SNR\rho T_{2,bulk}\} \tag{8.9}$$

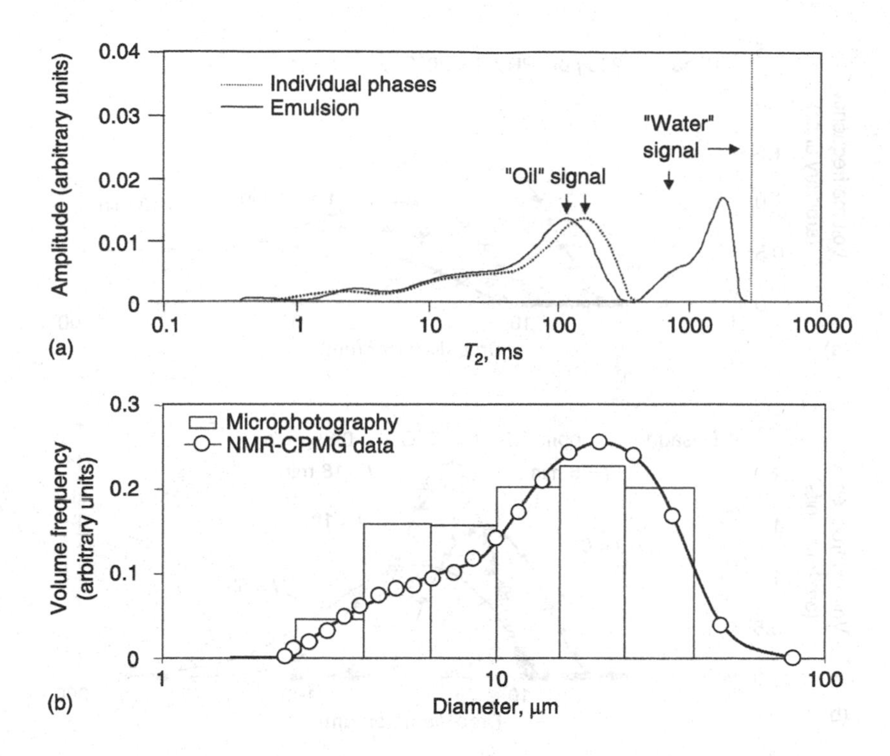

FIGURE 8.3 (a) T_2 distribution of a water-in-crude oil emulsion (W/O ratio = 30/70 (vol.)); (b) calculated drop size distribution from the T_2 distribution of the water phase, using $\rho = 0.88$ μm/s.

where *SNR* is the signal-to-noise ratio ($SNR = M_{xy}(t = 0)/v$, with v being the intrinsic noise of the measurement) and $T_{2,\min} = -2\tau/\ln(1 - \varepsilon)$, where ε is the fraction of the transverse component of magnetization of the fluid present in drops with diameter d_{MIN} that has relaxed when the first echo is acquired at time 2τ (Figure 8.1), i.e., the fraction of droplets with $d = d_{MIN}$ that would not be accounted for due to relaxation before the acquisition of the first spin-echo. The expression for d_{MIN} is valid whenever $T_{2,\min} \ll T_{2,bulk}$. Using the data reported in Ref. 47 ($D = 2.3 \times 10^{-9}$ m^2/s, $SNR = 99.7\%/0.3\% = 332$, $T_{2,bulk} = 2.8$ s, $2\tau = 315$ ms) with $\rho = 0.88$ μm/s and $\varepsilon = 0.1$, Equations 8.8 and 8.9 render $d_{MIN} = 16$ nm and $d_{MAX} = 580$ μm. These figures illustrate that, in principle, a very wide range of droplet sizes can be measured using the CPMG technique. In many practical cases, however, the minimum drop size may be limited by the ability to distinguish the water from the oil signal if they overlap.

The NMR-CPMG experiment is also useful for performing drop size distribution measurements in emulsions with limited stability. Figure 8.4 shows NMR-CPMG transient drop size distribution data for two emulsions of 5 wt% NaCl brine (30 vol%) in crude oil (70 vol%) that were treated with chemicals commonly used in demulsification operations [66]. Results shown in Figure 8.4(a) correspond to an emulsion sample treated with a nonylphenol formaldehyde resin. Results given in Figure 8.4(b) correspond to a sample of the same emulsion treated with a polyurethane. The data shown in Figure 8.4 depict a fast and steady growth of drop sizes, thus indicating that the nonylphenol formaldehyde resin induced rapid coalescence. In contrast, Figure 8.4(b) illustrates the inability of the polyurethane used in the experiment to induce significant coalescence. A shortcoming of this procedure is the fact that the surface relaxation

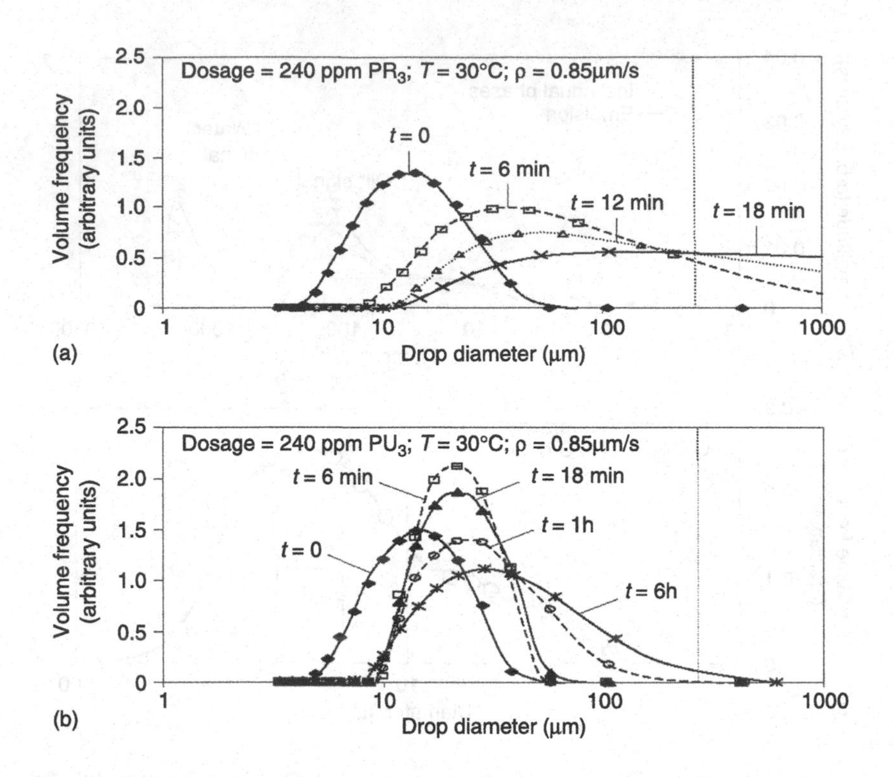

FIGURE 8.4 Drop size distributions of water-in-crude oil emulsions (W/O ratio = 30/70 (vol.)) undergoing phase separation as measured via NMR-CPMG: (a) emulsion treated with a nonylphenol formaldehyde resin (PR); (b) emulsion treated with a polyurethane (PU). Details on the experiment are given in Ref. 66.

is assumed constant throughout the phase separation process. Whereas the validity of such an assumption deserves consideration in future studies, the result is still useful because it provides a reference for the timeframes for which significant changes in the microstructure of the emulsion occur.

An interesting extension of the NMR-CMPG method consists of applying the sequence shown in Figure 8.1 in the presence of a magnetic field gradient (CPMG gradient or CPMG-g test). When a gradient of strength g is imposed, the steady magnetic field is not uniform in space and the Larmor frequency depends on the position of the nuclei [$\omega(\mathbf{r}) = \gamma \mathrm{B}(\mathbf{r})$] as mentioned earlier. The coil ("antenna") of the spectrometer is tuned to measure the response of the region where spins precess at a specific Larmor frequency ω_0. Therefore, only those spins are sensed by the spectrometer. This principle allows characterizing "slices" of the sample of interest. In this case, an additional contribution to the relaxation time T_2 is caused by the diffusion of the spins in the field gradient and Equation 8.5 becomes:

$$\frac{1}{T_{2,i}} = \frac{1}{T_{2,bulk}} + \rho\frac{3}{a_i} + \frac{(\gamma g t_E)^2 D}{12} \tag{8.10}$$

where $t_E = 2\tau$ is the time between consecutive 180° rf pulses (Figure 8.1) and γ is a physical constant called the gyromagnetic ratio that has a specific value for each nuclei ($\gamma = 2.67 \times 10^8$ rad $\mathrm{T}^{-1}\mathrm{s}^{-1}$ for ^{1}H).

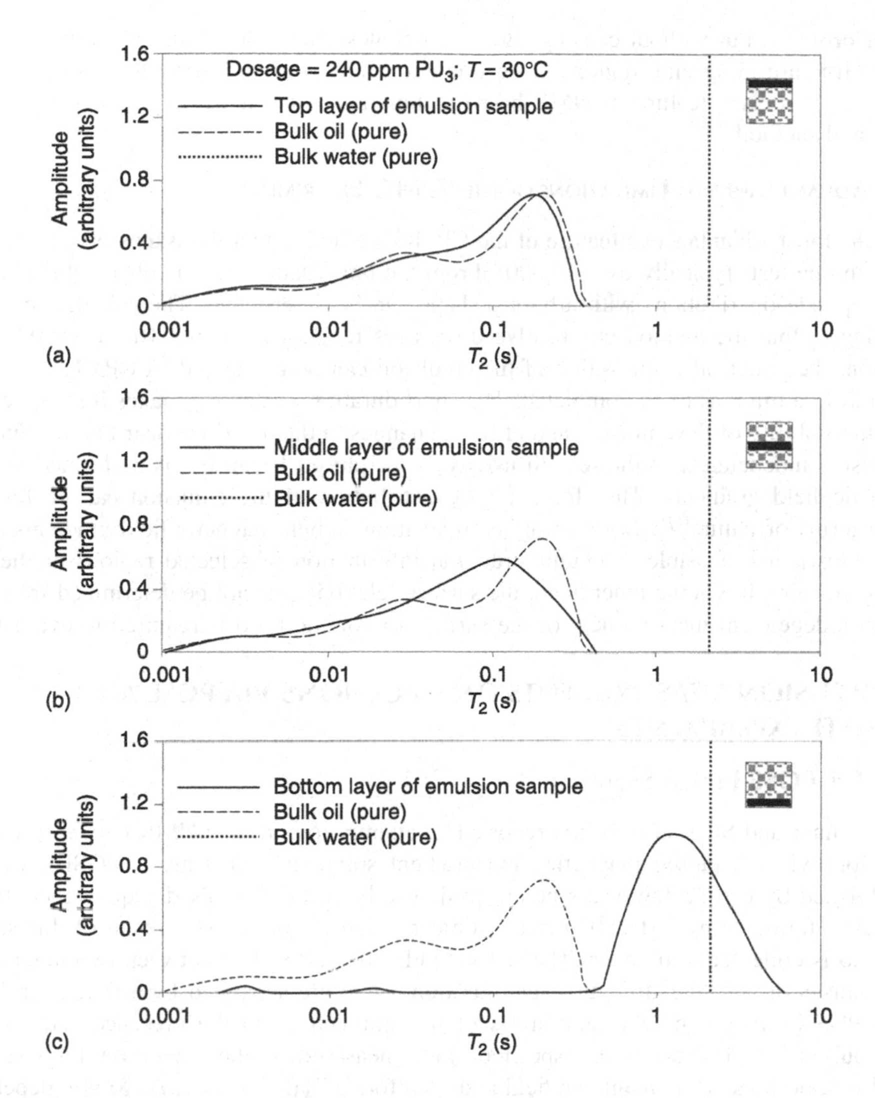

FIGURE 8.5 T_2 distributions of selected regions of the emulsion treated with a polyurethane. Data were obtained using the CMPG-*g* pulsed sequence.

Figure 8.5 shows the T_2 distributions that were obtained after applying the CPMG-*g* method on three regions (top, middle, and bottom) of the sample treated with the polyurethane for which data were reported in Figure 8.4(b). The T_2 distributions of the crude oil (dashed line) and of the brine (dotted line) are also shown as reference. Details on the experiment are provided in Ref. 66, the most important being that a clear layer of decanted water could not be observed at the bottom of the sample at the time that the experiment was performed. Nevertheless, the comparison of the T_2 distributions for "slices" of the emulsion with those of the individual phases clearly shows that the upper and middle regions of the sample had been deprived of water, which otherwise had accumulated at the bottom. These data, along with bottle tests and viscosity measurements, suggest that the polyurethane that was used in the experiment depicted in Figures 8.4(b) and 8.5 was able to induce flocculation and rapid sedimentation of the water droplets, but without causing significant coalescence. The ability to evaluate details of the microstructure in specific regions of opaque emulsions without having to destroy the sample (or aliquots of it) is a feature of NMR-based techniques that cannot be procured by any other experimental method.

8.3.4 Advantages and Limitations of the CPMG Experiment

Perhaps the most advantageous feature of the CPMG method is that thousands of spin-echoes are acquired in one test (typically over 10,000). From the large dataset that is obtained in this experiment, drop size distributions with arbitrary shape can be determined. The calculations reported above suggest that the method can resolve drop sizes ranging from $\sim$ 0.01 to $\sim$ 500 microns. In addition, the water/oil composition of the emulsion can be calculated. A CPMG test typically takes about five minutes to be completed. The short duration of the test makes it suitable to keep track of the stability of the emulsion and of rapid changes in the drop size distribution. Finally, the CPMG test is independent of the self-diffusivity of the phases because it is performed in absence of magnetic field gradients. Therefore, the T_2 distribution of the emulsion can be fully interpreted in terms of natural (bulk) and surface relaxation. When magnetic field gradients are used (CMPG-g test), it is feasible to obtain structural information of selected regions of the sample without perturbing it. On the other hand, the surface relaxivity cannot be determined from CPMG alone. An independent measurement of the surface-to-volume ratio is required to evaluate ρ.

8.4 DIFFUSION MEASUREMENTS ON EMULSIONS VIA PGSE AND PGSTE EXPERIMENTS

8.4.1 The PGSE Pulsed Sequence

In 1965, Tanner and Stejskal [67] first reported an attempt to correlate NMR data with drop sizes in emulsions via the pulsed magnetic field gradient spin-echo experiment (PGSE), which had been published by them a few years before [68]. The basic PGSE pulsed sequence consists of a rf 90° pulse, followed by a rf 180° pulse at time τ (Figure 8.6(a)). As a result of this sequence, a spin-echo is collected at time 2τ. The rf 180° pulse is sandwiched between two magnetic field gradient pulses of absolute strength g and duration δ that are separated by a time span Δ.

In a PGSE experiment, the amplitudes of the spin-echoes in the presence and absence of gradient pulses ($g > 0$ and $g = 0$, respectively) are measured. In the latter case, the spin-echo is acquired in a homogeneous magnetic field and therefore $M_{xy}(2\tau,\ g = 0,\ \Delta,\ \delta)$ is independent of the spatial distribution of spins in the sample. On the other hand, when $g > 0$ the first gradient pulse imposes an inhomogeneous magnetic field, thus causing a loss of coherence in the phases of the spins to an extent that depends on the position of the nuclei at the time the gradient is applied. In the absence of diffusion, the second gradient pulse would exactly reverse the phase shifts. However, since molecules diffuse and change their position during the *diffusion time* Δ, the refocusing is incomplete and the amplitude of the echo that is recorded at time 2τ [$M_{xy}(2\tau,\ g > 0,\ \Delta,\ \delta,\ D)$] is smaller than $M_{xy}(2\tau,\ g = 0,\ \Delta,\ \delta)$. For this reason, $0 < M_{xy}(2\tau,\ g,\ \Delta,\ \delta,\ D) < M_{xy}(2\tau,\ g = 0,\ \Delta,\ \delta)$. Whence,

$$R = \frac{M_{xy}(2\tau, g, \Delta, \delta, D)}{M_{xy}(2\tau, g = 0, \Delta, \delta)}; \quad 0 \leq R \leq 1 \tag{8.11}$$

R is termed the *spin-echo attenuation ratio*. In a typical PGSE experiment, attenuation ratios are measured changing systematically δ, Δ, or g.

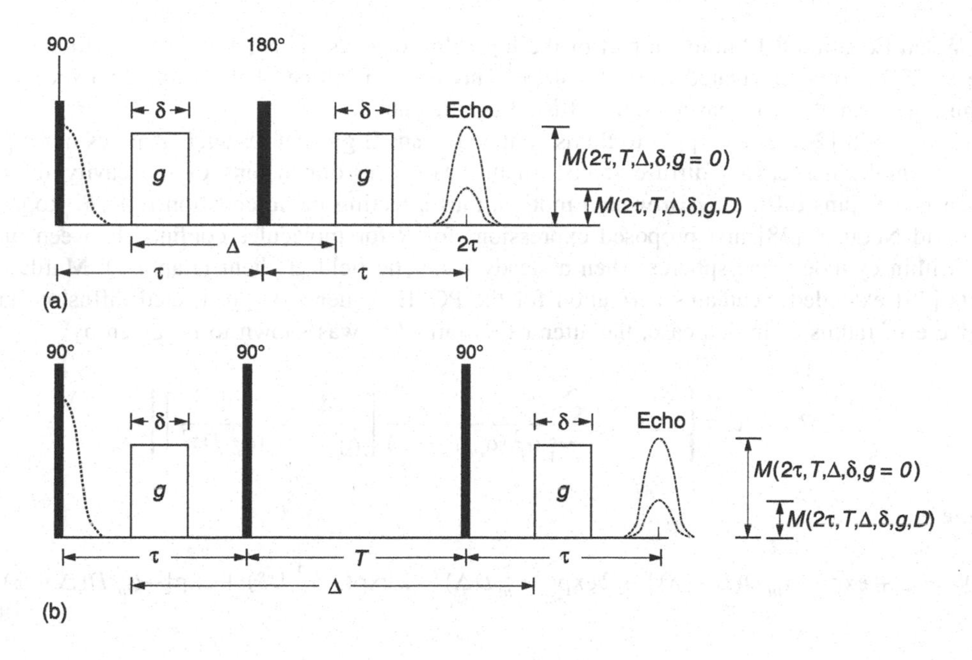

FIGURE 8.6 Basic (a) PGSE and (b) PGSTE pulsed sequences.

8.4.2 The PGSTE Pulsed Sequence

The stimulated spin-echo pulsed field gradient (PGSTE), which was also introduced by Tanner [69] (Figure 8.6(b)), is particularly useful for systems exhibiting significantly different longitudinal (T_1) and transverse (T_2) relaxation times ($T_1 > T_2$). This is often the case for systems exhibiting a large interfacial area, such as emulsions. By virtue of the second rf 90° pulse, the direction of the magnetization vector is shifted in a way that its initially transverse (xy) component is positioned in the longitudinal (z) direction, upon which relaxation of magnetization occurs at a slower rate (characteristic relaxation rate $= 1/T_1 < 1/T_2$). The third rf 90° pulse repositions M_{xy} in the transverse plane. As a result, a "stimulated" spin-echo occurs an interval after the third rf pulse equal to that between the first two pulses. This procedure allows increasing the diffusion time Δ while reducing the deleterious effect of extended relaxation upon the signal-to-noise ratio that otherwise would occur when increasing Δ for the 90° to 180° PGSE pulsed sequence. This feature has made the PGSTE sequence the NMR pulsed sequence of choice for a large number of experimental studies on emulsions (see, for example, Refs. 70–76). The definition for the attenuation ratio R given by Equation 8.11 also applies to PGSTE tests, with the acquisition time being $2\tau + T$ instead of 2τ. The mathematical description that follows is valid both for PGSE and PGSTE experiments, unless otherwise indicated.

8.4.3 Determination of Drop Sizes in Emulsions via PGSE and PGSTE

For isotropic bulk fluids in which molecules can diffuse freely (Fickian diffusion), the following expression holds [68]:

$$R_{bulk} = \exp\left[-\gamma^2 g^2 D\delta^2 \left(\Delta - \frac{\delta}{3}\right)\right] \tag{8.12}$$

When Equation 8.12 holds, a plot of the logarithm of R vs. $g^2\ \delta^2(\Delta - \delta/3)$ renders a straight line, and D can be calculated from the slope. This method is one of the very few experimental techniques available to measure self-diffusion coefficients.

Equation 8.12 does not apply to fluids confined in small geometries such as pores or droplets, because molecules cannot diffuse freely. In this case, the dimensions of the cavity limit the distance the spins diffuse by Brownian motion, thus affecting the attenuation ratio R. Robertson [77] and Neuman [78] first proposed expressions for R for molecules confined between planes and within cylinders and spheres when a steady magnetic field gradient is applied. Murday and Cotts [79] extended Neuman's derivation for the PGSE sequence for restricted diffusion within a sphere of radius a. In this case, the attenuation ratio R_{sp} was shown to be given by:

$$R_{sp} = \exp\left\{-2\gamma^2 g^2 \sum_{m=1}^{\infty} \frac{1}{\alpha_m^2(\alpha_m^2 a^2 - 2)}\left[\frac{2\delta}{\alpha_m^2 D} - \frac{\Psi}{(\alpha_m^2 D)^2}\right]\right\} \tag{8.13}$$

where

$$\Psi = 2 + \exp\lfloor -\alpha_m^2 D(\Delta - \delta)\rfloor - 2\exp(-\alpha_m^2 D\Delta) - 2\exp(-\alpha_m^2 D\delta) + \exp\lfloor -\alpha_m^2 D(\Delta + \delta)\rfloor \tag{8.14}$$

α_m is the mth positive root of the equation:

$$\alpha a \cdot J_{5/2}(\alpha a) - J_{3/2}(\alpha a) = 0 \tag{8.15}$$

and J_k is the Bessel function of the first kind, order k. Figure 8.7 shows theoretical calculations for R using Equations 8.13 to 8.15 (solid lines) for selected values of a. It is seen that the attenuation of R is less for smaller cavities.

Two limiting cases of Equation 8.13 are of interest. First, for very large spheres ($a \to \infty$), Equation 8.13 reduces to Equation 8.12 [$R_{sp}(a \to \infty) = R_{bulk}$]. Calculations for the attenuation ratio using Equation 8.12 are also reported in Figure 8.7 (squares) to show the agreement of this expression with Equation 8.13 for large values of a.

Second, for very small spheres, Equation 8.13 simplifies as follows:

$$R_{sp} = 1 - \frac{16}{175}\gamma^2 g^2 D^{-1} \delta a^4 \tag{8.16}$$

whence $R_{sp}(a \to 0) = 1$. In this case, the probability of the molecules to displace during the diffusion time Δ is reduced as $a \to 0$. For this reason, the loss of coherence in the phases of the spins caused by the magnetic field gradient diminishes and less attenuation of the spin-echo is observed. Figure 8.7 (inserted plot) also illustrates the agreement between Equation 8.16 (circles) and Equation 8.13 (solid lines) for small spheres.

In the derivation of Equation 8.13 it is assumed that the phase shifts of spins diffusing in a bounded region exhibit a normal (Gaussian) distribution. However, this Gaussian phase-distribution (GPD) approximation is exact only for spins undergoing free diffusion [78]. Balinov et al. [80] performed Brownian dynamic simulations for restricted diffusion in spheres of selected sizes at fixed g, Δ, and for various δ to calculate the *exact* attenuation ratio that would be observed in each case. Least-squares fits of these results were performed using Equation 8.13 and the sphere radius a as fitting parameter. The radii calculated with this expression differed by less than 5%

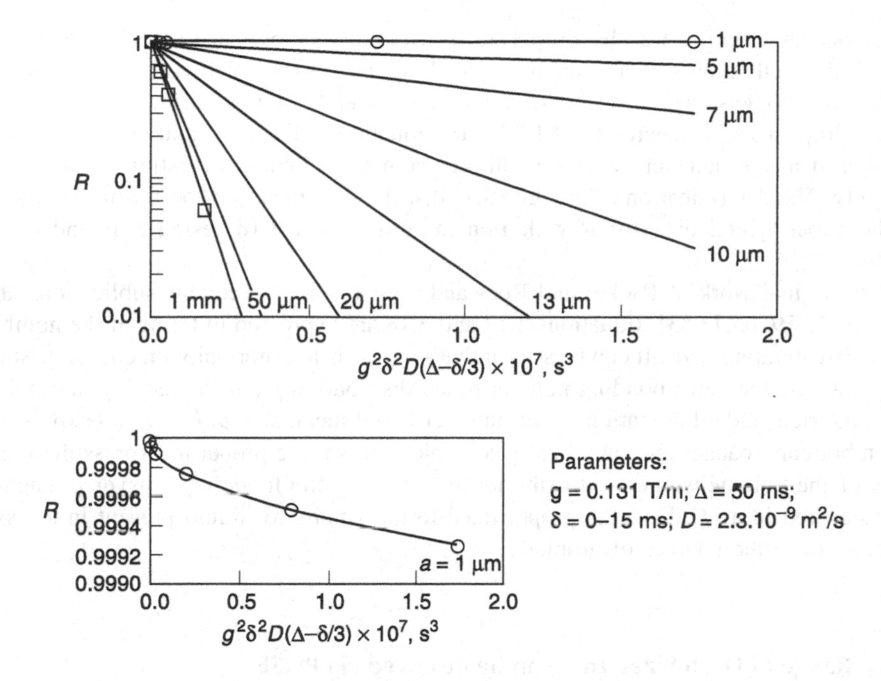

FIGURE 8.7 Theoretical calculations for the PFG attenuation ratio using Equation 8.13 (solid lines) and the limiting expressions for large values of a (Equation 8.12, squares) and for small values of a (Equation 8.16, circles; see also inserted plot for $a = 1\ \mu\text{m}$).

from the sizes set for the simulations. Whence it was concluded that the GPD approximation and Equation 8.13 are adequate to account for the decay of transverse magnetization of fluids confined in spheres.

For emulsion with a finite distribution of (spherical) droplet sizes, Packer and Rees [58] first proposed that the attenuation ratio of the drop phase (R_{DP}) can be calculated as the sum of the attenuation ratios $R_{sp}(a)$ that would be recorded for fluid confined in drops of radii a, weighted by the probability of finding drops with such sizes in the dispersion. This is:

$$R_{DP} = \int_0^{\infty} p_V(a) R_{sp}(a) da \bigg/ \int_0^{\infty} p_V(a) da \tag{8.17}$$

where $p_V(a)$ is the volume-weighted distribution of sizes, and $R_{sp}(a)$ is given by Equation 8.13. The task of determining $p_V(a)$ from the PGSE data is feasible, but requires a large number of measurements of R for which the duration of the test may become impractical as discussed later. Instead, few data are usually taken and an empirical form of $p_V(a)$ is assumed. The log–normal probability distribution function (p.d.f.),

$$p_V(a) = \frac{1}{2a\sigma(2\pi)^{1/2}} \exp\left\{-\left[\frac{\ln(2a) - \ln(d_{gV})}{2\sigma^2}\right]\right\} \tag{8.18}$$

is the classic assumption for the drop size distribution in absence of additional information, because it is well known that sequential break-up processes, such as grinding of solids or disruption of droplets under mechanical agitation, lead to a log–normal distribution of particle and drop sizes, respectively [81,82]. In Equation 8.18, d_{gV} and σ are the geometric volume-based mean diameter and the width or geometric standard deviation of the distribution, respectively. The determination of the drop size distribution consists of performing a least-squares fit of the experimental data for R with Equations 8.17 and 8.18, using d_{gV} and σ as fitting parameters.

In the original work of Packer and Rees and most of the subsequent publications about this method [26,37,59–65,73,83], Equations 8.17 and 8.18 are expressed in terms of the number-based drop size distribution $p_N(a)$. It can be shown that if $p_V(a)$ is log–normal with characteristic parameters d_{gV} and σ, the corresponding number-based distribution $p_N(a)$ is also log–normal with the same geometric standard deviation σ and number-based mean size $d_{gN} = d_{gV} \cdot \exp(-3\sigma^2)$ [84]. Although both approaches are numerically equivalent, it is more proper to express these equations in terms of the volume-weighted distribution because the amplitude of transverse magnetization that is measured in a PGSE test is proportional to the volume of liquid present in the system as drops, and not to the number of droplets.

8.4.3.1 Range of Drop Sizes That can be Resolved via PGSE

The maximum drop size that can be determined via PGSE is related to the one-dimension root-mean-square displacement of spins undergoing free (Fickian) self-diffusion in isotropic, isothermal media during the diffusion time Δ:

$$d_{MAX} \approx \sqrt{2}\langle x^2\rangle^{1/2} = \sqrt{2}\left\{\int_{-\infty}^{\infty} (x - x_0)^2 \frac{\exp[-(x - x_0)^2/4Dt]}{\sqrt{4\pi D_{DP} t}} dx\Bigg|_{t=\Delta}\right\}^{1/2} = 2\sqrt{D\Delta} \quad (8.19)$$

Most of the molecules confined in a droplet with diameter much larger than d_{MAX} would not "feel" any restriction in their diffusion path due to the presence of the water/oil interface. For this reason, such a drop would not be sized accurately. This expression for d_{MAX} is also obtained by requiring the characteristic diffusion time in the drops with size d_{MAX} to be comparable to Δ ($t_{D,MAX} = a_{MAX}^2/D \approx \Delta$). The factor $\sqrt{2}$ has been included in Equation 8.19 to assure consistency between these two approaches.

Equation 8.19 suggests that d_{MAX} can be increased at will with Δ. However, the diffusion time must be adjusted keeping in mind that $\Delta < (T_{2,bulk})_{DP}$, with $(T_{2,bulk})_{DP}$ being the characteristic bulk relaxation time of the drop phase. Otherwise, the data will be affected by natural (bulk) and surface relaxation of the magnetization, and therefore by a reduction in the signal-to-noise ratio.

Low self-diffusivities render small values for d_{MAX}. In addition, low-mobility molecules usually exhibit short relaxation times [57] and therefore short values of Δ must be chosen in such cases. In general, it is not possible to determine droplet sizes via PGSE for drop phases exhibiting self-diffusion coefficients below 10^{-12} m^2/s [37].

An expression for the minimum drop size d_{MIN} that are measurable from PGSE data can be obtained from Equation 8.16:

$$d_{MIN} = \left(175\lambda \frac{D_{DP}}{\gamma^2 g^2 \delta_{MAX}}\right)^{1/4} \tag{8.20}$$

where $\lambda = 1 - R_{sp}(a = d_{MIN}/2, \delta = \delta_{MAX})$ is the smallest reduction of the attenuation ratio that can be reliably detected experimentally. Plausible values for λ are $0.05 \geq \lambda \geq 0.01$. δ_{MAX} is the maximum duration that is chosen for the magnetic field gradient pulse.

Denkova et al. [74] considered the usefulness of Equations 8.19 and 8.20 when performing an evaluation of the precision of the PGSTE method to measure drop sizes in emulsions as applied by Goudappel et al. [73], by comparing NMR and video-enhanced optical microscopy measurements for 27 *o/w* emulsions. Measurements for the minimum drop sizes were in agreement with predictions from Equation 8.20. It was also noted that drop sizes five times larger than those dictated by Equation 8.19 could be characterized while underestimating d_{gV} by $\sim$20%.

8.4.3.2 Resolving for Drop Size Distributions with Arbitrary Shape in Short Timeframes

An important shortcoming of the original PGSE method as published by Packer and Rees [58] is the assumption that the distribution of drop sizes is invariably described by a log–normal probability distribution function. Drop sizes are often, but not always, distributed in a log–normal fashion [84], and therefore Equation 8.18 does not provide a satisfactory fit of the PGSE data in all cases [64,85,86].

Ambrosone et al. [85] developed a numerical procedure based on a solution of the Fredholm integral equation to resolve the distribution function without assuming an analytical type in advance. The method was shown to be able to reconstruct the distribution function of a bimodal emulsion from a simulated PGSE experiment. Hollingsworth and Johns [86] applied regularization techniques to resolve for the shape of monomodal and bimodal emulsions from synthetic and experimental PGSE data. The ability to retrieve the correct distribution function was found to depend heavily on the regularization method chosen to assess the optimum regularization parameter.

In practice, direct methods to resolve for the shape of the distribution function directly from a PGSE or PGSTE experimental set may be adequate in some, but not in all, cases, since the procedure is usually applied on a small dataset of attenuation ratios. In such cases, multiple solutions may provide satisfactory fits of the data. Very long diffusion experiments aimed at acquiring large dataset of R may not be practical either, since changes in the drop size distribution may occur within the timeframe of the experiment.

Recent developments in this area point to performing PGSE and PGSTE measurements in combination or as part of successive pulsed sequences to enlarge the dataset while shortening the duration of the test. For systems exhibiting significant surface relaxivity, Peña and Hirasaki [47] proposed defining the distribution function from a CPMG test as explaind in Section 8.3.3, and use such distribution for the analysis of the PGSE dataset. In this combined CPMG-PGSE method, the surface relaxation is varied through successive iterations until the drop size distribution that is determined via CPMG renders a satisfactory fit for the PGSE dataset. In this way, both the drop size distribution with arbitrary shape and the surface relaxation at the water–oil interfaces are resolved. The typical duration for a CPMG-PGSE test is 5 min in a low-frequency spectrometer.

Buckley et al. [87] applied a pulsed sequence (Difftrain) that uses successive stimulated echoes from a single excitation pulse to measure self-diffusion at varying diffusion times Δ to characterize drop size distributions in emulsions. The fact that this technique collects and retains all spectral information from the emulsion with a single acquisition allows collecting a significant number of echoes in very short timeframes (~15 spin-echoes/sec at the conditions reported in Ref. 87). Although regularization methods are still needed to retrieve the drop size distribution function from Difftrain data, they can be applied over much larger datasets, thus improving the ability to resolve for arbitrary distribution function of the emulsion's drop sizes. These authors showed the drop size distribution of an emulsion of PDMS in water, which was calculated from a Difftrain dataset that was collected in just 4 sec. Results compared favorably with data gathered using the conventional PGSTE sequence.

8.4.3.3 Effect of Surface Relaxation on Restricted Diffusion Measurements

The effect of surface relaxation on the amplitude of the spin-echo in a diffusion experiment is not accounted for in the method to size emulsion droplets discussed in Section 8.4.3, since Equation 8.13 was derived assuming $\rho = 0$. It can be anticipated that this effect is small in a PGSE experiment performed in the fast diffusion regime, because in such regime the relaxation due to diffusion of the spins is much more significant than surface relaxation (see Equation 8.7).

Dunn and Sun [49] considered the effect of surface relaxation on the attenuation ratio R_{sp} by changing the boundary condition in the derivation by Neuman et al. [78] for the solution of the diffusion propagator of the spins P (see Eq. 11 in Neuman's paper) from:

$$D\nabla P\big|_{r=a} = 0 \tag{8.21}$$

to:

$$D\nabla P + \rho P\big|_{r=a} = 0 \tag{8.22}$$

to obtain:

$$R_{sp,\rho} = \exp\left\{-2\gamma^2 g^2 \sum_{m=1}^{\infty} \frac{1}{\alpha_m^2(\alpha_m^2 a^2 - 2 + \rho^2 a^2/D^2 - \rho a/D)}\left[\frac{2\delta}{\alpha_m^2 D} - \frac{\Psi}{(\alpha_m^2 D)^2}\right]\right\} \tag{8.23}$$

where Ψ is given by Equation 8.14 as before and α_m is the mth positive root of the equation:

$$\alpha a \cdot J_{5/2}(\alpha a) - \left(1 + \frac{\rho a}{D}\right) J_{3/2}(\alpha a) = 0 \tag{8.24}$$

Equation 8.23 applies only for small ρ and large D so the probability of finding a spin anywhere in the drop is nearly uniform and the GPD approximation is satisfied (see discussion following Equation 8.16). Otherwise, significant surface relaxation would take place and a non-Gaussian distribution of the phase shifts of the spins near the water/oil interfaces would be observed. Equation 8.23 reduces to Equation 8.13 in the limit $\rho a/D \to 0$, as might be expected.

Codd and Callaghan [88] extended a matrix formalism developed by Callaghan [89] to interpret restricted diffusion data from pulsed sequences with gradient pulses of arbitrary shape, to take into account surface relaxation at the walls of spheres. These authors noted that when $\rho = 1$

to 10 μm/s and the magnetization data is collected from pores of the order of 10 μm, the effect of surface relaxation on the attenuation ratio can be neglected. This assessment can be confirmed in a straightforward manner via Equation 8.23.

8.4.4 A Generalized Theory for the Time-Resolved Attenuation Ratio of Emulsions

Packer and Rees [58] pointed out that the procedure described in Section 8.4.3 is useful to determine the drop size distribution whenever the spin-echo is originated solely from the drop phase of the emulsion. This is

$$R_{EMUL} = R_{DP} \tag{8.25}$$

where R_{EMUL} and R_{DP} are the attenuation ratio that is measured for the emulsion and for the drop phase, respectively. This assumption limits the applicability of the method to emulsions for which the signal from the continuous phase is suppressed (i.e., emulsions of oil in D_2O). Equation 8.25 is also valid if the transverse magnetization of the continuous phase has relaxed completely and the natural (bulk) relaxation of the drop phase is small at the time the spin-echo is acquired. Namely,

$$(T_{2,bulk})_{CP} \ll 2\tau \ll (T_{2,bulk})_{DP} \tag{8.26}$$

where $(T_{2,bulk})_{CP}$ is the characteristic bulk relaxation time of the continuous phase. Equation 8.26 is often satisfied by emulsions of water in viscous oils, but not by emulsions of viscous oils in water. This explains why the PGSE method was often applied during the last 30 years for the determination of drop sizes in *w/o* emulsions such as margarine, butter, and low-calorie spreads [59,60,62,83,90] and water-in-crude-oil emulsions [61], but not as often for the characterization of *o/w* emulsions [73,74].

The impossibility of resolving for the drop size distributions, when the contribution of the magnetization of the continuous phase to the spin-echo is significant, was overcome with the introduction of the Fourier-transform PGSE method (FT-PGSE). Excellent reviews on this method and its applications in the characterization of water/oil/surfactant systems have been published by Stilbs [24] and Söderman and Stilbs [25]. In the FT-PGSE procedure, the second half of the spin-echo that is generated in the PGSE sequence is Fourier-transformed, and the individual contributions of the components to the spin-echo appear as separate peaks in the frequency domain due to differences in self-diffusivity. Lönnqvist et al. [26] applied this technique to resolve for the individual signal of water and oil in simple (*w/o*) and multiple (*w/o/w*) emulsions. Ambrosone et al. [90] used the method to isolate the water signal from margarine and emulsions of water in olive oil.

Better resolution of the Fourier spectrum is attained as the strength of the magnetic field is increased. For this reason, high-resolution magnets with frequencies typically above 80 MHz are used in FT-PGSE studies. Unfortunately, individual peaks cannot be resolved satisfactorily at the frequencies at which low-resolution NMR spectrometers operate (∼ 2 to 20 MHz). Therefore, the determination of the individual contributions of water and oil to the spin-echo that is obtained from an emulsion in the time domain is relevant, particularly for industrial applications in which high-resolution spectrometers cannot be afforded.

A generalized expression for the time-resolved attenuation ratio of emulsions was reported recently [47]:

$$R_{EMUL} = (1-\kappa)R_{DP} + \kappa R_{CP}; \quad 0 \le \kappa \le 1 \tag{8.27}$$

where R_{CP} is the time-resolved attenuation ratios of the continuous phase, respectively. κ is a parameter associated with the natural relaxation of the transverse component of the magnetization of both phases as shown later. R_{DP} is given by Equation 8.17 as before, and $R_{sp}(a)$ is calculated using Equation 8.13. Diffusion in the continuous phase may occur as bulk diffusion ($R_{CP} \sim R_{bulk,CP}$), particularly for dilute emulsions. In such a case, R_{CP} can be calculated using Equation 8.12. The case in which restricted diffusion in the continuous phase due to the presence of droplets significantly affects R_{CP} has also been considered [47,91].

In general, κ is given by:

$$\kappa = \left[1 + \frac{M_{xy,DP}(t_{SE}, g=0)}{M_{xy,CP}(t_{SE}, g=0)}\right]^{-1} \tag{8.28}$$

where t_{SE} is the time for maximal amplitude of the spin-echo, and $M_{xy,DP}(t_{SE}, g=0)$ and $M_{xy,CP}(t_{SE}, g=0)$ are the magnitudes of the transverse component of the magnetization vector at time t_{SE} when $g=0$ for the drop and continuous phase, respectively. For the basic PGSE sequence (Figure 8.6(a)), $t_{SE} = 2\tau$ and Equation 8.28 becomes:

$$\kappa = \left[1 + \frac{\sum (f_i)_{DP} \exp[-2\tau/(T_{2,i})_{DP}]}{\sum (f_i)_{CP} \exp[-2\tau/(T_{2,i})_{CP}]}\right]^{-1} \simeq \left[1 + \frac{\phi_{DP}}{\phi_{CP}}\frac{HI_{DP}}{HI_{CP}}\frac{\exp[-2\tau/(T_{2,bulk})_{DP}]}{\exp[-2\tau/(T_{2,bulk})_{CP}]}\right]^{-1} \tag{8.29}$$

where the $T_{2,i} - f_i$ values for each phase are determined from the T_2 distribution of the emulsion via CPMG as explained above. For the PGSTE sequence (Figure 8.6(b)), $t_{SE} = 2\tau + T$ and κ becomes:

$$\begin{aligned}\kappa &= \left[1 + \frac{\sum (f_i)_{DP} \exp[-2\tau/(T_{2,i})_{DP} - T/(T_{1,i})_{DP}]}{\sum (f_i)_{CP} \exp[-2\tau/(T_{2,i})_{CP} - T/(T_{1,i})_{CP}]}\right]^{-1} \\ &\cong \left[1 + \frac{\phi_{DP}}{\phi_{CP}}\frac{HI_{DP}}{HI_{CP}}\frac{\exp[-2\tau/(T_{2,bulk})_{DP} - T/(T_{1,bulk})_{DP}]}{\exp[-2\tau/(T_{2,bulk})_{CP} - T/(T_{1,bulk})_{CP}]}\right]^{-1}\end{aligned} \tag{8.30}$$

assuming that the ratio $T_{1,i}/T_{2,i} = T_{1,bulk}/T_{2,bulk}$ remains constant. Surface relaxation effects are neglected for the approximations made in Equations 8.29 and 8.30.

According to Equation 8.28, $\kappa \to 0$ when the signal of the continuous phase is suppressed, or when it has relaxed completely at the time the spin-echo is acquired. If so, Equation 8.27 reduces to Equation 8.25. It is inferred from this analysis that the classic method to interpret NMR PGSE data in the time domain to calculate drop sizes is valid only whenever $\kappa \to 0$. Neglecting the effect of κ may lead to significant errors in the interpretation of PGSE data in the time domain as shown below.

Figure 8.8 shows (circles) experimental data for a sample containing water (10 vol%) and a crude oil (90 vol%) in contact as bulk fluids, and for a *w/o* emulsion (squares) made from the same water/crude oil mixture [47]. The emulsion was ultrasonicated until the drop sizes were

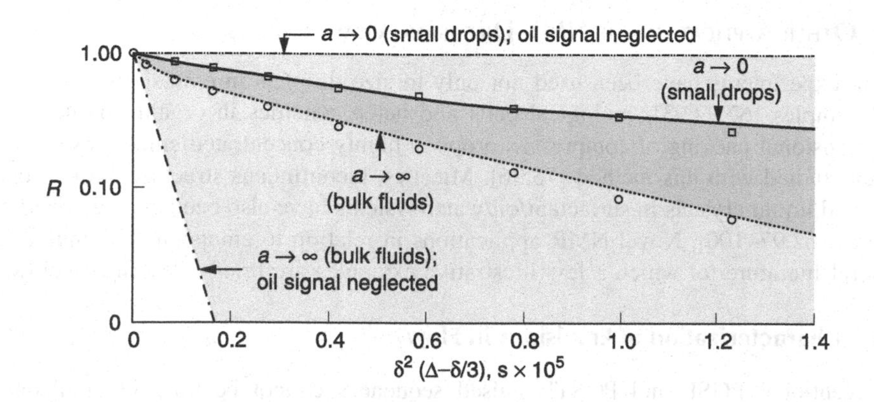

FIGURE 8.8 Effect of the transverse magnetization of the continuous phase on the PGSE response of mixtures of water and crude oil, contacted as bulk fluids and emulsified.

below the resolution limit of the PGSE experiment, which was $d_{MIN} \sim 3$ μm, according to Equation 8.20. The mean drop size was followed via microphotography, and the final emulsion exhibited a narrow distribution of droplets with sizes close to 0.5 μm. Since water is dispersed as droplets, Equation 8.27 becomes $R_{EMUL} = (1 - \kappa)R_W + \kappa R_O$. In this case, $\kappa = 0.755$ was calculated from Equation 8.29, which indicates that 75.5% of the attenuation ratio of the mixture was determined by the oil phase.

The dotted curve in Figure 8.8 shows the predicted behavior for the sample before emulsification using Equation 8.27 with $R_{DP} = R_{bulk,DP}$ and $R_{CP} = R_{bulk,CP}$. The predicted behavior for the attenuation ratio of the emulsion via Equation 8.27 is also shown (continuous line). R_O was calculated using Equation 8.12 and considering the distribution of diffusion coefficients in the oil phase as described in Ref. 47, whereas $R_W \sim 1$ according to Equation 8.16. The excellent agreement between experiments and calculations shows that Equation 8.27 is valid both for mixtures of bulk fluids and emulsions.

The dash–dot and dash–dot–dot lines in Figure 8.8 stand for the attenuation profiles that would be obtained for the same system before and after emulsification, respectively, if the contribution of the oil signal to the spin-echo amplitude is neglected. The area of the plot between these two lines has been shaded in light tone to indicate the range of conditions in which the attenuation ratios for emulsions with any given drop size distribution could be found if the oil signal were indeed negligible, i.e., if Equation 8.25 were correct. If Equation 8.27 holds instead as experiments suggest, such conditions are restricted to the area shaded in dark tone. Clearly, neglecting the signal from the oil phase could lead to significant error in the prediction of the attenuation profile for this system. The trends shown in Figure 8.8 have been confirmed in experiments with other water/oil mixtures [47].

It is worth noting that the parameters of a PGSE or PGSTE experiment should be chosen in order to *minimize* the effect of the continuous phase on the spin-echo and therefore on the attenuation ratio of the emulsion, while maintaining a satisfactory signal-to-noise ratio. The idea is to broaden the range of conditions at which attenuation ratios can be obtained (that is, to expand the extension of the dark-shaded area in Figure 8.8), so the uncertainty in the drop size distribution that is determined from the PGSE diminishes. Therefore, Equation 8.27 can be used as a tool to predict limiting attenuation profiles and optimize the selection of parameters for testing.

8.4.5 Other Applications of NMR Diffusion Measurements

Diffusion experiments have been used not only to size droplets in emulsions, but also pores in mineral samples [67,92,93], biological cells and heterogeneities in organic tissue [94,95]. The three-dimensional packing of compressed drops in highly concentrated emulsions ($\phi_{DP} \sim 1$) has also been studied with this method [75,96]. Micelles, bicontinuous structures in microemulsions, vesicles and liquid crystals in surfactant/oil/water/systems have also been characterized with these techniques [62,97–100]. Novel NMR applications in relation to emulsions continue to appear in the general literature, of which a few illustrative examples are briefly described below.

8.4.5.1 Characterization of Emulsions in Flow

The conventional PGSE and PGSTE pulsed sequences cannot be used in emulsions in flow because both self-diffusion and velocity contribute to the attenuation of magnetization in the presence of field gradients. A flow-compensating PGSTE pulsed sequence was introduced by Johns and Gladden [65] and used to characterize xylene-in-water emulsions in laminar flow. In essence, this sequence halves the diffusion time Δ by adding two rf 90° pulses sandwiched between two additional gradient pulses to the classic PGSTE sequence. The flow effect is compensated whenever the assumption that the displacement of each individual molecule during the initial diffusion time $\Delta/2$ is equivalent to the displacement of such molecule due to the second $\Delta/2$ time period is valid. Therefore, only emulsions in laminar flow exhibiting negligible diffusion in the direction transverse to the flow are suited for evaluation. Figure 8.9(a) shows a comparison of the drop size distributions that were measured in a quiescent emulsion using confocal microscopy, the conventional PGSTE sequence, and the flow-compensating PGSTE sequence. Excellent agreement was found among results from the three techniques, thus showing that the modification to the pulsed sequence does not affect the ability to quantify for drop sizes. Figure 8.9(b) shows results for the drop size distribution of the emulsion under flow at two different superficial velocities. It is seen that an increase in velocity resulted in narrower drop size distributions with smaller mean sizes. The effect was attributed to breakage of larger drops into smaller drops as the superficial velocity was increased.

8.4.5.2 Kinetics of Emulsion Freezing

NMR has also been used to characterize emulsions undergoing freezing and freeze/thaw cycling. For such application, NMR proves to be particularly useful since it allows identifying changes in the physical state not only of the water and oil phases, but also of components within each of such phases during cooling/heating processes.

Hindmarsh et al. [71] followed the freezing behavior of drops containing emulsions of hydrocarbons in water and water/sucrose mixtures using the PGSTE technique. Figure 8.10(a) shows the transient behavior of the intensity of the NMR signal for the water, sucrose, and dodecane that were present in an emulsion sample of 20 wt% dodecane-in-20 wt% sucrose solution at −40°C. The plots indicate that nucleation of water, sucrose, and dodecane started at about 35, 65, and 80 sec after subjecting the emulsion sample to cooling, respectively. Interestingly, the NMR data captures the fact that an inflexion in the water curve appeared once the nucleation of sucrose started, while an inflexion in the sucrose curve appears when the nucleation for dodecane began. These phenomena are due to a reduction in the freezing rate whenever latent heat is released by a component for which nucleation begins. Figure 8.10(b) shows the drop size distribution of the same emulsion described above, first unfrozen and then cooled down to temperatures at which freezing of the drop phase (dodecane) would not occur. The inserted plot reports the same data

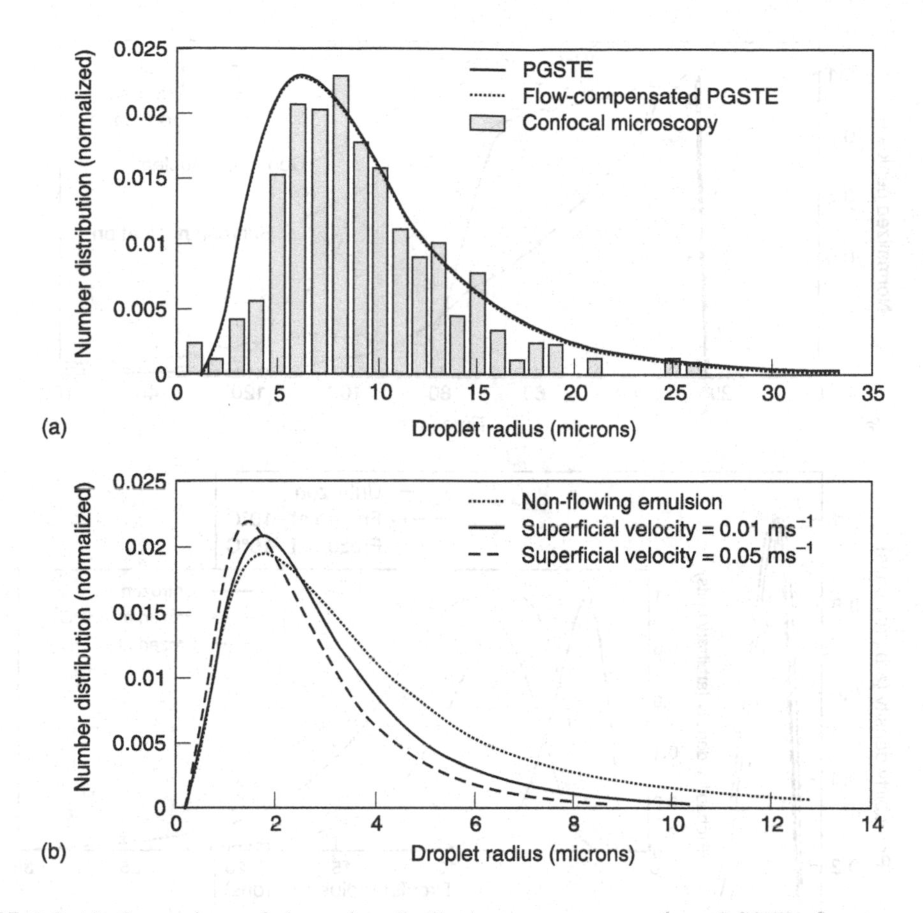

FIGURE 8.9 (a) Comparison of drop size distribution measurements from PGSTE, flow-compensated PGSTE, and optical microscopy for a quiescent water-in-oil emulsion; (b) flow-compensated PGSTE measurements for the distribution of drop sizes of emulsions in flow. (Adapted from Ref. 65.)

for a dodecane-in-water emulsion for which sucrose was not added. These data suggest that the presence of sucrose inhibits the coalescence and growth of droplet sizes when freezing the emulsion. This assessment confirms previous studies reporting a stabilizing effect of water-soluble carbohydrates in *o/w* emulsions subject to freezing [101,102].

8.4.6 Advantages and Limitations of the PGSE and PGSTE Experiments

In the PGSE and PGSTE methods, the contributions of the water and oil phases to the attenuation ratio R can be resolved independently. Also, surface relaxation has a negligible effect on R in the fast diffusion regime and therefore the PGSE and PGSTE data can be interpreted solely in terms of self-diffusivity and natural (bulk) relaxation. As a result, an independent measurement of the drop size distribution, and therefore of the surface-to-volume ratio of the drop phase in the emulsion, can be obtained with these procedures.

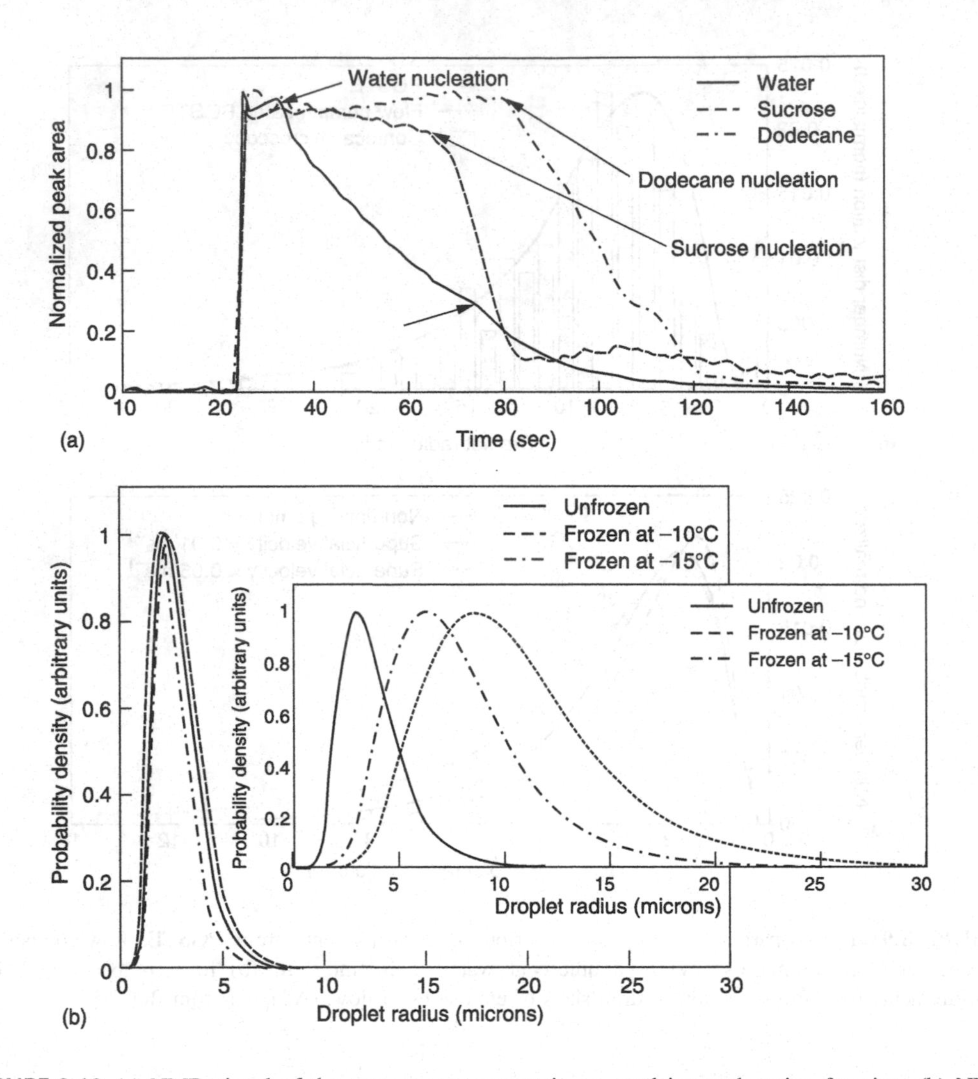

FIGURE 8.10 (a) NMR signal of the components present in an emulsion undergoing freezing; (b) NMR-resolved drop size distributions of freezing emulsions. (Adapted from Ref. 71.)

Nevertheless, the range of drop sizes that can be determined precisely with these methods is narrow (~1 to 50 μm). Furthermore, the PGSE and PGSTE experiments are slow: acquiring a dataset comparable to that of a single CPMG test would take several days. For this reason, in a typical PGSE experiment only few (~10 to 20) attenuation ratios are measured in 5 to 20 min. Due to lack of data, the drop size distribution is resolved assuming in advance a p.d.f. to describe it. Recent studies have unveiled the possibility of combining diffusion and relaxation experiments to overcome these limitations [47,87].

REFERENCES

1. Friberg, S.E., K. Larsson, and J. Sjöblom, Eds. *Food Emulsions*. 4th Edition. 2003, Marcel Dekker, Inc.: New York.
2. McClements, D.J., *Food Emulsions: Principles, Practices and Techniques*. 2nd Edition. 2004, CRC Press: Boca Raton, USA.

3. Rieger, M.M. and L.D. Rhein, Eds. *Surfactants in Cosmetics*. 2nd Edition. 1997, Marcel Dekker, Inc.: New York.
4. Nielloud, F. and G. Marti-Mestres, Eds. *Pharmaceutical Emulsions and Suspensions*. 2000, Marcel Dekker, Inc.: New York.
5. Turner, G.P.A. and J. Bentley, *Introduction to Paint Chemistry and Principles of Paint Technology*. 4th Edition. 1997, CRC Press: Boca Raton, USA.
6. Foy, C.L. and D.W. Pritchard, Eds. *Pesticide Formulation and Adjuvant Technology*. 1996, CRC Press: Boca Raton, USA.
7. Becher, P., *Emulsions: Theory and Practice*. 3rd Edition. 2001, Oxford University Press: New York.
8. Muncy, H.W., *Asphalt Emulsions*. ASTM Special Technical Publications # 1079. 1990, ASTM International: West Conshohoker, USA.
9. Schramm, L.L., *Emulsions: Fundamentals and Applications in the Petroleum Industry*. Advances in Chemistry. Vol. #231. 1992, American Chemical Society: Washington D.C.
10. Sjöblom, J., N. Aske, I.H. Auflem, O. Brandal, T. Havre, O. Saether, A. Westvik, E.E. Johnsen, and H. Kallevik, Our Current Understanding of Water-in-Crude Oil Emulsions. Recent Characterization Techniques and High Pressure Performance. *Advances in Colloid and Interface Science*, (2003). **100–102**: pp. 399–473.
11. Hartland, S. and S. Jeelani, Gravity settlers, in *Liquid–liquid Extraction Equipment*, J.C. Godfrey and M.J. Slater, Editors. 1994, John Wiley & Sons: New York.
12. Otsubo, Y. and R.K. Prud'homme, Effect of drop size distribution on the flow behavior of oil-in-water emulsions. *Rheologica Acta*, (1994). **33**(4): pp. 303–306.
13. Ramírez, M., J. Bullón, J. Andérez, I. Mira, and J.L. Salager, Drop size distribution and its effect on O/W emulsion viscosity. *Journal of Dispersion Science and Technology*, (2002). **23**(1–3): pp. 309–321.
14. Pal, R., Effect of droplet size on the rheology of emulsions. *AIChE Journal*, (1996). **42**(11): pp. 3181–3190.
15. Walstra, P., *Emulsion stability*, in *Encyclopedia of Emulsion Technology*, Volume 4, P. Becher, Editor. 1996, Marcel Dekker, Inc.: New York. Chapter 1.
16. Peña, A.A. and C.A. Miller, Transient behavior of polydisperse emulsions undergoing mass transfer. *Industrial and Engineering Chemistry Research*, (2002). **41**(25): pp. 6284–6296.
17. McClements, D.J., Theoretical prediction of emulsion color. *Advances in Colloid and Interface Science*, (2002). **97**: pp. 63–89.
18. Kilcast, D. and S. Clegg, Sensory perception of creaminess and its relationship with food structure. *Food Quality and Preferences*, (2002). **13**: pp. 609–623.
19. Lissant, K., *Demulsification: Industrial Applications*. 1983, Marcel Dekker, Inc.: New York.
20. Orr, C., *Determination of Particle Size*, in *Encyclopedia of Emulsion Technology*, Vol. 3, Chapter 3, P. Becher, Editor. 1988, Marcel Dekker, Inc.: New York.
21. Sjöblom, J., Ed. *Encyclopedic Handbook of Emulsion Technology*. 2001, Marcel Dekker, Inc.: New York. pp. 736.
22. Coupland, J.N. and D.J. McClements, *Analysis of Droplet Characteristics Using Low-Intensity Ultrasound*, in *Food Emulsions*, S.E. Friberg, K. Larsson, and J. Sjöblom, Editors. 2003, Marcel Dekker, Inc.: New York.
23. Callaghan, P., *Principles of Nuclear Magnetic Resonance Microscopy*. 1991, Oxford University Press: New York. pp. 492.
24. Stilbs, P., Fourier transform pulsed-gradient spin-echo studies of molecular diffusion. *Progress in NMR Spectroscopy*, (1987). **19**: pp. 1–45.
25. Söderman, O. and P. Stilbs, NMR studies of complex surfactant systems. *Progress in NMR Spectroscopy*, (1994). **26**: pp. 445–482.
26. Lönnqvist, I., B. Hakansson, B. Balinov, and O. Söderman, NMR self-diffusion studies of the water and the oil components in a W/O/W emulsion. *Journal of Colloid and Interface Science*, (1997). **192**(1): pp. 66–73.
27. Parhami, P. and B.M. Fung, Fluorine-19 relaxation study of perfluorochemicals as oxygen carriers. *Journal of Physical Chemistry*, (1983). **87**: pp. 1928–1931.

28. Schnur, G., R. Kimmich, and R. Lietzenmayer, Hydrogen/fluorine retuning tomography. Applications to ^{1}H image-guided volume-selective ^{19}F spectroscopy and relaxometry of perfluorocarbon emulsions in tissue. *Magnetic Resonance in Medicine*, (1990). **13**(3): pp. 478–489.
29. Chiba, K. and M. Tada, Study of the emulsion stability and headgroup motion of phosphatidylcholine and lysophosphatidylcholine by carbon-13 and phosphorus-31. *Agric. Biol. Chem.*, (1989). **53**: pp. 995–1001.
30. Rotenberg, M., M. Rubin, A. Bor, D. Meyuhas, Y. Talmon, and D. Lichtenberg, Physico-chemical characterization of intralipid emulsions. *Biochim. Biophys. Acta*, (1991). **1086**(3): pp. 265–272.
31. Drew, J., A. Liodakis, R. Chan, H. Du, M. Sadek, R. Brownlee, and W.H. Sawyer, Preparation of lipid emulsions by pressure extrusion. *Biochem. Int.*, (1990). **22**(6): pp. 983–992.
32. Férézou, J., T.L. Nguyen, C. Leray, T. Hajri, A. Frey, Y. Cabaret, J. Courtieu, C. Lutton, and A.C. Bach, Lipid composition and structure of commercial parenteral emulsions. *Biochim. Biophys. Acta*, (1994). **1213**(2): pp. 149–158.
33. Westesen, K. and T. Wehler, Investigation of the particle size of a model intravenous emulsion. *J. Pharm. Sci.*, (1993). **82**(12): pp. 1237–1244.
34. ter Beek, L.C., K. Ketelaars, D.C. McGain, P.E. Smulders, P. Walstra, and M.A. Hemminga, Nuclear magnetic resonance study of the conformation and dynamics of beta-casein at the oil/water interface in emulsions. *Biophys. J.*, (1996). **70**(5): pp. 2396–2402.
35. Wolber, J., I. Rowland, M. Leach, and A. Bifone, perfluorocarbon emulsions as intravenous delivery media for hyperpolarized xenon. *Magnetic Resonance in Medicine*, (1999). **41**(3): pp. 442–449.
36. Möller, H.E., M.S. Chawla, X.J. Chen, B. Driehuys, L.W. Hedlund, C.T. Wheeler, and G.A. Johnson, Magnetic resonance angiography with hyperpolarized 129-Xe dissolved in a lipid emulsion. *Magnetic Resonance in Medicine*, (1999). **41**(5): pp. 1058–1064.
37. Balinov, B. and O. Söderman, *Emulsions – the NMR perspective*, in *Encyclopedic Handbook of Emulsion Technology*, J. Sjöblom, Editor. 2001, Marcel Dekker, Inc.: New York.
38. Håkansson, B., O. Söderman, and B. Balinov, *Nuclear Magnetic Resonance of Emulsions*, in *Encyclopedia of Surface and Colloid Science*, A.T. Hubbard, Editor. 2002, Marcel Dekker, Inc.: New York. pp. 3684–3699.
39. Carr, H.Y. and E.M. Purcell, Effects of diffusion on free precession in nuclear magnetic resonance experiments. *Physical Review*, (1954). **94**(3): pp. 630–638.
40. Meiboom, S. and D. Gill, Modified spin-echo method for measuring nuclear relaxation times. *The Review of Scientific Instruments*, (1959). **29**(8): pp. 688–691.
41. Tikhonov, A.N. and V.Y. Arsenin, *Solution of Ill-posed Problems*. 1977, Winston & Sons: Washington D.C. pp. 258.
42. Dunn, K.J., G.A. LaTorraca, J.L. Warner, and D.J. Bergman. *On the Calculation and Interpretation of NMR Relaxation Times Distribution, SPE Paper 28367*. In: 69th Annual SPE Technical Conference and Exhibition. 25–28 September 1994. New Orleans, USA.
43. LaTorraca, G.A., K.J. Dunn, P.R. Webber, and R.M. Carlson, Low-field NMR determinations of the properties of heavy oils and water-in-oil emulsions. *Magnetic Resonance Imaging*, (1998). **16**(5–6): pp. 659–662.
44. Cannon, D.E., C. Cao Minh, and R.L. Kleinberg. *Quantitative NMR interpretation*. In: SPE Annual Technical Conference and Exhibition. SPE Paper 49010, September 27–30, 1998. New Orleans, USA.
45. Zhang, Q., S.W. Lo, C.C. Huang, and G.J. Hirasaki. *Some exceptions to the default NMR rock and fluid properties*. In: SPWLA 39th Annual Logging Symposium. May 25–28, 1998. Keystone, Colorado.
46. Kleinberg, R.L. and H.J. Vinegar, NMR properties of reservoir fluids. *The Log Analyst*, (1996). **37**(6): pp. 20–32.
47. Peña, A.A. and G.J. Hirasaki, Enhanced characterization of oilfield emulsions via NMR diffusion and transverse relaxation experiments. *Advances in Colloid and Interface Science*, (2003). **105**: pp. 103–150.
48. Allsopp, K., I. Wright, D. Lastockin, K. Mirotchnik, and A. Kantzas, Determination of oil and water compositions of oil/water emulsions using low field NMR relaxometry. *Journal of Canadian Petroleum Technology*, (2001). **40**(7): pp. 58–61.

49. Dunn, K.J. and B. Sun, Personal communication. 2002.
50. Hills, B.P., H.R. Tang, P. Manoj, and C. Destruel, NMR diffusometry of oil-in-water emulsions. *Magnetic Resonance Imaging*, (2001). **19**(3–4): pp. 449–451.
51. Clark, B. and R.L. Kleinberg, Physics in oil exploration. *Physics Today*, (2002). **April**: pp. 48–53.
52. Kenyon, W.E., Nuclear magnetic resonance as a petrophysical measurement. *Nuclear Geophysics*, (1992). **6**(2): pp. 153–171.
53. Matteson, A., J.P. Tomanic, M.M. Herron, D.F. Allen, and W.E. Kenyon, NMR relaxation of clay/brine mixtures. *SPE Reservoir Evaluation & Engineering*, (2000). **3**(5): pp. 408–413.
54. Logan, W.D., J.P. Horkowitz, R. Laronga, and D.W. Cromwell, Practical application of NMR logging in carbonate reservoirs. *SPE Reservoir Evaluation & Engineering*, (1998). **1**(5): pp. 438–448.
55. Brownstein, K.R. and C.E. Tarr, Importance of classical diffusion in NMR studies of water in biological cells. *Physical Review A*, (1979). **19**(6): pp. 2446–2453.
56. Flaum, M., J. Chen, and G.J. Hirasaki, NMR diffusion editing for D-T_2 maps: Application to recognition of wettability change. *Petrophysics*, (In press).
57. Freedman, R., S.W. Lo, M. Flaum, G.J. Hirasaki, A. Matteson, and A. Sezginer, A new NMR method of fluid characterization in reservoir rocks: experimental confirmation and simulation results. *SPE Journal*, (2001). **December**: pp. 452–464.
58. Packer, K.J. and C. Rees, Pulsed NMR studies of restricted diffusion. I. droplet size distribution in emulsions. *Journal of Colloid and Interface Science*, (1972). **40**(2): pp. 206–218.
59. Van Den Enden, J.C., D. Waddington, H. Vanaalst, C.G. Vankralingen, and K.J. Packer, Rapid determination of water droplet size distributions by PFG-NMR. *Journal of Colloid and Interface Science*, (1990). **140**(1): pp. 105–113.
60. Balinov, B., O. Söderman, and T. Warnheim, Determination of water droplet size in margarines and low-calorie spreads by nuclear-magnetic-resonance self-diffusion. *Journal of the American Oil Chemists Society*, (1994). **71**(5): pp. 513–518.
61. Balinov, B., O. Urdahl, O. Söderman, and J. Sjöblom, Characterization of water-in-crude-oil emulsions by the NMR self-diffusion technique. *Colloids and Surfaces A – Physicochemical and Engineering Aspects*, (1994). **82**(2): pp. 173–181.
62. Söderman, O., I. Lönnqvist, and B. Balinov, *NMR Self-diffusion Studies of Emulsion Systems. Droplet Sizes and Microstructure of the Continuous Phase*, in *Emulsions – a Fundamental and Practical Approach*, J. Sjöblom, Editor. 1992, Kluwer Academic Publishers: The Netherlands. pp. 239–258.
63. Fourel, I., J.P. Guillement, and D. Le Botlan, Determination of water droplet size distributions by low-resolution PFG-NMR 1. "Liquid" emulsions. *Journal of Colloid and Interface Science*, (1994). **164**(1): pp. 48–53.
64. Lee, H.Y., M.J. McCarthy, and S.R. Dungan, Experimental characterization of emulsion formation and coalescence by nuclear magnetic resonance restricted diffusion techniques. *Journal of the American Oil Chemists' Society*, (1998). **75**(4): pp. 463–475.
65. Johns, M.L. and L.F. Gladden, Sizing of emulsion droplets under flow using flow-compensating NMR-PFG techniques. *Journal of Magnetic Resonance*, (2002). **154**(1): pp. 142–145.
66. Peña, A.A., G.J. Hirasaki, and C.A. Miller, Chemically induced destabilization of water-in-crude oil emulsions. *Industrial and Engineering Chemistry Research*, (2005), **44**(15): pp. 1139–1149.
67. Tanner, J.E. and E.O. Stejskal, Restricted self-diffusion of protons in colloidal systems by the pulsed-gradient, spin-echo method. *The Journal of Chemical Physics*, (1968). **49**(4): pp. 1768–1777.
68. Stejskal, E.O. and J.E. Tanner, Spin diffusion measurements: spin echoes in the presence of a time-dependent field gradient. *The Journal of Chemical Physics*, (1965). **42**(1): pp. 288–292.
69. Tanner, J.E., Use of the stimulated echo in NMR diffusion studies. *The Journal of Chemical Physics*, (1970). **52**(5): pp. 2523–2526.
70. Hedin, N. and I. Furó, Ostwald Ripening of an Emulsion Monitored by PGSE NMR. *Langmuir*, (2001). **17**(16): pp. 4746–4752.
71. Hindmarsh, J.P., K.G. Hollingsworth, D.I. Wilson, and M.L. Johns, An NMR study of the freezing of emulsion-containing drops. *Journal of Colloid and Interface Science*, (2004). **275**: pp. 165–171.
72. Garasanin, T. and T. Cosgrove, NMR Self-Diffusion Studies on PDMS oil-in-water emulsion. *Langmuir*, (2002). **18**: pp. 10298–10304.

73. Goudappel, G.J.W., J.P.M. van Duynhoven, and M.M.W. Mooren, Measurement of oil droplet size distributions in food oil/water emulsions by time domain pulsed field gradient NMR. *Journal of Colloid and Interface Science*, (2001). **239**: pp. 535–542.
74. Denkova, P.S., S. Tcholakova, N.D. Denkov, K.D. Danov, B. Campbell, C. Shawl, and D. Kim, Evaluation of the precision of drop-size determination in oil/water emulsions by low-resolution NMR spectroscopy. *Langmuir*, (2004). **20**: pp. 11402–11413.
75. Håkansson, B., R. Pons, and O. Söderman, Structure determination of a highly concentrated W/O emulsion using pulsed-field-gradient spin-echo nuclear magnetic resonance "diffusion diffractograms." *Langmuir*, (1999). **15**(4): pp. 988–991.
76. Mezzenga, R., B.M. Folmer, and E. Hughes, Design of double emulsions by osmotic pressure tailoring. *Langmuir*, (2004). **20**: pp. 3574–3582.
77. Robertson, B., Spin-echo decay of spins diffusing in a bounded region. *Physical Review*, (1966). **151**(1): pp. 273–277.
78. Neuman, C.H., Spin echo of spins diffusing in a bounded medium. *The Journal of Chemical Physics*, (1974). **60**(11): pp. 4508–4511.
79. Murday, J.S. and R.M. Cotts, Self-diffusion coefficient of liquid lithium. *The Journal of Chemical Physics*, (1968). **48**(11): pp. 4938–4945.
80. Balinov, B., B. Jönsson, P. Linse, and O. Söderman, The NMR self-diffusion method applied to restricted diffusion – simulation of echo attenuation from molecules in spheres and between planes. *Journal of Magnetic Resonance Series A*, (1993). **104**(1): pp. 17–25.
81. Epstein, B., Logarithmico-normal distribution in breakage of solids. *Industrial and Engineering Chemistry*, (1948). **40**(12): pp. 2289–2290.
82. Salager, J.L., M. Pérez-Sénchez, M. Ramírez-Gouveia, J.M. Andérez, and M.I. Briceño-Rivas, Stirring-formulation coupling in emulsions. *Récent Progrès en Génie des Procédés*, (1997). **11**(52): pp. 123–130.
83. Fourel, I., J.P. Guillement, and D. Le Botlan, Determination of water droplet size distributions by low-resolution PFG-NMR. 2. Solid emulsions. *Journal of Colloid and Interface Science*, (1995). **169**(1): pp. 119–124.
84. Orr, C., *Emulsion Droplet Size Data*, in *Encyclopedia of Emulsion Technology*, 1, P. Becher, Editor. 1983, Marcel Dekker, Inc.: New York. pp. 369–404.
85. Ambrosone, L., G. Colafemmina, M. Giustini, G. Palazzo, and A. Ceglie, Size distribution in emulsions. *Progress in Colloid and Polymer Science*, (1999). **112**: pp. 86–88.
86. Hollingsworth, K.G. and M.L. Johns, Measurement of emulsion droplet sizes using PFG NMR and regularization methods. *Journal of Colloid and Interface Science*, (2003). **258**: pp. 383–389.
87. Buckley, C., K.G. Hollingsworth, A.J. Sederman, D.J. Holland, M.L. Johns, and L.F. Gladden, Applications of fast diffusion measurement using Difftrain. *Journal of Magnetic Resonance*, (2003). **161**: pp. 112–117.
88. Codd, S.L. and P.T. Callaghan, Spin echo analysis of restricted diffusion under generalized gradient waveforms: Planar, cylindrical and spherical pores with wall relaxivity. *Journal of Magnetic Resonance*, (1999). **137**: pp. 358–372.
89. Callaghan, P., A simple matrix formalism for spin echo analysis of restricted diffusion under generalized gradient waveforms. *Journal of Magnetic Resonance*, (1997). **129**: pp. 74–84.
90. Ambrosone, L., A. Ceglie, G. Colafemmina, and G. Palazzo, NMR studies of food emulsions: the dispersed phase self-diffusion coefficient calculated by the least variance method. *Progress in Colloid and Polymer Science*, (2000). **115**: pp. 161–165.
91. Liu, E.-H., P. Callaghan, and K.M. McGrath, Bicontinous and closed-cell foam emulsions as continuum structures in an oil in water emulsion system. *Langmuir*, (2003). **19**: pp. 7249–7258.
92. Bergman, D.J. and K.J. Dunn, Theory of diffusion in a porous-medium with applications to pulsed-field-gradient NMR. *Physical Review B*, (1994). **50**(13): pp. 9153–9156.
93. Hürlimann, M.D., K.G. Helmer, L.L. Latour, and C.H. Sotak, Restricted diffusion in sedimentary-rocks – Determination of surface-area-to-volume ratio and surface relaxivity. *Journal of Magnetic Resonance Series A*, (1994). **111**(2): pp. 169–178.
94. Topgaard, D. and O. Söderman, Diffusion of water absorbed in cellulose fibers studied with ^{1}H-NMR. *Langmuir*, (2001). **17**(9): pp. 2694–2702.

95. Ngwa, W., O. Geier, L. Naji, J. Schiller, and K. Arnold, Cation diffusion in cartilage measured by pulsed field gradient NMR. *European Biophysics Journal*, (2002). **31**: pp. 73–80.
96. Balinov, B., O. Söderman, and J.C. Ravey, Diffraction-like effects observed in the PGSE experiment when applied to a highly concentrated water/oil emulsion. *Journal of Physical Chemistry*, (1994). **98**(2): pp. 393–395.
97. Söderman, O. and U. Olsson, Dynamics of amphiphilic systems studied using NMR relaxation and pulsed field gradient experiments. *Current Opinion in Colloid & Interface Science*, (1997). **2**(2): pp. 131–136.
98. Giustini, M., G. Palazzo, A. Ceglie, J. Eastoe, A. Bumujdad, and R.K. Heenan, Studies of cationic and nonionic surfactant mixed microemulsions by small-angle neutron scattering and pulsed field gradient NMR. *Progress in Colloid and Polymer Science*, (2000). **115**: pp. 25–30.
99. Söderman, O. and M. Nydén, NMR in microemulsions. NMR translational diffusion studies of a model microemulsion. *Colloids and Surfaces a – Physicochemical and Engineering Aspects*, (1999). **158**(1–2): pp. 273–280.
100. Wennerström, H., O. Söderman, U. Olsson, and B. Lindman, Macroemulsions versus microemulsions. *Colloids and Surfaces A – Physicochemical and Engineering Aspects*, (1997). **123**: pp. 13–26.
101. Thiebaud, M., E.M. Dumay, and J.C. Cheffel, Pressure-shift freezing of o/w emulsions: influence of fructose and sodium alginate on undercooling, nucleation, freezing kinetics and ice crystal size distribution. *Food Hydrocolloids*, (2002). **16**: pp. 527–545.
102. Flores, A.A. and H.D. Goff, Ice crystal size distributions in dynamically frozen model solutions and ice cream as affected by stabilizers. *Journal of Dairy Science*, (1999). **82**(7): pp. 1399–1407.

9 Ultrasound for Characterizing Emulsions and Microemulsions

A.S. Dukhin and P.J. Goetz

CONTENTS

Abstract

Ultrasound-based techniques (acoustics and electroacoustics) offer a unique opportunity to characterize concentrated emulsions and microemulsions in their natural state, without dilution. Elimination of the dilution protocol is crucial for an adequate characterization of liquid dispersions as dilution can change the thermodynamic equilibrium in these systems.

In this review, we present a short review of theoretical basis for the mentioned ultrasound techniques. Emphasis is placed on the mechanism of thermal losses, which is related to the thermodynamic properties of liquids and dominates ultrasound attenuation of many emulsions, microemulsions, and latex systems.

There are already many examples when this theoretical model allows us to achieve a successful characterization of various liquid-based systems. It is incorporated into the software of commercially available instruments made by Dispersion Technology Inc. In this review, we present several of these examples.

The first application is characterization of various dairy products. We show that acoustic attenuation technique is able to provide reliable information on the milk fat droplet size in milks, as well as a droplet size of water in butter. In addition this method might be suitable for characterizing fat content of dairy products.

The second application presents our recent observation of the evolution of the water-in-kerosene system from emulsion to mini-emulsion state. We show how ultrasound-based techniques allow us to monitor the kinetics of droplet size variation and reveal the reason for this variation. We show that the droplet size could be controlled by ion exchange between water droplets and the bulk of nonpolar liquid.

The third application presents characterization of microemulsions. We show how concentration of surfactant determines the water droplet size distribution in hexadecane. These results are in a very good agreement with independent measurements by X-ray and neutron scattering.

9.1 INTRODUCTION

The widespread acceptance and commercialization of acoustic spectroscopy has been slow to develop. This technique has been overlooked by many in academia and industry in the past, but has recently been showing increased levels of acceptance. This powerful method of characterizing concentrated heterogeneous systems has all the capabilities for being successful.

The first hardware for measuring acoustic properties of liquids was developed more than 50 years ago at MIT [1] by Pellam and Galt. The first acoustic theory for heterogeneous systems was created by Sewell 90 years ago [2]. Epstein and Carhart formulated the general principles of the acoustic theory 50 years ago [3]. There is a long list of applications and experiments using acoustic spectroscopy presented in recent publications [4–6].

Acoustics is able to provide reliable particle size information for concentrated dispersions without any dilution. There are examples when acoustics yields size information at volume fractions above 40%. This in-situ characterization of concentrated systems makes the acoustic method very useful and unique in this capability compared to alternate methods, including light scattering, where dilution is required.

Acoustics is also able to deal with low dispersed phase volume fractions and in some systems can characterize down to below 0.1% vol. This flexibility for concentration range provides an overlap with classic methods for dilute systems. In the overlap range, acoustics size characterization has been found to have excellent agreement with these other techniques [6].

Many people have perceived acoustics to have a high degree of complexity. The operating principles are in fact quite straightforward. The acoustic spectrometer generates sound pulses that pass through a sample system and are then measured by a receiver. The passage through the sample system causes the sound energy to change in intensity and phase. The acoustic instrument measures the sound energy losses (attenuation) and the sound speed. The sound attenuates due to the interaction with the particles and liquid in the sample system. Acoustic spectrometers generally operate with sound in the frequency range of 1 to 100 MHz. This is a much higher sound frequency than the upper limit of our hearing, which is only 0.02 MHz.

While the operating principles are relatively simple, the analysis of the attenuation data to obtain particle size distributions does involve a degree of complexity in fitting experimental results to theoretical models based on various acoustic loss mechanisms. The advent of high-speed computers and the refinement of these theoretical models have made the inherent complexity of this analysis of little consequence. In comparison, many other particle sizing techniques such as photon correlation spectroscopy also rely on similar levels of complexity in analyzing experimental results.

Acoustics is not only a particle sizing technique, but also provides information about the microstructure of the dispersed system. The acoustic spectrometer can be considered as a micro-rheometer. In acoustics, stresses are applied in the same way as regular rheometers, but over a very short distance on the micron scale. In this way, the microstructure of the dispersed system can be sensed. Currently, this feature of the acoustics is only beginning to be exploited, but it is certainly very promising.

In addition to characterizing special heterogeneity and structural properties, ultrasound offers an opportunity for studying electric surface properties of concentrated emulsions. There is a related field that is usually referred to as "electroacoustics" [7–9]. Electroacoustics can provide information on the zeta potential and surface charge of the droplets. This relatively new technique is more complex than acoustics because an additional electric field is involved. As a result, both hardware and theory become more complicated. There are even two different versions of electroacoustics depending on what field is used as a driving force. Electrokinetic sonic amplitude (ESA) involves the generation of sound energy caused by the driving force of an applied electric field. Colloid vibration current (CVC) is a phenomenon when sound energy is applied to a system. This results in electric field or current created by the vibration of the droplets' electric double layers.

Coming back to acoustics, its lack of widespread acceptance may be related to the fact that it yields too much, sometimes overwhelming, information. Instead of dealing with interpretation of the acoustic spectra it is often easier to dilute the system of interest and apply light-based techniques. It was often naïvely assumed that the dilution had not affected the emulsion characteristics. Lately, many researchers are coming to the realization that emulsion systems need to be analyzed in their natural concentrated form, and that dilution destroys many useful and important properties.

We are optimistic about the future of acoustics in colloid science. It is amazing what this technique can do especially in combination with electroacoustics for characterizing electric surface properties. We hope that this review will provide additional information concerning the power and opportunities related to these sound-based techniques.

9.2 THEORETICAL BACKGROUND

There are six known mechanisms of ultrasound interaction with a dispersed system:

- Viscous
- Thermal
- Scattering
- Intrinsic
- Structural
- Electrokinetic

Here we give only short qualitative descriptions omitting complicated mathematical models.

Viscous losses of the acoustic energy occur due to the shear waves generated by the particle oscillating in the acoustic pressure field. These shear waves appear because of the difference in the densities of the particles and medium. This density contrast causes the particle motion with respect to the medium. As a result, the liquid layers in the particle vicinity slide relative to each other. This sliding non-stationary motion of the liquid near the particle is referred to as the "shear wave." Viscous losses are dominant for small rigid particles with sizes below 3 microns, such as oxides, pigments, paints, ceramics, cement, and graphite.

The reason for the *thermal losses* is the temperature gradients generated near the particle surface. These temperature gradients are due to the thermodynamic coupling between pressure and temperature. This mechanism is dominant for soft particles, including emulsion droplets and latex beads.

The mechanism of the *scattering losses* is quite different than the viscous and thermal losses. Acoustic scattering does not produce dissipation of acoustic energy. This scattering mechanism is similar to light scattering. Particles simply redirect a part of the acoustic energy flow and as a result this portion of the sound does not reach the sound transducer. This mechanism is important for larger particles (>3 micron) and high frequency (>10 MHz).

The *intrinsic losses* of the acoustic energy occur due to the interaction of the sound wave with the materials of the particles and medium as homogeneous phases on a molecular level.

Structural losses are caused by the oscillation of a network of particles that are interconnected. Thus, this mechanism is specific for the given type of structured system.

Electrokinetic losses are caused by the oscillation of charged particles in an acoustic field that leads to the generation of an alternating electrical field, and consequently to alternating electric current. As a result, a part of the acoustic energy is transformed into electric energy and then irreversibly to heat.

Only the first four loss mechanisms (viscous, thermal, scattering, and intrinsic) make a significant contribution to the overall attenuation spectra in most cases. Structural losses are significant only in structured systems that require a quite different theoretical framework. These four mechanisms form the basis for acoustic spectroscopy.

The contribution of electrokinetic losses to the total sound attenuation is almost always negligibly small [10] and will be neglected. This opens an opportunity to separate acoustic spectroscopy from electroacoustic spectroscopy because the acoustic attenuation spectrum is independent of the electric properties of the dispersed system. Following this distinction between acoustics and electroacoustics, the corresponding theories will be considered separately.

9.2.1 Theory of Acoustics

The most well known acoustic theory for heterogeneous systems was developed by Epstein and Carhart [3]. Later Allegra and Hawley [11] made some important modifications. The resulting general theory has abbreviated to ECAH theory. This theory takes into account the four most important mechanisms (viscous, thermal, scattering, and intrinsic). This theory describes attenuation for a monodisperse system of spherical particles and is valid only for dilute systems.

The term "monodisperse" assumes that all of the particles have the same diameter. Extensions of the ECAH theory to include polydispersity have typically assumed a simple linear superposition of the attenuation for each size fraction. The term "spherical" is used to denote that all calculations are performed assuming that each particle can be adequately represented as a sphere.

Most importantly, the term "dilute" is used to indicate that there is no consideration of particle–particle interactions. This fundamental limitation normally restricts the application of the resultant theory to dispersions with a volume fraction of less than a few volume percent. However, there is some evidence that the ECAH theory, in some very specific situations, does nevertheless provide a correct interpretation of experimental data, even for volume fractions as large as 30%.

Allegra and Hawley provided the earliest demonstration of this ability of the ECAH theory. They observed almost perfect correlation between experiment and dilute case ECAH theory for several systems: a 20% by volume toluene emulsion; a 10% by volume hexadecane emulsion; and a 10% by volume polystyrene latex. Similar work with emulsions by McClements [12,13] has provided similar results. The recent work by Holmes, Challis, and Wedlock [14,15] shows good agreement between ECAH theory and experiments even for 30% by volume polystyrene latex.

A surprising absence of particle–particle interaction was observed with neoprene latex [16]. This experiment showed that attenuation is linear function of the volume fraction up to 30% for this particular system. This linearity is an indication that each particle fraction contributes to the total attenuation independently of other fractions, and is a superposition of individual contributions. Superposition works only when particle–particle interaction is insignificant.

It is important to note that the surprising validity of the dilute ECAH theory for moderately concentrated systems has only been demonstrated in systems where the "thermal losses" were dominant, such as emulsions and latex systems. In contrast, the solid rutile dispersion exhibits non-linearity of the attenuation above 10% by volume [6].

The difference between the "viscous depth" and the "thermal depth" provides an answer to the observed differences between emulsions and solid particle dispersions. These parameters characterize the penetration of the shear wave and thermal wave correspondingly into the liquid. Particles oscillating in the sound wave generate these waves, which damp in the particle vicinity. The characteristic distance for the shear wave amplitude to decay is the "viscous depth" δ_v. The corresponding distance for the thermal wave is the "thermal depth" δ_t. The following expressions give these parameters values in the dilute systems:

$$\delta_v = \sqrt{\frac{2\nu}{\omega}} \tag{9.1}$$

$$\delta_t = \sqrt{\frac{2\tau}{\omega\rho C_p}} \tag{9.2}$$

where ν is the kinematic viscosity, ω is the frequency, ρ is the density, τ is heat conductance, C_p is a heat capacity at constant pressure.

The relationship between δ_v and δ_t has been considered before. For instance, McClements plots "thermal depth" and "viscous depth" versus frequency [4]. It is easy to show that "viscous depth" is 2.6 times more than "thermal depth" in aqueous dispersions [16]. As a result, the particle viscous layers overlap at the lower volume fraction more than the particle thermal layers. Overlap of the boundary layers is the measure of the corresponding particle–particle interaction. There is no particle interaction when corresponding boundary layers are sufficiently separated.

Thus, an increase in the dispersed volume fraction for a given frequency first leads to the overlap of the viscous layers because they extend further into the liquid. Thermal layers overlap at higher volume fractions. This means that the particle hydrodynamic interaction becomes more important at the lower volume fractions than the particle thermodynamic interaction.

The 2.6 times difference between δ_v and δ_t leads to a large difference in the volume fractions corresponding to the beginning of the boundary layers' overlap. The dilute case theory is valid for the volume fractions smaller than these critical volume fractions φ_v and φ_t. These critical volume fractions, φ_v and φ_t, are functions of the frequency and particle size. These parameters are conventionally defined from the condition that the shortest distance between particle surfaces is equal to $2\delta_v$ or $2\delta_t$. This definition gives the following expression for the ratio of the critical volume fractions in aqueous dispersions:

$$\frac{\varphi_v}{\varphi_t} = \left(\frac{a\sqrt{\pi f} + 2.6^{-1}}{a\sqrt{\pi f} + 1}\right)^3 \tag{9.3}$$

where a is particle radius in microns, and f is the frequency in MHz.

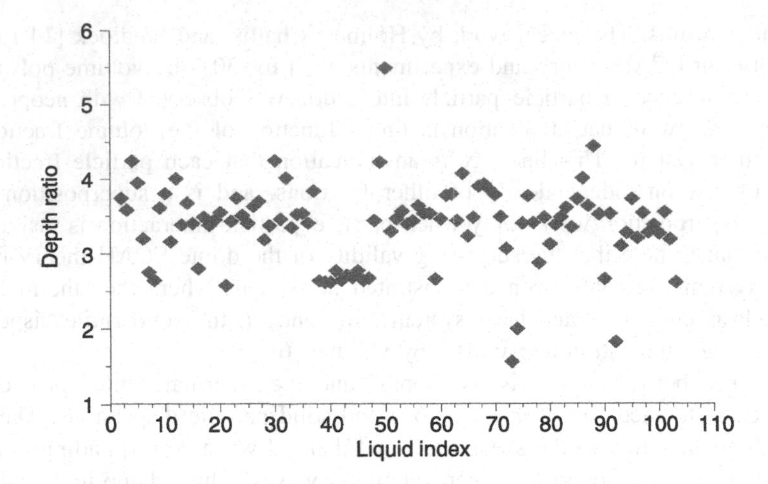

FIGURE 9.1 Ratio of viscous and thermal depth for various liquids.

The ratio of the critical volume fractions depends on the frequency. For instance for neoprene latex, the critical "thermal" volume fraction is 10 times higher than the critical "viscous" volume fraction for 1 MHz and only three times higher for 100 MHz.

It is interesting that this important feature of the "thermal losses" works for almost all liquids. We have more than 100 liquids with their properties in our database. The core of this database is the well-known paper by Anson and Chivers [17]. We can introduce a parameter referred to as "depth ratio":

$$\text{depth ratio} = \frac{\delta_\nu}{\delta_t}$$

This parameter is 2.6 for water, as mentioned before. Figure 9.1 shows values of this parameter for all liquids from our database relative to the viscous depth of water. It is seen that this parameter is even larger for many liquids.

Therefore "thermal losses" are much less sensitive to the particle–particle interaction than "viscous losses" for almost all known liquids. It makes ECAH theory valid in a much wider range of emulsion volume fractions than one would expect.

There is one more fortunate fact for ECAH theory that follows from the values of the liquid's thermal properties. In general, ECAH theory requires information about three thermodynamic properties: thermal conductivity τ, heat capacity C_p, and thermal expansion β. It turns out that τ and C_p are almost the same for all liquids except water. This reduces the number of required parameters to one – thermal expansion. This parameter plays the same role in "thermal losses" as density in "viscous losses."

ECAH theory has a great disadvantage of being mathematically complex. It cannot be generalized for particle–particle interactions. This is not important as we have found for emulsions, but may be important for latex systems, and is certainly very important for the high density contrast systems. There are two ways to simplify this theory using a restriction on the frequency and particle size. The first is the so-called "long wave requirement" [11] which requires the wave length of the sound wave λ to be larger than particle radius a. This "long wave requirement" restricts particle size for a given set of frequencies. Our experience shows that particle size must be below 10 microns for the frequency range from 1 to 100 MHz. This restriction is helpful for characterizing small particles.

The long wave requirement provides a sufficient simplification of the theory for implementing particle–particle interaction. Work has been done by Dukhin and Goetz [18] on the basis of the "coupled phase model" [19,20]. This new theory [18] works up to 40% vol even for heavy materials, including rutile.

There is another approach to acoustics, which employs a "short wave requirement" [21]. This approach works only for large particles above 10 microns and requires limited input data about the sample. This theory may provide an important advantage in the case of emulsions and latex systems when the thermal expansion is not known.

There is opportunity in the future to create a mixed theory that could use a polynomial fit merging together "short" and "long" wave range theories. Such combined theory will be able to cover a complete particle size range from nanometers to millimeters for concentrated systems.

There are two recent developments in the theory of acoustics, which deserve to be mentioned here. The first one is a theory of acoustics for flocculated emulsions [22]. It is based on ECAH theory but it uses in addition an "effective medium" approach for calculating thermal properties of the flocs. The success of this idea is related to the feature of the thermal losses that allows for insignificant particle–particle interactions even at high volume fractions. This mechanism of acoustic energy dissipation does not require relative motion of the particle and liquid. Spherical symmetrical oscillation is the major term in these kinds of losses. This provides the opportunity to replace the floc with an imaginary particle assuming a proper choice of the thermal properties.

Another significant recent development is associated with Temkin, who, in his recent papers [23,24], offers a new approach to the acoustic theory. Instead of assuming a model dispersion consisting of spherical particles in a Newtonian liquid, he suggests that the thermodynamic approach be explored as far as possible. This new theory operates with notions of particle velocities and temperature fluctuations. This very promising theory yields some unusual results [23,24]. It has not been yet used, as far as we know, in commercially available instruments.

9.2.2 Theory of Electroacoustics

Whereas acoustic spectroscopy describes the combined effect of the six separate loss mechanisms, electroacoustic spectroscopy, as it is presently formulated, emphasizes only one of these interaction mechanisms, the electrokinetic losses.

In acoustic spectroscopy sound is utilized as both the excitation and the measured variable, and therefore there is but one basic implementation. In contrast, electroacoustic spectroscopy deals with the interaction of electric and acoustic fields and therefore there are two possible implementations. One can apply a sound field and measure the resultant electric current, which is referred to as the colloid vibration current (CVC), or conversely one can apply an electric field and measure the resultant acoustic field, which is referred to as the electronic sonic amplitude (ESA).

First let us consider the measurement of CVC. When the density of the particles ρ_p differs from that of the medium ρ_m, the particles move relative to the medium under the influence of an acoustic wave. This motion causes a displacement of the internal and external parts of the double layer (DL) (see Figure 9.2). This phenomenon is usually referred to as a polarization of the DL [7]. This displacement of opposite charges gives rise to a dipole moment. The superposition of the electric fields of these induced dipole moments over the collection of particles gives rise to a macroscopic electric field, which is referred to as the colloid vibration potential (CVP). This potential in turn creates electric current, the CVC. Thus, the fourth mechanism of particle interaction with sound leads to the transformation of part of the acoustic energy to the

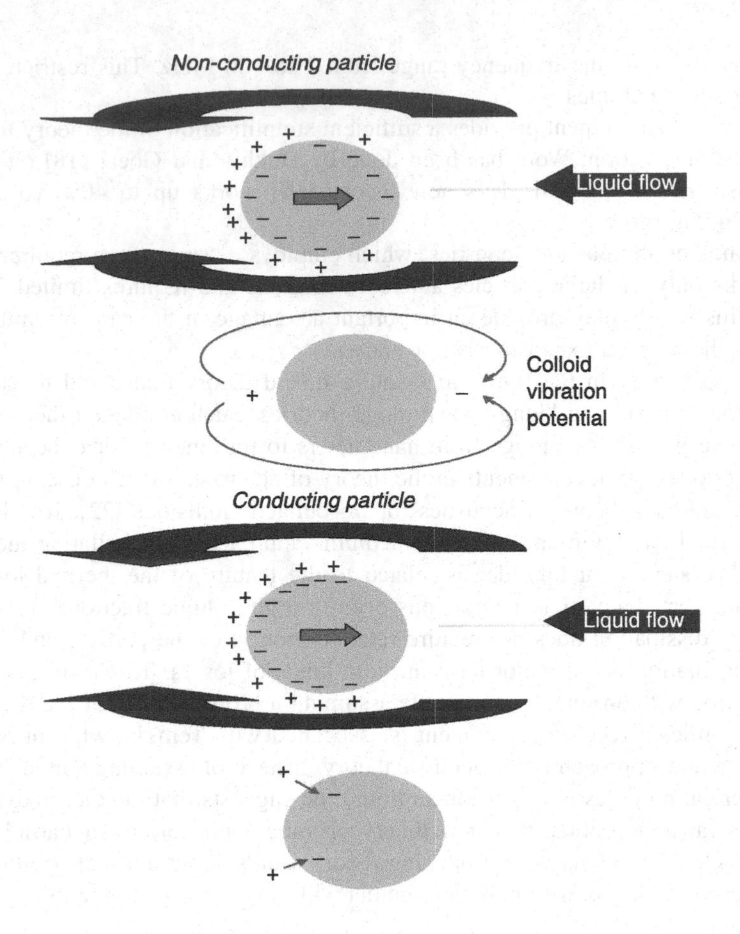

FIGURE 9.2 Illustration of the exterior DL polarization by the liquid flow relative to the particle surface induced with ultrasound.

electric energy. This electric energy may then be dissipated if the opportunity exists for the flow of electric current.

Now let us consider the measurement of ESA, which occurs when an alternating electric field is applied to the disperse system [9]. If the zeta potential of the particle is greater than zero, then the oscillating electrophoretic motion of the charged dispersed particles generates a sound wave.

Both electroacoustic parameters, CVC and ESA, can be experimentally measured. The CVC or ESA spectrum is the experimental output from electroacoustic spectroscopy. Both of these spectra contain information about zeta potential and PSD. However, only one of the electroacoustic spectra is required because both of them contain essentially the same information about the dispersed system.

The conversion electroacoustic spectra into the PSD requires theoretical model of the electroacoustic phenomena. This conversion procedure is much more complicated for electroacoustics comparing to the acoustics. The reason for the additional problems relates to the additional field involved in the characterization: electric field. The theory becomes much more complicated because of this additional field.

For some time O'Brien's theory [25,26] has been considered as a basis for electroacoustics. It provides a widely accepted expression for the dynamic electrophoretic mobility μ_d (Equation 5.28

in Ref. 4); that is:

$$\mu_d = \frac{2\varepsilon_0\varepsilon_m \varsigma(\rho_p - \rho_s)\rho_m}{3\eta(\rho_p - \rho_m)\rho_s} G(s,\varphi)(1 + F(Du,\omega',\varphi)) \tag{9.4}$$

where ε_0 and ε_m are dielectric permittivities of the vacuum and liquid, ζ is the electrokinetic potential, η is dynamic viscosity, ρ_m, ρ_p, and ρ_s are the densities of liquid, particle, and dispersion, $\omega' = \frac{\omega}{\omega_{MW}}$, $\omega_{MW} = \frac{K_m}{\varepsilon_0\,\varepsilon_m}$, K_m is conductivity of the media, ω is frequency of ultrasound, and $Du = \kappa^{\sigma}/K_m a$ is the Dukhin number that reflects the contribution of the surface conductivity κ^{σ}.

Function G presents the contribution of hydrodynamic effects, whereas function F reflects electrodynamic aspects of the electroacoustic phenomena.

This initial theory by O'Brien does not take into account hydrodynamic and electrodynamic interactions. Later, two more theories were developed for oil-in-water emulsions. The first, by O'Brien, is presented in a review, Ref. 9. The second is by Dukhin et al. (see Refs. 27–29, and Chapter 5 in Ref. 6).

Application of these theories to water-in-oil emulsions is complicated by the conductivity of the water droplets. In order to apply this theory to water-in-oil emulsions we should simply introduce the conductivity of the particles in the expression for function F. A water-in-oil emulsion can be considered as conducting particles in a non-conducting media. In order to achieve this we should simply use a more general expression for the Dukhin number:

$$Du = \frac{K_p}{K_m} + \frac{\kappa^{\sigma}}{K_m} \tag{9.5}$$

where K_p is conductivity of water droplets.

This modified Dukhin number can be used in the general expression for the F function:

$$F(Du,\omega',\varphi) = \frac{(1-2Du)(1-\varphi) + j\,\omega'(1-\frac{\varepsilon_p}{\varepsilon_m})(1-\varphi)}{2(1+Du+\varphi(0.5-Du)) + j\omega'(2+\frac{\varepsilon_p}{\varepsilon_m}+\varphi(1-\frac{\varepsilon_p}{\varepsilon_m}))} \tag{9.6}$$

The properties of the water-in-oil emulsion allow us to approximate the value of this function. It is known that the conductivity of water is much higher than the conductivity of nonpolar liquids. This means that

$$Du \approx \frac{K_p}{K_m} \gg 1 \tag{9.7}$$

In addition, the dielectric permittivity of water is about 40 times higher than the permittivity of nonpolar liquids:

$$\frac{\varepsilon_p}{\varepsilon_m} \gg 1 \tag{9.8}$$

Using these two strong inequalities we can neglect 1 in comparison to the relevant parameters in Equation 9.3. As a result we obtain the following approximate value for the function F in water-in-oil emulsions:

$$F(Du,\omega',\varphi)\approx\frac{1-\varphi}{1-\varphi}\frac{-2Du-j\,\omega'\dfrac{\varepsilon_p}{\varepsilon_m}}{2Du+j\omega'\dfrac{\varepsilon_p}{\varepsilon_m}}=-1 \tag{9.9}$$

This approximate value is actually quite accurate due to the large conductivity and permittivity of water.

If we use this value for function F in Equation 9.4 for the dynamic mobility, we would come to the interesting result:

$$\mu_d\left(\frac{K_p}{K_m}\gg 1,\ \frac{\varepsilon_p}{\varepsilon_m}\gg 1\right)=0 \tag{9.10}$$

for thin isolated DLs only.

This means simply that water-in-oil emulsions with isolated and thin DLs should not exhibit any electroacoustic effect.

It is possible to give a simple explanation to this unexpected conclusion. In order to do this we compare the electric field structure induced by the relative particle–liquid motion for non-conducting and conducting particles, as illustrated in Figure 9.2.

It is known that ultrasound generates particle motion relative to the liquid due to the density contrast. In drawing Figure 9.2 we assume that the particles move from left to right. This causes a liquid motion, relative to the particle surface, which is illustrated with arrows above and below the particle.

This motion of the liquid drags ions within the diffuse layer towards the left pole of the particle. This redistribution of the diffuse layer leads to the different results for conducting and non-conducting particle.

In the case of non-conducting particle, the surface charge cannot move. It retains its spherical symmetry. As a result the particle gains an excess negative charge at the right hand side pole and excessive positive charge at the left hand side pole. It gains a dipole moment. This dipole moment generates an external electric field, which gives rise to the colloid vibration current.

In the case of a conducting particle, the charge carriers inside the particle can follow the counter-ions of the DL that are being dragged to the left hand side pole. This means that the surface charge, or charge associated with the particle, loses its internal spherical symmetry as well as the charge of the diffuse layer. Most importantly, each element of the surface, as well as the adjacent diffuse layer, remains electroneutral. No external electric field appears and electroacoustic effect is practically zero.

This does not mean that electroacoustics is useless for water-in-oil emulsions. It is known that existing electroacoustic theories for both ESA and CVC effects do not take account of overlapping of the double layers (DLs). There is only one exception, the recently created Shilov's theory [30].

The assumption of isolated, not-overlapped, double layers requires either a large droplet radius a or a sufficiently high ionic strength leading to a shorter Debye length κ^{-1}. Figure 9.3 illustrates approximately the range of volume fractions φ, where assumption of isolated DLs is valid for given κa, following Ref. 6. It is seen that water-in-oil emulsions should have overlapped DLs at very low volume fractions due to the large κ^{-1}. In the water-in-oil emulsions with overlapped DLs the nature of electroacoustic effect is quite different.

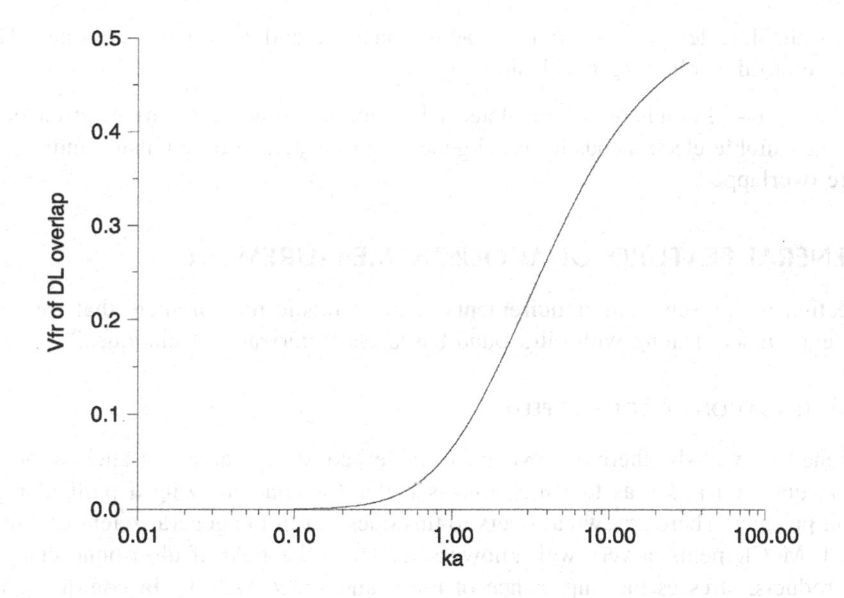

FIGURE 9.3 Estimate of the volume fraction of the overlap of the electric double layer.

Homogeneous distribution of the counter-ions eliminates polarization charges induced on the left hand side pole of the particles by the liquid motion. The electroacoustic effect is related simply to the displacement current of the oscillating motion of the particle surface charges. Shilov's theory [30] describes this effect. It yields the following expression for the dynamic mobility:

$$\mu_d = \frac{2\sigma a}{3} \frac{\rho_m}{\rho_s \eta \Omega + i\omega (1-\varphi) \dfrac{2a^2}{9} \rho_p \rho_m} \tag{9.11}$$

where σ is surface charge density, Ω is a hydrodynamic drag coefficient, and ω is the ultrasound frequency.

It is seen that there are no electrodynamic parameters for either the particles or the media involved in this expression. This leads us to the conclusion that, in the case of overlapped DLs, water-in-oil emulsions generate an electroacoustic signal in a similar manner as non-conducting particles.

According to Shilov's theory, in the case of overlapped DLs, we can calculate the surface charge of the particles from the dynamic mobility knowing nothing about the relevant ions or even the ionic strength. We know of only one other electrokinetic theory with a similar level of simplicity – Smoluchowski's theory.

This general nature of Shilov's theory is especially important for nonpolar liquids. The calculation of zeta potential requires information on the ions, because it includes Debye length

$$\sigma = \frac{1}{3} \frac{RT}{F} \frac{1-\varphi}{\varphi} \varepsilon_0 \varepsilon_m a \kappa^2 \sinh \frac{F\zeta}{RT} \tag{9.12}$$

where T is absolute temperature, F is Faraday constant, and R is a gas constant. This simple analysis leads to the following conclusions:

1. A water-in-oil emulsion with isolated DLs should not generate any electroacoustic signal
2. A measurable electroacoustic signal generated by water-in-oil emulsion indicates that DLs are overlapped

9.3 GENERAL FEATURES OF ACOUSTIC MEASUREMENT

In this section we present some requirements of the acoustic measurement that we could derive from our experience dealing with ultrasound-based characterization techniques.

9.3.1 ATTENUATION OR SOUND SPEED?

As mentioned previously, there are two measurable acoustic parameters: attenuation and sound speed. The question arises as to which one is better for characterizing a particular effect in a given food product. There are two answers to this question in the general scientific literature. For example, J. McClements, a very well known scientist in the field of ultrasound characterization of food products, stresses the importance of the sound speed [12,13]. In contrast, our group at Dispersion Technology Inc. relies mostly on attenuation. It is clear that this question deserves special consideration.

We have performed several simple tests for answering this question. First, we have measured sound speed of water varying concentration of several simple chemicals. It turns out that sound speed is very sensitive to the chemical composition (Figure 9.4). It changes roughly 100 m/sec per 1 mol/l of the simple salt. Keeping in mind that precision of the sound speed measurement is roughly 1 cm/sec, we could conclude that it is possible to monitor variation of the water chemical composition with precision of 0.0001 mol/l. Sound speed is also very sensitive to the temperature. For water it changes 2.4 m/sec per degree Celsius.

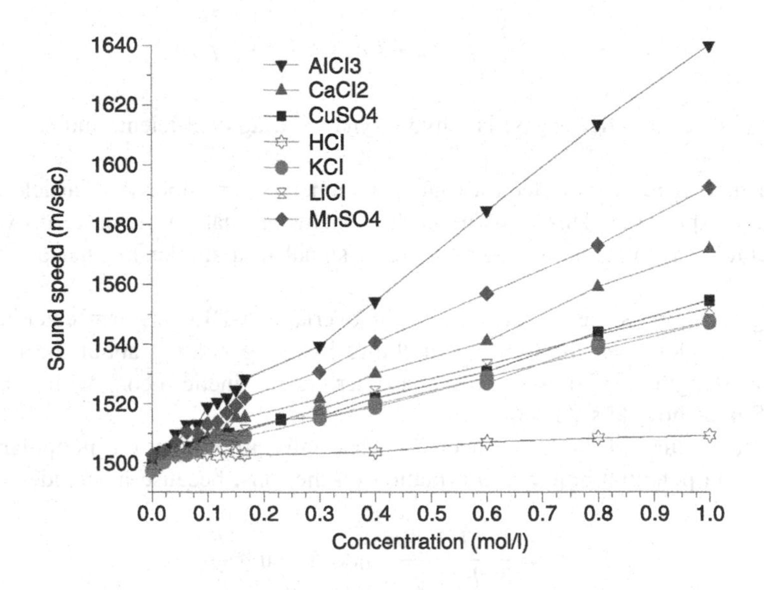

FIGURE 9.4 Sound speed of water versus concentration of various simple chemicals.

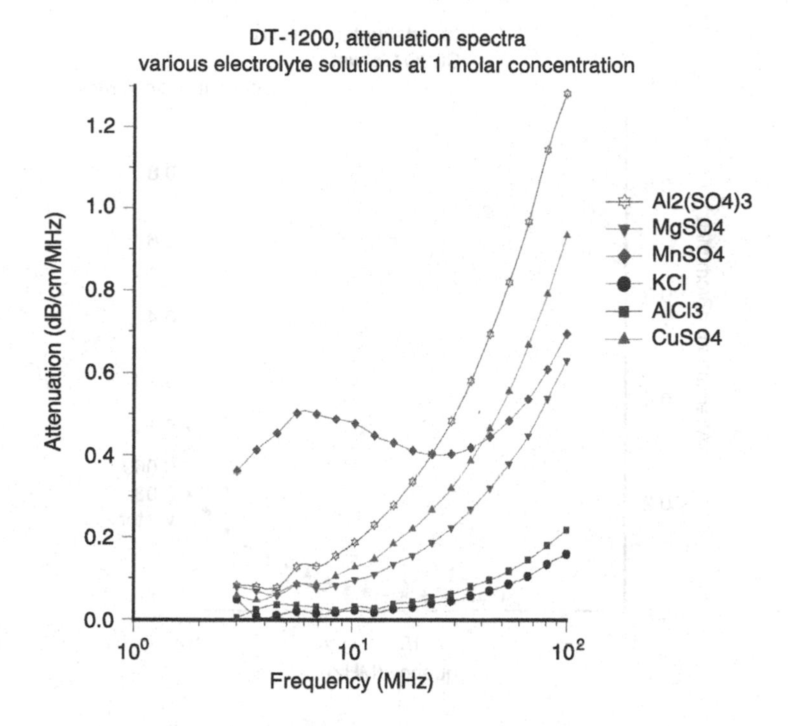

FIGURE 9.5 Attenuation of water versus concentration of various simple chemicals.

Attenuation in turn is much less sensitive to chemical composition. Figure 9.5 shows attenuation spectra of water at 1 mol/l concentration of similar chemical additives. It is seen that even at this very high concentration it changes significantly only for rather charged ions. This influence is much less pronounced at lower concentration, which we could expect in food products. Figure 9.6 shows that even for 2:2 valency salt it changes 0.1 dB/cm/MHz only at 0.14 mol/l. It might be generally concluded that the impact of chemical variation on attenuation is negligible if concentration varies less than 0.1 mol/l.

Attenuation is not very sensitive to the molecular and ionic composition, but it changes dramatically with composition of the dispersed phase. In other words, attenuation reflects variation occurring on the scale of colloid sizes, from nanometers to microns, whereas sound speed is more representative for effects that originate on the molecular scale of angstroms.

This simple analysis indicates that sound speed is the better parameter for investigating effects on the molecular scale, whereas attenuation is more suitable for characterizing effects related to the heterogeneity and phase composition of the particular system.

9.3.2 Ranges of Values

The next point of interest is the potential range of attenuation and sound speed in food products. Sound speed varies within the range 1000–2000 m/sec for practically all food products. For water-based food products it would be close to 1500 m/sec, which is sound speed of water at room temperature.

Attenuation varies in much wider dynamic range. That is why it is normally expressed on a logarithmic scale in decibels (dB).

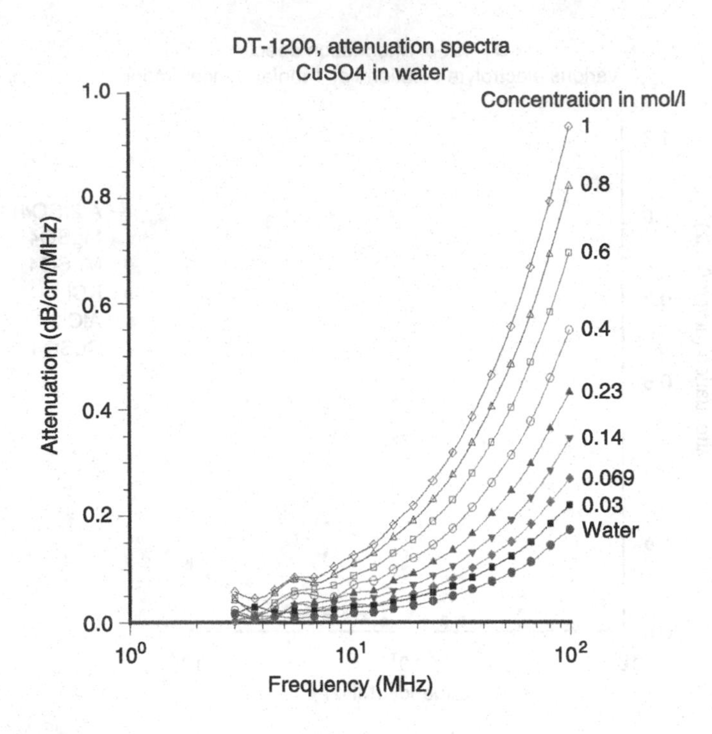

FIGURE 9.6 Attenuation of water versus concentration of $CuSO_4$.

Water is the least attenuating liquid. Its attenuation expressed in units of dB/cm/MHz increases linearly with frequency reaching 0.2 at 100 MHz.

According to our experience, an acoustic spectrometer must be able to cover a dynamic range from water up to 20 dB/cm/MHz for a complete frequency range from 1 to at least 100 MHz.

9.3.3 Precision

It is also possible to derive some conclusions regarding precision of the acoustic measurement because it depends directly on the precision of monitoring and maintaining other parameters.

For instance, it is practically impossible to maintain a uniform temperature of the sample and instrument with variations less than 0.01 °C. We know that a variation in temperature of 1 °C leads to the sound speed variation of water 2.4 m/sec. Following this number, a variation of 0.01 °C corresponds to the variation of sound speed of about 1 cm/sec. This is about 10^{-5} of the absolute sound speed.

Higher precision of the sound speed measurement makes no sense because uncontrolled temperature variations would mask it.

It is possible to make a similar estimate of the sound speed measurement precision using chemical composition variations. It is known that maintaining dissolved ion and molecule composition in water with precision higher than 10^{-4} mol/l is practically impossible. Adsorption of various chemical species from the atmosphere would create larger uncontrolled variations. Figure 9.4 indicates that sound speed changes about 100 m/sec per mol/l of the simple salts. This leads us to the sound speed precision of 1 cm/sec for the smallest controllable variations of the chemical composition. It is interesting that this level of precision is practically identical to the one calculated from the temperature stability requirement.

Unfortunately, it is impossible to estimate precision of attenuation measurement using similar analysis. As we mentioned above, attenuation is much less sensitive to variations in chemical composition and temperature. Attenuation depends mostly on the composition and state of the dispersed phase, particles, and droplets. As a result precision of its measurement must be related somehow to the particles.

We have done this using resolution of the attenuation measurement for characterizing content of the large particles (see Chapter 8 in Ref. 6). It turns out that in order to resolve a single 1 micron particle on the background of the 100,000 particles with size 100 nm, attenuation must be measured with precision of 0.01 dB/cm/MHz.

9.3.4 Frequency Range

The frequency range might be different depending on the purpose of the measurement. For droplet size distribution, the wider the frequency range, the more information about the system can be extracted. It turns out that measurement within frequency range from 1 to 100 MHz makes it possible to characterize sizes as small as 5 nm, as it will be shown with water-in-heptane microemulsion. Restricting the frequency range to the lower frequencies would limit dramatically the small size limit. It would also restrict the ability to characterize details of the particle size distribution. The frequency range specified above makes it possible to characterize bimodality, which would be impossible if high frequencies are not included in the measurement.

At the same time, low frequency attenuation below 10 MHz could be affected by bubbles. This is especially important for process control, when consistent removal of bubbles might present a problem.

This factor might be especially important for fat content measurement. It requires just one frequency, which simplifies electronics and software. Position of this frequency is open for discussion. From the cost viewpoint it is highly desirable to make this frequency as low as possible. This would reduce transducer and electronics costs. However, low frequency attenuation is highly affected by bubbles. In addition, low frequency attenuation is very sensitive to the droplet size. Elimination or at least minimization of these two factors (bubbles and droplet size) would require frequency above 10 MHz. Frequency in the range 40–50 MHz might be optimal.

9.4 APPLICATIONS

There are many instances of successful characterization of the particle size distribution and zeta potential of emulsion droplets. There are two quite representative reviews of these experiments published by McClements [4] (acoustics) and Hunter [9] (electroacoustics). We present here some more recent results that have not been published in any reviews.

9.4.1 Dairy Products: Droplet Sizing

We use dairy products from two different sources. The first is a local A&P supermarket. We purchased various milks and butters. Containers of these products present table of product composition, including fat, proteins, and sugar content. We use some of these data in our analysis. In addition we have two samples of the whole milk supplied by the Department of Food Science at Cornell University (courtesy of Dr Carmen Moraru). One of the samples is homogenized, whereas the other is unhomogenized. These samples have well defined compositions: 3.9%

fat, 3.25% proteins, and 4.6% lactose. These two samples allow us to verify the particle size calculation procedure.

There is no special sample preparation involved. Milks and butters are measured "as is," with no dilution. Milk is simply poured into the DT-100 sample chamber with a volume between 20 ml and 100 ml depending on the instrument setup. There is no mixing or pumping applied for measuring attenuation of the stable products. DT-100 has a built-in magnetic stirrer, which can be used for mixing unstable samples. It is used for measuring sound speed during milk spoiling processes.

Butter is pushed into the chamber at the temperature around 25 °C when it becomes a soft gel-like substance. Measurement must be performed when the gap between the transducer and receiver closes. It is opposite to the normal mode of DT-1200 functioning.

We use unsalted butter for extracting pure milk fat. This is necessary for measuring the intrinsic acoustic properties of the fat. Heating the butter leads eventually to its phase separation. We use a middle phase that is supposed to be a pure milk fat with no water, proteins, or other constituents.

Attenuation frequency spectra could be considered as experimental data. Figure 9.7 presents attenuation spectra for general dairy products. The legends on this figure indicate the type of the dairy product, relative fat content according to the manufacturer label, date and time of the measurement. This figure shows only one curve for each product. Actually, each product is measured several times for controlling reproducibility. Figure 9.8 shows multiple measurements for the low fat products.

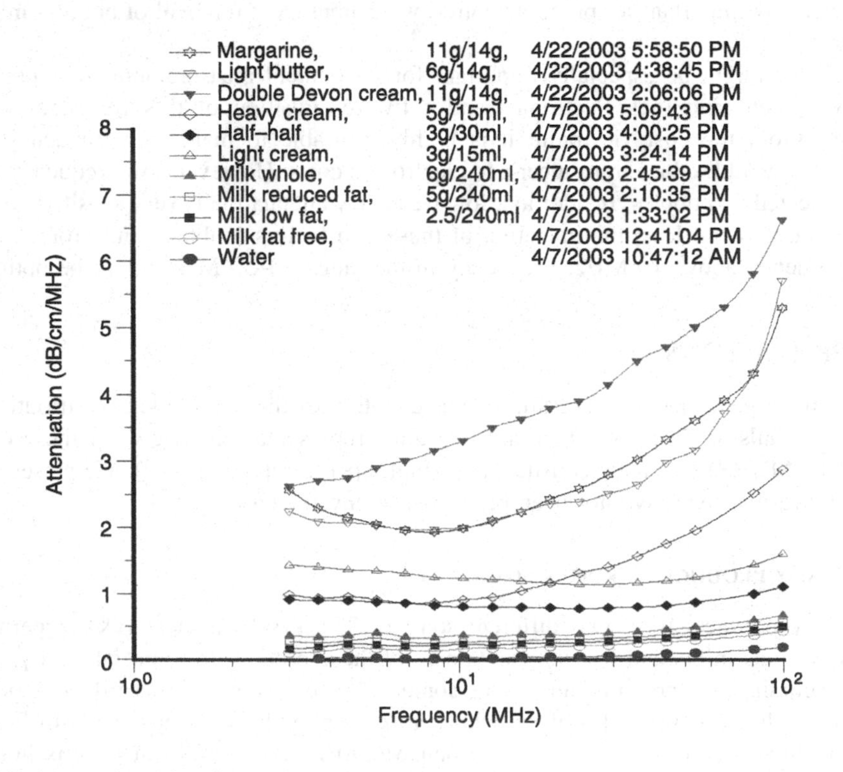

FIGURE 9.7 Attenuation of various dairy products.

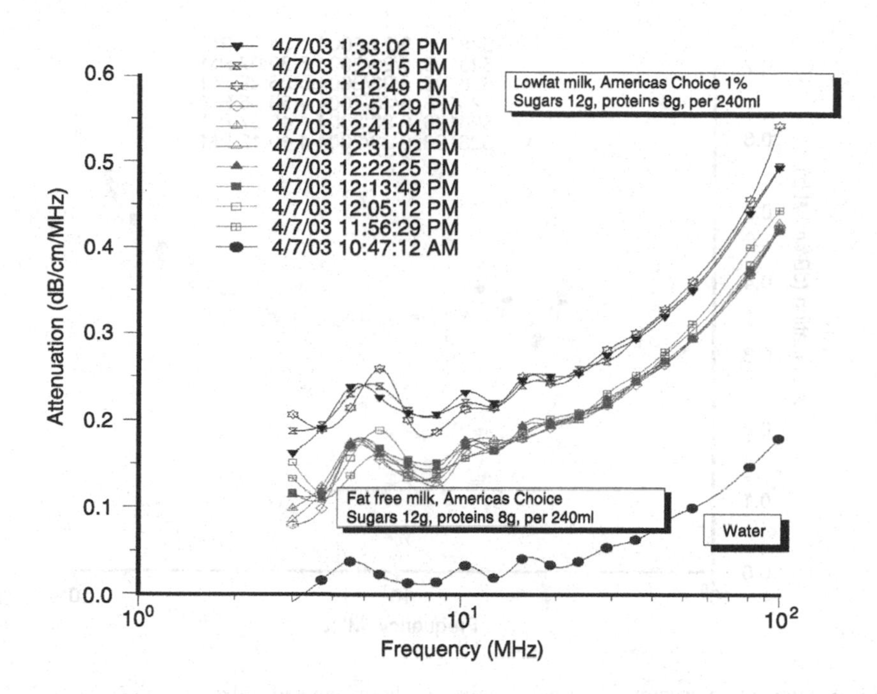

FIGURE 9.8 Precision test with low fat content milks.

Figure 9.9 provides attenuation spectra for the samples from Cornell University. These samples are identical in composition but different in terms of water droplet size distributions. A distinctive difference in attenuation spectra of these samples is a clear proof that acoustic spectroscopy can be used for particle sizing in dairy systems. Attenuation spectra within frequency range from 3 to 100 MHz contain information on the droplet size distribution. We show below how this information could be extracted.

Calculation of the droplet size distribution and fat content requires certain information on the milk fat properties. That is why we measured attenuation spectra of the pure milk fat. It is shown in Figure 9.10. It turns out that acoustic properties of the milk fat are very temperature dependent. In order to incorporate this temperature dependence into the size calculation we measured dependence of the fat attenuation on the temperature. It is almost linear, as shown in Figure 9.11.

Experimental data contains information for calculating droplet size distribution and/or fat content of these samples. Calculation of the droplet size distribution requires some input parameters for each phase. This set of the input parameters depends strongly on the properties of the system.

Dairy products are emulsions, oil-in-water for milks and water-in-oil for butters. The densities of fat and water are very close. This eliminates a viscous dissipation mechanism of the ultrasound attenuation. In addition, droplet size of these systems usually does not exceed 10 microns. In combination with the low density contrast this practically eliminates scattering effect.

This simple analysis leads to the conclusion that ultrasound attenuation in the dairy products occurs mostly due to the thermal dissipation and intrinsic attenuation of the water and fat phases. We can describe these systems as "soft particles" using terminology from Ref. 6. Therefore, we need a set of thermodynamic properties for each phase, such as thermal conductivity τ (J/m sec K), heat capacity C_p (J/kg K), and thermal expansion β (10^{-4} K^{-1}). In addition

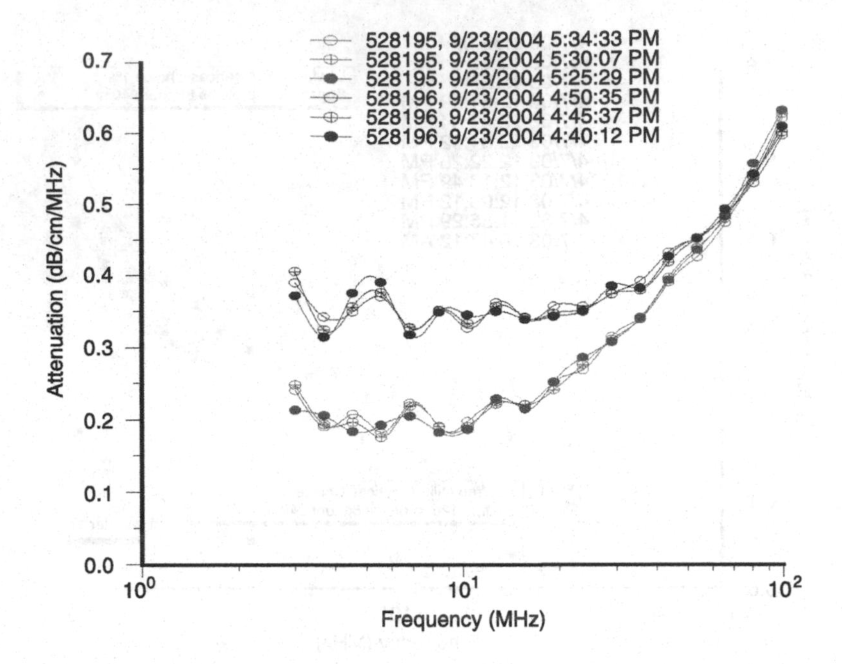

FIGURE 9.9 Attenuation spectra of the whole milk and homogenized milk.

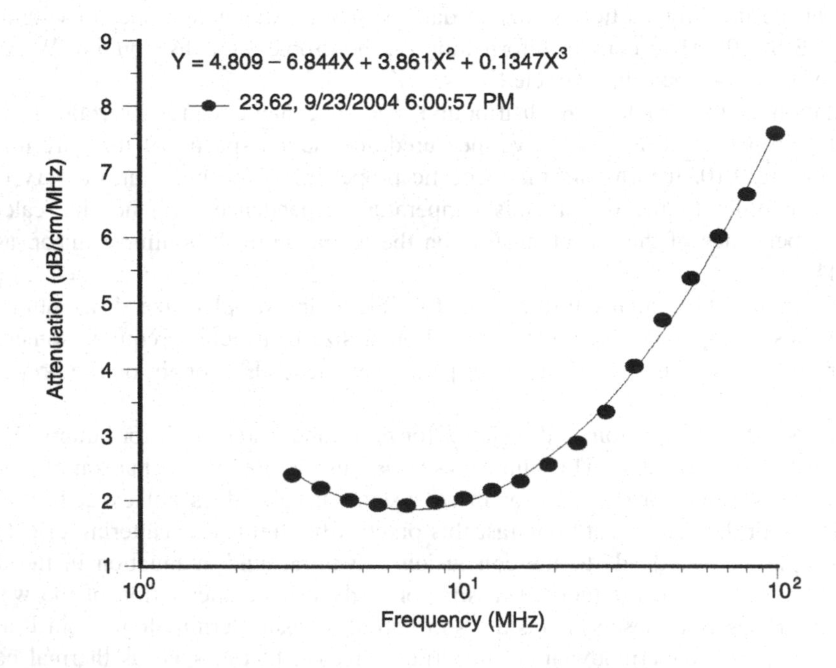

FIGURE 9.10 Intrinsic attenuation of the milk fat at room temperature.

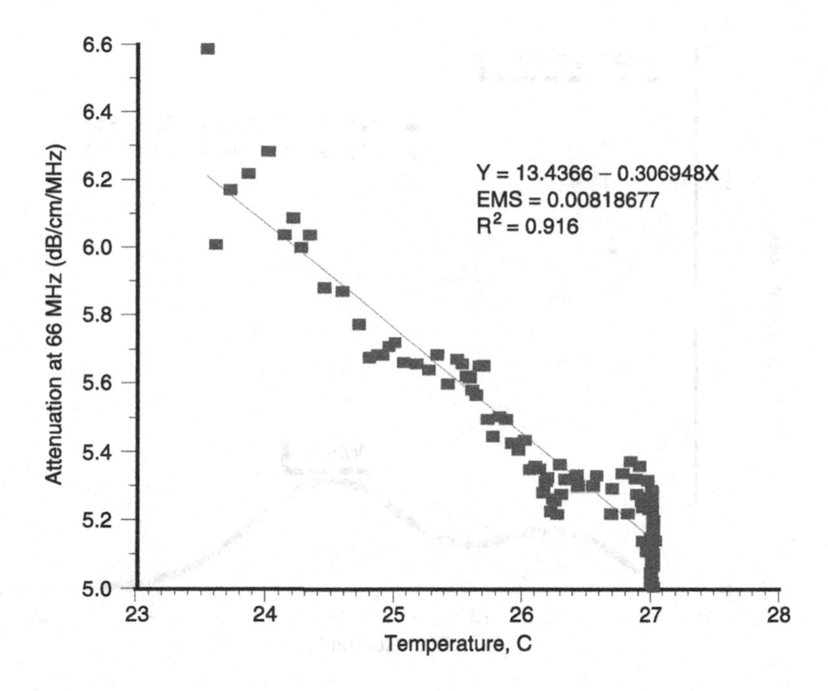

FIGURE 9.11 Temperature dependence of the milk fat attenuation at 66 MHz.

TABLE 9.1
Thermodynamic Properties of the Milk Fat and Water Phase Around Room Temperature

	Thermal conductivity (J/m s K)	Heat capacity (J/kg K)	Thermal expansion (10^{-4} K^{-1})	Attenuation (dB/cm/MHz) – frequency (MHz) function
Water	0.61	4.18	2.07	$\alpha = 4.809 - 6.844\omega + 3.861\omega^2 + 0.147\omega^3$
Milk fat	0.17	2.1 (at 40 °C)	6.2	$\alpha = 0.002\omega$

we need the intrinsic attenuation of the fat and water phases. Attenuation of the fat might be frequency (ω) dependent. That is why we should specify a function $\alpha(\omega)$ that describes this dependence. Thermodynamic properties and intrinsic attenuation of the milk fat and water are summarized in Table 9.1.

It is important to mention here that we neglect for the time being influence of proteins, sugar, and other additives on the properties of the water phase. This assumption must be verified later. We also neglect temperature dependence of all of these parameters. These two assumptions would create some uncertainty in the calculated droplet size distribution.

Milk is an obvious candidate for verifying particle sizing capability of ultrasound-based techniques. It is known that homogenization leads to substantial reduction of the fat droplet size. Attenuation spectra presented in Figure 9.9 reflects this change. Figure 9.12 shows droplet size distributions calculated from these attenuation spectra using input parameters presented in Table 9.1.

It is interesting to compare contributions of the different mechanisms to the ultrasound attenuation. Figure 9.13 shows these contributions calculated for the size distributions from Figure 9.12.

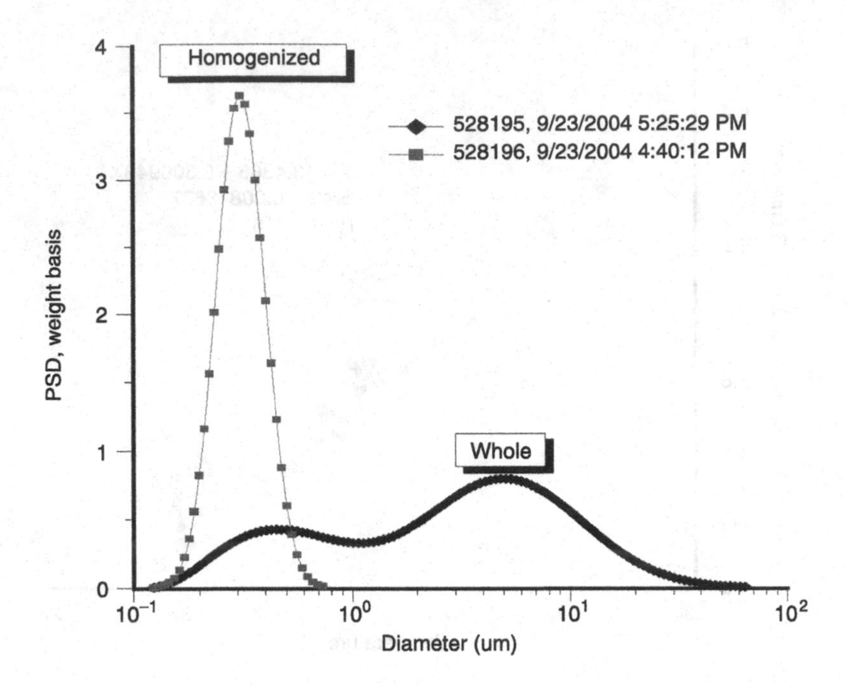

FIGURE 9.12 Droplet size distributions of the fat in whole and homogenized milks.

It is seen that at low frequency thermal attenuation and intrinsic attenuation are comparable. At the high frequency intrinsic attenuation dominates. This conclusion would have a profound effect for fat content calculation. At this point we could conclude that low frequency range below 10 MHz is the most important for particle sizing.

The more interesting feature of acoustic attenuation characterization procedure is that it allows characterization of droplet size distribution even in gel like systems, like butter, with no dilution or melting. It is possible because thermal attenuation effect is thermodynamic in nature; it does not require relative droplet–liquid motion.

Figure 9.14 shows size of the water droplets in light butter. The second graph on Figure 9.14 presents theoretical attenuations with intrinsic and thermal contributions. Intrinsic contribution is much higher in this case because it corresponds to the pure fat, not to the water. It dominates high frequency spectra even more than in the case of milks.

These calculations use the same thermodynamic properties of the fat as in the case of the milks.

9.4.2 Dairy Products: Fat Content

Fat content is a very important quality control parameter of the dairy products. Attenuation measurement offers a simple way to characterize it with no special treatment of the sample. Figure 9.7 shows attenuation spectra of the various dairy products. It is clear that attenuation increases with increasing fat content. This dependence could be used for calculating fat content.

In previous sections we showed that total ultrasound attenuation consists of thermal dissipation and intrinsic attenuation. Thermal dissipation is droplet size dependent, whereas intrinsic attenuation is not sensitive to the variation of the droplet size.

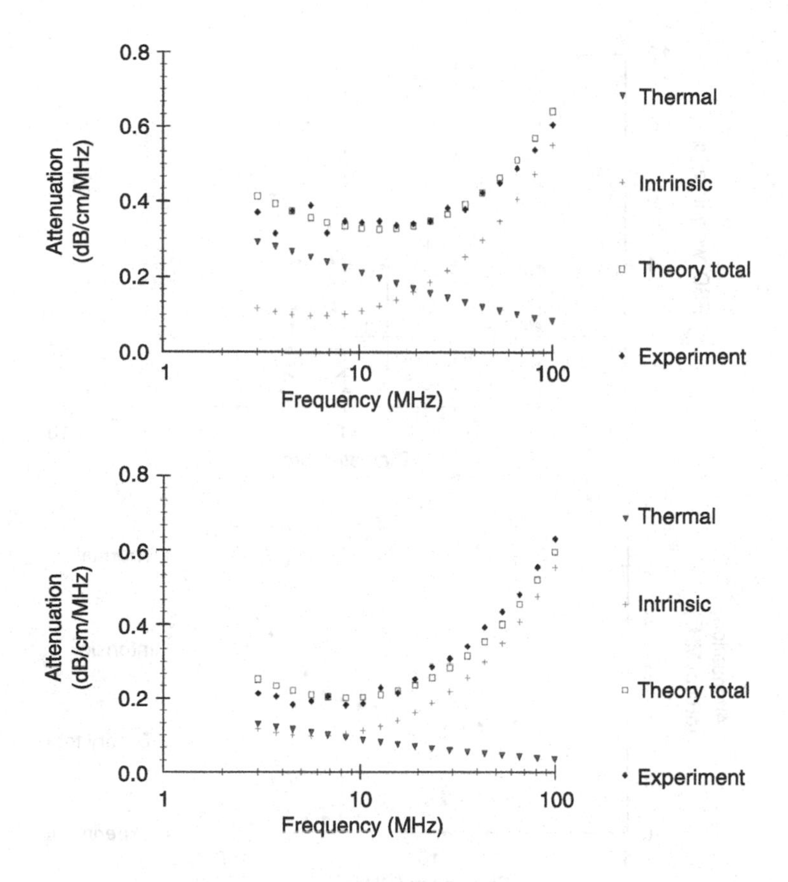

FIGURE 9.13 Experimental and theoretical attenuations for whole milk and homogenized milk.

Dependence on the droplet size would complicate characterization of the fat content. In order to minimize this factor we should use attenuation at the frequency range where intrinsic effect dominates. Figures 9.13 and 9.14 indicate that it is high frequency range, above at least 40 MHz.

Figure 9.15 shows attenuation at 44 MHz plotted versus fat weight fraction. We use data from Figure 9.7 for generating this graph. It is seen that there is practically linear correlation between attenuation and fat content. This might be used as a simple and direct way of determining this parameter.

We might expect a high reproducibility of this procedure because attenuation spectra measured for different brands of the same dairy product are practically the same. Figure 9.16 shows this for two different half–half milks.

9.4.3 Microemulsion

The mixture of heptane with water and AOT is a classic three components system. It has been widely studied due to a number of interesting features it exhibits. This system forms stable reverse microemulsions (water-in-oil) without the complication introduced by additional cosurfactant. Such a cosurfactant (usually alcohol) is required by many other reverse microemulsion systems. This simplification makes the alkane/water/AOT system a model for studying reverse microemulsions.

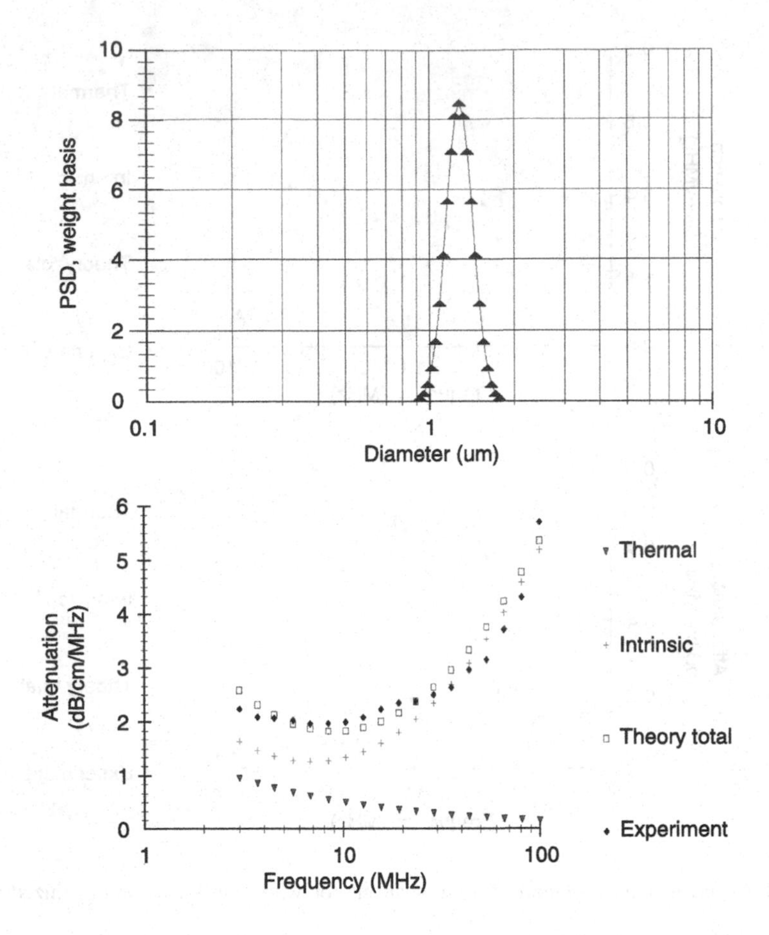

FIGURE 9.14 Droplet size distribution of the water droplets in light butter and corresponding attenuation spectra.

There have been many studies devoted to characterization of these practically important systems. Reverse emulsion droplets have been used as chemical micro reactors to produce nano size inorganic and polymer particles with special properties that are not found in the bulk form. These microemulsion systems have also been a topic of research for biological systems and the AOT head groups have been found to influence the conformation of proteins and increase enzyme activity. The unique environment created in the small water pools of swollen reverse micelles allows for increased chemical reactivity. The increase in surface area with decrease in size of the droplets also can significantly increase reactivity by allowing greater contact of immiscible reactants.

There have been many attempts to measure the droplet size of this microemulsion. Several different techniques were used: PCS [28–33], classic light scattering [30,31,34], SANS [35–37], SAXS [29,38,39], ultra-centrifugation [32,34,40], and viscosity [29,32,34]. It was observed that the heptane/water/AOT microemulsions have water pools with diameters ranging from 2 to 30 nm. The water drops are encapsulated by the AOT surfactant so that virtually all of the AOT is located at the interface shell. The size of the water droplets can be conveniently altered by

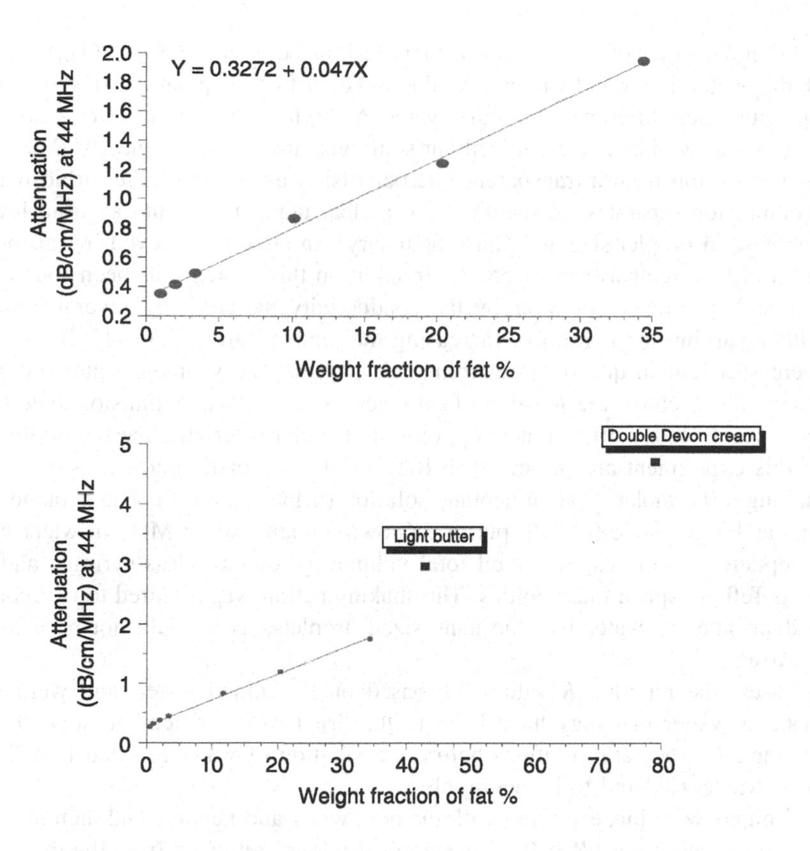

FIGURE 9.15 Attenuation of the oil-in-water dairy products at the single frequency of 44 MHz versus fat content.

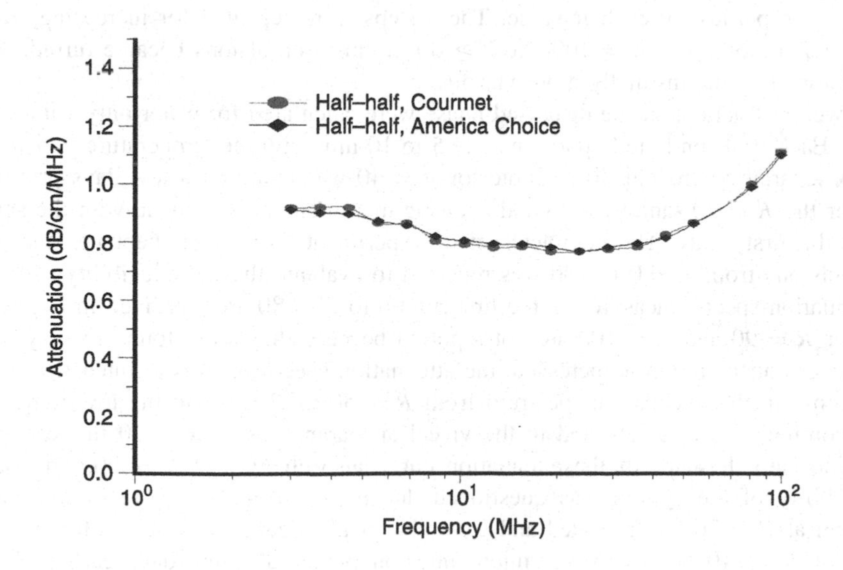

FIGURE 9.16 Attenuation of the half–half milk from different brands.

adjusting the molar ratios of water to surfactant, designated as R ([H_2O]/[AOT]). At low R values ($R \leq 10$) the water is strongly bound to the AOT surfactant polar head groups and exhibits unique characteristics different from bulk water. At higher water ratios ($R > 20$) free water is predominant in the swollen reverse micellular solutions, and at approximately $R = 60$, the system undergoes a transition from a transparent microemulsion into an unstable turbid macroemulsion. This macroemulsion separates on standing into a clear upper phase and a turbid lower phase.

The increase in droplet size and phase boundary can also be achieved by raising the temperature up to a critical temperature of 55 °C. In addition this system has been found to exhibit an electrical percolation threshold whereby the conductivity increases by several orders of magnitude by either varying the R ratio or increasing the temperature [37,38,41,42]. Despite all these efforts, there still remain questions regarding the polydispersity of the water droplets, and few studies are available above the R value of 60 where a turbid macroemulsion state exists.

Acoustic spectroscopy offers a new opportunity for characterizing these complicated systems. Details of this experiment are presented in Ref. 43. The reverse microemulsions were prepared by first making a 0.1 molar AOT in heptane solution (6.1% wt. AOT). The heptane was obtained from Sigma as HPLC grade (> 99% purity). Known amounts of 18 MΩ cm water were added to the AOT–heptane solution using a 1 ml total volume, graduated glass syringe, and then shaken for 30 sec in Teflon capped glass bottles. The shaking action was required to overcome an energy barrier to distribute the water into the nano-sized droplets, as it could not be achieved using a magnetic stirrer.

In all cases, the reported R values are based on the added water, and were not corrected for any residual water that may have been in the dried AOT or heptane solvent. Karl Fischer analysis of the AOT–heptane solutions before the addition of water resulted in a R value of 0.4. This amount was considered to be negligible.

Measurements were made starting with the pure water and heptane and then the AOT–heptane sample with no added water ($R = 0$). The sample fluid was removed from the instrument cell and placed in a glass bottle with a Teflon cap. Additional water was titrated and the microemulsion was shaken for 30 sec before being placed back into the instrument cell. The sample cell contained a cover to prevent evaporation of the solvents. The samples were visually inspected for clarity and rheological properties for each R value. These steps were repeated for increasing water weight fraction or R ratios up to $R = 100$. At $R \geq 60$ the microemulsions became turbid. At $R > 80$, the emulsions became distinctly more viscous.

The weight fractions of the dispersed phase were calculated for water only without including the AOT. Each trial run lasted approximately 5 to 10 min with the temperature varied from 25 to 27 °C. A separate microemulsion sample for $R = 40$ was made up a few days prior to the first study. For the $R = 70$ sample, a second acoustic measurement was made with the same sample used for the first study. The complete set of experiments for water, heptane, and the reverse microemulsions from $R = 0$ to 100 was repeated to evaluate the reproducibility.

Attenuation spectra measured in the first run up to $R = 80$ are presented in Figure 9.17. The results for $R = 90$ and $R = 100$ are not reported because they were found to vary appreciably. As the water concentration is increased, the attenuation spectrum rises in intensity and there is a distinct jump in the attenuation spectrum from $R = 50$ to $R = 60$ in the low frequency range. This discontinuity is also reflected in the visual appearance as at $R = 60$ the system becomes turbid. The smooth shape of the attenuation curve also changes at $R > 60$. The stability and reproducibility of the system was questioned due to the irregular nature of the curve so the experiment at $R = 70$ was repeated and gave almost identical results. An additional experiment was run at $R = 40$ for a separate microemulsion prepared a few days earlier. This showed excellent agreement with the results for freshly titrated microemulsion.

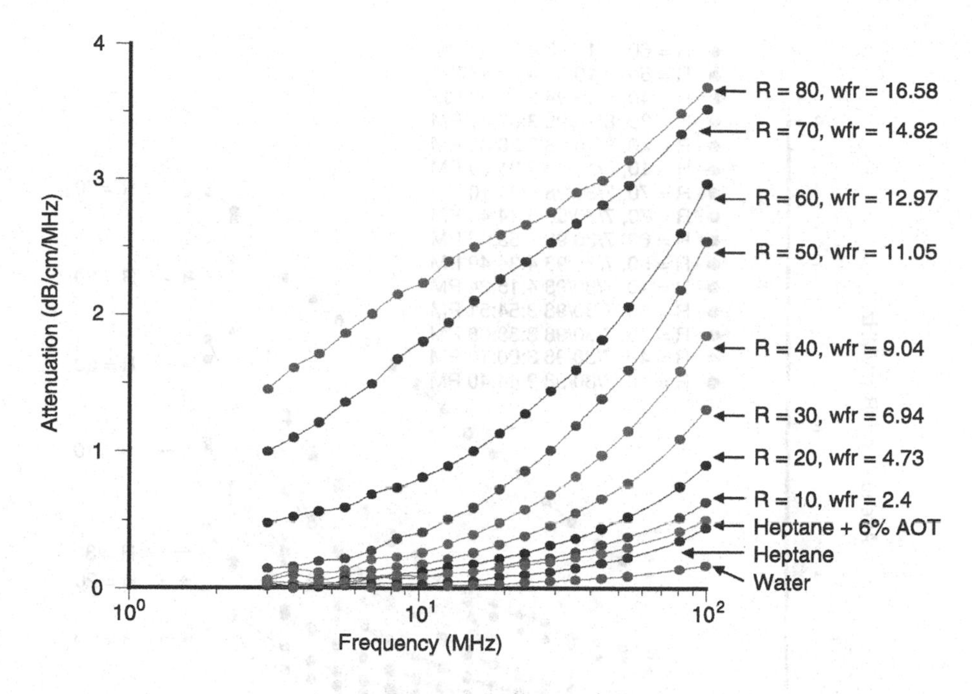

FIGURE 9.17 Acoustic attenuation spectra measured for water/AOT/heptane system for different water to AOT ratios *R*.

For *R* values > 70, an increase in the viscosity and a decrease in the reproducibility of the attenuation measurement were observed. This could be due to the failure of the model for this system as a collection of separate droplets at high R values.

A second set of experiments was run to check the reproducibility. The results of both sets of experiments up to $R = 60$ are given in Figure 9.18. It can be seen that the error related to the reproducibility is much smaller than the difference between attenuation spectra for the different *R* values. This demonstrates that the variation of attenuation reflects changes in the sample properties of water weight fraction and droplet size. The sound attenuation at *R* values above 60 were not as reproducible, but did give the same form of a bimodal distribution as the best fit for the experimental data.

The two lowest attenuation curves correspond to the attenuation in the two pure liquids: water and heptane. This attenuation is associated with oscillation of liquid molecules in the sound field. If these two liquids are soluble in each other, the total attenuation of the mixture would lie between these two lowest attenuation curves. But it can be seen that the attenuation of the mixture is much higher than that of the pure liquids. The increase in attenuation, therefore, is due to this heterogeneity of the water in the heptane system. The extra attenuation is caused by motion of droplets, not separate molecules. The scale factor (size of droplets) corresponding to this attenuation is much higher than that for pure liquids (size of molecules).

The current system contains a third component – AOT. A question arises on the contribution of AOT to the measured attenuation. In order to answer this question, measurements were done on a mixture of 6.1% wt AOT in heptane ($R = 0$). It is the third smallest attenuation curve on Figure 9.17. It is seen that attenuation increases somewhat due to AOT. However, this increase is less than the extra attenuation produced by water droplets. The small increase in attenuation

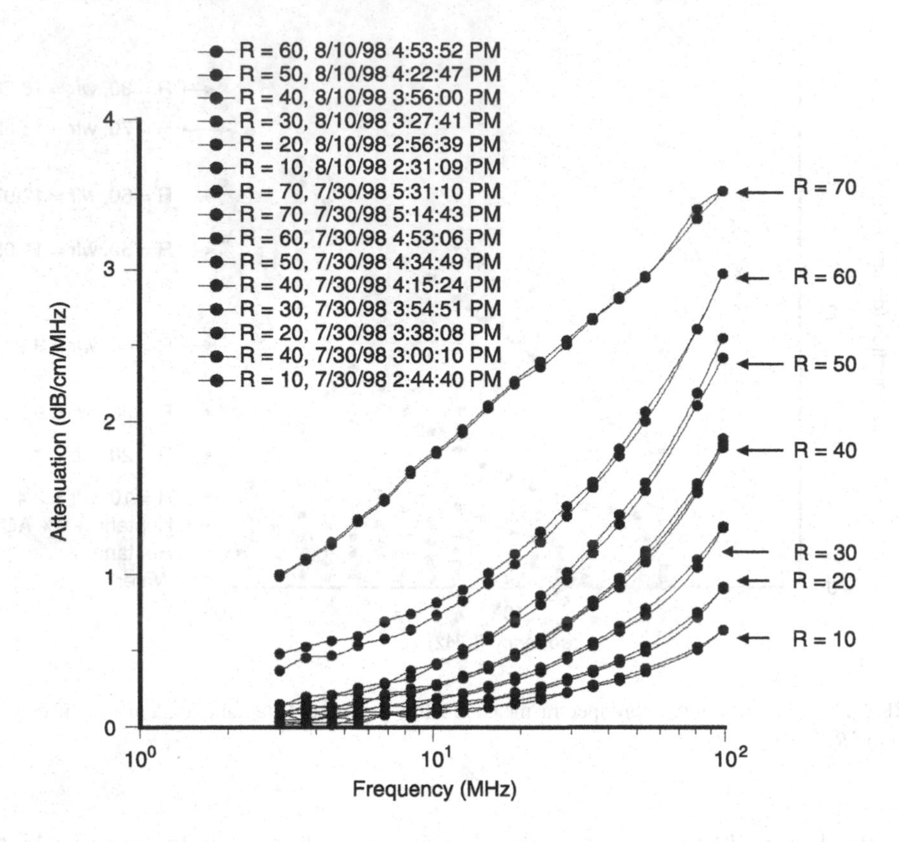

FIGURE 9.18 Reproducibility test of the attenuation measurement.

is attributed to AOT micelles. Unfortunately thermal properties of the AOT as a liquid phase are not known and the size of these micelles could not be calculated.

The droplet size distributions corresponding to the measured attenuation spectra are presented in Figure 9.19. It can be seen that the distribution becomes bimodal for $R \geq 60$ that coincides with the onset of turbidity. It is to be noted that such a conclusion could not easily be arrived at with other techniques. However, Figure 9.19 illustrates a peculiarity of this system that can be compared with independent data from literature: mean particle size increases with R almost in a linear fashion. This dependence becomes apparent when mean size is plotted as a function of R as in Figure 9.20.

It is seen that mean particle size measured using acoustic spectroscopy is in good agreement with those obtained independently using the neutron scattering (SANS) and X-ray scattering (SAXS) techniques [29,35] for R values ranging from 20 to 60. A simple theory based on equi-partition of water and surfactant can reasonably explain the observed linear dependence.

At $R = 10$ the acoustic method gave a slightly larger diameter than expected. This could be due to the constrained state of the "bound water" in the swollen reverse micelles. The water under these conditions may exhibit different thermal properties than the bulk water used in the particle size calculations. Also at the low R values ($R \leq 10$ or $\leq 2.4\%$ water), the attenuation spectrum is not very large as compared to the background heptane signal. Contribution of droplets to attenuation spectrum then may become too low to be reliably distinguished from the background signal coming from heptane molecules and AOT micelles.

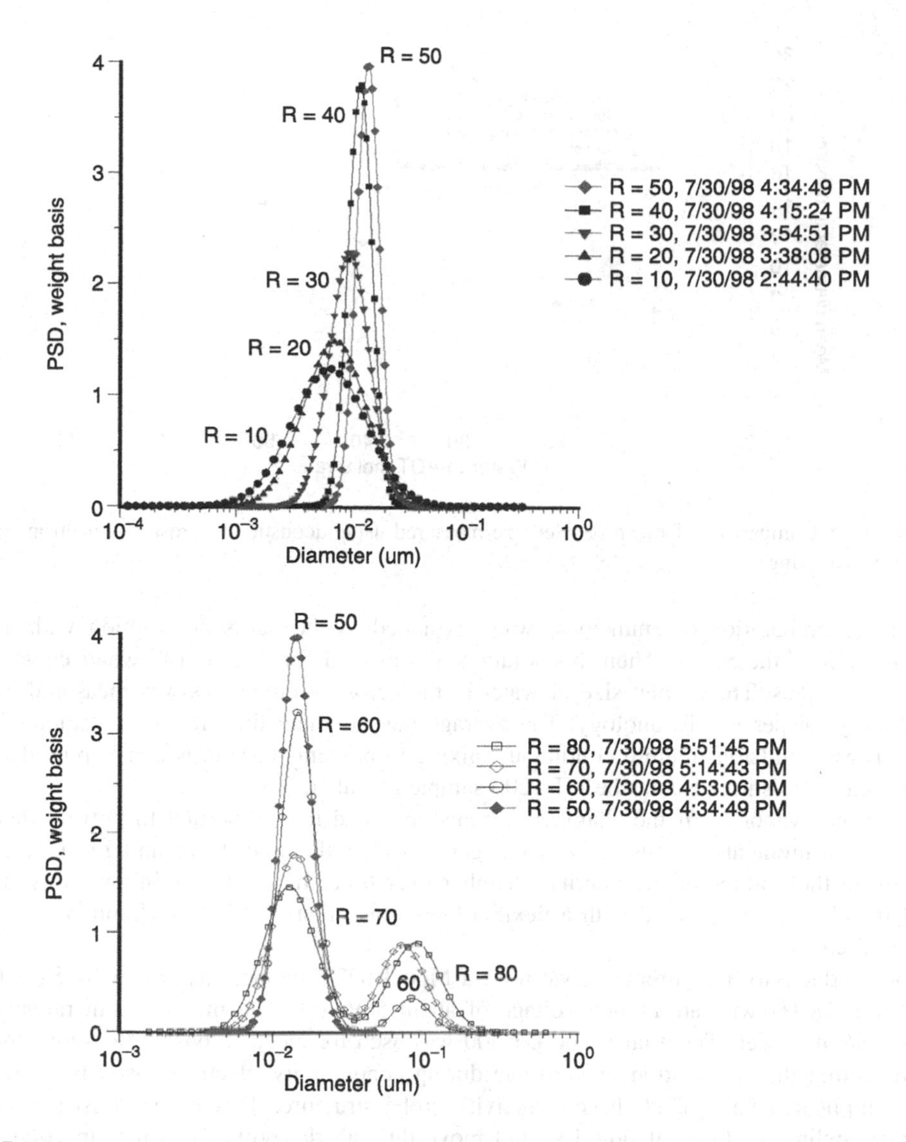

FIGURE 9.19 Drop size distribution for varying *R* [H_2O]/[AOT] from 10 to 50 and from 50 to 80.

9.4.4 Evolution of Water-in-Oil Emulsion Controlled by Droplet–Bulk Ion Exchange

This section describes interesting observations made while working with water-in-kerosene emulsion stabilized with nonionic surfactant. This emulsion exhibits time evolution of the acoustic and electric properties. These results have been published in *Colloids and Surfaces* [44].

We used the surfactant sorbitan mono-oleate, known also as SPAN 80 (Fluka), at surfactant contents of 0.5%, 1%, and 5% by weight relative to kerosene. The system with 1% is the most convenient for illustrating the observed effects. We used kerosene of different origins available at hardware stores. The origin of the kerosene had practically no influence on the results of measurements. We used a value of 2 for the dielectric constant of kerosene because it is not clear how impurities would affect this number. We used distilled water and water with 0.01 M KCl.

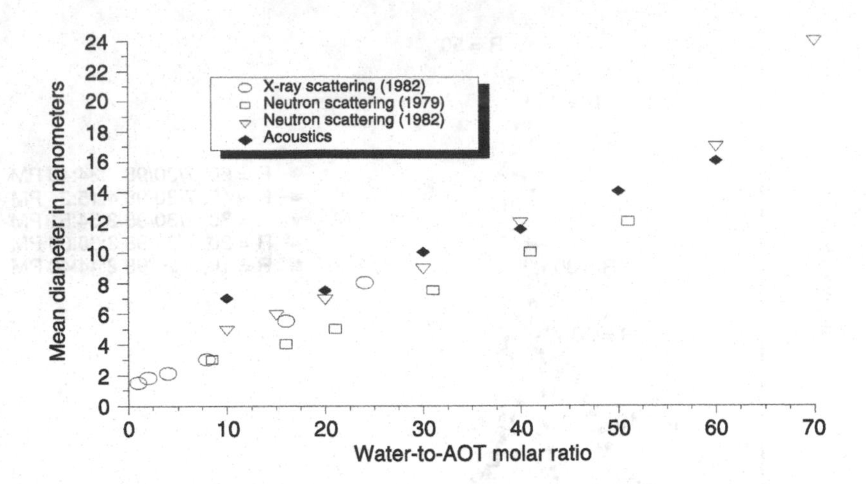

FIGURE 9.20 Comparison of mean droplet size measured using acoustic spectroscopy, neutron scattering, and X-ray scattering.

For the preparation of emulsions, water is added to the kerosene solution with a certain concentration of the SPAN. Then this solution is sonicated for 2 min. The water content is 5% vol in all samples. The droplet size of water in the kerosene emulsions was measured using the DT-1200 by Dispersion Technology. The average measurement time for these samples is about 6 min. These emulsions require continuous mixing to prevent settling, which is provided by the built-in magnetic stirrer bar in the DT-1200 sample chamber.

Kerosene evaporates if the chamber remains open and it is important to prevent this during acoustic measurements because a variable gap between the transducer and receiver causes a variation in the volume of the internal chamber over time. In order to minimize this effect the top of the chamber is covered with a flexible latex glove that could expand and contract during the measurement.

For conductivity measurements we used a Model 627 Conductivity Meter by Scientifica. It operates at 18 Hz with an applied voltage of about 5 V rms. The measurement range is 20 to 20,000 pS/cm. A Zeta Potential Probe DT-300 was used for electroacoustic measurement.

Preventing the evaporation of kerosene during conductivity–electroacoustic measurement is more complicated because of the conductivity probe structure. This probe consists of the two co-centric cylinders. Flow of liquid should move through the probe, which is inserted into the DT-1200 chamber from the top. A connecting cable causes a problem when we attempt to create an airtight seal. In addition, kerosene vapor affects the properties of the sealant. As a result we can only keep the chamber sealed for a restricted amount of time, which is shorter than in the case of acoustic measurement.

We tested reproducibility of the observed effect by repeating the same experiment several times. Altogether we made 4313 measurements from December 2003 through April 2004. We present some of these repeated runs below.

The attenuation frequency spectrum is the raw data for calculating the droplet size distribution. Figure 9.21 shows the attenuation curve as it evolves in time. Figure 9.22 illustrates just trends using a simpler two-dimensional presentation. It indicates that low frequency attenuation begins to decay as soon as sonication stops. The high frequency attenuation begins to increase with a 10-h delay. This increase is the most striking feature of the process. It is quite reproducible, as shown in Figure 9.23, and occurs at different water and surfactant contents.

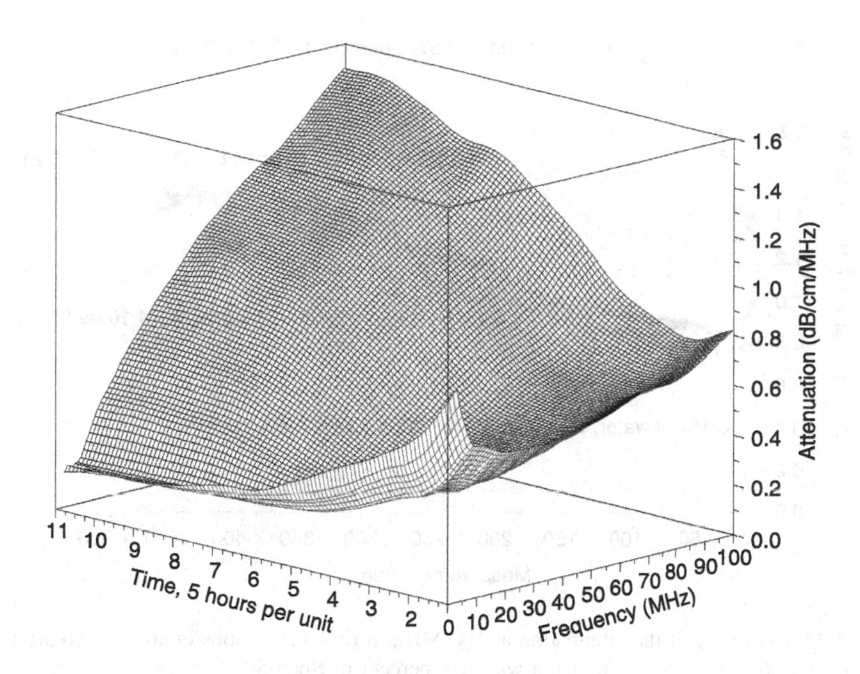

FIGURE 9.21 Attenuation frequency spectra evolution in time.

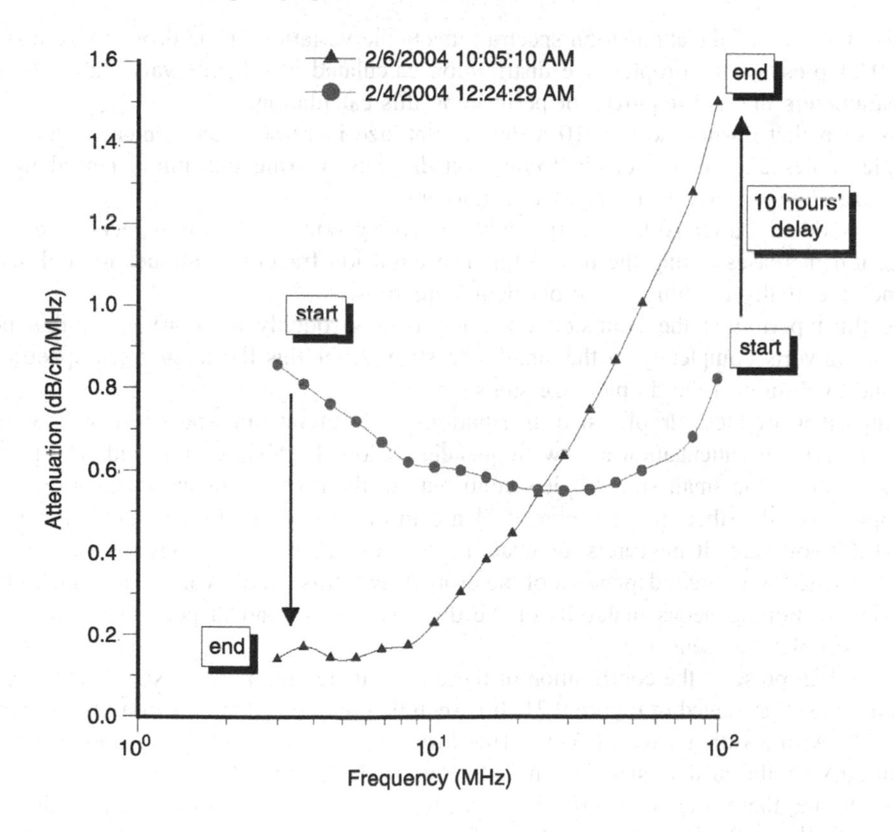

FIGURE 9.22 Attenuation frequency spectra at the beginning and at the end of experiment. Arrows show trends in attenuation evolution.

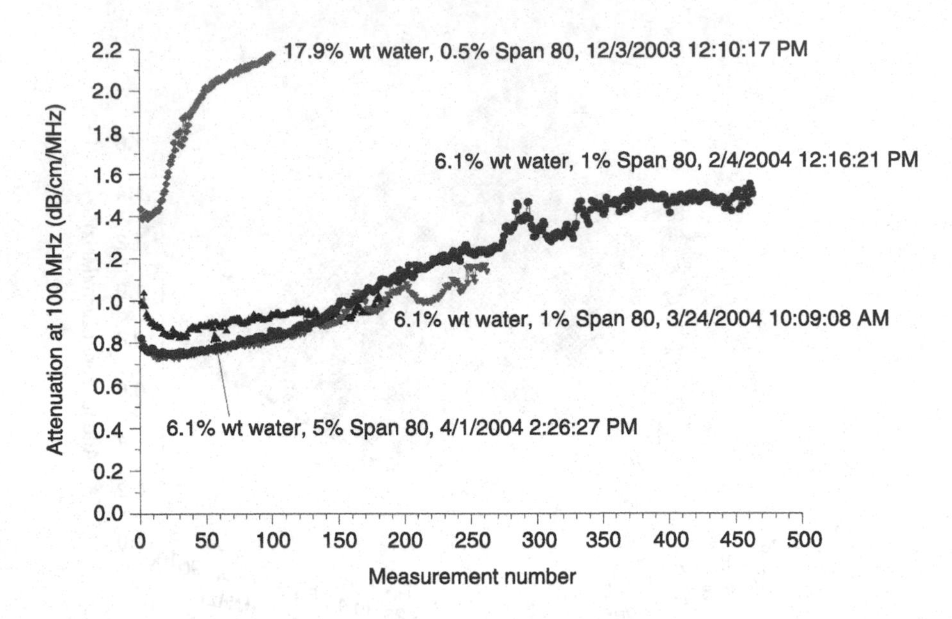

FIGURE 9.23 Evolution of the attenuation at 100 MHz in time for various emulsions. Reproducibility test with 6.1% wt water-in-kerosene emulsion with one percent of Span 80.

This evolution of the attenuation spectra reflects the variation of the droplet size distribution. Figure 9.24 presents the droplet size distribution calculated at 5-h intervals. Table 9.2 presents input parameters that are required for performing this calculation.

It is seen that during the first 10 h the droplet size increases monotonically, corresponding to simple coalescence of the original emulsion droplets. During this initial period the median droplet size changes roughly from 0.4 to 2 microns.

Then, suddenly, after 10 h a fraction with an average size of 25 nm appears. The content of this fraction increases during the next 30 h. The emulsion fraction continues to coalesce during this time, eventually reaching a size of about 7 microns.

The third period in the emulsion evolution begins roughly after 40 h. At this point the emulsion converts completely to the small size state. After this the attenuation spectra become stable and evolution of the droplet size stops.

Comparison of these droplet size distributions with attenuation spectra reveals a very simple correlation. The attenuation at low frequencies below 10 MHz corresponds to the emulsion droplets, whereas the small size droplets contribute to the high frequency attenuation.

Droplet size distributions in Figure 9.24 are the result of automatic calculation performed by DT-1200 software. It interprets attenuation as a combination of intrinsic losses and thermal losses. Intrinsic losses are independent of the droplet size. It is simple volume averaged ultrasound attenuation in homogeneous materials of the dispersed phase and dispersion medium. Thermal losses are droplet size sensitive.

Figure 9.25 presents the contribution of these two attenuation mechanisms to the last attenuation from the set presented in Figure 9.21. It is seen that theoretical attenuation fits the experiment pretty well, with a fitting error of 5.6%. This fitting corresponds to the log–normal droplet size distribution with the median size 17 nm and standard deviation 0.22.

In principle, there is another possibility for high frequency attenuation interpretation. It might be related to the scattering losses, instead of the thermal losses. In this case calculation would

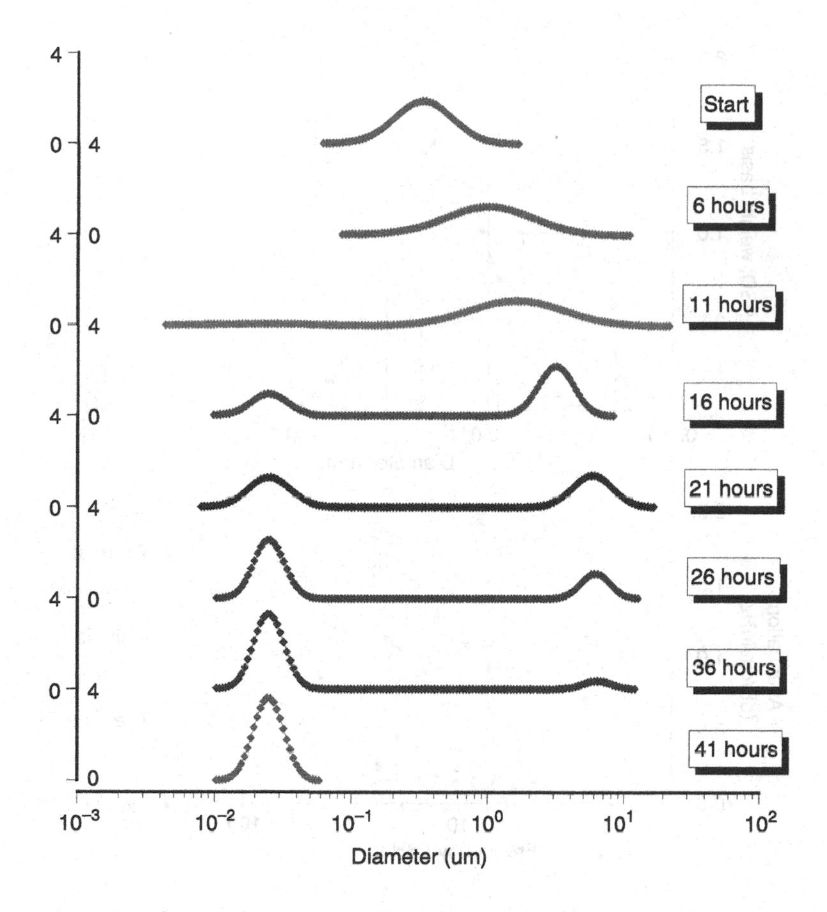

FIGURE 9.24 Evolution of the droplet size distribution in time.

TABLE 9.2
Thermodynamic Properties of the Kerosene and Water Phases Around Room Temperature

	Density (g/cm^3)	Thermal conductivity (J/m s K)	Heat capacity (J/kg K)	Thermal expansion (10^{-4} K^{-1})	Attenuation (dB/cm/MHz) – frequency (MHz)
Water	0.997	0.61	4.18	2.07	$\alpha = 0.002\omega$
Kerosene	0.81	0.15	2.85	10.6	$\alpha = 0.0106\omega$

yield a larger droplet size with increasing high frequency attenuation. It is important to clarify why DT-1200 software ignored this option and selected thermal effect over scattering. We could answer this question by forcing a searching routine in the DT-1200 software into the particular range of large sizes above 5 microns.

The best solution in this range and corresponding theoretical fitting are shown in Figure 9.26. The best median droplet size is 11.7 microns. This droplet size distribution yields theoretical attenuation that fits experimental data with error of 21.6%. This error is much larger than in the case of the thermal attenuation assumption (5.6%). This is the reason why DT-1200 software

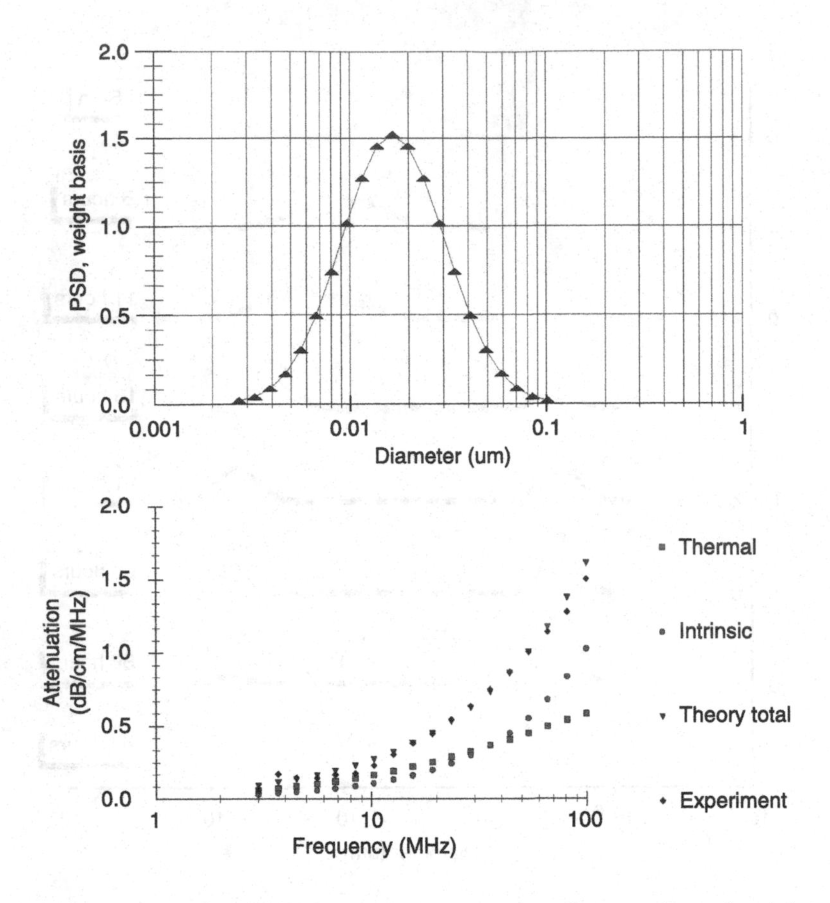

FIGURE 9.25 Droplet size of the final mini-emulsion and corresponding experimental and theoretical attenuation spectra. Droplet size does not reflect small amount of large droplets because attenuation spectra does not contain enough information for its determination.

selected small droplet size with a corresponding thermal mechanism over larger droplet sizes with a scattering mechanism.

At the same time, a combination of intrinsic and thermal attenuation for only small droplets does not provide a perfect fit. It is our experience that a fitting error of 5.6% is still quite large. We think that it happens because of the presence of a small fraction of large droplets. This could explain the leveling of the attenuation curve at low frequency as shown in Figure 9.3 for the curve marked "end." Instead of becoming practically 0 at low frequency it becomes stable at the level of about 0.1 dB/cm/MHz when the frequency is below 10 MHz. It is a clear indication of the larger emulsion droplets. Unfortunately, this attenuation is not sufficient to determine the size and amount of these droplets. That is why DT-1200 software has simply ignored them. The inability to perfectly fit this part of the attenuation spectra determines the relatively large fitting error of 5.6%.

We would like to stress the importance of being able to collect attenuation data over a wide frequency range. The absence of frequency data below 20 MHz would simply miss the variation of the emulsion fraction. Similarly, the absence of high frequency data would miss the variation due to the small size fraction.

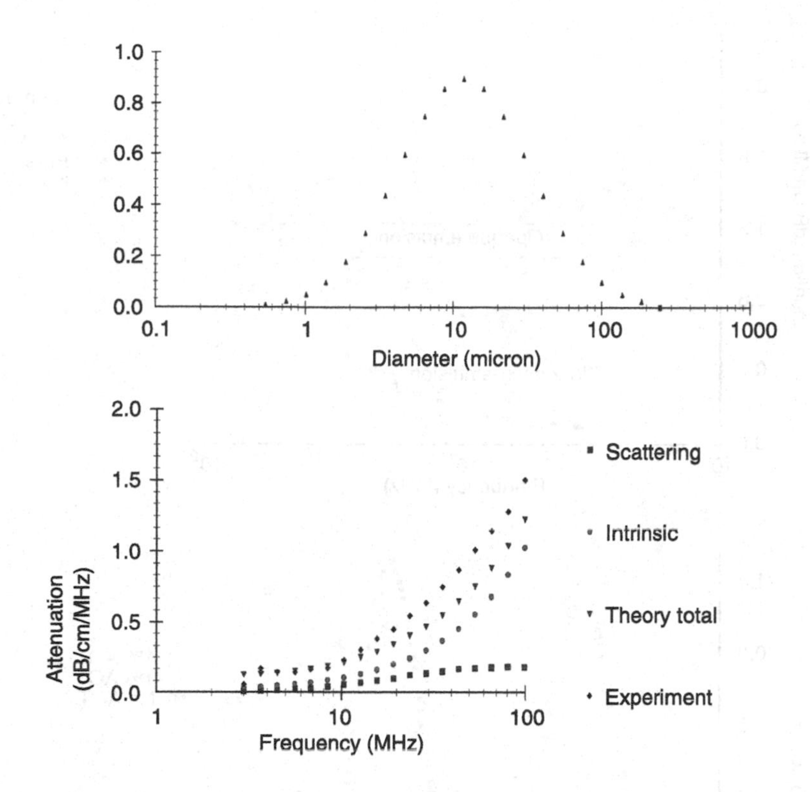

FIGURE 9.26 The best droplet size distribution assuming scattering mechanism of ultrasound attenuation. Corresponding theoretical attenuation that fail to fit experimental data.

At this point in thc discussion we do not know the nature of the small droplets: they might be either microemulsion or mini-cmulsion droplets. Sedimentation analysis and image analysis are used to answer this question. In addition, the reason for the strange kinetic behavior is not known. Next we will consider electroacoustic measurements to help obtain these answers, because it yields information about the surface properties of the water droplets.

The evolution of the emulsion described above occurs under conditions of continuous stirring. We use a magnetic stirrer to prevent sedimentation of the original emulsion. If we turn the stirrer off after the system reaches a steady state, the water droplets sediment. If the system was a thermodynamically microemulsion, we would observe no sedimentation effect. Hence sedimentation points toward a thermodynamically unstable mini-emulsion. This mini-emulsion turns into a sedimenting emulsion when the stirrer is turned off. At the same time there is no phase separation. If we turn the stirrer on again, the system exhibits the same acoustic properties as before sedimentation occurred with the same small droplet size. These observations indicate that we are dealing with an unstable mini-emulsion, which undergoes coalescence after stirring is turned off.

We performed two tests to confirm this conclusion. First we compared acoustic properties of this unstable system with acoustic properties of another water-in-oil emulsion, which becomes definitely microemulsion after adding surfactant. It is water-in-car oil with AOT as stabilizer. Water in pure car oil builds up an emulsion when sonicated, even without any additional surfactant. We prepared such an emulsion at 5% vol by applying 1 min of sonication. The final

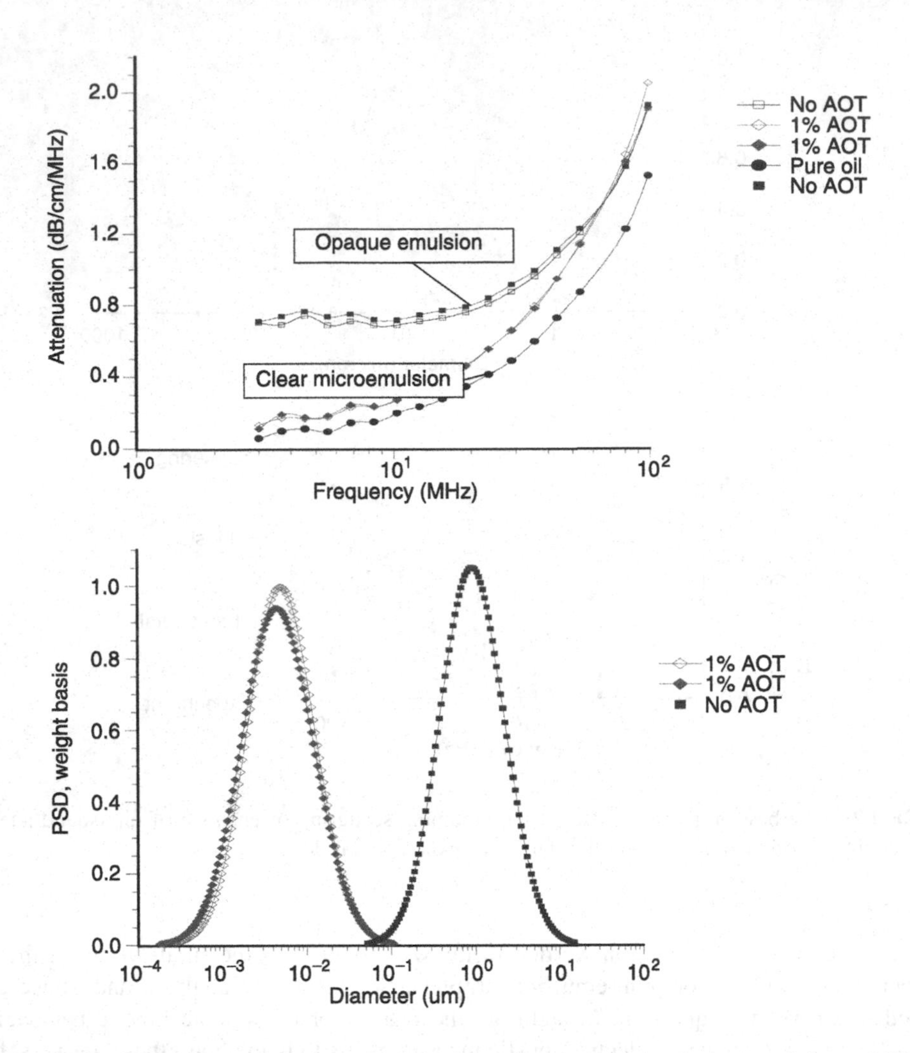

FIGURE 9.27 Attenuation spectra of 5% vol water-in-car-oil emulsion and microemulsion stabilized with 1% AOT. Corresponding droplet size distributions.

emulsion has a white color, which is specific for emulsions. It indicates very high turbidity, which is usually associated with droplet sizes of the order of microns. The attenuation spectra and corresponding droplet size for this emulsion are shown in Figure 9.27. The median size is about 2 microns, as expected.

At the second stage of this experiment we added 1% of AOT to this emulsion. After 1 min of sonication this liquid becomes transparent with a color practically identical to the color of the original car oil. This simple observation is definite proof that AOT converts emulsion into microemulsion. Figure 9.27 shows that acoustic properties have changed dramatically as well. Attenuation becomes much smaller and the median droplet size becomes about 10 nm. This experiment shows that there is a clear correlation between acoustics and optical properties of the stable emulsion–microemulsion transition.

We could now compare attenuation spectra of water-in-kerosene emulsion (Figure 9.22) with water-in-car-oil emulsion (Figure 9.27). There is clear similarity between these curves. We know that the low attenuation curve in the case of water-in-car-oil corresponds to the stable and optically transparent microemulsion. This could be used as justification for our earlier conclusion that lower attenuation at low frequency is indication of the smaller droplet sizes in the case of water-in-kerosene emulsion at the final stages of its evolution.

Unfortunately, the optical test in the case of the water-in-kerosene emulsion is not as definite as in the case of the water-in-car-oil emulsion. The water-in-kerosene emulsion remains opaque independently on variation of the acoustic properties. However, it is not sufficient to claim that there are no small mini- or microemulsion droplets in the system at the end of its evolution. A small amount of larger droplets could be responsible for the high turbidity.

In order to verify this hypothesis we performed a microscopic analysis of the final system after 40 h of stirring. We used a dark field microscope with a digital camera to capture images continuously for 1 h. Figure 9.28 shows five snapshots of the emulsion image at intervals of 10 min. It is clearly seen that the size of the emulsion droplets grows with time. However, the large droplets of micron size do not coalesce. The growth that we observe is apparently related to

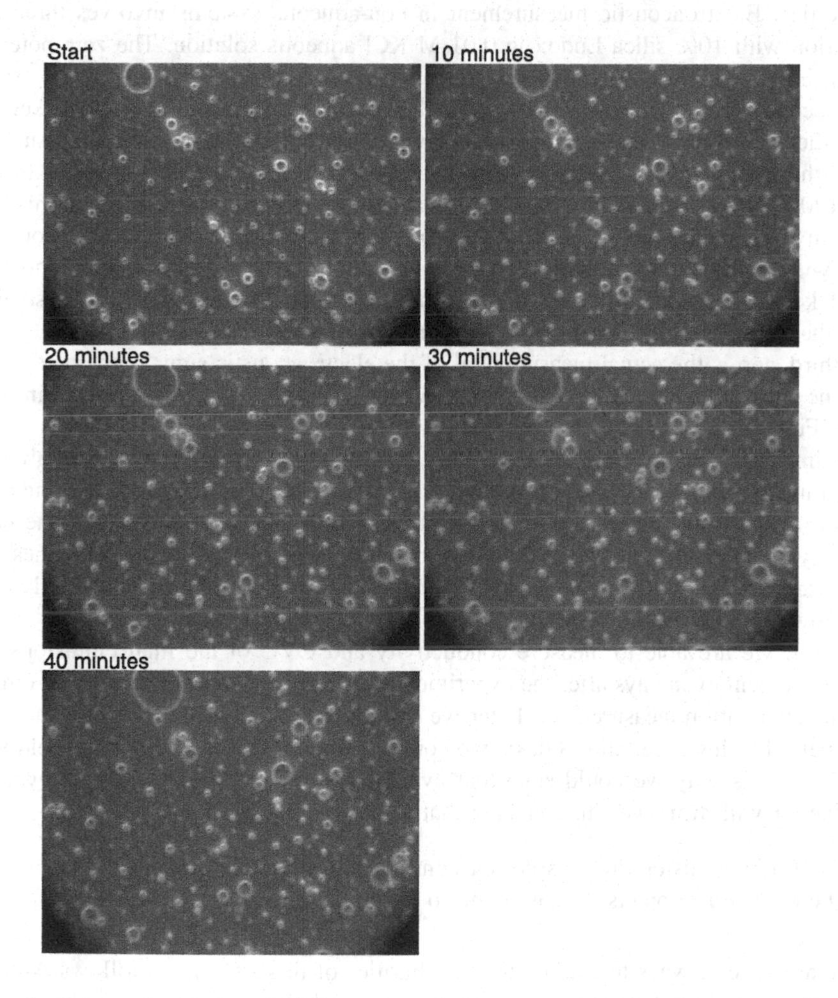

FIGURE 9.28 Emulsion evolution after stirrer is turned off. The width of each picture is approximately 100 microns.

the capturing of the very small droplets by the larger ones. These small droplets look like bright dots in the dark field. Microscopy does not allow estimate of their sizes, which are clearly less than a micron.

This microscopic test confirmed that the system contains a number of very small droplets, which slowly coalesce in larger droplets with a higher degree of aggregative stability. Stirring breaks quickly these large droplets into the smaller ones. These small mini-emulsion droplets dominate the acoustic attenuation, which is measured under stirring conditions.

At the same time it confirms presence of larger droplets with sizes on micron scale. This is the fraction that contributes 0.1 dB/cm/MHz to the low frequency attenuation in Figure 9.22, curve "end." Apparently volume fraction of these large droplets is not sufficient for being reflected in the droplet size distributions calculated from the attenuation spectra.

The presence of this small fraction of larger droplets could explain the high turbidity of the final mini-emulsion. Unfortunately, we could not estimate the amount of these larger droplets either from acoustics or from optical images. In order to detemine the cause of this interesting evolution we use electroacoustic and conductivity measurements for monitoring variation of electric properties. Electroacoustic measurement in non-aqueous systems involves three steps. First is calibration with 10% silica Ludox in 0.01 M KCl aqueous solution. The zeta potential of this dispersion is −38 mV, which is the basis for calibration.

The second step is the measurement of the electroacoustic signal of pure kerosene. This gives us the value of the background or ion vibration signal. The presence of surfactant does not affect this background signal, as shown with measuring kerosene that contains the surfactant. The DT-1200 software allows us to save this value and to subsequently subtract this background signal from further measurements. It is a vector subtraction because the electroacoustic signal is a vector with a certain magnitude and phase. As a simple test that this subtraction works, we measured kerosene again, this time making the background subtraction. This test gives us the value of the noise level of the electroacoustic measurement.

The third step is the actual measurement of the electroacoustic signal generated by the water-in-kerosene emulsion. This signal substantially exceeded the noise level, as shown in Figures 9.29 and 9.30. Figure 9.30 illustrates reproducibility of this effect.

With time, the CVC magnitude gradually decreases. Unfortunately, we are not able to measure CVC continuously during 40 h with the existing experimental setup. We mentioned before that the measuring chamber must be airtight for preventing kerosene evaporation. The cable of the conductivity probe makes this hard to achieve. Simple sealant that we use becomes affected by kerosene vapor. That is why conductivity and CVC measurement are conducted only during 20 h, as it is shown in Figures 9.9 and 9.10.

However, we are able to measure conductivity and CVC of the final emulsions as a single point measurement even days after the experiment finished. For instance, we stored emulsion after 40 h of the attenuation measurement. Later we could use this emulsion for CVC and conductivity measurement. For instance, after 8 days two consequent measurements of CVC yield values 1318 and 1973. That is why we could state that eventually CVC reaches the noise level. From this CVC behavior with time we can conclude that:

- The microemulsion droplets do not contribute to the CVC signal
- The emulsion droplets do contribute to the CVC signal

There are several ways to explain the peculiarities of this system, as follows. We can use the CVC values at the initial stage of coalescence for calculating the surface charge σ of the water droplets. According to Shilov's theory, this calculation requires certain input parameters, such as

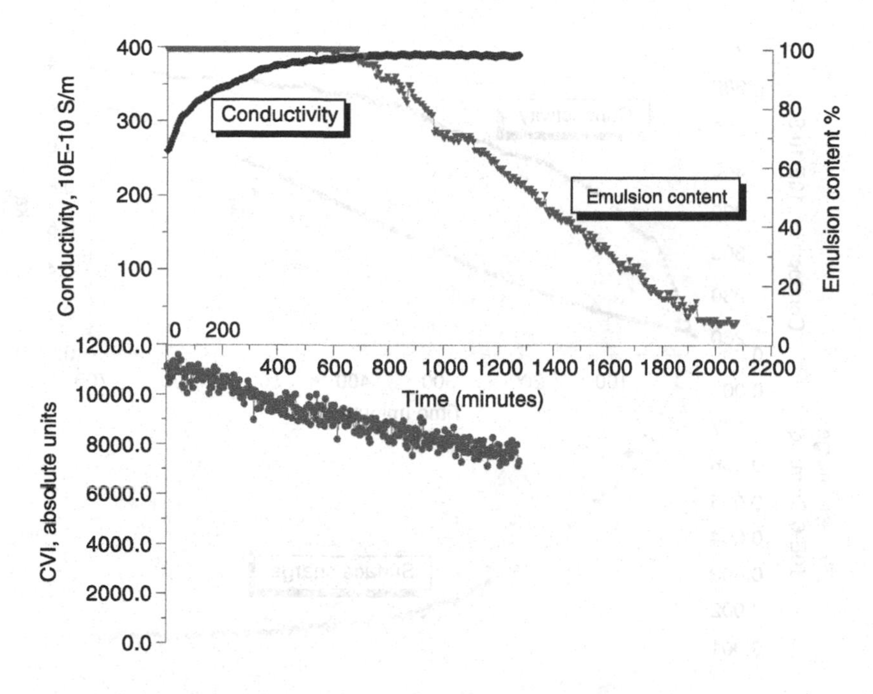

FIGURE 9.29 Correlation between emulsion droplet content and measured conductivity and electroacoustic signal.

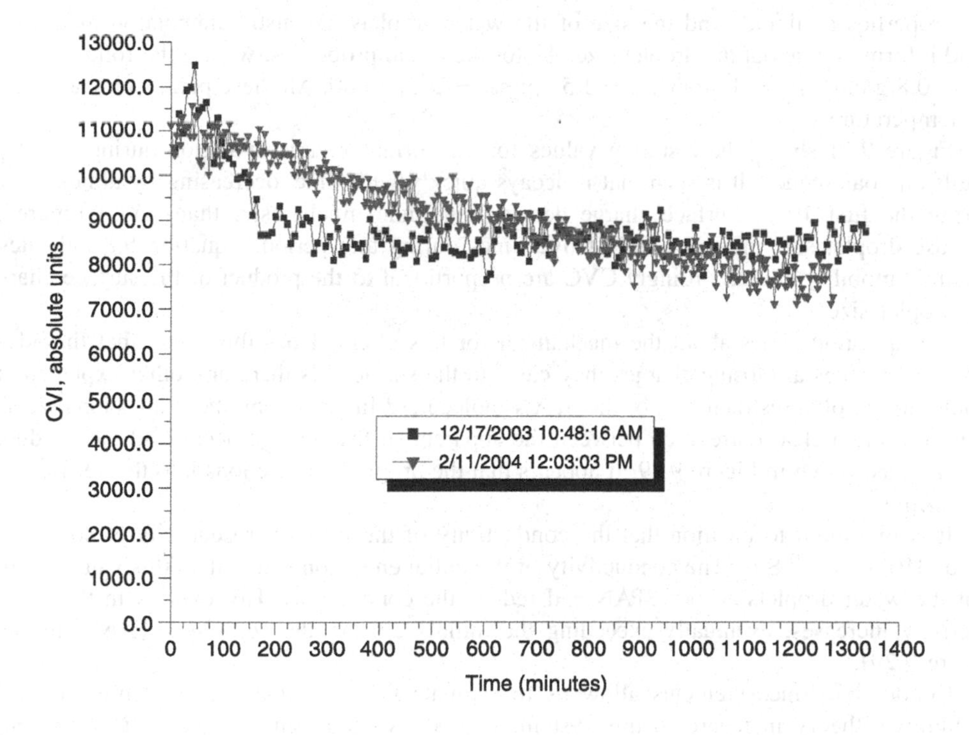

FIGURE 9.30 Illustration of reproducibility of the effect and electroacoustic measurement.

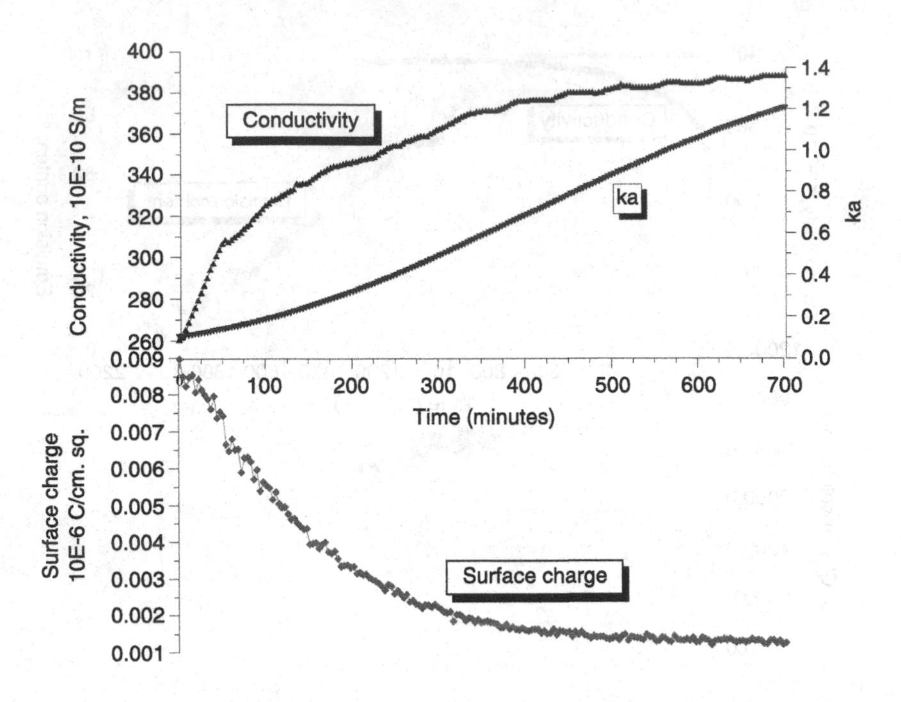

FIGURE 9.31 Correlation between measured conductivity and calculated apparent surface charge density.

the properties of liquids and the size of the water droplets. Acoustic attenuation measurements yield information about the droplet size. As for the liquid properties, we use the following values: $\rho_m = 0.8$ g/cm^3; $\rho_p = 1$ g/cm^3; $\eta = 1.5$ cp; $\varepsilon_m = 2$; $\varepsilon_p = 80$. All these parameters are corrected for temperature.

Figure 9.31 shows the resulting values for the surface charge evolution during the stage of emulsion coalescence. It is seen that it decays quickly with time, decreasing by almost 10 times during the first 10 h. Surface charge decays with time much faster than CVC (Figure 9.31) because droplet size increases almost five times during this period. Equation 9.8 indicates that dynamic mobility and accordingly CVC are proportional to the product of the surface charge by the droplet size.

The question arises about the mechanism for this effect. Does this mean that the adsorbed SPAN molecules are losing charges they carry to the surface? Is there any other explanation that would not involve restructuring of the SPAN molecules? In answering these questions we should keep in mind a clear correlation between the apparent surface charge decay and the conductivity increase, as shown in Figure 9.29. It appears that the droplets release ions into the kerosene while coalescing.

It is important to mention that the conductivity of the initial kerosene–SPAN 80 solution is about 310×10^{-10} S/m. The conductivity of the initial emulsion is less than this value. It appears that the water droplets adsorb SPAN and reduce the conductivity. However, with time the conductivity increases, eventually exceeding the initial value in the kerosene–SPAN solution (see Figure 9.29).

Conductivity measurements allow us to estimate the κa value and determine the validity of Shilov's theory in regard to this system. It is known that Shilov's theory is valid only for overlapped DLs. Figure 9.3 shows the relationship between the volume fraction and κa for overlapped DLs. A more detailed description is given in Ref. 6. It is seen that at a volume fraction of 5% the DLs do overlap if $\kappa a < 1$. This is the range of κa values when Shilov's theory can be applied.

Figure 9.31 presents κa values calculated using the measured conductivity and the particle size computed using attenuation spectroscopy. We also assumed that there are simple small ions in the kerosene phase with an approximate diffusion coefficient of 10^{-5} cm^2/sec. If this assumption is valid, κa satisfies the condition of overlapped DLs during the complete initial period of emulsion coalescence. This is a justification for using Shilov's theory for calculating the surface charge.

We have established so far that the evolution of the water-in-kerosene emulsion with SPAN 80 as emulsifier proceeds in three distinct periods.

Period 1: the first 10 h. The water droplets are coalescing. The apparent surface charge of the droplets decays. The conductivity increases, eventually exceeding the initial value in the kerosene–SPAN solution.

Period 2: between 10 and 40 h. The mini-emulsion fraction appears and grows. The emulsion droplets continue to coalesce. The CVC continues to decay. The rate of conductivity increase is much smaller.

Period 3: after 40 h. All parameters are stable. There are no emulsion droplets. The water is in mini- and possibly microemulsion droplets. These droplets do not generate any CVC signal.

It is quite possible that there are multiple explanations of this set of facts. Here we just suggest one of them. We do not claim that it is the only one. We want to be absolutely clear that it is just one of, perhaps, several possible models. Our model is based on the assumption that the water–kerosene interface with adsorbed SPAN remains unchanged. We assume that the SPAN molecules stabilized this interface almost instantly during initial sonication and brought it to thermodynamic equilibrium. However, it is only a local equilibrium for each small element of the interface.

The part of the SPAN molecules in the bulk of the solution is responsible for the conductivity of the original kerosene SPAN solution. It means that they carry charge, either as individual molecules, or as micelles. It is also possible that at least a fraction of the adsorbed SPAN molecules is charged and brings this charge to the surface. We assume that the electric charge of the water–kerosene interface that is associated with adsorbed SPAN molecules remains constant, time independent. Local thermodynamic equilibrium is the justification for this assumption.

We attribute the observed evolution of the parameters to ion exchange between the interior of the water droplets and the bulk of the kerosene solution. Initially the water droplets have an electro-neutral interior. Ions are trapped inside of the large water droplet at about equal amounts. The surface charge induced by adsorbed SPAN molecules is screened with an external diffuse layer in the kerosene solution. Figure 9.32 illustrates this structure.

The external diffuse layers are very thick due to the low ionic strength. They are overlapped. This gives rise to the CVC signal. We could estimate the amount of energy W_{ext} required to charge this external DL to a certain level of surface charge Q as:

$$W_{ext} = \frac{Q^2}{2C_{ext}} = \frac{Q^2}{\varepsilon_p \varepsilon_0 S \kappa_{ext}} \tag{9.13}$$

where S is the surface area, κ_{ext} is the Debye parameter for external DLs.

Obviously, there is another possibility to screen the surface charge induced by SPAN adsorption. It could be done with screening the diffuse layer inside of the particle. This second opportunity would also take a certain amount of energy, which we could estimate with the

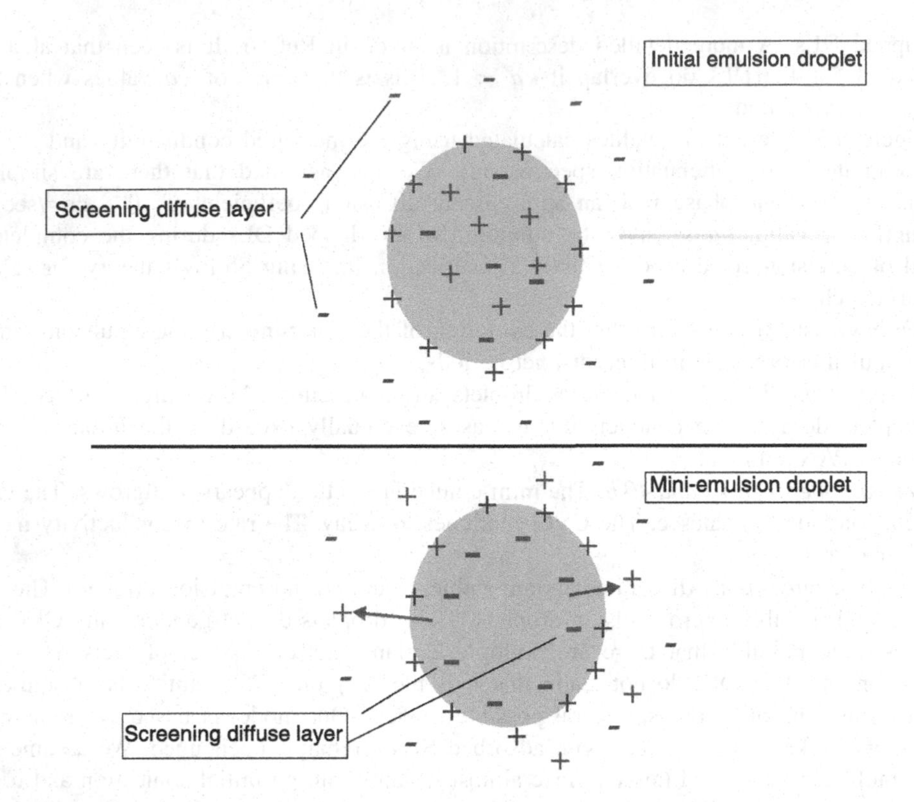

FIGURE 9.32 Illustration of the double layer restructuring due to the ion exchange between interior and exterior of the water droplet.

following equation:

$$W_{in} = \frac{Q^2}{2C_{in}} = \frac{Q^2}{\varepsilon_p \varepsilon_0 S \kappa_{in}} \tag{9.14}$$

Comparison of these two charging energies yields the following approximate result:

$$\frac{W_{in}}{W_{ext}} = \frac{\varepsilon_m \kappa_{ext}}{\varepsilon_p \kappa_{in}} = \sqrt{\frac{\varepsilon_m K_m}{\varepsilon_p K_m}} = \sqrt{\frac{K_m}{40 K_p}} \approx 0.001 \tag{9.15}$$

where we have assumed that the conductivity of water is about $10^6/40$ times larger than conductivity of kerosene, the effective diffusion coefficients D_{eff} being the same, and that the Debye parameter can be given with the following equation:

$$\kappa^2 \approx \frac{K}{\varepsilon_0 \varepsilon D_{eff}} \tag{9.16}$$

This is certainly a very approximate estimate. However, it clearly indicates that charging of the interior DL to a certain electric charge is orders of magnitude more energy efficient than the exterior one.

It is important to mention that charging up to a certain level of the electric potential, instead of the electric charge, would lead to the opposite conclusion. We believe that in this case we are dealing with the constant surface charge case because it is related to the adsorption of thermodynamically determined amount of the surfactant.

The above mentioned difference in energy for charging the DLs creates a driving force for co-ions to leave the interior of the droplet. This simply means that there is a gradient of co-ion electrochemical potential that drives co-ions from the droplet out. This statistical process is slow because the ion should become solvated by surfactant molecules on the kerosene–water interface. It is known [44] that only sufficiently large ions could exist in nonpolar liquids. This slow leakage of co-ions from the droplet generates an internal electric charge that would screen the adsorbed surface charge. Eventually this would lead to the complete collapsing of the external DL. The surface charge of the adsorbed SPAN would be completely compensated by the interior charge of the water droplet. Figure 9.12 illustrates this final stage of the transition.

This model could explain all observed features of the emulsion–mini-emulsion transition. First, it explains why the conductivity increases during the coalescence period and eventually significantly exceeds the conductivity level of the original kerosene–SPAN solution (see Figure 9.29). This excess conductivity comes from the counter-ions released from the water droplets into the kerosene.

An alternative model that could potentially explain some of the observed facts with restructuring of the adsorbed SPAN molecules would not explain the excess conductivity. This is a serious argument supporting the "ion exchange model."

The "ion exchange model" also explains the lack of the CVC signal for the microemulsion. It happens because the surface charge of the adsorbed SPAN molecules is completely screened with the internal charge of the microemulsion droplet. Consequently it looks from the outside as being electro-neutral.

In this model redistributing screening electric charge from the outside diffuse layer to the inside one is a limiting factor of the emulsion droplet break up into much smaller mini- and possibly microemulsion droplets. Apparently this break up occurs when a sufficient critical amount of energy is released due to the collapsing external DL.

Another potential factor controlling this break-up, droplet size, does not seem to be important. We could conclude this based on the observation that the size of the emulsion droplets continues to grow during period 2. If break-up was related to the droplet size, we would observe growth only up to a certain maximum value of the size. This is obviously not the case.

This break-up is not exactly spontaneous because it strongly depends on the presence of stirring. However, stirring with a magnetic stirrer bar in the DT-1200 chamber is not vigorous. It is clear that reduction of the surface tension is a most important factor. In this sense the observed phenomenon is similar to the "negative interfacial tension" mechanism of spontaneous emulsification. It is also possible that the combination of stirring and reduction of the surface tension intensifies "interfacial turbulence."

Finally, we would like to repeat that the suggested mechanism explains all the experimental facts, but we could not claim that it is the only one to do so. There might be some other explanations of the phenomenon presented in this paper.

REFERENCES

1. Pellam, J.R. and Galt, J.K. "Ultrasonic propagation in liquids: Application of pulse technique to velocity and absorption measurement at 15 megacycles", *J. Chem. Phys.* **14**(10), 608–613 (1946).

2. Sewell, C.T.J. "The extinction of sound in a viscous atmosphere by small obstacles of cylindrical and spherical form", *Phil. Trans. Roy. Soc., Lond.* **210**, 239–270 (1910).
3. Epstein, P.S. and Carhart, R.R. "The absorption of sound in suspensions and emulsions", *J. Acoust. Soc. Amer.* **25**(3), 553–565 (1953).
4. McClements, D.J. "Ultrasonic characterization of emulsions and suspensions", *Adv. Colloid Interface Sci.* **37**, 33–72 (1991).
5. "Ultrasonic and Dielectric Characterization Techniques for Suspended Particulates", V.A. Hackley and J. Texter (Eds), The American Chemical Society, Ohio (1998).
6. Dukhin, A.S. and Goetz, P.J. "Ultrasound for Characterizing Colloids. Particle Sizing, Zeta Potential, Rheology", New York, Elsevier (2002).
7. Lyklema, J. "Fundamentals of Interface and Colloid Science", Volume 1, Academic Press (1993).
8. Hunter, R.J. "Foundations of Colloid Science", Oxford University Press, Oxford (1989).
9. Hunter, R.J. "Review. Recent developments in the electroacoustic characterization of colloidal suspensions and emulsions", *Colloids and Surfaces* **141**, 37–65 (1998).
10. Strout, T.A. "Attenuation of sound in high-concentration suspensions: development and application of an oscillatory cell model", Thesis, University of Maine (1991).
11. Allegra, J.R. and Hawley, S.A. "Attenuation of sound in suspensions and emulsions: theory and experiments", *J. Acoust. Soc. Amer.* **51**, 1545–1564 (1972).
12. McClements, D.J. "Ultrasonic determination of depletion flocculation in oil-in-water emulsions containing a nonionic surfactant", *Colloids and Surfaces* **90**, 25–35 (1994).
13. McClements, D.J. "Comparison of multiple scattering theories with experimental measurements in emulsions", *J Acoust. Soc. Amer.* **91**(2), 849–854 (1992).
14. Holmes, A.K., Challis, R.E. and Wedlock, D.J. "A wide-bandwidth study of ultrasound velocity and attenuation in suspensions: Comparison of theory with experimental measurements", *J. Colloid Interface Sci.* **156**, 261–269 (1993).
15. Holmes, A.K., Challis, R.E. and Wedlock, D.J. "A wide-bandwidth ultrasonic study of suspensions: The variation of velocity and attenuation with particle size", *J. Colloid Interface Sci.* **168**, 339–348 (1994).
16. Dukhin, A.S., Goetz, P.J. and Hamlet, C.W. "Acoustic spectroscopy for concentrated polydisperse colloids with low density contrast", *Langmuir* **12**(21), 4998–5004 (1996).
17. Anson, L.W. and Chivers, R.C. "Thermal effects in the attenuation of ultrasound in dilute suspensions for low values of acoustic radius", *Ultrasonic* **28**, 16–25 (1990).
18. Dukhin, A.S. and Goetz, P.J. "Acoustic spectroscopy for concentrated polydisperse colloids with high density contrast", *Langmuir* **12**(21), 4987–4997 (1996).
19. Harker, A.H. and Temple, J.A.G. "Velocity and attenuation of ultrasound in suspensions of particles in fluids", *J. Phys. D., Appl. Phys.* **21**, 1576–1588 (1988).
20. Gibson, R.L. and Toksoz, M.N. "Viscous attenuation of acoustic waves in suspensions", *J. Acoust. Soc. Amer.* **85**, 1925–1934 (1989).
21. Riebel, U., et al. "The fundamentals of particle size analysis by means of ultrasonic spectrometry", *Part. Part. Syst. Charact.* **6**, 135–143 (1989).
22. Chanamai, R., Coupland, J.N. and McClements, D.J. "Effect of temperature on the ultrasonic properties of oil-in-water emulsions", *Colloids and Surfaces* **139**, 241–250 (1998).
23. Temkin, S. "Sound speed in suspensions in thermodynamic equilibrium", *Phys. Fluids* **4**(11), 2399–2409 (1992).
24. Temkin, S. "Sound propagation in dilute suspensions of rigid particles", *The Journal of the Acoustical Society of America.* **103**(2), 838–849 (February 1998).
25. O'Brien, R.W. "Electro-acoustic effects in a dilute suspension of spherical particles", *J. Fluid Mech.* **190**, 71–86 (1988).
26. O'Brien, R.W. "Determination of particle size and electric charge", US Patent 5,059,909 (22 October 1991).
27. Dukhin, A.S., Ohshima, H., Shilov, V.N. and Goetz, P.J. "Electroacoustics for concentrated dispersions", *Langmuir* **15**(10), 3445–3451 (1999).
28. Dukhin, A.S., Shilov, V.N, Ohshima, H. and Goetz, P.J. "Electroacoustics phenomena in concentrated dispersions. New theory and CVI experiment", *Langmuir* **15**(20), 6692–6706 (1999).

29. Dukhin, A.S., Shilov, V.N, Ohshima, H. and Goetz, P.J. "Electroacoustics phenomena in concentrated dispersions. Effect of the surface conductivity", *Langmuir* **16**, 2615–2620 (2000).
30. Shilov, V.N., Borkovskaja, Yu.B. and Dukhin, A.S. "Electroacoustic theory for concentrated colloids with arbitrary κa. Nano-colloids. Non-aqueous colloids", *J. Colloid Interface Sci.* **277**, 347–358 (2004).
31. Dukhin, A.S. and Goetz, P.J. "Characterization of concentrated dispersions with several dispersed phases by means of acoustic spectroscopy", *Langmuir* **16**(20), 7597–7604 (2000).
32. Dukhin, A.S., Goetz, P.J. and Truesdail, S.T. "Surfactant titration of kaolin slurries using ζ-potential probe", *Langmuir* **17**, 964–968 (2001).
33. Crupi, V., Maisano, G., Majolino, D., Ponterio, R., Villari, V. and Caponetti, E. *J. Mol. Struct.* **383**, 171 (1996).
34. Bedwell, B. and Gulari, E. In: "Solution Behavior of Surfactants" (K.L. Mittal, Ed.), Vol. 2., Plenum Press, New York (1982).
35. Gulari, E., Bedwell, B. and Alkhafaji, S. *J. Colloid Interface Sci.* **77**(1), 202 (1980).
36. Zulauf, M. and Eicke, H.-F., *J. Phys. Chem.* **83**(4), 480 (1979).
37. Eicke, H.-F. In: "Microemulsions" (I.D. Rob, Ed.), p. 10, Plenum Press, New York (1982).
38. Nicholson, J.D., Doherty, J.V. and Clarke, J.H.R. In: "Microemulsions" (I.D. Rob, Ed.), p. 33, Plenum Press, New York (1982).
39. Eicke, H.-F. and Rehak, J. *Helv. Chim. Acta* **59**(8), 2883 (1976).
40. Radiman, S., Fountain, L.E., Toprakcioglu, C., de Vallera, A. and Chieux, P. *Progr. Coll. Polym. Sci.* **81**, 54 (1990).
41. Fletcher, P.D.I., Robinson, B.H., Bermejo-Barrera, F., Oakenfull, D.G., Dore, J.C. and Steytler, D.C. In: "Microemulsions" (I.D. Rob, Ed.), p. 221, Plenum Press, New York (1982).
42. Cabos, P.C. and Delord, P. *J. App. Cryst.* **12**, 502 (1979).
43. Dukhin, A.S. and Goetz, P.J. "New developments in acoustic and electroacoustic spectroscopy for characterizing concentrated dispersions", *Colloids and Surfaces* **192**, 267–306 (2001).
44. Dukhin, A.S. and Goetz, P.J. "Evolution of water-in-oil emulsion controlled by droplet-bulk ion exchange: acoustic, electroacoustic, conductivity and image analysis", *Colloids and Surfaces* (2004).

10 Environmental Emulsions: A Practical Approach

Merete Ø. Moldestad, Frode Leirvik, Øistein Johansen, Per S. Daling, and Alun Lewis

CONTENTS

10.1 INTRODUCTION

When oils are spilt at sea different weathering processes alter the properties of the oil as a function of time and weather conditions. Evaporation of volatiles, water-in-oil emulsification (w/o emulsions), and natural dispersion (o/w emulsions) are important weathering processes taking place when oil is spilled at sea. The emulsification of water in the oil contributes to a change in properties of the spilled oil and increases the total volume of pollutant. The natural dispersion contributes to a removal of the oil from the sea surface into the water column. The behavior of spilled crude oils and refined oil products depends on the ambient conditions (e.g., temperature, sea-state, currents) and on the chemical composition of the oil. Large variations in oil properties cause them to behave differently when spilled at sea. The Gullfaks crude spilled at the *Braer* incident in the Shetlands has a low content of waxes and asphaltenes, which are important compounds for stabilizing water-in-oil (w/o) emulsions formed on the sea surface. This, combined with heavy sea state, resulted in consequences that were much less severe than might have been expected (1). Almost all of the 84,000 tons of the spilled Gullfaks crude oil was naturally dispersed into the water column as o/w-emulsions (2). In the *Amoco Cadiz* and the *Metula* spills, persistent "chocolate mousse" has contaminated the shorelines for years after the initial spill (3,4). Knowledge about weathering behavior of crude oils and fuel oils is therefore of importance for environmental risk assessment of a spill, for contingency planning, response analysis, net environmental benefit analysis (NEBA), and for rapid and right decision making in the case of an oil spill.

10.2 WEATHERING OF OIL ON THE SEA SURFACE

When a crude oil is spilt at sea a number of natural processes take place, which change the volume and the chemical properties of the oil. These natural processes are evaporation, water-in-oil (w/o) emulsification, oil-in-water (o/w) dispersion, dissolution of oil components into the water column, spreading, sedimentation, oxidation, and biodegradation. A common term for all of these natural processes is weathering. The relative contribution of each process varies during the duration of the spill. Figure 10.1 illustrates the various weathering processes taking place.

The behavior of spilled crude oils and refined oil products depends on:

- The physicochemical properties of the spilled oil and its propensity to disperse into the water column or to form stable water-in-oil (w/o) emulsions on the sea surface
- The release conditions (the rate and amount of oil spilled, surface release or underwater release, presence of gas etc.)
- The ambient conditions (e.g., temperature, sea-state, currents)

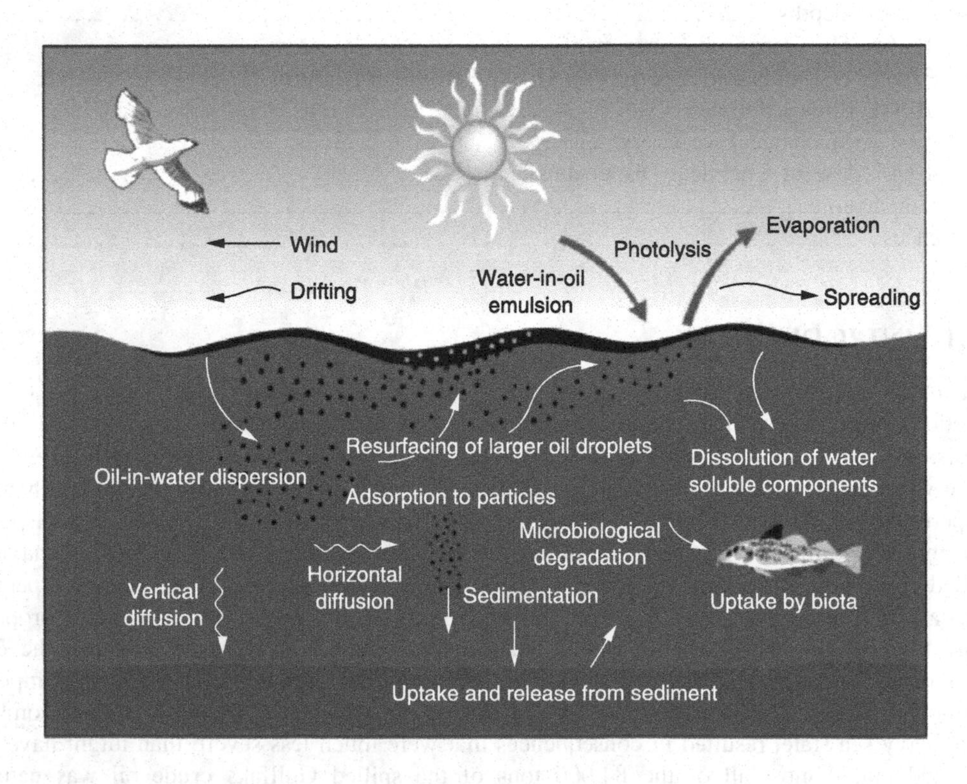

FIGURE 10.1 The weathering processes that take place when oil is spilt on the sea surface.

Evaporation, natural dispersion, and emulsification are the most important processes affecting the fate of the oil on the sea surface.

The weathering properties of crude oils and fuel oils are studied in the laboratory using a "stepwise" weathering procedure of the "fresh" crude oil including evaporation and water-in-oil emulsification. The laboratory data from various weathered oil samples generated in the laboratories form the basis for the input in SINTEF Oil Weathering Model (OWM) for prediction of the weathering behavior of oils at different weather conditions. A description of the SINTEF weathering methodology including small- and meso-scale laboratory testing of oils and modeling of weathering behavior of oils is given by Daling et al. (5), Johansen (6), Hokstad et al. (7), and Daling and Strøm (8).

10.2.1 Evaporation

Evaporation is one of the natural processes removing the volatile components from the spilled oil on the sea surface. The evaporation process starts immediately after the oil is spilled and the evaporation rate decreases throughout the duration of the oil spill. The relative amount evaporated depends on the chemical composition of the oil in addition to the prevailing weather conditions, sea temperature, and the oil slick thickness. The rate of evaporation will therefore vary for different oil types. Light refinery products (e.g., gasoline and kerosene) may evaporate completely after a few hours/days on the sea surface. Condensates and lighter crude oils can lose 50% or more of their original volume during the first days after an oil spill (9). Predicted evaporation for some residual fuel oils and Norwegian crude oils are shown in Figure 10.2.

The most significant changes caused by evaporation is that the loss of volatile and semi-volatile compounds increases the relative amount of higher molecular weight compounds (boiling point above 250 to 300° C). The chemical and physical properties of the remaining oil change

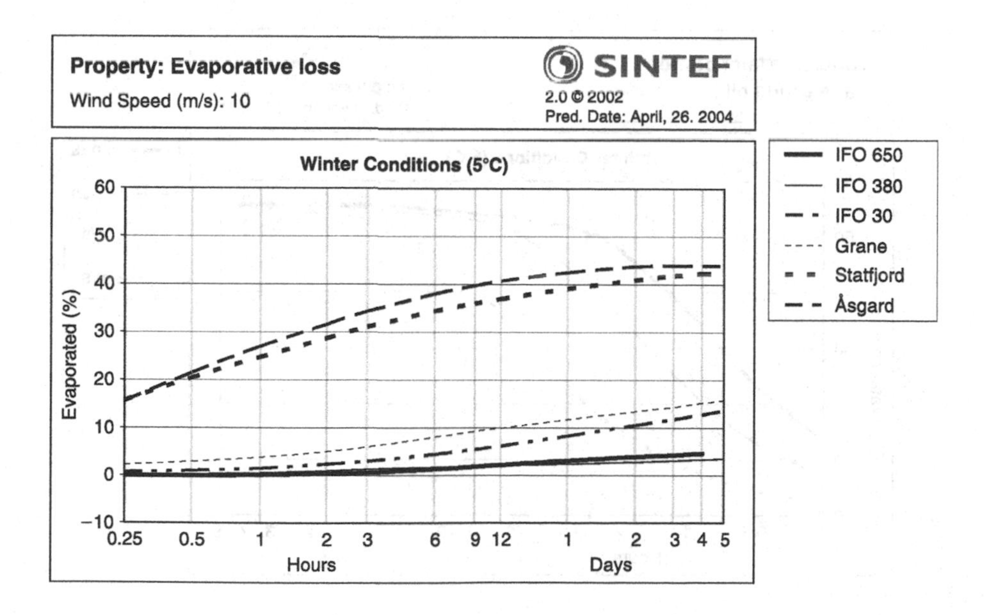

FIGURE 10.2 Predicted evaporative loss with the SINTEF OWM on the basis of laboratory studies for some residual fuel oils and Norwegian crude oils.

due to the evaporative loss. For example the density, viscosity, pour point, and the relative wax and asphaltene contents of the oil residue will increase with the increased amount of evaporation.

10.2.2 Water-in-Oil Emulsification

The formation of water-in-oil emulsions significantly affects the behavior and clean-up of oil spilled at sea. As a result of emulsification, the total emulsion volume may increase to as much as six times the original spilled oil volume depending on the properties of the oil.

Formation of w/o emulsions takes place in the presence of breaking waves (i.e., wind speeds of above 5 m/s), but a slow water uptake can also occur during calmer weather or lower sea states. The maximum water uptake will vary for different crude oils. Tests performed at SINTEF have shown that the maximum water uptake is fairly independent of the prevailing weather conditions, although the emulsification rate depends highly on the weather conditions. Predicted maximum water content and emulsification rate for Grane crude oil at different wind speeds are shown in Figure 10.3.

The physical and chemical properties of the continuous phase of the emulsion (oil) govern the stability of the emulsion formed, and thus the emulsification rate and maximum water content. As different oils have different physical and chemical properties the kinetics of the emulsion formation will be highly oil dependent. Very heavy fuel oils like the oils spilled by the *Erika, Prestige*, and the *Baltic Carrier* form water-in-oil emulsions slowly. However, many spilled crude oils will rapidly form w/o emulsions when spilled at sea (e.g., Ref. 10). Such emulsions will initially have low viscosities, will be unstable, and will tend to revert to the oil residue and water if they are removed from the mixing action of the sea. Predicted water uptake for some residual fuel oils and Norwegian crude oils under similar conditions (10 m/s wind and 5° C) are shown in Figure 10.4.

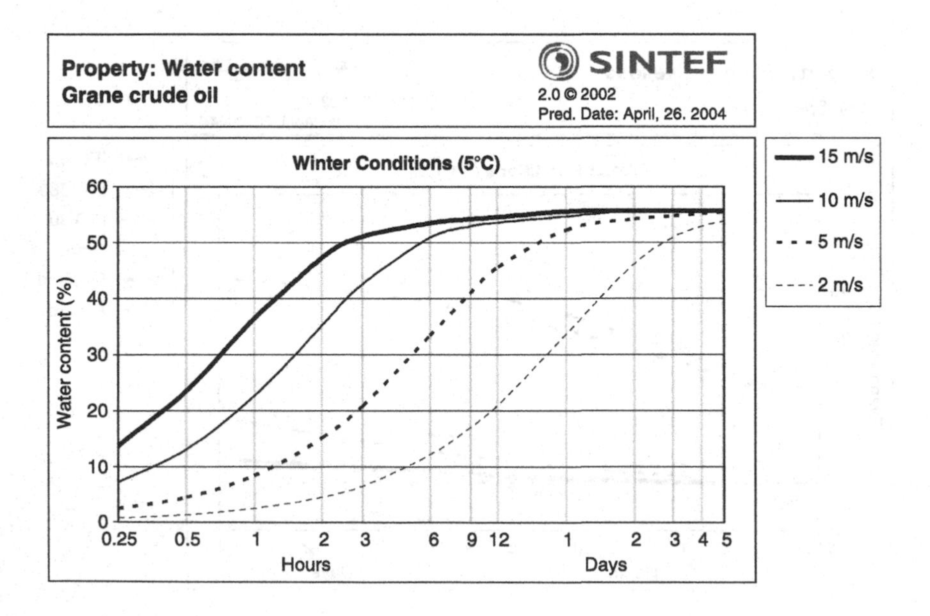

FIGURE 10.3 Predicted emulsification of Grane crude oil with SINTEF OWM on the basis of laboratory data at different wind speeds.

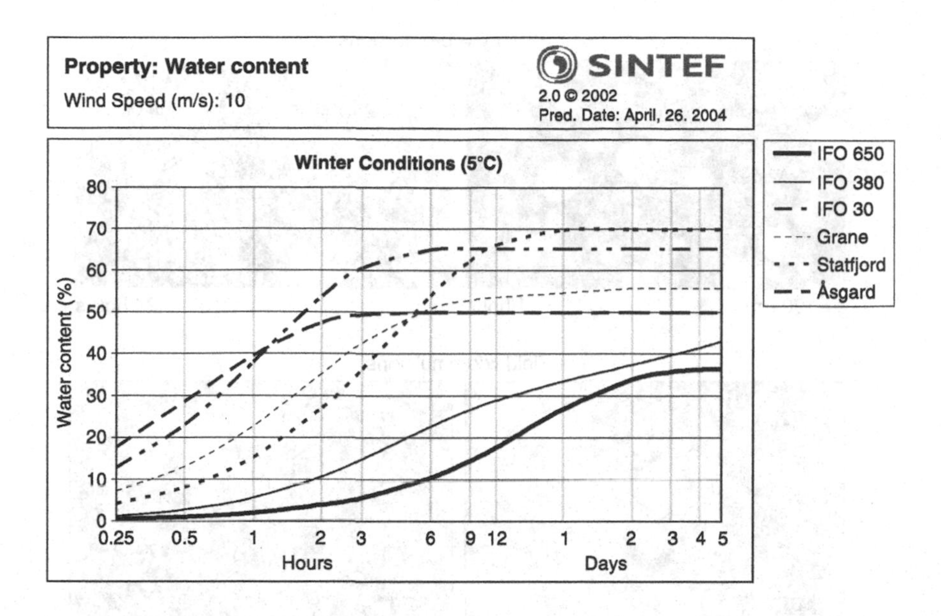

FIGURE 10.4 Predicted water uptake with SINTEF OWM on basis of laboratory data for some oils.

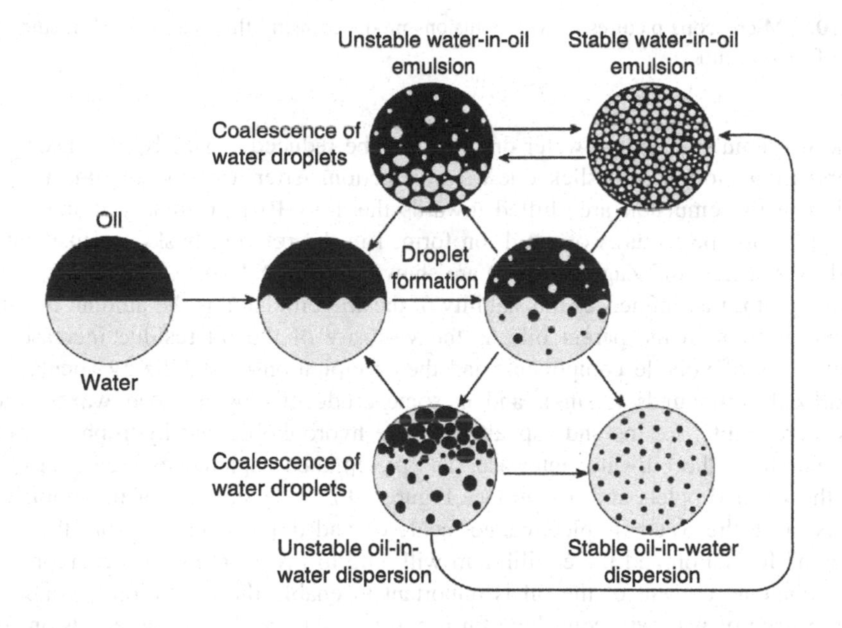

FIGURE 10.5 Mixing of oil and water at sea.

Unstable emulsions are simply mixtures of water droplets in oil and the w/o emulsion present on the surface will be the result of the dynamic equilibrium of emulsion formation and emulsion breakdown (see Figure 10.5).

The stability of the w/o emulsion depends on the water droplet size in the emulsion. Not all of the water droplets in the emulsion are stable. The largest droplets may coalesce and settle

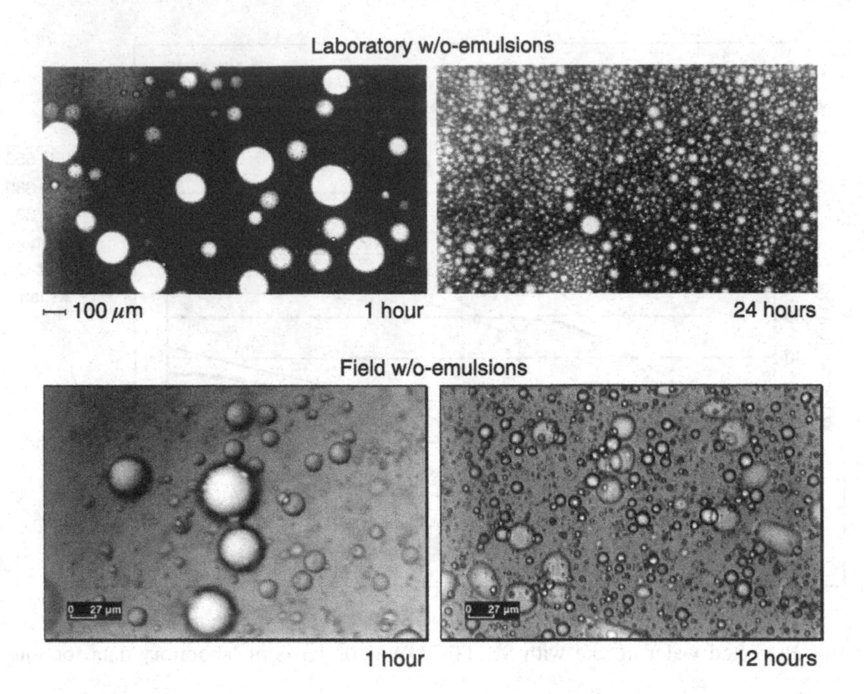

FIGURE 10.6 Microscope pictures of w/o emulsions produced using the rotating flask method showing the influence of mixing time.

out of the w/o emulsion. Large water droplets may be reduced in size by the flexing, stretching, and compressing motion of a slick due to wave action. After a period of time the droplet size distribution in the emulsion are shifted towards the 1 to 10 μm area, yielding a more stable emulsion. Microscope pictures of emulsion formed in the rotating flasks (emulsification method described by Mackay and Zagorski (11)) are shown in Figure 10.6.

Another factor that influences the stability of the w/o emulsion is the amount of surface-active components present in the parent oil. As the viscosity of the oil residue increases due to the evaporative loss of volatile components and the precipitation of stabilizing agents (asphaltenes, photo-oxidized compounds (resins), and in some crude oils precipitated waxes) the emulsion becomes more stable. Resins and asphaltenes have hydrophobic and hydrophilic properties and will concentrate at the oil/water interface. The precipitated asphaltenes create an elastic "skin" between the water droplets and the oil (see Figure 10.7). The stability of the emulsion will then increase because the water droplets cannot coalesce and drain so easily from the emulsion due to this "skin" formation and the equilibrium will tend to favor emulsion formation.

The asphaltene content of the oil is important to enable the formation of stable emulsions, and the presence of wax will contribute further to the stabilization of the emulsion formed. Oils low in asphaltenes but high on wax form emulsions that are stabilized by the rheological strength of the continuous phase (oil) due to precipitated wax rather than the chemical stabilization of asphaltenes and resins. Emulsions from such waxy oils are constituted by larger droplets and are generally less stable than asphaltene stabilized emulsions. Stability and stability classes of w/o emulsions formed during oil spills are also discussed by others (12,13).

The Gullfaks crude spilled at the *Braer* incident in the Shetlands is a biodegraded crude oil with a very low content of waxes and asphaltenes. As the oil lacked stabilizing compounds

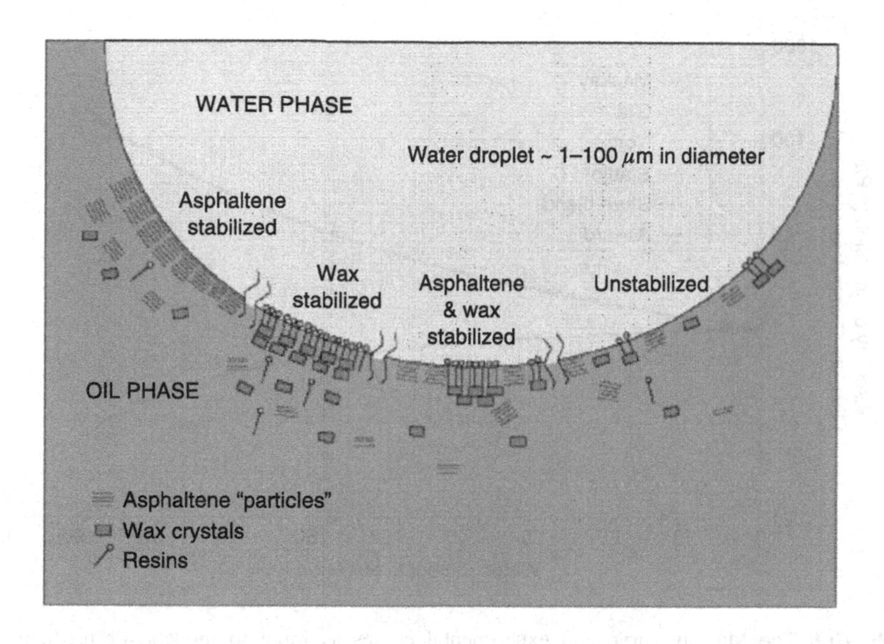

FIGURE 10.7 Stabilization of the interfacial layer between the water and oil in a w/o emulsion by wax and asphaltenes.

TABLE 10.1
Wax and Asphaltene Content in Some Selected Crude Oils from The North Sea

Crude Oil	Wax (wt%)	Asphaltenes* (wt%)	SARA Analysis**			
			Asphaltenes (wt%)	Resins (wt%)	Aromatics (wt%)	Saturated (wt%)
Forties		0,2				
Grane	3.2	1.4	3	37	38	22
Gullfaks	1.6	<0.1	0.8	7	53	39
Statfjord	4.2	0.1	1.5	11	29	59
Åsgard	4.6	0.05				

*The asphaltenes are those that are insoluble in hexane (IP 143).

**Iatroscan.

the emulsion formed were unstable and had a low viscosity. This, combined with the weather conditions that prevailed at the time of the spill, resulted in almost all of the 84,000 tons of the spilled Gullfaks crude oil being naturally dispersed (14). During the *Sea Empress* spill of Forties Blend crude oil (another North Sea crude oil) a significant amount of the surface oil was converted into w/o emulsions (2). Forties crude oil is a paraffinic crude oil with a relative high wax content and a low content of asphaltenes.

Wax and asphaltene content and SARA-analysis (Saturated, Aromatics, Resins, Asphaltines) for some crude oils are shown in Table 10.1. As seen in Table 10.1 from the SARA analysis, the relative composition of the crude oils is very different. Gullfaks has low content of stabilizing

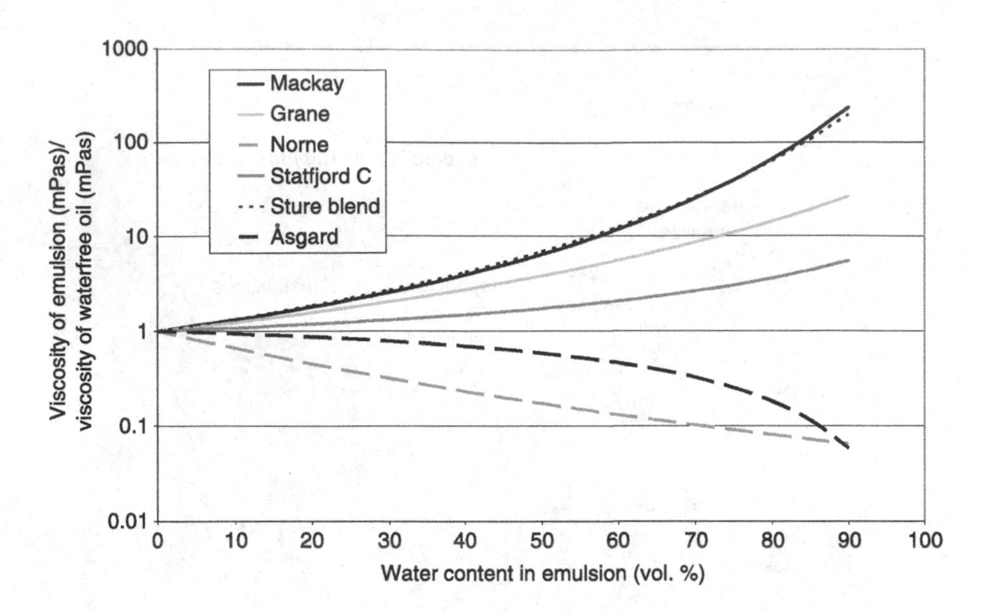

FIGURE 10.8 The Mackay curve and experimental curves as input to models for predicting emulsion viscosities for emulsions formed at sea.

compounds and a higher aromatic content. The paraffinic Statfjord, which has similarities to Forties crude oils, has a higher content of stabilizing agents and saturates.

As oil emulsifies on the sea surface, the viscosity will increase with increasing water content. A widely used approach for predicting w/o emulsion viscosity is the Mooney equation which gives the viscosity ratio as a function of water content. The viscosity ratio is defined as the w/o emulsion viscosity divided by the parent oil viscosity. Mackay (15) has studied viscosity of w/o emulsions and determined the typical values for the constants in the equation for emulsions formed when oil is spilt at sea. Figure 10.8 shows the "general correlation equation" based on the Mackay relationships between dispersed water phase volume and viscosity used in many models to predict the viscosity of emulsion based on the viscosity of parent oil and water content. Specific laboratory measurements of the viscosity of oil residues and the emulsions generated from them in the rotating flask method are used to "tune" this correlation equation for specific oil being considered. This will lead to more reliable predictions of the emulsion viscosity for individual oils, compared to using only a "general" correlation equation, because the variation in water droplet size distribution is considered.

The maximum water content of w/o emulsions is different (Figure 10.9) due to various chemical and physical properties of oils. Weathering studies of oils have shown that the maximum water content decreases as the viscosity of the residue increases.

Predicted emulsion viscosities are shown for some crude oils and fuel oils in Figure 10.10 and shows very high viscosity for the heavy fuel oils (100,000 to 300,000 mPas after some days at sea). The viscosity of the crude oils after some days at sea varies from 10,000 to 70,000 mPas. This shows that crude oils and fuels may behave very differently when spilt at sea and that various crude oils behave differently at sea due to differences in chemical and physical properties of the oils.

The weathering processes change the appearance of the oils as seen in Figure 10.11. The pictures are taken after 3 days' weathering in a meso-scale flume. The colors of the emulsified

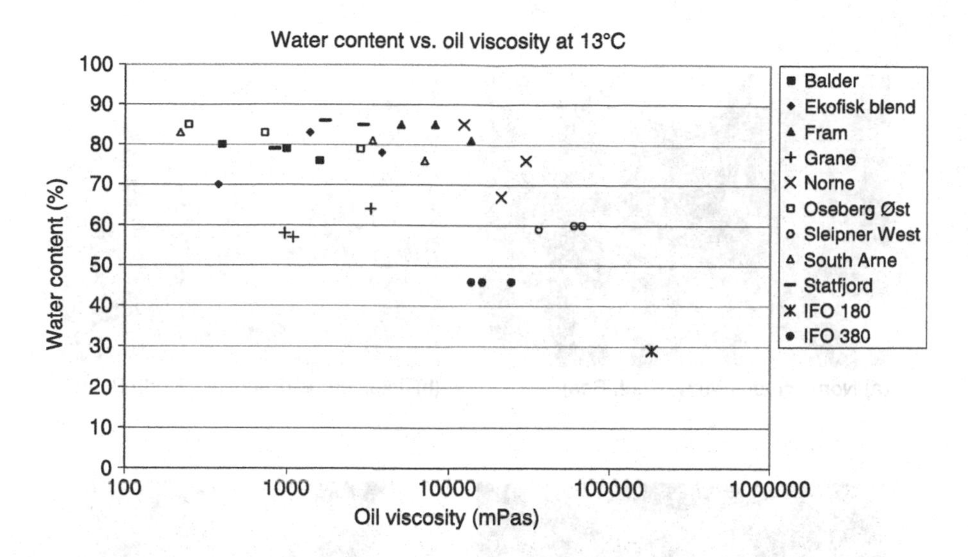

FIGURE 10.9 Water content in emulsion versus oil residue viscosity.

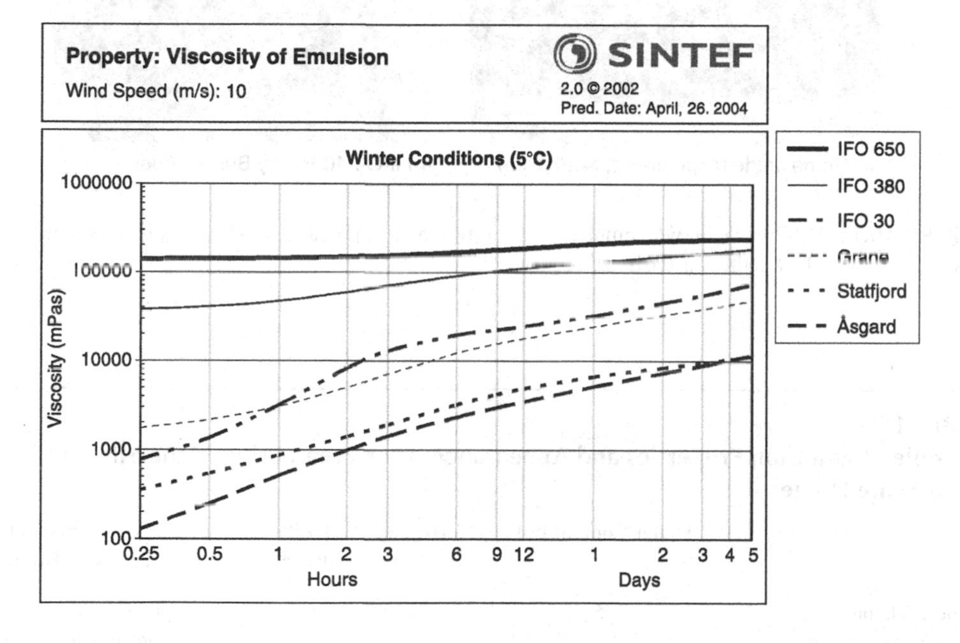

FIGURE 10.10 Predicted viscosities of a selection of crude oils and fuel oils.

oils vary from the yellow/brown, paraffinic Jotun to the dark brown, solidified waxy Norne and the black lumps of the asphaltenic Grane. Some oils form continuous slick while others form lumps. Table 10.2 shows the water content and viscosity of these emulsions. The viscosity varies from approximately 2400 to 23,000 mPas for these oils, and the water content from 35 to 70%.

(a) Norne crude (Waxy, North Sea)

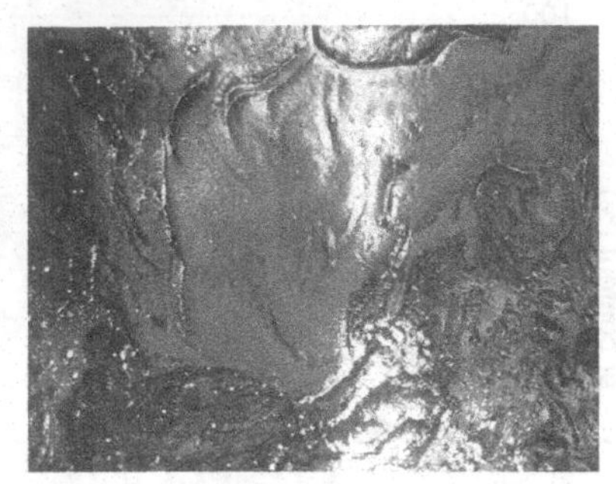

(b) Jotun crude (Paraffinic, North Sea)

(c) Grane crude (Asphaltenic, North Sea)

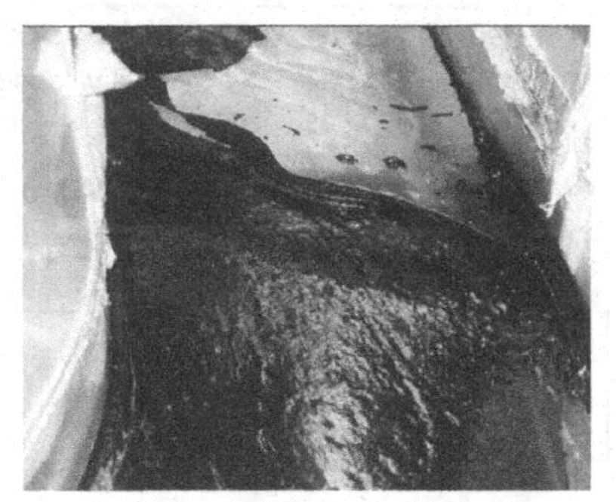

(d) IFO-240 Heavy Bunker Fuel

FIGURE 10.11 Difference in w/o emulsion properties and appearance after 3 days' weathering in the SINTEF meso-scale weathering basin.

TABLE 10.2
Example of Emulsion Properties and Appearance After Weathering in the SINTEF Meso-Scale Flume

Oil	Water Content (%)	Dynamic Viscosity at 10 sec^{-1} (mPa sec)	Appearance of the Emulsion Formed
Norne crude oil (waxy, Norwegian Sea)	35	3000 (5 °C)	Solidified oil lumps, turn slowly into brown color
Jotun crude oil (paraffinic, North Sea)	58	2430 (13 °C)	Yellow/light brown
Grane crude oil (asphaltenic, North Sea)	70	11,000 (13 °C)	Dark, viscous, sticky lumps
IFO-240 Heavy Fuel Oil (Esso refinery)	50	23,000 (13 °C)	Dark emulsion

10.2.3 Natural Dispersion

Natural dispersion is caused by the action of breaking waves. As a breaking wave collapses on an oil slick, oil droplets of various sizes are detached from the principal slick and entrained into the water masses (16). The oil droplets can remain dispersed, resurface, and coalesce with the principal slick, or resurface and form a thin film trailing behind the principal slick. This depends on the size of the droplets, the submergence depth, and the height (or energy) of the breaking wave. If the droplets coalesce with the principal slick, water may become trapped within the reformed slick and a stable water-in-oil emulsion may be formed by successive breaking events. Oil slick dispersion (o/w emulsification) and w/o emulsification comprises therefore a series of batch process, the outcome of each determined by a combination of oil properties (oil chemistry, density, interfacial tension, and viscosity) and hydrodynamics (breaking wave frequency, submergence depth, and intensity of mixing associated with each breaking wave). The intermittent nature of the breaking events may also be important, especially for viscous oils, as the duration of the event may be too short to form stable droplet size distributions corresponding to the prevailing intensity of mixing.

According to Shaw (16), the key variables that determine the size distribution of the detached oil droplets are oil–water interfacial tension, oil or emulsion density (relative to the density of sea water), oil or emulsion viscosity, and the prevailing forces in the environment (shear stress). Others (17) maintain that turbulent dissipation rate is a more relevant measure of the splitting force, but none of these parameters are easy to relate to the actual sea state. It may thus be more practical to use wave energy as a measure of the splitting force, as in the empirical correlations derived by Delvigne and Sweeney (17). The same authors neglected oil–water interfacial tension in their experimental study, assuming that oil viscosity was the most representative property of the oil in this context. This may be justified since the variation in the interfacial tension is limited among fresh crude oil, and the relative importance of the interfacial tension seems to diminish with increasing oil viscosity (16). However, addition of surfactants to the oil–water will reduce the interfacial tension significantly, an effect that is exploited in oil spill countermeasures based on chemical dispersion.

Delvigne and Sweeney found that breaking waves entrain oil droplets to a depth corresponding to the height of the breaking wave. They also found that the number distribution of the droplets that were formed by breaking waves could be described by the same power law relationship, $N(d) \sim d^{-2.3}$, independent of the applied wave energy and the oil properties. The oil mass entrained per unit area in droplets of a given size range was independent of the surface oil film thickness, but increased almost in proportion to the height of the breaking wave. Experiments with oil weathered to various degrees showed that the entrained oil mass decreased with increasing oil viscosity of the oil. In a later study, Delvigne and Hulsen (18) found that the entrained oil mass was practically independent of oil viscosity for viscosities up to 100 cSt, but started to decrease considerably as viscosities increased above this range. The reduction in oil droplet entrainment with increasing viscosity implies that the natural dissipation rate will diminish as the oil remaining in the surface slick is subjected to weathering and emulsification.

These findings have been used to derive algorithms that are used in SINTEF's Oil Weathering Model for prediction of the dissipation rate of surface oil depending on oil type, degree of weathering, and sea state (19). The main assumption in that context is that droplets smaller than a certain size will stay submerged in the water masses, while larger droplets will resurface and rejoin the primary slick within a short time span. The entrainment rate of droplets smaller than the size limit is calculated from sea state and oil/emulsion viscosity with relations derived from Delvigne's studies. Examples of predictions are shown in Figure 10.12.

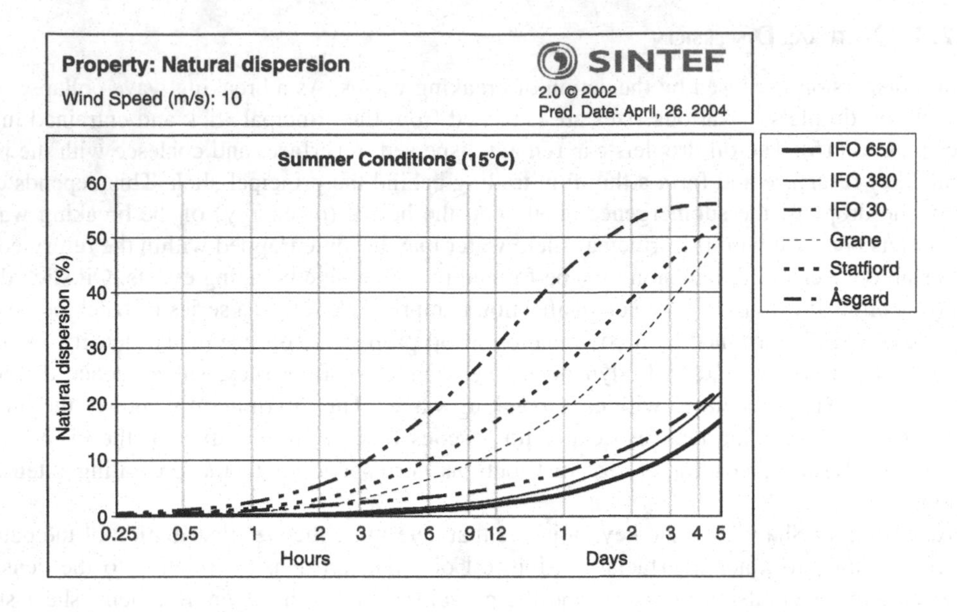

FIGURE 10.12 Predictions of natural dispersion for different crude oils and oil products.

10.3 FULL-SCALE FIELD TRIALS IN NORWAY

Emulsification and behavior of spilled oils have been studied during series of experimental releases with various oils simulating various release conditions. These studies involved both surface releases (1994, 1995, and 1996), simulated sub-sea pipeline leaks (release of oil and no gas, 1994), and simulated sub-sea blowouts (oil with gas) released both from 100 m depth (1996) and 850 m depth (DeepSpill June 2000). Three of the field series were carried out in the Frigg area in the North Sea, while the DeepSpill experiment was performed in the Helland Hansen area in the Norwegian Sea. The field trials have been performed in cooperation between spill response organizations (NOFO), individual oil companies, and governmental pollution control agencies (SFT and MMS). A summary report of the field trials in Norway has been prepared (20).

10.3.1 The NOFO 1994 Field Trial: Surface Release

One of the main objectives of the 1994 field trials was to verify the results from laboratory studies and model predictions of oil weathering (evaporation, natural dispersion, and emulsification) up to 2 days at sea. Two oil slicks (2×20 m^3 of Sture Blend North Sea crude) were released on the surface. The sea state was quite rough with wind speed varying between 8 and 12 m/s and about 2.5 m significant wave heights. A control slick ("Charlie") was followed with extensive surface monitoring for 32 h before dispersants treatment. The other slick ("Tango") was treated twice with dispersant after 3 and 7 h weathering at sea.

As soon as the oil was released on the surface, the slicks rapidly formed an elongated shape as some of the oil was temporarily dispersed and then re-surfaced away from the thicker areas (in front of the slick) and spread out again as sheen. Measurements showed a very wide variation

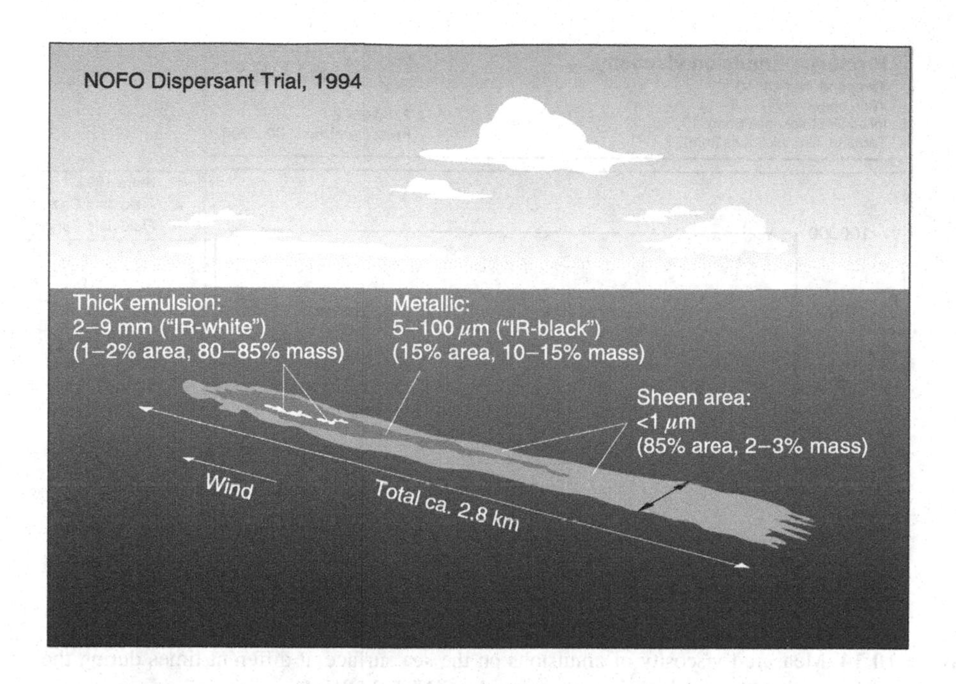

FIGURE 10.13 Schematic drawing of the distribution of oil slick thickness after 3 h at sea.

in the slick thickness, with three main categories observed:

- Thick w/o-emulsion (2 to 9 mm) in the front part of the slick
- Metallic (up to about 0.1 mm, also called IR-black)
- Sheen (up to about 1 to 5 μm)

This correlation between visual appearance and ground truth film thickness measurements was in good agreement with the Bonn Agreement Oil Appearance Code (BAOAC), OTSOPA (21).

An estimate of the distribution in area and mass of these categories based on remote sensing and film thickness measurement is presented schematically in Figure 10.13. The weathering behavior and physical properties of the thick emulsion that formed on the surface were in accordance with results from previous laboratory studies.

The weathering data were used to calibrate and verify the SINTEF Oil Weathering Model (8). Figure 10.14 shows the measured viscosity of the emulsions sampled from the sea surface, at different times during the field trial. The viscosity of the fresh crude oil is approximately 10 mPas increasing to 9000 mPas after 28 h of weathering on the sea surface.

Figure 10.15 gives an estimate of amount of oil on the surface and the lifetime of the two slicks based on aerial surface monitoring. Further details concerning operational aspects, monitoring, analytical methods, and conclusions from this sea trial have been published in several reports, e.g., Lewis et al. (10).

10.3.2 The NOFO 1995 Field Trial: Surface and Underwater "Pipeline" Releases

One of the aims of the NOFO field trial in 1995 (22) was to study the behavior of oil released from sub-sea pipeline leaks. A total of 25 m^3 of Troll Crude was released over a period of 20 min

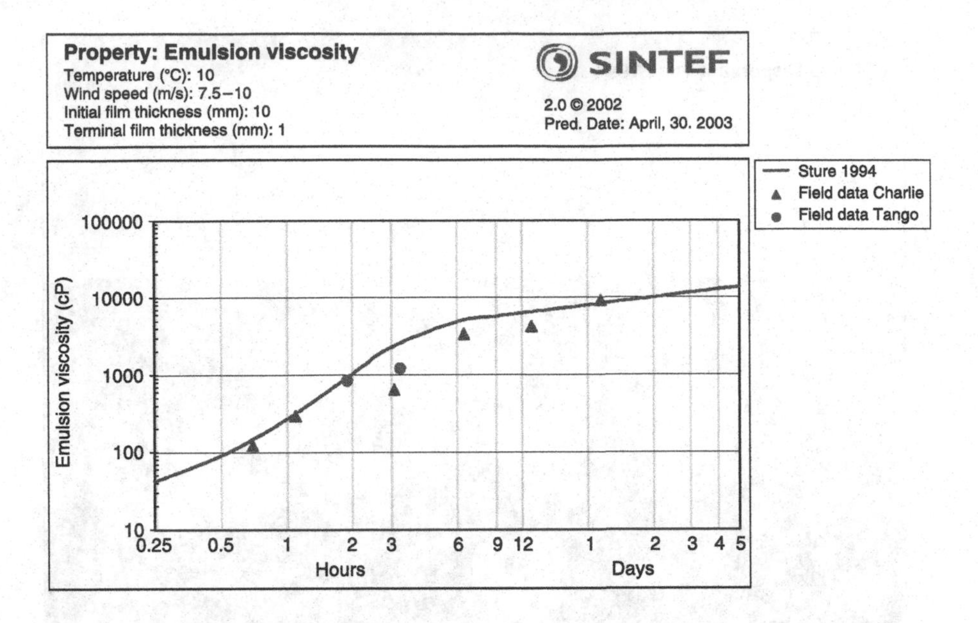

FIGURE 10.14 Measured viscosity of emulsions on the sea surface at different times during the field trial. The line shows predicted emulsion viscosity from the SINTEF OWM.

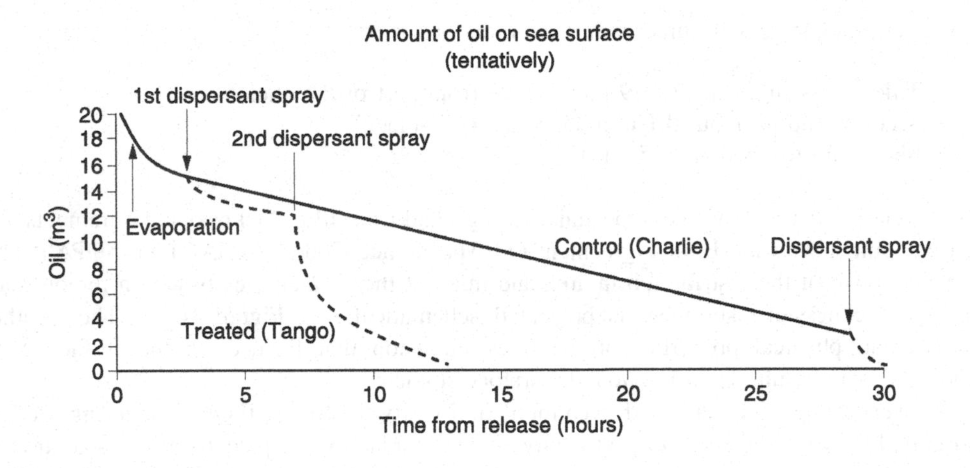

FIGURE 10.15 Lifetime of surface oil slick; slick treated with dispersant ("Tango") versus control slick ("Charlie").

from 100 m depth (slick called "Uniform"). A reference spill (surface slick "Sierra") was released using the same conditions as in the 1994 trials. The release and monitoring arrangement of the sub-sea release is illustrated in Figure 10.16. Due to relatively low exit velocities (about 2 m/s), relative large oil droplets were formed (typically 2 to 6 mm in diameter) when released from the sub-sea installation. An oil slick started to form at the surface about 10 min after the start of the release. Oil samples taken as the oil appeared on the surface indicated that no w/o emulsion

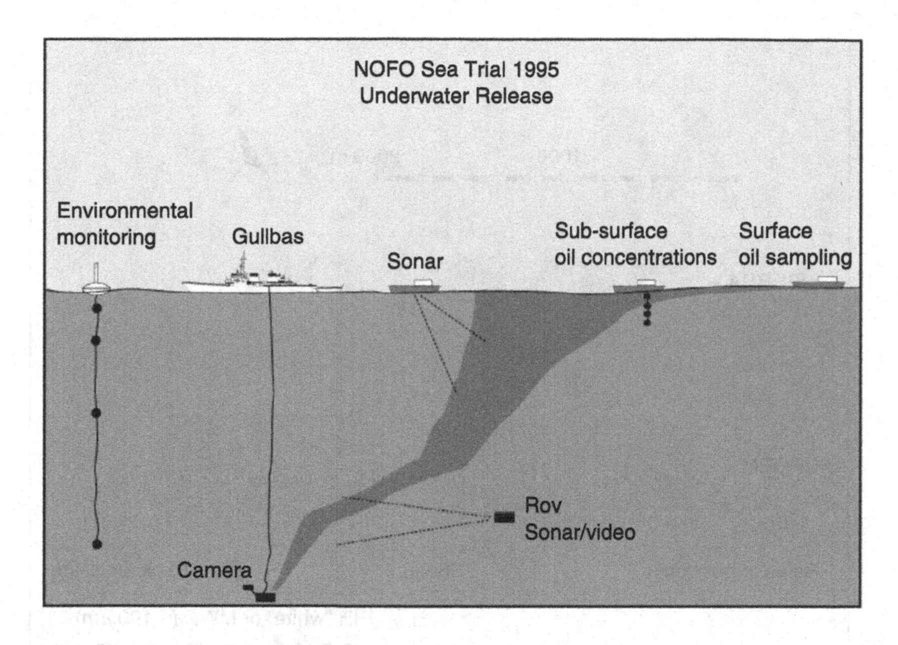

FIGURE 10.16 Schematic illustration of the sub-sea release arrangement, the plume created and monitoring strategy.

was formed in the plume, but that the emulsion was subsequently formed after the oil appeared on the sea surface. The weathering properties of the thick emulsion formed on the surface were similar to the properties of the emulsion generated in the surface release ("Sierra"). Subsequent analysis of oil samples from both slicks showed good agreement with model predictions based on laboratory input data of Troll crude oil (8).

Observations of the two oil slicks 4 h after the releases indicated that the slick formed from the sub-sea release was slightly larger due to the larger spreading at the start of the release (Figure 10.17). Further details both concerning operational aspects, monitoring, analytical methods, and conclusions from this sea trial are described in several reports including Brandvik et al. (22).

10.3.3 The NOFO 1996 Field Trial: Simulated Sub-Sea Blowout from Moderate Depths

One of the aims of the NOFO field trial in 1996 (23) was to increase the knowledge of the behavior of the oil in sub-sea blowouts. This was done by simulating a blowout released from 100 m depth with a realistic gas-to-oil ratio (GOR = 65) and a release rate of 1 m^3 Troll crude oil per minute, with a total of 45 m^3. For safety reasons, pressurized air was used in place of natural gas. The release method and monitoring arrangement were similar to the 1995 sub-sea release (see Figure 10.16). However, due to a much higher exit velocity (about 20 m/s), very small oil droplets were formed that were carried to the surface by the gas bubble plume. The first oil appeared on the surface about 2 min after the start of the release. The high initial spreading of the oil on the surface (see picture in Figure 10.18) was in accordance with calculations using the blowout model developed by Fanneløp and Sjøen (24). The horizontal spreading of the plume near the surface resulted in a very thin and homogenous oil film on the surface. Some small tendency

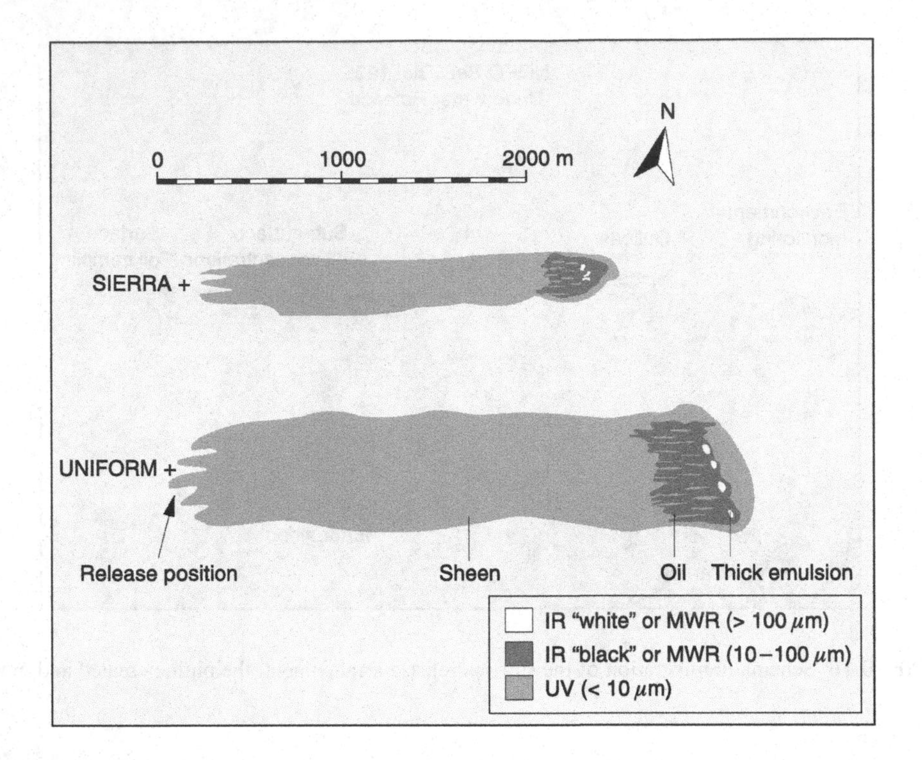

FIGURE 10.17 Composite sketch of remote sensing imagery showing the slick dimensions of the surface release (Sierra) and the underwater release (Uniform) after 4 hours of weathering at sea.

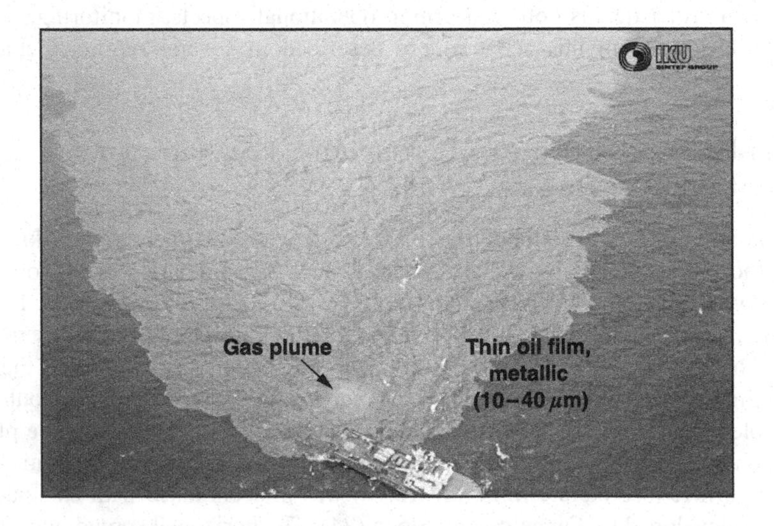

FIGURE 10.18 Photo of the surface slick taken during the simulated sub-sea blowout (NOFO-1996 trial).

to Langmuir cells could also be observed by IR images, in a thickness range of 10 to 40 μm. This thickness appeared, however, to be too thin for any emulsion to form on the surface, and the oil slick dissipated naturally within a few hours. Further details concerning operational aspects, monitoring, analytical methods, and conclusions from this sea trial are described in several reports (e.g., Refs. 23 and 25).

10.3.4 The DeepSpill Experiment

The DeepSpill experiment was conducted in the Norwegian Sea in June 2000, and included releases of oil and natural gas from 850 m depth. The main objective was to provide data for verification of numerical models for simulating accidental releases in deep water. Three experiments were performed, one with gas only, one with gas and marine diesel (simulating a non-emulsifying oil), and one with gas and crude oil (Sture Blend). Each release lasted for 1 h, with gas rates of 1 Sm^3/sec and oil rates of 1 m^3/min. The oil and gas was pumped from the discharge vessel through separate coiled steel tubing lines down to a discharge platform deployed on the seabed. The gas was transported to the site in liquefied state (LNG), and was transformed into gas in a seawater-heated evaporator during the release (26).

Met-ocean data were measured continuously during the field trial with wind sensors and a downward facing acoustic Doppler current profiler (ADCP) mounted under the research vessel, supplemented with an upward facing ADCP mounted on the seabed. The wind speed during the experiments varied between 9 and 14 m/sec. Ocean currents that were measured in the range of 10 to 20 cm/sec showed a significant variation with depth. Video cameras mounted on remote operated vehicles (ROVs) provided close-up pictures of gas bubbles and oil droplets, and ship mounted echo-sounders provided images of the plumes of water, gas bubbles, and oil droplets that were rising through the water column.

In both experiments with oil, the first oil appeared on the surface 1 h after the start of the release, while the main surfacing took place around 2 to 4 h after release. The water in the plume was trapped below the perennial thermocline (below 500 m depth), while gas bubbles and oil droplets were then separated from the trapped water and continued towards the surface with the rise velocity of individual bubbles and droplets (see Figure 10.19). Observations during the pure gas release (without oil) showed no presence of the gas bubbles above 150 m depth, indicating that the gas bubbles had been dissolved completely in seawater at this level. However, the oil droplets continued to the surface and formed a surface slick that developed gradually over time. In the crude oil experiment, the oil film thickness in the surface slick was measured in the range from 200 to 400 μm of non-emulsified oil (see Figure 10.20(a)). This thickness was sufficient for the formation of patches of emulsified oil during the coming hours on the sea surface (see Figure 10.20(b)). The water uptake rate and the time development of the viscosity of the thick emulsion was in good agreement with model predictions using the SINTEF OWM based on laboratory input data of the Sture Blend crude oil (see Figure 10.21). The emulsion became stable after about 5 h weathering on the sea surface. The viscosity of the fresh crude oil is approximately 10 mPas increasing to 8000 mPas after 8 h of weathering on the sea surface.

The last aerial survey that was made about 8 h after the start of the crude oil release depicted a slick about 8 km in length and 1 km in width. In the marine diesel experiment, where no emulsion was formed, the maximum slick size was observed about 5 h after the start of the release, with a length in the order of 1.5 km and a width of 500 m. No remains of the slick could be detected during the last surveillance flight about 3 h later. The difference in behavior between emulsifying and non-emulsifying oils is illustrated schematically in Figure 10.22.

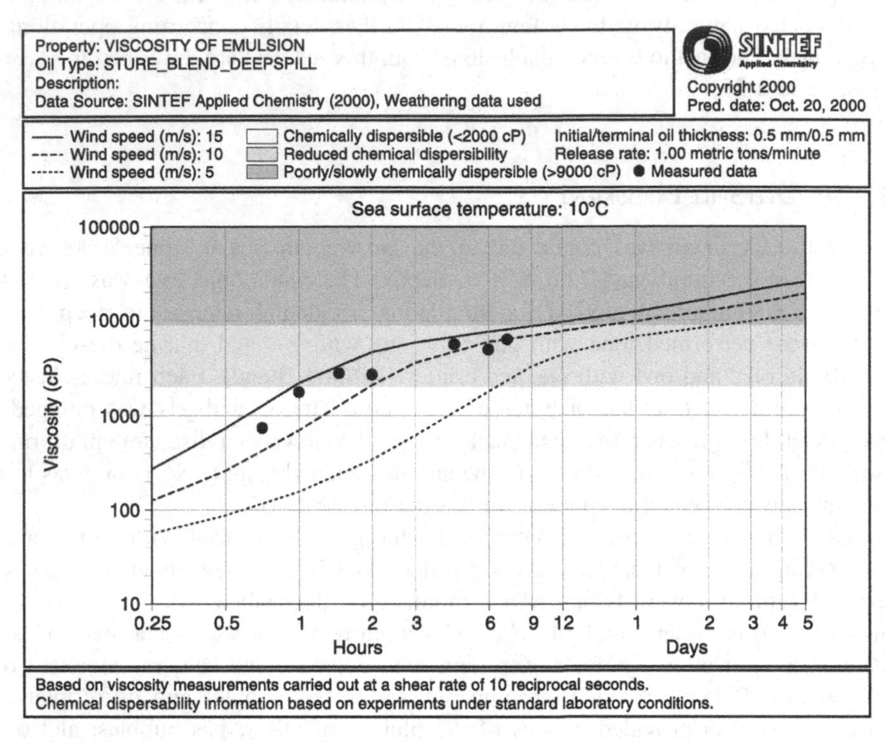

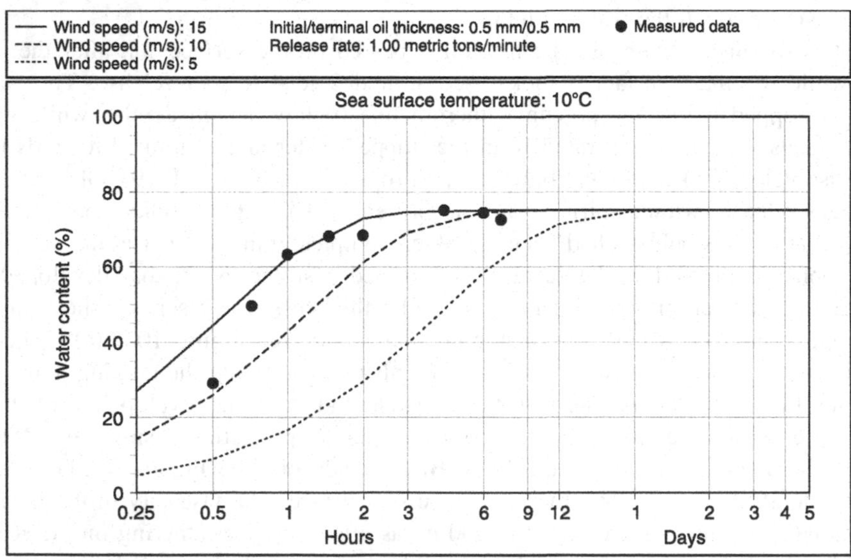

FIGURE 10.19 Comparison between predicted and measured values of viscosity of emulsion (top) and water content after surfacing of the oil (bottom) (DeepSpill 2000).

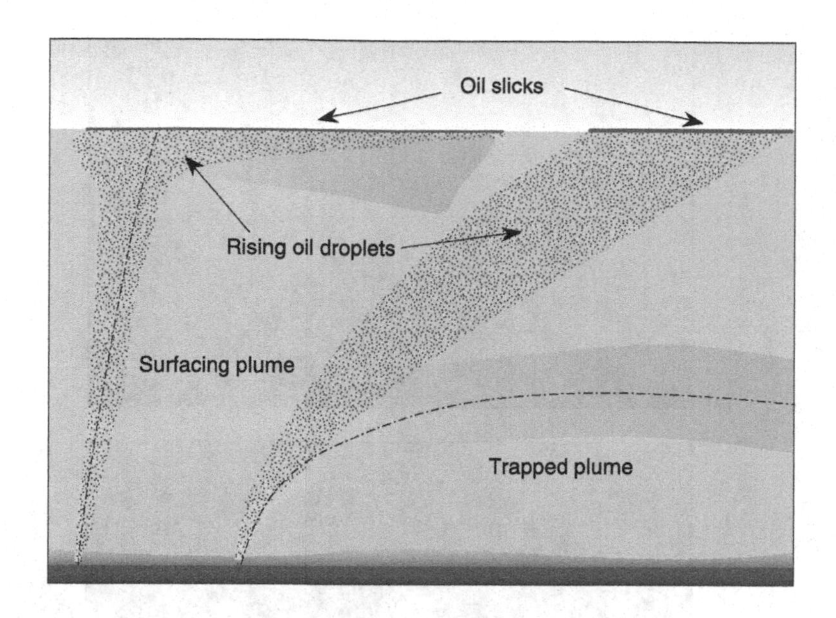

FIGURE 10.20 Schematic drawing of sub-sea blow-outs from medium depths (top) and deep waters (bottom).

Further details both concerning operational aspects, monitoring, analytical methods, and conclusions from this sea trial are described by Johansen et al. (26).

10.4 EFFECTS OF EMULSION PROPERTIES ON OIL SPILL COUNTERMEASURES

One of the main reasons for studying the properties of emulsions formed on the sea surface is its effect on oil spill countermeasures. Good knowledge about the specific oil properties is crucial in connection with net environmental benefit analysis (NEBA), contingency planning, and decision making in spill situations. Some examples of the direct impact of emulsion properties in oil spill response operations are described below.

10.4.1 Slick Lifetime

One of the first tasks in an oil spill response operation is to assess the lifetime of the slick on the sea surface. If the oil's estimated lifetime is short, mobilization of oil spill countermeasures may not be needed.

The two processes contributing to the removal of oil from the sea surface are evaporation and natural dispersion. For most crude oils natural dispersion is the main contributor to removing oil from the sea surface. Figure 10.23 shows predictions of the lifetime of slicks of different oils on the sea surface. A slick's lifetime on the sea surface is very dependent on the release conditions and weather conditions. The example below is only representative to one specific set of release conditions (i.e., thick enough to form emulsion on the surface), temperature, and wind speed.

The relatively light Åsgard crude has a limited lifetime on the sea surface while the heavy bunker fuel oils are persistent to natural dispersion and can survive for weeks and months on the sea surface.

FIGURE 10.21 (a) Aerial photo of the surface slick taken after the crude oil release. (b) Photo taken during surface sampling from emulsified patches (1 to 1.5 mm thick) generated during 3 h weathering on the surface (DeepSpill 2000).

As illustrated in Figure 10.13, the thickness within an oil slick varies greatly in a surface release spill. Figure 10.24 shows the predicted lifetime of a Norwegian crude oil as a function of oil film thickness and wind conditions. The predictions show how important it is to give priority to the thick w/o emulsion within the slick during a response operation.

10.4.2 Mechanical Recovery

Past experiences from Norwegian field trials have shown that the effectiveness of many mechanical clean-up operations is reduced due to a high degree of leakage of the confined oil or w/o emulsion from the boom systems (especially in high current). This leakage is especially pronounced if the viscosity of the oil or the w/o emulsion is lower than 1000 mPas (at a shear rate of 10 s^{-1}) (27). It is therefore often recommended to let the oil weather to a viscosity of minimum 1000 mPas before recovery in order to optimize the recovery operation.

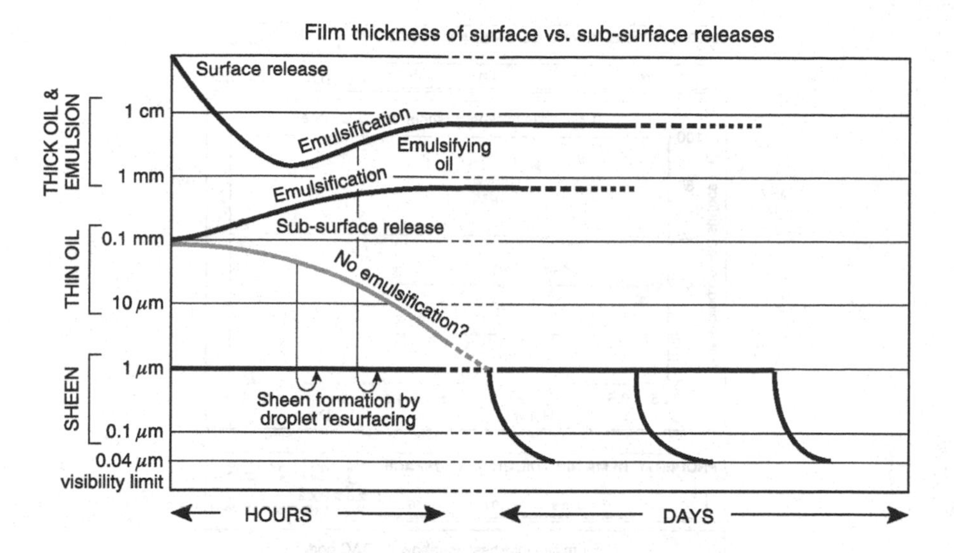

FIGURE 10.22 Schematic illustration of the development in film thickness, breakdown, and relative lifetime on the sea surface of an emulsifying surface release, versus emulsifying and non-emulsifying underwater releases.

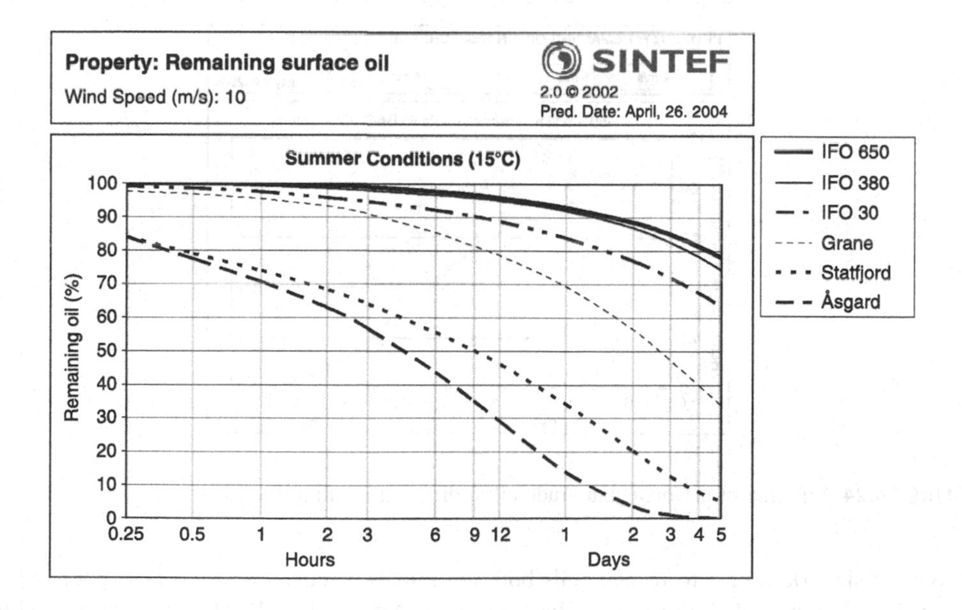

FIGURE 10.23 Predicted lifetime of different oils at the sea surface. The oil is removed from the surface by evaporation and natural dispersion.

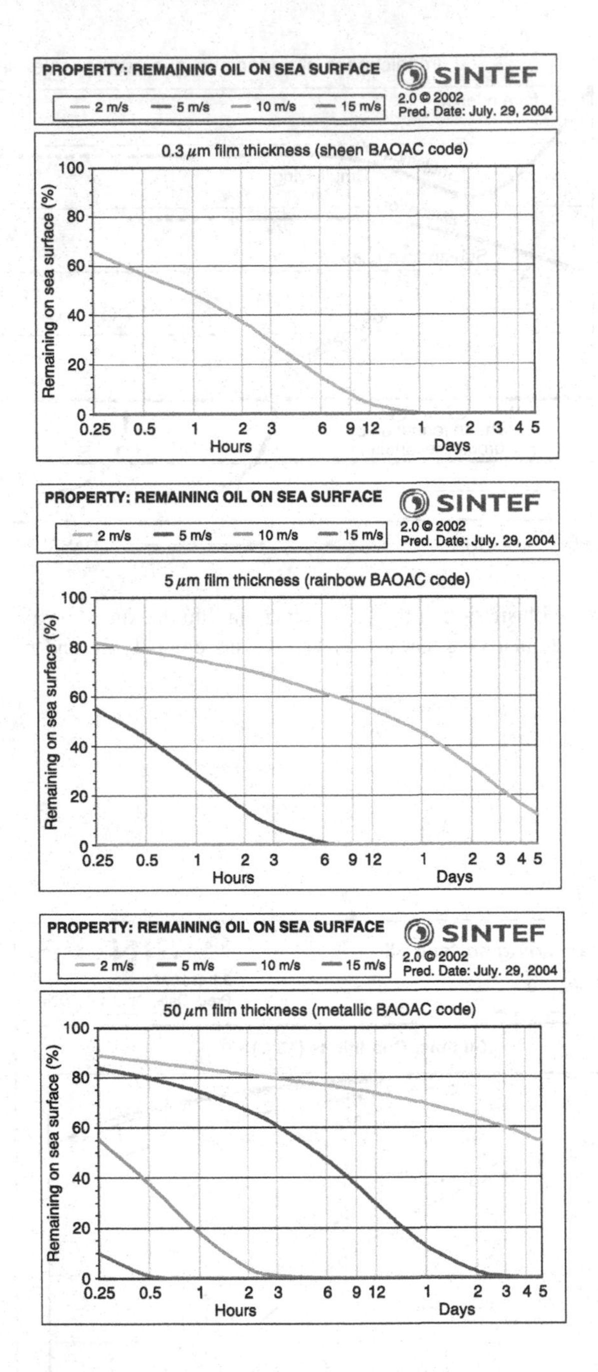

FIGURE 10.24 Lifetime of a Norwegian crude oil at different oil film thickness.

When using skimmers to recover oils both too high a viscosity and too high pour point may reduce the efficiency of the skimmer. Recent tests performed by SINTEF using a weir skimmer showed that the efficiency may be reduced for semi-solid and solidified oils (i.e., oils with a high wax content and pour point values higher than 10 to 15° C above ambient sea temperature) and for oils with viscosities above approximately 20,000 mPas (28). For oils with viscosities

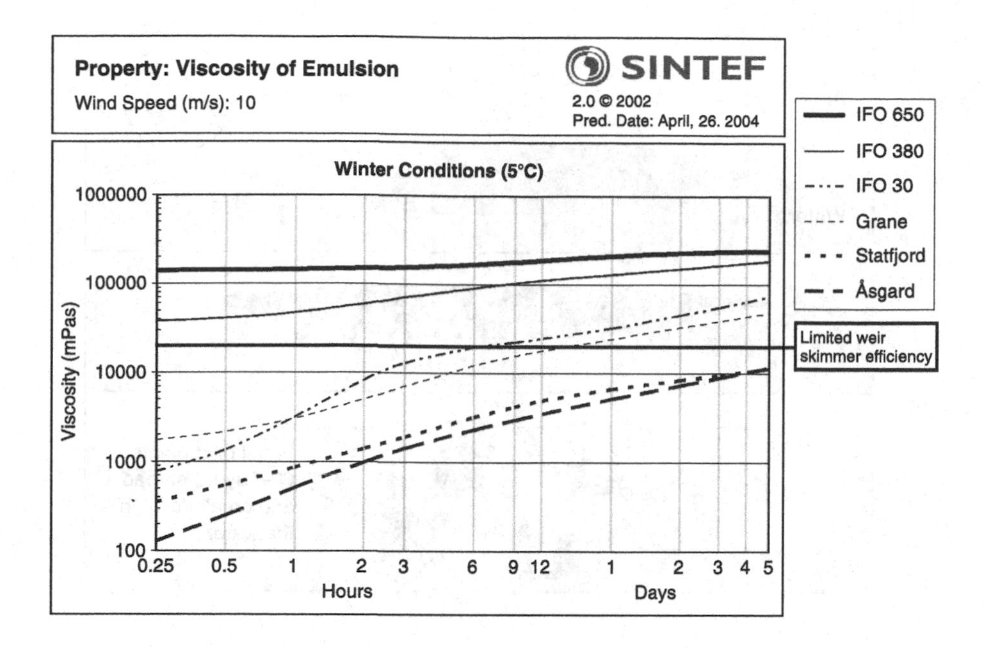

FIGURE 10.25 Predicted viscosities of a selection of crude oils and fuel oils.

above these values, or with high yield stress values, skimmer equipment requiring the oil to passively spread towards the skimmer may render useless, and special equipment for high viscous oil recovery may be needed.

Figure 10.25 shows variations in emulsion viscosities between different oils with weathering on the sea surface. As for limitations in skimmer efficiency some of the oils will always have viscosities above the critical limit, some will obtain critical viscosities with time on the sea surface, and some will never cause problems in a mechanical recovery operation even after days of weathering on the sea surface. Knowledge about the viscosity of the emulsions formed is crucial when choosing skimmer equipment both in contingency planning and in the case of a spill situation.

10.4.3 Use of Chemical Dispersants

One of the means in Norwegian oil-spill contingency for combating oil slicks is the application of oil spill dispersants. Dispersants work in the same way as a household soap, stabilizing the formation of small oil droplets, enhancing the mixing of oil into the water. The purpose of using oil spill dispersants is to remove spilled oil from the surface and dilute it into the bulk of the water column as droplets at a faster rate than occurs naturally. The appropriate use of dispersants could prevent subsequent shoreline pollution or damage to other sensitive areas/resources. A schematic picture of how dispersants work is shown in Figure 10.26.

Effective use of dispersants depends on chemical interactions between the added chemicals and naturally occurring components within the oil (e.g., wax, asphaltenes, resins). Oils have different chemical composition, causing variations in their ability to interact with the dispersants. This gives a great variation in the potential for use of dispersants between different oils. Viscosity and pour point will also be important to the effectiveness of chemical dispersants. A high viscosity or high pour point of the oil will give poor mixing between oil and chemical dispersant, and will also cause the oil to resist being mixed into the water by wave energy.

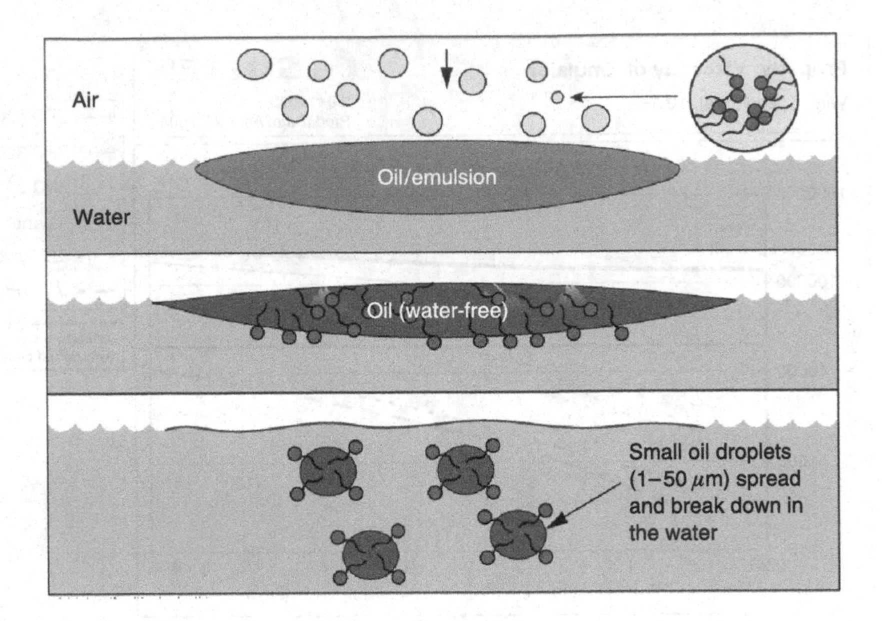

FIGURE 10.26 Dispersion of oil into water.

The methodology for defining the "window of opportunity" for use of dispersants is described in detail by Daling and Strøm (8). Laboratory data on weathering properties and dispersibility is used as input to the OWM for prediction of the "window of opportunity." Figure 10.27 shows the "window of opportunity" divided into categories (easily, reduced, and poorly dispersible) for a selection of Norwegian crude oils with weathering on the sea surface.

Figure 10.27 shows examples of how the dispersibility varies with time on the sea surface for different Norwegian crude oils. As the oils hold different chemical properties, the potential for use of dispersants is not directly correlating with the viscosity of the emulsion. It is important that the predicted "window of opportunity" for use of dispersants is taken into account in oil-spill contingency planning and operational decision making in spill situations.

10.5 CONCLUSIONS

Laboratory experiments and full scale field experiments have been performed during the last decades in order to obtain more knowledge about the w/o emulsification and o/w dispersion of oils at sea. A broad range of oils has been investigated in the laboratory both in bench- and meso-scale. Field trials have been performed to verify laboratory results, for calibration of models, and to study the influence of different release conditions. The major findings from these studies are:

- When oils are spilt at sea different weathering processes alter the properties of the oil as a function of time and weather conditions. Water-in-oil emulsification (w/o) and natural dispersion (o/w) are two important weathering processes taking place when oil is spilt at sea.
- Laboratory studies of a wide spectrum of oils have revealed that both the physicochemical properties of the oils and the release conditions that control the initial film thickness are fundamental parameters for the rate of emulsion formation and for the rheological properties of the emulsion formed.

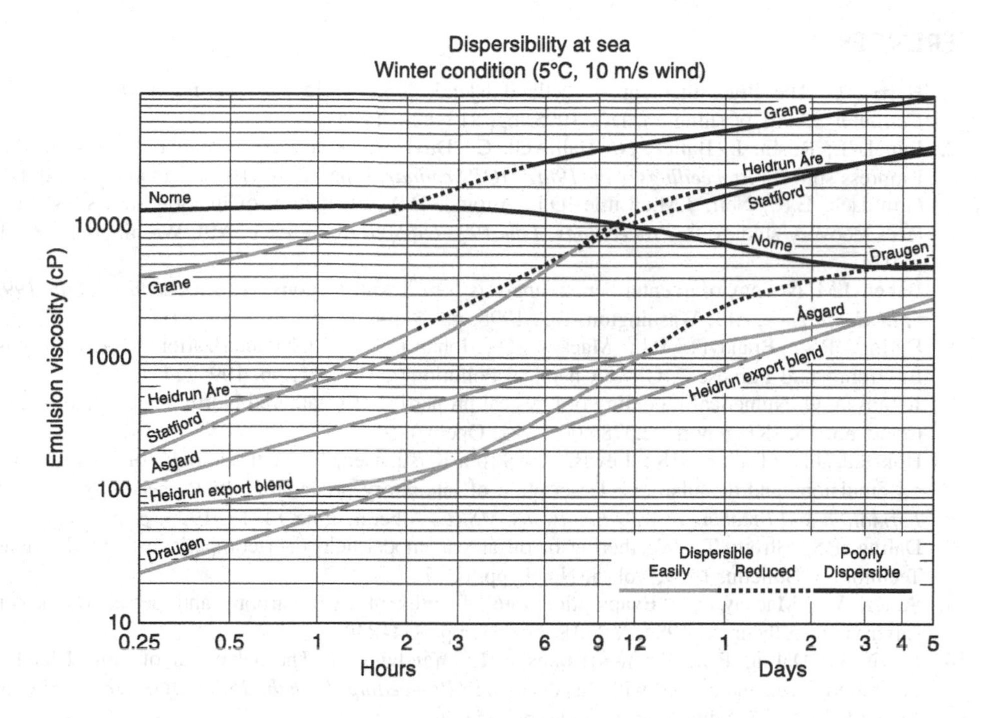

FIGURE 10.27 Window of opportunity for the use of chemical dispersants for a selection of Norwegian crude oils.

- Knowledge about weathering behavior of crude oils and fuel oils is of importance for contingency planning, response analysis, net environmental benefit analysis (NEBA), and for rapid and appropriate decision making in the case of an oil spill.
- The appearance and rheology of oil slicks at the sea surface can be very different for different oils and degrees of weathering. Continuous oil slicks may be observed at an early stage after release, but weathering processes will soon transform the slicks into a patchy appearance consisting of broken fragments of highly emulsified oil.
- There is a generally good agreement between analysis of oil samples taken during experimental field trials and model prediction using the SINTEF Oil Weathering Model based on oil specific laboratory data.
- None of the three underwater releases (1995, 1996, and 2000) showed any emulsion formation in the rising plumes of oil droplets. The emulsification took place on the sea surface (when the initial oil film thickness was sufficient).
- Simulated underwater blowouts from moderate depths (<300 m) showed that the gas-bubble plume will come to the surface, bringing entrained water with it. The rapid surface spreading of this entrained water will cause the surfacing oil to spread into a thin oil film. This film may be too thin for emulsification to take place unless further concentration of the oil into thicker narrow bands (windrows) occurs with the development of wind-driven Langmuir circulation cells.
- With blowouts in deep waters (>500 m), the plume may be trapped in the water column, and the rising oil droplets will surface within a more limited area. This may lead to initial oil film thicknesses that are sufficient for emulsion formation.

REFERENCES

1. Harris, C. The Braer incident – Shetland Island, January. In *Proceedings of the 1995 Oil Spill Conference*, API Washington, DC, 1995, pp. 813–820, 1995.
2. Lunel, T.; Rusin, J.; Bailey, N.; Halliwell, C.; Duvis, L. A successful at sea response to the Sea Empress spill. In *Proceedings of the 19th AMOP Seminar*, June 12–14, 1996, Canada, pp. 1499–1520.
3. Grundach, E.R.; Neff, J.M.; Little, D.I.; Aurand, D.A. Evaluation of historic spill sites for long-term recovery studies. In *Proceedings of the 1995 Oil Spill Conference*, API, Washington, DC, 1995, p. 974.
4. Baker, J.M. Net environmental benefit analysis for oil spill response. In *Proceedings of the 1995 Oil Spill Conference*, API, Washington, DC, 1995, pp. 611–614.
5. Daling, P.S.; Brandvik, P.J.; Mackay, D.; Johansen, Ø. "Characterization of Crude Oils for Environmental Purposes". Oil & Chemical Pollution 7, 1990/91, pp. 199–224.
6. Johansen, Ø. Numerical modeling of physical properties of weathered North Sea crude oils. DIWO-report no. 15. IKU-report 02.0786.00/15/91. Open. 1991.
7. Hokstad, J.N.; Daling, P.S.; Lewis, A.; Strøm-Kristiansen, T. Methodology for testing water-in-oil emulsions and demulsifiers: Description of laboratory procedures. In *Proceedings Workshop on Formation and Breaking of W/O Emulsions. MSRC*, Alberta, June 14–15, 1993, p. 24.
8. Daling, P.S.; Strøm, T. "Weathering of oil at sea; model/field data comparisons". Spill Science & Technology Bulletin, 1999, Vol. 5, No. 1, pp. 63–74.
9. Stiver, W.; Mackay, D. "Evaporation rate of spills of hydrocarbons and petroleum mixtures." Environ. Sci. Technol., 1984, Vol. 18, No. 11, pp. 834–840.
10. Lewis, A.; Daling, P.S.; Strøm-Kristiansen, T.; Brandvik, P.J. The behaviour of Sture Blend crude oil spilled at sea and treated with dispersants. In *Proceedings from the 18th AMOP Technical Seminar*, June 14–15, 1995, Edmonton, Canada, pp. 453–469.
11. Mackay, D.; Zagorski, W. Studies of water-in-oil emulsions. Report EE-34: Environment Canada, Ottawa, Ontario, 1982.
12. Fingas, M.F.; Fieldhouse, B.F.; Lane, J.; Mullin, J.V. 2000. Studies of water-in-oil emulsions: Long-term stability, oil properties, and emulsions formed at sea. In *Proceedings of the 1997 AMOP Technical Seminar*, Environment Canada No. 23a, pp. 145–160.
13. *Encyclopedic Handbook of Emulsion Technology*. Johan Sjøblom (editor). Marcel Dekker Inc., New York, 2001, pp. 1–736.
14. Ritchie, W.; et al. (Ecological Steering Group on the oil spill in Shetland). An interim report on survey and monitoring, May 1993, Edinburgh, The Scottish Office.
15. Mackay, D.; Buist, I.; Mascarenhas, R.; Paterson, S. Oil spill processes and models. Department of Chemical Engineering, University of Toronto, Toronto, Ontario, Publ. No. EE-8.
16. Shaw, J.M. "A microscopic view of oil slick break-up and emulsion formation in breaking waves". Spill Science & Technology Bulletin, 2003, Vol. 8, pp. 491–501.
17. Delvigne, G.A.L.; Sweeney, C.E. "Natural dispersion of oil". Oil and Chemical Pollution, 1988, Vol 4, pp. 281–310.
18. Delvigne, G.A.L.; Hulsen, L.J.M. Simplified laboratory measurement of oil dispersion coefficient – application in computations of natural oil dispersion. In *Proceedings of the 17th Arctic and Marine Oil Spill Program (AMOP) Technical Seminar*, Vol. 1, pp. 173–187. Environment Canada, 1994.
19. Daling, P.S.; Aamo, O.M.; Lewis, A.; Strøm-Kristiansen, T. SINTEF Oil Weathering Model – predicting oil's properties at sea. In *Proceedings from 1997 International Oil Spill Conference*, Fort Lauderdale, Florida, April 2–10, pp. 297–307.
20. Moldestad, M.Ø.; Brandvik, P.J.; Daling, P.S. Development of data sets from experimental oil spills for OWM algorithm and model testing and validation. SINTEF report: STF66 A04025, 2004.
21. OTSOPA. Bonn Agreement Aerial Surveillance Handbook. OTSOPA, 2004.
22. Brandvik, P.J.; Lewis, A; Strøm-Kristiansen, T; Hokstad, J.N.; Daling, P.S. 1996. NOFO 1996. Oil on water exercise – Operational testing of Response 3000D Helibucket. IKU report no. 41.5164.00/01/96. p. 53.
23. Rye, H.; Brandvik, P.J.; Strøm, T. "Subsurface Blowouts: results from field experiments". Spill Science and Technology Bulletin, 1996, Vol. 4, No. 4, pp. 239–256.

24. Fanneløp, T.K. and Sjøen, K. "Hydrodynamics of underwater blowouts". Norwegian Maritime Research, 1980, No. 4, pp. 17–33.
25. Rye, H.; Brandvik, P.J.; Strøm-Kristiansen, T.; Lewis, A; Daling, P.S. NOFO 1996 Oil on water exercise – Simulated blow-out, releasing oil and gas at 106 meters depth, 1996.
26. Johansen, Ø.; Rye, H; Melbye, A.G.; Jensen, H.V.; Serigstad, B.; Knutsen, T. DeepSpill JIP – Experimental Discharges of Gas and Oil at Helland Hansen - June 2000, Technical Report. SINTEF Report STF66 F01082, SINTEF Applied Chemistry, Trondheim, Norway, 2001, p. 159.
27. Nordvik, A.B.; Daling, P; Engelhardt, F.R. Problems in the interpretation of spill response technology studies. In: *Proceedings of the 15th AMOP Technical Seminar*, June 10–12, Edmonton, Alberta, Canada, pp. 211–217.
28. Leirvik, F.; Moldestad, M.; Johansen, Ø. Kartlegging av voksrike råoljers tilflytsevne til skimmere. SINTEF Report. 2001.

11 Bitumen Emulsions

Per Redelius and Jean Walter

CONTENTS

11.1 INTRODUCTION

Even if most of us are not familiar with bitumen, we all know the "black" roads on which we drive every day. The majority of road surfaces are black because the binding agent used to manufacture the surfacing course is bitumen mixed with crushed rock aggregate. Road surfaces can also come in gray to white color, in which case another alternative binder has been used: Portland cement concrete.

Bitumen is the residue from crude oil distillation. It is a black visco-elastic solid at ambient temperature that turns into a viscous liquid as temperature increases. Bitumen presents unique adhesive and waterproofing properties that make this material particularly ideal to bind together aggregate.

To prepare an asphalt mix bitumen can simply be brought to a high enough temperature (usually above 150 °C) to reach a sufficiently low viscosity to be directly applied to dried and preheated aggregate material. Among its numerous properties, bitumen can also be emulsified using specific surfactants. Mixing can then be carried out at ambient temperature and these techniques do not require that the materials be first dried.

Road surfacing was invented centuries ago, possibly by the Romans who used large stone slabs to build permanent roads. The Roman road-construction method was abandoned at the end of the Roman Empire and very little improvement took place during the following centuries in terms of road technology. Roads were either not surfaced, surfaced with wood or rock cobbles in urban areas, or "surfaced" with hand crushed rock usually bound with soil in rural areas until the end of the 19th century. It is only at the end of the 19th century and in the early 20th century that the first applications of a binder, initially coal tar rather than bitumen, was used to stabilize crushed rock and slag aggregate. Road history literature refers to an early use of tar in the city of Auch, France, to surface the "Place Sallinis" in 1854 (1) and a patent was placed on the application of tar to stabilize slag by Purnell Hooley around 1901 in the UK (2). At the time, coal tar was sprayed hot onto a crushed rock mat to bind the stones together. The use of bitumen in road surfacing is more recent and related to the parallel development of the petroleum and car industries at the beginning of the 20th century.

References to the use of anionic bitumen emulsions for dust control go back to 1904 in the Ardennes race track in France. As cars became popular, road dust generated by vehicles became a real nuisance and the need for even and lasting wearing surfaces became more and more urgent. Despite an early start, emulsion remained secondary to hot sprayed tar and bitumen. One hypothesis to explain the development of hot processes over emulsions is that tar was initially more abundant, particularly in Europe. As tar cannot be emulsified, it was used in hot process and the transition from tar to bitumen may have given preference to the hot process over emulsions.

Tar, then combinations of tar and bitumen, was sprayed on roads throughout the Western world until the Second World War. A new breakthrough in the U.S.A. took place at that time: the invention of pre-mixed bituminous concrete (asphalt) where special equipment had been developed to blend in a controlled way crushed aggregate with hot bitumen. This approach was a real success and led to the generalized use of bituminous concrete in the U.S.A. and everywhere in the world where car and truck traffic developed.

Meanwhile, emulsions, initially almost entirely of anionic type, have continued to progress. Hugh Alan MacKay filed a patent in 1922 on bitumen emulsion (3). This patent is regarded as the start of emulsion technology. It is reported in the literature (1) that by the end of 1926, the production of emulsion in the U.K., Germany, Austria, Denmark, and India reached 150,000 tons, which is not negligible. France was also an important producer with reference to 6000 tons of emulsion produced in 1925. Then, in 1951, cationic emulsions appeared, opening the emulsion market to high performance processes, in the sense that anionic emulsions break by evaporation of water, which takes time, while cationic emulsions present a chemical breaking behavior leading to much shorter delays between construction and trafficking as well as higher performance.

Aside from the advantage of working at ambient temperature, the interest of emulsifying bitumen is that, as a result of the surfactants used, bitumen maintains its adhesive properties toward minerals in the presence of water. This is particularly interesting in road construction, as crushed rock aggregates are stock piled in the open air and are therefore moist. Using bitumen in emulsion form eliminates the need to dry the aggregate, an essential step when working with molten bitumen. Drying aggregate requires a complex and energy intensive plant.

This chapter will first review briefly the nature of bitumen, the specificity of surfactants available to manufacture bitumen emulsion and the emulsification techniques. We will then cover important issues such as bitumen emulsion stability and breaking mechanisms in the different type of construction applications. Focus will be on road construction where most of the current emulsion developments are taking place.

11.2 BITUMEN

Bitumen is a semi-solid material, which can be produced from certain crude oils by distillation. It can also be found in nature as "natural asphalt." It consists of a mixture of hydrocarbons of different sizes containing small amounts of heteroatoms like sulfur, nitrogen, and oxygen, as well as traces of metals like vanadium and nickel. Bitumen behaves as a visco-elastic thermoplastic solid at ambient temperature and turns into a viscous liquid at high temperature. It presents unique adhesive and waterproofing properties which make it ideal to manufacture asphalt for road construction as well as a wide range of industrial application, from waterproofing in construction to sound dampening in the car industry.

The term "bitumen" is not completely unambiguous since it has been given different meanings in different parts of the world. In Europe the term "bitumen" is defined as above, while in Canada, for example, the term "bitumen" is used for heavy crude oils. In the U.S.A. the term "asphalt" is used instead of bitumen. Sometimes bitumen is confused with tar, which is a product of completely different origin. Tar is produced by dry distillation of coal or wood.

The most common process for production of bitumen is by distillation under vacuum of properly selected crude oils. There are, however, a limited number of crude oils which permit direct distillation to proper bitumen grades suitable for production of road asphalt. Although the reserves of such crude oils is very large worldwide, they are not primarily produced since they contain too small amounts of fuel, which is the most important and profitable product for refiners. A vacuum residue which is not directly suitable as a bitumen binder may be further processed by extraction or oxidation. The primary product from an extraction process using propane as solvent is heavy oil. There will also be a hard residue of bitumen remaining after the extraction process that is sometimes referred to as propane bitumen. Propane bitumen may be diluted with a distillate to give bitumen fulfilling the specification for a road binder. A process sometimes used to increase the stiffness of bitumen is oxidation. This can be achieved by blowing air through hot bitumen. The oxidation makes the bitumen harder at high temperatures while partly maintaining the soft properties at low temperatures.

The true chemical nature of bitumen is not completely known. Most books and papers on bitumen chemistry teach that bitumen is a colloidal dispersion of asphaltenes in maltenes. The dispersion is stabilized with resins. This statement is based on the well-known fact that when bitumen is diluted with certain hydrocarbon liquids, n-alkanes, a precipitate appears. If the hydrocarbon liquid is n-heptane, the precipitate is called "asphaltenes." It has been proposed that the "asphaltenes" are present in the bitumen in the form of micelles. The first author to introduce this concept was Nelensteyn in 1924 (4). The model was later refined by Pfeiffer and Saal (5). Other models, which question the existence of micelles, have also been proposed, for example by Park and Mansoori (6), and later as a result from the SHRP development program in the U.S.A. (7). Recent research has shown that the asphaltenes are soluble in the maltenes and thus no micelles can exist in the bitumen (8). To further investigate the chemistry of bitumen the maltenes are separated into different fractions using chromatographic techniques. A large number of different procedures have been described, for example gel permeation chromatography (GPC), ion exchange chromatography (IEC), high pressure liquid chromatography (HPLC), and others (9). The most common procedure is probably the SARA separation, which divides bitumen into four generic groups. In the first step the asphaltenes are precipitated by n-heptane, followed by separation of the maltenes with respect to polarity in three fractions: saturates, aromatics, and resins (10).

The functional properties of bitumen are usually related to its use as binder in asphalt for roads. Thus the most common properties are related to the rheology of bitumen. Since the road construction area is very conservative and bitumen has been used for about 100 years, most tests are empirical and have been used for a long time. Two of the most common tests are penetration at 25 °C and softening point Ring&Ball. The penetration gives a measure of the stiffness of the bitumen at most common service temperatures of a road, while the Ring&Ball gives the stiffness close to the highest expected temperature in the road. In Europe bitumens are graded according to their penetration at 25 °C, for example 50/70, where the two numbers give the highest and lowest limit for the particular grade. It is also common, particularly in the U.S.A., to use viscosity gradation based on viscosity at 60 °C. Bitumen is, however, a visco-elastic material with a complex rheology and thus can not be completely described by simple penetration and softening point. The development of modern and reliable rheometers, for example the dynamic shear rheometer (DSR), has made it possible to describe the full rheology of bitumen.

11.3 EMULSIFIERS

Most types of emulsifier have been used for emulsification of bitumen. In the early days anionic emulsifiers were the most common, but since the 1950s cationic emulsifiers have taken over more and more. One of the advantages of cationic emulsifiers is their good adhesion to acidic stone materials like granite, which also are good construction materials in roads. Another advantage is the "active breaking" against the surface of acidic stones. The anionic emulsions have maintained some importance for certain industrial applications as well as for road construction using basic stones like limestone. Non-ionic emulsions and clay emulsions have occasionally been used for certain applications but are currently not produced in any substantial amounts.

11.3.1 Anionic Emulsifiers

During the early days of bitumen emulsions, anionic emulsifiers were the most common types. In principle, any type of long chain fatty acid or mixture of fatty acids can be used for emulsification of bitumen. The first type of emulsifier, which was reported for emulsification of bitumen, is oleic acid (Figure 11.1) potassium or sodium soap. In later years they were replaced by rosin soaps, for example Vinsol® resin and tall oil soaps. Tall oil is distilled from a byproduct of the paper pulp industry making use of the so-called "sulfate process." The main constituents of tall oil are oleic acid, linoleic acid, and abietic acid. In 1960 about 95% of all bitumen emulsions used in the U.S.A. was of the anionic type (11).

In spite of the dominance for cationic emulsifiers for bitumen many new anionic emulsifiers, tailor-made for bitumen, are available on the market. They are mainly derivatives of the same byproducts from the paper pulp industry, tall oil, tall oil rosin or lignin, as used in the early days.

A typical formulation of an anionic bitumen emulsion is:

Bitumen 160/220 65%
Water 34.1%
Tall oil (crude) 0.4%
Vinsol® resin 0.4%
Sodium hydroxide 0.09%

In the first manufacturing step the fatty acid is dispersed in water followed by addition of sodium hydroxide (or potassium hydroxide). Hot bitumen is then added to the soap solution under vigorous mixing. In full-scale production a colloid mill is usually used.

One disadvantage with carboxylic acid soaps is that most calcium salts of carboxylic fatty acids are insoluble in water. Since calcium ions are very common in tap water or water from natural wells, it is very important to keep careful control of the water quality when working with anionic emulsions based on fatty acids. Calcium salts of fatty acids remain in the soap solution as lumps, which might block filters and pipes.

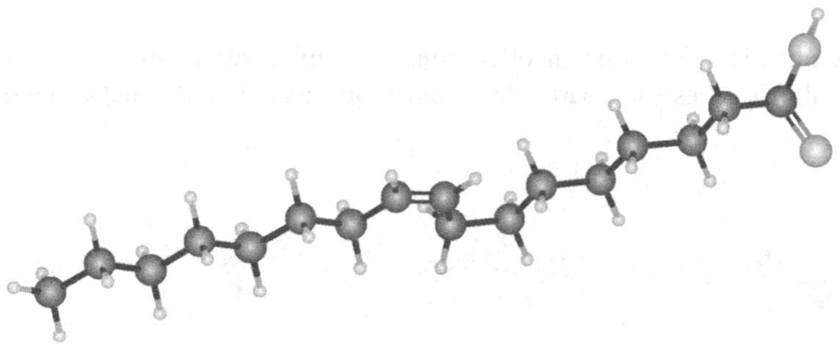

FIGURE 11.1 Oleic acid. Black is carbon, light gray is oxygen atoms.

One way of overcoming the disadvantage with calcium sensitivity is to use anionic emulsifier based on sulfonic acids or sulfates. These emulsifiers are very effective for emulsification of bitumen and they do not form insoluble calcium salts.

11.3.2 Cationic Emulsifiers

During the last 30 years the cationic emulsifiers have dominated in bitumen emulsions for road applications. The main advantage is the good adhesion to acidic stone materials which usually is preferred for road constructions due to their good mechanical properties.

A large variety of commercially available cationic emulsifiers are suitable for emulsification of bitumen. The most common is of the general type "fatty diamine." In these emulsifiers the hydrocarbon part originates from fatty acids where the acid part has been replaced by some type of diamine or polyamine. The alkyl part commonly originates from fatty acids from tallow, which are more or less hydrogenated to increase their stability. The mixture of fatty acids varies slightly depending on origin of the tallow but a common figure for the hydrogenated tallow mixture is:

Tetradecanoic acid (myristic acid) 4%
Hexadecanoic acid (palmitic acid) 30%
Octadecanoic acid (stearic acid) 60%
cis-9-Octadecanoic acid (oleic acid) 2%

Also the hydrocarbon part of tall oil and lignin from the paper and pulp industry are used to produce cationic emulsifiers. Tall oil also consists of a mixture of fatty acids while lignin has a completely different structure.

Perhaps the most common emulsifier for bitumen emulsification is *N*-tallow alkyltrimethylene-diamine (Figure 11.2) but a variety of emulsifiers exist with different types of amino groups. For example, tallow amine ethoxylates, tallow polyamines, tallow amido amines, quaternized tallow diamines, imidazolines. Although alkyl chains originating from tallow are the most common for bitumen emulsifiers there are also a large variety of synthetic alkyl chains used to manufacture emulsifiers for bitumen.

A common formulation of a cationic bitumen emulsion is:

Water 33.6%
N-tallow-1,3-propanediamine 0.2%
Hydrochloric acid (32%) 0.2%
Bitumen 66%

Commercial emulsifiers for bitumen often consist of mixtures of more than one emulsifier to give the optimal properties for a particular application. Recently, automatic plants where bitumen

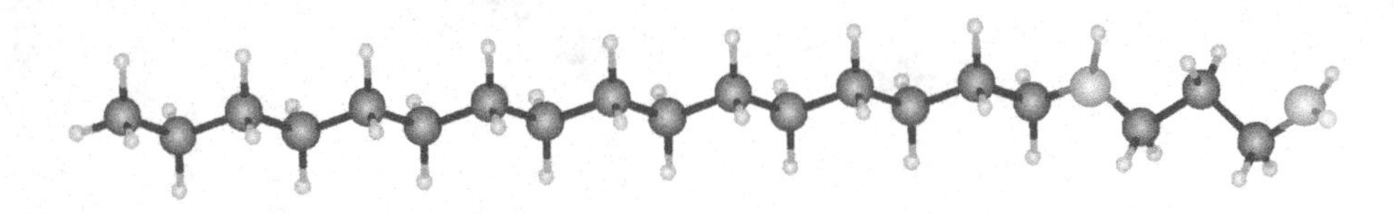

FIGURE 11.2 *N*-hexadecanoic-1,3-propanediamine. Black is carbon, light gray is nitrogen atoms.

emulsions are prepared in one continuous process have been developed. This requires emulsifiers which can be pumped at ambient temperature and thus the development of liquid emulsifiers has been important. This can be achieved by oxyalkylation of the *N*-tallow-diamines. One example is oxypropylation with a polypropylenoxide containing 1 to 4 propylenoxid units (12).

Before the fatty amines can be used as emulsifier they have to be acidified (except the quaternary amines). This is usually done with hydrochloric acid, making a soap solution in water. Sometimes other acids have been used, such as acetic acid or phosphoric acid. The latter acid should be used with care because, generally, fatty-amine phosphates are insoluble in water. The emulsion is produced by pumping the soap solution together with hot bitumen in a colloid mill.

11.3.3 Non-Ionic Emulsifiers

Non-ionic emulsifiers can also be used for emulsification of bitumen. These emulsifiers become less immobilized when in contact with soil and stone material which make the emulsions very stable. They have thus frequently been used in applications where the bitumen emulsion is supposed to penetrate through soil, dust, and even clay. The advantage is that the bitumen makes the soil less susceptible for water.

Typical emulsifiers used for bitumen belong to the group of alkyl ethoxylates. The previous most common emulsifier from this group is the nonyl phenol ethoxylates. This emulsifier has been suspected, however, of being harmful for the aquatic environment and is not recommended for bitumen emulsification any more, although up to the end of 2004 there has been no legislation restricting its use.

Non-ionic emulsifiers have also been used for emulsification of very heavy crude oils (Orimulsion) to facilitate pumping, storing, and transportation of the crude oil. These crude oils have such a high viscosity that they cannot be handled with conventional equipment for crude oil transportation (pipelines and tankers), not even at higher temperature. As emulsions they can be used, however, as fuel for industrial boilers without any further refining.

11.3.4 Clay Stabilizers

Bitumen can also be emulsified using colloidal clays as stabilizers. Most common are clays of the montmorillonite, kaolinite, and attapulgite type. One of the most common clays of the montmorillonite type is bentonite.

The most effective bitumen emulsifiers are clays which are readily wet in water and disperse as very fine particles (<1 μm) having colloidal properties and which adhere to the surface of the bitumen with a contact angle less than 90°. If enough clay particles attach to the surface, the surface charge of the clay will create a repulsive energy, which prevents coalescence of the bitumen drops.

Clay emulsions are mainly used for house building purposes as protective coatings for roofs, metals, and masonry. They have also been used as adhesives for laminates and tiles as well as flooring mastics.

11.4 MANUFACTURING AND HANDLING

Bitumen emulsions are generally of the oil-in-water type of emulsions; thus most emulsifiers for bitumen emulsification are more soluble in the water phase than in the bitumen. Therefore it would be an advantage for the emulsification process if the emulsifier was added to the bitumen phase, since the migration of the emulsifier into the water phase will help the emulsification process. In the production of bitumen emulsions, however, the bitumen is usually at temperatures between 120 °C and 180 °C to keep the viscosity at a suitable level for emulsification. At these temperatures most emulsifiers are not stable, or at least have a very short lifetime. Thus the common practice in bitumen emulsification is to mix the emulsifier into the water phase. The demand on thermal stability is then less, but it still requires the emulsifier to be stable up to 100 °C for several days.

11.4.1 Colloid Mills

The general process of emulsification involves controlled mixing of water, emulsifier, and bitumen. With the aid of high shear action, the bitumen is broken up into minute particles and dispersed in the water phase. The dispersion may be accomplished by use of simple mixers, centrifugal pumps, homogenizers, or colloid mills. The selection of the type of equipment is dependent on the expected volume of manufacture as well as plant investment considerations. For the manufacture of road emulsions, colloid mills are most commonly used. A colloid mill consists of a rapidly revolving conical disc, or rotor, which fits closely in a stationary part known as a stator. The bitumen, water, and emulsifier are forced through the narrow clearance between the rotor and the stator thus effecting dispersion (Figure 11.3). The capacity is usually between 10 and 40 ton/h.

The equipment for manufacturing anionic emulsions and cationic emulsions is principally very similar, so for the simplicity of the following discussion we will focus on cationic emulsions.

The first step in the emulsification process is to acidify the emulsifier. Most diamines and particularly *N*-tallow alkyltrimethylendiamine are supplied as slightly basic products. It is a paste-like semi-solid at room temperature. The *N*-tallow alkyltrimethylendiamine is dispersed in warm water followed by addition of hydrochloric acid until a pH of 2 to 4 is reached. By this process the diamine is protonized and the amine functionalities will carry a positive charge. In this state the *N*-tallow alkyltrimethylendiamine is a very effective emulsifier for bitumen. The manufacture

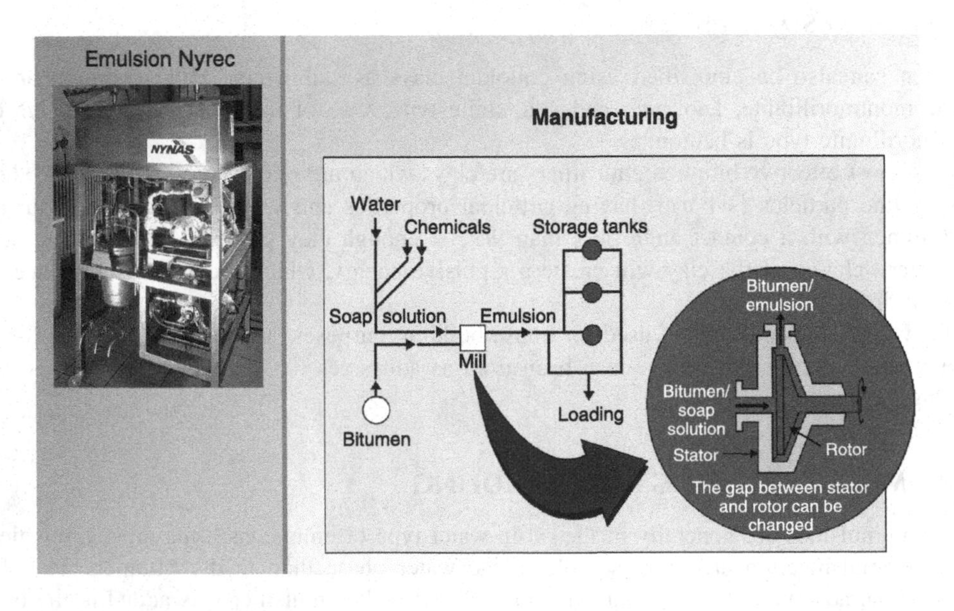

FIGURE 11.3 Schematic of a bitumen emulsion plant. Left: Laboratory plant.

of the soap solution is usually a batch process while the production of the emulsion in the colloid mill preferentially is made continuously. Therefore several batches of soap solution have to be made in parallel to maintain a continuous production in the mill.

During the last 20 years the development has been toward fully continuous emulsification plants, including the acidification process. The emulsifier, the water, and the acid are then mixed in line to make the soap solution. Sometimes a small container is located on the line before the mill to give enough time for the emulsifier to react with the acid, but this reaction is fast and no extensive reaction time is necessary. The key is to have access to special liquid emulsifiers or to install a heating system to handle the emulsifiers in a liquid state. Modern emulsion plants are built as fully mobile plants that can easily be transported in containers and be set up close to the work site.

Another important factor to take into consideration is that bitumen is too viscous at ambient temperature to be emulsified. Bitumen must therefore be brought to high temperature (above 100 °C) to reach a sufficiently low viscosity for emulsification to be possible. It is commonly agreed that the bitumen must be at a temperature corresponding to a viscosity of 200 cP or less for emulsification. The temperature of the bitumen and the aqueous phase should therefore be adjusted so that the resulting emulsion temperature does not exceed 100 °C. In some cases when hard bitumen (pen 50 and below) and polymer modified bitumen are emulsified it is not possible to keep the resulting temperature below 100 °C. If this is the case, a cooler must be fitted on the output of the emulsion mill to bring the emulsion below the boiling point of water.

Calculation of expected emulsion temperature can be achieved using the following equation:

$$(T_b - T_e)^*SH_b{}^*b = (T_e - T_w)^*SH_w{}^*w$$

where

T_b = bitumen temperature
T_w = water temperature
T_e = emulsion temperature
b = bitumen content, %
w = water content, %
SH_b = specific heat of bitumen (1.9 kJ/kg K)
SH_w = specific heat of water (4.2 kJ/kg K)

Note that the milling operation in itself generates an emulsion temperature increase of approximately 10 °C.

Sometimes it is desirable to add minor amounts (<2%) of light petroleum naphtha (kerosene) to the emulsion. The kerosene improves the wetting to stone surfaces as well as improves the coalescence once the emulsion is broken. The kerosene could either be added by an extra feeding line into the bitumen line before emulsification, or alternatively it could be added directly to the finished emulsion under gentle mixing.

11.4.2 Other Industrial Emulsification Methods

Colloid mills are not ideal tools, however. Many emulsion formulators would very much like to be in full control of their emulsion particle size, which is far from being the case with colloid mills. Some mills have a possibility of adjusting rotor speed or rotor/stator gap supposedly to help control emulsion particle size. However, all colloid mills seem to produce polydisperse bitumen emulsions with particle size ranging from 1 to 100 microns. Several properties, in particular viscosity and breaking characteristics, are affected by the particle size distribution of the emulsion and work is under way to develop potential methods to achieve monodispersed emulsions of adjustable median diameter (13).

Although homogenizers in association with a standard colloid mill would be an option the authors are not aware at the time this book was written that such a device has been used industrially for production of bitumen emulsions.

The only industrial alternative to the colloid mill seems to be the so-called SMEP process for static mixer emulsification process (14). This emulsification method has been applied at industrial scale in Europe. It is claimed that it is possible to reach monodispersed emulsions of controlled median diameter with this method.

11.4.3 Stability

Bitumen emulsion stability is a critical issue due to the high viscosity of bitumen at ambient temperature. Once a bitumen emulsion is broken it is impossible to bring it back to the emulsified state again. An emulsion sufficiently unstable to cause coalescence and to break will cause major problems in the industry: storage tanks, pipes, pumps, and filter blockage which can be cleaned only by melting back the bitumen to liquid stage, i.e., above 150 °C; or through solvents. On the other hand, the aim of the bitumen is to act as glue between stones, which only can be achieved once the emulsion is broken. It is therefore necessary to formulate the emulsion to provide stability for storage and transportation but sufficiently unstable so it breaks once in the right place in the application.

11.4.3.1 Different Aspects of Bitumen Emulsion Stability

As many as five types of bitumen emulsion stability are recognized: chemical, storage, freezing, mechanical, and mixing. Some of these types are related to each other, whereas often there is no relationship between others. In this discussion we will focus on storage stability which is mainly due to sedimentation or creaming, and chemical stability which is more related to true breaking and coalescence of the emulsion. In the first case it is possible to restore the emulsion to its original state by simple stirring, but in the second case it is necessary to re-emulsify it.

Another aspect of stability is the emulsion's ability to break on contact with stone materials. Depending upon the type of application it can be desirable to have a very quick breaking, for example with surface dressing, slurry surfacing, or various types of sealing. In other applications it is desirable to have a very slow breaking, for example with mixing grade asphalt stored for later use for patching or pot hole filling. It is, however, common to all emulsions that once they are in place on the road they should break and cure as quickly as possible.

11.4.3.2 Storage Stability

Storage stability relates to "static" storage of the emulsion. Storage stability is usually solved in the industry by maintaining the emulsion under slow agitation in the storage tank. Agitation may, however, promote breaking if too intense. Bitumen emulsions can not always be stored in tanks with agitation, but are sometimes supplied in drums to smaller work sites. In these cases it is important to select emulsions with good storage stability or to have a procedure to stir or roll the drums before using the emulsion.

11.4.3.3 Properties that Control Storage Stability

11.4.3.3.1 Bitumen Density

Creaming and sedimentation are mostly related to the density difference between the bitumen (from 0.95 to 1.10) and the water. Since the density of bitumen and water are very close and since the thermal expansion coefficients are different for water and bitumen it is often possible to find a temperature where the density is practically identical. This temperature is usually found around 70 °C. If the density of the bitumen is very high it can be an advantage to increase the density of the water by adding a salt, for example calcium chloride. It should be noted that the high temperature and high salt content decrease the chemical stability of the emulsion. High temperature facilitates coalescence by decreasing the viscosity of the bitumen while high salt content decreases the electrical double layer which decreases the repulsive force between bitumen particles.

11.4.3.3.2 Bitumen Content

Bitumen emulsions containing high concentrations of bitumen (67 to 70%) generally have fewer problems with storage stability than emulsions with low bitumen content (40 to 60%). The reason is that in the 70% emulsions so much space is filled by the bitumen phase that there is practically no room for sedimentation or creaming to take place. For the 50% emulsions, however, the distance between the bitumen drops is so large that only Brownian movements can keep the emulsion stable, and these movements are too weak for particles larger than 10 μm; thus it is almost impossible to avoid sedimentation or creaming. Unfortunately, it is very common to supply 50% emulsions in drums which are stored at ambient temperature and thus severe separation always has to be expected, so these emulsions should be carefully stirred before use.

11.4.3.3.3 Freezing

Freezing generally destroys oil-in-water emulsions, and this is also the case with bitumen emulsions. They should never be stored at temperatures below the freezing point of the continuous phase. A bitumen emulsion which is suspected of having been frozen should not be used for road applications. If, for some reason, bitumen emulsions are to be used below the freezing point, antifreeze liquids should be added. Examples are ethanol or ethylene glycol. None of these is soluble in bitumen and therefore will not destroy the bitumen properties.

11.4.3.3.4 Maximum Particle Size

If the emulsion contains particles which are larger than approximately 10 μm it becomes more sensitive to mechanical stress. This results in increased risk for breaking during pumping, transport, and stirring of the emulsion. The expected storage time for such emulsions is also shorter due to Oswald ripening that consumes the smallest particles which coalesce into the largest ones.

11.4.3.4 Chemical Stability

With chemical stability we refer to the stability against breaking and coalescence. There are several factors which determine the chemical stability. Some of these are related to the formulation such as emulsifier type and amount, pH, and bitumen content. Other factors are related to handling of the emulsion, such as mechanical stress, pumping, stirring, and transportation. Another aspect of emulsion stability is stability in contact with stone material. This is further discussed in Section 11.6.

11.4.3.4.1 Emulsifier Type and Content

Surfactant type and concentration in the water phase is a very important factor in bitumen emulsion stability. Generally speaking an increase of emulsifier content and an increase in acidity (lower pH) increase the stability of the emulsion. This statement is true within reasonable limits, such as an emulsifier content between 0.1 and 3% and from pH 2 to 5, but it is not generally valid. Some cationic emulsifiers are of the quartenary amines type. These emulsifiers do not lose their effect as emulsifiers when the pH is changed, and are thus not sensitive to pH changes.

It is usually easy to make very stable bitumen emulsions which can be stored for months and even years but many bitumen emulsions are formulated to have certain breaking characteristics when coming into contact with stone material, and are thus formulated to be on the borderline of stability. The selection of emulsifier type and amount is usually done on an empirical basis, with respect to the type of bitumen, type of emulsion mill, and type of stone material expected.

11.4.3.4.2 Mixing of Bitumen Emulsions

Mixing of bitumen emulsions made with emulsifiers of opposite charge, for example anionic and cationic, is not possible. Emulsions made with identical emulsifier can generally be mixed. In more unclear cases, mixing should be avoided since the exact effect of mixing can not easily be predicted. One possibility of using emulsion mixing is to have emulsions with the same type of emulsifier but different grades of bitumen. By mixing the emulsions, any bitumen grade between the two can be achieved in the final application when the emulsion is fully broken and the bitumen droplets have coalesced. Another advantage is the lower stiffness and the better coating of the stones in a cold asphalt mix prepared with mixed emulsions (15).

11.5 PROPERTIES OF BITUMEN EMULSIONS

Properties of bitumen emulsions for roads are specified in many standards. Each country usually has a national standard, although there are many similarities among the different countries. It is obvious that the national standardization bodies have to some extent copied each other. In Europe a harmonized standard will soon be launched which will make all national standards in Europe obsolete. Examples of standards on bitumen emulsions are ASTM D-977 in the U.S.A., EN 13808 in Europe and JIS K 2208 in Japan. All standards specify only very basic properties and it is usually not possible to determine from the specification whether the emulsion will perform well in a particular road application or not. The performance as a construction material can only be determined from the properties of the binder as retained after full breaking and coalescence of the emulsion. The most common properties of bitumen emulsions specified are given in Sections 11.5.1 to 11.5.10.

11.5.1 Charge on Emulsion Particles

This is a test to distinguish anionic and cationic emulsions. It is included in most specifications for bitumen emulsions, but is rarely performed in any laboratory. Usually the type of emulsion is known, or as is the case in many countries, only cationic emulsions are used.

11.5.2 Viscosity

The viscosity of the bitumen emulsion is important for pumping and transportation. In some applications, for example surface dressing, bitumen emulsion is sprayed on the road followed by a thin layer of stone material. In this case the viscosity is critical. It should be low enough to permit even spraying but at the same time high enough to prevent run-off, once it is sprayed on the road.

In most standards the viscosity is determined by some kind of efflux viscometer, for example in the European EN 12846 which makes use of a STV (standard tar viscometer). The value given is an efflux time rather than a viscosity and should more correctly be named pseudoviscosity. It has been proposed that modern rheometers are used to specify the true viscosity of bitumen emulsions. There are, however, problems with the non-Newtonian behavior and further complications due to breaking of the emulsion against metal surfaces in the rheometers which so far have prevented standardization.

The viscosity is mainly controlled by the binder content, but also by type of emulsifier, particle size distribution, and salt content in the bitumen. The latter might give an osmotic effect which will drag water inside the bitumen drops, resulting in expansion of the drops and a decrease in continuous water phase which will be seen as an increase in viscosity (16). The effect could be overcome by adding small amounts of a salt, for example calcium chloride, to the water phase. The increased salt content in the water phase will decrease the osmotic pressure for the water into the bitumen droplets.

11.5.3 Breaking Behavior

This test is used to classify bitumen emulsions into slow breaking, medium breaking, and fast breaking. Most tests for breaking behavior make use of some kind of mineral filler additive which is added to the emulsion until it breaks. The amount of additive is a measure of the stability. The European specification EN 13075 makes use of standard filler which is added until the emulsion is completely broken. The breaking speed in real applications may, however, be completely different since the reaction against fines varies a lot with different minerals.

11.5.4 Mixing Stability with Cement

A special test used for identification of so-called "over stabilized emulsions" is the mixing stability with cement. Since Portland cement is very reactive against most emulsifiers and has very high specific surface area, it breaks most cationic and anionic emulsions immediately. Over-stabilized emulsions are characterized by very high stability and should not break even after mixing with cement. These emulsions are generally formulated with nonionic emulsifiers or sometimes with anionic emulsifiers to make them very stable even in contact with soil and clay. They are mainly used for soil stabilization and to make soil less water sensitive when the soil contains relatively large amounts of clay and fines. The presence of clay also causes very stable emulsions to break.

11.5.5 Residue on a Sieve

This test is mainly used for quality control. The test determines the amount of bitumen which is present in the emulsion as larger particles or lumps. The test consists of simple pouring of the emulsion through a sieve with relatively low mesh size. The amount of material which is collected on the sieve should not exceed a certain value. In the European specification there are several options with two mesh sizes (0.500 mm and 0.160 mm) and different levels of acceptance for retained material on the sieve from 0.1 to 0.5%.

A special case of the sieve test is a storage stability test, where the emulsion is stored for several days (usually 7) followed by a second sieve test. If the amount of material collected on the sieve is higher than the specified amount the emulsion is deemed not to be stable enough to sustain storage for a certain time. In this test the collected material is considered to be broken and it is thus not possible to restore the emulsion by simple stirring.

11.5.6 Settling Tendency

In this test the settling or the creaming tendency of bitumen emulsion is determined. The test is part of the European standards and the principle is to hold the emulsion for 7 days in a cylinder and then measure the binder content in the top and bottom of the cylinder. If the binder content has increased in the top part of the cylinder, the emulsion is "creaming," while an increase of binder content at the bottom of the cylinder indicates a tendency of the emulsion for "sedimentation." This property is particularly important for emulsions containing less than 60% bitumen and which are stored in drums or tanks without stirring. An emulsion that has settled can usually be restored by careful stirring provided that no breaking has occurred.

11.5.7 Coating and Adhesion

These are two of the very few tests which are related to performance of the bitumen emulsion in road applications. Most coating tests are empirical and make use of a mixing procedure with stone material followed by a visual inspection of the degree of coating. If the inspection takes place after full breaking and curing of the emulsion it is sometimes referred to as adhesion. For practical purpose it is not advisable to specify a coating test with standard stones since the behavior is very much dependent on the mineralogy of the stone. The test should preferably be performed with the local stone material and is consequently more a test of the stone material rather than the emulsion.

11.5.8 Recovery of Bitumen

Strictly speaking this is not a property of the emulsion, but in most standards for bitumen emulsion a process for recovery of the bitumen from the emulsion is included. This is to give the user a standard test to recover the material that gives the functional properties to the finished asphalt. In the European standard EN 13808 two methods have been included. The first method consists of a distillation procedure as specified in ASTM D-244 or EN 1431. In this test the water and solvents, if present, are distilled and collected in such a way that the solvent content can easily be determined. The bitumen content is determined by weighing and the recovered bitumen can also be subjected to further testing. The high temperature used during the distillation (260 °C) may, however, have slightly changed the properties of the binder. Caution should then be applied to the validity of further testing on recovered bitumen. This is particularly important when testing emulsions of polymer modified binders which are more sensitive to high temperatures. When further testing of polymer modified binder is needed the use of a milder test to recover the binder without too much heating should be considered. In EN 13074, the second European method for recovery of binder, the water is evaporated at maximum 50 °C. In this method, a residue of solvent might still remain in the binder, changing its original properties.

11.5.9 Binder Content

A basic property of a bitumen emulsion is the bitumen content. The bitumen content may be calculated based on the result from any of the methods for recovery of bitumen above. If, however, the emulsion contains solvent with high volatility (kerosene) the two methods will give different values. In the U.S.A. it has been decided to use the residue after distillation as the binder content. In Europe it has been decided to consider everything which is not water as "binder content." Therefore a method for determination of water content according to Dean and Stark has been approved as European standard (EN 1428). Thus if solvent is present in the emulsion it will be considered as part of the binder. In most cases the solvent content is low (<2%) and the difference between the two methods for recovery of binder will be of minor importance for the practitioner.

11.5.10 Particle Size

One of the most important factors determining the properties of a bitumen emulsion is the particle size and the size distribution of the bitumen droplets. This is never specified, but many producers of bitumen emulsion include measurement of particle size in their quality control system. The most common technique is laser scattering and many practical commercial instruments are available on the market. Particle size distribution is partly determined by the formulation of the emulsion, but the manufacturing conditions are also of vital importance for the size distribution. Some examples are:

- The temperature difference between bitumen and aqueous phases
- The viscosity of the bitumen
- The gap between the rotor and stator in the emulsion mill
- The rotation speed of the mill
- The contacting configuration between the soap and the bitumen

In fact, it is known that different emulsion mills give products with different properties although identical formulations of bitumen emulsions are applied. An attempt to relate particle size to other properties is illustrated in Figure 11.4 (17).

11.6 BREAKING AND ADHESION

The main purpose of bitumen in asphalt is to act as a glue to bind the stones together. The emulsification of bitumen should be considered as a way of transportation of the binder (bitumen) to the right place between the stones in the asphalt. Once the bitumen is in the right place it should separate from the water followed by wetting the stone surface and gluing the stones together. We may call this process the "breaking of the emulsion" although in reality it consists of a number of different processes. It is, however, not enough to make the water separate from the bitumen; we also need a good adhesion to the surface of the stone. This is also a complication since the surface of the stone is generally more hydrophilic than hydrophobic. This means that the affinity for water is higher than the affinity for the bitumen. Thus the full strength of the adhesion between the bitumen and the stone surface will not be achieved until the stone surface is completely dry, which sometimes can take a very long time. Here a proper selection of surfactant could improve the situation, since certain cationic emulsifiers could displace the water from the stone surface and promote good adhesion between the bitumen and the stone. In these aspects

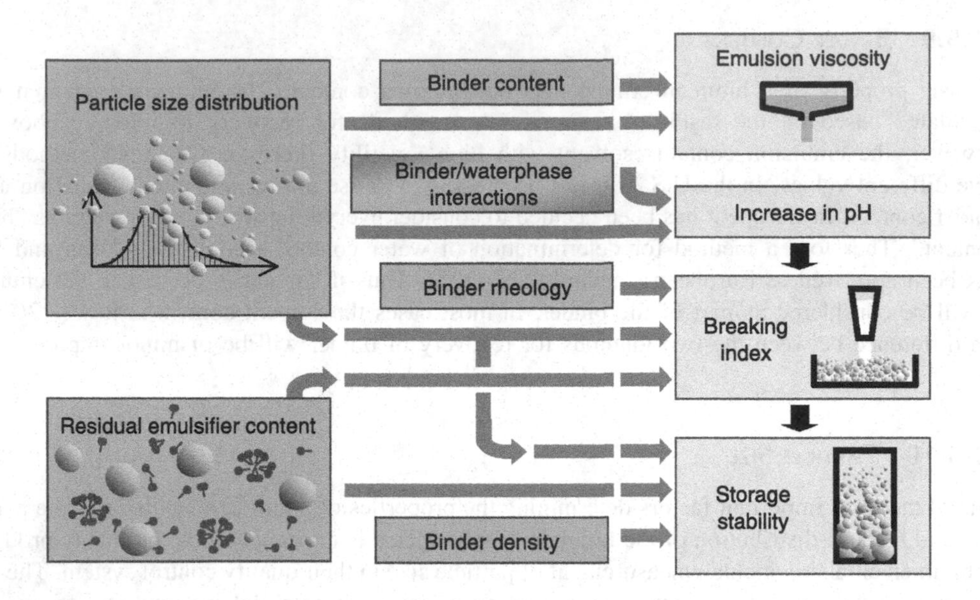

FIGURE 11.4 Proposed relation between the formulation of a bitumen emulsion and its properties (17).

cationic emulsifiers are preferred for acidic stone surfaces, while anionic emulsifiers are preferred for basic stones.

Depending on type of application the desired breaking process is different. In some applications a fast breaking is preferred while in others a delayed process is preferred. The conditions could also be very different, as for example in some applications the stone material could be dry or wet as well as clean or dusty. For example in surface dressing applications (see below) the breaking should be as fast as possible as well as having good adhesion of the stones. In cold asphalt mix applications on the other hand the breaking process must be delayed to permit enough time for transportation, laying, and compaction after the aggregates have been mixed with the bitumen emulsion. After compaction the breaking should take place as quickly as possible to achieve good strength in the new road to be able to sustain traffic as soon as possible, preferably within hours rather than days.

The breaking mechanism is mainly of two types, breaking under influence of added stone material or breaking due to evaporation of the water (drying). In the case of pH sensitive emulsifiers it is also possible to cause breaking by changing the pH of the emulsion. This will be further discussed under "breaking additives." It is preferred to select an emulsifier that adsorbs strongly to the stone surface because then the breaking will mainly be controlled by the addition of the aggregate. The breaking mechanism itself is rather complicated and can be divided into different steps. The exact nature of these steps is not completely known in detail but basically they consist of the following:

- Destabilization of the emulsion due to adsorption of emulsifiers on the added stone material
- Agglomeration of the emulsion particles forming a solid but nontacky and brittle material (cheesy state)
- Coalescence of the bitumen droplets to give back the tackiness and stiffness of the bitumen
- Complete evaporation of water giving full adhesion to the aggregate surface and full strength of the asphalt

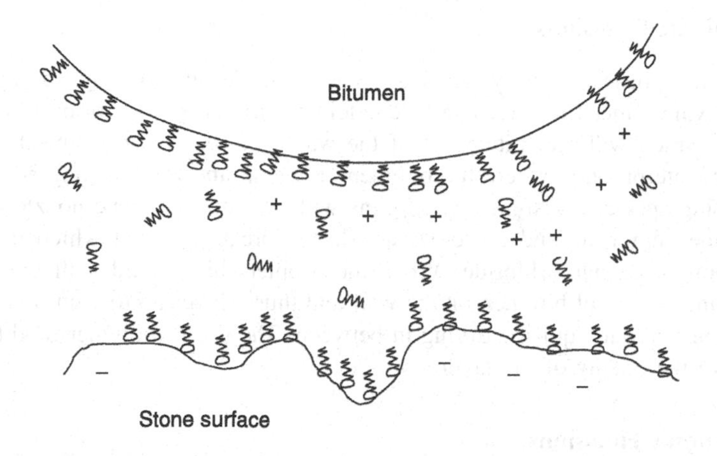

FIGURE 11.5 Situation when bitumen emulsion has been contacted with stone material.

A hypothetical process for fast breaking bitumen emulsions is illustrated in Figure 11.5. It is assumed that the concentration of emulsifier is below the critical micelle concentration (CMC) in the water phase and that the surface of the stone material is negatively charged. The emulsifiers are adsorbed on the surface of the stone, and consequently if the surface of the stone material is large enough the emulsion becomes destabilized. At the same time the layer of emulsifier sticking to the stone will make the surface hydrophobic, promoting the wetting of the aggregate by the hydrophobic bitumen. At the same time the emulsifier will act as an anti-stripping agent.

If the emulsifier content in the emulsion is increased to a concentration higher than the CMC, the excess emulsifier will form micelles in the water phase. When mixing this type of emulsion with stone material, the excess emulsifier might coat the stones without destabilization of the emulsion. The breaking of the emulsion then becomes much slower and may even not take place until the water has evaporated.

It is evident that the breaking speed is dependent on the amount of emulsifier, but the amount of stone surface is equally important. The aggregates used in road construction are usually crushed stones, which may contain various amounts of fines. The fines have a very high specific surface area (surface per gram) and the breaking will thus be very sensitive to the filler content in the stone material. The third important step in the breaking process is coalescence, which is heavily dependent on the viscosity of the bitumen. At low temperatures the bitumen droplets can be considered more as solids rather than liquids. Under these conditions the coalescence can take a very long time. A common way of improving the coalescence is to add a small amount of light petroleum liquid, for example kerosene. If the coalescence is not complete the binder may stay in a state where full strength has not been achieved (cheesy state) for days and even weeks. When the road is opened for traffic the kneading action will easily make the coalescence complete.

11.6.1 Breaking Additives

One way of increasing the flexibility of bitumen emulsions is to have good breaking additives to control the breaking rather than to be dependent on the fines in the stone material or evaporation of the water. Due to the different nature of cationic and anionic emulsions, different approaches have to be applied for breaking of the different types of emulsions.

11.6.1.1 Anionic Emulsions

Anionic emulsions based on fatty acids can easily be destabilized by adding a salt solution containing a divalent metal ion, for example calcium ions. They will form an insoluble salt with the fatty acid, which will precipitate out of the water phase. This principle has been exploited in systems for waterproofing where the bitumen emulsion and the breaking additive are sprayed on surfaces using specially designed spray guns with two nozzles. One nozzle sprays a polymer modified bitumen emulsion and the other nozzle the breaking agent which usually consists of a water solution of calcium chloride. When the droplets hit a solid wall the emulsion breaks instantly forming a layer of bitumen on the wall and thus a layer up to 1 cm thick can be sprayed in one operation without requiring drying in between. The use of polymer modified bitumen will further improve the quality of the layer.

11.6.1.2 Cationic Emulsions

Cationic emulsions may be destabilized by increasing the pH, which usually leads to destabilization and consecutive breaking of the emulsion. The reason for the breaking is that the amine group loses its positive charge and becomes a less effective emulsifier. This technique may be used by spraying an alkaline water solution on top of a bitumen emulsion. The breaking will start, however, on the surface forming a bitumen skin on top of unbroken emulsion. This will prevent further evaporation of water, resulting in a delayed breaking instead of the desired accelerated breaking.

Several techniques have been presented to solve this problem. One solution is to use two spray bars spraying the emulsion and the additive simultaneously when the stones are applied (18). The technique is limited, however, to surface dressings where the bitumen emulsion is sprayed on top of an already existing road. But in cold mix applications the stone material and the emulsion have to be mixed before laying and compaction. In these applications it is very important that the emulsion does not break immediately, since the workability of the mix has to be good enough to permit laying and compaction before breaking takes place. The breaking additive thus has to be given a delayed effect. One solution is to encapsulate the breaking additive, as has been presented in some patents (19–21). Another technique, which has successfully been applied in several road constructions in Sweden, is to emulsify a basic water-soluble material in oil making a W/O emulsion (22–24) (Figure 11.6). If this W/O breaking additive emulsion is added to the O/W bitumen emulsion the situation illustrated in Figure 11.7 is obtained.

The alkaline breaking additive is dispersed in oil, preventing the alkaline material to neutralize the acid in the bitumen emulsion. After a certain time the breaking additive diffuses into the water phase of the bitumen emulsion causing neutralization of the acid. When the cationic emulsifier loses its positive charge it also loses its ability as emulsifier and the emulsion breaks. The time for breaking is mainly determined by the viscosity of the oil and the agitation of the emulsion.

11.7 APPLICATIONS

The use of bitumen emulsions may be divided into two main construction areas: roads and other industrial applications. The largest volumes of bitumen emulsions are by far those used in road applications. It is also in this area that research and development is most intense as a result of environmental constraints towards less solvent, better work environment, and lower energy consumption. The phrase "industrial applications" covers a variety of different applications of which waterproofing can be considered as one of the most important. Road applications using

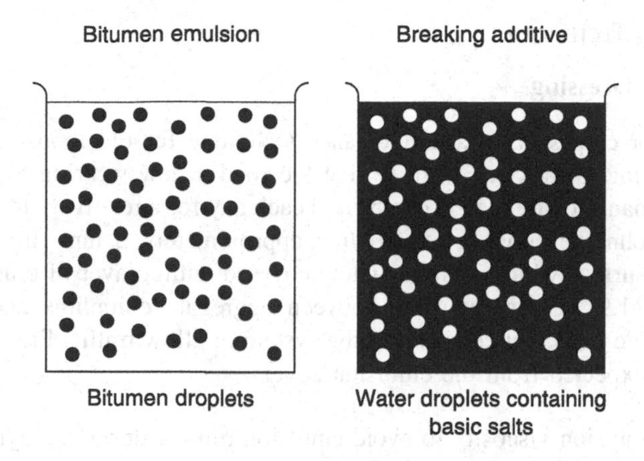

FIGURE 11.6 Bitumen O/W emulsion and W/O breaking additive.

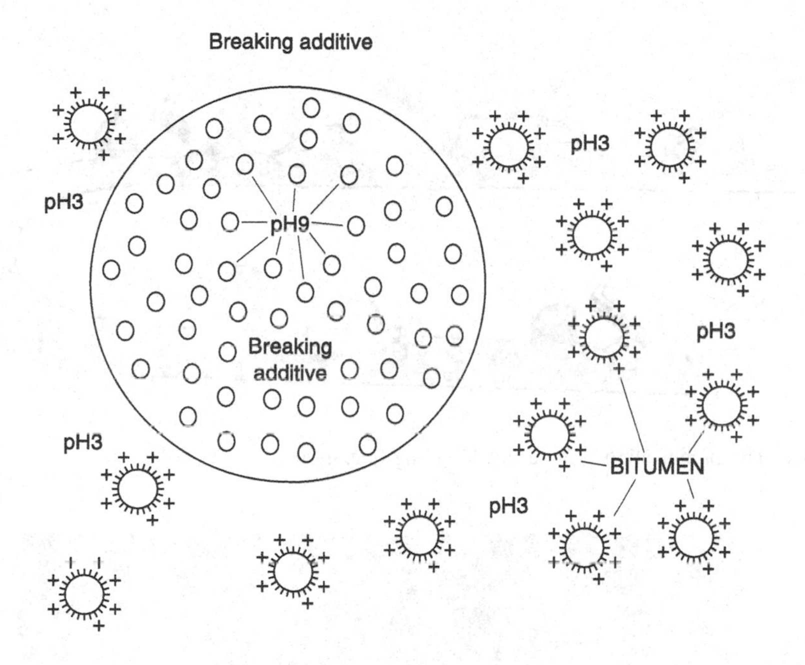

FIGURE 11.7 Bitumen emulsion immediately after addition of breaking additive.

bitumen emulsion can be classified into two main techniques: sprayed applications and cold mix asphalt. There are several basic differences in the requirements of emulsions for spraying and mixing. Spraying emulsions are usually formulated to break rather fast on the road and since they are going to be sprayed, viscosity is also usually an important property. Mixing grade emulsions are usually formulated to be rather stable to permit contact with stone material without premature breaking.

11.7.1 Spraying Techniques

11.7.1.1 Surface Dressing

Surface dressing, or chip seal, is a maintenance technique for old roads which can restore the surface by sealing microcracks as well as give the road a new wearing layer. The technique is attractive because many square meters can be laid each day for a relatively low price (Figure 11.8). In principle the technique consists of applying approximately 2 mm film of emulsion on the existing pavement surface which is immediately covered with a layer of crushed rock of selected gradation (Figure 11.9). Grain interlock between aggregate chippings and adhesive effect of bitumen must provide sufficient cohesion to the system to allow traffic directly after construction. The performances expected from the emulsion cover:

- Sufficient emulsion viscosity to avoid emulsion run-off during spraying on sloped surface (road crown, grade).
- Good reactivity, which is a combination of break and adhesion as soon as the emulsion is in contact with the mineral aggregate. Reactivity is a phenomenon occurring before

FIGURE 11.8 The organization of a surface dressing operation.

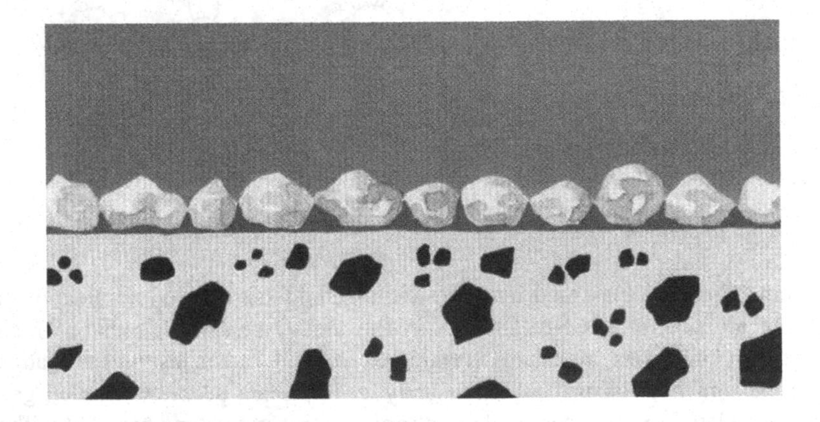

FIGURE 11.9 Cross-section of a single layer surface dressing.

the emulsion has had time to lose water and is related to the adsorption of the emulsifier on the surface of the stones. At this point it is important to note that the emulsion must not break too fast, since there should be enough time to wet the stone before breaking takes place.

- Fast coalescence. It is expected from the emulsion film that it loses water by evaporation quickly. Coalescence and reactivity are the two main phenomena involved in the breaking of the emulsion in this technique.

Emulsions for surface dressing usually have high binder content (68 to 71%) in order to have a high viscosity and a minimum amount of water to evaporate. They are also generally formulated to be very fast breaking. Formulations making use of polymer modified binders are becoming more and more popular and make up a considerable part of the use today.

Surface dressings come in several different variations. Some include double application of emulsion and stone material, sometimes referred to as sandwich technique. In such applications the requirement on high viscosity of the emulsion is not so strong. It is also common to consecutively apply two different gradations of stone material, first large stones, and then smaller stones expected to "lock in" the large stones. This system is sometimes referred to as "racked in system."

11.7.1.1.1 Evaporation Filtration Test

This is a test which has been proven useful for characterization of surface dressing emulsion. In surface dressing applications where a thin film of bitumen emulsion is sprayed on the road and covered with aggregate chipping, two phenomena contribute to the break of the emulsion:

- Break in contact with minerals, referred to as reactivity
- Break due to increasing emulsion binder content as water evaporates until the emulsion becomes unstable and breaks

The evaporation filtration test has been developed to follow the second phenomenon (25). This test follows the evolution with time of thin emulsion films placed in standard dishes. The evaporation rate is obtained by weighing the samples at regular time intervals. Coalescence is estimated by washing the film with soap solution whilst stirring with a glass rod through a 160-micron sieve. Coalesced binder will either adhere to the dish or be retained on the sieve. Noncoalesced emulsion is washed away.

The details of the test are as follows. A complete test requires a total of six samples. Approximately 13.58 g of emulsion is placed in a 100 mm diameter dish (emulsion film thickness 1.5 mm) and placed in an environmental chamber at 40 °C. Each sample is weighed regularly until it reaches 75, 80, 85, 90, and 95% binder content. When the desired binder content is reached, the time is recorded and the sample is washed. Usually a test takes approximately 2 to 4 h. The sixth sample is kept in the environmental chamber for 24 h and its binder content measured as an indication of how much water remains trapped in the fully coalesced film.

Results are expressed in terms of coalescence versus binder content and time versus binder content (evaporation rate). Figure 11.10 presents a typical result for a surface dressing emulsion. The rapid change in coalescence between 75 and 85% binder indicates that phase inversion may be occurring without any real effect on evaporation. When 86% binder is reached the change in evaporation rate is quite dramatic indicating that water is now trapped in the system and finding it more difficult to escape.

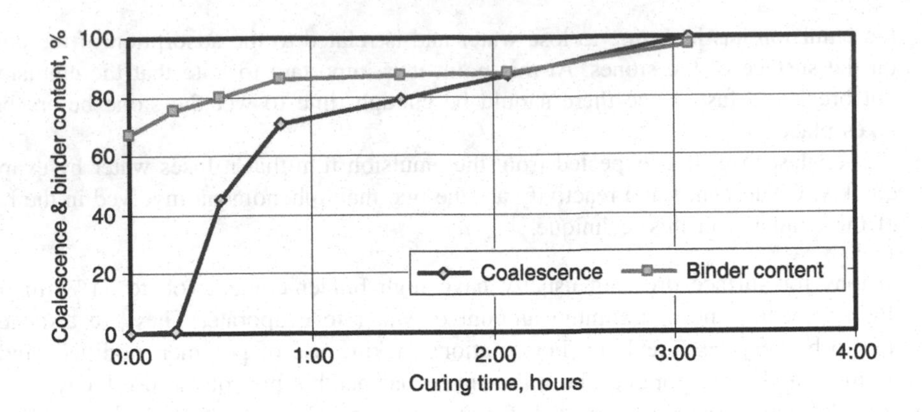

FIGURE 11.10 Evaporation filtration test result for a typical surface dressing emulsion.

11.7.1.2 Tack Coat (or Bond Coat)

Tack coats are a thin emulsion film sprayed onto the existing pavement surface before applying a new asphalt layer. The purpose of tack coats is to get good adhesion between successive pavement layers. This is of fundamental importance for the strength and the durability of roads consisting of several asphalt layers and even more important when thin layers are applied as wearing courses. As this new layer is usually applied hot (minimum of 140 °C), breaking of the tack coat emulsion is expected to occur during compaction under the heat of the new layer. The major characteristics expected from a tack coat emulsion are storage stability and proper viscosity. Viscosity must be sufficiently low to insure good wetting of the existing surface and also have ability to penetrate through remaining layers of dust or fines on the old surface.

11.7.2 Techniques for Cold Mix Asphalt

Potentially cold mix asphalt can replace hot mix asphalt in most constructions. There are, however, a number of limitations with cold mix asphalt which so far have limited their extensive use. Part of the limitation is conservatism by road constructors and specifiers, but there are also technical limitations which need to be overcome. Three of the major factors which are particularly important when working with emulsion mixes are:

- Workability (stiffness of the mix)
- Void content in the stone material
- Breaking and curing combined with release of water

Workability is related to the stiffness of the uncompacted cold mix. A good workability is needed to secure a smooth paving and proper compaction of the mix. The main reason for poor workability is that the emulsion starts breaking immediately when contacted with the stone material. This is particularly difficult to avoid if the stone material contains large amounts of fines. There are many ways of achieving good workability, but many solutions are also associated with some disadvantages. One of the most common solutions is to make the binder soft enough to permit laying and compaction, even if broken. This can be done by using a soft binder or alternatively using a solvent which will soften the binder during mixing, laying, and compaction and then eventually evaporate and give full strength to the pavement. Another solution is to use

TABLE 11.1
Emulsion Cold Mix Workability

Broken Systems		Nonbroken System,
Soft Binder	Solvent Fluxed Binder	Slow Set Emulsion
• Soft paving • Flexible paving • Self-healing • Easy to lay • Easy to make	• High VOC emission • Improve strength with time	• Long curing time • Good strength after curing • Very sensitive after laying

a very stable emulsion which does not break in contact with the stones. The disadvantage is that it may take a very long time to achieve full strength of the road. Table 11.1 summarizes the different approaches for achieving good workability on cold mix asphalt.

The second factor is void content. Crushed stone material used for road construction consists of a mixture of aggregates with different size or different gradations. The amount of space between the mineral aggregates is called the "volume of voids in mineral aggregate." In dense graded hot mix asphalt the void content in the aggregate should be slightly higher than the volume of the bitumen. The air void in the compacted asphalt sample should be as small as possible, but the air voids must never be completely filled with bitumen, since this will cause the asphalt to be unstable and very sensitive to deformation and bleeding. This makes it impossible to use the same gradation with emulsion mixes since we must always allow for additional 35% water from the emulsion. This makes it very difficult to make a truly dense graded cold asphalt mix using bitumen emulsions.

The third factor is breaking and curing. After paving and compaction of the cold mix asphalt, breaking and curing should take place as quickly as possible. This contradicts the desire to have as stable emulsion as possible to have good workability. A further complication is the necessity to remove the water as soon as possible. In wet climatic conditions this could take a very long time. In open graded mixes water could drain out of the paving but in dense graded mixes the water has to evaporate out of the paving unless a water consuming additive like cement is added.

The choice of emulsion type for a certain application is heavily dependent on the amount of fines in the stone material since the fines are very reactive towards the emulsion. Too much fines can cause premature breaking, while too little fines could cause "run off" of the emulsion from the mix. Thus we may classify different applications according to the type and amount of fines (defined as particles less than 80 microns) present in the minerals to be coated (Table 11.2).

11.7.2.1 Open Graded Emulsion Mixes

In this technique an open gradation of stone material is selected. This means that there is plenty of space between the stones to allow for binder and water. Figure 11.11 illustrates the internal structure of this type of mix. The open gradation usually contains very little fines making premature breaking less probable and thus the mix workability is usually good. On the other hand the retention of the emulsion in the mix becomes very critical. A high amount of binder is desired to have good durability and good strength. If, however, too much emulsion is added it will simply run off the aggregates and leave just a thin layer of binder on the stone surface. This is sometimes referred to as "run off." The retention of emulsion can be improved by thickening additives. Open graded mixes are mainly used as base courses since the open structure makes the asphalt

TABLE 11.2
Application and Emulsion Type with Respect to Filler Content

	Construction Method			
	Sprayed	**Mixed**		
Type of fines	None	Crushed aggregate		Silt/clay
Fines content	<1%	<2%	2 to 10%	≤10%
Application	• Surface dressing • Tack coats	• Cold recycling • Open graded emulsion mixes	• Dense graded emulsion mixes • Slurry seal • Microsurfacing	Soil stabilization
Emulsion type	Cationic/anionic Formulated to pass storage and transportation requirements Fast break upon contact with road and chipping required	Cationic/anionic Sufficiently stable to allow coating of aggregate	Cationic/anionic Sufficiently stable to allow coating of fines Quick set after construction usually required	Anionic High stability required to allow mixing in presence of silt/clay particles

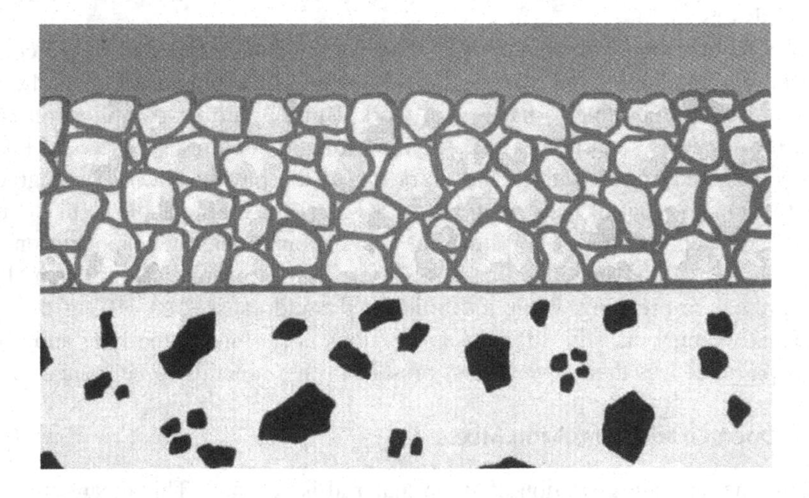

FIGURE 11.11 Cross-section of an open graded emulsion mix layer (the top part).

sensitive to aging and penetration of water and thus less suitable for top layers. If open graded mix is used as top layer it is advisable to seal the asphalt with a surface dressing or a slurry seal to guarantee durability. The emulsion used must be able to coat the aggregate and have a break time after mixing adapted to the construction methods. When mixing and construction are carried out with little delay, emulsion break times similar to surface dressing can be required in order to have a quick cohesion build-up after compaction. Requirements for this type of technique include:

- Aggregate must be moist for best results (about 2 to 6% water content).
- Emulsion must have sufficient viscosity to avoid emulsion run-off after mixing (particularly true in the case of open graded emulsion mixes, where thick bitumen film is needed).
- Emulsion binder content depends on the moisture content of the aggregate; it is best to work with low bitumen content (60 to 65%) when working with relatively dry aggregate. If the aggregate is wet, it is possible to work with higher binder content emulsions (65 to 70%).

As in the case of surface dressing, reactivity is an important phenomenon. The mix, however, does not depend as strongly as surface dressing on coalescence, since compaction has an important role in the break of the emulsion. An example of paving so-called semi-dense cold mix asphalt is illustrated in Figure 11.12.

FIGURE 11.12 Paving of semi-dense cold mix asphalt.

11.7.2.2 Dense Graded Emulsion Mixes

Cold dense graded mixes may be compared to hot mixes. They are characterized by a high fine content and a dense gradation leaving minimum amount of voids. This is the most challenging type of mix for the emulsion formulator in a sense that he must both design for fines coating and coarse aggregate coating. In most cases, the fine coating criterion is critical and the emulsion is formulated for high stability. However, by increasing stability, reactivity is decreased and there may be difficulties achieving 100% coating of the larger aggregate particles. This has been solved by coating the fine fraction and the coarse fraction separately in the cold mix plant, followed by mixing the two fractions before laying.

In any case stone gradations used in emulsion mixes can not be identical to the gradations used in hot mixes. There must always be space for at least 35% water present in the emulsion. In reality there is usually even more water in the mix, coming from natural moisture in the aggregate piles. It might also be necessary to add extra water to secure even coating and good workability. If more liquid is added, than the available voids, the emulsion will be squeezed out at compaction, and the mix will contain less binder than designed. The emulsion water on the surface will also cause a disaster by sticking to vehicles, persons, and tools around the work site. To give enough space for the water in the cold mix we need to use a gradation with more voids than normally used in hot mix design. The result is that when the emulsion is fully cured and the water is evaporated, we have more voids in the asphalt than with the corresponding hot mix. All this taken together makes it impossible to produce a true dense graded cold mix asphalt, but experience has proven that cold mix asphalt with void content of 7% will perform as well as a hot mix asphalt with void content of 3%.

So far, many commercial dense graded cold laid mixes have been made with very soft binder or hard binder being fluxed with solvent to become soft enough to permit paving and compaction even if the emulsion is broken. These soft systems are also subjected to after compaction which will make the void content more comparable to hot mix asphalt but also sensitive to deformations.

11.7.2.3 Slurry Seal and Microsurfacing

Slurry seal and microsurfacing are special types of dense graded mixes and are often viewed as the top of the scale of complexity in bitumen emulsion techniques. In the U.S.A. they are frequently used to seal and improve skid resistance on highways as well as for maintenance of local roads in urban areas with very little traffic. Sealing on top of an old road can extend the lifetime of the road by 10 years or more. While ordinarily dense mixes form a relatively stiff mix before compaction, slurry seal and, to a large extent, microsurfacing form a slurry composed of:

- Fine aggregate gradation (usually up to 5 mm for slurry seal and 10 mm for microsurfacing)
- Water
- Break control additives dispersed in the water
- Emulsion
- Cement

Slurry seals and microsurfacing are always applied directly after blending in one continuous operation, with machines which are specially designed to lay microsurfacing. They are equipped with an individual dosing system to adjust the balance between the components listed above. A skilled technician may adjust the formulation to have more or less immediate breaking once the material is laid on the road. The breaking speed is very much dependent on the type of aggregate and the type of emulsion used. Thus the formulation of the slurry should be individually optimized for each work site using proper selection of additives and cement. Microsurfacing systems make use of particular types of emulsifiers, of which di-quartenary fatty amine is one example, but many other excellent types are available on the market. The breaking in such systems is a true chemical break in which both chemical nature of the aggregate and the emulsifier play a major role. Slurry seal is applied in thin layers not exceeding 5 mm while microsurfacing systems usually build thicker layers and use polymer modified binders. Once the emulsion breaks, it is expected that the system rejects the water. Compaction and traffic help in this process. Set is completed once the mix is dry.

11.7.2.4 Recycling

Recycling of old asphalt pavements is becoming more and more important. Today most old asphalt is recycled. Cold recycling is preferred because most of the operations can be made close to the work site without extensive transportation and heating. The mix is composed of reclaimed asphalt pavement (RAP) issued from cold milling of asphalt materials which are mixed with a suitable bitumen emulsion. The RAP for cold recycling already contains 4 to 6% aged bitumen and usually comes in relatively open gradations with very little fines. One of the difficulties with cold recycling is the lack of control over the material entering the process. The existing binder can be of different types and the stone material can be of different type of mineral as well as different gradations. Sometimes small amount of virgin stone material (up to 20%) is added to adjust the gradation and to improve the quality.

The selection of binder follows generally two lines. One is that a soft binder should be used to rejuvenate the old binder. The viscosity of the added binder is selected to give a suitable stiffness when the new and the old binder have been completely mixed. The other line considers the old binder as a part of the stone, and the new added binder should have the desired stiffness for the road. Experience has shown that using hard binder gives much better initial strength (modulus) to the new road.

Although it is sometimes expected that coated stones and lack of fines guarantee low reactivity of the RAP, this is not always the case. To secure an even coating of all RAP by the new binder, the emulsion should be rather stable. The addition of new binder has a maximum of approximately 3%. Too much binder gives a risk for permanent deformation (rutting) or bleeding, while too little binder decreases the durability of the road considerably.

11.7.3 Other Road Applications

Bitumen emulsions are also used in many other road applications, but it is outside the scope of this chapter to go into details of all these, but just to mention a few:

- Soil stabilization
- Penetration macadam
- Grave emulsion (particularly France)
- Prime coat
- Dust binding

11.7.4 Industrial Applications

Bitumen emulsions have also proven to be very useful in many applications other than road building. In most cases anionic emulsions or clay emulsions are used in these applications although sometimes cationic emulsions are also used. The producers of cationic emulsions for road applications generally do not like to produce anionic or clay emulsions due to the consequences if the two types are accidentally mixed. Instead the emulsions for industrial applications are produced in small plants specially designed for the particular product. Some examples of applications are:

- Paint
- Roofing mastic
- Protective cover for freshly seeded areas
- Mulch treatment
- Metal and masonry coating
- Spray coating for water proofing
- Grout mastic for vibration dampening
- Adhesives
- Tile cement
- Laminant
- Fiber board sizing

11.8 THE FUTURE FOR BITUMEN EMULSIONS

Even 80 years ago it was predicted that within 10 years bitumen emulsions would dominate as binders in the road construction industry. It has not happened yet, although we are convinced that sooner or later they will, although we do not expect this to happen within 10 or even 20 years. On the contrary, during the last 10 years we have seen a slight decline in the use of bitumen emulsions, reflecting the general trend of declining road construction and maintenance work on public roads. The ratio of emulsions to hot mix asphalt has been maintained, however, or even slightly increased during the same period.

Today bitumen emulsions dominate in some construction applications, such as where spraying is required (surface dressing in roads, waterproofing in construction), because it is possible with emulsions to work at ambient temperature on wet surfaces and reduce almost entirely the need for solvents. However, usage of bitumen emulsions remains secondary to hot bitumen for several technical and economic reasons. On the technical side, emulsion behavior is more complex to master than hot bitumen. Engineers generally feel more comfortable with hot bitumen. In the hot process, there is no water to eliminate and hardening is achieved simply by cooling instead of being achieved through the coalescence and setting phenomena as in the case of emulsions. On the economics side, the cost of emulsification always needs to be justified. The economical advantages of emulsions are real, however, and related to the savings in energy from working directly with moist minerals and therefore avoiding the drying stage of minerals. Also, emulsions give the possibility of working on site with lighter equipment than in the case of hot techniques.

Continuous research into emulsion and emulsion mixes is still on-going, identifying more and more areas where emulsions present clear advantages or where molten bitumen would not be applicable, such as most on-site techniques. Finally, the need to reduce the use of volatile solvents is an important reason for using emulsions in sprayed applications.

During recent years bitumen emulsion technology has sometimes been presented as the "green" alternative to hot mix asphalt. A comparison between hot mix asphalt and cold mix asphalt is, however, rather complicated and several factors should be taken into consideration. Cold mix has without doubt several advantages over hot mix, but we should be careful about presenting cold mix as the "green" alternative without any reservations. Let us take a look at some factors related to cold mix as an environmentally preferred product.

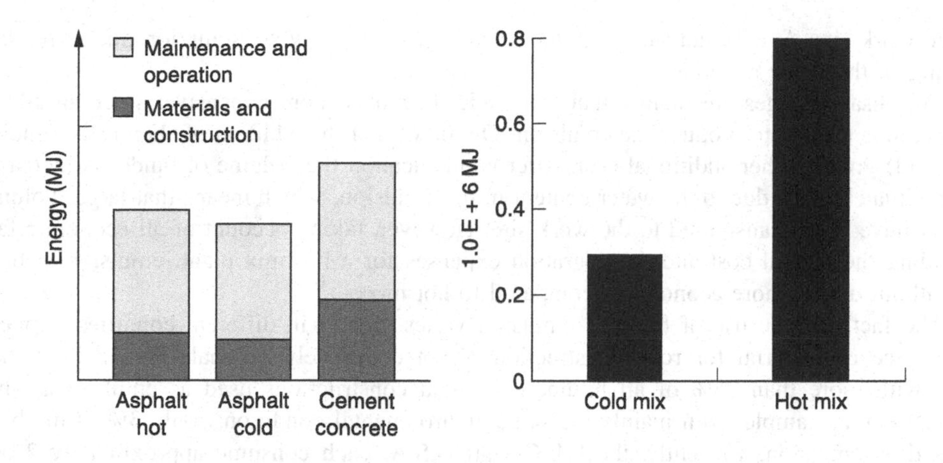

FIGURE 11.13 Comparison of energy consumption using hot mix asphalt and cold mix asphalt. The right diagram refers to production only.

Cold mix technology can lead to energy savings. Since this technology does not require heating of the stone material, which makes up 94% of the asphalt, there is considerable saving in energy. Cold mix technology also makes use of less sophisticated and cheaper production equipment. Usually cold asphalt mix plants are mobile and can easily be placed close to the work site, considerably reducing transportation. If all these factors are taken together, cold mix gives a considerable saving in energy compared to hot mix (Figure 11.13).

Consider also the health and safety aspect. A historical success story showing the advantage with bitumen emulsions is the reduction in volatile organic compounds (VOCs) when surface dressings previously produced with cut back bitumen (bitumen and kerosene) were replaced by bitumen emulsions. This change took place between 1975 and 1990 and considerably reduced the amount of VOCs released to the atmosphere. It is also clear that the working environment when laying cold mix asphalt completely eliminates all problems with fuming from hot bitumen and also reduces the risk of burns from hot asphalt.

Everything is not "green," however, with bitumen emulsions. There is still a large number of formulations, containing small amounts of kerosene or other solvents, which eventually evaporates to the atmosphere, although not to the extent of cutbacks in the 1970s. There has also been concern that the emulsifier may be washed out from the asphalt by rain and could possibly contaminate ground water, particularly since most cationic emulsifiers are considered as hazardous to the environment. On the other hand, investigations have shown that after breaking of the emulsion the cationic emulsifiers are irreversibly captured by the soil and would not be released until they are completely degraded by micro-organisms (26).

The main reason why road construction and maintenance technology are based on hot mix asphalt is mainly due to the fact that most countries, particularly in Europe, have an infrastructure adapted for hot mix asphalt. The available specifications and practices are also more adapted to hot mix asphalt rather than cold mix asphalt. There is an excess of hot mix plants covering the most densely populated areas of the Western countries. These plants represent considerable capital employed and it is important for the owners to use the plants for production of hot mix asphalt.

The most important advantages with cold mix technology over hot mix are the inexpensive equipment required for production, giving high flexibility for production of cold mix asphalt close to the work site. Another advantage is the relatively low energy consumption due to the lack of heating of the stone material.

The disadvantages are mainly that the binder becomes more expensive, since an additional operation is needed to produce the emulsion. On top of that the addition of chemicals (emulsifiers and acid) gives further additional cost. After emulsification the volume of binder is increased by approximately 30% due to the water content in the emulsion, which means that larger volumes of binder have to be transported to the work site. However, taking account of all economic factors, including the capital cost and the operation expenses for a hot mix plant, emulsion technology is, without doubt, more economical compared to hot mix.

The fact that the use of bitumen emulsion varies greatly in different countries implies that the choice of material for road construction is not completely rational. France is the largest user, with more than 25% of all bitumen for road constructions used as emulsions, while in Germany, for example, with mainly the same environmental conditions, only 3% of the bitumen is used as emulsion. Currently, the E.E.C. and U.S.A. each consume approximately 2 million tons of emulsion, representing on average, respectively, 12% and 7% of the annual bitumen consumption. There is, however, no doubt that bitumen emulsions will become more and more important, particularly for maintenance of our road network in the future.

REFERENCES

1. *Bitumen Emulsions, General Information Applications*; Syndicat des Fabricants d'Emulsion Routieres de Bitume (SFERB), Paris, 1991.
2. Reader, W.J. *Macadam, the McAdam family and the Turnpike Roads 1798–1861*; W. Heinemann: London, 1980.
3. MacKay, H.A. Bitumen emulsion. Patent 202.021, 1922.
4. Nellensteyn, F.J. *Relation of the Micelle to the Medium in Asphalt*; Inst. Petrol. Technol., 1928; Vol 14, pp 134–138.
5. Pfeiffer, J.P.; Saal R.N.J. *Asphaltic Bitumen as Colloidal System*; Phys. Chem., 1940; Vol 44, pp 139–149.
6. Park, S.J.; Mansoori, G.A. *Aggregation and Deposition of Heavy Organics in Petroleum Crudes*; Energy Sources, 1988; Vol 10, pp 109–125.
7. *Binder Characterization and Evaluation*, volume 1: *Physical Characterization*; Strategic Highway Research Program SHRP-A-367; 1994.
8. Redelius, P.G. *Bitumen Solubility Model Using Hansen Solubility Parameter*; Energy & Fuels, 2004; Vol 18, pp 1087–1092.
9. *Binder Characterization and Evaluation*, volume 2: *Chemistry*; Strategic Highway Research Program SHRP-A-368; 1994.
10. Altgelt, K.H.; Jewell, D.M.; Latham, D.R.; Selucky, M.L. In *Chromatographic Science Series*, Ed. K.H. Altgelt; Marcel Dekker, New York, 1979; pp 194–196.
11. *Bituminous materials: Asphalt, Tars and Pitches*; Ed. A.J. Hoiberg; Vol II: Asphalts, Part One.
12. CECA S.A. *Emulsifiants perfectionnes à base de diamines grasses oxypropylees et les emulsions de liants hydrocarbons obtenues en utilisant ces emulsifiants.* Patent application EP 0051030.
13. Poirier, J.E. *Custom-made drops: Emulsion high fashion*; Revue Generale des Routes et Aerodromes, September 2002.
14. ESSO Societe anonyme Francaise. Bitumen Emulsion. PCT WO92/02586.
15. CECA S.A. *Procédé pour la confection d'enrobés bitumineux denses à froid stockable et de graves emulsions.* Patent application EP 0589740.
16. James, A.; Furlong, S.; Kalinowsky, E.; Thompson, E. *Water Enclosed Within the Droplets of Bitumen Emulsion and its Relation to Emulsion Viscosity Changes During Storage.* 2nd World Congress on Emulsion, Bordeaux, France 1997. Paper 2-4/009.
17. Eckmann, B.; Van Nieuwenhuyze, K.; Tanghe, T.; Verlhac, P. *Prediction of Emulsion Properties from Binder/Emulsifier Characteristics*; 2nd Eurasphalt & Eurobitume Congress, Barcelona; 2000.

18. COLAS S.A. *Process for obtaining surface coatings of bitumen.* UK Patent GB 2 167,975.
19. Kenobel AB. *Sätt att bilda en beläggning av bitumen och en bitumenhaltig composition* (in Swedish). Swedish patent application SE 453 760.
20. Societe Anonyme d'Application des Derives de l'Asphalte SAADA. *Process for breaking an emulsion.* UK Patent application GB 2 208,270.
21. COLAS. *Bitumen emulsion, method for their preparation and their use for the maintenance or the making of road surfaces.* Patent EP 0,864,611.
22. AB Nynäs Petroleum. *Bitumen emulsion, process for its preparation, breaking additive for use therein and the use of said bitumen emulsion.* European Patent 0491107.
23. Redelius, Per. *Breaking Control of Bitumen Emulsions*; 5th Eurobitume Congress, Stockholm, 16–18 June 1993; Vol 1A, pp 353–357.
24. Redelius, Per. *A Novel System for Delayed Breaking Control of Bitumen Emulsion.* First World Congress on Emulsions, 19–22 October 1993, Paris; Vol 1, Paper 1–22, 147.
25. Marchal, J.-L. *Evaporation–Filtration Test for Emulsion Inversion*; Asphalt Emulsions, ASTM STP 1079. Ed. H.W. Muncy, American Society for Testing and Materials, Philadelphia, 1990; pp 106–111.
26. Campbell, D.; James, A.; Redelius, P.; Thorstensson, B.-A. *Chemical Emissions from Asphalt Emulsion Applications*; AEMA 27th Annual Conference, 12–15 March 2000, Florida, USA.

12 Modern Characterization Techniques for Crude Oils, Their Emulsions, and Functionalized Surfaces

Johan Sjöblom, Gisle Øye, Wilhelm R. Glomm, Andreas Hannisdal, Magne Knag, Øystein Brandal, Marit-Helen Ese, Pål V. Hemmingsen, Trond E. Havre, Hans-Jörg Oschmann, and Harald Kallevik

CONTENTS

12.1 INTRODUCTION

Crude oil exploitation offshore has advanced to the point, at least on the Norwegian Continental Shelf, where the era of large fields with a high quantity and quality is over. Since the possibility of finding highly productive formations is small the corresponding strategy is focused on an improved exploitation of the existing large fields and to tie in, in an efficient way, small fields. The most important thing when focusing on existing large fields is to increase the recovery rate. This, however, can be connected with typical flow assurance problems. At older fields the co-production of water is in most cases substantial. It is well known that many wells co-produce water to an extent of 50 to 70%. This production profile will cause problems with high water volumes in the separator and emulsions close to the inversion point.

To tie in small fields with larger production streams means that the fluids that are to be mixed must be compatible. If one encounters precipitation of organic matter (for instance asphaltenes) the whole production line and the regularity will be endangered.

In parallel with these efforts the trend to build up a sub-sea processing network will emerge. Also in this case the nature of the fluids will be decisive for the design and dimensions of the process unit. Since this unit will function on a sea depth of 300 to 400 m (in the future about 1000 m) it is obvious that maintenance operations are very costly and should be rare.

A central question when solving the design of the process unit is the nature of the fluids that are to be processed. Hence fluid characterization will be a key technology for success in the future offshore processing.

In this chapter we present modern instrumentation for such a fluid characterization. We cover the basic theory behind the measurements together with recent examples. We have also included some examples on characterization of functionalized solid surfaces.

At the Ugelstad laboratory, as at many other modern laboratories, we aim at doing good analytical chemical characterization of components in the crude oil as well as mapping the physicochemical properties of these components.

12.2 ANALYTICAL CHEMICAL CHARACTERIZATION

12.2.1 SARA Fractionation by High Performance Liquid Chromatography

Crude oils have a complex nature and it is not possible to determine the individual molecular constituents. Compositional studies are usually done by fractionation into predefined chemical families. The SARA group type fractionation separates the crude oil into the following classes: saturates, aromatics, resins, and asphaltenes. The definition of the SARA fractions is well accepted and will not be presented in detail here. Speight [1] gives a thorough introduction to the chemistry and structure of crude oil. In short, saturates are defined as the saturated hydrocarbons ranging from straight-chain paraffins to cycloparaffins (napthenes) while the aromatic fraction includes hydrocarbons containing one or more aromatic nuclei which may be substituted with napthenes or paraffins. Asphaltenes are defined as the solubility class of crude oils that precipitate in the presence of aliphatic solvents (here n-hexane or n-pentane) while the resin fraction is defined as the fraction soluble in light alkanes but insoluble in liquid propane [1]. Asphaltenes and resins are known as large, polar, polynuclear molecules consisting of condensed aromatic ring systems and heteroatoms like sulfur, nitrogen, and oxygen. Asphaltenes and resins have gained increased interest with the knowledge of the large effect of these constituents in particular on the overall performance of heavy crude oils.

Although the SARA fractionation method is just a rough sorting of the crude oil constituents, it provides an important classification of crude oils. As will be seen from other examples of applications in this chapter, SARA characteristics are frequently used when discussing different crude oil properties. The SARA fractionation method has found great utility in combination with high-performance liquid chromatography (HPLC) [2–6]. HPLC is an analytical technique for the separation and determination of organic and inorganic solutes in any sample, especially biological, pharmaceutical, food, environmental, and industrial. In a normal-phase liquid chromatographic process a nonpolar mobile phase permeates through a porous solid stationary phase (polar) usually in the form of small uniform particles, packed into a cylindrical column. The sample is injected into the mobile phase, travels through the column, and is retained by the stationary phase mainly depending on polarity. Adsorption interactions between sample components, the mobile phase, and the stationary phase can be manipulated by the choice of mobile and stationary phases and flow conditions.

As an example of the SARA fractionation method we will discuss the procedure from a study by Hannisdal and co-workers [7]. The ability of vibrational spectroscopy to predict SARA components in heavy and particle rich crude oil will be discussed in the following section.

In this study, 20 crude oil samples at ambient temperature and pressure were received from exploration sites on the Norwegian Continental Shelf and sites located in Brazil, France, the South China Sea, the Atlantic Ocean, and the Gulf of Mexico. These oils were quantitatively fractionated into saturates, aromatics, resins, and asphaltenes (SARA) by asphaltene precipitation in n-hexane and preparative HPLC. Here, the focus will be on the HPLC procedure.

The HPLC system used in this study was built up from the modules shown schematically in Figure 12.1. Two columns were used: one 21.2×250 mm column packed with unbonded silica 15 μm and one 21.2×50 mm amino (10 μm) column. Dichloromethane (99.8%) and n-hexane (95%) were used as mobile phases. Generally, preparative separations of organic samples favor the use of unmodified silica as column packing material [8]. However, experiments on the preparative silica column showed that dichloromethane did not have the necessary solvent strength to elute the most polar components in heavy crude oils. These resins were irreversibly adsorbed onto the silica packing material. Instead an amino column, weaker in column strength, was used for this study in a multidimensional column setup shown in Figure 12.2. The two

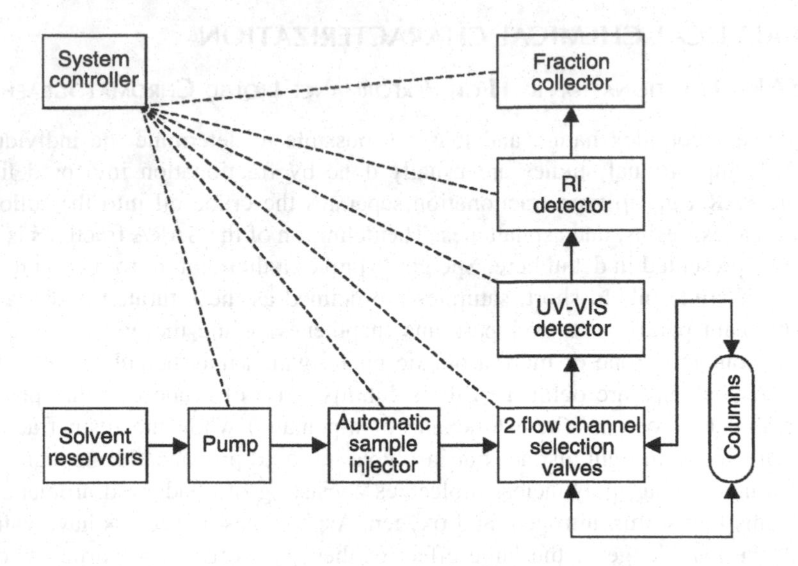

FIGURE 12.1 Schematic representation of the HPLC system with main modules.

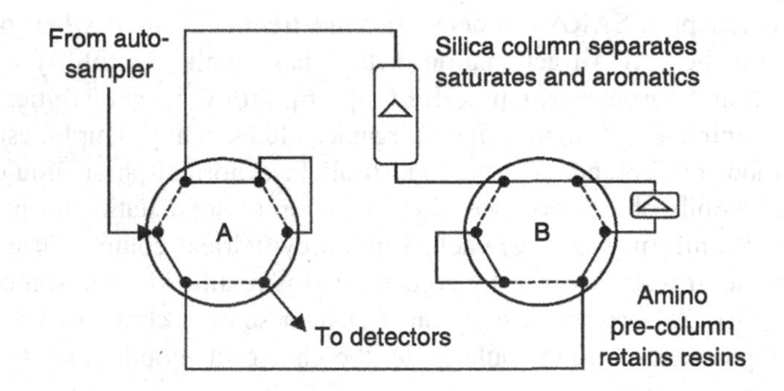

FIGURE 12.2 Scheme of the flow channel selection valves. Valve A controls the flow direction through columns (normal/backflush) while valve B includes or excludes the amino pre-column.

selection valves shown in the figure allow an individual reloading of the two columns. Valve A controls the flow direction through columns (normal/backflush) while valve B includes or excludes the pre-column. The new setup reduced the time of analysis from 70 to 26 min.

A sample of 5 ml asphaltene-free crude oil in n-hexane (corresponding to 0.6 g crude oil) was injected into the 20 ml/min flow (isocratic) of filtrated and degassed n-hexane. Crude oil components travel through the columns and are retained mainly depending on polarity. Saturates, having no retention on the columns, are collected from RI signals (3 to 5 min). After 15 min, all aromatics have left the amino pre-column but some condensed polycyclic aromatic ring systems are not yet eluted from the silica column. These are collected from the preparative column by dichloromethane backflush (<23 min) while the resin fraction is desorbed from the amino pre-column by dichloromethane backflush, 23 to 26 min after sample injection. This procedure provides the chromatogram in Figure 12.3 where the three fractions are well resolved. The SAR

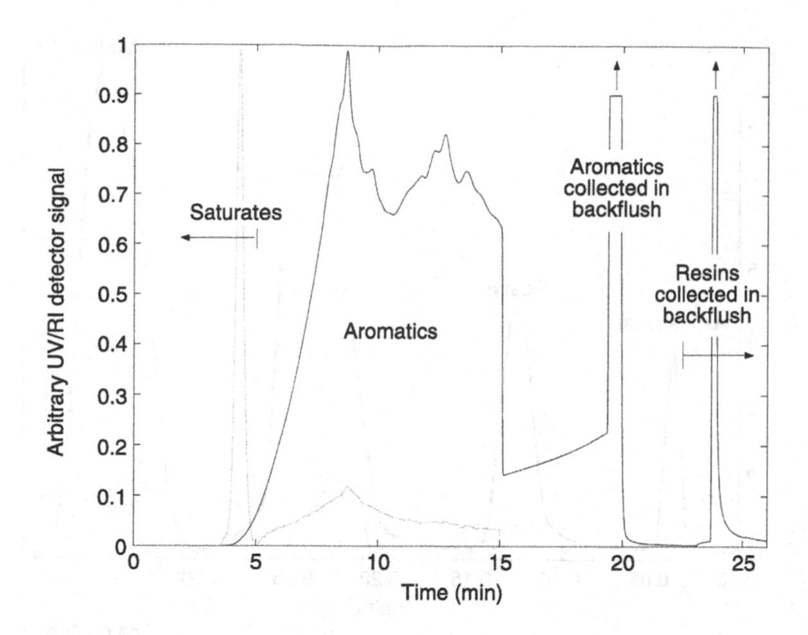

FIGURE 12.3 Typical chromatogram from preparative HPLC analysis. The solid line represents the RI signal while the broken line represents the UV chromatogram at 254 nm. Peaks are cut for better visualization.

weight fractions are determined gravimetrically after controlled evaporation of solvents in an N_2 atmosphere.

This example shows how SARA values are determined using a procedure for asphaltene preprecipitation followed by automated preparative high performance liquid chromatography. The discussed HPLC method had a sample capacity corresponding to 0.6 g of heavy crude oil. Since the determination of the SARA components is done gravimetrically, larger sample capacity will improve the accuracy of the procedure significantly. Moreover, the large sample capacity allows further characterization of the individual SARA fractions.

12.2.2 SARA Fractionation by Automated TLC-FID

In this section we will describe how the SARA fractions can be separated and quantified by means of thin layer chromatography (TLC) in combination with an automated flame ionization detector (FID). This method is of great benefit when there is a need to run a larger number of samples on a routine basis.

The sample to be analyzed is diluted and spotted on quartz rods that are coated with sintered silica particles. The four SARA fractions are then separated to various positions on the rods. This is done by successively emersing the rods into three development tanks containing solvents of various polarity; nonpolar solvent for saturates and increasing polarity for the aromatics and resin fractions. In the first tank n-heptane is eluated almost to the top of the rod and thereby separating the saturates from the rest of the sample. The rods are then taken out and dried before being placed in the second tank containing 80% toluene and 20% n-heptane. Here the aromatics fraction is separated as the solvent moves up the rod. The solvent is eluated to a lower height on the rod than in the first tank so that the aromatic fraction does not mix with the already separated saturate fraction. In the last development tank 95% dichloromethane and 5% methanol is used

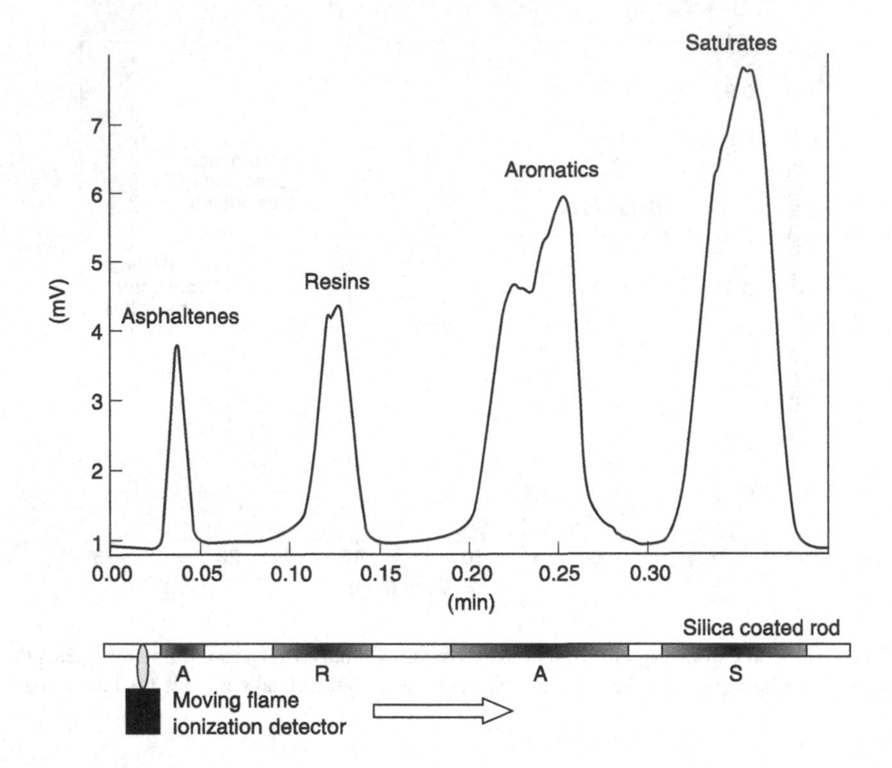

FIGURE 12.4 After being prepared in the three development tanks, the SARA fractions can be found at various positions on the silica coated rod. The flame ionization detector moves along the rod and the resulting signal is plotted as a function of its position. The SARA fractions are quantified by integrating the corresponding peak.

to separate the resins from the asphaltenes, which will be irreversibly adsorbed to the silica rod and will stay at the same position where the sample was first spotted.

The amount of the various fractions is determined by a moving flame ionization detector (FID), which moves along the entire length of the rods. Figure 12.4 illustrates a typical chromatogram that is obtained from such analysis where the FID signal is plotted as a function of the position on the rod. The SARA fractions are quantified by integrating each peak and assuming the same response factor for all four fractions.

The SARA fraction as determined by the TLC-FID method is described in the IP 469/01 standard [9]. Given the differences in the principle for the TLC-FID method compared to the HPLC or the clay-gel absorption method [10], it is obvious that there will be some differences in the obtained results. The magnitude of the differences depends on the nature of the crude oil. Fan and Buckley [11] pointed out that volatile components will evaporate during the TLC procedure; thus underestimating the saturates and aromatics. This fact makes the TLC-FID procedure less suitable for crude oils with high amount of volatiles. Bharati and co-workers [12,13] have developed a North Sea oil based standard for the TLC-FID method. The use of this standard improves the accuracy of the obtained data, and data comparison between laboratories becomes more reliable.

The TLC-FID instrumentation is known in the industry under the name "Iatroscan." This instrument is gaining increased popularity due to the fast and easy way of performing SARA analysis. The amount of sample needed is only 100 mg or less, which is beneficial in cases where the amount of sample is limited.

Another way of determining the SARA fractions is by means of near-infrared spectroscopy in combination with multivariate analysis. This method is described in the following section.

12.2.3 Near-Infrared Spectroscopy

Over the last 30 years near-infrared spectroscopy (NIR) has been increasingly used as an analytical tool, particularly by the food and agricultural industries, but also by the textiles, polymers, and petroleum industries. The increasing popularity is due to four principal advantages of the method: efficiency, simplicity, multiplicity of analysis from a single spectrum, and the nonconsumption of the samples. The rapid development of advanced and user-friendly software for multivariate analysis further enhances the usability of NIR. Optical fibers can be used to carry the light from the light source to the point of measurement and back to the light detector. This makes NIR applicable on-line in many processes.

The near-infrared region is found between the visible and middle infrared regions (MIR) of the electromagnetic spectrum. The absorption spectra in the NIR range from 780 to 2500 nm (12820 to 4000 cm^{-1}) consist of overtones and combinations of the fundamental molecular vibration bands, which are primarily due to stretches of hydrogen in C-H, N-H, S-H, and O-H bonds. This makes NIR an excellent choice for hydrocarbon analysis, where functional groups such as methylenic, olefinic, and aromatic C-H give rise to various C-H stretching vibrations that are mainly independent of the rest of the molecule. In addition to molecular absorption, the NIR spectra are dependent upon several physical parameters, where the most prominent is scattering from particles. As the particle size changes it causes a change in the amount of radiation scattered by the sample, and this is reflected in the NIR spectra as a shift of the baseline. For slightly lossy dielectric spheres in the Rayleigh limit ($r/\lambda \leq 0.05$, where λ is the wavelength of incident light), the scattering and absorption processes contribute separately to the extinction coefficient. That is:

$$\sigma_{tot} = \sigma_{sc} + \sigma_{abs} \tag{12.1}$$

where σ_{tot}, σ_{sc}, and σ_{abs} are, respectively, the total, scattering, and absorption cross-sections. The ratio of scattering to absorption scales with r^3, indicating the importance of particle size on the total light extinction. The relation between optical density (OD), light intensity (I), particle diameter (N), and particle cross-section (σ_{tot}) is given as:

$$\text{OD} = \log(I_0/I) = 0.434N\sigma_{tot} \tag{12.2}$$

where I_0 and I are the intensities of incident and transmitted light. The effect of multiple scattering is not accounted for in this equation. Details on light scattering in the near-infrared region can be found in the literature [14,15].

12.2.3.1 Applications

The applications of near-infrared spectroscopy in the oil industry are numerous and much work has been done in our group within this area. Generally, these applications of NIR spectroscopy can be divided into two groups:

- For prediction of physicochemical properties
- For studies of the degree of aggregation and/or aggregation size

To give an impression of the diversity of applications, we will briefly present some examples of recent work using NIR spectroscopy.

12.2.3.1.1 Prediction of SARA Values Using NIR in Combination with Multivariate Analysis

Since the SARA fractionation method presented in Section 12.2.1 is time consuming and requires expensive laboratory equipment, and some attempts have been made recently to find alternative analytical options. Aske and co-workers have shown the ability of NIR spectroscopy in combination with partial least squares (PLS) regression to predict SARA components in lighter crude oils and condensates [16]. Subsequently, Hannisdal and co-workers carried out a similar study, where the range of application was extended to particle rich and heavy crude oils [7]. Regression models were built for each SARA component from NIR data to predict the amount of SARA components. These models successfully fitted the experimental data from NIR analyzes and showed good predictive ability for the crude oil composition. This is a typical example of the ability of NIR to reflect chemical properties through vibrational characteristics as well as electronic absorption from asphaltenes.

12.2.3.1.2 Prediction of Solubility Parameters

In another study recently performed in our group [17] NIR and MIR spectra were correlated to the Hildebrand [18] and Hansen solubility parameters [19–21], using multivariate data analysis. Models were built from NIR and MIR spectra of different solvents and solvent mixtures. Table 12.1 shows the results from the correlation of solubility parameters to NIR spectra. This work will be useful in order to predict solubility parameters for crude oils and asphaltenes, and ultimately asphaltene stability.

12.2.3.1.3 Emulsion Stability Correlated to Physicochemical Parameters and NIR

In a study by Aske and co-workers [22], emulsion stability as measured by the critical electric field method ($E_{critical}$ presented later) was correlated to physicochemical properties and NIR spectra of 21 crude oils. Analysis of the significance of the regression coefficients of the model revealed that the NIR spectra, among some other factors, were closely correlated to the measured emulsion stability. Based on this, an attempt was made to predict emulsion stability exclusively

TABLE 12.1
Results from PLS Modeling and Predictions Based on the Correlation of Solubility Parameters to Unmodified NIR Spectra

Quality parameters	Hildebrand (Total)	Dispersive parameter	Hydrogen bonding	Polar contribution
Principal components	4	4	2	4
Correlation validation	0.92	0.88	0.95	0.95
Correlation prediction	0.92	0.53	0.91	0.78
RMSEV ($MPa^{1/2}$)	1.8	0.6	1.7	1.1
RMSEP ($MPa^{1/2}$)	2.6	1.1	3.5	3.8
Explained *Y*-variance (%)	87	81	90	89
Explained *X*-variance (%)	80	84	67	82

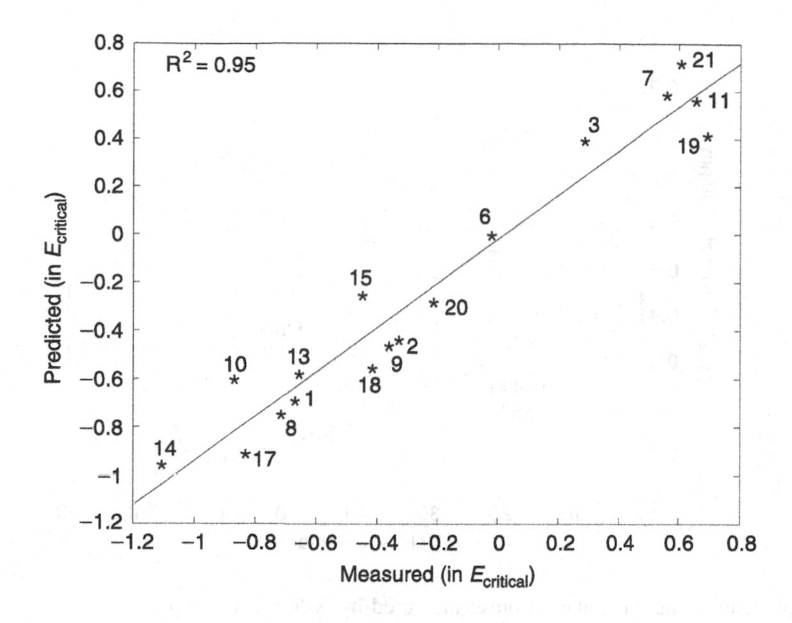

FIGURE 12.5 Predicted vs. measured plot from the PLS regression of emulsion stability (E_{crit}) based on the NIR spectra. (From Aske, N.; Kallevik, H., Sjöblom, J. *J. Petrol. Sci. Engin.* **2002**, *36*, 1–17.)

from NIR spectra. The predicted to measured emulsion stability is presented in Figure 12.5. NIR spectra contain information on both the aggregation state of asphaltenes and on chemical composition. This may explain the reason for its good predictive power for the $E_{critical}$ values.

12.2.3.1.4 High-Pressure Study of Asphaltene Aggregation

Asphaltenes tend to self-associate depending on solvent, pressure, and temperature conditions. Generally, aliphatic solvent conditions and pressure reductions will increase asphaltene aggregation size. Under increasingly unfavorable solvent condition, asphaltenes aggregate and eventually precipitate as large asphaltene flocculates. The asphaltene precipitation onset gives important information about the solubility of the asphaltenes in a given hydrocarbon system. Such measurements are carried out by titrating the system with an n-alkane, while measuring some physical property (e.g., light absorbance, viscosity, electrical conductivity), which shows a distinctive discontinuity at the precipitation onset [23]. NIR is an excellent and popular method for measuring the asphaltene precipitation onset [24–27]. As shown in Figure 12.6, the optical density of the system decreases as a consequence of dilution, until the precipitation threshold is reached. At that point the apparent absorbance increases due to light scattering of the precipitating asphaltenes. Aske and co-workers [28] have investigated the asphaltene aggregation behavior from crude oils and model systems under high pressure. Model systems and the crude oil were pressurized to 100 and 300 bar respectively. The systems were then depressurized in steps, and the resulting NIR spectra were recorded at each pressure level and analyzed with multivariate analysis. The asphaltene precipitation onset pressure was identified from increased optical density due to light scattering. By reducing the pressure even further, the bubble point was detected from the subtle drop in absorbance due to gas evolution. The reversibility of the asphaltene aggregation was studied for the crude oil and a model system by repressurizing the systems stepwise from the bubble point to the original pressure. The systems were then left to equilibrate for several hours.

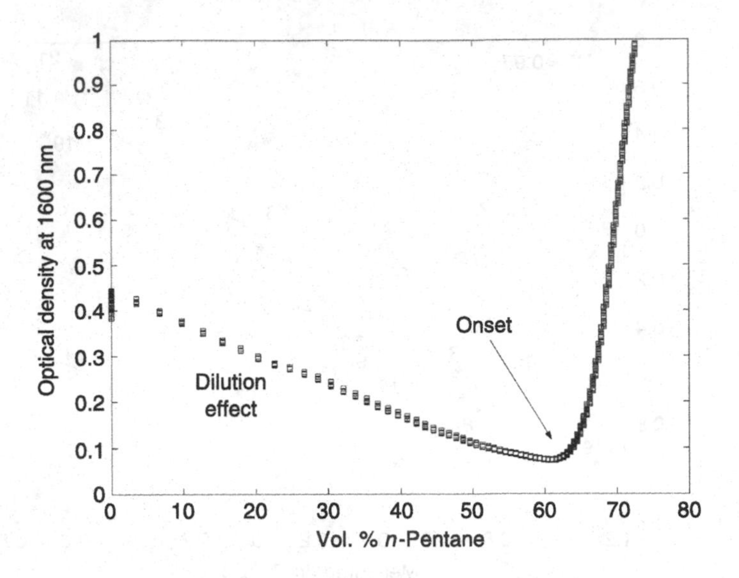

FIGURE 12.6 Asphaltene precipitation onset measured by NIR spectroscopy.

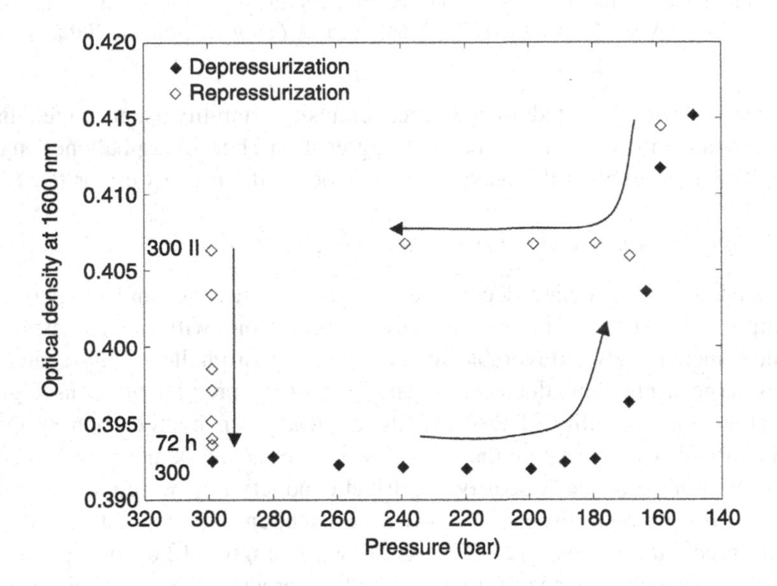

FIGURE 12.7 Asphaltene aggregation and reversibility of crude oil studied by high-pressure NIR. (From Aske, N.; Kallevik, H.; Johnsen, E. E., Sjöblom, J. *Energy & Fuels* **2002**, *16*, 1287–1295.)

Here, we will focus on the part of the study dealing with repressurization. Figures 12.7 and 12.8 show the recorded optical density at 1600 nm of the crude oil and the model system during the pressure cycle.

For the crude oil, considerable redissolution of the aggregates was seen when increasing the pressure from 150 to 170 bar, as reflected by the decreased optical density. After 72 h the aggregate size had practically returned to its original state. For the model system, the aggregates seemed to redissolve steadily during the depressurizing, but came to a stop after approximately

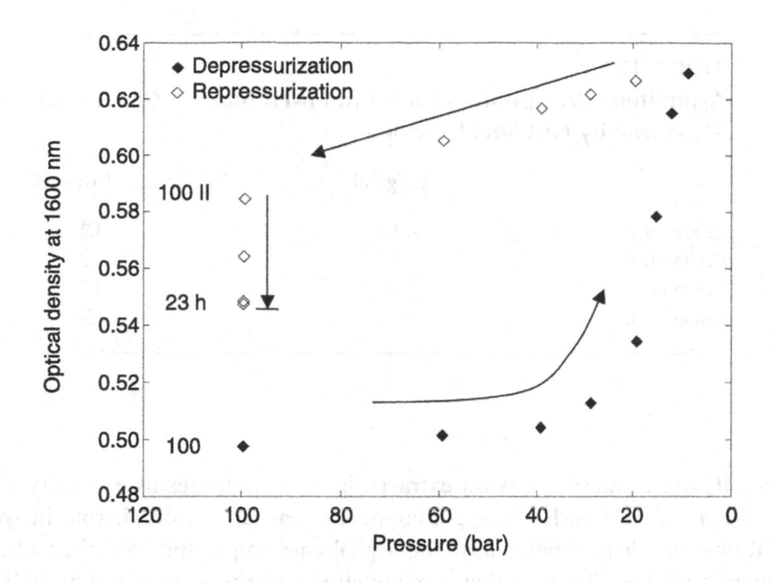

FIGURE 12.8 Asphaltene aggregation and reversibility of model system studied by high-pressure NIR. (From Aske, N.; Kallevik, H., Johnsen, E. E.; Sjöblom, J. *Energy & Fuels* **2002**, *16*, 1287–1295. Model system: 1.2 wt % asphaltenes; 35 wt % pentane in toluene.)

23 h at the original pressure of 100 bar. While the asphaltene aggregation of the crude oil was more or less completely reversible, the asphaltene aggregation in the model system was only partially reversible. The authors explained the different behavior of the crude oil and the model system to be caused by the lack of dispersing resins during the repressurization of the model system. This study is a typical example of the power of NIR spectroscopy to monitor physical changes in crude oil systems.

12.2.3.1.5 NIR Study on the Dispersive Effect of Amphiphiles and Napthenic Acids on Asphaltenes

Auflem and co-workers [29] showed the ability of NIR to follow the disintegration of asphaltene aggregates upon addition of chemicals or indigenous crude oil components. In order to hinder asphaltene deposition, the petroleum industry injects large volumes of chemicals into reservoirs and pipelines. These chemicals are supposed to imitate the indigenous resin fraction, by dispersing the asphaltenes in the hydrocarbon mixture as discussed earlier in this chapter. In addition to the direct problems concerning asphaltene deposition in process equipment, the stability of water-in-oil emulsions will be strongly dependent on the asphaltene aggregation size and state [30,31]. The asphaltene aggregation state is monitored with NIR spectroscopy after addition of chemicals known to efficiently disperse asphaltenes through different interaction mechanisms depending on the additive. The results showed that additives, which are efficient in replacing hydrogen bonds in asphaltene aggregates, possess dispersive power and can serve as inhibitors. Commercial blends gave the best results.

12.2.3.1.6 Asphaltene Destabilization

Recently, in a study concerning the potential for hydrate plugging, crude oils washed with a strong alkaline solution (pH 14) showed much higher water-in-oil emulsion stability than the

TABLE 12.2
Asphaltene Precipitation Onset (ml n-Heptane/g Crude Oil) Measured by NIR Spectroscopy

	Original	pH14 washed
Crude oil A	7.0	4.2
Crude oil B	3.2	2.3
Crude oil C	3.9	3.2
Crude oil D	5.0	1.7

original crude oils [32]. The pH 14 wash extracts the most polar resins, typically napthenic acids and phenols. As mentioned earlier, these compounds play an important role in solubilizing the asphaltenes. Removing them should cause the asphaltenes to precipitate earlier when titrating the crude oil with an n-alkane. The asphaltene precipitation onset was measured by NIR spectroscopy. As shown in Table 12.2, the asphaltenes in the pH 14 washed crude oils precipitated *earlier* than the asphaltenes in the original crude oils. NIR can be used for the initial screening of a large number of chemicals for asphaltene inhibition.

12.3 PHYSICOCHEMICAL PROPERTIES

12.3.1 Critical Electric Field as a Measure of Emulsion Stability

Electro-coalescers are commonly used in the oil industry to enhance the separation of water from crude oil. The electro-coalescer applies an electric field of 1 to 10 kV cm^{-1} through the flowing water-in-oil emulsion to enhance the flocculation and coalescence of the dispersed water droplets [33]. Basically, the water droplets increase their sizes through the electro-coalescer, and thereby the water sedimentation rate is increased [34,35]. There is a variety of factors influencing the electrically induced coalescence, such as the dielectric properties of the dispersed and the continuous phase, the volume fraction of the dispersed phase, conductivity, size distribution of the dispersed droplets, electrode geometry, electric field intensity, the nature of the electric field (AC or DC), etc. [36–38].

An electric field cell containing a small volume of sample has recently been developed for the determination of emulsion stability [22]. The method is similar to the one employed by Kallevik et al. [39]. The cell consists of two brass plates separated by a 0.5 mm Teflon plate (Figure 12.9). The Teflon plate has a 10 mm diameter hole in the center so that a small sample of emulsion can be injected between the two brass plates. The brass plates are connected to a power supply (Agilent Model 6634B). The power supply can increase the voltage stepwise and measure the current passing through the emulsion. A sudden increase of the current will indicate that the electric field has broken the emulsion and that there will be a free transport of ions between the electrodes. The corresponding electric field is the E-critical value.

Figure 12.10 illustrates how emulsions behave under the influence of the applied field. When no field is applied, the water droplets are randomly distributed according to Stokes law of sedimentation and Brownian motion. In addition, some degree of droplet flocculation may be present. As the field increases, the droplets line up between the electrodes due to polarization of the aqueous

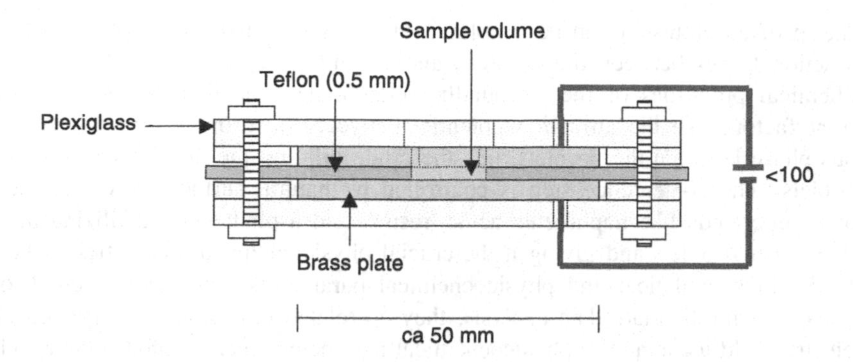

FIGURE 12.9 The critical electric field cell.

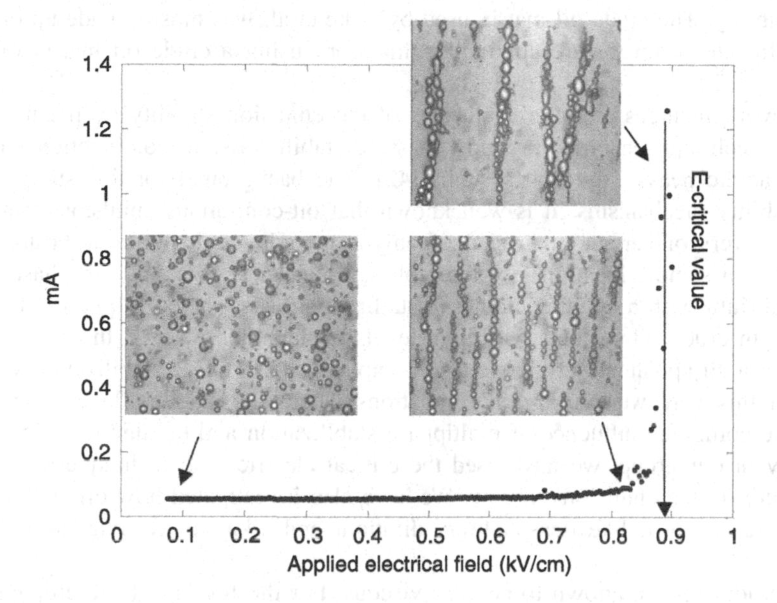

FIGURE 12.10 Behavior of w/o emulsions in an external electric field.

droplets containing electrolyte. Finally, at some point, the applied electric field causes irreversible rupture of the interfacial films between the droplets [33]. The droplets coalesce, resulting in a water-continuous bridge between the electrodes. The ions of the water phase contribute to a sudden increase in the conductivity. The emulsion stability is hence defined as the strength of the field when the conductivity suddenly rises.

12.3.1.1 Applications

The critical electric field technique can only be used for emulsions where the continuous phase is an insulator and the dispersed phase has conducting capacity. Hence it can be used for measuring the stability of water-in-oil emulsions, but not for oil-in-water emulsions.

One question that often arises when facing an emulsion problem is: what factors are causing the stabilization of the emulsion? This is an important question in order to select the right remedy.

The break-up of an emulsion can be simplified in two steps: first the droplets flocculate due to weak attraction forces between the droplets, and secondly they coalesce which depends on the physicochemical properties of the surrounding membrane. The flocculation step is controlled by different factors like the attraction potential between the droplets and the viscosity of the continuous phase between the droplets, i.e., the crude oil, and the droplet sizes of the dispersed water droplets. The coalescence step is controlled by natural surfactants within the crude oil, indigenous compounds like naphthenic acids, resins, and asphaltenes, stabilizing the interfacial film on the water droplets and giving it the crucial physicochemical properties. Aske et al. [40] investigated which analytical and physicochemical parameters contribute to emulsion stability properties. Using multivariate data analysis, they correlated emulsion stability measured by the critical electric field technique to physicochemical parameters such as molecular weight, density, viscosity, interfacial tension, interfacial elasticity, total acid number (TAN), and SARA fractions for a crude oil matrix consisting of 21 samples. They concluded that the asphaltene content, state of asphaltene aggregation, and interfacial elasticity were the most important factors with regard to emulsion stability. The crude oil matrix used by Aske et al. was mostly made up of conventional crude oils. In our group we are continuing this work using a crude oil matrix of heavy crude oils [41].

Earlier work includes also the evaluation of the emulsion stability by means of the critical electric field technique in order to study D-phase stabilization in combination with asphaltene particles in acidic heavy crude oil systems [42]. The background for this study is to compare different stability mechanisms. It is well known that oil-continuous emulsions can be stabilized by multiple layers of surfactant instead of only a monolayer. In an equilibrium situation this corresponds to a sample location in a three-phase area where two solution phases (L1 and L2) are in equilibrium with a lamellar liquid crystalline phase (so-called D phase). This situation is of relevance in crude oil systems with high levels of naphthenic acids. In order to simulate the situation in a high aphaltenic crude we also combined the D-phase stabilization with asphaltene particles. In this way we could create situations where the stability level rose several times reflecting the combined influence of multiphase stabilization and organic particles.

Recently in our group, we have used the critical electric field technique to study the effect of temperature on the emulsion stability. We have also investigated how different kinds of modifications on crude oils, like deasphalting, dilution, and alkaline washing, affect the emulsion stability.

Heavy crude oils are known to be very viscous. For the line up of the droplets between the electrodes in the critical electric field cell, the viscosity of the continuous phase should be one of the time-limiting factors. In order to test how the emulsion stability is affected by changing the viscosity, five crude oils were diluted in different ratios with a heptane–toluene mixture (70:30 vol%) [41]. The diluted crude oils were then mixed with 30% water containing 3.5 wt% NaCl and the emulsion stability was measured by the critical electric field cell. Figure 12.11 shows that, generally, for the five crude oils, the emulsion stability decreases as the viscosity is decreased by dilution. However, as the crude oils are diluted, the concentration of surface active compounds like resins and asphaltenes is also decreasing, which will also affect the emulsion stability. Likewise, the interfacial tension and interfacial elasticity will also change as a result of the dilution. The most interesting information from Figure 12.11 is what happens to crude oils 3 and 4. At high viscosity their corresponding emulsions show high stability, but as we increase the dilution (decreasing the viscosity), the stability drops, but levels off at a certain level, independently of the viscosity. In this region, where the E-critical value is fairly stable, the line up of the water droplets cannot be the limiting step for the break-up of the emulsion.

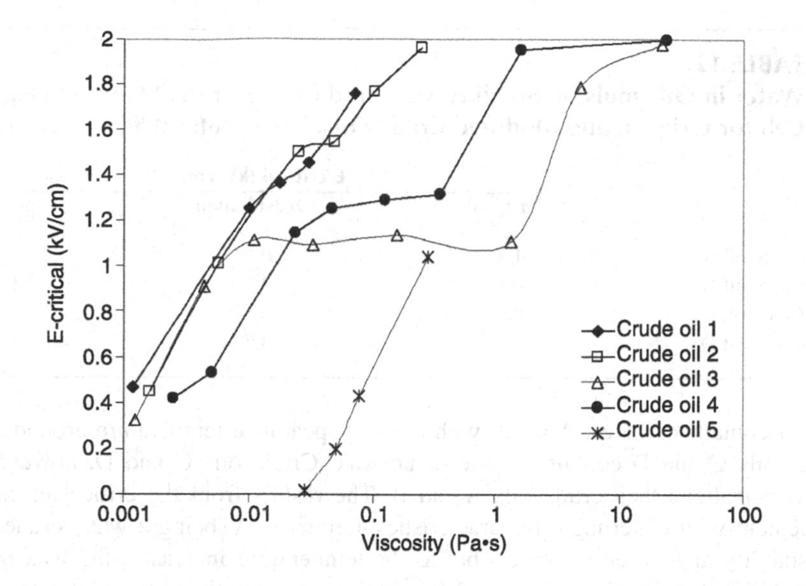

FIGURE 12.11 Emulsion stability, measured by means of critical electric field technique, as a function of the viscosity of the crude oil, for five different oils. The viscosity was changed by dilution of the different crude oils by a heptane–toluene mixture (70:30 vol%).

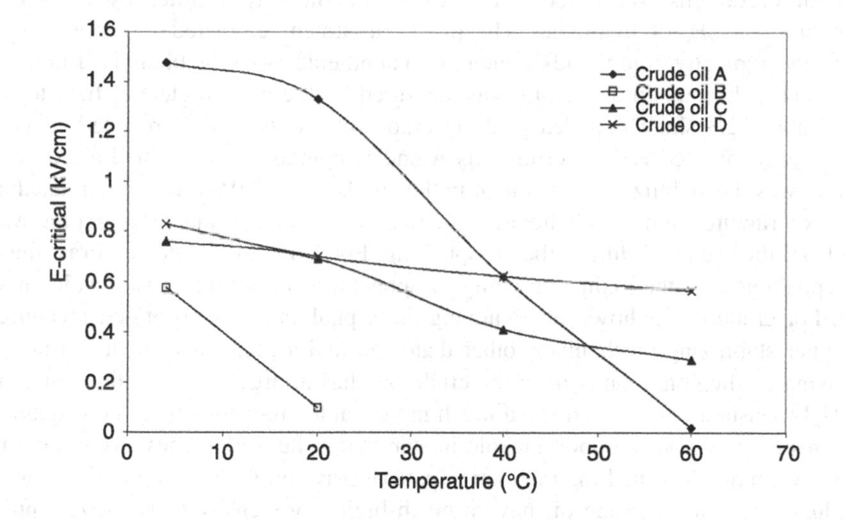

FIGURE 12.12 Emulsion stability, represented by the critical electric field, E-critical, as a function of temperature for four different crude oils.

Another important aspect of emulsions is how they are affected by temperature. Figure 12.12 shows the emulsion stability, by means of the critical electric field, as a function of the temperature for four crude oil systems [32]. Prior to making the emulsions, the crude oil and water were heated/cooled to the appropriate temperature. The critical electric field cell was also kept at the same temperature as the emulsions. The results show that there is a big difference in the temperature dependence on the emulsion stability. Viscosity measurements showed that crude

TABLE 12.3
Water in Oil Emulsion Stability Measured by the Critical Electric Field Cell for Original and Modified Crude Oils. Watercut: 30%. T = 20 °C

	E-critical (kV/cm)		
	Original	**Deasphalted**	**pH 14**
Crude oil A	1.3	0.03	1.9
Crude oil B	0.1	0	1.8
Crude oil C	0.7	0	3.0
Crude oil D	0.7	0.7	1.9

oils A and B contained waxes, A most, with a wax appearance temperature around 20 to 25 °C, while crude oils C and D contained little or no wax. Crude oils C and D, however, contained much more asphaltene than crude oils A and B. The results from the critical electric field cell are in agreement with the crude oil characteristics. Crude oil A, being a waxy crude, shows high emulsion stability at low temperatures, but as the temperature increases, the wax melts and the emulsion stability drops rapidly. Crude oil B contains less wax than crude oil A, and shows less stability than crude oil A. For crude oils C and D, the emulsion stability decreases less than for the two waxy crude oils, and the emulsions are most likely stabilized by asphaltenes, resins, and/or inorganic fines.

The four crude oils were modified in two different ways; either by deasphalting using n-pentane or by a pH 14 treatment. The pH 14 treatment extracted the most polar resins in the crude oils, typically compounds containing naphthenic acids or phenols. Emulsion stability of the original and modified crude oils was measured by the critical electric field technique. The results in Table 12.3 show that deasphalting crude oils A, B, and C resulted in very unstable emulsions. As mentioned earlier, crude oils A and B contained wax, and Figure 12.12 showed that the wax was the stabilizing component in the emulsions. FTIR of the precipitated asphaltenes showed wax structures in the asphaltenes from crude oils A and B, meaning that the wax was also removed from the crude oil during the deasphalting. For crude oil C one can draw the conclusion that the asphaltenes are the main stabilizing component, as removing those results in an unstable emulsion. For crude oil D, however, removing the asphaltenes has no effect, meaning that there must be other stabilizing mechanisms other than wax and asphaltenes for this crude.

Removing acidic compounds from the crude oils has a large effect on the emulsion stability. All the pH 14 washed crude oils formed much more stable emulsions than the original crude oils. The polar resins play a very important role in stabilizing the asphaltenes. By removing the most polar resins we have disturbed the interaction pattern between the resins and the asphaltenes, and the asphaltenes left in the crude oil have a much higher propensity to stabilize emulsions. This is in accordance with several studies [29,31,43].

12.3.2 Determination of the Rheological Characteristics of a W/O Interface

Interfacial rheology is the measurement of the elasticity and viscosity of interfacial films that form on liquid surfaces. These properties play an important role for the stability of water-in-crude-oil emulsions and the mechanism of demulsification. It has been shown that when indigenous crude oil components adsorb onto an interface between oil and water they can protect the interface against coalescence [30,31]. In short, when the interface is stressed, the uneven distribution of interfacially active components will create tension gradients on the interface that will oppose the strain, i.e., the Gibbs–Marangoni effect. The interface will hold elastic properties protecting the interface from coalescence. Determination of the rheological properties of an interface can be undertaken with dilational stress where the interfacial resistance to changes in area is monitored. Measurements are made by simultaneously analyzing the interfacial tension and surface area of a drop which is periodically perturbed. A fresh oil droplet is created from a needle immersed in water and its volume is modulated by a syringe pump to give a sinusoidal variation of the surface area of the drop.

$$\Delta A = A - A_0 = A_a \sin(\omega t) \tag{12.3}$$

A_a is the area amplitude and A_0 is the equilibrium area. The drop shape remains constant. The principle of the method is to monitor the change in interfacial tension as a result of processes trying to re-establish equilibrium of the interface. The dynamic interfacial tension is estimated using axisymmetric drop shape analysis as presented in Section 12.3.4.

$$\Delta\gamma = \gamma - \gamma_0 = \gamma_a \sin(\omega t + \theta) = \gamma_a \sin(\omega t)\cos\theta + \gamma_a \cos(\omega t)\sin\theta \tag{12.4}$$

Here γ_a is the tension amplitude and γ_0 is the equilibrium interfacial tension. The phase angle θ is shown in Figure 12.13 and can be estimated by a phase comparison between the variation in area and the response in tension. The interfacial dilational modulus E describes the response of an interface to local compression and expansion and is defined as the interfacial tension increment per unit fractional change in the interfacial area [40].

$$E = \frac{d\gamma}{d \ln A} \tag{12.5}$$

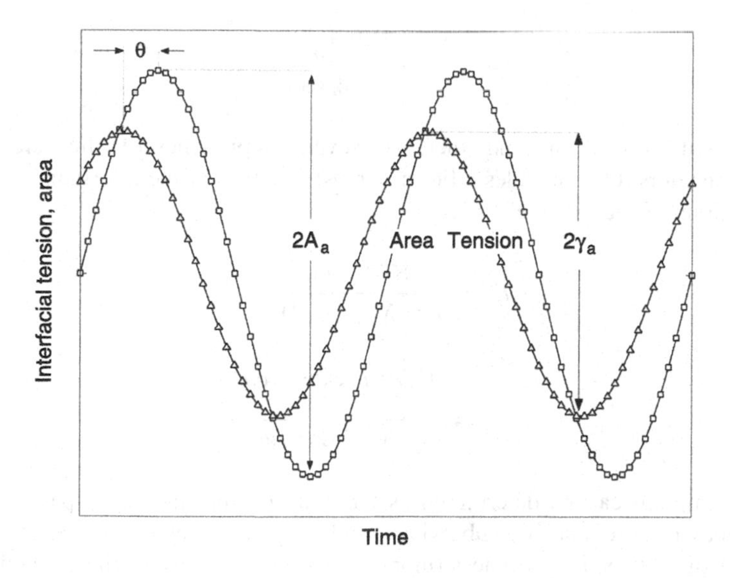

FIGURE 12.13 Time dependent relationship between area and interfacial tension during a dilational rheology experiment.

When analyzing the relaxation processes at or near the interface between oil and water, we expect the rheological properties of the interface to be viscoelastic rather than pure elastic. The interfacial dilational modulus is a complex function of the angular frequency of oscillation ω [44].

$$E^* = E_d + i\omega\eta_d = E' + iE'' \tag{12.6}$$

E_d is known as the interfacial dilational elasticity whereas the interfacial dilational viscosity η_d corresponds to the dynamic viscosity in an ordinary bulk oscillatory rheology experiment. The complex dilational modulus E^* may be written as the sum of two contributions. An elastic component accounting for the recoverable energy stored in the interface (the storage modulus E') and a viscous component accounting for the energy lost through relaxation processes (the loss modulus E''). The storage modulus is in phase with the modulation and the loss modulus is 90° out of phase. By using the Gibbs definition of elasticity (Equation 12.5) with insertion of the variation in drop area (Equation 12.3) and the corresponding variation in interfacial tension (Equation 12.4), the complex interfacial dilational modulus E^* can be expressed as:

$$\begin{aligned} E^* &= \frac{\Delta\gamma}{\Delta \ln A} = \frac{\Delta\gamma}{\Delta A/A_0} \\ &= \frac{\gamma_a \sin(\omega t)\cos\theta + \gamma_a \cos(\omega t)\sin\theta}{A_a \sin(\omega t)/A_0} \\ &= \frac{\gamma_a}{A_a/A_0}\left(\cos\theta + \frac{\cos(\omega t)}{\sin(\omega t)}\sin\theta\right) \\ &= |E|\cos\theta + i|E|\sin\theta \end{aligned} \tag{12.7}$$

where

$$|E| = \frac{\gamma_a}{A_a/A_0}$$

With the analyzed "movie" in hand, there are several approaches for the determination of the rheological parameters. One includes a Fourier transformation of the interfacial tension waveform inside subdivisions of the "movie" [45].

$$\begin{aligned} E^* &= \frac{\mathrm{FT}\Delta\gamma(t)}{\mathrm{FT}(\Delta A/A_0)(t)} \\ &= \frac{\int_0^\infty \Delta\gamma(t)\exp(-i\omega t)dt}{(\Delta A/A_0)i\omega} \end{aligned} \tag{12.8}$$

The E' and E'' moduli can be determined as the real and the imaginary part of Equation 12.8. Figure 12.13 may represent such a subdivision including two cycles. The duration of each subdivision is a tradeoff between smoothness (more cycles results in less variance) and time resolution (less cycles gives quicker response). The frequency component corresponding to the modulation can be easily picked out from the spectrum and other frequencies (noise) rejected.

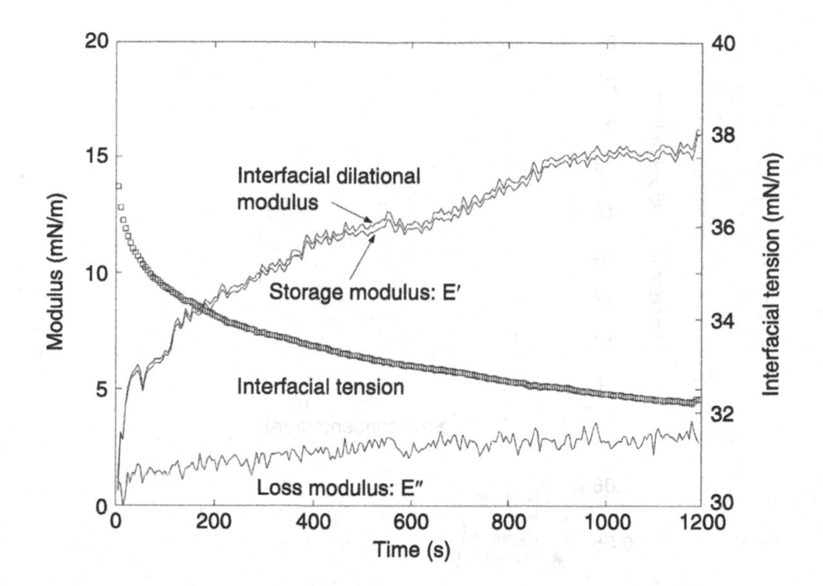

FIGURE 12.14 Dynamic pendant drop experiment of diluted (heptane/toluene equal volumes) Brazilian heavy crude oil immersed in water (3.5 wt% NaCl). Frequency of oscillations is 0.1 Hz. $\theta \sim 12°$.

Figure 12.14 shows the interfacial tension response of a sinusoidal oscillation of a diluted Brazilian oil against water (3.5 wt% NaCl). Each point represents a Fourier transformed subdivision of the interfacial tension profile. The interfacial tension decays from the initial tension of a pure interface to 32 mN/m, 20 min after drop formation. Storage and the loss moduli are also included. The loss modulus stays essentially constant while the storage modulus increases greatly when the interface is organized with interfacially active crude oil components. The interface is mainly elastic (close fit between E' and E) but it also holds some viscous properties reflected by a non-zero E'' and a phase angle: $\theta \sim 12°$.

12.3.2.1 Applications

Aske and co-workers investigated the interfacial rheology of water–crude oil systems [40]. An oscillating pendant drop apparatus was used in the systematic characterization of 21 crude oil and condensates with respect to interfacial elasticity. The elasticity modulus of the interface after 1000 sec was reported. These experiments were supplemented by near-infrared spectroscopy (optical density at 1600 nm) to keep track of the asphaltene aggregation state in the oil phase.

Results presented in this example are the elasticity modulus with corresponding optical density of one of the crude oils, a West African crude oil with 1 wt% asphaltenes. The crude oil was diluted to 3 different concentration levels (ml oil/ml solvent) with solvents of heptane and toluene. The effect of crude oil concentration and solvent composition will be discussed with reference to Figure 12.15, showing the elasticity modulus of the oil–water interface and the optical density of the oil phase.

In poor solvents (high heptane content) the interfacial elasticity decreased with increasing oil concentration. In good solvents the interfacial elasticity generally increased with oil concentration. From the optical density the formation of larger aggregates is seen to be much more dominating in poorer solvents. The lowering of the interfacial elasticity appeared to be caused by the formation

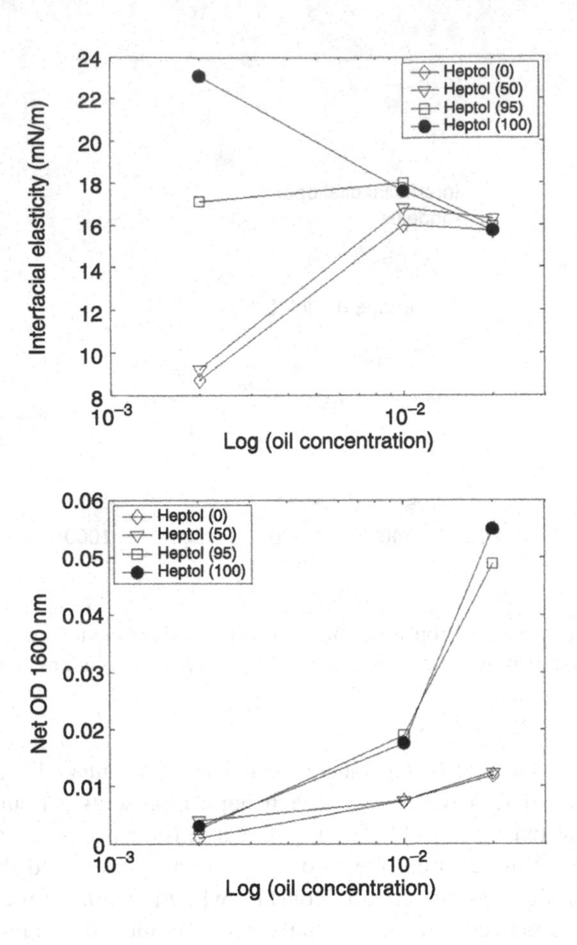

FIGURE 12.15 Elasticity modulus of an interface between diluted crude oil and water (3.5 wt% NaCl) (top). Corresponding optical density of the oil phase is also reported (bottom). The number in the legend indicates the heptane content of the solvent (vol%).

of larger asphaltene aggregates. At the lowest oil concentration, the highest elasticity of the interface was obtained in the poorest crude oil surfactant solvents. In such solvents the interfacial activity of resins and asphaltenes were high, and asphaltenes appeared to be in a favorable aggregation state. At higher oil concentrations this effect of interfacial activity seemed to be opposed by the formation of asphaltene aggregates.

This study by Aske and co-workers [40] demonstrates that measurements of interfacial rheological properties of crude oil systems are highly dependent both on oil concentration and of the solvent used for dilution. The results are consistent with the asphaltene aggregation model proposed by Kilpatrick and co-workers [30,31]. They found that precipitated asphaltenes had a lower ability to form elastic interfaces. This was attributed to lower interfacial activity of such precipitated aggregates and possibly high amounts of defects in films of precipitated material. It should be mentioned that the measurements by Kilpatrick et al. were performed according to other measurement principles.

12.3.3 The Langmuir Technique

12.3.3.1 Background

In studies of insoluble monolayers on liquid surfaces, the parameter of greatest interest and the one usually measured is the difference in surface tension between a clean or pure liquid surface and one covered with a film. This quantity is called the surface pressure and denoted by a Greek pi (Π):

$$\Pi = \gamma - \gamma_0 \tag{12.9}$$

where γ is the surface tension in the absence of a film and γ_0 the tension of the film-covered surface. The surface tension of water is 73.05 mN/m at 18 °C and atmospheric pressure [46]. Hence the maximum possible surface pressure for a monolayer on a water surface at this temperature is 73.05 mN/m.

The Langmuir technique is used in order to characterize film properties of surface-active materials. The instrumentation consists of a shallow rectangular container (trough) in which the liquid subphase is added, whereupon the film is spread. The barriers, for manipulation of the film, rest across the edges of the trough (Figure 12.16(a)). The surface pressure is measured by means of the Wilhelmy method [47–49]. Modification of the trough design has made it possible to carry out the same kind of experiments on a liquid/liquid interface, i.e., the oil/water interface. In this case the trough is a "double trough" (Figure 12.16(b)) where the barriers contain holes to allow the flow of the light phase as the compression of the interphase proceeds. The Wilhelmy plate is first placed in the aqueous phase, and then the oil phase is added until the plate is totally immersed.

The most common way of presenting the data obtained from the Langmuir technique is a plot of surface pressure as a function of surface area. The measurements are carried out at a constant temperature and are known as surface pressure/area isotherms. The film is compressed at constant rate by the moving barriers while the surface pressure is continuously monitored. As the surface area is reduced from its initial high value a series of different regions indicating different states of the monolayer is observed. Analogously to bulk matter these states are characterized as gas-like, liquid-like, and solid-like [50,51]. At small surface areas the monolayer will collapse;

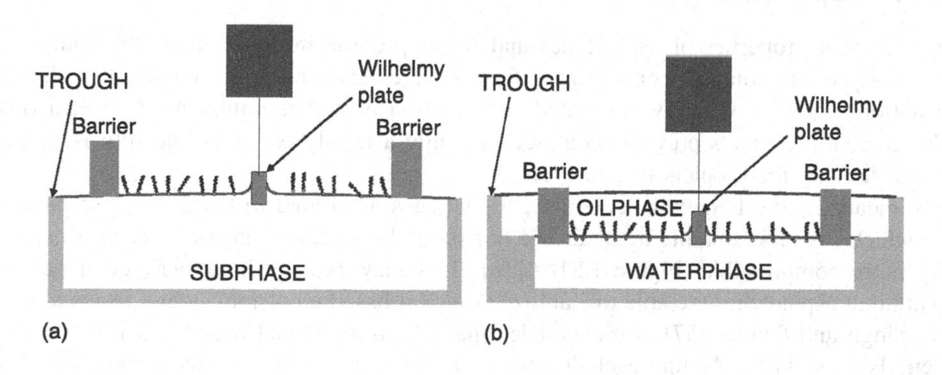

FIGURE 12.16 Illustration of the different Langmuir troughs. (a) Liquid–air surface trough. (b) Liquid–liquid interfacial trough.

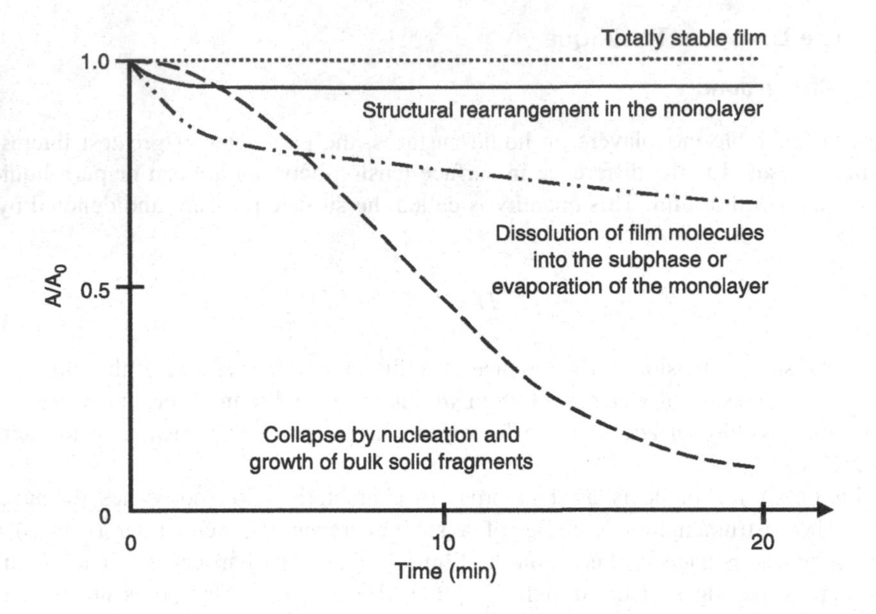

FIGURE 12.17 Relaxation curves at constant surface pressure for monolayers of different stability.

the collapse pressure is the highest surface pressure to which a monolayer can be compressed without a detectable movement of the film molecules and formation of a new phase. The variation and complexity of the monolayer behavior is a result of the different intermolecular interactions in the film and between the film and the subphase. This implies that the changes in interaction forces are related to the packing of the molecules in the two-dimensional plane [48,49].

In order to be able to interpret a surface pressure/area isotherm, the film stability has to be characterized. The stability of a monolayer is studied by measuring the area loss at constant surface pressure or the decrease in surface pressure at constant area. The different shapes of the relative area isotherms in Figure 12.17 illustrate the most common destabilization mechanisms [52–55].

12.3.3.2 Applications

Studies of film properties of asphaltenes and resins provide information on the compressibility and stability of monolayers consisting of these indigenous crude oil components. This kind of information might be virtually correlated to formation of stable emulsions. Hence, a rigid film on the emulsion droplets prevents coalescence while a highly compressible film is more easily ruptured, leaving the droplets free to coalesce.

By means of the Langmuir technique, asphaltenes are found to build up close-packed rigid films, which give rise to quite high surface pressures. Resin films, on the other hand, are considerably more compressible (Figure 12.18) [56]. This may explain the experimental observations showing that asphaltenes are able to stabilize crude oil based emulsions, while resins alone fail to do so. Singh and Pandey [57] also concluded that a high interfacial pressure correlated with high w/o emulsion stability. Adding asphaltenes and resins together into a mixed film, the properties gradually change from a rigid to a compressible structure as the resin content is increased. The more hydrophilic resin fraction starts to dominate the film properties due to the higher affinity towards the surface.

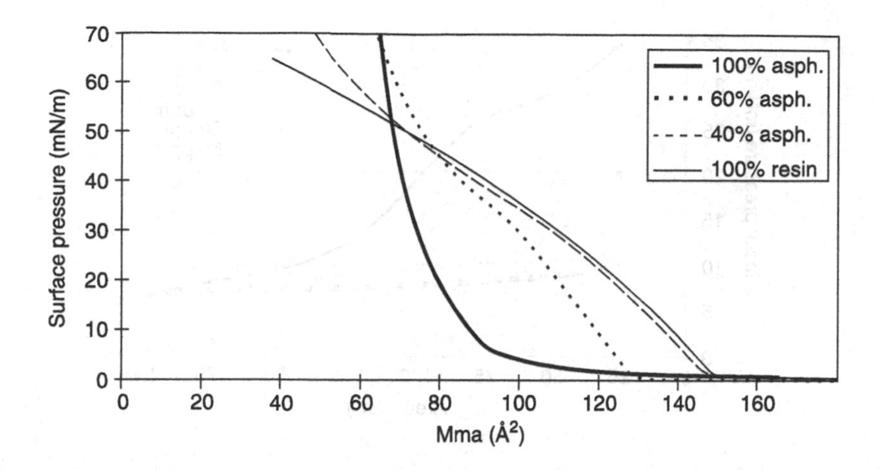

FIGURE 12.18 *Π*-A isotherms of asphaltene/resin mixtures spread from pure toluene on pure water (bulk concentration = 4 mg/ml).

The influences of chemical additives on asphaltene films on water surface and at oil/water interface have also been studied by means of the Langmuir technique. This was done in order to view the interaction between demulsifiers added and asphaltenes, and to show the importance of this on emulsion stability.

Highly compressible resin films alone will not stabilize a crude oil emulsion. Related to this, demulsifiers, which form films of low rigidity and high compressibility, should be most efficient. When used as demulsifiers, the efficiency depends on the ability of the chemicals to interact with and modify the film built up by asphaltene particles.

Addition of demulsifiers of high molecular weight in the asphaltene film gave the isotherms in Figure 12.19 [58]. Chemical G is highly effective with respect to increased compressibility together with a reduced rigidity. The effect of this kind of manipulation of the asphaltene film is similar to the effects observed when resins are mixed together with asphaltenes (Figure 12.18).

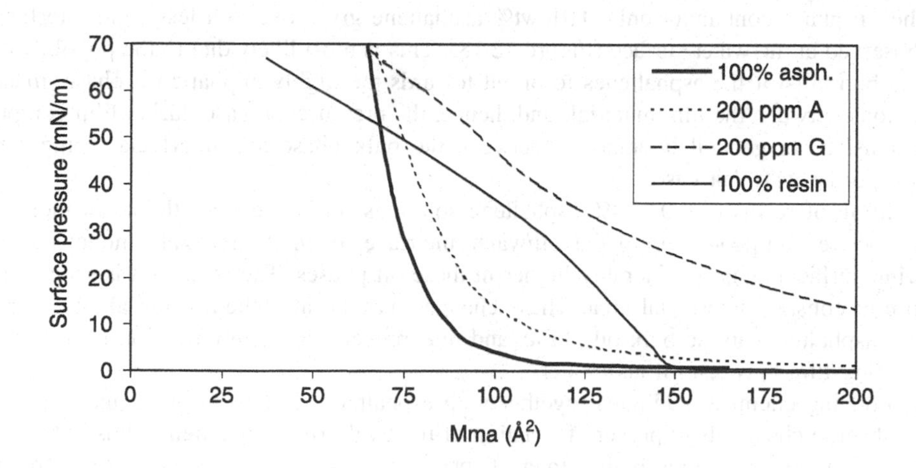

FIGURE 12.19 *Π*-A isotherms of mixed monolayers of asphaltenes and two different demulsifers on pure water. The mixed monolayers are compared with the pure asphaltene and the pure resin film.

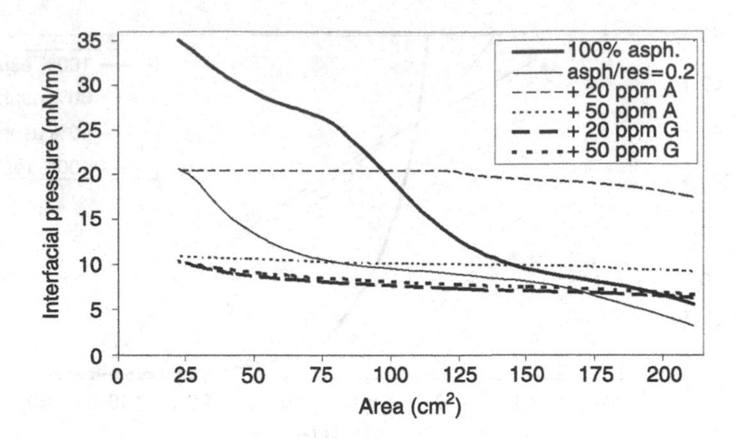

FIGURE 12.20 Interfacial pressure isotherms of films formed between water and oil containing different ratios of asphaltenes and resins or asphaltenes and demulsifiers.

However, the concentration needed to achieve the same effects is considerably lower when demulsifiers are used instead of resins. Demulsifier A has a quite small influence on a film of asphaltenes.

From film studies above one can conclude that the best candidates for emulsion breaking should be G. However, the efficiency depends not only on the direct influence of chemical additives within the film, but also on the ability of demulsifiers to reach the w/o interface in an emulsion (diffusion through the fluid). This is a critical step regarding the effective concentration of demulsifiers at the interface. These aspects make it difficult to undertake a direct comparison between influence of demulsifier in Langmuir surface films, where all demulsifier molecules are implanted in the film (2-D), and in real emulsions with a hydrocarbon environment (3-D).

In order to represent more realistic emulsion conditions, Langmuir interfacial films adsorbed at the oil/water interface were analyzed [58]. The isotherms given in Figure 12.20 illustrate some of the film properties of naturally occurring crude oil components adsorbed at the o/w interface.

The oil phase containing only 0.01 wt% asphaltene gives rise to a less rigid interfacial film than observed at the water surface (Figure 12.18). This is most likely due to the possibility for the hydrocarbon tails of the asphaltenes to orient towards the highly aliphatic oil phase, making the interactions between the film material, and, hence, the pressure increase during film compression, less extensive. In general interactions between the bulk phase and interfacial components are different to the water/air case.

Addition of resins to 0.01 wt% asphaltene solutions further reduces the adsorption of interfacially active components onto the oil/water interface, even if the total amount of naturally occurring surfactants is considerably higher in these oil phases. The reduction is seen as reduced pressure at constant interfacial area. These changes may be attributed to the ability of resins to disperse asphaltenes in the bulk oil phase, and thus prevent this heavy fraction from building up a stabilizing film between oil and water.

Introducing chemical additives together with asphaltenes in the oil phase may highlight the ability of these chemicals to prevent formation of relatively rigid asphaltene films at the oil/water interface. For concentrations higher than 20 ppm of chemical A there is no pressure increase during the compression. Hence, the film that is formed at the interface is highly compressible. So instead of increasing the pressure, the components will build up a multilayer, or the film

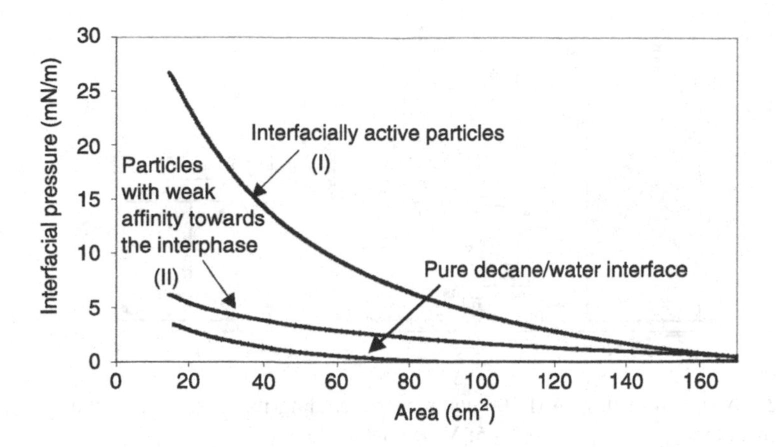

FIGURE 12.21 Interfacial pressure isotherms of films formed between water containing different types of particles and an oil phase of pure decane.

may dissolve under the influence of compression. An increased inhibitor concentration reduces the interfacial pressure, but has no influence on the film behavior. The reduced pressure is probably due to a more complete cover of inhibitor at the interface. That is, fewer components from the asphaltene fraction are adsorbed together with the chemical additive when the inhibitor concentration becomes high enough.

The results obtained upon addition of demulsifier G are similar to those of A. However, G clearly increases the compressibility of the film even at low concentration. The difference between 20 and 50 ppm is quite small, so it is reasonable to believe that maximal efficiency resulting from the competing adsorption in a system like this is reached already at a concentration of 20 ppm of the oil phase. With 20 ppm or more G present, only small amounts of asphaltene will reach the interface.

The results obtained from the Langmuir interfacial film studies are important in explaining why certain chemicals are more effective as inhibitors than as demulsifiers. Obviously the inhibitor/asphaltene interaction is so strong in the bulk oil phase that the interfacial structures being gradually built up will no longer possess properties required to stabilize w/o emulsions.

Interfacial activity of particles might also be characterized by means of the Langmuir technique. Figure 12.21 illustrates the film properties of some naturally occurring reservoir particles adsorbed at the o/w interface (the particle sizes are the same for both fractions). The oil phase consists of pure decane while 1 wt% particles are added to the water phase. One of the particle fractions (type I) forms a relatively rigid film at the interface, while the other (type II) is more hydrophilic and remains dispersed in the aqueous phase. Studies of these particles' ability to stabilize emulsions have shown that type I particles are highly efficient emulsifiers, while type II particles will not form stable emulsion [59].

12.3.4 Pendant Drop Instrumentation/Interfacial Reactions

Interfacial tension (IFT) is a measurement of the cohesive energy present at an interface between two phases and is the most used parameter to characterize the dynamic properties of liquid adsorption layers. For a liquid in contact with a vapour, the molecules in the bulk phase are symmetrically surrounded by other molecules so that the molecular interactions in all directions

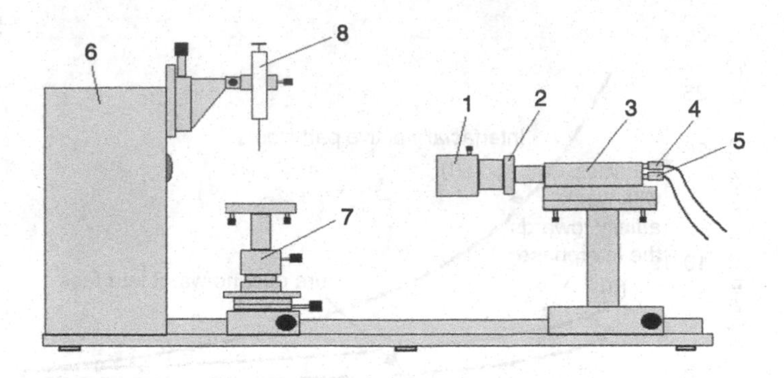

FIGURE 12.22 A sketch of the CAM 200 equipment including its main components:

1: Camera lens zoom
2: Lens aperture adjustment
3: CCD camera
4: Light synchronizing cable
5: Video out cable
6: Light source and interface unit
7: Stage for sample
8: Syringe

are identical and the sum of all interacting forces is zero. At the liquid–vapour interface, however, the molecules are mainly affected by interacting forces from the liquid side, causing a net force towards the liquid bulk. The liquid molecules at the interface are thus "drawn" towards the bulk phase, which in turn cause a minimizing of interfacial area.

Hence, it is clear that the IFT increases with increasing intermolecular interactions. Thus, any factor which reduces the strength of these interactions, like increasing the temperature or introducing interfacially active molecules, will lower the IFT.

Several experimental techniques have been developed with the object of measuring IFT. This includes the technique of capillary rise, Wilhelmy's plate method, ring- and rotating drop tensiometry, the method of drop counting etc. In this section, the focus will be on the pendant drop technique, where liquid boundary tension is determined from the shape of drops without any direct contact.

The CAM 200 equipment is an optical instrument for contact angle and IFT measurements. The instrument includes a CCD video camera with telecentric optics, a frame grabber, and a LED based background light source. The resolution is 512 × 512 pixels and the frame interval is between 40 msec and 1000 sec. The LEDs are housed in a reflective sphere which integrates their light and directs it towards the sample. The light is also strobed and monochromatic, and all these features help to assure sharp images. Figure 12.22 shows the instrument including all its main components.

The pendant drop technique has a number of advantages compared to other techniques developed to determine IFT. Firstly, very accurate (± 0.15% or less) boundary tension measurements can be made [60]. Furthermore, the measurements might be made rapidly and, since no direct contact exists, successive measurements may be performed without disturbing the interface.

The first studies involving the pendant drop technique were performed by measuring characteristic drop diameters and interpreting them to various tables [61]. Later, computer technology made it possible to obtain direct measurements through drop-shape analysis by coordinating the data to suitable mathematical equations. Reviews of the theory behind the pendant drop technique have frequently been given in the literature [60,62–64].

A relation between drop shape and IFT might be given by introducing the classical Young–Laplace equation:

$$\Delta P = \gamma\left(\frac{1}{R_1} + \frac{1}{R_2}\right) \tag{12.10}$$

where ΔP is the pressure difference across the drop interface, γ is the IFT and R_1 and R_2 are the radii of curvature. In the absence of external forces other than gravity, the difference in pressure is linear to the ascending forces:

$$\Delta P = \Delta P_0 + (\Delta\rho)\, gz \tag{12.11}$$

where ΔP_0 is the pressure difference at a selected datum plane, $\Delta\rho$ is the difference in the densities of the two bulk phases, g is the gravity, and z is the vertical distance between the datum plane and a given point. A sketch of a pendant drop including symbols and dimensions is shown in Figure 12.23.

By combining Equations 12.10 and 12.11, and by introducing the geometrical correlations according to Figure 12.23, the following relation appears:

$$\frac{d\theta}{dS} = \frac{2}{R_0} + \frac{(\Delta\rho)\, g}{\gamma} z - \frac{\sin\theta}{x} \tag{12.12}$$

In most cases it is appropriate to transform x, z, and S into the dimensionless coordinates x', z', and S' by dividing by R_0. This results in the following three dimensionless first-order differential equations:

$$\frac{dx'}{dS'} = \cos\theta, \qquad \frac{dz'}{dS'} = \sin\theta, \qquad \frac{d\theta}{dS'} = 2 + \alpha \cdot z' - \frac{\sin\theta}{x'} \tag{12.13}$$

where the term $\alpha\left(= \frac{(\Delta\rho)\, g \cdot R_0^2}{\gamma}\right)$ constitutes the shape factor of the pendant drop.

Thus, for any pendant drop where the densities of the two liquids in contact are known, the IFT might be calculated by iterating the above equations simultaneously.

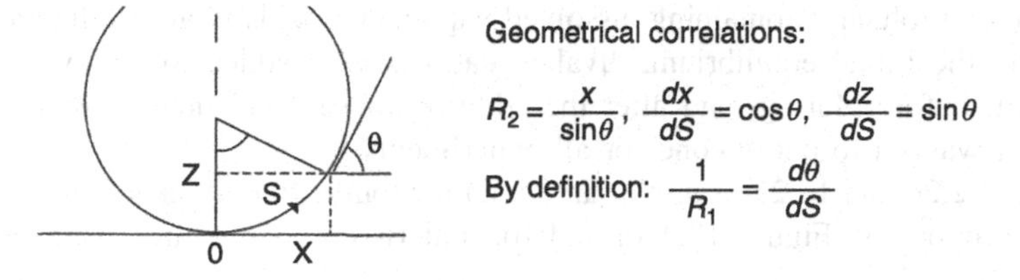

Geometrical correlations:
$R_2 = \frac{x}{\sin\theta}$, $\frac{dx}{dS} = \cos\theta$, $\frac{dz}{dS} = \sin\theta$
By definition: $\frac{1}{R_1} = \frac{d\theta}{dS}$
Z
S
θ
0
X

FIGURE 12.23 A sketch of a pendant drop including some geometrical considerations.

12.3.4.1 Applications

The CAM 200 equipment has recently been utilized to capture interactions between dissociated naphthenic acids and divalent cations across water–oil interfaces [65,66]. The work is related to crude oil production, where this process is assumed to be the origin of several problems, like stabilization of colloidal structures as well as formation of metal naphthenates through chemical reactions.

The naturally occurring naphthenic acids are complex mixtures consisting of aromatic and saturated rings connected by aliphatic chains [67,68]. The acids show polydispersity in stoichiometry and molecular weight [68–71] and due to the complex distribution of different structures, their physicochemical behavior is quite different from normal fatty acids. From an operational point of view, the naphthenic acids are causing several problems. Due to their amphiphilic nature, they may accumulate at interfaces and stabilize water/oil emulsions [72–74], which in turn causes enhanced separation problems. Dissociated carboxylic groups may also react stoichiometrically with metal cations present in the coproduced water to form metal naphthenates. Deposition of metal naphthenates in topside facilities like oil/water separators and desalters is becoming a serious problem in a number of fields with highly acidic crude, including fields in West Africa and in the North Sea [75,77,78].

Basically, there are two main approaches to how the naphthenic acids may react with cations to form metal naphthenate. The first is to consider bulk reactions between water soluble compounds and cations in the water phase. This was done by Havre [79], for example, who investigated the reaction between synthetic naphthenic acids and Ca^{2+} at high pH by using near-infrared spectroscopy and by correlating changes in optical density to the growth of naphthenate particles. At normal operational conditions, however, only the low-molecular weight napthenic acids show significant solubility in water. The major part, constituting larger molecules, will preferably be oil soluble. According to this, and by considering the huge interfaces created when water is emulsified with the crude during the transportation from the reservoir to the topside, it is obvious that the contact and the interactions between the compounds *across* the interface may lead to a second approach, based on interfacial reactions.

The interest in interfacial behavior of carboxylic acids in combination with cations is not new and the topic has during the last decades been a subject of several studies. In that regard, a number of studies is carried out in order to establish correlations between pH and IFT [80–84] and to investigate effects of counter-ion binding on film stability [85–89]. However, few studies have been focused directly on interfacial reactions as a mean to understand naphthenate formation. The aim of the present work is thus to improve the understanding of how naphthenic acids of various structures may react across interfaces with different divalent cations. The experimental approach has been to correlate changes in dynamic IFT to plausible reaction mechanisms. In the following, some of the results obtained from the pendant drop study are discussed. The discussion is limited to the effect of adding divalent cations.

The experiments were executed by forming a pendant drop of a model oil (9:1 ratio of n-hexadecane and toluene) containing dissolved naphthenic acids in an alkaline aqueous solution. After reaching the initial equilibrium, divalent cations were added to the system. The IFT was measured both before, during, and after the addition of salt and then plotted versus time. The frame interval was set to one second for all experiments.

In Figures 12.24 and 12.25 examples are given for four different naphthenic acids, comprising two model compounds, Figure 12.24(a and b), and two acid mixtures from crude distillation,

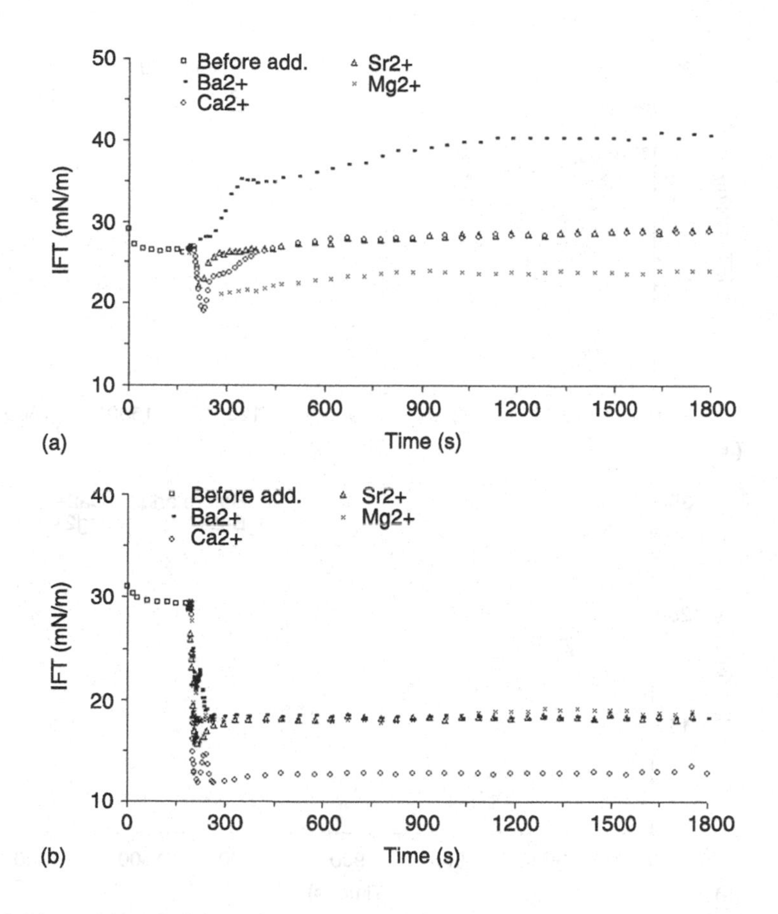

FIGURE 12.24 Plots of interfacial tension versus time for two different model naphthenic acids in combination with various divalent cations. (a) *p*-(n-dodecyl) benzoic acid and (b) 4-n-dodecyl-cyclohexane carboxylic acid.

Figure 12.25(a and b). Table 12.4 shows an overview of all the structures whereas the experimental matrix is given by Table 12.5.

As indicated by the plots, the curves show quite different progress, depending on acid structure and type of added salt. The effect of adding different salts is more significant for the model acid comprising an aromatic ring (Figure 12.25(a)), than for the saturated model acid (Figure 12.25(b)). In a recent paper this was discussed as a result of dealing with compounds with different polarity [66]. Due to the aromatic ring, the unsaturated structures will penetrate deeper into the interface, enabling them to react chemically with the cations on the aqueous side of the interface. The dissimilarity in curve shapes is a result of adding cations with different sizes. The fact that the degree of hydration increases with decreasing ionic size will most likely influence the density of the cations at the interface. Ba^{2+}, which is less hydrated, will thus mainly form stoichiometric 2:1 complexes with the naphthenic acids. This will reduce the interfacial activity which is observed by an increase in IFT. Mg^{2+}, on the other hand, is highly hydrated and the cations will preferably be located in the water bulk, causing longer distance between the interacting compounds.

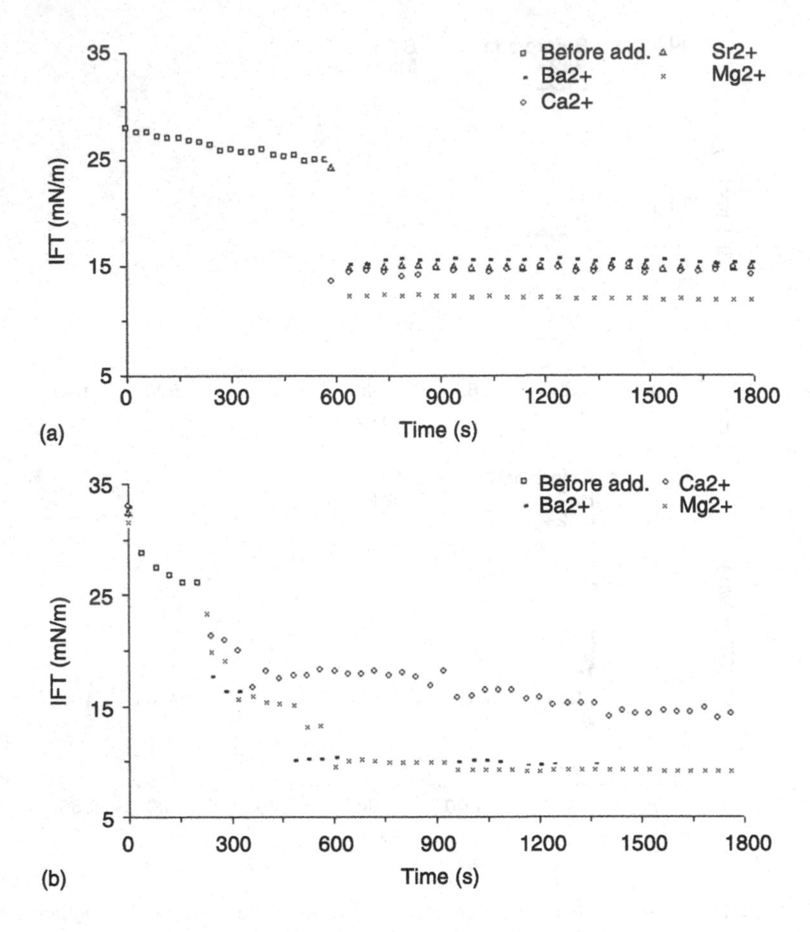

FIGURE 12.25 Plots of interfacial tension versus time for two different indigenous naphthenic acid mixtures in combination with various divalent cations. (a) Fluka naphthenic acid mixture and (b) naphthenic acid mixture from a North Sea crude.

The permanent decline in IFT, which reflects coverage of more interfacial active compounds, is thus a result of electrostatic attracting forces exerted by Mg^{2+} upon the carboxylate groups, or formation of positively charged monoacid complexes which remain at the interface due to high interfacial activity. If considering the latter, the question is whether the reaction with Mg^{2+} will propagate through two steps or if the second reaction step is very slow.

Contrary to the systems involving the aromatic naphthenic acid, all the plots of the saturated structure show permanent lowering of the IFT (Figure 12.24(b)). This was explained in a similar way as for Mg^{2+} in the former case: due to the hydrophobic saturated ring, the acid penetrates less into the interfacial layer and the distance between the interacting compounds across the interface might be too far to get a 2:1 formation of acid and cation. Now, since this is caused by the acid nature, all cations will affect the IFT in a similar way and the dominating mechanism should be electrostatic attractions.

Figure 12.25(a) shows the results from some experiments performed on the Fluka naphthenic acid, which is a commercial acid mixture. Due to high water solubility at pH 9.0, the experiments were carried out at pH 8.0. As indicated, addition of salt caused a sudden decline in IFT which

TABLE 12.4
An overview of the naphthenic acids utilized in the experiments

Name	M (g/mol)	Naphthenic acid structure	Source
p-(n-dodecyl) benzoic acid	290.4	CO_2H C_{12}	Chiron AS
4-n-dodecyl-cyclohexane carboxylic acid	296.5	CO_2H C_{12}	Chiron AS
Fluka naphthenic acid	250*	Mixture	Fluka
Naphthenic acids from a North Sea crude	200	Mixture	Statoil ASA

*Average values calculated from total acid number (TAN).

TABLE 12.5
The experimental matrix

Figure	pH	Acid conc. [M]	Conc. Ratio [M^{2+}/Ac]
12-24a	9.0	10^{-4}	3/1
12-24b	9.0	10^{-4}	3/1
12-25a	8.0	10^{-1}	3/1
12-25b	9.0	10^{-2}	1/1

happened within 1 sec since no frame has been captured between the upper and lower state. The decline is also permanent for all systems, similar to the case involving the saturated model naphthenic acid. Analysis by FT-IR and NMR has also proved that the Fluka mixture mainly consists of saturated structures. Based on earlier statements, it is fair to suppose that the main mechanism also in this case constitutes electrostatic attractions.

In Figure 12.25(b), one example is given for measurements performed on the acid mixture from the North Sea crude. After addition of salt, the IFT shows a stair-formed progress until it equilibrates 10 to 15 units below the starting value. The stair-formed curve is most likely a result of a combination of two processes. Firstly, since the mixture is polydisperse, small molecules will reach the interface at an earlier stage than the larger ones due to higher diffusion rate. Gradually, however, the smallest molecules will be replaced by larger molecules as they cross the interface due to high water solubility. A stepwise adsorption of monomers to the interface might thus occur. Secondly, for polydisperse mixtures, more dynamic processes are taking place at the interface than for monodisperse surfactants and normally longer time is required to get the system equilibrated. The sequential drops in IFT might thus refer to temporary states before the real equilibrium is reached as the curves flatten out. At this final state, the interface is covered by a film comprising more interfacially active molecules in combination with the divalent cations.

12.3.5 Quartz Crystal Microbalance

The quartz crystal microbalance (QCM) has been known for several decades and a variety of devices are available. Common for all these instruments are that they utilize an electric field to oscillate a crystal, and find the resonance frequency. Adsorption of a compound increases the mass of the crystal and thereby lowers the resonance frequency. The principle is shown in Figure 12.26, and the lateral displacement is typically 1 to 2 nm [90]. The odd numbered harmonics can also be recorded, and in the case of a rigid and thin adsorbate, they will give the equivalent response as the fundamental frequency. As the overtones have smaller amplitude, they are changed more for the adsorbed mass close to the surface rather than a thick layer. Sauerbrey's equation [91] is used when the adsorbate is rigid (i.e., low dissipation values, described below) and a thin layer. Then the harmonic oscillations will give a linear relation between change in mass Δm and change in resonance frequency Δf,

$$\Delta m = -\frac{\rho_q \cdot \upsilon_q \cdot \Delta f}{2 f_0^2 \cdot n} = -C\frac{\Delta f}{n} \tag{12.14}$$

Here ρ_q, υ_q, and f_0 are the specific density, shear wave velocity, and fundamental resonance frequency, respectively, and are constant for each quartz crystal. The factor n is one for the fundamental frequency, three for the third harmonic oscillation, five for the fifth harmonics, etc. Newer instruments also measure the dissipated energy, the inverse of the Q-factor.

$$D = \frac{1}{Q} = \frac{E_{Dissipated}}{2\pi E_{Stored}} \tag{12.15}$$

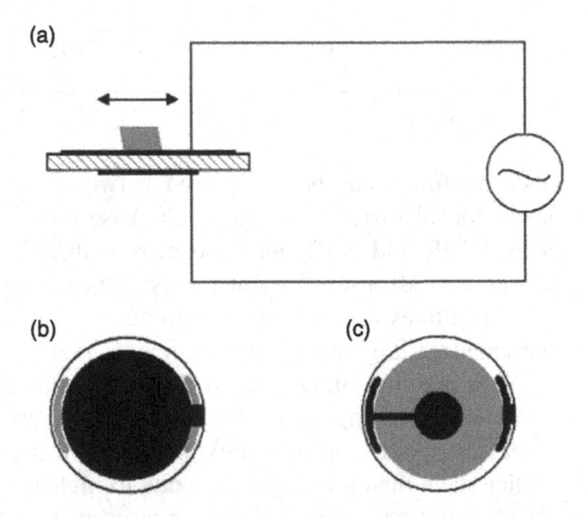

FIGURE 12.26 The heart of the QCM is the AT-cut quartz crystal sandwiched between two gold electrodes. (a) The crystal is given a mechanical lateral oscillation by an oscillating potential difference, and its resonance frequency is dependent on the mass. (b) The larger gold electrode is facing the solution investigated, and the area where the adsorption is measured is slightly larger than (c) the smaller counter electrode. Crystals used in Q-sense and KSV instruments are 1.4 cm in diameter, has electrode thickness of 100 nm, 5 MHz resonance frequency, and the factor C in Sauerbrey's equation 17.7 ng cm^{-2} Hz^{-1}.

The more viscous the adsorbed layer is the more energy is lost from the oscillation. Q-Sense QCM-D instrument measures the damping of the oscillation when the electrical circuit driving the oscillation is opened. KSV-Z500 uses impedance measurements to find the resonance frequencies and the Q-factor. Other QCM devices are made by Elchema, EG&G, Maxtek, and Universal.

The QCM technique is very sensitive, and measures data every second or even faster depending on the model. The sensitivity in liquid is 1 ng cm^{-2} for adsorption, and a D value of 4×10^{-8} compared to D value around 1×10^{-6} in air (Q-Sense QCM-D). This makes it suitable for surfactant adsorption that traditionally is measured by depletion method.

One of the advantages with measuring adsorption with QCM is that crystals can be coated with different materials, and a variety of surfaces (SiO_2, stainless steels, polystyrene, titanium, and many metals) are available from suppliers. Thin metal oxide films, polymer coating, and organic films can be deposited with a spin coater, and compounds like cellulose can be deposited with the Langmuir–Blodgett technique.

12.3.5.1 Applications

Corrosion inhibitors are usually investigated in corrosion experiments, using weight loss of coupons or electrochemical methods as a measure of their protecting properties. Little is known beyond that they adsorb to steel surfaces. We have used QCM to study adsorption processes of a corrosion inhibitor (quaternary ammonium) onto iron and cementite, mimicking the steel walls of a oil pipe [92]. From the kinetics it is seen that the adsorption and desorption is a fast process and that maximum adsorption is achieved around the critical micelle concentration. The adsorbed mass is less than what is needed for a monolayer/bilayer, hence it is likely that the adsorbed corrosion inhibitor organize on the surface like discrete aggregates. Varying the chain length of the quaternary ammonium, or adding salt to the system, changed the concentration for maximum adsorption. However, this is due to changing the critical micelle concentration. Corrosion inhibitors are often dissolved in alcohols when injected into oil pipes, therefore we also investigated the effect of alcohols on the adsorption. The typical solvent methyl ethylene glycol did not change the level of the adsorbed amount of quaternary ammonium. The effect of long chain alcohols was also investigated as they behave like cosurfactants and can alter the geometry of the adsorbate. At concentrations of 25 mM hexanol with quaternary ammonium, the QCM gave a sudden increase in adsorbed mass, corresponding to a bilayer. When the quaternary ammonium was removed from the bulk solution, the adsorbed mass slowly decreased, indicating a more stable adsorbate.

It is important that the surfactant is added in moderate concentrations in the QCM experiments to be certain that it does not change the viscosity and density of the bulk solution. The effect of carbon chain lengths on the adsorption of the model corrosion inhibitor (alkyltrimethylammonium bromide) on cementite is presented in Figure 12.27. The surfactant adsorption increases with concentration up to the critical micelle concentration. The long-chain homologues, with carbon chain C_{14} and upwards, flatten out completely at this concentration, while the shorter compounds seem to give higher adsorption. As the CMC increases with the shorter carbon chain lengths, these observations are attributed to the bulk effect. Thus, there are two contributions to the change in frequency, the actual adsorption of mass, and the changes in the bulk solution. A comparative study of adsorption of nonionic surfactants on silica from benzene applying QCM and depletion methods has been reported [93]. Both methods agree very well for the series of poly (ethylene glycol) ethers at low concentrations. However, large deviation is observed for the nonyl phenol ethoxylate, and this is explained by the interactions between the benzene solvent and the phenyl in the adsorbed surfactant.

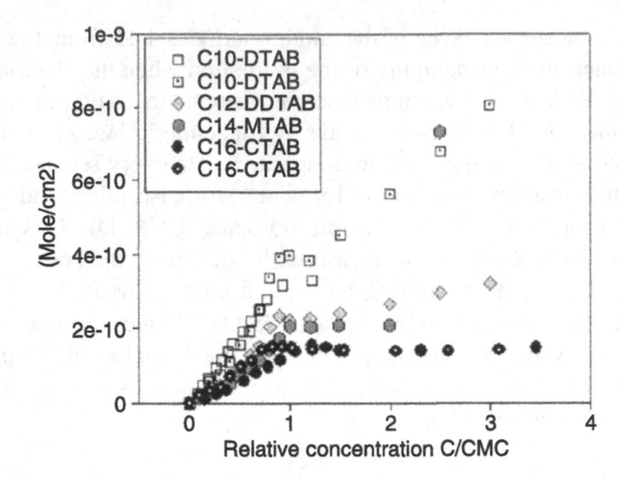

FIGURE 12.27 Adsorption of the cationic surfactant alkyltrimethylammonium bromide on a cementite surface, measured with QCM. The concentration along the *x* axis is normalized with respect to the critical micelle concentration for the sake of comparison. As the shorter homologs have higher CMC, it is likely that the frequency change is due to both adsorption and the changed bulk properties. While the C16TAB flattens out at CMC, the C10TAB continues to increase.

Different solvents can be used in the QCM adsorption measurements, and unlike optical methods like ellipsometry, the solution does not need to be transparent. The adsorption of asphaltenes and resins from organic solutions has been investigated with this technique. As explained in previous sections, asphaltene precipitation may cause problems in the oil production chain, all the way from reservoir to refinery. The solvents and presence of indigenous oil-soluble surfactants like resins can affect the precipitation and adsorption to surfaces, and the phenomena has been investigated with the QCM method and a hydrophilic surface [94]. It was found that the adsorbed amount of resins from a heptane solution corresponds to a rigidly attached monolayer. The amount adsorbed decreased with an increasing amount of toluene and was virtually zero in pure toluene. Asphaltenes adsorb in large quantities and the mass and dissipation show that it is adsorbed as aggregates on the surface. The aggregates were firmly attached, and could not be removed by adding resins. The system of asphaltenes and resins in mixture was markedly different from the adsorption of the pure compounds as the compounds associate in the liquid bulk, and it was concluded that the preformed resin aggregate adsorbed to the surface.

The majority of QCM experiments have been carried out on biological systems, using gold or a modified surface. The effect of hydrophilic surfaces is that lipid vesicles can adsorb as monolayer, bilayer, or intact vesicles [95]. A close study of the adsorption on a silica surface [96], which gives the bilayer structure, is complemented by SPR and AFM studies, and together they give a complete picture of the process on silica surface. First, lipid vesicles are adsorbed, until a certain coverage where the vesicles collapse and a bilayer without any visible defects are formed. The adsorbed mass from SPR and QCM agrees well after the vesicle collapse on the surface, and the water inside the vesicles is released. The lipid vesicles and bilayer are a good starting point for further investigations of reactions. For example a fluid biotin containing phospholipid bilayer supported on an SiO_2 surface immobilized with the tetrameric protein streptavidin to give a rigid layer. Then a mixed sequence 15-mer biotin-PNA and DNA with identical base pair sequences were linked to streptavidin, and the difference in adsorption of fully complementary DNA was detectable [97]. It was also possible to differentiate adsorption of complementary DNA and a singly mismatched DNA.

12.3.6 Atomic Force Microscopy

Atomic force microscopy (AFM), invented in 1986 [98], is a type of the scanning probe microscope, and measures the attracting/repulsive forces arising when two bodies are close to each other. The force is measured by using a cantilever with a certain spring constant (0.01 to 100 N m^{-1}) that is pulled towards the investigated surface in case of attraction, or pushed up in case of repulsion from the investigated surface. On the downside of the cantilever is a well defined tip (most often Si_3N_4 or Si), and on the reverse side is a reflective surface. The cantilever displacement is most often measured with the laser beam deflection method [99,100]. A laser beam is reflected from the rear side of the cantilever, and the cantilever displacement is detected as deflected beam hitting a split diode photodetector. A sketch of the main components in an AFM is given in Figure 12.28, and the typical deflection-distance curve in Figure 12.29. AFM is used for measuring force between two bodies approaching–retracting and for imaging of a surface by letting the probe scan the surface in the *xy* plane in contact, non-contact, and tapping mode. The strength of AFM lies in its extreme resolution and its operating conditions. At optimal conditions, large atoms on the surface can be seen in the images, such as graphite [101], mica [102], boron nitride [103], and NaCl (001) [104].

Various modulation techniques can be used when AFM experiments are carried out. *Force spectroscopy*, that is moving the cantilever in the *z* direction with no movement in the *xy* plane, is used for measuring different surface mechanical properties. The first direct measurements in air and water were using tips with complex or unknown geometry. With the development of the colloidal probe, where a microsphere is glued to the AFM tip, a better force measurement is possible. The spherical geometry allows a comparison with theoretical predictions. When the spring constant of the cantilever is known, the raw data from a force measurement can be

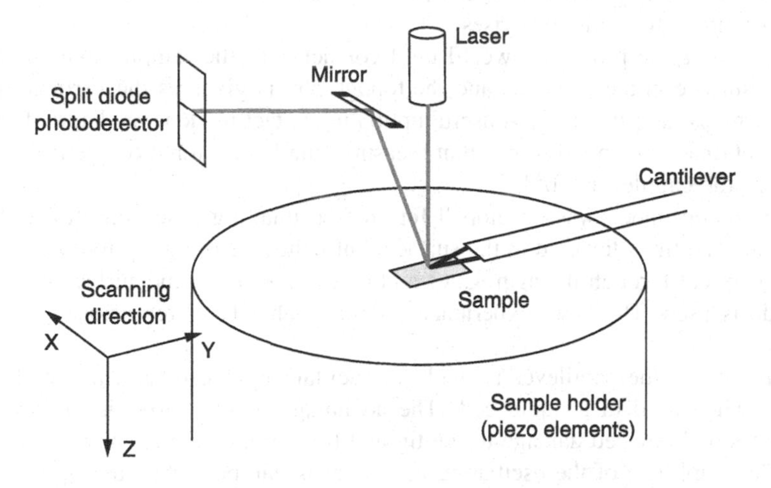

FIGURE 12.28 AFM main components. The deflection of the cantilever changes the reflected laser beam which is detected by a split diode photodetector (divided in two or four parts). The sample holder is built around a cylindrical piezo element with several electrodes, which controls the *xy* movements used for scanning images and the *z*-direction for height control.

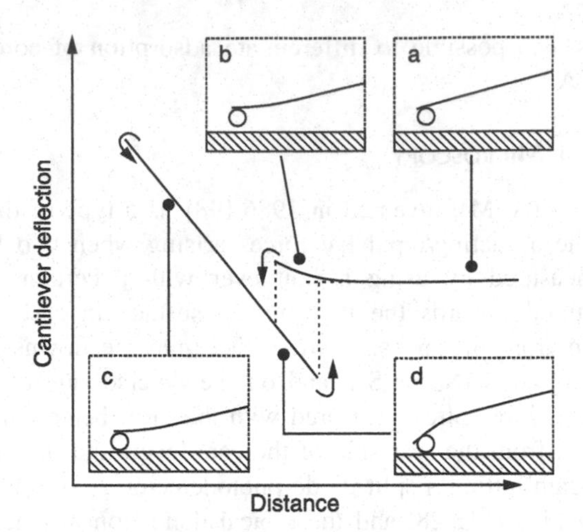

FIGURE 12.29 A typical deflection–distance curve. (a) Far apart the cantilever is not interacting with the surface and the zero-line is obtained. (b) As the cantilever approaches the surface the cantilever is deflected by coulombic forces before the jump-to-contact, noted by the dotted line. (c) For hard surface a linear compression–decompression is observed with a slope proportional to the spring constant of the cantilever. (d) When the cantilever is withdrawn from the surface the adhesive forces will hold it in contact with the sample until jump-off-contact.

converted by using the method of Neumeister and Ducker [105] to a force versus tip separation curve. Forces that are measured with the AFM are capillary, Coulomb, van der Waals, double layer, solvation, hydration, hydrophobic specific, and steric forces [106]. The high sensitivity of soft cantilevers can measure forces at the pico-newton level and can be used to measure intermolecular forces, interactions between single molecules or single molecules and surfaces. In such experiments a single molecule is fixed to the cantilever tip with covalent bonding, and the force is measured as the cantilever rises.

In *contact mode* the probe is lowered until contact with the sample. Imaging takes place by scanning the surface in the *xy* plane, and the topography is given as the vertical displacement of the cantilever gives the *z* direction. A useful option in contact mode is the lateral force microscopy (LFM), or frictional force microscopy, that measures the lateral force on the tip causing buckling and torsion of the cantilever [105].

The *non-contact* mode was developed for surface imaging based on attractive forces rather than repulsion. The tip is lowered to the surface, but halted before jump-to-contact. A sharp AFM probe can easily cut through the aggregates and record images of the surface beneath if ordinary contact mode is used. The force experienced is the double layer repulsion between the tip and the sample.

In *tapping mode* the cantilever is made to oscillate by piezoelements on the cantilever or by acoustics when used in a liquid cell. The advantage of this mode is the absence of lateral torsional forces and reduced damage to the tip and the sample, and is suitable for imaging softer materials. The amplitude of the oscillating cantilever is damped when the tip is close enough to interact with the surface. A PID regulator regulates the *z* direction to keep amplitude constant through the scan.

In addition to the above techniques, *phase imaging* is a powerful imaging technique. It is based on the tapping mode and utilizes the phase lag between the periodic signal driving the oscillation and actual oscillation of the cantilever. Images from phase shift give the contrasts due to surfaces with different hardness, hydrophilicity, magnetic fields, and other surface properties.

12.3.6.1 Applications

Atomic force microscopy features many techniques of great interest in colloid and surface science. It has the great advantage over other techniques that it can work in gases and liquids, the natural environment for colloidal systems. Characteristic for colloidal particles is the extremely large surface area, and the interface reactions and attraction/repulsion of particles are therefore one of the main interests. The three different regions of a force–distance curve (contact line, non-contact region, and zero line) contain different information. In the contact region an elastic material will give a straight line with no hysteresis, while a plastic surface will not follow a retracting cantilever and therefore give loading–unloading hysteresis. There are different theories for the interpretation of the contact region. Hertz theory from 1881 [107] considers an elastic sphere on a flat surface and applies in some cases, but does not take into account neither surface forces or adhesion. The theory gives the reduced Young modulus expressed by the Young modulus and Poisson ratios of the sample and the tip. Sneddon's theory [108] considers the case when the indenter is not deformed while the surface takes the deformation. This gives an expression for the force exerted by the tip/sample contact and the deformation of the surface which is solved numerically. Three theories, Bradley, Derjaguin–Müller–Toporov, and Johnson–Kendall–Roberts, take into account the effect of surface energies on the deformation on the contact deformation. Maugis [109] has the most complete theory, which applies for materials ranging from large rigid spheres with high surface energies to small compliant bodies with low surface energies. Above the linear contact line in Figure 12.29 we find the non-contact region. This contains information about the repulsive and attractive forces between two bodies. The approach curve is increasing until the jump-to-contact is observed when the gradient of the tip–sample interactions exceeds the elastic constant of the spring. The maximum value of the attractive force is equal to the jump-to-contact cantilever deflection times the spring constant of the cantilever. Knowing the tip shape, it is possible to evaluate the force. The Hamaker constant can be calculated in two ways: (1) fitting one of the force laws to the attractive part of the force laws, or (2) to use the jump to contact force, and calculate it from the relation to van der Waals force and the geometry. In a review of force spectroscopy, Cappella and Dietler [106] have evaluated the Hamaker constants determined experimentally with theoretically derived values. They remark that the first measurements done did not take the capillary force into account, and give too high values. Measurements in water give significantly lower values, and even though the experimental error is large they often show a valid trend with surface energies. Upon withdrawal of the cantilever a jump-off-contact takes place when the cantilever spring constant is larger than the gradient of the tip–sample interaction. At this point the product of the cantilever deflection and the spring constant equals the adhesive forces. The jump-off-contact is related to the surface energies and geometry. The pull-off force gives many adhesive material properties, but is not an accurate method for determining the surface energies.

12.3.6.2 Atomic Force Microscopy Imaging and Colloidal Systems

12.3.6.2.1 Topography of a Substrate

AFM imaging reveals the structure of adsorbate on solid surfaces. A surface that appears planar to the naked eye can still be rough on the micro- or nano scale. The imaging technique determines the surface rugosity/roughness in micro- and nano range.

12.3.6.2.2 Adsorbate on a Surface

AFM can also be used for determining coverage and structure of an adsorbate on a surface, the adsorbate being a self-assembly structure, a Langmuir–Blodgett (L–B) film, biopolymers (proteins, DNA, cells, viruses etc.), and other substances. Most of the samples need to be imaged under water to eliminate the capillary forces. Mica is the most common substrate for adsorption on hydrophilic surfaces, as it has a well defined charge density and planar surface. Graphite is chosen for adsorption on hydrophobic surfaces.

Self-assembly structures are very soft, and therefore difficult to image. The non-contact AFM is suitable as the tip is not in contact with the surface. The first direct imaging of "hemimicelles," analogous to bulk micelle, was CTAB on graphite [110]. An overview of AFM imaging of cationic surfactants [111] shows the effect of counter-ion, carbon chain length, and cosurfactant on surfaces like mica and graphite.

The Langmuir–Blodgett technique is suitable for depositing water-insoluble compounds to a substrate, where the compound (e.g., lipids, particles, biopolymers) is packed on a water surface to a known surface pressure and area per unit. Then it is transferred to a solid surface by dipping a substrate into the solution with sufficiently low speed. It can be used to find structure and eventual defects. L–B films of cadmium arachidate on mica [112] show that the method is suitable for studying the quality of a deposited L–B film. The technique has also been used for investigating the effect of resins and asphaltene inhibitors on the precipitation of colloidal asphaltene [113]. In this study monolayers of asphaltenes and resins were transferred onto mica substrates using the L–B technique, and the topography analyzed by means of AFM.

This work shows the structural change in the monolayer when the composition of the film was gradually changed from pure asphaltenes to pure resins. Pictures of pure asphaltene show a closed-packed layer of round disks or rod formed units. Addition of resins will change this rigid structure towards a more open network with regions completely uncovered by film material. Pure resins build up a layer with an open structure, i.e., more like a fractal pattern.

Another observation is the growth of the size of the individual film units upon addition of resins. This indicates interactions between asphaltenes and resins, providing aggregates of larger dimensions than observed for the pure fractions. Small and moderate amounts of resins give rise to a more polydisperse distribution of the film material, while a further increase in the resin content (i.e., 60 wt% resins) reduces the polydispersity, i.e., the monolayer becomes more uniform in component size when one of the pure fractions dominates the film properties.

AFM pictures of asphaltene films containing 100 ppm of different high molecular weight demulsifiers/inhibitors were also investigated and it was demonstrated that these components introduce similar effect on the film as the structural changes initiated by the resins. These results indicate that the observed changes in the film, i.e., opening of the structure and increased size of the film components, are qualitatively essential in order to reduce the emulsion stability.

AFM images of samples of supported planar lipid-protein membranes and actin filaments have been obtained by applying a very low force to the cantilever [114].

Biomaterial is often dissolved into a very dilute aqueous solution and either sprayed on mica in a small amount, or a droplet is left on for a while before the excess is blown off with an inert gas. In some cases pretreatment of the mica, or counter-ion in the solution, is necessary to get sufficient bonding to the surface; for instance mica is soaked in magnesium acetate 2 h and rinsed with water prior to DNA (aq) addition [115]. Another method of fixing material to the surface is to use molecules with thiol or silane groups that bind to the surface and another group that binds to the desired particle. Determining particle size distribution of nano-particles is possible if no aggregation takes place and being aware of the convolution of the image. Convolution of

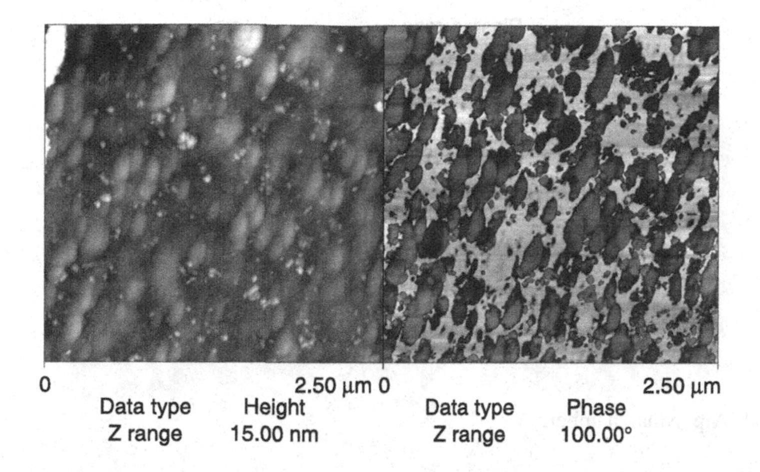

FIGURE 12.30 AFM image of an iron surface exposed to solution of sodium oleate and poly-DADMAC. The topography (left) only give the structure of the sputtered iron, and the thin film of organic material is not visible. The phase image (right) shows that the iron (dark region) is partly covered with a softer organic material (bright region).

an image will make particles larger as the tip on the cantilever has a certain size. Dealing with very small particles, the width in the image is larger than the true dimension. Determining size of nano-particles of Pd, Au, CdS, and ZnS based on the height is in good agreement with TEM with an average of ~3 nm, while the radius in the *xy* plane is about 50 nm [116]. The same was observed earlier [117] for particles of 37 ± 5 nm (TEM) to be 95 ± 12 nm in AFM.

Contrast between different regions can be enhanced with phase imaging. It is often a much more sensitive measurement technique than the amplitude detection. The phase shift is due to variations in composition, adhesion, friction, viscoelasticity, and electric and magnetic properties. The phase imaging is not only getting better contrast. It can get out information that is not visible in the topography image at all. Figure 12.30 shows the topography image and the simultaneously recorded phase image of a surfactant and polymer adsorbed on an iron surface. The two regions with different hardness are distinguishable in the image to the right. Tapping on iron gives little phase shift, while the tapping of surfactant and polyelectrolyte gives a significant phase shift, and appears as bright regions. In wood pulp fibers the technique can be used to show how the lignin component is distributed on the cellulose surface. Phase imaging can also be used to visualize the distribution of hard particles in a softer matrix, like in composite materials [118,119], or different crystalline phases in polymers.

12.3.7 Plasmachemical Surface Modification

The plasma state can be defined as a gaseous mixture of freely moving charged particles which overall are electrically neutral. More specifically, the plasma is a blend of electrons, ions, molecular fragments, radicals, excited states, photons, and neutral atoms and molecules. The degree of ionization in the plasma depends highly on how the plasma is generated. Here the focus will be on so-called cold discharges. These are low-pressure, non-equilibrium plasmas where the degree of ionization typically is less than 10%, and direct current (DC), radio frequency (RF), or microwave (MW) power sources are used to generate and maintain the plasma.

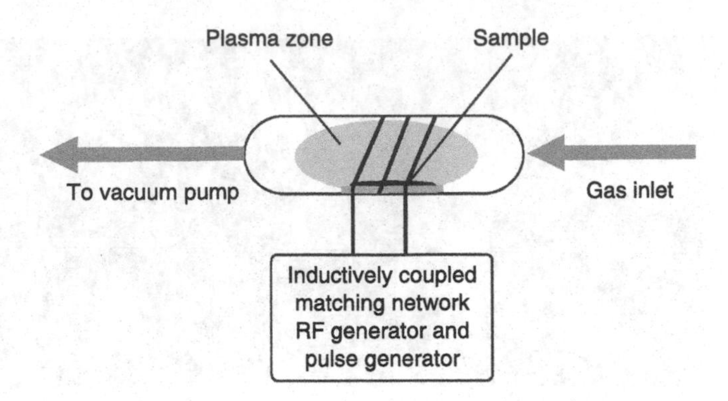

FIGURE 12.31 A plasma chamber.

The schematic set-up of an inductively coupled RF plasma chamber is shown in Figure 12.31. The main parts are the glass chamber where the samples are placed, the inlet for gas or vapor on one side and the outlet to a vacuum pump, via a nitrogen cold trap, on the other side. A copper coil is wound around the glass chamber, and connected to a 13.56 MHz RF generator, via a matching network. The purpose of the matching network is to match the output impedance from the RF generator with the impedance in the chamber. A pulse generator can also be connected to the RF generator if pulsed plasmas are desired.

When a sample is placed in the chamber and a steady state, low-pressure flow is obtained through the system, the RF generator and matching network is turned on in order to create an electrical field. A plasma state is then generated by acceleration of free electrons in the electromagnetic field, and the kinetic energy of these will reach a level where ionization, excitation, and molecular fragmentation will occur. Since the electron mass is much smaller than the mass of molecular and atomic species, the kinetic energy of the electrons is high compared to the ions and free radicals (i.e., the plasma is not in equilibrium). Typical energies of the electrons are between 0.5 and 5 eV, and this is intense enough to dissociate chemical bonds and form radicals in most organic structures. Interactions between the sample and plasma species will then entail plasmachemical modification of the sample surface. The non-equilibrium energy distribution in the plasma enables chemical reactions to proceed via nonthermodynamic channels at approximately room temperature, while a corresponding thermally activated reaction could require temperatures as high as 10,000 °C.

The most important variables in the system described here are the power into the chamber, the pressure in the chamber prior to ignition of the plasma, and the time the sample are exposed to the plasma state. Furthermore, the type of surface modification the samples undergo depends on the properties of the gas or vapor going into the chamber. In polymerizing plasmas, molecular fragmentation leads to reorganization of species and formation of macromolecular structures, so-called plasma polymers, which will deposit as thin films on the sample. Properties such as film thickness and functional groups can be controlled by the experimental variables. In nonpolymerizing plasmas fragmentation and implantation of functional groups occur directly.

Etching occurs in plasma systems with high particle energies, and since material is removed from the sample surface, the process is typically used for cleaning of surfaces. Finally, plasma activated grafting is a two-step process where an inert plasma is used to activate the substrate, and subsequently a reactive vapor is led above the sample in order to modify the surface.

12.3.7.1 Applications

Much research has been carried out on plasmachemical modification of flat substrates [120], but very little has been done on particles. However, a study of how the level of amine loading on porous polymer beads is affected by the physical properties of the particles has been carried out by Øye et al. [121–124]. A range of highly cross-linked polystyrene beads with different particle sizes and pore structures were modified by allylamine plasma. Allylamine is polymerizing in the plasma state, and the external surface of all the particle types were readily covered by a thin film. Raman spectroscopy chemical mapping and scanning electron microscopy of cross-sectioned beads revealed that the smallest beads (15 μm in diameter) had a uniform distribution of amine groups throughout the particles. However, partial filling of the pore structure by the plasma polymer was also observed. For larger beads, there was an accumulation of amine groups towards the outer edge of the particles. The effective penetration of the plasma polymer was found to be 3 to 4 μm, and also in this case it partially filled the pores. An alternative functionalization method was tried out in order to reduce the pore filling effects. This entailed activation of the porous polymer beads by inert argon plasma, and subsequent grafting by diaminopropane. In this case a more homogenous amine distribution was obtained for larger beads as well.

The ultimate goal was to utilize the plasma modified particles in solid-phase synthesis of organic compounds, and several synthesis schemes were carried out. Generally, it was found that the low available amine loading of the modified beads, in comparison with commercial beads, resulted in limited amounts of recovered material. However, the produced yields were comparable to those obtained for commercial bead systems. Furthermore, a scavenging process using benzoyl chloride as a model electrophile was carried out. These processes are normally fast, and the fastest rate was obtained for the plasma enhanced diaminopropane grafted beads. The allyl amine modified material gave slower scavenging rates, also compared to a commercial material.

Plasmachemical modification by water plasma has also been used to surface oxidize carbon nanofibers (CNFs) [125]. CNFs have attracted large interest for use as catalyst supports. The surface chemistry is essential for this application, as oxygen moieties on the surface will improve the fiber–metal precursor interactions during impregnation or deposition precipitation of the materials. It was shown that in this study that the ratio between acidic and basic oxygen surface sites could be tailored by varying the conditions in the plasma chamber.

12.3.8 Contact Angles

The primary focus of contact angle studies is in assessing the wetting characteristics of solid/liquid interactions. A sessile drop of liquid on a solid surface will be affected by two kinds of forces, namely gravity forces and surface forces. In most cases, the gravity forces can be neglected for small droplets and the behavior is determined by surface forces.

Contact angle is a quantitative measure of the wetting of the solid by the liquid. The less the contact angle, the higher is the degree of wetting. If it is greater than 90° it is said to be nonwetting. On the other hand, a contact angle of zero represents complete wetting.

A correlation between contact angle and surface forces, given in terms of the interfacial tensions between the involving phases, may be given by the Young equation:

$$\gamma_{sg} = \gamma_{sl} + \gamma_{lg} \cos\theta \tag{12.16}$$

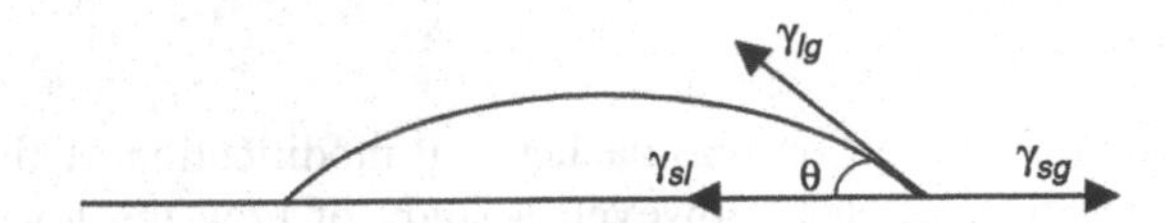

FIGURE 12.32 A sketch of a sessile drop, showing the contact angle θ and the vectors of interfacial tensions between the various phases.

Here, γ_{sg}, γ_{sl}, and γ_{lg} are the interfacial tensions between solid–gas, solid–liquid, and liquid–gas, respectively, whereas θ is the contact angle (Figure 12.32). The corresponding spreading coefficient is defined according to:

$$S = \gamma_{sg} - (\gamma_{sl} + \gamma_{lg}) \quad (12.17)$$

Contact angle is commonly used as the most direct measure of wetting. Other experimental parameters, like work of adhesion, work of cohesion, work of spreading, and wetting tension, may be derived directly from contact angle and surface tension results. A detailed theoretical review of the theory of contact angle is given by Good [126].

Several experimental techniques are today available for contact angle measurements [127]. The CAM 200 equipment is based on goniometry, one of the most traditional techniques for solid–liquid systems. In conventional goniometry the contact angle can be assessed directly by measuring the angle formed between the solid and the tangent to the drop surface. Consequently, the assignment of the tangent line is the limited factor in reproducibility, and significant errors might occur, especially if the instrument is operated by multiple users.

For CAM 200 this potential problem is avoided by utilizing computer analysis of the drop shape to generate consistent contact angle data. The Young–Laplace equation (equation 12.10) is then fitted to the drop profile and the tangent might be assigned in the intersection between the curve and the baseline.

12.3.9 Zeta Potential

Surface charges will arise when particles are suspended in a solution. The most common ways these occur is by ionization of chemical groups on the surface, differential loss of ions from the crystal lattice, or by adsorption of charged species from the surrounding solution. The presence of ions in the liquid will also be influenced by the surface charges, as the concentration of counter-ions will increase close to the charged solid–liquid interface. The consequence of this is that an electrical double layer is built up around the particles [128]. This layer is commonly divided into an inner layer, the so-called Stern layer, where the ions are adsorbed onto the charged particle surface and a diffuse layer, often called the Gouy–Chapman layer, containing more loosely bound ions. Factors that will affect the thickness of the electrical double layer are pH, ion strength, and concentration of additives such as surfactants. Furthermore, when particles move, by diffusion or in a gravitational or induced field, a shear plane exists outside which the ions in the double layer will no longer be influenced by the movements of the particles. The potential at this boundary is defined as the zeta potential, and several experimental techniques can be used to measure it. Electro-osmosis, streaming potential, and sedimentation potential are treated in more detail elsewhere, while the focus in the following will be on electrophoresis.

The result of applying an electrical field across a suspension of charged colloidal particles is that the particles move towards the oppositely charged electrode. The constant velocity that is obtained in the electric field with equilibrium with the opposing viscous forces is called the electrophoretic mobility. The zeta potential and the electrophoretic mobility are linked by the Henry equation:

$$u_E = \frac{2\varepsilon\zeta f(\kappa a)}{3\eta} \tag{12.18}$$

where

u_E = electrophoretic mobility
ε = dielectric constant of the solvent
ζ = zeta potential
$f(\kappa a)$ = Henry's function
η = viscosity of solvent

Henry's function is effectively the ratio of the particle radius to the electrical double layer thickness, as κ and a represent the thickness of the electrical double layer and the radius of particles, respectively. For large values of κa, Henry's function approaches 1.5. This is called the Smoluchowski approximation, and the relation between the zeta potential and electrophoretic mobility is straightforward within this limit, i.e., when the particles are much larger than the double layer thickness. Henry's function approaches 1 for small values of κa, and this is the Hückel approximation. Also, in this case, the relationship between the zeta potential and electrophoretic mobility is very simple, and the particles are much smaller than the double layer thickness. Both these approximations are normally available in instruments used for measuring the electrophoretic mobility.

In the Malvern Zetasizer system the electrophoretic mobility is measured by laser Doppler velocimetry (LDV). In this method a He–Ne laser at 633 nm is first split, and subsequently made to cross in the capillary cell containing the sample. Young's interference fringes of known spacing are created at the crossing point. As the particles move through this pattern by an applied electric field, the scattered light undergoes similar fluctuations, and the frequency of fluctuations is related to the velocity of the particles. The scattered light is collected by an avalanche photodiode and the signals transferred to a digital correlator. Analysis of the correlation function (described in detail in the following section) gives a frequency spectrum that is utilized to calculate the electrophoretic mobility and subsequently the zeta potential.

12.3.9.1 Applications

The film properties of reservoir particles were described in the Langmuir section. Figure 12.33 shows the zeta potential as a function of pH for the same particles. The hydrophilic Type II particles have a negative zeta potential over the entire pH range, while the Type I particles have an isoelectric point at pH 6.84. The difference is due to adsorbed species on the Type I particles.

The zeta potential as a function of pH was also measured for plasma oxidized carbon nanofibers [125]. It was found that the ratio between acidic and basic surface groups was dependent on the power of the water plasma. Furthermore, zeta potential has been used to study the electrokinetic properties of oil-in-water emulsions [129] and dispersed asphaltenes [130,131].

12.3.10 Dynamic Light Scattering

Dynamic light scattering (DLS) is a means of measuring particle size and particle size distribution of dilute suspensions in the range from ~2 nm to ~2 μm. For particles in this size range, the only

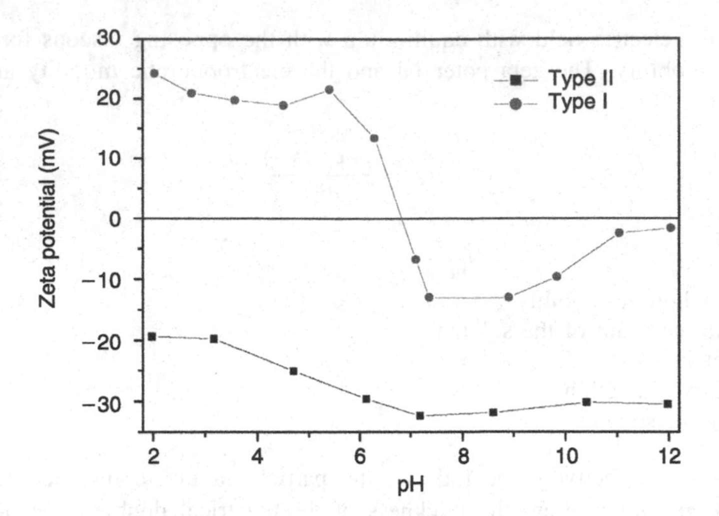

FIGURE 12.33 Zeta potential of reservoir particles.

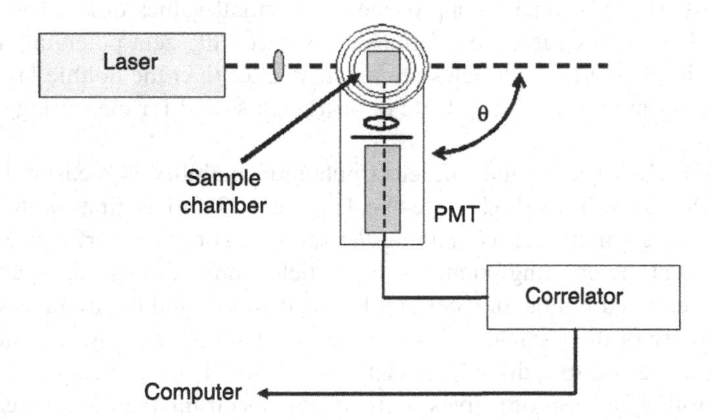

FIGURE 12.34 Instrumental set-up for a DLS experiment. The scattered light is collected at an angle θ (usually 90°), before passing through a photomultiplier tube (PMT) to the computer.

major competing technique is electron microscopy. However, electron microscopy experiments are often slow, sample preparation can be tedious, and it is only cost- and time effective to measure a small amount of sample, leading to poor statistics and the risk of not displaying a representative part of the sample. DLS has the advantage of being rapid, of requiring little or no sample preparation, and of having a solid statistical foundation through measuring a large number of particles per experiment.

In dynamic light scattering, a laser beam (usually a He–Ne laser, λ = 633 nm) is utilized to probe a small volume of particle suspension, as shown in Figure 12.34.

As the particles undergo Brownian motion, interference between scattered light produces a time-dependent intensity fluctuation at the detector, where the intensity i(s,t) fluctuates around an

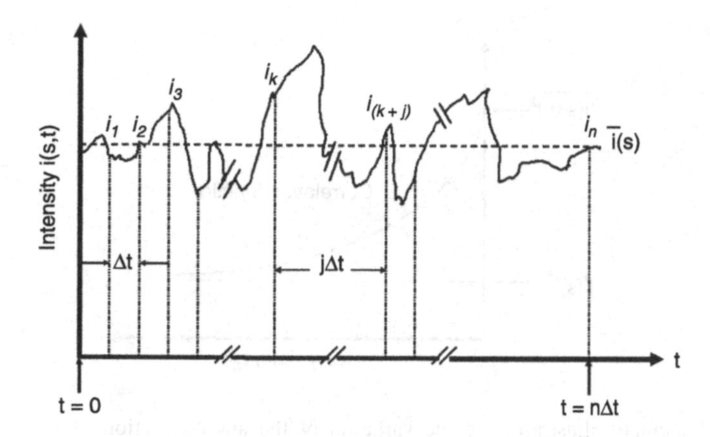

FIGURE 12.35 Schematic illustration of the intensity of scattered light with time (From Hiemenz PC, Rajagopalan R. *Principles of Colloid and Surface Chemistry*; 3rd ed.; Marcel Dekker, Inc: New York, 1997. [201].)

average value $\bar{i}(s)$ as illustrated in Figure 12.35.

$$\bar{i}(s) = \lim_{t_n \to \infty} \frac{1}{t_n} \int_0^{t_n} i(s,t)dt \approx \lim_{n \to \infty} \sum_{j=1}^{n} i(s, j\Delta t) \tag{12.19}$$

where j is the number of discrete time intervals Δt, and s is the scattering vector, a function of the laser output wavelength λ and the angle of the detector relative to the incident beam θ:

$$\begin{aligned} s &= \frac{4\pi}{\lambda}\sin\left(\frac{\theta}{2}\right) \xrightarrow[\lambda=633\ l\text{nm}]{\theta=\pi/2} s \\ &= \frac{4\pi}{633\text{nm}}\sin\left(\frac{\pi}{4}\right) \end{aligned} \tag{12.20}$$

This temporal fluctuation is then modeled as the decay between its highest value $\overline{[i(s,0)]^2}$ and its lowest asymptotic value $\left[\bar{i}(s)\right]^2$ as a function of delay time t_d between measurements to yield the autocorrelation function $C(s,t_d)$ as shown in Figure 12.36 [132,133]:

$$\begin{aligned} C(s, t_d) &= \lim_{t_n \to \infty} \int_0^{t_n} i(s,t)i(s, t+t_d)dt \\ &= \overline{i(s,0)i(s,t_d)} \\ &\approx \lim_{n \to \infty} \sum_{k=0}^{n} i(s, k\Delta t)i(s, (k+j)\Delta t) \end{aligned} \tag{12.21}$$

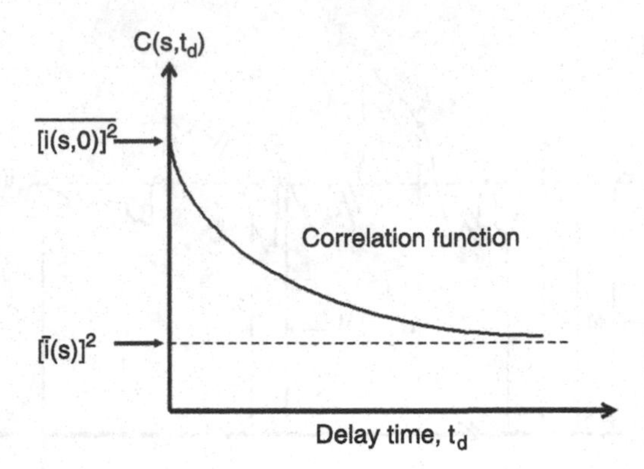

FIGURE 12.36 Schematic illustration of the variation of the autocorrelation $C(s,t_d)$ with delay time t_d. (From Hiemenz PC, Rajagopalan R. *Principles of Colloid and Surface Chemistry*; 3rd ed.; Marcel Dekker, Inc: New York, 1997.)

By applying the Siegert relation, the ratio of the autocorrelation function $C(s,t_d)$ to its asymptotic value $\left[\bar{i}(s)\right]^2$ can be found:

$$\begin{aligned}\frac{C(s, t_d)}{\left[\bar{i}(s)\right]^2} &= g_2(s, t_d)\\ &= 1 + \xi\, |g_1(s, t_d)|^2\end{aligned} \tag{12.22}$$

where ξ is an instrumental constant approximately equal to unity.

For a dilute suspension of (non-interacting) monodisperse spherical particles, $g_1(s,t_d)$ reduces to:

$$g_1(s, t_d) = e^{-s^2 D t_d} \tag{12.23}$$

where D is the self-diffusion coefficient of the particle, and s is the magnitude of the scattering vector as defined above. Application of the Stokes–Einstein equation yields the hydrodynamic particle radius R_H (see Figure 12.37) from the diffusion coefficient:

$$D = \frac{kT}{6\pi \eta R_H} \tag{12.24}$$

where η is the viscosity of the suspending fluid.

In polydisperse systems, the overall decay of $g_1(s,t_d)$ is determined by the decay rate s^2D corresponding to each particle size, and can be written as a weighted average of all possible decays:

$$g_1(s, t_d) = \lim_{n\to\infty} \sum_{j=1}^{n} w_j \left(s^2 D\right) e^{-s^2 D_j t_d} \tag{12.25}$$

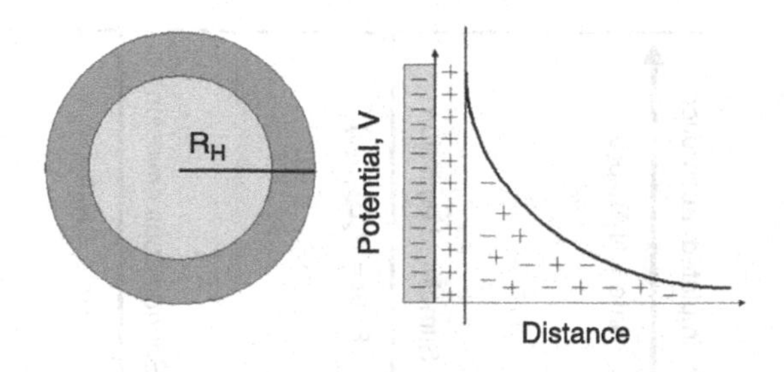

FIGURE 12.37 Illustration of the hydrodynamic radius R_H of a spherical particle with an overall negative surface charge.

where $w_j(s^2D)$ is a weighting function determined by the population in size range j. As real systems rarely, if ever, are truly monodisperse, numerical algorithms such as cumulant expansion (Z-average) [134], CONTIN [135,136], NNLS [137,138] etc. are required in order to properly fit Equation 12.25. What algorithm to use depends on what prior knowledge exists of the system to be studied, and on what model best fits the data set (yields the highest correlation coefficient).

12.3.10.1 Applications

In addition to its routine application in studies of polymers [139,140], food emulsions [141], and drug delivery vectors [142], dynamic light scattering has been widely used for studying coagulation rates in colloidal systems [143–147], including aqueous suspensions of gold nanoparticles. Wilcoxon and co-workers have studied the aggregation behavior – kinetics and structure – of gold sols with respect to addition of functional ligands (such as dyes), changes in pH, and addition of salt [147]. Here, the large scattering cross section of gold colloids enables detection of low concentrations, but the strong extinctions may also lead to distortion of the relaxation time, thus making accurate particle size measurements difficult, a problem which is compounded by the aggregation process. Dynamic light scattering has also been used for determination of Ostwald ripening rates and destabilization phenomena in aqueous emulsions [143]. Furthermore, aggregation and precipitation of asphaltenes have been studied by this technique [148–151].

12.3.11 UV–Visible Spectroscopy

The recent interest in nanotechnology has spurred resurgence in the use of steady-state absorption and emission spectroscopy in colloid and surface chemistry. In contrast to the corresponding bulk materials, material properties in the colloidal domain, including optical, are size-dependent. Metal colloids such as gold and silver nanoparticles and quantum dots exhibit strongly size-dependent extinctions in the visible frequency spectrum. This allows for observation of the flocculation behavior of the sol by monitoring the absorption bands of the system. Moreover, it is possible to look at changes in the dielectric function on the gold surface from adsorption of chromophores and emitters, as will be described later in this section.

Figure 12.38 shows a simple two-state diagram which represents the three processes considered in a quantum mechanical treatment of the radiation field: (stimulated) absorption, stimulated emission, and spontaneous emission, represented by the phenomenological rate constants B_{12}, B_{21}, and A_{21}, respectively.

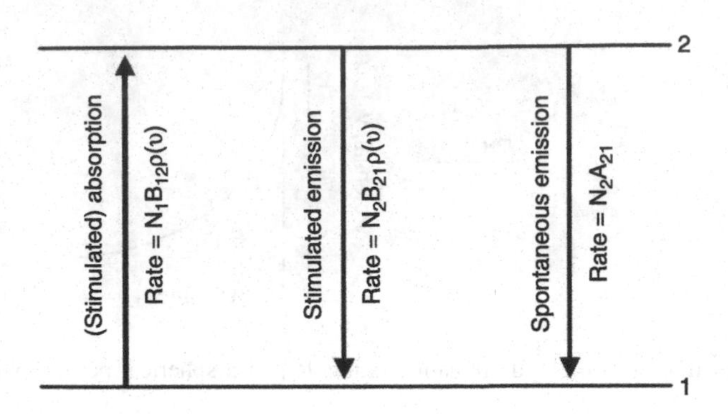

FIGURE 12.38 Einstein coefficients for absorption and emission.

The rate of transitions out of a state is proportional to the population (number of molecules) in that state. Designating the populations of the lower and upper levels by N_1 and N_2 and letting $\rho(\nu)$ represent the frequency dependent energy density of the radiation, the rates of upward (W_{12}) and downward (W_{21}) transitions can be expressed as follows:

$$W_{12} = N_1 B_{12} \rho(\nu) \tag{12.26}$$

$$W_{21} = N_2 B_{21} \rho(\nu) + N_2 A_{21} \tag{12.27}$$

To maintain equilibrium, the rates of upward and downward transitions must be balanced:

$$N_1 B_{12} \rho(\nu) = N_2 B_{21} \rho(\nu) + N_2 A_{21} \tag{12.28}$$

In linear spectroscopy experiments, the Boltzmann populations of the two states are unperturbed [152]:

$$\frac{N_2}{N_1} = \frac{g_2}{g_1} \exp\left[\frac{-(E_2 - E_1)}{kT}\right] = \frac{g_2}{g_1} \exp\left[\frac{-h\nu}{kT}\right] \tag{12.29}$$

Here, the possibility of degeneracy is allowed for; g_i is the number of states at the energy level E_i. Assuming that the energy density is that of a black body:

$$\rho(\nu) = \frac{8\pi h \nu^3}{c^3} \frac{1}{e^{h\nu/kT} - 1} \tag{12.30}$$

Combining Equations 12.28, 12.29, and 12.30 and solving for the energy density gives the relationships between the Einstein coefficients:

$$g_1 B_{12} = g_2 B_{21} \tag{12.31}$$

$$\frac{A_{21}}{B_{21}} = \frac{8\pi \nu^3}{c^3} \tag{12.32}$$

Although Equation 12.32 demonstrates the frequency dependency of the ratio of the rate constants for spontaneous to stimulated, A_{21} and B_{21} are not directly comparable as they are not dimensionally equivalent. Thus, a better comparison is of the rates A_{21} with $B_{21}\rho(\nu)$, the ratio of which is $A_{21}/B_{21}\rho(\nu) = e^{h\nu/kT} - 1$. It is clear from this ratio that when $h\nu \gg kT$, as is usually the case for luminescence experiments, the rate of spontaneous emission greatly exceeds that of stimulated emission. Thus, in systems at equilibrium at room temperature, the spontaneous emission of light at optical frequencies is greatly favored over stimulated emission. The properties of the two types of radiation are quite different; in the case of stimulated emission, the stimulated photon has the same properties as the incident radiation, resulting in emission which is collimated and coherent, whereas spontaneous emission, such as ordinary fluorescence, is emitted in all directions with random phase, and is red-shifted as compared to the incident radiation.

Although many principles presented thus far are quite general, the focus of this section is on applications in electronic spectroscopy or, more specifically, UV–vis absorption. The intensity of absorption by a sample varies with the path length x of the sample in accord with Beer–Lambert's law [153]:

$$\log\frac{I}{I_0} = -\varepsilon Cx \tag{12.33}$$

where I_0 is the incident intensity at a particular wavelength, I is the intensity after passage through a sample of length x (in cm), and C is the molar concentration (in mol/L) of the absorbing species. The quantity ε (or $\varepsilon(\nu)$) is called the molar absorptivity (formerly, and still widely, referred to as the extinction coefficient), and is generally expressed in L mol^{-1} cm^{-1}. The dimensionless product $A = \varepsilon Cx$ is called the absorbance of the sample, and the ratio I/I_0 is the transmittance T. These two quantities are related as follows:

$$\log T = -A \tag{12.34}$$

Hence, the absorbance is indirectly measured experimentally by determining the ratio of the incident to emerging intensities and applying Equation 12.34.

Information gathered from UV–vis absorption is not necessarily limited to information about the concentration and frequency of electronic transitions of a molecule or solid. By applying the Fermi Golden Rule (FGR) transition rate for (stimulated) absorption

$$B_{if} = \frac{|\mu_{if}|^2}{6\varepsilon_0\hbar^2} \tag{12.35}$$

between two states i and f, where μ_{if} is the transition dipole moment between the ground (i) and first excited (f) electronic states respectively, it is even possible to find the absolute value of the given transition moment through use of the integrated molar absorptivity [152]:

$$|\mu_{if}|^2 = \frac{(6\varepsilon_0\hbar^2)\,2303c}{N_A hn}\int_{band}\frac{\varepsilon(\nu)}{\nu}d\nu \tag{12.36}$$

where n is the refractive index of the medium, N_A is Avogadro's number, and the integer 2303 is a conversion factor.

The intensity of emission $f \rightarrow i$ is also proportional to $|\mu_{if}|^2$ and to A_{fi}, although there are several reasons why this is not a practical way to determine the radiative lifetime (to be discussed in the next section) or the transition dipole moment. One reason is that fluorescence intensities, unlike absorption intensities, are difficult to quantify, due to the emitted radiation being spread in all directions with random phase. The signal observed will depend strongly on the angle subtended by the detector, so it is sensitive to experimental conditions. One does not usually have the advantage of being able to compare a sample and a reference beam as in absorption spectroscopy.

12.3.11.1 Applications

It is well known that the optical signature of gold nanoparticles is strongly dependent on particle size and interparticle distance [154,155]. Thus, by monitoring the position and full width at half maximum (FWHM) of the plasmon band, information about the stability and flocculation behavior of the suspension can be collected. When the interparticular distance is substantially greater than the average particle diameter, the nanoparticle suspension appears red, but as the interparticle distance decreases to less than approximately the average particle diameter, the color shifts to blue or purple, depending on the level of flocculation and the particle concentration. The changes in peak position and in FWHM of the plasmon band as observed in UV–visible spectroscopy have been used to obtain experimental conditions where surface modification of the gold nanoparticles do not induce aggregation on the time scale of the experiment [156], to determine the critical coagulation concentration of gold nanoparticles capped with various stabilizing agents [157], and to monitor photoinduced fusion of gold colloids [158,159]. Moreover, the changes in optical signature have been widely used in colorimetric detection schemes for DNA-linked gold nanoparticle assemblies [160–167]. Here, the color changes from red to blue/purple accompanying changes in the flocculation behavior has been utilized to detect and monitor the programming of assemblies of two- and three-dimensional architectures, and to detect and quantify hybridization of gold nanoparticle-immobilized oligonucleotides. Figure 12.39 illustrates the principle behind the programmed materials synthesis of DNA-linked gold nanoparticle assemblies. The strong visible extinctions of gold nanoparticles have also been used for characterizing the presence, location, and properties of gold nanoparticles dispersed in mesoporous solids [168–173]. *In situ* synthesis of gold nanoparticles within ordered mesoporous materials can be monitored with UV–visible spectroscopy, as shown in Figure 12.40. The presence of a plasmon band indicates the presence

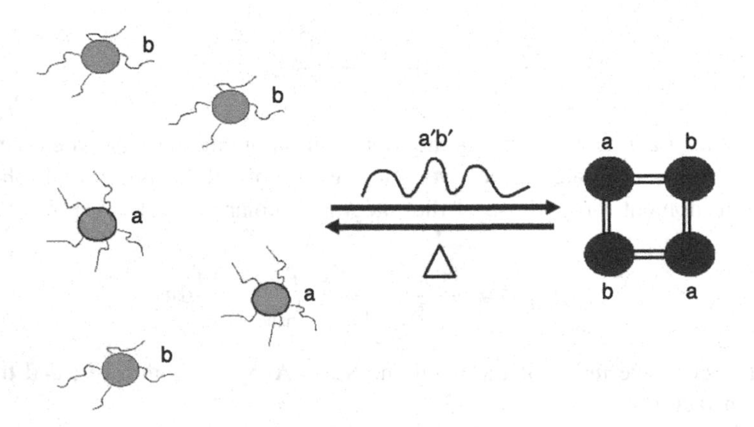

FIGURE 12.39 Programmed materials synthesis and accompanying changes in optical signature of DNA-capped gold nanoparticles. Two probe species (a and b) are linked by a complementary DNA target strand (a′b′).

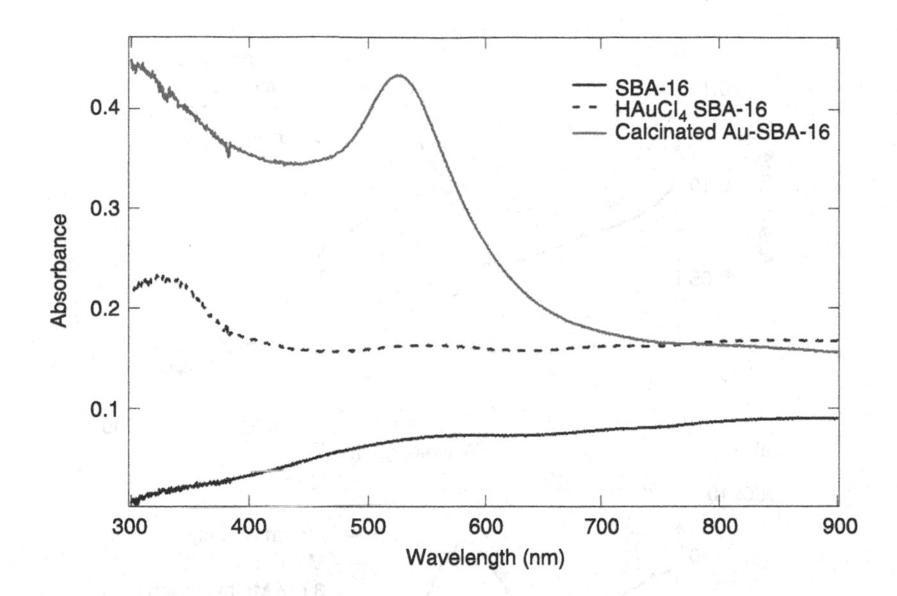

FIGURE 12.40 UV–visible spectrum of gold nanoparticles synthesized within the ordered mesoporous material SBA-16. When only the gold nanoparticle precursor ($HAuCl_4$) is present, no plasmon band is observed in the material. Following thermal reduction, a distinct absorption band centered at ~520 nm appears – the gold nanoparticle plasmon band.

of gold nanoparticles with a diameter ≥3 nm [174]. Shifts in the peak position of the plasmon band can be attributed to changes in the dielectric function on the gold nanoparticle surface. Thus, changes in the spectral lineshape of gold nanoparticles yield information about the chemical environment of the colloids. Shifts in the peak position of the plasmon band λ_{max} towards lower frequencies (red shifts) have been attributed to interfacial charge transfer reactions between the nanoparticles and the pore walls for gold nanoparticles dispersed in mesoporous materials [171]. Conversely, shifts in the λ_{max} towards higher frequencies (blue shifts) of gold [175] and silver [176] nanoparticles dispersed on porous supports have been ascribed to coherent coupling of the plasmon resonances of closely spaced colloidal entities. This coupling would be facilitated by gold (or silver) nanoparticles residing within the pore system of the ordered mesoporous material rather than on the external surface.

UV–visible spectroscopy can also be used for the study of changes in the dielectric function on the nanoparticle surface as a result of adsorption of dyes and chromophores and emitters on noble metal colloids [156,177]. Franzen and co-workers have studied the optical properties of dye molecules adsorbed on single gold and silver nanoparticles, using different particle sizes, solvent conditions, and passivating layers [156]. The dye molecules were similar to those reported in studies of surface-enhanced Raman scattering (SERS). From the observed spectral changes of the adsorbates, the dyes studied could be divided into two classes according to the nature of the interaction between the adsorbate (dye) and the substrate (gold or silver nanoparticles), and the position of the absorption band of the dye relative to the plasmon frequency ω_p of the nanoparticles (see Figure 12.41). Dyes where the absorption band occurred at higher frequencies than the plasmon frequency, and where the adsorption to the nanoparticles most likely involved chemical bonding, showed little or no changes in the absorption spectrum following adsorption.

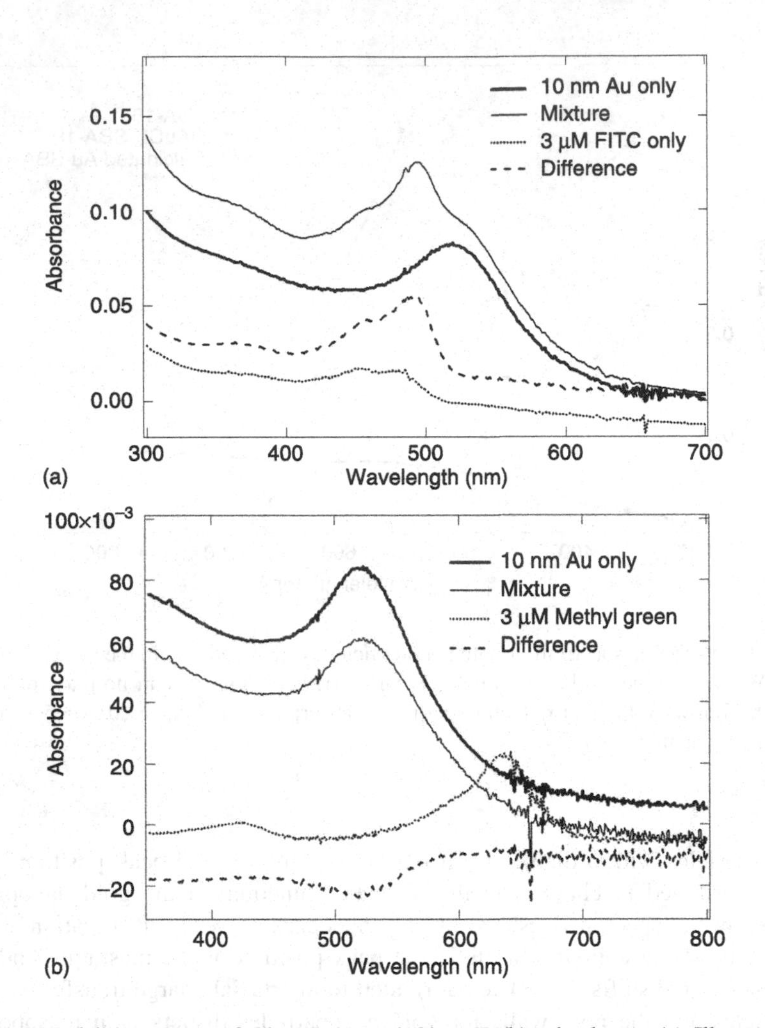

FIGURE 12.41 UV–visible spectra of Class I (a) and Class II (b) adsorbates. (a) Illustration of Class I adsorbate behavior. UV–visible spectra of F5ITC (5 μM) adsorbed to 10 nm citrate stabilized Au nanoparticles in a 50/50 H_2O/MeOH solvent. (b) Illustration of Class II adsorbate behavior. UV–visible spectrum of methyl green (3 μM) adsorbed to 10 nm citrate stabilized Au nanoparticles in a 50/50 H_2O/MeOH solvent. The difference spectra were obtained by subtracting the nanoparticle-only spectrum from the spectrum of the adsorbate–nanoparticle mix.

This set of dyes was denoted Class I adsorbates. However, dyes with absorption bands occurring at lower frequencies than ω_p, and where the adsorption most likely occurred through electrostatic interactions, showed significant bleaching of the absorption spectrum following adsorption. This set of adsorbates was denoted Class II. As the changes in the optical signature of the adsorbates depend on the size, surface charge, and nature of the nanoparticles, this effect could be called colloidochromism. This is illustrated in Figure 12.41(a), where the absorbance spectrum of a pH sensitive Class I adsorbate, fluorescein-5-isothiocyanate (F5ITC), is enhanced by a factor of ~5. This enhancement can be ascribed to either (1) a change in the pKa of the dye upon adsorption due to the interaction with the dielectric function on the gold surface or (2) a response to the local pH on colloidal gold surfaces being different to the solution pH.

12.3.12 Time-Correlated Single Photon Counting Spectroscopy

Emission experiments are typically performed in sample geometries that are large relative to the size of the emitters and relative to the absorption and emission wavelengths. In this arrangement, the emitters radiate into free space [178]. These free-space conditions serve as the basis for our knowledge and intuition about emission, i.e., fluorescence and phosphorescence. However, the presence of nearby metallic surfaces, defined as highly conducting particles of a surface and not metal ions or oxides, can alter the free-space condition, the result of which could be dramatic spectral changes as compared to those observable in the absence of metal surfaces. These changes are due to interactions of the excited-state emitters with free electrons in the metal, which polarize the metal and impose a reactive field on the emitter [179–183], the observable effect of which is an alteration of the radiative decay rates and thus the observed lifetimes of the excited state of an emitter. This observable effect is usually referred to as "quenching," as the presence of a metal surface often provides additional nonradiative pathways for the emitter [184–188], thus altering the measured lifetime of the excited state τ_{obs}. Time-resolved spectroscopies such as TCSPC make it possible to do direct measurements of the excited-state lifetimes of emitters.

Under free-space conditions, the quantum yield F_{rad}, as well as the observed lifetime of the excited/radiative state τ_{obs}, of an emitter are determined by the rates of radiative decay, k_{rad}, and nonradiative decay, k_{nr}, to the ground state. The quantum yield is defined as [152]:

$$\Phi_{rad} = \frac{k_{rad}}{k_{rad}+k_{nr}} = \frac{k_{rad}}{k_{obs}} \tag{12.37}$$

The observed lifetime in the absence of metallic surfaces is given by:

$$\tau_{obs} = (k_{rad} + k_{nr})^{-1} = (k_{obs})^{-1} \tag{12.38}$$

The natural lifetime [189] of the radiative excited state of an emitter (τ_N) is the inverse of the radiative decay rate ($\tau_N = k_{rad}^{-1}$), which is the lifetime if $k_{nr} = 0$ or $F_{rad} = 1$.

When the emitter is associated with a quencher, i.e., a metallic surface such as a nanoparticle, an additional nonradiative process with a rate $Q = k_q[Qu]$ is introduced, which competes with emission. Here, k_q represents the rate constant for the quenching process, and [Qu] represents the concentration of the quencher. The quantum yield $\Phi_{rad,Q}$ and observed lifetime $\tau_{obs,Q}$ are now given by:

$$\Phi_{rad,Q} = \frac{k_{rad}}{k_{rad} + k_{nr} + Q} = \frac{k_{rad}}{k_{obs,Q}} \tag{12.39}$$

$$\tau_{obs,Q} = (k_{rad} + k_{nr} + Q)^{-1} = \left(k_{obs,Q}\right)^{-1} \tag{12.40}$$

For mixed systems, such as a suspension of a quencher (like gold colloids) with an emitter (i.e., a fluorophore-labeled protein, peptide, or other macromolecule) where the concentration of emitter exceeds the number of adsorption sites on the quencher, the total decay rate k_{tot} as measured by TCSPC can be resolved into two components: (1) a decay rate representing the fraction of emitter associated with the quencher $k_{obs,Q}$, and (2) a decay rate representing the fraction of emitter under free-space conditions (not associated with the quencher) k_{obs} as given by Equations 12.39 and 12.37, respectively, where the corresponding lifetimes are given by Equations 12.40 and 12.38. In order to separate these time components and quantify the concentrations of quenched and free emitter, the measured decay k_{tot} can be normalized and fitted to a sum of exponentials so as to maximize the correlation coefficient of the fit, R^2:

$$Y(t) = \sum_{i=1}^{N} A_i e^{-k_i t} \tag{12.41}$$

where N is the total number of luminescent components, and A_i and k_i represent the amplitude/fraction/population and rate constant for each component, respectively. The decay rates are identified as either quenched ($k_{obs,Q}$) or free (k_{obs}) by comparison with the decay rate in a sample containing only free emitter. For systems obeying Stern–Volmer type quenching, $k_{obs,Q} \gg k_{obs}$, and thus $\tau_{obs,Q} \ll \tau_{obs}$. The concentrations of quenched and free emitter can now be determined by multiplying the total emitter concentration by the appropriate amplitude/fraction A_i associated with each decay rate/lifetime. By comparing the concentration of surface-bound emitter with the concentration of quencher, i.e., Au nanoparticle, the ratio of emitter (or labeled macromolecule) to quencher can be determined.

12.3.12.1 Applications

One application for time-correlated single photon counting spectroscopy has been studies of energy migration dynamics in light-harvesting antenna polymer systems derivatized with Ru(II) and Os(II) polypyridyl complexes [190–192]. For mixed Ru(II) and Os(II) systems, efficient energy transfer is expected to occur, based on the high degree of overlap between the emission spectrum of the donor – Ru(II)* complexes – and the absorption spectrum of the acceptor – Os(II) complexes. Because the Os(II) excited state lies 0.36 eV below the Ru(II) excited state, the Os sites are considered to be deep traps that terminate the energy migration process [190,191]. In the development of molecular assemblies for studies in energy conversion, an important issue is the coupling of light absorption to electron transfer indirectly by use of intervening energy transfer in an antenna array. Another critical factor in the development of these nanoscale assemblies is the ability to control the spatial arrangement of the molecular components, especially if intermolecular energy-transfer and charge-transfer processes are at the core of the material's function. Spatial organization can be achieved through the design of covalently bonded supramolecules in which molecular subunits are linked together so that their relative geometries (i.e., separations and orientations) are well defined [193–198]. Derivatized polymers are attractive choices for positioning of molecular constituents because they offer flexibility and simplicity in the design of multicomponent assemblies. The functional capabilities of light/harvesting systems can be quantified on the basis of the efficiency with which they conduct excited-state energy. For efficient conduction to occur, the time scale for energy transfer must be fast compared to the lifetime of the excited state. For example, if the time scale for an excited state to hop to an adjacent site is 10 times faster than its lifetime τ_N, energy transfer will occur with 90% efficiency. Thus, efficient energy transfer is best achieved by combining fast energy transfer between monomer units with long-lived excited states.

The Papanikolas group has studied the photophysical properties of Ru(II) and Os(II) polypyridyl complexes linked to polystyrene through use of TCSPC and steady-state spectroscopies, and by Monte Carlo simulations [190,191]. A series of supramolecular assemblies consisting of 20 Ru(II) and/or Os(II) polypyridyl coordination complexes with varying Ru(II)/Os(II) ratios linked together through a polystyrene backbone as shown in Figure 12.42 were synthesized and investigated. Here, the Ru(II) complexes can act as efficient "antennas" for collecting visible light and sensitizing the lower energy Os(II) sites on the polymeric backbone. Energy transfer

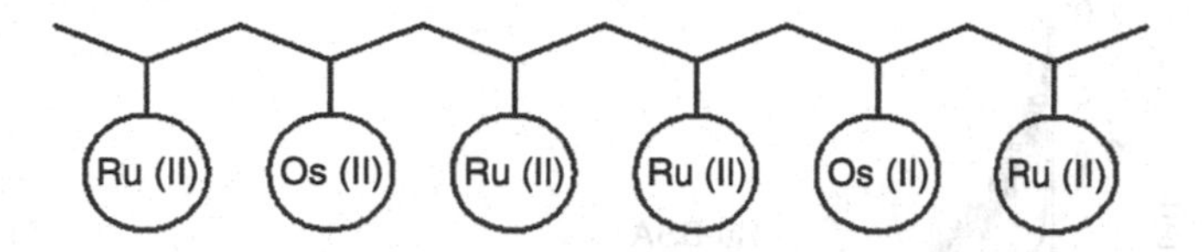

FIGURE 12.42 Illustration of the Ru(II)/Os(II) complexes linked together through a polymer backbone.

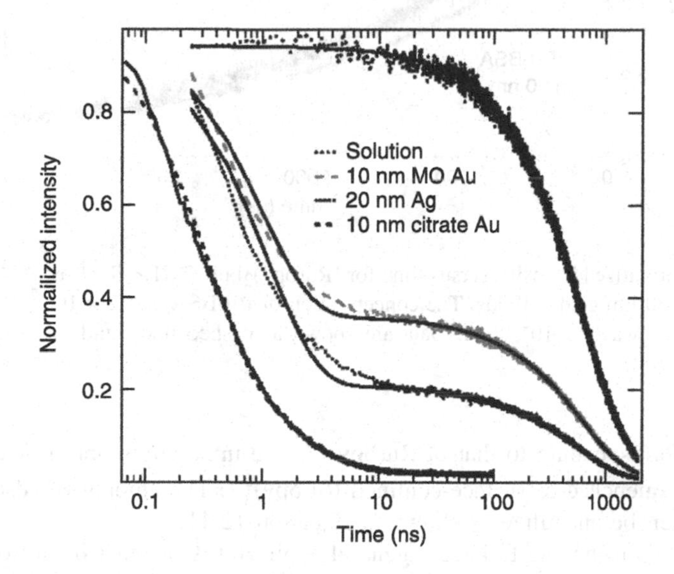

FIGURE 12.43 Time-resolved normalized emission from $Ru(bpy)_3^{2+}$ in solution and adsorbed onto various colloids, including 10 nm mercaptooctanoate (MO) stabilized Au colloids (3.1×10^{-9} M), 20 nm citate stabilized Ag colloids, and 10 nm citrate stabilized Au colloids (4.7×10^{-9} M). The solution concentration of $Ru(bpy)_3^{2+}$ was kept at $\sim 6 \times 10^{-6}$ M. The excitation wavelength was 423 nm, and emission was detected at 630 nm. Experimental data is shown as a dashed or dotted line, and the fit is shown as a solid line.

efficiency was found to be dependent on the degree of Os(II) loading, however; with an average loading of 11 Ru(II) sites per 5 Os(II) sites, the energy transfer Ru(II)* → Os(II) was determined to occur with an efficiency of 95% independent of excitation wavelength from 420 to 500 nm.

Glomm et al. [177] have described the application of Ru(II) tris bipyridine ($Ru(bpy)_3^{2+}$) and Os(II) tris bipyridine ($Os(bpy)_3^{2+}$) as phosphorescent labels for the quantification of surface binding of molecules to gold and silver nanoparticles. The fraction of $Ru(bpy)_3^{2+}$ and $Os(bpy)_3^{2+}$ in solution can be distinguished from the surface-confined fraction by the relative lifetimes and integrated emission yields as determined by TCSPC. Complementary steady-state measurements confirmed surface attachment of the phosphorescent label molecules. Figure 12.43 shows the time-resolved normalized emission from $Ru(bpy)_3^{2+}$ in solution and adsorbed onto gold and silver colloids. Here, it is evident that the observed emission traces for mixed $Ru(bpy)_3^{2+}$ and metal colloid systems consists of two or more decay rates, where at least one of the excited states has a significantly shorter lifetime than that of $Ru(bpy)_3^{2+}$ free in solution (Figure 12.43). In other words, the traces of mixed $Ru(bpy)_3^{2+}$ and metal colloid systems contain one component

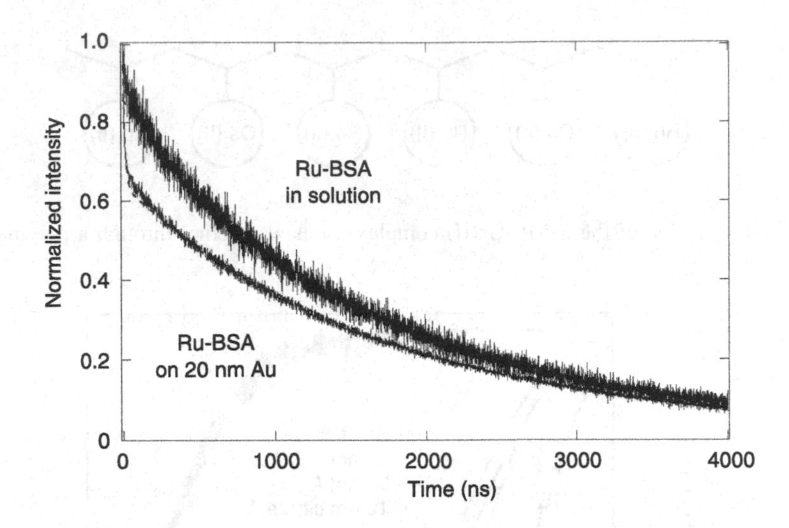

FIGURE 12.44 Normalized intensity versus time for $[Ru(bpy)_2bpy\text{-}C_6H_{12}\text{-}S]^{2+}$-BSA (Ru-BSA) in solution and adsorbed onto 20 nm gold colloids. The concentration of Ru-BSA was 2×10^{-7} M, and the 20 nm Au colloid concentration was 4×10^{-10} M. Data are shown as dashed lines, and the fits are shown as solid lines.

with a lifetime corresponding to that of $Ru(bpy)_3^{2+}$ free in solution, and at least one component corresponding to quenched or surface-confined $Ru(bpy)_3^{2+}$. The amount of adsorbed luminescent molecules can then be quantified as shown in Equation 12.41.

In order for this technique to have a general applicability, it must be not only able to detect and quantify surface attachment of the label molecules themselves (i.e., $Ru(bpy)_3^{2+}$); it must also be able to distinguish between adsorbed and free states of bigger constructs, such as fluorophore-labeled proteins, peptides, and other macromolecules. However, it is important to note that when using constructs wherein luminescent labels are incorporated, the emission quantum yield of the construct will differ from that of the luminescent label molecule by itself. This technique has been proven effective for detecting and quantifying adsorption of Rhodamine-B labeled adenoviral receptor-mediated endocytosis (RME) peptide (CKKKKKKSEDEYPYVPN) on 5 nm and 20 nm gold colloids [199]. Further scale-ups of the technique have enabled quantification of Ru(bpy) labeled bovine serum albumin (BSA) adsorbed on gold nanoparticles [157,199]. For a 20 nm gold nanoparticle, monolayer coverage of BSA was determined to be 160 ± 8 protein molecules (see Figure 12.44). Multifunctional gold nanoparticle–peptide complexes where BSA functions partly as a stabilizer for the gold nanoparticles, and partly as a scaffold for covalent attachment of RME and nuclear localization signal (NLS) peptides to the gold–protein bioconstructs for use in targeted drug delivery, have been extensively studied by the Franzen and Feldheim research groups [142,157,199,200]. In order to preserve the functionality of the functional and targeting ligands, as well as to quantify the efficacy of the delivery vector in a given chemical and/or biological environment such as a specific cell line, detailed information about the interaction between each active component (proteins, peptides, oligonucleotides) with the substrate, such as obtained through use of TCSPC, is absolutely crucial.

TCSPC provides a rapid and powerful tool for direct measurement and quantification of surface-confined molecules [157,177,199]. Compared to steady state emission measurements, TCSPC has the advantage that one can distinguish a component that contributes only a few percent to the total emission signal using the lifetime of that component. While steady-state emission measurements require the preparation of a series of samples with varying quencher concentrations and a reference in order to obtain Stern–Volmer plots, the method described here only requires a single sample plus a reference.

ACKNOWLEDGMENTS

In preparing this chapter the authors would like to acknowledge the support of the industrial members of the Ugelstad Laboratory at NTNU in Trondheim. In addition we would also like to acknowledge The Research Council of Norway (NFR) for financial support to instrumentation and PhD candidates.

REFERENCES

1. Speight, J. G. *The Chemistry and Technology of Petroleum*; 3rd ed.; Marcel Dekker, Inc.: New York, 1998.
2. Ali, M. A.; Nofal, W. A. *Fuel Science & Technology International* **1994**, *12*, 21–33.
3. Fan, T. G.; Buckley, J. S. *Energy & Fuels* **2002**, *16*, 1571–1575.
4. Hammami, A.; Ferworn, K. A.; Nighswander, J. A.; Overa, S.; Stange, E. *Petroleum Science and Technology* **1998**, *16*, 227–249.
5. Radke, M.; Willsch, H.; Welte, D. H. *Analytical Chemistry* **1980**, *52*, 406–411.
6. Suatoni, J. C.; Swab, R. E. *Journal of Chromatographic Science* **1976**, *14*, 535–537.
7. Hannisdal, A.; Hemmingsen, P. V.; Sjöblom, J. *Industrial & Engineering Chemistry Research* **2005**, *44*, 1349–1357.
8. Snyder, L. R.; Kirkland, J. J.; Glajch, J. L. *Practical HPLC Method Development*; 2nd ed.; John Wiley and Sons, Inc.: New York, 1997.
9. The Institute of Petroleum Standards Method, IP 469/01.
10. ASTM Standard 2001, D2007-01a.
11. Fan, T.; Buckley, J. S. *Energy Fuels* **2002**, *16*, 1571–1575.
12. Bharati, S.; Røstum, G. A.; Løberg, R. *Org Geochem* **1994**, *22*, 835–862.
13. Bharati, S.; Patience, R.; Mills, N.; Hanesand, T. *Org Geochem* **1997**, *26*, 49–57.
14. Mullins, O. C. *Analytical Chemistry* **1990**, *62*, 508–514.
15. Mullins, O. C.; Joshi, N. B.; Groenzin, H.; Daigle, T.; Crowell, C.; Joseph, M. T.; Jamaluddin, A. *Applied Spectroscopy* **2000**, *54*, 624–629.
16. Aske, N.; Kallevik, H.; Sjöblom, J. *Energy & Fuels* **2001**, *15*, 1304–1312.
17. Fossen, M.; Hemmingsen, P. V.; Sjöblom, J. *Journal of Dispersion Science and Technology* **2005**, *26*, 227–241.
18. Hildebrand, J. H.; Scott, R. L. *The Solubility of Nonelectrolytes*; 3rd ed.; Reinhold Publishing Corporation: New York, 1950.
19. Hansen, C. M. *Journal of Paint Technology* **1967**, *39*, 104–117.
20. Hansen, C. M. *Journal of Paint Technology* **1967**, *39*, 505–510.
21. Hansen, C. M. *Journal of Paint Technology* **1967**, *39*, 511–514.
22. Aske, N.; Kallevik, H.; Sjöblom, J. *Journal of Petroleum Science and Engineering* **2002**, *36*, 1–17.
23. Donaggio, F.; Correra, S.; Lockhart, T. P. *Petroleum Science and Technology* **2001**, *19*, 129–142.
24. Oh, K.; Ring, T. A.; Deo, M. D. *Journal of Colloid Interface Science* **2004**, *271*, 212–219.
25. Rogel, E.; León, O.; Espidel, Y.; González, Y. *Society of Petroleum Engineers Production & Facilities* **2001**, SPE no. 72050.
26. Rogel, E.; León, O.; Contreras, E.; Carbognani, L.; Torres, G.; Espidel, Y.; Zambrano, A. *Energy & Fuels* **2003**, *17*, 1583–1590.
27. Hu, Y. F.; Guo, T. M. *Fluid Phase Equilibria* **2001**, *192*, 13–25.
28. Aske, N.; Kallevik, H.; Johnsen, E. E.; Sjöblom, J. *Energy & Fuels* **2002**, *16*, 1287–1295.

29. Auflem, I. H.; Havre, T. E.; Sjöblom, J. *Colloid and Polymer Science* **2002**, *280*, 695–700.
30. McLean, J. D.; Kilpatrick, P. K. *Journal of Colloid and Interface Science* **1997**, *196*, 23–34.
31. McLean, J. D.; Kilpatrick, P. K. *Journal of Colloid and Interface Science* **1997**, *189*, 242–253.
32. Hemmingsen, P. V.; Li, X.; Peytavy, J. L.; Sjöblom, J. *Journal of Dispersion Science and Technology*, Submitted.
33. Chen, T. Y.; Mohammed, R. A.; Bailey, A. I.; Luckham, P. F.; Taylor, S. E. *Colloids Surf.* **1994**, *83*, 273–284.
34. Williams, T. J.; Bailey, A. G. *IEEE Trans. Ind. Appl.* **1986**, *IA-22*, 536–541.
35. Atten, P. *J. Electrostatics* **1993**, *30*, 259–270.
36. Førdedal, H.; University of Bergen: Norway, 1995.
37. Eow, J. S.; Ghadiri, M.; Sharif, A. O.; Williams, T. J. *Chem. Eng. J* **2001**, *84*, 173–192.
38. Eow, J. S.; Ghadiri, M. *Chem. Eng. J* **2002**, *85*, 357.
39. Kallevik, H.; Kvalheim, O. M.; Sjöblom, J. *J. Colloid Interface Sci.* **2000**, *225*, 494.
40. Aske, N.; Orr, R.; Sjöblom, J.; Kallevik, H.; Oye, G. *Journal of Dispersion Science and Technology* **2004**, *25*, 263–275.
41. Hemmingsen, P. V.; Silset, A.; Hannisdal, A.; Sjöblom, J. *Journal of Dispersion Science and Technology* **2005**, *26*, 615–627.
42. Havre, T. E.; Sjöblom, J. *Coll. Surf. A. Physiochem. Eng. Asp.* **2003**, *228*, 131–142.
43. Spiecker, P. M.; Gawrys, K. L.; Trail, C. B.; Kilpatrick, P. K. *Colloids Surfaces A* **2003**, *220*, 9–27.
44. Myrvold, R.; Hansen, F. K. *Journal of Colloid and Interface Science* **1998**, *207*, 97–105.
45. Wang, Y. Y.; Zhang, L.; Sun, T. L.; Jiayong, Y. *Journal of Colloid and Interface Science* **2004**, *270*, 163–170.
46. West, R. C.; Astle, M. J., Eds. *CRC Handbook of Chemistry and Physics*; 63 ed.; CRC Press, Inc.: Boca Raton, Florida, 1984.
47. Petty, M. C.; Barlow, W. A. In *Langmuir–Blodgett Films*; Roberts, G., Ed.; Plenum Press: New York, 1990, p 93.
48. Gaines Jr., G. L. *Insoluble Monolayers at Liquid–Gas Interfaces*; Wiley: New York, 1966.
49. Birdi, K. S. *Lipid and Biopolymer Monolayers at Liquid Interfaces*; Plenum Press: New York, 1989.
50. Harkins, W. D. *The Physical Chemistry of Surface Films*; Reinhold Publishing Corporation: New York, 1952.
51. Adam, N. K. *Physics and Chemistry of Surfaces*; 3rd ed.; Oxford University Press: London, 1941.
52. Smith, R. D.; Berg, J. C. *Journal of Colloid and Interface Science* **1980**, *74*, 273–286.
53. Pezron, E.; Claesson, P. M.; Berg, J. M.; Vollhardt, D. *Journal of Colloid and Interface Science* **1990**, *138*, 245–254.
54. Tomoaiacotisel, M.; Zsako, J.; Chifu, E.; Cadenhead, D. A. *Langmuir* **1990**, *6*, 191–197.
55. Honig, E. P.; Hengst, J. H. T.; Denengel, D. *Journal of Colloid and Interface Science* **1973**, *45*, 92–102.
56. Ese, M. H.; Yang, X.; Sjöblom, J. *Colloid and Polymer Science* **1998**, *276*, 800–809.
57. Singh, B. P.; Pandey, B. P. *Indian Journal of Technology* **1991**, *29*, 443–447.
58. Ese, M. H.; Galet, L.; Clausse, D.; Sjöblom, J. *Journal of Colloid and Interface Science* **1999**, *220*, 293–301.
59. Ese, M. H.; Selsbak, C. M.; Hannisdal, A.; Sjöblom, J. *J. Dispersion Sci. Technol.* **2005**, *26*, 145–154.
60. Ambwani, D. S.; Fort, T., Jr. *Surface and Colloid Science* **1979**, *11*, 93–119.
61. Stauffer, C. E. *Journal of Physical Chemistry* **1965**, *69(6)*, 1933–1938.
62. Andreas, J. M.; Hauser, E. A.; Tucker, W. B. *Journal of Physical Chemistry* **1938**, *42*, 1001–1019.
63. Rotenberg, Y.; Boruvka, L.; Neumann, A. W. *Journal of Colloid and Interface Science* **1983**, *93*, 169–183.
64. Boucher, E. A.; Evans, M. J. B.; Jones, T. G. J. *Advances in Colloid and Interface Science* **1987**, *27(1–2)*, 43–79.
65. Brandal, Ø.; Sjöblom, J.; Øye, G. *Journal of Dispersion and Science Technology* **2004**, *25(3)*, 367–374.
66. Brandal, Ø.; Hanneseth, A-M.; Sjöblom, J. *Colloid and Polymer Science* **2005** (Accepted).

67. Brient, J. A.; Wessner, P. J.; Doyle, M. N. In *Encyclopedia of Chemical Technology*; 4th ed.; Kirk-Othmer, Ed.; John Wiley & Sons: New York, 1995; Vol. 16, pp 1017–1029.
68. Seifert, W. K. *Fortschritte der Chemie Organischer Naturstoffe* **1975**, *32*, 1–49.
69. Tomczyk, N. A.; Winans, R. E.; Shinn, J. H.; Robinson, R. C. *10.1021/ef0201228* **2001**, *15*, 1498–1504.
70. Fan, T.-P. *10.1021/ef0201228* **1991**, *5*, 371–375.
71. Koike, L.; Reboucas, L. M. C.; Reis, F. d. A.; Marsaioli, A. J.; Richnow, H. H.; Michaelis, W. *Org Geochem* **1992**, *18*, 851–860.
72. Ovalles, C.; Carcia, M. d. C.; Lujano, D.; Aular, W.; Barmúdez, R.; Cotte, E. *Fuel* **1998**, *77*, 121–126.
73. Márquez, M. L. In *AICHE Spring National Meeting, Session T6005*: Houston, Texas, 1999; Vol. 99sp 56d.
74. Pathak, A. K.; Kumar, T. In *Proceedings of PETROTECH-95, Technology Trends in Oil Industry*: New Dehli, 1995, pp 217–224.
75. Poggesi, G.; Hurtevent, C.; Buchart, D. In *SPE Oilfield Scale Symposium*: Aberdeen, UK, 2002; Vol. SPE74649, pp 1–6.
76. Gallup, D. L.; Smith, P. C.; Chipponeri, J.; Abuyazid, A.; Mulyono, D. In *SPE International Conference on Health, Safety and Environment in Oil and Gas Exploration and Production*: Kuala Lumpur, Malaysia, 2002; Vol. SPE73960, pp 1–16.
77. Vindstad, J. E.; Bye, A. S.; Grande, K. V.; Hustad, B. M.; Hustvedt, E.; Nergård, B. In *5th SPE Oilfield Scale Symposium*: Aberdeen, UK, 2003; Vol. SPE80375.
78. Dyer, S. J.; Graham, G. M.; Heriot-Watt, C. A. In *5th SPE International Symposium on Oilfield Scale*: Aberdeen, UK, 2003; Vol. SPE 80395.
79. Havre, T. E. *Colloid and Polymer Science* **2004**, *282(3)*, 270–279.
80. Hartridge, H.; Peters, R. A. *Proc R Soc A* **1922**, *101*, 348–367.
81. Danielli, J. F. *Proc R Soc A* **1937**, *122*, 155–174.
82. Chifu, E.; Salajan, M.; Demeter-Vodnar, I.; Tomoaia-Cotisel, M. *Revue Roumaine de Chimie* **1987**, *32*, 683–691.
83. Rudin, J.; Wasan, D. T. *Colloids Surf* **1992**, *68*, 67–79.
84. Cratin, P. D. *J Dispersion Sci Technol* **1993**, *14*, 559–602.
85. Wolstenholme, G. A.; Schulman, J. H. *Trans. Faraday Soc.* **1950**, *46*, 475–487.
86. Goddard, E. D. *Journal of Colloid and Interface Science* **1967**, *24(3)*, 297–309.
87. Veale, G.; Peterson, I. R. *Journal of Colloid and Interface Science* **1985**, *103(1)*, 178–189.
88. Yazdanian, M.; Yu, H.; Zografi, G.; Kim, M. W. *Langmuir* **1992**, *8*, 630–636.
89. Avila, L. V. N.; Saraiva, S. M.; Oliveira, J. F. *Colloids and Surfaces, A: Physicochemical and Engineering Aspects* **1999**, *154(1–2)*, 209–217.
90. Marx, K. A. *Biomacromolecules* **2003**, *4*, 1099–1120.
91. Sauerbrey, G. *Zeitschrift fuer Physik* **1959** *155*, 206–222.
92. Knag, M.; Sjöblom, J.; Øye, G.; Gulbrandsen, E. *Colloids and Surfaces A* **2004**, *250*, 269–278.
93. Caruso, F.; Rinia, H. A.; Furlong, D. N. *Langmuir* **1996**, *12*, 2145–2152.
94. Ekholm, P.; Blomberg, E.; Claesson, P.; Auflem, I. H.; Sjöblom, J.; Kornfeldt, A. *Journal of Colloid and Interface Science* **2002**, *247*, 342–350.
95. Keller, C. A.; Kasemo, B. *Biophys. J.* **1998**, *75*, 1397–1402.
96. Keller, C. A.; Glasmastar, K.; Zhdanov, V. P.; Kasemo, B. *Physical Review Letters* **2000**, *84*, 5443–5446.
97. Hook, F.; Ray, A.; Norden, B.; Kasemo, B. *Langmuir* **2001**, *17*, 8305–8312.
98. Binnig, G.; Quate, C. F.; Gerber, C. *Physical Review Letters* **1986**, *56*, 930–933.
99. Meyer, G.; Amer, N. M. *Applied Physics Letters* **1988**, *53*, 1045–1047.
100. Alexander, S.; Hellemans, L.; Marti, O.; Schneir, J.; Elings, V.; Hansma, P. K.; Longmire, M.; Gurley, J. *J. Appl. Phys.* **1989**, *65*, 164–167.
101. Binnig, G.; Gerber, C.; Stoll, E.; Albrecht, T. R.; Quate, C. F. *Europhysics Letters* **1987**, *3*, 1281–1286.
102. Drake, B.; Prater, C. B.; Weisenhorn, A. L.; Gould, S. A. C.; Albrecht, T. R.; Quate, C. F.; Cannell, D. S.; Hansma, H. G.; Hansma, P. K. *Science* **1989**, *243*, 1586–1589.

103. Albrecht, T. R.; Quate, C. F. *J. Appl. Phys.* **1987**, *62*, 2599–2602.
104. Meyer, G.; Amer, N. M. *Applied Physics Letters* **1990**, *56*, 2100–2101.
105. Neumeister, J. M.; Ducker, W. A. *Rev. Sci. Instrum.* **1994**, *65*, 2527–2531.
106. Cappella, B.; Dietler, G. *Surface Science Reports* **1999**, *34*, 1.
107. Hertz, H.; Reine, J. *Angew. Math.* **1881**, *92*, 156.
108. Sneddon, J. N. *Int. J. Eng. Sci.* **1965**, *3*, 47.
109. Maugis, D. *Journal of Colloid and Interface Science* **1992**, *150*, 243–269.
110. Manne, S.; Cleveland, J. P.; Gaub, H. E.; Stucky, G. D.; Hansma, P. K. *Langmuir* **1994**, *10*, 4409–4413.
111. Atkin, R.; Craig, V. S. J.; Wanless, E. J.; Biggs, S. *Advances in Colloid and Interface Science* **2003**, *103*, 219–304.
112. Hansma, H. G.; Gould, S. A. C.; Hansma, P. K.; Gaub, H. E.; Longo, M. L.; Zasadzinski, J. A. N. *Langmuir* **1991**, *7*, 1051–1054.
113. Ese, M.-H.; Sjöblom, J.; Djuve, J.; Pugh, R. *Colloid and Polymer Science* **2000**, *278*, 532–538.
114. Weisenhorn, A. L.; Drake, B.; Prater, C. B.; Gould, S. A. C.; Hansma, P. K.; Ohnesorge, F.; Egger, M.; Heyn, S. P.; Gaub, H. E. *Biophys. J.* **1990**, *58*, 1251–1258.
115. Hansma, H. G.; Sinsheimer, R. L.; Li, M. Q.; Hansma, P. K. *Nucleic Acids Res.* **1992**, *20*, 3585–3590.
116. Sato, H.; Ohtsu, T.; Komasawa, I. *Journal of Colloid and Interface Science* **2000**, *230*, 200–204.
117. Grabar, K. C.; Brown, K. R.; Keating, C. D.; Stranick, S. J.; Tang, S. L.; Natan, M. J. *Analytical Chemistry* **1997**, *69*, 471–477.
118. Wang, K.; Wu, J. S.; Zeng, H. M. *Composites Science and Technology* **2001**, *61*, 1529–1538.
119. Ray, S.; Bhowmick, A. K.; Bandyopadhyay, S. *Rubber Chemistry and Technology* **2003**, *76*, 1091–1105.
120. Denes, F. S.; Manolache, S. *Prog. Polym. Sci.* **2004**, *29*, 815–885.
121. Øye, G.; Roucoules, V.; Cameron, A. M.; Oates, L. J.; Cameron, N. R.; Steel, P. G.; Badyal, J. P. S.; Davis, B. G.; Coe, D.; Cox, R. A. *Langmuir* **2002**, *18*, 8996–8999.
122. Øye, G.; Roucoules, V.; Oates, L. J.; Cameron, A. M.; Cameron, N. R.; Steel, P. G.; Badyal, J. P. S.; Davis, B. G.; Coe, D. M.; Cox, R. A. *Journal of Physical Chemistry B* **2003**, *107*, 3496–3499.
123. Badyal, J. P.; Cameron, A. M.; Cameron, N. R.; Oates, L. J.; Øye, G.; Steel, P. G.; Davis, B. G.; Coe, D. M.; Cox, R. A. *Polymer* **2004**, *45*, 2185–2192.
124. Badyal, J. P.; Cameron, A. M.; Cameron, N. R.; Coe, D. M.; Cox, R.; Davis, B. G.; Oates, L. J.; Øye, G.; Spanos, C.; Steel, P. G. *Chemical Communications* **2004**, *12*, 1402–1403.
125. Kvande, I.; Øye, G.; Ochoa-Fernández, E.; Hammer, N.; Rønning, M.; Holmen, A.; Sjöblom, J.; Chen, D. *Nano Letters* (Submitted).
126. Good, R. J. *Journal of Adhesion Science and Technology* **1992**, *6(12)*, 1269–1302.
127. Neumann, A. W.; Good, R. J. *Surface and Colloid Science* **1979**, *11*, 31–91.
128. Hunter, R. J. *Introduction to Modern Colloid Science*; Oxford Science Publications: Oxford, 1993.
129. Barchini, R.; Saville, D. A. *Langmuir* **1996**, *12*, 1442–1445.
130. Neves, G. B. M.; Dos Anjos de Sousa, M.; Travalloni-Louvisse, A. M.; Lucas, E. F.; Gonzalez, G. *Petroleum Science and Technology* **2001**, *19*, 35–43.
131. Leon, O.; Rogel, E.; Torres, G.; Lucas, A. *Petroleum Science and Technology* **2000**, *18*, 913.
132. Berne, B. J.; Pecora, R. *Dynamic Light Scattering: With Applications to Chemistry, Biology and Physics*; Dover Publications: New York, 2000.
133. Van de Hulst, H. C. *Light Scattering by Small Particles*; Dover: New York, 1981.
134. Koppel, D. E. *Journal of Chemical Physics* **1972**, *57*, 4814.
135. Provencher, S. W. *Computer Physics Communications* **1982**, *27*, 213–227.
136. Provencher, S. W. *Computer Physics Communications* **1982**, *27*, 229–242.
137. Lawson, C. L.; Hanson, R. J. In *Solving Least Square Problems*; Prentice Hall: Englewood Cliffs, NJ, 1974.
138. Lawson, C. L.; Hanson, R. J. *Solving Least Squares Problems*; SIAM: Philadelphia, 1995.
139. Wagner, J.; Hartl, W.; Hempelmann, R. *Langmuir* **2000**, *16*, 4080–4085.
140. Sedlak, M. *Langmuir* **1999**, *15*, 4045–4051.

141. Dalgleish, D. G.; Hallett, F. R. *Food Research International* **1995**, *28*, 181–193.
142. Tkachenko, A. G.; Xie, H.; Coleman, D.; Glomm, W.; Ryan, J.; Anderson, M. F.; Franzen, S.; Feldheim, D. L. *Journal of the American Chemical Society* **2003**, *125*, 4700–4701.
143. De Smet, Y.; Deriemaeker, L.; Parloo, E.; Finsy, R. *Langmuir* **1999**, *15*, 2327–2332.
144. Holthoff, H.; Egelhaaf, S. U.; Borkovec, M.; Schurtenberger, P.; Sticher, H. *Langmuir* **1996**, *12*, 5541–5549.
145. Holthoff, H.; Borkovec, M.; Schurtenberger, P. *Physical Review E* **1997**, *56*, 6945–6953.
146. Holthoff, H.; Schmitt, A.; FernandezBarbero, A.; Borkovec, M.; CabrerizoVilchez, M. A.; Schurtenberger, P.; HidalgoAlvarez, R. *Journal of Colloid and Interface Science* **1997**, *192*, 463–470.
147. Wilcoxon, J. P.; Martin, J. E.; Schaefer, D. W. *Physical Review A* **1989**, *39*, 2675–2688.
148. Anisimov, M. A.; Yudin, I. K.; Nikitin, V.; Nikolaenko, G.; Chernoutsan, A.; Toulhoat, H.; Frot, D.; Briolant, Y. *Journal of Physical Chemistry* **1995**, *99*, 9576–9580.
149. Yudin, I. K.; Nikolaenko, G. L. *NATO ASI Series, Series 3: High Technology* **1997**, *40*, 341–352.
150. Yudin, I. K.; Nikolaenko, G. L.; Gorodetskii, E. E.; Kosov, V. I.; Melikyan, V. R.; Markhashov, E. L.; Frot, D.; Briolant, Y. *Journal of Petroleum Science & Engineering* **1998**, *20*, 297–301.
151. Yudin, I. K.; Nikolaenko, G. L.; Gorodetskii, E. E.; Markhashov, E. L.; Frot, D.; Briolant, Y.; Agayan, V. A.; Anisimov, M. A. *Petroleum Science and Technology* **1998**, *16*, 395–414.
152. McHale, J. L. *Molecular Spectroscopy*; 1st ed.; Prentice-Hall Inc.: Upper Saddle River, New Jersey, 1999.
153. Atkins, P. W. *Physical Chemistry*; 5th ed.; Oxford University Press: Oxford, 1994.
154. Brust, M.; Kiely, C. J. *Colloids and Surfaces a – Physicochemical and Engineering Aspects* **2002**, *202*, 175–186.
155. Kreibig, U.; Genzel, L. *Surface Science* **1985**, *156*, 678–700.
156. Franzen, S.; Folmer, J. C. W.; Glomm, W. R.; O'Neal, R. *Journal of Physical Chemistry A* **2002**, *106*, 6533–6540.
157. Xie, H.; Tkachenko, A. G.; Glomm, W. R.; Ryan, J. A.; Brennaman, M. K.; Papanikolas, J. M.; Franzen, S.; Feldheim, D. L. *Analytical Chemistry* **2003**, *75*, 5797–5805.
158. Chandrasekharan, N.; Kamat, P. V.; Hu, J. Q.; Jones, G. *Journal of Physical Chemistry B* **2000**, *104*, 11103–11109.
159. Dawson, A.; Kamat, P. V. *Journal of Physical Chemistry B* **2001**, *105*, 960–966.
160. Elghanian, R.; Storhoff, J. J.; Mucic, R. C.; Letsinger, R. L.; Mirkin, C. A. *Science* **1997**, *277*, 1078–1081.
161. Mirkin, C. A.; Letsinger, R. L.; Mucic, R. C.; Storhoff, J. J. *Nature* **1996**, *382*, 607–609.
162. Mirkin, C. A. *Inorganic Chemistry* **2000**, *39*, 2258–2272.
163. Mirkin, C. A. *Mrs Bulletin* **2000**, *25*, 43–54.
164. Reynolds, R. A.; Mirkin, C. A.; Letsinger, R. L. *Journal of the American Chemical Society* **2000**, *122*, 3795–3796.
165. Storhoff, J. J.; Elghanian, R.; Mucic, R. C.; Mirkin, C. A.; Letsinger, R. L. *Journal of the American Chemical Society* **1998**, *120*, 1959–1964.
166. Storhoff, J. J.; Mirkin, C. A. *Chemical Reviews* **1999**, *99*, 1849–1862.
167. Storhoff, J. J.; Lazarides, A. A.; Mucic, R. C.; Mirkin, C. A.; Letsinger, R. L.; Schatz, G. C. *Journal of the American Chemical Society* **2000**, *122*, 4640–4650.
168. Bronstein, L. M. In *Colloid Chemistry 1*; Springer-Verlag Berlin: Berlin, 2003; Vol. 226, pp 55–89.
169. Chen, W.; Cai, W. P.; Liang, C. H.; Zhang, L. D. *Materials Research Bulletin* **2001**, *36*, 335–342.
170. Chen, W.; Cai, W. P.; Zhang, L.; Wang, G. Z.; Zhang, L. D. *Journal of Colloid and Interface Science* **2001**, *238*, 291–295.
171. Chen, W.; Cai, W. P.; Wang, G. Z.; Zhang, L. *Applied Surface Science* **2001**, *174*, 51–54.
172. Chen, W.; Cai, W. P.; Zhang, Z. P.; Zhang, L. *Chemistry Letters* **2001**, 152–153.
173. Glomm, W. R.; Øye, G.; Walmsley, J.; Sjöblom, J. *Journal of Dispersion Science and Technology* **2005**, *26*, 729–744.
174. Alvarez, M. M.; Khoury, J. T.; Schaaff, T. G.; Shafigullin, M. N.; Vezmar, I.; Whetten, R. L. *Journal of Physical Chemistry B* **1997**, *101*, 3706–3712.

175. Cury, L. A.; Ladeira, L. O.; Righi, A. *Synthetic Metals* **2003**, *139*, 283–286.
176. Chumanov, G.; Sokolov, K.; Cotton, T. M. *Journal of Physical Chemistry* **1996**, *100*, 5166–5168.
177. Glomm, W. R.; Anthireya, S. J.; Brennaman, M. K.; Papanikolas, J. M.; Franzen, S. *J. Phys. Chem* (Submitted).
178. Lakowicz, J. R. *Principles of Fluorescence Spectroscopy*; Kluwer Academic/Plenum: New York, 1999.
179. Ford, G. W.; Weber, W. H. *Phys. Rep.* **1984**, *113*, 195–287.
180. Chance, R. R.; Prock, A.; Silbey, R. *Adv. Chem. Phys.* **1978**, *37*, 1–65.
181. Barnes, W. L. *J. Mod. Opt.* **1998**, *45*, 661–699.
182. Gersten, J. I.; Nitzan, A. *Surf. Sci.* **1985**, *158*, 165–189.
183. Kummerlen, J.; Leitner, A.; Brunner, H.; Aussenegg, F. R.; Wokaun, A. *Mol. Phys.* **1993**, *80*, 1031–1046.
184. Gersten, J. I.; Nitzan, A. *J. Chem. Phys.* **1981**, *75*, 1139–1152.
185. Gersten, J. I.; Nitzan, A. *Chem. Phys. Lett.* **1984**, *104*, 31–37.
186. Hua, X. M.; Gersten, J. I.; Nitzan, A. *J. Chem. Phys.* **1985**, *83*, 3650–3659.
187. Lakowicz, J. R. *Analytical Biochemistry* **2001**, *298*, 1–24.
188. Lakowicz, J. R.; Shen, Y.; D'Auria, S.; Malicka, J.; Fang, J.; Gryczynski, Z.; Gryczynski, I. *Analytical Biochemistry* **2002**, *301*, 261–277.
189. Strickler, S. J.; Berg, R. A. *J. Chem. Phys.* **1962**, *37*, 814–822.
190. Fleming, C. N.; Maxwell, K. A.; DeSimone, J. M.; Meyer, T. J.; Papanikolas, J. M. *Journal of the American Chemical Society* **2001**, *123*, 10336–10347.
191. Fleming, C. N.; Dupray, L. M.; Papanikolas, J. M.; Meyer, T. J. *Journal of Physical Chemistry A* **2002**, *106*, 2328–2334.
192. Brennaman, M. K.; Alstrum-Acevedo, J. H.; Fleming, C. N.; Jang, P.; Meyer, T. J.; Papanikolas, J. M. *Journal of the American Chemical Society* **2002**, *124*, 15094–15098.
193. Kleverlaan, C. J.; Indelli, M. T.; Bignozzi, C. A.; Pavanin, L.; Scandola, F.; Hasselman, G. M.; Meyer, G. J. *Journal of the American Chemical Society* **2000**, *122*, 2840–2849.
194. Bignozzi, C. A.; Schoonover, J. R.; Scandola, F. In *Molecular Level Artificial Photosynthetic Materials*; John Wiley & Sons Inc: New York, 1997; Vol. 44, pp 1–95.
195. Bignozzi, C. A.; Argazzi, R.; Kleverlaan, C. J. *Chemical Society Reviews* **2000**, *29*, 87–96.
196. Gust, D.; Moore, T. A.; Moore, A. L. *Accounts of Chemical Research* **2001**, *34*, 40–48.
197. Hissler, M.; McGarrah, J. E.; Connick, W. B.; Geiger, D. K.; Cummings, S. D.; Eisenberg, R. *Coordination Chemistry Reviews* **2000**, *208*, 115–137.
198. Slate, C. A.; Striplin, D. R.; Moss, J. A.; Chen, P. Y.; Erickson, B. W.; Meyer, T. J. *Journal of the American Chemical Society* **1998**, *120*, 4885–4886.
199. Tkachenko, A. G.; Xie, H.; Ryan, J.; Glomm, W. R.; Franzen, S.; Feldheim, D. L. In *Bionanotechnology Protocols*; Rosenthal, S., Wright, D., Eds.; Humana Press, Inc.: New York, **2005**, pp. 85–99.
200. Tkachenko, A. G.; Xie, H.; Liu, Y. L.; Coleman, D.; Ryan, J.; Glomm, W. R.; Shipton, M. K.; Franzen, S.; Feldheim, D. L. *Bioconjugate Chemistry* **2004**, *15*, 482–490.
201. Hiemenz, P. C.; Rajagopalan, R. *Principles of Colloid and Surface Chemistry*; 3rd ed.; Marcel Dekker, Inc: New York, 1997.

13 Production Issues of Acidic Petroleum Crude Oils

Christian Hurtevent, Guy Rousseau, Maurice Bourrel, and Benjamin Brocart

CONTENTS

13.1 INTRODUCTION

13.1.1 Acidic Oil Origin and Related Production Problems

Many oil fields recently discovered in deep offshore in Angola, Congo, and Nigeria produce oil with a high acid content. However, this observation is not limited to West Africa; some fields in the North Sea and in Venezuela have the same characteristics (Table 13.1).

Naphthenic acids in petroleum are considered to be a class of biological markers [1–3] closely linked to the degree of biodegradation of the fields. Naphthenic acids are predominantly found in immature, biodegraded, heavy crudes [4]. The alteration of petroleum by living microorganisms, which may occur, for example, when meteoric water infiltrates an accumulation [5], significantly increases the density of the crude and, at the same time, decreases the level of paraffinic components. It is highly probable, therefore, that acidic crudes contain low levels of paraffins and have higher densities than nonacidic crudes. The correlation is quite good for all wells on the same field or for all fields located in the same block (Table 13.2).

In oil and gas production, acidic crudes cause scale to form inside tubing or in surface installations. The scale is often a mixture of calcium soaps associated with other minerals such

TABLE 13.1
Total Acid Numbers (TANs) of Crudes from Different Fields

Crude	Country	TAN (mg KOH/g)
CAMELIA	Angola	1.90
DALIA 2	Angola	2.37
ORQUIDEA	Angola	3.73
MOHO	Congo	0.87
BILONDO	Congo	1.80
UKOT	Nigeria	1.01
AFIA	Nigeria	1.20
IME	Nigeria	2.08
ALBA	North Sea	1.83
CAPTAIN	North Sea	2.32
HEIDRUN	North Sea	2.60
LAGOTRECO	Venezuela	1.18
MEREY	Venezuela	1.24
LAGUNA	Venezuela	4.10

TABLE 13.2
Paraffin Level, API Degree, and TAN Values for Acidic Crudes from Angola Block 17

Crude	%Paraffin	°API	TAN (mg KOH/g)
CAMELIA	0.49	23.02	1.90
DALIA 3	1.33	23.68	1.37
TULIPA	1.61	24.14	1.11
LIRIO	3.47	32.38	0.54
GIRASSOL	5.34	30.60	0.38
CRAVO	5.90	32.88	0.35

as calcium carbonate, clays, iron carbonate, etc. Such type of scales have been found in several countries, for example:

- Nigeria, on the Afia field
- Indonesia, on the Attaka field
- Great Britain, on the Blake field
- Norway, on the Heidrun field
- Angola, on the Kuito field
- China, on the EDC field
- Cameroon, on the Kita and Asoma fields

It has been discovered only recently [6] that such calcium soap scale is composed primarily of the calcium salt of a tetra-acid ($M_W \sim 1230$). Identified by Statoil and named ARN, this acid is the main component of all the scale already analyzed.

In the case of the Kuito field in Angola, 135 tons of scale had to be removed mechanically from two separator strings, the cleaning operation took 1 month, incurring considerable expenses and production losses.

On the Heidrun field in Norway, before an appropriate inhibitor was formulated and injected, 500 kg of scale a day were generated in the separators. In 1985, when production started on Kita in Cameroon, 80 m^3 of scale comprising a mixture of napthenates and calcium carbonate had to be removed from the electric desalters.

Among the fields that have experienced problems due to the formation of calcium soap scale, most of them have a high total acid number (TAN) relative to their API degree (Figure 13.1).

This particular characteristic can also be found in the Usan reservoir in Nigeria. In this case, the high TAN relative to the API can be explained by the multiple hydrocarbon formation and migration phases in the reservoir. These migration and formation phases at the scale of reservoir geology can be explained simplistically as a mixture of several oils with different characteristics.

In reservoirs like Usan, biological tracers have also been found, indicating high biodegradation, whereas elsewhere, a high API degree and a large quantity of paraffins were recorded, indicating at the scale of reservoir geology that this reservoir contains at least two different hydrocarbons. Many types of crude producing calcium soap scale fit into the latter case.

Depending on the nature of the naphthenic acids present in the oils, the formation of stable emulsions, associated with the very strong surfactant capacity of the naphthenate group, may also be observed in the separators. These emulsions, which cause separation problems, usually generate discharge waters having a high hydrocarbon and naphthenate content. Here, the naphthenic acids

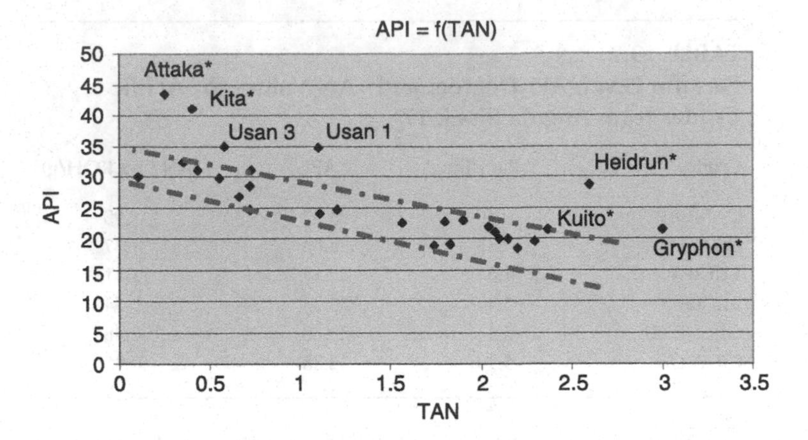

FIGURE 13.1 Relation between TAN and API for different fields. * These fields gave calcium soap scales.

responsible for such emulsions are not the ARN, responsible for scale, but rather monoacids with lower molecular weights (~200 to 400).

Refineries are concerned by the potential corrosivity of acid crudes at high temperatures. It is for this reason that these crudes are downgraded on the basis of their TAN value. The downgrade may be estimated using the following formula:

$$\text{Downgrade in \$/Bbls} = (\text{TAN} - 0.6)/2$$

where TAN is expressed in mg KOH/g.

13.1.2 Challenges Associated with Acidic Crudes

In the context of a new development project or assistance for an operator who is facing problems of the above type, the following points must be addressed.

13.1.2.1 Prediction of the Risks of Scale and/or Emulsion

For the producer, this essentially means being able to predict the risk of formation of emulsion and/or scale in the pipes or surface installations. The tools developed so far are insufficient and the models commercially available in this domain do not apply, since they imply a balance between the aqueous phase and the organic phase, and also since the parameters to be taken into account have not all been identified.

The acidity of a crude by itself is not a sufficient criterion; some weakly acidic oils in Cameroon or in Indonesia [7] may form stable emulsions while other highly acidic crudes can be treated with no problem. It is the actual structure of the naphthenic acids that may explain these differences in behaviors, whence the importance of characterizing the naphthenic acids of a crude.

Identifying the entire set of reactions that occur in each phase and at the water–oil contact, and characterizing the naphthenic acids, particularly their surfactant capacity, are the two first steps that allow us to model the balances and therefore to more accurately evaluate the potential risks associated with the treatment of acid oils.

Moreover, although it has been possible to identify the ARN in many deposits, it is still very difficult to identify this molecule in oils. Initial tests conducted by adding specific amounts of ARN to an oil showed that the methods of analysis described later on in this paper have a detection sensitivity of less than 50 ppm. This molecule is therefore present in oils at concentrations of a few ppm and must be concentrated (through deposit preparation, for example) to be identified. Recently, we have been able to find it in sludge prepared in our laboratory (unpublished result).

13.1.2.2 Prevention of Deposits and/or Emulsions

In some cases, like Kuito, modifying the separation process was sufficient to substantially decrease the mass of scale.

Some dispersants produce good results, especially with the Heidrun crude.

13.1.3 Theoretical Background on Calco-Carbonic and Naphthenic Equilibria

13.1.3.1 Calco-Carbonic Equilibrium

Reservoir waters are naturally saturated with carbon dioxide which produces carbonic acid, H_2CO_3. At bottom hole conditions, water is at equilibrium with the rock and the gas of the reservoir. On the other hand, during production, water is drained to the surface and undergoes significant pressure and temperature variations in the tubing, downstream from the choke at the well head, in the exchangers or in the separators.

The successive pressure drops lead to degassing of the carbon dioxide with an increase in the pH value of the produced water. This pH increase induces the formation and precipitation of calcium carbonate which has very low solubility, thereby generating undesirable deposits which plug the chokes or fill the separators. The chemical relation between carbon dioxide, pH value, and calcite solubility can be expressed as follows:

$$\mathbf{CO_2 + H_2O \Leftrightarrow HCO_3^- + H^+}$$

$$\mathbf{Ca^{2+} + HCO_3^- \Leftrightarrow CaCO_3 + H^+}$$

13.1.3.2 Naphthenate Equilibrium

The so-called "naphthenic acids" are mainly carboxylic acids with saturated cyclic structures, represented by a general formula $C_nH_{2n-z}O_2$, where n indicates the carbon number and z specifies a homologous series, from 0 for saturated acyclic acids to 8 in tetracyclic acids [8].

Naphthenic acids of low molecular weight contain alkylated cyclopentane carboxylic acids with smaller amounts of cyclohexane derivatives:

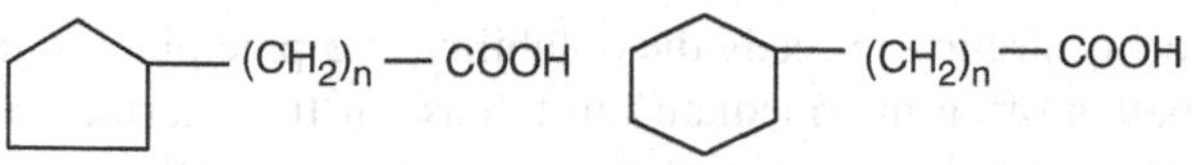

(cyclopentane)—$(CH_2)_n$—COOH   (cyclohexane)—$(CH_2)_n$—COOH

One thousand five hundred acids were identified in a single California crude [9] with boiling points ranging from 250 to 350 °C. They are fully soluble in organic solvents and fairly insoluble

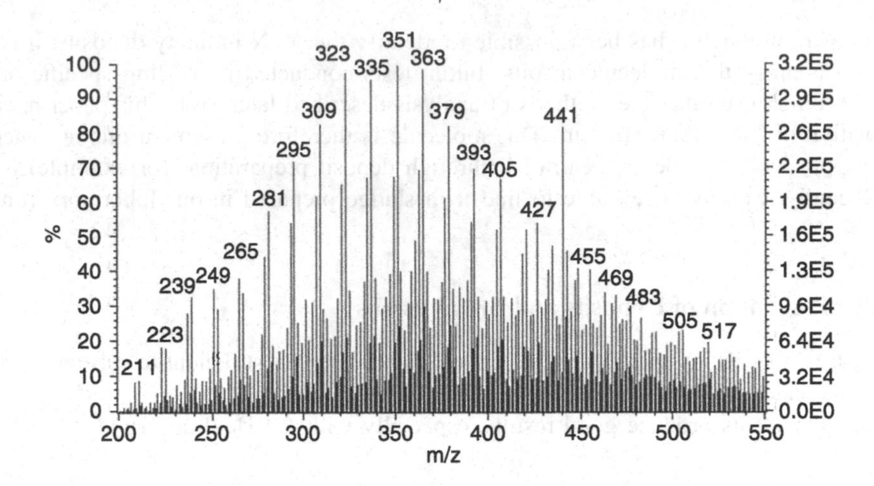

FIGURE 13.2 Dalia 3 acid characterization.

in water. Figure 13.2 shows the results of mass spectrometry analysis of the naphthenic acids extracted from Dalia 3 by anionic resin.

The naphthenic acids of the crude will follow acid–base equilibrium reactions in a multiphase system, which imply (1) acid partitioning between oil and water phases, characterized by a partition coefficient, and (2) acid dissociation, described by a dissociation constant (pKa). These phenomena are incorporated in a model presented in Section 13.4.

A number of studies have attempted to determine the pKa [10,11] of the various families of naphthenic acids – a complicated task in view of the number of different acids present in a given crude. However, this pKa value does not take into account the very low solubility of these organic acids in the water phase. At a given pH, the total amount of dissociated acids depends not only on the pKa value, but also on their partitioning coefficient. The higher the pH value, the heavier the naphthenates formed, even if the pKa values of light and heavy acids are close. Therefore, the potential capacity for a naphthenic acid to create stable emulsions depends on its partitioning coefficient, which may be correlated to its molecular weight. In a report published in 1969, Seifert and Howells state that the average molecular weights of interfacially active acids range from 300 to 400 [35].

The cations in reservoir water can react with $RCOO^-$ naphthenate groups to form salts, commonly named soaps, essentially sodium and calcium naphthenates, which dissolve in either water or oil, depending on their affinity for one phase or the other.

Sodium naphthenates having a low molecular weight tend to dissolve into the water phase. Calcium, as a divalent cation, is associated with two naphthenate groups, yielding lipophilic calcium naphthenates partitioning preferentially into the oil phase or adsorbing at the oil/water interface [12,13].

When the soap concentration exceeds the solubility, the precipitation of a solid deposit or in some cases the formation of an intermediate third phase at the interface between the oil and the water phases is observed. Sludge containing mainly sodium naphthenates has been observed on the Serang field [7] in Indonesia.

Contrary to field observations, in lab experiments we never observed solid deposits. We have only observed the formation of sludge of sodium and/or calcium naphthenates, possibly due to the formation of mesophases [14–16]. Interestingly, ARN has been found at least in one case, in sludges formed from the Usan crude.

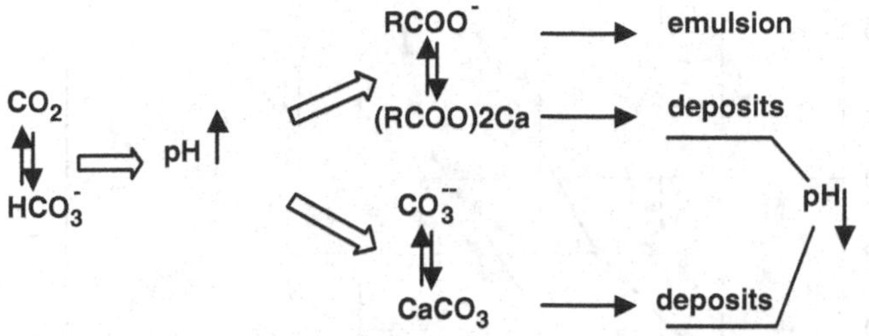

The release of CO_2 due to a shift in the calco-carbonic balance causes the pH value of the reservoir water to increase. This is followed by a competition between the formation of sodium or calcium naphthenate and the precipitation of calcium carbonate. The result is the potential formation of calcium carbonate deposit and/or emulsion stabilized by sodium/calcium naphthenates, and/or ARN deposits associated with a decrease in the pH value, if the water is not buffered by bicarbonate.

In the absence of bicarbonate, an increase in the pH value due to CO_2 degassing produces naphthenates and, at the same time, the dissociation of the naphthenic acids releases protons which act against the increase in pH, and thus impedes the formation of naphthenates. In the opposite case, naphthenates are produced as long as some bicarbonate is available to buffer the medium.

Thus, the species which mainly contribute to naphthenate formation are $RCOO^-$, Na^+, Ca^{2+}, and HCO_3^-.

From this, it may be assumed that the following criteria should be taken into account when predicting the formation of stable emulsion and/or scale in the processing of an acidic crude:

- For water, its pH value at process conditions and the level of bicarbonate and calcium content at reservoir conditions
- For oil, the amount of available naphthenic acids

Carbonates produce scale, whereas naphthenates can form emulsions at the water–oil contact, and ARN can form scale. None of the models on the market factor-in this potential competition in acidic crudes. As a result, the precipitable quantities of calcium carbonate are systematically overestimated.

13.1.4 Emulsion Stabilization by Naphthenates

The stability of water in crude oil emulsions is described in Section 13.3. The general conclusion of these tests is that the emulsion stability is strongly dependent on pH.

On a given oil field, the stabilization of emulsions by naphthenates depends on the capacity of naphthenic acids to form naphthenates at the lowest possible pH. Effectively, in the treatment of acidic oils, the pH increases throughout the entire process when the pressure decreases.

We have looked at the quantity of acid likely to react with the aqueous phase as a function of pH, for several crude oils. The measurements were carried out on a mixture of 90% oil and 10% water containing increasing quantities of caustic soda. The results are given as final pH, after the transfer of acids into the water as naphthenates, versus the quantity of naphthenates formed.

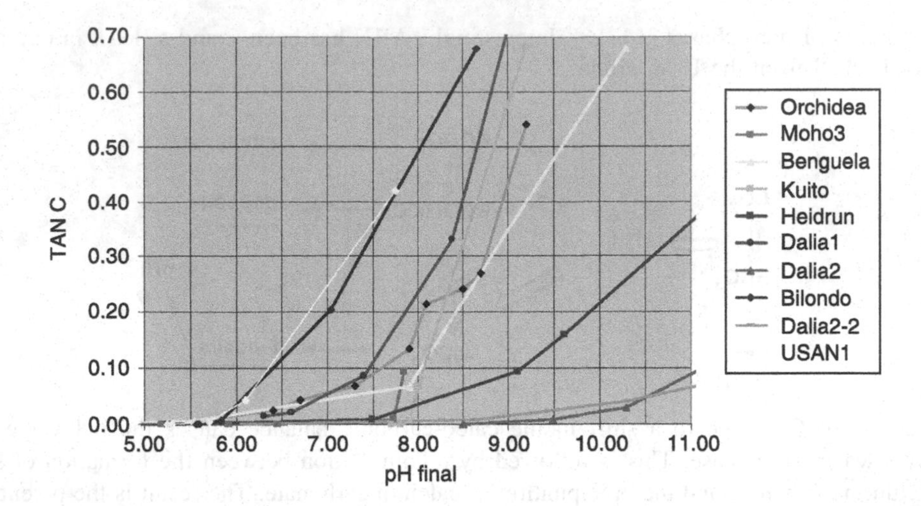

FIGURE 13.3 Naphthenate formation as a function of pH.

These naphthenates are expressed as the TAN (TANc) equivalent to the quantity of naphthenic acids that have come out of the oil (see Section 13.3 for more details).

The curves in Figure 13.3 show that some crude oils form naphthenates from pH as low as 6, whereas others only begin to form them at a pH value greater than 8 to 9. As naphthenates can be powerful surfactants, the risk of forming emulsions is especially greater whenever the transfer of acids occurs at a low pH.

In a petroleum installation, the gas associated with the produced oil generally contains a given proportion of CO_2. When the pressure decreases, CO_2 degassing promotes an increase in the pH value, simulated by soda addition in Figure 13.3. The formation of naphthenates will increase as the pressure in the installations decreases. One of the important ideas generated from this observation is that emulsion problems can be reduced if the water is separated from the oil at the highest possible pressure. As naphthenates are soluble in water, they will also have a strong tendency to carry oil into the water and therefore cause significant downgrading of the produced waters.

These naphthenates dissolved in the water are not included in the 30 to 40 ppm stated in the standards for determining hydrocarbon levels in produced waters. Effectively, naphthenates, being polar compounds, are retained by the silica generally used in methods for determining hydrocarbon levels. These compounds, which are almost nonexistent in water in a process occurring under pressure, may represent several hundreds of ppm when a process at atmospheric pressure is used. Though these compounds are not counted as hydrocarbons, the use of a process under pressure for oil/water separation is highly recommended for strictly environmental reasons. Certain regulations, onshore in particular, impose threshold values of total organic carbon (TOC). This property of crude oils concerns not only those crude oils qualified as acidic, but also all crude oils having a significant TAN. The same observations have been made on crude oils having a TAN of 0.3. Transfer of naphthenic acids from the oil phase as naphthenates in the water phase is described in Section 13.4 below, on modeling.

13.2 CHARACTERIZATION OF NAPHTHENIC ACIDS IN CRUDE OILS

The difference in behavior observed during the processing of acidic crudes mainly arises from the difference of the nature of the acids; which is why it is expected that the isolation and characterization of naphthenic acids could be of prime importance for the prediction of the associated risks. The standard methods used to analyze naphthenic acids are discussed hereafter with a special interest to mass spectrometry.

13.2.1 Naphthenic Acids Quantification in Crude Oils

There are several methods for the determination of naphthenic acid content, which, unfortunately, yield different results.

13.2.1.1 ASTM Procedure for Total Acid Number

Assessment and predictability of corrosion by naphthenic acids are generally based on the determination of total acid number (TAN) and total sulfur in the crude oil [17]. Acid numbers are expressed numerically as mg KOH required to neutralize the acidity in 1 g of sample. However, the rules of thumb used in most refineries to predict corrosion based on TAN, sulfur species, based on previous experiences indicate a limited applicability of these characteristics since it has been discovered that crudes having a high TAN could be noncorrosive. Thus, high vacuum units, alloyed to deal only with sulfidation, have successfully processed feed stocks having a TAN of 4 mg KOH/g for more than 37 years [18]. More recent research has begun to highlight deficiencies in relying upon this method to quantify the oil acidity; ASTM D974 is a colorimetric titration method and has a reproducibility of 15%. ASTM D664 is a potentiometric titration method with reproducibility of 20% depending on the end point, type of oil, and titration mode. Inorganic acids, esters, phenolic compounds, salts, and additives such as inhibitors and detergents interfere with both methods. In fact, all compounds that react with potassium hydroxide (including hydrogen sulfide) are counted as acids; this is the case with calcium or iron naphthenates. Consequently, TAN is no longer regarded as such a reliable indicator of acidity [19].

13.2.1.2 FT-IR Spectroscopy

The measure of the absorbance at 1710 cm^{-1} by FT-IR spectroscopy is another method used to quantify specifically the acidity of crude oils [20]. An example of spectrum is shown in Figure 13.4.

Naphthenates can also be evaluated using the same method by quantification of the ester carbonyl band near 1600 cm^{-1}. The results are expressed respectively as neutralization and saponification values. The procedure requires calibration with known compounds in an oil representative of the samples to be analyzed.

We found a good correlation between the concentration of acids determined by FT-IR and TAN values: the higher the absorbance at 1710 cm^{-1}, the higher the TAN. This is illustrated in Figure 13.5 where the surface of the peak measured at the specified wavelength is plotted versus TAN for several crude oils. The FT-IR method is more reliable than TAN regarding the quantification of acidity, but is not usable for crudes having a low acidity.

In MOBIL analytical method 1463-89, acids are first separated from the crude oil sample using an amine bonded silica gel Sep Pak cartridge. Then the naphthenic acids are eluted from the cartridge with acetone and methanol and finally analyzed by IR spectroscopy.

Unfortunately, the instrument is calibrated with a purchased mixture of crude naphthenic acids, which are likely not to have the same composition and molecular weight as the sample. Thus, the weight percent can be used for comparison purpose only.

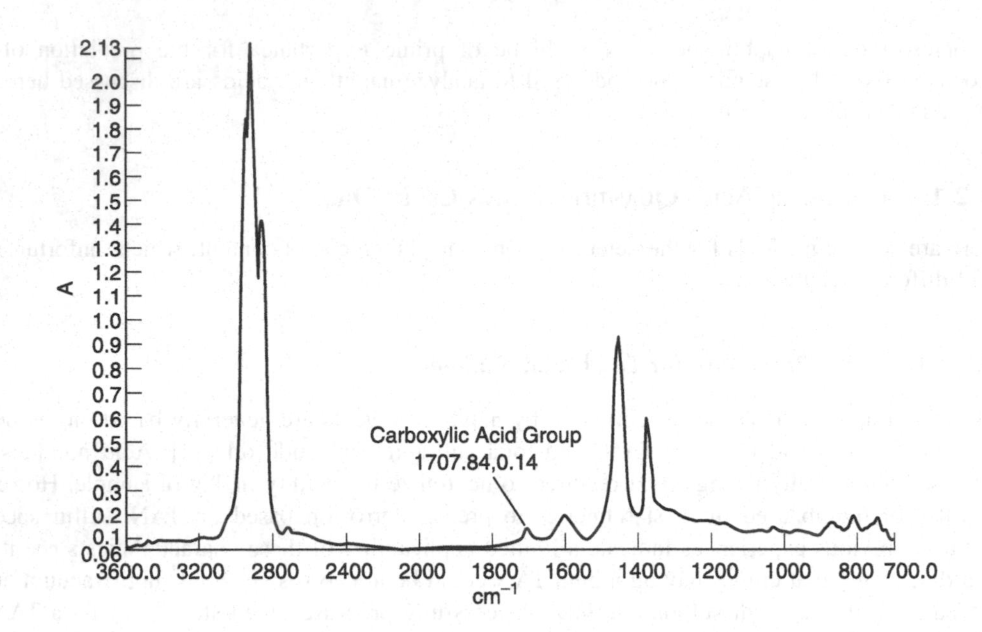

FIGURE 13.4 FT-IR spectrum of crude oil.

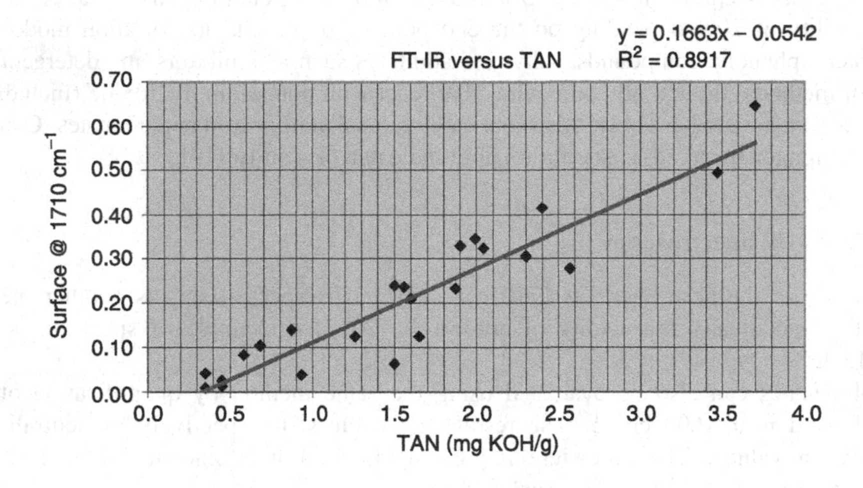

FIGURE 13.5 Comparison between TAN values and acidity quantified by FT-IR.

Another variation of the chromatographic method is the naphthenic acid number (NAN) or naphthenic acid titration (NAT) whereby the sample is extracted by chromatography and then titrated per ASTM D664 [19].

13.2.1.3 Gas Chromatography

Naphthenic acids comprise a highly complex mixture containing hundreds of compounds that are impossible to separate into individual components, even by high resolution chromatography.

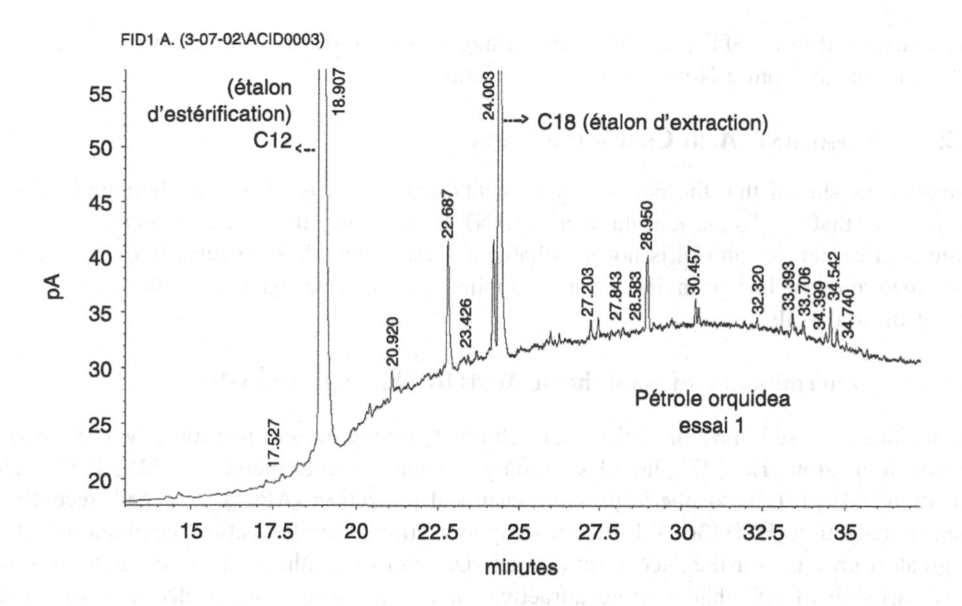

FIGURE 13.6 GC spectrum of esterified naphthenic acids after extraction from Orquidea crude oil.

The most described technique is based on a liquid–liquid or solid–liquid extraction of naphthenic acids from the crude oil sample followed by the derivatization of the isolated fractions as methyl esters [21,22]. Acids are analyzed by gas chromatography with a cold on-column injector and a flame ionization detector using HT-5 capillary column.

The percentage of acids in the crude is quantified by measuring the total GC-FID chromatogram areas above the baseline of the blank (hexane) analysis through the internal standard (C12), assuming a response factor of 1. Chromatograms of the extracted and esterified acids from crudes present a strong background noise, as illustrated in Figure 13.6. The peaks seen at approximately 19 and 25 min correspond, respectively, to the lauric and octadecanoic acids used as internal and extraction standards [9].

13.2.2 Naphthenic Acid Extraction from Crude Oil

One of the key points in naphthenic acid characterization is to use the suitable procedure to extract all the acids from the crude. The different methods are grouped into the categories of: solid–liquid extraction, liquid–liquid extraction, and chromatography.

Acid-IER (ion exchange resin) is the preferred method; carboxylic acids in the crude are selectively extracted by an alkaline resin. They are adsorbed onto the resin as naphthenates $RCOO^-$ that are later desorbed during the regeneration of the resin by sodium hydroxide. The acids are recovered after acidification and dichloromethane back-extraction.

AMBERLYST® IRA 900 or QAE Sephadex A-25, strong negative ion exchangers, are both suitable for such an extraction. The adsorption of the acids on the resin can be done either in a batch process or by flowing the crude through a column. The amount of needed resin roughly corresponds to 10 equivalents of base per equivalent of acid determined by TAN measurement.

The extraction yield can be very different depending on the used procedure or on the nature of the crude. Results from 60% to 90% have been found with the same crude, just by running the extraction at a different temperature or for a longer time. A suitable procedure has been optimized and is described in an SPE paper [23]. According to STATOIL, it would allow recovery of up to 97% of the acids from a Norwegian crude oil sample.

13.2.3 Naphthenic Acid Characterization

Research has shown that the corrosivity of naphthenic acids is related to their molecular mass [24–26] and that the "total acid number" (TAN), traditionally used as an indicator of the naphthenic acid content of an oil, is not as reliable as first believed. With regards to these concerns, mass spectrometry has been increasingly applied to the investigation of the naphthenic acid content of crude oils.

13.2.3.1 Determination of Naphthenic Acids by Mass Spectrometry

The techniques used have included: gas chromatography mass spectrometry (GC-MS) [27], electron ionization (EI) [27], liquid secondary ion mass spectrometry (LSIMS) [28], chemical ionization (CI) [29], atmospheric pressure chemical ionization (APCI) [29], and, recently, electrospray ionization (ESI) [30–32]. Electrospray ionization is the ionization technique which holds the greatest promise for the successful characterization of naphthenic acids in crude oil samples.

A variation of ESI that is more attractive for this purpose is nano electrospray ionization, also known as "nanospray" [33]. Nanospray entails the use of a fine glass needle, coated in a conductive layer such as gold and/or palladium, into which the sample is loaded. No pumping is required and the ions are formed and extracted through the use of a strong electrical field alone.

Fast atom bombardment mass spectroscopy (FAB-MS) [2] uses the negative ion detection mode; it detects the $(M - 1)^-$ ion in carboxylic acids, including high molecular species with no fragmentation. FAB-MS determines the molecular weight distribution of the naphthenic acids extracted from the oil. Component information, based on carbon number and acid group type distributions, is obtained clearly from the FAB spectrum. The results show that this group type distribution provides a fingerprint that can be a valuable tool to predict the risks associated with crude oils processing.

FAB-MS gives all the masses of the acids extracted from the crude ranging from 150 to 650, centered on 350. However, acids having the same nominal molecular weights but different formulas, such as $C_{14}H_{26}O_2$ and $C_{15}H_{14}O_2$, are not completely resolved under the operating instrument resolution. Typical LC/MS ESI negative mode spectra are presented in Figures 13.7 and 13.8.

13.2.3.2 Determination of Naphthenate Deposits by Mass Spectrometry

Recently, the STATOIL Company has identified the dominating naphthenic acid in naphthenate deposits [6]; it would be a 4-protic acid with a molecular weight in the range of 1227 to 1235 g/mol, called ARN, acid. The presence of that acid would be a universal characteristic of oilfield naphthenate deposits. It is not possible by LC-MS to determine the detailed structure of ARN, but its formula would correspond to $C_{80}H_{142}O_8$. This acid has been detected by FTMS and LC-MS in naphthenate deposits from West Africa dissolved in a blend of one aromatic solvent and acetic acid (Figure 13.9).

The first part of the mass spectrum is representative of the naphthenic acids commonly found in acidic crudes, but one can see a peak with the highest relative intensity at $m/z' = 1230$ corresponding to the acids specific of deposits. The presence of a doubly charged ion at $m/z' = 614.5$ confirms the fact that naphthenate deposits are multiacids. The 4-protic acid claimed by STATOIL could be seen by high resolution MS as it would become possible to see other charged

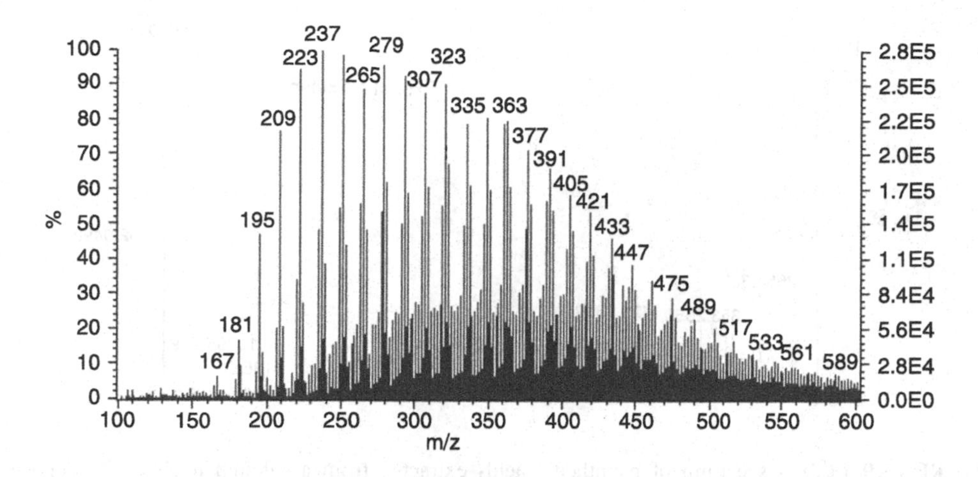

FIGURE 13.7 LC/MS spectrum of Orquidea crude oil from West Africa.

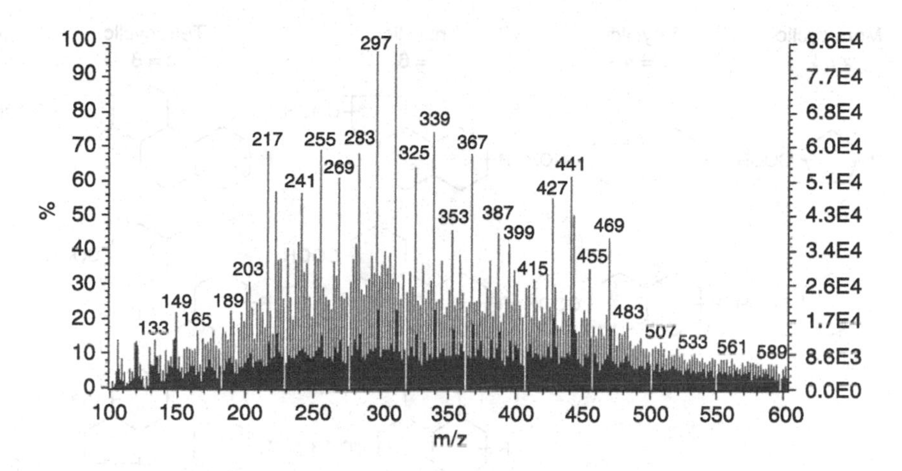

FIGURE 13.8 LC/MS spectrum of Dalia 2 crude oil from West Africa.

ions at m/z′ = 410 and m/z′ = 307.5. However, this new discovery demonstrates the great interest of mass spectrometry to predict a potential risk of naphthenate deposit.

13.2.3.3 Characterization of Naphthenic Acids by FTMS

The most interesting analytical methods for the characterization of naphthenic acids in crude oils are based on the extraction of the acids followed by soft ionization mass spectrometry techniques.

The composite mass spectrum provides an envelope that follows the molecular weight distribution of the mixture. The naphthenic homologs are represented by a general formula $C_nH2_{n-z}O_2$, where n indicates the carbon number and z specifies a homologous series. z is equal to 0 for saturated aliphatic carboxylic acids; it increases by 2 with the increase of hydrogen deficiency resulting from the formation of rings or double bonds in the molecule. Some typical naphthenic structures, where z = 2 to 8, are shown in Figure 13.10.

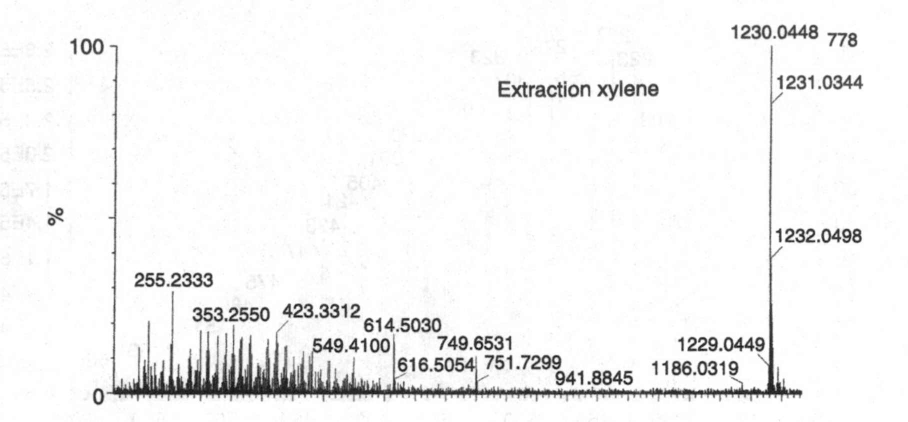

FIGURE 13.9 LC/MS spectrum of naphthenic acids extracted from a calcium naphthenate deposit from West Africa.

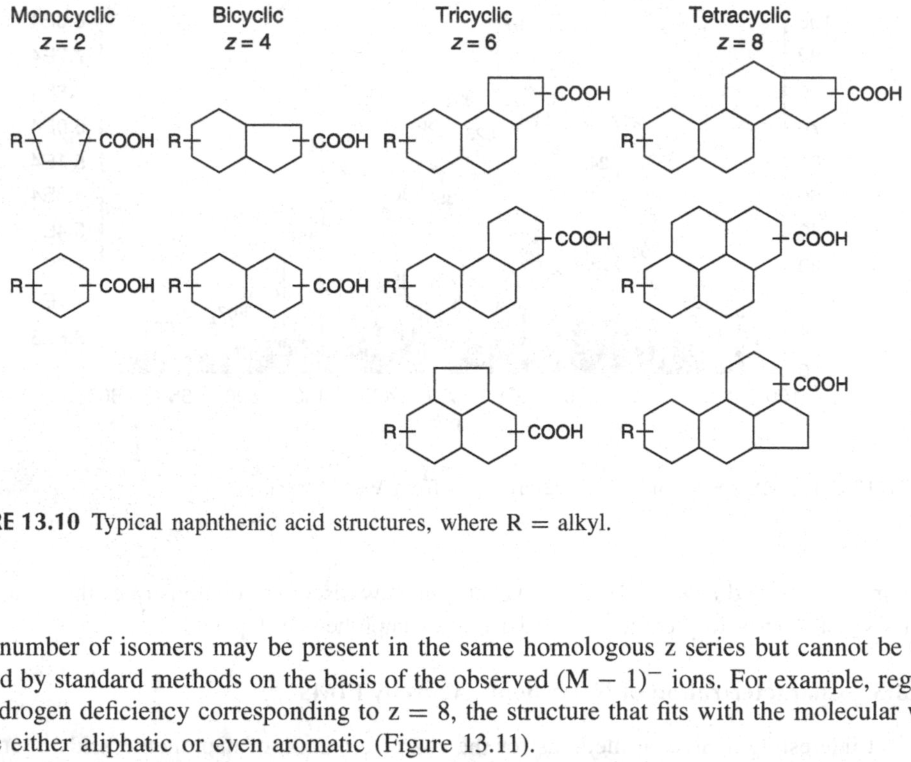

Monocyclic $z = 2$
Bicyclic $z = 4$
Tricyclic $z = 6$
Tetracyclic $z = 8$

FIGURE 13.10 Typical naphthenic acid structures, where R = alkyl.

A number of isomers may be present in the same homologous z series but cannot be distinguished by standard methods on the basis of the observed $(M - 1)^-$ ions. For example, regarding the hydrogen deficiency corresponding to $z = 8$, the structure that fits with the molecular weight can be either aliphatic or even aromatic (Figure 13.11).

Some isomers can be identified by using Fourier transform ion cyclotron resonance mass spectrometry (FT-ICR MS or FTMS) technique, regarding its inherent high mass accuracy and high resolution [34]. A 9.4 T FT-ICR mass spectrometer was used in order to illustrate the advantage of routine high resolution, permitting the distinction to be made between nominally isobaric acid species. Nanospray has been selected as the ionization technique to combine the advantages of reduced sample fragmentation, the inherent sensitivity of nanospray, and minimization of the risks of instrument contamination.

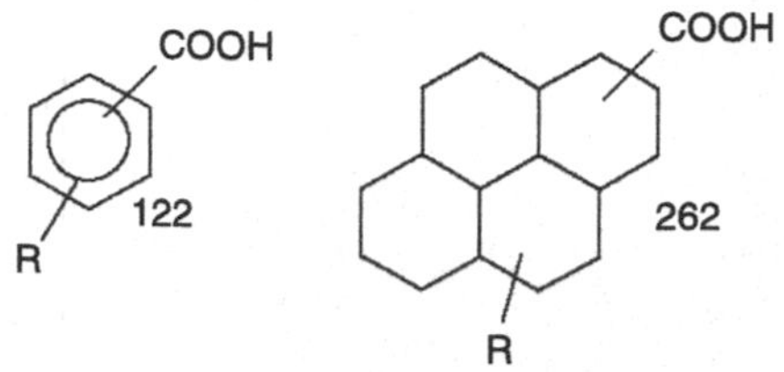

FIGURE 13.11 Possible naphthenic acid structures corresponding to a deficiency of eight hydrogen atoms.

Ten crude oil samples were analyzed using the negative-ion mode to determine the presence of different naphthenic acid species. Spectra obtained in the positive-ion mode contained approximately three times the number of signals, as species other than the acids could be observed, and were thus more complex with regard to the data analysis stage.

Once the spectrum had been acquired, the resulting data processed with XMASS was imported to a spread sheet and sorted into corresponding z homologs and carbon contents, comparing the theoretical m/z ratio with the experimental value. Finally, graphs of the absolute intensities were plotted as a function of carbon content for each of the z homologs.

FTMS gives the exact mass of the extracted acids up to 8 decimals for molecular weights ranging from 300 to 650, centered on 450. Doublets were observed which revealed the presence of naphthenic acids with a high degree of hydrogen deficiency. Such signals would not be observable when using instruments such as quadrupoles and some time-of-flight instruments, which are frequently used for routine analyses. Eighteen families of acids could be identified, including diacids; however, it is not yet possible to determine whether aromatic rings are present simply from the empirical formula. Examples of FTMS spectrum and subsequent analysis of BILONDO crude from West Africa are shown in Figures 13.12 and 13.13.

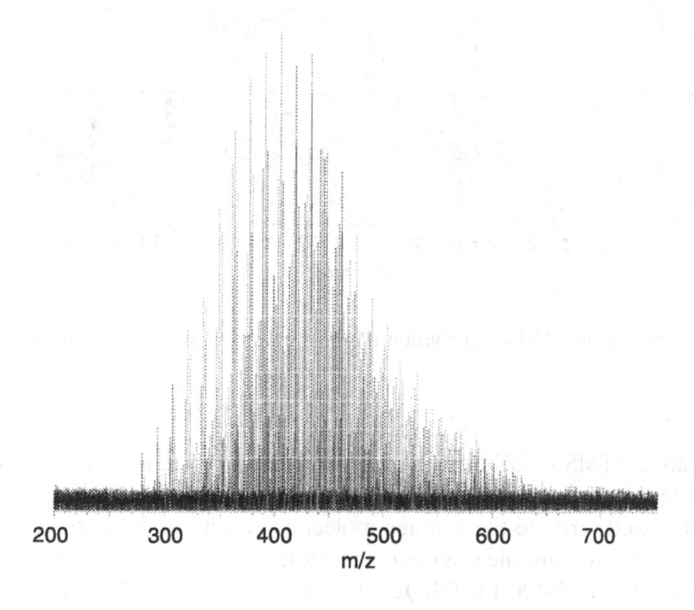

FIGURE 13.12 FTMS spectrum of BILONDO crude.

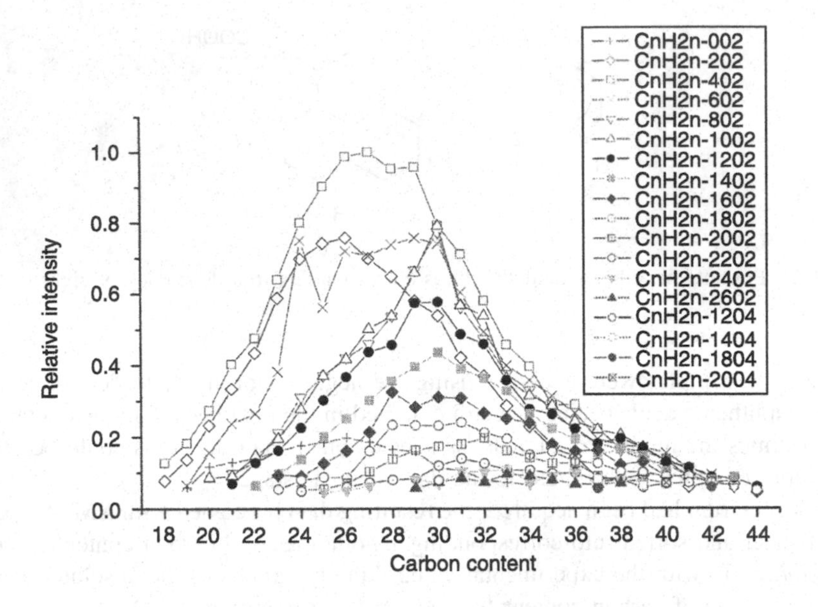

FIGURE 13.13 Distribution of naphthenic acids extracted from BILONDO crude and analyzed by FTMS.

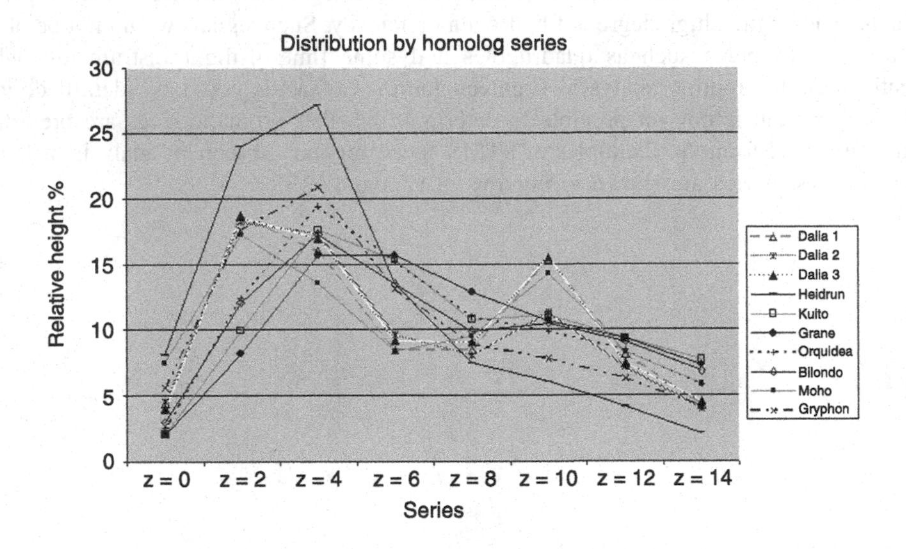

FIGURE 13.14 Comparative FTMS distribution of acids extracted from 10 crude oils.

The comparative FTMS distribution by homolog series of the naphthenic acids extracted from ten different crude oils is illustrated in Figure 13.14 for the first homolog series. From the results, we have been able to determine the average molecular weight of the extracted acids and observe significant differences: for instance, an average molecular weight of 380 for HEIDRUN crude, 440 for DALIA crudes, and 490 for ORQUIDEA crude. Seifert and co-workers claimed that the number average molecular weights of the interfacially active acids range from 300 to 400 [35].

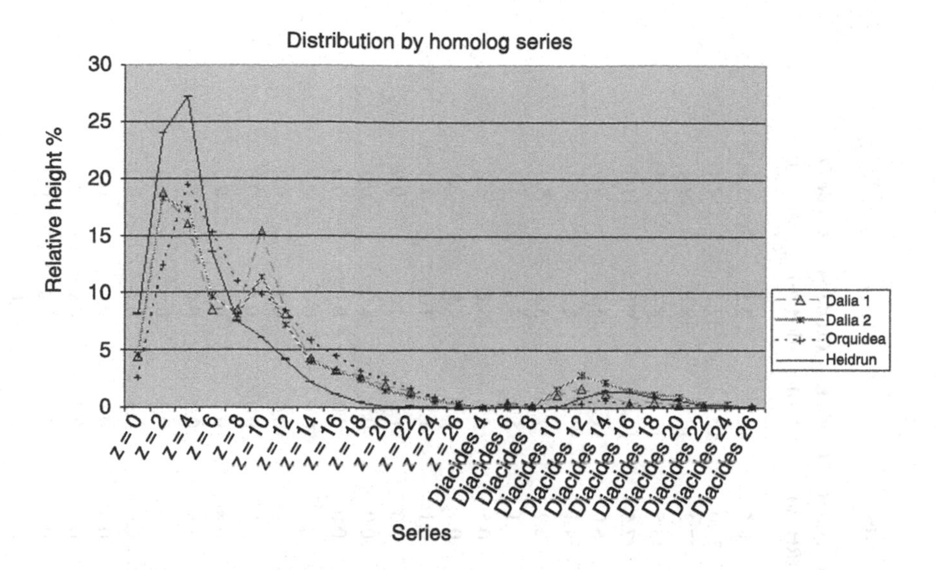

FIGURE 13.15 Distribution of the naphthenic acids determined by FTMS on acid extracted from various crude oils.

Regarding the distribution of the acids, we observed significant differences; all comparative data are given in Table 13.3. They are highlighted in Figure 13.15 for HEIDRUN, DALIA, and ORQUIDEA crudes, as they are deeply investigated in Section 13.3. HEIDRUN crude, known to form naphthenate deposits, has the highest amount of acids from z series 0, 2, and 4. DALIA 1 crude has the highest level of acids from z series 10 whereas DALIA 2 showed the highest level of diacides.

It is not possible at this stage to find the connection between naphthenic acid distribution and the risks associated with crude oils processing; however, by providing information about the composition and range of naphthenic acids present in a crude oil, information is obtained which can be used as a fingerprint that would identify a particular oil sample and link it to a particular oil field. On top of that, MS-MS technique associated with FTMS could be useful to identify aromatics, as well as to determine the nature of ARN responsible for naphthenate deposits.

13.3 PHYSICO-CHEMICAL BEHAVIOR OF ACIDIC CRUDE OILS

13.3.1 Materials and Experimental Techniques

A number of acidic crude oils have been studied, most of them coming from the deep off-shore fields of block 17 operated by Total in Angola. We present here first the results obtained with three crudes representative of the variety of behaviors observed with acidic crude oils [36,37]. Their characteristics are given in Table 13.4. The TAN has been measured according to ASTM D664 for the crudes and their cuts. The distribution of the acid species among the various cuts is different from one crude to another: Dalia 1 acids are concentrated in the 150 to 230 cut, contrary to Orquidea. Dalia 2 exhibits a high acid content in the heaviest cut (550+).

Emulsions were prepared by mixing brine (20 ml) and crude oil (180 ml) with a whisking system in a 500 ml beaker. The rate and application time of the whisk were chosen according to the nature of the oil in order to obtain about 80% of decanted water after 8 h storage at 40 °C

TABLE 13.3
Distribution of Naphthenic Acids Determined by FTMS for Various Crude Oils

Series	Crude B (RH %)	Crude D1 (RH %)	Crude D2 (RH %)	Crude D3 (RH %)	Crude Ga (RH %)	Crude Gy (RH %)	Crude H (RH %)	Crude K (RH %)	Crude M (RH %)	Crude O (RH %)
Z = 0	3.0	4.4	4.5	4.0	2.1	5.6	8.1	2.1	7.5	2.5
Z = 2	12.1	18.7	18.2	18.6	8.2	17.6	24.0	10.0	17.3	12.4
Z = 4	17.3	16.0	17.3	17.0	15.7	20.9	27.2	17.6	13.6	19.4
Z = 6	13.5	8.5	9.6	9.2	15.6	13.1	13.5	15.3	8.4	15.3
Z = 8	9.9	8.5	8.1	9.1	12.9	8.9	7.5	10.8	9.5	11.0
Z = 10	10.5	15.4	11.3	15.5	10.7	7.8	6.0	11.1	14.3	9.9
Z = 12	9.1	8.2	7.1	7.4	9.3	6.3	4.1	9.3	8.3	8.4
Z = 14	6.8	4.3	4.0	4.5	7.3	4.2	2.2	7.7	5.9	5.8
Z = 16	5.3	3.2	3.1	3.6	5.3	2.9	1.1	5.4	4.4	4.5
Z = 18	4.1	2.8	2.5	2.5	4.0	2.2	0.5	3.9	3.2	3.2
Z = 20	2.9	2.0	1.5	1.7	3.1	2.0	0.1	2.5	1.9	2.5
Z = 22	2.1	1.4	1.1	1.1	2.4	1.1	0.1	1.5	1.7	1.7
Z = 24	1.7	0.8	0.5	1.0	1.9	0.7	0.0	1.2	1.5	0.9
Z = 26	1.1	0.2	0.3	0.1	1.5	0.3	0.0	0.5	1.0	0.3
Diacides 4	0.0	0.0	0.0	0.2	0.0	1.7	0.0	0.0	0.0	0.0
Diacides 6	0.0	0.4	0.0	0.7	0.0	2.4	0.1	0.0	0.0	0.3
Diacides 8	0.0	0.2	0.0	0.6	0.0	0.7	0.0	0.0	0.1	0.2
Diacides 10	0.0	1.0	1.6	0.8	0.0	0.5	0.0	0.0	0.5	0.0
Diacides 12	0.1	1.6	2.8	1.7	0.0	0.5	0.8	0.0	0.7	0.3
Diacides 14	0.2	1.2	2.2	0.3	0.0	0.5	1.4	0.3	0.0	0.7
Diacides 16	0.1	0.3	1.5	0.2	0.0	0.1	1.3	0.3	0.1	0.3
Diacides 18	0.2	0.5	1.1	0.0	0.0	0.1	0.9	0.3	0.1	0.0
Diacides 20	0.1	0.3	1.0	0.1	0.0	0.0	0.7	0.1	0.0	0.1
Diacides 22	0.0	0.1	0.3	0.0	0.0	0.1	0.3	0.0	0.1	0.1
Diacides 24	0.0	0.0	0.3	0.0	0.0	0.0	0.2	0.0	0.0	0.1
Diacides 26	0.0	0.1	0.0	0.0	0.0	0.0	0.0	0.0	0.0	0.0

RH = Relative Height

TABLE 13.4
Distribution of the Total Acid Number (TAN) Within the Different Crude Cuts

TAN (mg KOH/g)	DALIA 1	DALIA 2	ORQUI
Crude	1.4	2.5	3.4
150 to 230 cut	1.11	0.43	0.13
230 to 350 cut	2.28	2.70	3.73
350 to 400 cut	1.62	2.64	7.39
400 to 550 cut	0.35	0.58	1.25
550+ cut	0.40	1.05	0.20

for the least stable system (deionized water at pH = 5.5): 500 rpm during 10 s for Dalia 1 and Orquidea, 750 rpm during 30 s for Dalia 2. Before whisking, water was added according to the following procedure: aqueous solutions at given pH were prepared by adding solid NaOH to 200 ml of distilled water. 10 ml was poured in the 500 ml beaker. 180 ml of crude oil was then added and finally 10 ml of brine prepared by adding NaCl or $CaCl_2$ to distilled water. Amounts of NaOH, $CaCl_2$, and NaCl were calculated in order to reach the targeted pH value and sodium and/or calcium concentration for a water volume of 20 ml. Such a method was performed in order to prevent any precipitation by direct contact of OH^- with calcium chloride. The same procedure was carried out even in the absence of calcium chloride.

The 200 ml volume was then divided in two 100 ml bottle testers (ASTM D96) placed in a controlled temperature bath at 40 °C. The percentage of decanted water was followed over time. After 3 days, when decantation was too slow, phase separation was forced by centrifugation. On decanted water samples, pH was measured (pH_f), and found different from initial pH (pH_i). Surface tension measurements were performed on recovered water samples as a function of time with a drop tensiometer (IT Concept). Surface tensions at 4 min are reported. Similar experiments have been carried out on model systems, where the oil is n-heptane and the organic acids are the octanoic acid from Aldrich and naphthenic acid from Fluka (a mixture of average molecular weight = 230).

13.3.2 Results and Discussion: Crude Oils

The stability of water in crude oil emulsion was studied first as a function of pH. Figure 13.16 typically shows the fraction of non-separated water of Dalia 1 emulsions as a function of time. For each system, the initial (pH_i) and final pH (pH_f) are indicated.

Various systems are then compared by plotting the percentage of non-decanted water after 48 h against the final pH. Due to the consumption of NaOH by the organic acids of the crude, pH_f is generally lower than pH_i. We have reported in Figure 13.17 the evolution of pH_f as a function of pH_i for the three presented crudes. It is interesting to observe that pH_f varies only slightly over a wide range of initial pH_i, that is, over a wide range of NaOH concentration, and then increases for the highest pH_i. This behavior will be discussed at the light of the results obtained on model systems (Section 13.4.2). It is interesting also to notice that, for Dalia 1 and Dalia 2, the equilibrium pH is higher than the initial pH for pH_i = 5.3. This surprising result will be explained later in the theoretical section (Section 13.4.2). It is attributed to the presence of sodium and/or calcium naphthenates in the oil phase prior to contact with the aqueous phase. Dalia 2, especially, is known to contain around 150 ppm of calcium.

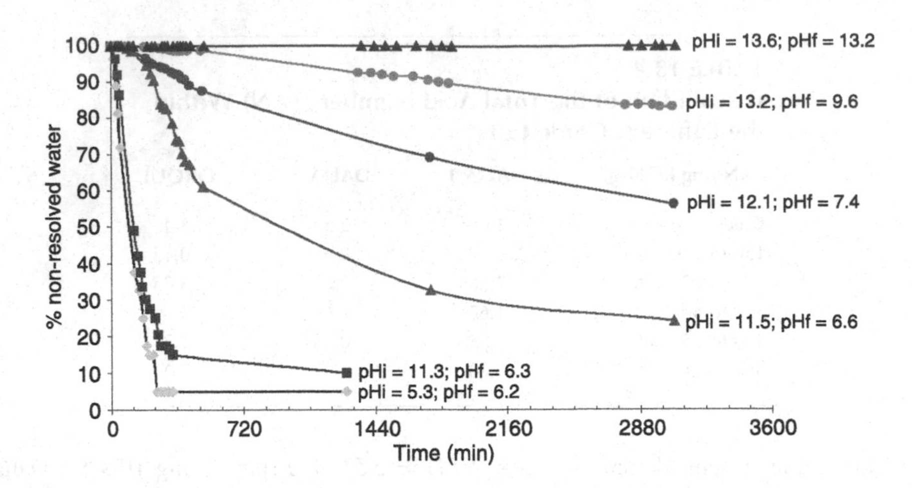

FIGURE 13.16 Stability of Dalia 1 crude emulsions.

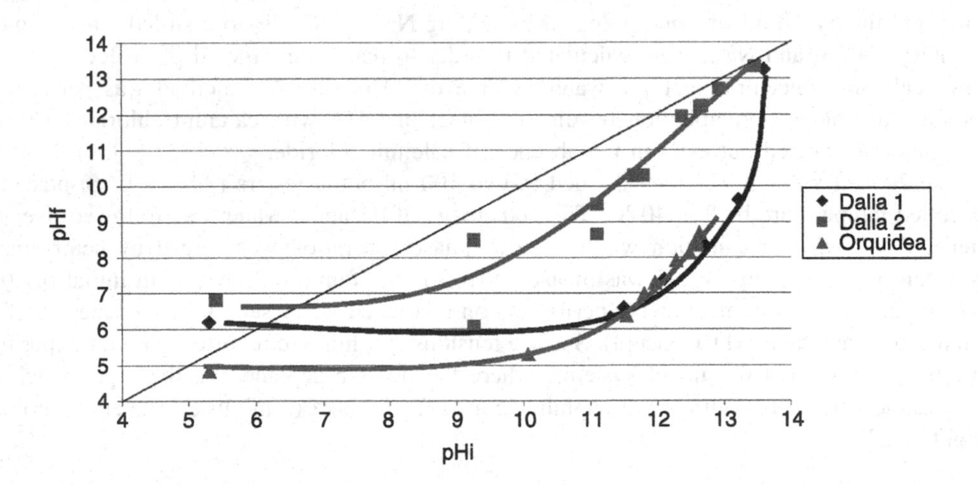

FIGURE 13.17 Equilibrium pH (pH_f) versus initial pH (pH_i) of water (10%) contacted with three different crudes (90%).

From pH_i and pH_f values, the amount of OH^- consumed has been calculated and expressed in TAN units: TAN_C. The ratio TAN_C/TAN represents the fraction of naphthenic acid originally present in the crude which have been neutralized by OH^- during emulsification.

13.3.2.1 Dalia 1 Results

The evolution of TAN_C/TAN with the final pH_f is reported in Figure 13.18 for the Dalia 1 crude. It increases progressively from $pH_f = 6.2$ to $pH_f = 13.2$. That is, at this pH, all the acids are neutralized by NaOH.

In a complementary way of evaluating the fraction of naphthenic acids neutralized by the base in each bottle test, it is interesting to calculate the fraction of the base introduced in the

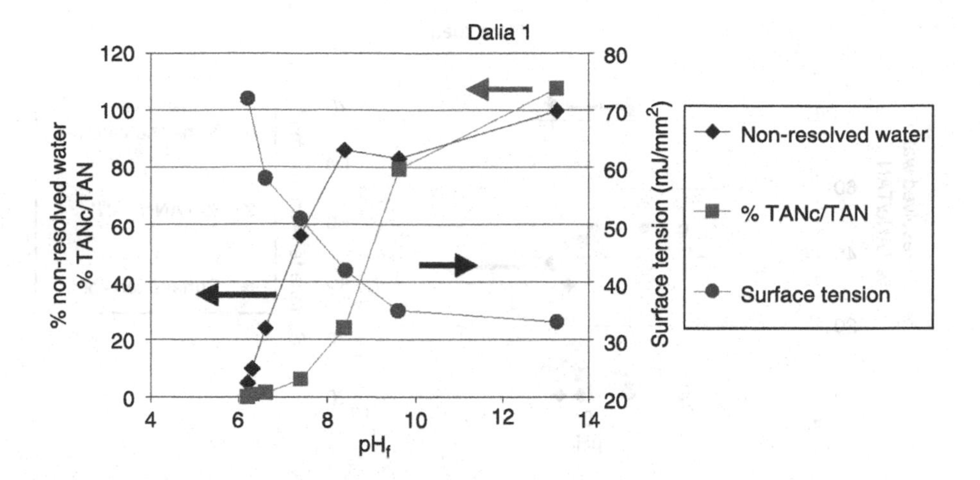

FIGURE 13.18 Dalia 1: emulsion stability, fraction of consumed TAN, and surface tension of the separated aqueous phase as a function of final pH.

system which has been consumed by the naphthenic acid. Similarly as above, the initial NaOH concentration has been expressed in TAN units = TAN_o, for "TAN offered." TAN_C/TAN_o is thus the fraction of initial NaOH neutralized by the naphthenic acid. In the present case of Dalia 1, all the base is consumed until $pH_f \sim 9.8$. Then, NaOH is in excess with respect to the naphthenic acid content, which explains the increase in pH_f.

Figure 13.18 also displays the surface tension of the decanted water as a function of the final pH. The value of the surface tension is linked to the presence of surface active species: a decrease in surface tension reveals an increase in concentration of surface active molecules in water. At $pH_f = 6.2$, the surface tension of the separated aqueous phase is 71 mN/m, i.e., very close to that of pure water: in the present case, no surface active species of the crude dissolve in the water. The undissociated naphthenic acids, which are known to display surface activity [38,39], do not partition into the water phase, as seen also from the pH_f value, despite the elevated TAN value of the lightest cut of the Dalia 1 crude (Table 13.4). Conversely, the basic species (sodium or calcium naphthenates?) which partition into water at this pH do not exhibit any surface activity. When pH_f increases, the water surface tension decreases markedly, indicating an increased partitioning of the naphthenic acids into water, more specifically of their salts. At the same time, the emulsion becomes more and more stable, accompanying the increase in TAN_C/TAN, which demonstrates further the surfactant character of the salts of the Dalia 1 naphthenic acids.

13.3.2.2 Orquidea Results

For Orquidea (Figure 13.19), although the same general features as for Dalia 1 are displayed, significant quantitative differences are observed. TAN_C/TAN, contrary to Dalia 1, varies only slightly when pH_f increases up to 8.5 roughly. Beyond that value, a sharp raise is observed, but, at $pH_f = 9.2$, only 25% of the naphthenic acids initially present in the crude have been neutralized. (At that point, TAN_C/TAN_o indicates that all the NaOH introduced in the system has been consumed.)

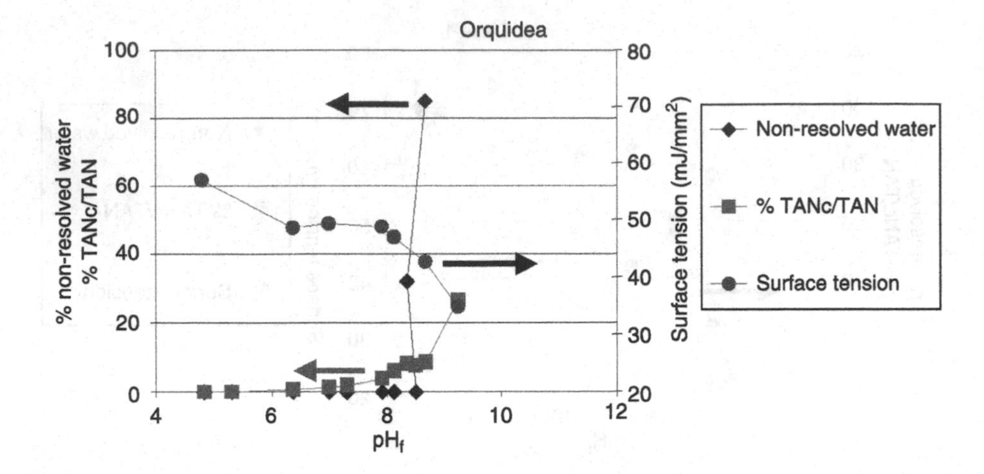

FIGURE 13.19 Orquidea: emulsion stability, fraction of consumed TAN, and surface tension of the separated aqueous phase as a function of final pH.

The surface tension of the extracted water phase is only 60 mN/m at $pH_f = 6.2$, which shows that some surface active species already partition into water at this low pH. Upon increasing pH_f to 8.5, roughly, the surface tension decreases only slightly, although the value of TAN_C/TAN_o indicates that all the base introduced in the system is neutralized by the crude. This suggests that the organic acids involved for $pH_f < 8.5$ are not highly active at the water surface. For $pH_f >$ 8.5, the surface tension of water decreases significantly, indicating an increase of the partitioning of more surface active species.

Similarly, the emulsion stability curve exhibits an abrupt transition around $pH_f = 8.5$, where the emulsions become very stable. This is consistent with the transitions observed on TAN_C/TAN and on water surface tension, supporting the idea that two main types of naphthenic acids are present in the Orquidea crude: the first ones, likely of very low molecular weight, are soluble in water and display low surface activity, at least as regards the stabilization of water-in-oil emulsions. It is known from literature that the pK_a of such compounds is low, on the order of 4.7, which means that at $pH_f \sim 6.2$, there is a significant proportion of dissociated species. The other type of naphthenic acids, likely of higher molecular weight, displays the expected surface activity and strongly stabilizes the water-in-oil emulsion.

When looking at the distribution of TAN values among the various cuts (Table 13.4), Orquidea has clearly a stronger acidity in the heavier cuts compared to Dalia 1, which is coherent with the above interpretation. Regarding the light cut (150 to 230) it is, however, less obvious to correlate its TAN value to the observed phase behavior, unless it is speculated that organic acids might be present in even higher cut.

13.3.2.3 Dalia 2 Results

The results are reported in Figure 13.20. Again, as for Orquidea, TAN_C/TAN remains small when pH_f increases, but up to much higher pH_f value: 12 compared to 8.5 for Orquidea. Then a sharp increase is observed but again, at $pH_f = 13.3$, only 22% of the naphthenic acids of the crude are neutralized, although the calculation of TAN_C/TAN_o reveals that, at that point, 75% of the base is still available. This raises the question of the quality of the contact of the naphthenic acids of

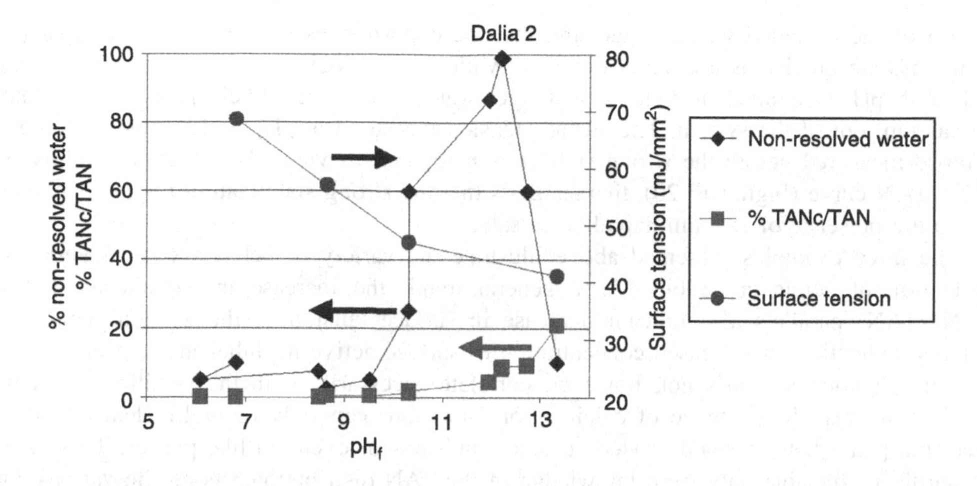

FIGURE 13.20 Dalia 2: emulsion stability, fraction of consumed TAN, and surface tension of the separated aqueous phase as a function of final pH.

the crude with the aqueous base, and thus the question of the initial agitation of the system. The experiment has thus been repeated, with a vigorous initial mixing of the fluids. The result should be approximately the same, indicating that some likely heavy naphthenic acids do not participate in the acid–base equilibrium under the conditions of the experiment (although they are counted in the TAN measurement, i.e., in the presence of methanol).

The surface tension of the aqueous phase starts, at $pH_f = 6.8$, with a value close to that of pure water (72 mN/m), as for Dalia 1, but then exhibits a more progressive decrease when pHf increases. Contrary to the previous cases of Dalia 1 and Orquidea, this strong increase in emulsion stability does not coincide with a large increase in naphthenate formation, as seen from TAN_C/TAN: the neutralization of only roughly 2% of the naphthenic acids is sufficient to produce a very stable emulsion, indicating likely that the naphthenates produced in these conditions ($pH_f \sim 10.2$) are particularly well adapted to Dalia crude (water-in-oil) emulsification. Above $pH_f = 10.2$, around $pH_f = 12$, three very stable emulsions are obtained, presenting a particular behavior: after some phase separation was achieved, a gel-like phase was observed at the water–oil interface.

At that point, it is important to mention that around 150 ppm calcium are present in the Dalia 2 crude. The calcium salts of long chain carboxylic acids are known to be highly hydrophobic and water insoluble, forming colloidal precipitates, or liquid crystal mesophases, or oil soluble complexes. This is further discussed below. In the present case, it is believed that the observed gel-like phase is due to the structure of calcium naphthenates, which, by adsorption at the water–oil interface, could also be responsible for the high emulsion stability of the three systems at $pH_f \sim 12$. Such compounds may form at lower pH_f, such as $pH_f = 10.2$, but in much smaller concentrations so that gel-like phases are not visible, but, because of their high hydrophobicity, they may be at the origin of the strong increase in emulsion stability observed at that pH_f.

Surface tension measurements performed on the centrifuged water from the very stable systems obtained at $pH_f = 12$ could not give reliable results: after a few seconds, some particles appeared at the water–air interface, modified the shape of the drop, thus confirming the presence of a third phase. Beyond $pH_f = 12.5$, it was again possible to measure properly the surface tension of the decanted water, suggesting that the deposit does not form any more. At $pH_f > 12.5$, unstable emulsions are again obtained while the gel-like phase is no more observed. At such high pH, calcium hydroxide is no longer water soluble and likely precipitates, removing the calcium out of the system. The surface tension measured at $pH_f = 12.7$ and 13.3 are close to those measured before the strong stabilization regime. In view also of the continuity of the TAN_C/TAN curve (Figure 13.20), this suggests that the strong stabilization regime is essentially due to the presence of calcium naphthenate salts.

The three examples presented above illustrate the variety of behaviors encountered in the production of acidic crude oils. As a general trend, the increase in naphthenate formation (TAN_C/TAN) parallels a significant decrease in surface tension of the aqueous phase of the systems, indicating an increased concentration of surface active naphthenates in water.

This phenomenon does not, however, correlate necessarily with the stabilization of water-in-oil emulsions: the presence of calcium ions, or more generally of multivalent cations, may affect the phase behavior and produce coarse emulsions or even gel-like phases. These features are hardly predictable only from knowledge of the TAN distribution among the various cuts of the crude.

From a practical standpoint, the preparation of a series of bottle tests at various pH values according to the procedure presented above seems to be relevant for evaluating the potential risk of coarse emulsion or gel formation of a given crude upon pH increase. From this point of view, the Dalia 1 crude, which exhibits, under the conditions of the experiments, emulsion formation at the lowest pH, seems to be the most "dangerous" crude among the three investigated. It is interesting that it displays the lowest TAN. Indeed, the presence of multivalent cations is of paramount importance and is investigated below.

13.3.2.4 Effect of the Presence of Salt

From a general point of view, as will be discussed in more detail here, the stability of emulsions is related to the relative interactions of the surfactant with oil and water [40,41]. This property has been used to optimize demulsifier molecules or mixtures for destabilizing emulsions of "regular" crude oils [42]. Indeed, the salinity of water has a strong effect on the interactions of surfactant with water, particularly in the case of ionic surfactants [43–45]: the higher the ionic strength, the lower the interactions with water (and the lower their solubility in water). Therefore, it is expected that, for a given water–oil–surfactant system, changing the brine salinity can produce or destabilize an emulsion. An example is provided in Figure 13.21 where the system contains a naphthenic crude, water, and a demulsifier [42].

It can be seen that a maximum in coalescence rate, i.e., a minimum in emulsion stability, occurs roughly at 60 g/l NaCl for demulsifier I_1. Hence, depending on the starting point, increasing the salinity may stabilize or destabilize an emulsion. Indeed, the effect of multivalent cations is much stronger than monovalent, particularly in the case of carboxylic acids: calcium soaps are known to be sparingly soluble in water. To study the effect of salt, systems presenting an intermediate decantation rate (typically 30% of decanted water after 8 h and 50% after 48 h) were selected.

13.3.2.4.1 Dalia 1

The effect of sodium chloride and calcium chloride has been investigated on the system obtained for $pH_i = 12.1$ ($pH_f = 7.4$) in the absence of salt. Figure 13.22 shows the evolution of the

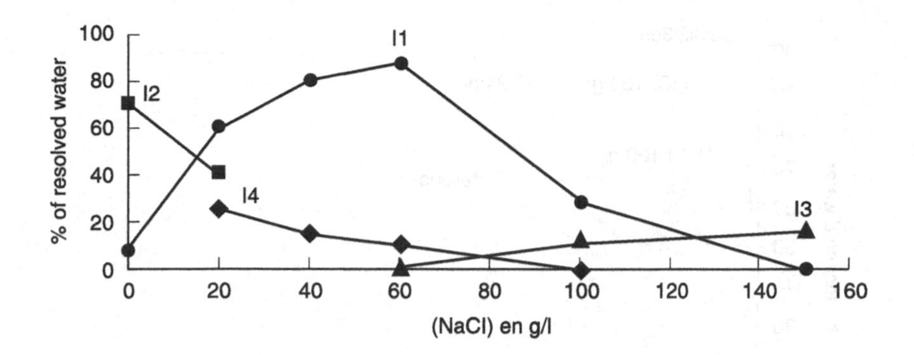

FIGURE 13.21 Effect of brine salinity on the stability of water in crude oil emulsions, in the presence of various demulsifiers (I1 to I4).

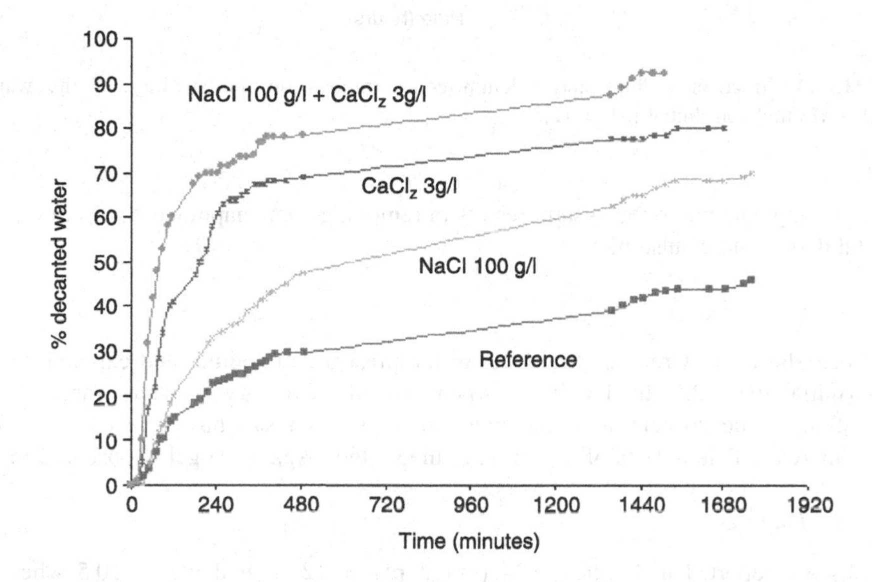

FIGURE 13.22 Effect of sodium and calcium concentrations on the stability of the water in oil (DALIA 1) emulsion. Initial pH = 12.1.

decanted water as a function of time for systems containing various amounts of salt. In all cases the presence of sodium or calcium destabilizes the emulsion.

The destabilizing effect is much more important with calcium than with sodium since only 3 g/l of $CaCl_2$ more than doubles the decantation rate of emulsion after 8 h, while 100 g/l of NaCl are required to increase it by 50%.

It is likely that NaCl destabilizes the emulsion through a modification of the relative interactions of the naphthenates with water and oil, as explained above. In the case of $CaCl_2$, it has been verified that, after phase separation, most of the calcium is located in the crude oil phase. At the same time, the surface tension of the water phase increases [46]. No gel is observed. This shows that the calcium salts of the Dalia 1 naphthenic acids are soluble in the crude. Indeed, in

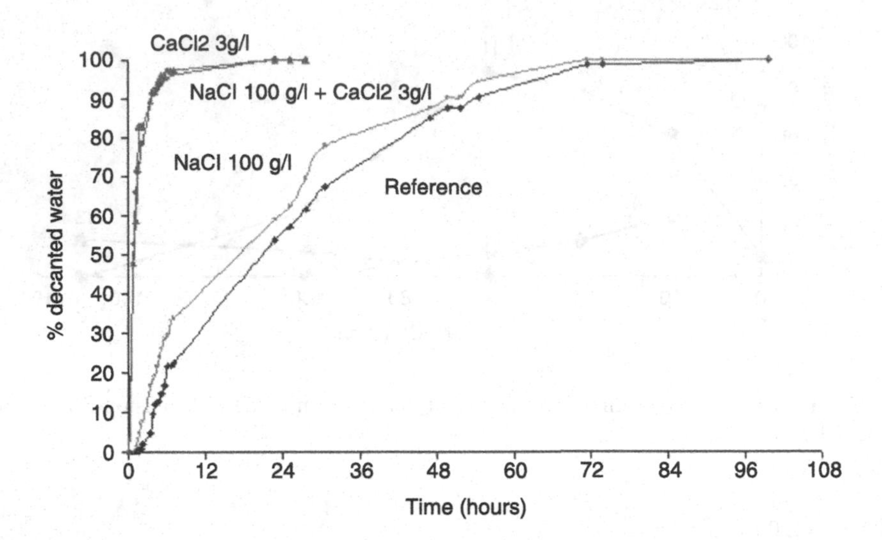

FIGURE 13.23 Effect of sodium and calcium concentrations on the stability of the water in oil (ORQUIDEA) emulsion. Initial pH = 12.5.

that case, adding calcium to the system results in removing some naphthenates from the interface, thus destabilizing the emulsion.

13.3.2.4.2 Orquidea

Figure 13.23 shows for Orquidea the effect of the presence of sodium and calcium on emulsion stability (initial pH: 12.5; final pH: 7.7 when no salt). Contrary to what was observed with Dalia 1 at the same concentration the presence of sodium salt has little effect on emulsion stability, whereas calcium destabilizes strongly the system. Again, no gel is observed in this case.

13.3.2.4.3 Dalia 2

The results are reported in Figure 13.24 (initial pH = 12.5; final pH = 10.5 when no salt). This case is interesting since, as it has been mentioned above, the Dalia 2 crude contains some calcium.

Contrary to what happened with Dalia 1 and Orquidea, the addition of calcium salt has a very small effect on the stability of the emulsion. Some solid gel-like deposit is visible at the interface, similar to that described above for the Dalia 2 behavior (Figure 13.20). The lower pH_f used here (10.5) likely explains why a lower gel concentration, and thus lower emulsion stability, is observed, compared to the systems at $pH_f = 12.5$ in Figure 13.20. Nevertheless, the stabilizing character of the structures formed from Dalia 2 calcium naphthenates is demonstrated.

As regards NaCl, it displays a strong destabilizing effect, even stronger than what was observed with Dalia 1. The surface tension of the aqueous phase extracted from different systems has been measured. The results are reported in Table 13.5. It can be seen that in the presence of NaCl, the surface tension of the aqueous phase is lower than that of the reference system. It shows that the Dalia 2 sodium naphthenates become more surface active at the water–air interface in the presence of NaCl, although they have been measured in lower concentration in the water phase ("salting out" effect). This clearly demonstrates the modification by NaCl of the

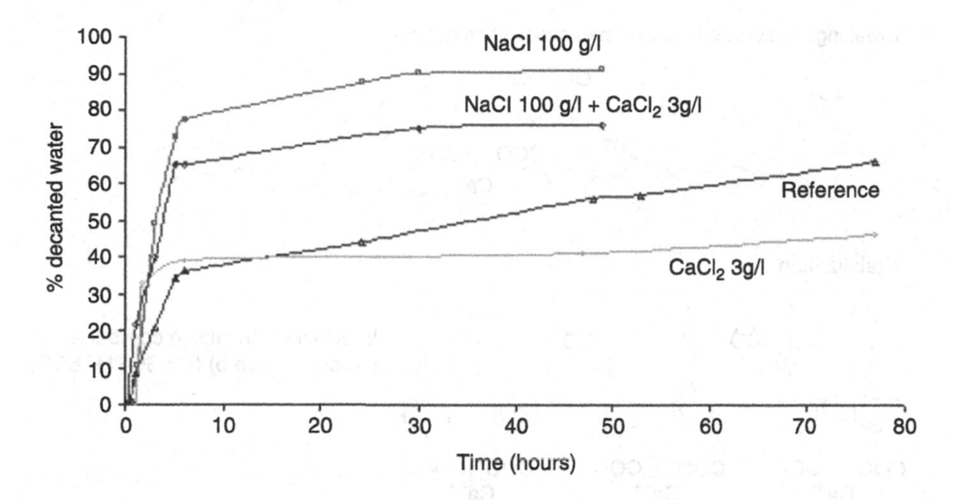

FIGURE 13.24 Effect of sodium and calcium concentrations on the stability of the water in oil (Dalia 2) emulsion. Initial pH = 12.5.

TABLE 13.5
Surface Tension of Aqueous Phase. Dalia 2: $pH_i = 12.5$; $pH_f = 10.5$

Salt Concentration (g/l)		Surface Tension (mN/m)	Decanted Water after 8 h (%)
NaCl	$CaCl_2$		
0	0	50	40
100	0	42	90
0	3	51	36
100	3	42	78

interactions of the sodium naphthenates with water. However, this makes them less effective at the water–Dalia 2 interface, because of disequilibrated interactions with oil relative to water.

It is interesting to compare the behavior of the calcium naphthenates of Dalia 1 and Orquidea on the one hand and of Dalia 2 on the other. As it has been shown, the first ones readily dissolve in the oil phase and have no contribution to emulsion stabilization, while the second form structured mesophases (as evidenced by observation under polarized light), displaying viscoelastic behavior and some interfacial activity since they may provide strong emulsion stabilization. Emulsion stabilization by lamellar liquid crystalline phases has been demonstrated in the past [47–49], such mesophases providing interfacial rigidity, thus preventing droplet coalescence.

In Figure 13.25, we propose schematically two types of structural arrangements of the calcium naphthenates which could correspond to the types of above-described behaviors. Type a is a more or less "linear" association of the naphthenates linked by the calcium ion, where the organic chains are solvated by the crude. In type b, the "comb-shape" structure is promoted by strong lateral interactions between the naphthenate molecules and by the structuring presence of water. Indeed, the occurrence of type a or b structure depends heavily on the interactions of the naphthenates

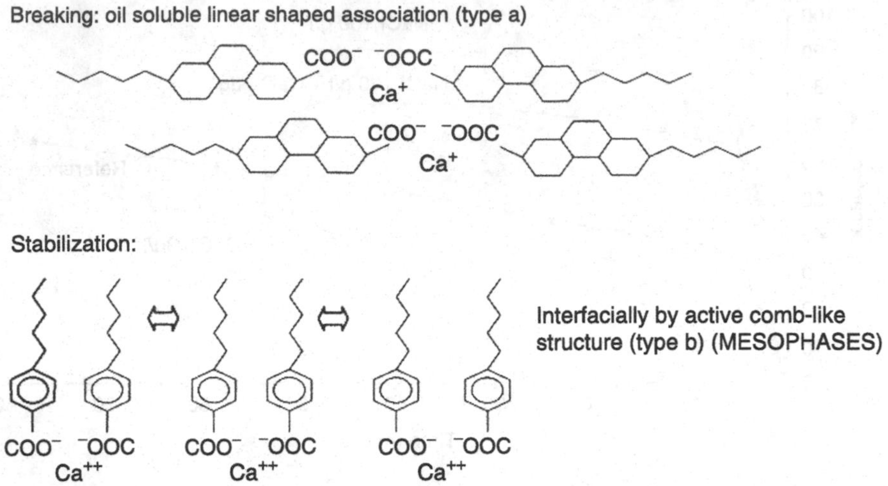

FIGURE 13.25 Hypothetical structures for various types of calcium naphthenates.

with the oil ("solvent quality" of the crude), as well as on the lateral interactions between the naphthenate molecules themselves, which obviously involves steric considerations.

13.4 THEORETICAL ASPECTS

13.4.1 Background to Phase Behavior Modeling

A number of factors contribute to emulsion stabilization, with respect to coalescence, which have to be taken into account for possible modeling. Obviously, the relative interactions of the surface active agents with oil and water are of paramount importance. They are in fact the base of the old Bancroft rule and of the hydrophile–lipophile balance (HLB) concept [50]. More recently, the relationship between microemulsion formulation and emulsion formulation has been established [40]. For optimum microemulsion formulation, it is essential that the interactions of the surfactant with oil and water be equal [44,45]. Under these conditions, at low surfactant concentrations, a third phase containing the surfactant is observed in equilibrium with excess oil and water. Such a system has been found coalescing extremely rapidly, as shown in Figure 13.26.

In that case, the modification of interactions has been achieved through salinity change. At 5 g/l NaCl, phase separation is extremely fast, and, at equilibrium, three phases are observed. On both sides of this system, a maximum in emulsion stability is exhibited corresponding to oil-in-water emulsions at low salinity, where the surfactant interactions with water are higher than those with oil, and to oil-in-water emulsions at high salinity, where it is the reverse. *The general character of this pattern must be recognized: a change in oil type, for example, or in surfactant type, or in temperature, or in any other parameter affecting the interactions at the oil/water interface, will shift the whole diagram towards higher or lower salinities, but the relative positions of the transitions will be maintained.*

This property has been shown to apply also to the case of crude oils, as shown in Figure 13.21 above, and exploited for demulsifier optimization [42], the demulsifier being a surface active molecule able to appropriately modify the interfacial interactions. Indeed, as it has been seen

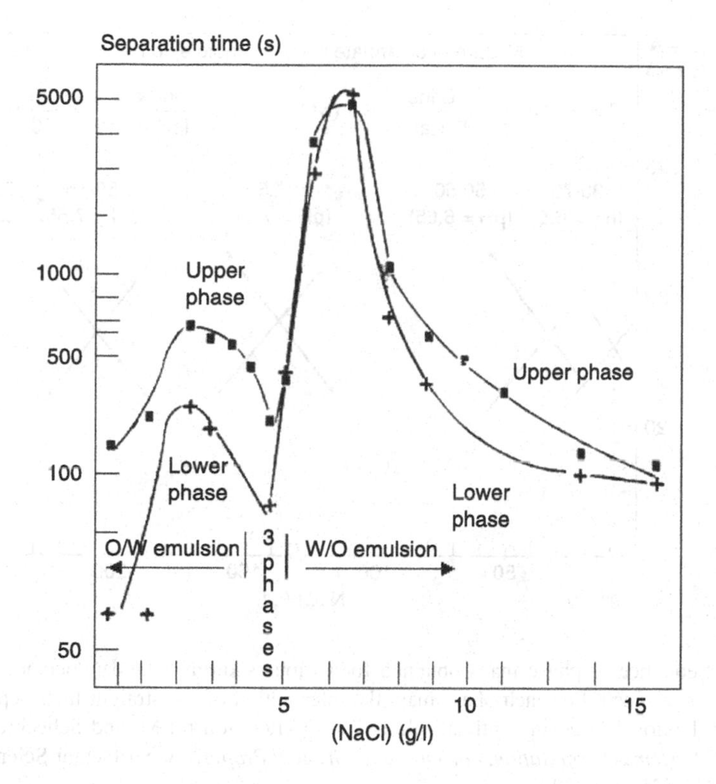

FIGURE 13.26 Time required for half of the considered phase to clear as a function of salinity. Water/hexane = 1; 1.5% petroleum sulfonate (TRS 10-80); 2.5% 2. butanol. Room temperature.

above with the Dalia 2 case, liquid crystalline phases may contribute greatly to the stabilization of emulsions, provided they display an amphiphilic character promoting their adsorption at the oil–water interface.

In the case of organic carboxylic acids, the ionization degree of the molecule, which of course depends on pH, is essential for determining its interaction with water at the water–oil interface. It must be recognized, however, that the acid from RCOOH, where R represents the organic moiety of the molecule, also displays an amphiphilic character. Indeed, this species is much less hydrophilic than the ionized from $RCOO^-$. A change in pH, which changes the ratio of the two species, depending on the pK_a, allows the ratio of the hydrophile to lipophile interactions at the interface to be varied continuously, and thus promote the expected transitions. Alternatively, this can be achieved by mixing the salt and acid species in various ratios, thus determining the pH. This is illustrated in Figure 13.27, which shows the phase behavior of various mixtures of octanoic acid and sodium octanoate in contact with decane and water [45]. Isobutanol has been added to the system to prevent liquid crystalline phase appearing.

For a given salt/acid ratio, for example 75/25 (wt) for which the pH is found to be 7.0, the water salinity (NaCl) is varied and the phase behavior is observed at various surfactant + alcohol (S + A) concentrations. A phase map is thus obtained which comprises four regions: two 2-phase regions on the left and right part of the diagram, a single-phase region (microemulsion regime), and a 3-phase region underneath the previous one, where a surfactant-containing phase is in equilibrium with both excess oil and water phases. High separation rates are observed in

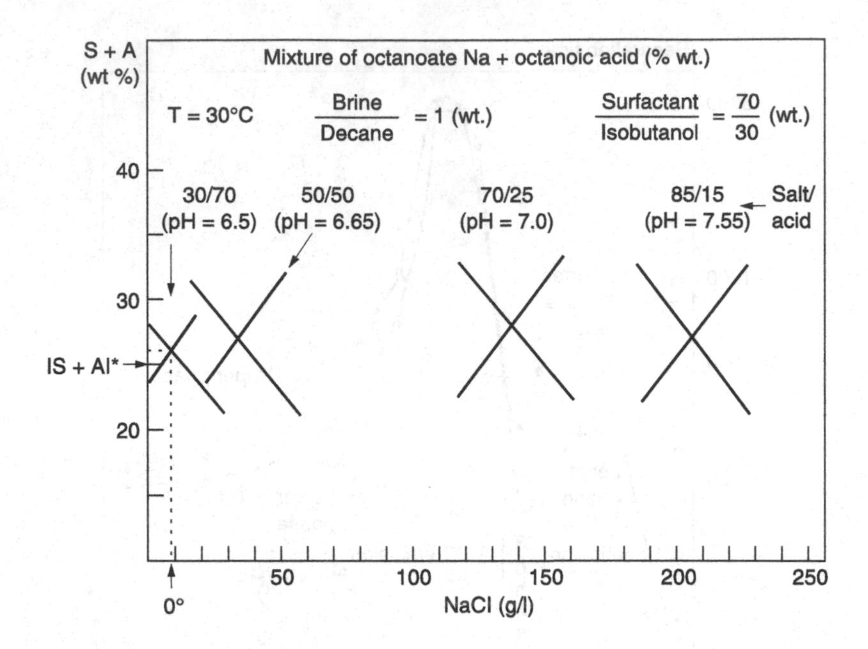

FIGURE 13.27 Sequence of phase maps obtained for various sodium octanoate/octanoic acid ratios when the brine salinity is scanned. For each phase map, the intersection of the straight lines separating regions of different phase behavior defines an "optimal salinity" s*. (After Bourrel M. and Schechter RS. *Microemulsions and Related Systems: Formulation, Solvency and Physical Properties*. Surfactant Science Series, Vol 30. Marcel Dekker, New York, 1988.)

this region. The salinity at which the amount of surfactant + alcohol required to reach the single phase region goes through a minimum is called "optimal salinity" s*. The corresponding S + A concentration is noted (S + A)*. At that point, the surfactant (acid + salt forms) interactions with oil and water are equilibrated. A maximum in emulsion stability for water-in-oil emulsions is expected to occur at salinity above s*. It is interesting to note the change in optimal salinity with the sodium octanoate/octanoic acid ratio. Indeed, modifying this ratio does not change the interactions with the oil phase, but changes the interactions with water: increasing the salt/acid ratio entails a strong increase of the interactions with water. To compensate and keep them constant in order to recover the optimal point, the NaCl concentration must be increased since it is well known that the solubility of ionic surfactant in water decreases when the salinity increases ("salting out" effect). This is exactly what is observed in Figure 13.27, demonstrating the interfacial activity of the octanoic acid. This is further emphasized by Figure 13.28 where the optimal salinity has been plotted against pH and compared to the ionization degree α calculated according to:

$$\alpha = \frac{10^{(\mathrm{pH}-\mathrm{pK_a})}}{1 + 10^{(\mathrm{pH}-\mathrm{pK_a})}} \tag{13.1}$$

which in principle is valid only for dilute solutions of weak electrolytes. α is the ratio of ionized molecules to the total number of ionized + non-ionized molecules.

The calculation has been carried out taking $\mathrm{pK_a} = 6.65$, a value which has been experimentally determined from the neutralization curve of the octanoic acid by its sodium salt in the case

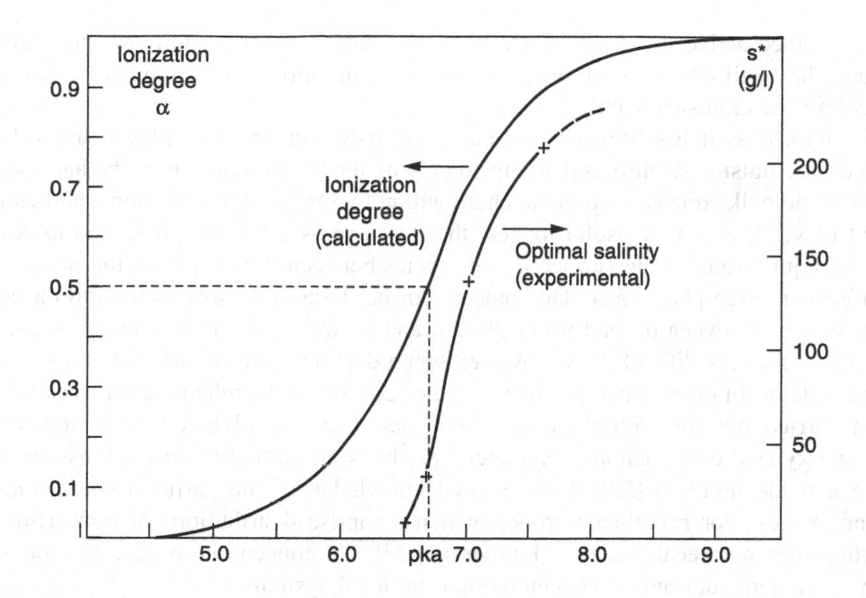

FIGURE 13.28 Correlation between the degree of ionization and optimal salinity s* of Figure 13.27 for carboxylated surfactants. (After Bourrel M. and Schechter RS. *Microemulsions and Related Systems: Formulation, Solvency and Physical Properties.* Surfactant Science Series, Vol 30. Marcel Dekker, New York, 1988.)

of brine–decane surfactant + alcohol systems, that is to say in microemulsion systems. It was found to be different from that obtained from highly dilute solutions: $pK_a = 4.89$ at 20 °C. This difference is probably due to the fact that in such highly dilute aqueous solutions, the acid and salt molecules are in monomeric forms, whereas in microemulsions, emulsions, or concentrated aqueous micellar solutions, it is known that the dissociation equilibrium of an electrolyte is affected by the presence of the interface [51,52]. This is even accentuated with brines displaying high ionic strength, due to the double-layer compression effect. In other words, the local pH at the interface is not identical to that measured commonly in the bulk. Additionally, the pK_a of acetic acid has been measured in mixtures of water + ethanol (50/50) and found to be 5.5 instead of 4.7 in pure water. Similarly, for octanoic acid, the pK_a is 6.05 in water–ethanol mixtures, compared to 4.9 in pure water.

Nevertheless, it is interesting to observe in Figure 13.28 that α calculated by Equation 13.1, which is oversimplified, displays the same sigmoidal shape as the optimal salinity when plotted against pH. This gives more support to the conclusion that octanoic acid must be considered as a surfactant when interpreting emulsion phase behavior.

To summarize, the stability with respect to coalescence of water in acidic crude oil emulsion is ruled by the interactions of the naphthenic species *present* at the interface with water, oil, and between themselves. The parameters affecting these interactions are:

- Water side of the interface: pH, salinity (with a dramatic effect of multivalent cations), temperature, naphthenate structure, pK_a
- Oil side of interface: oil type ("solvency quality"), temperature, naphthenate structure, naphthenic acids structure

Indeed, if surface active molecules, such as demulsifiers, corrosion inhibitors etc., are added to the system, they will adsorb at the water–oil interface and modify the above interaction energies, thus affecting the emulsion stability.

It is obvious, as it has been pointed out previously, that the role played by surface active compounds in emulsion stability is determined only by the fraction present at the interface. Species dissolved in the bulk aqueous or organic phase will not participate in emulsion stabilization. From this point of view, it is very useful to treat the interface as a pseudo-phase and to consider the partitioning equilibrium of the surface active species between the oil phase, the water phase, and the interfacial pseudo-phase [52]. This indeed can be described using two partition coefficients of the surfactant: between oil and water phases, and between oil and interface, for example.

This can be a very difficult task, however, when dealing with mixtures of surfactants such as those encountered in commercial products, and, of course, in naphthenic crudes. Good modeling has been carried out for multiphase microemulsion behavior obtained with isooctane, water, and polyethoxylated octylphenols (commercial products presenting the usual Poisson distribution of ethylene oxide number) [53]. This required knowledge of the partition coefficients of each surfactant species. For naphthenic crudes, which comprise distributions of both acidic and salt forms, this does not seem realistic. Furthermore, if the concentrations are not too small, the partition coefficients depend on concentration (non-ideal systems).

Finally, it must be recognized that the area of the interface (i.e., the "volume" of the interfacial pseudophase) depends on the mixing conditions of the system: when dealing with a partly emulsified system, the composition of the resolved water and oil phases may depend on the total area of oil–water interface in the residual emulsion, that is on the initial particle size distribution of the emulsion. Illustrations will be given below. Indeed, for a completely separated system, the interfacial area is small and the phenomenon can be ignored [39].

13.4.2 The Model

The main features observed during the experiments carried out on crude oils are:

- At low initial pH, the final pH is almost constant
- At higher initial pH, the final pH increases dramatically, with a simultaneous stabilization of water in crude oil emulsions

The basic hypothesis explaining the results is that the naphthenic acids of the crude react with the basic in the water phase. They impose the final pH, with respect to their pKa value and concentration in the phases. The sharp increase in final pH at higher initial pH is due to the massive neutralization of the naphthenic acids, and the concomitant emulsion formation due to the surface activity of the dissociated form of the acids. This interpretation can be further detailed by using the simple description presented below.

Let us consider a single naphthenic acid AH, at concentration C_{TAN} initially in the oil phase. Sodium naphthenates A^- could also be present in the oil phase at concentration C_{TBN}. The oil phase (volume V_o) is contacted through vigorous agitation with the water phase (volume V_w), the initial pH of which is fixed by addition of either (H_3O^+,Cl^-) or (Na^+,OH^-) in a known concentration. Naphthenic acid can react when contacted with soda according to:

$$AH + OH^- \leftrightarrows A^- + H_2O$$

Naphthenates can react with H_3O^+ in the water phase according to:

$$A^- + H_3O^+ \leftrightarrows AH + H_2O$$

At equilibrium, we therefore have six unknown concentrations:

- Concentrations in the oil phase: AH_o, A_o^-
- Concentrations in the water phase: AH_w, A_w^-, h, OH^-,

where h is the final proton concentration, i.e., the final equilibrium pH. To calculate these concentrations, we have to simultaneously solve six equations, which are:

- Mass balance:

$$V_o\left(C_{TAN} + C_{TBN}\right) = V_o\left(AH_o + A_o^-\right) + V_w\left(AH_w + A_w^-\right)$$

- Electroneutrality:

$$V_w\left(A_w^- + OH^-\right) = V_w\left(h + H_i\right) + V_o\left(C_{TBN} - A_o^-\right)$$

where H_i is either the initial OH^- concentration OH_i or the initial H_3O^+ concentration h_i, the last term on the right side being linked to the number of Na^+ associated to naphthenates transferred from the oil phase to the water phase.

For dissociations in water:

$$K_e = h \cdot OH^-$$

and in acid:

$$K_a = \frac{h \cdot A_w^-}{AH_w}$$

For partitioning we assume a constant partitioning coefficient with respect to concentration and water–oil ratio for both AH and A^-:

$$K_S^{AH} = \frac{AH_o}{AH_w}$$

for AH, and

$$K_S^{A^-} = \frac{A_o^-}{A_w^-}$$

for A^-. From this set of equations, the following equation in the variable h can be derived:

$$\begin{aligned} h^3 + h^2 &\left[K_a' + H_i + \frac{V_o}{V_w}C_{TBN}\right] \\ &+ h \cdot \left[K_a' \cdot \left(H_i - \frac{V_o}{V_w}C_{TAN}\right) - K_e\right] \\ &- K_a' \cdot K_e = 0 \end{aligned} \tag{13.2}$$

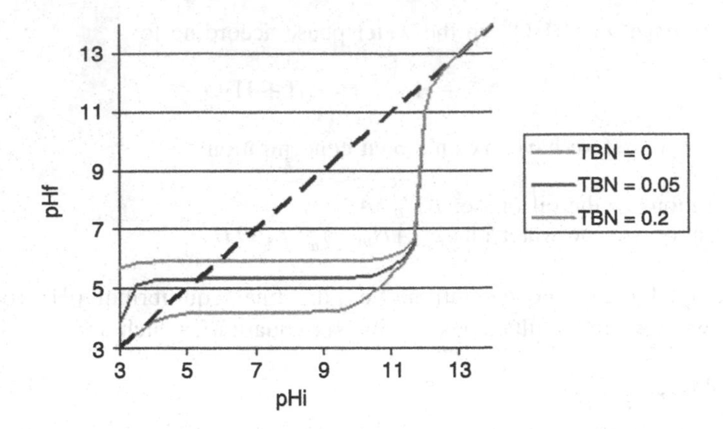

FIGURE 13.29 Calculated final pH as a function of the initial pH for an apparent dissociation constant value of 5.10 to 7 and TAN = 0.5. The TBN value is varied from 0 to 0.2.

where

$$K'_a = \frac{1 + \dfrac{V_o}{V_w} K_s^{A^-}}{1 + \dfrac{V_o}{V_w} K_s^{AH}} \cdot K_a$$

This equation is similar to that obtained in the case of an acid dissolved in a water phase, the dissociation constant of which would be K'_a instead of K_a. The partitioning coefficient of the acid and its conjugate base are therefore incorporated in an apparent dissociation constant of the acid.

Typical calculation results expressed as final pH as a function of the initial pH are shown in Figure 13.29. For this calculation, the apparent dissociation constant K'_a is set to 5×10^{-7}. The TAN of the oil phase is 0.5, and the total base number (TBN), which measures the concentration in sodium naphthenates in the oil phase, is varied from 0 to 0.2. The TBN value has a tremendous effect on the final pH value at low initial pH. Specifically, the final pH can be higher than the initial pH even with a very low and non-detectable concentration of naphthenates in the oil phase. This is important information to correctly interpret experimental curves such as those shown in Figure 13.17.

The physics behind this model is similar to that developed by Havre et al. [39], but has been detailed in order to describe our experimental conditions. Indeed, it has been developed to extract dissociation constant from bottle test experiments, without dosing the acid in the water or oil phase.

It should be pointed out that this modeling does not take account of the fraction of acids at the interface. The main reason for this simplification is that we are considering totally separated systems in which the interfacial area is negligible. The likely micellization of the acids is also discarded, as we are considering diluted systems only.

13.4.3 Application to Model Systems

To test these representations of the complex crude oil/water system, model systems made of carboxylic acid in heptane as an oil phase were used. These systems were studied using the bottle-test procedure described above for the crude oils.

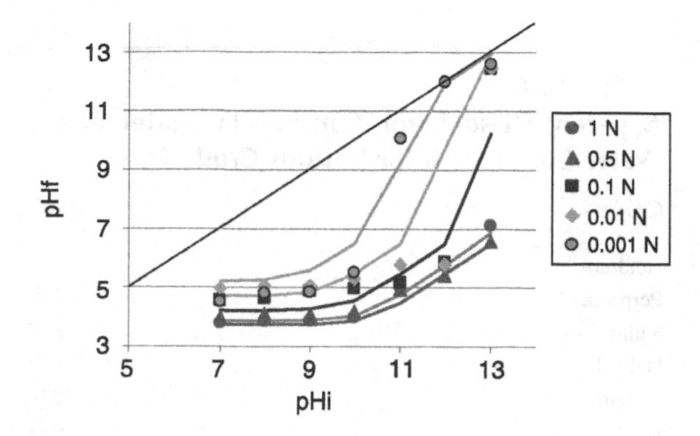

FIGURE 13.30 Final pH vs. initial pH for octanoic acid as a function of initial acid concentration in the oil phase. Solid lines are calculated using Equation 13.2 with pKa = 4.9, $K_S^{AH} = 360$, and $K_S^{A^-} = 0$.

13.4.3.1 Materials and Method

Octanoic acid is used as a model carboxylic acid in this study. Dissolving octanoic acid in heptane produces a model acidic crude, which can be used in bottle-test. Solutions at various concentrations of octanoic acid in heptane were prepared, and placed in contact with aqueous solutions at various initial pH. The pH is fixed by addition of soda in distilled water. The studied acid concentrations were: 1 N, 0.5 N, 0.01 N, and 0.001 N. For each concentration, the initial pH considered was 7, 8, 9, 10, 11, 12, and 13. The water–oil ratio (WOR) in the experiments is 1. The aqueous and oil solutions are prepared in 20 ml tubes that are manually shaken three times. No stable emulsions are observed in these experiments. The final pH is measured on the water phase after 24 h of decantation.

13.4.3.2 Results

Figure 13.30 gives the variation of the measured final pH as a function of the initial pH and the concentration. At a given total concentration, we observe an initial plateau followed by a sudden increase in the final pH. It is seen that the initial plateau depends strongly on the initial concentration of the acid in the oil phase. As expected, the final pH increases when the concentration decreases. The magnitude of the increase at high initial pH also strongly depends on the concentration. When little acid is available to neutralize the soda, the pH increases very rapidly.

Fitting of the data using Equation 13.2 is also reported in Figure 13.30. The generally admitted value of 4.9 is used for the pKa of the octanoic acid, and the partitioning coefficient of the protonated acid is fixed at 360. The partitioning coefficient of the dissociated acid is considered to be 0. The agreement between calculated and experimental values is satisfactory, especially for the higher octanoic acid concentrations.

13.4.4 Application to Crude Oils

13.4.4.1 Determination of Apparent Dissociation Constants

The model described above has been applied successfully to crude oils. Fitting the experimental data at a fixed water/oil ratio can give values of the apparent dissociation constant, which incorporates the partitioning effect, for various crude oils and provides a numerical basis for

TABLE 13.6
Apparent Dissociation Constant Evaluated at a Water Cut of 10% for Various Crude Oils

Crude	pKa′
Heidrun	6.16
Perpetua	7.85
Kuito	8.70
Dalia 1	8.86
Orquidea	9.00
BAP	10.00
Ekoundou	10.45
Dalia 2	13.00

comparison. Values of the apparent dissociation constant determined at a water cut of 10% are reported in Table 13.6. A wide range of values is observed, from a pKa′ value of 6.16 for the Heidrun crude to 13 for the Dalia 2 fluid. The majority of crude tested range from pKa′ = 8 to pKa′ = 9.

13.4.4.2 Relationship to Emulsion Stability

Fitting the final pH values from the bottle test experiments can only give access to the apparent dissociation constant of the crude. To estimate the actual dissociation constant and the partitioning coefficients, more data are needed.

In the case of the Dalia 1 fluid, UV measurements have been performed on the water phase extracted from the bottle tests. It is assumed that UV absorption is mainly sensitive to the dissolved naphthenic acids and naphthenates, thus providing a good indication of their concentration and therefore on the partition coefficient. Consequently, fitting the UV data with the calculated total dissolved acid concentration provides pKa and K_S^{AH} values. First, a fit of the final pH is obtained with an apparent dissociation constant value of pKa′ = 8.86 (Figure 13.31). Then the UV data are used to determine pKa = 7.3 and $K_S^{AH} = 4$ (Figure 13.32). It should be pointed out that the actual TAN value is used in the calculation, and that a small TBN value of 0.001 is introduced to recover a higher final than initial pH value under acidic conditions. The agreement between experimental and calculated values is excellent.

Having determined the dissociation constant and the partition coefficient of the acid, it is possible to calculate the dissociation coefficient of the acid, defined as the percentage of the acid dissociated in the water phase (Equation 13.1).

A good correlation is observed between the dissociation coefficient and non-resolved water fraction in the bottle test experiment, as shown in Figure 13.33. These results suggest that the naphthenates are indeed responsible in this case for the stabilization of the water in crude oil emulsion. It should be pointed out that the stabilization of the water in crude oil emulsion is an interfacial property which is here described by a model considering only bulk phase properties (partitioning and dissociation).

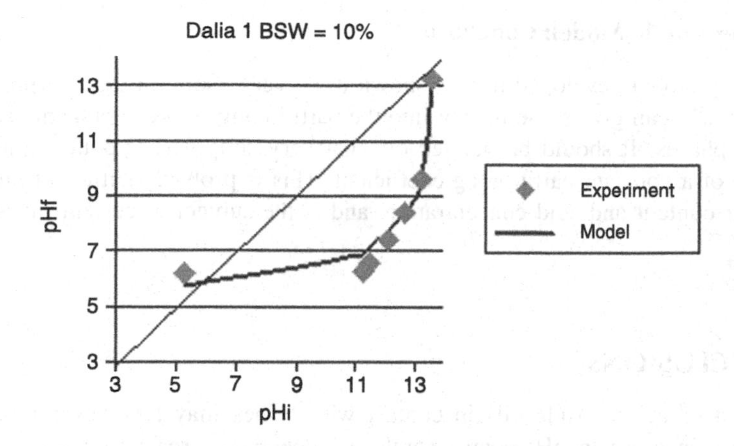

FIGURE 13.31 Final pH vs. initial pH and fitting by the numerical model for Dalia 1. pKa′ = 8.86.

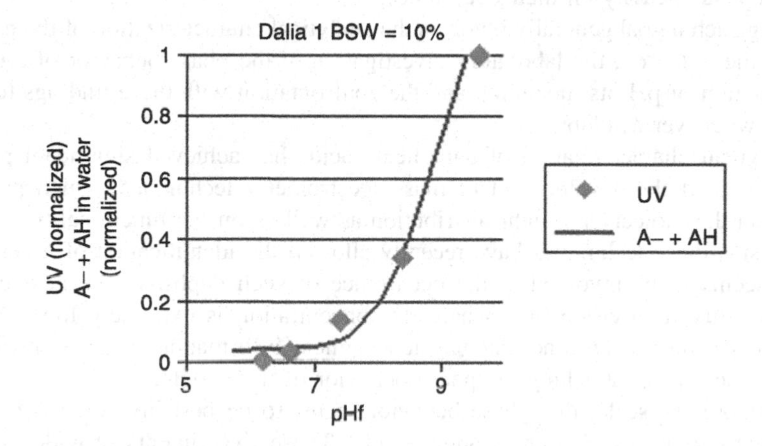

FIGURE 13.32 UV measurements vs. final pH and calculated total acid concentration in water. pKa = 7.3 and $K_S^{AH} = 4$.

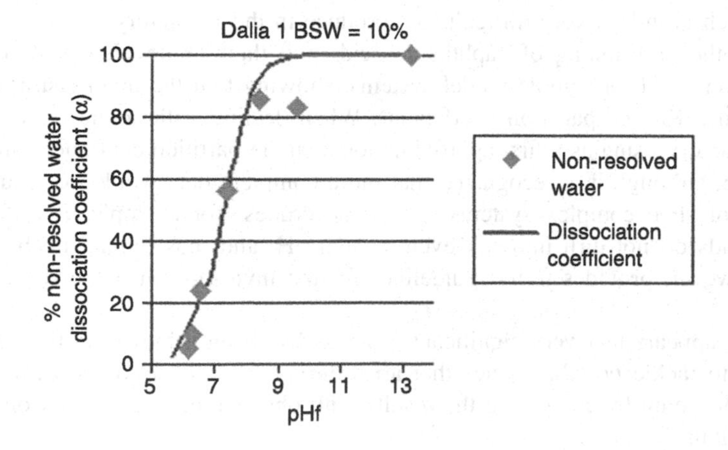

FIGURE 13.33 Non-resolved water and dissociation coefficient vs. final pH for the Dalia 1 case.

13.4.4.3 The Simple Model: Conclusion

The very simple model developed here is shown to be very useful in the interpretation of bottle-test experiments. It can give some insight into the partitioning of the naphthenic acid between the water and oil phases. It should be pointed out that very simplistic hypotheses are used, among which the use of a constant partitioning coefficient. This is probably sufficient only for a limited range of water content and acid concentration, and is the subject of current investigations.

13.5 CONCLUSIONS

The production of acidic crude oils in contact with brines may face several issues, whenever submitted to an increase in pH: coarse emulsion appearance, scale formation, and poor water quality. Predicting such phenomena is a real challenge for oil companies since the design of field equipment depends heavily on their occurrence.

Achieving such a goal generally involves the analytical characterization of the naphthenic acid contained in the crude oil, the laboratory investigation of the phase behavior of crude–brine systems as a function of pH, its modeling, and the confrontation with these findings to the observed field results, whenever available.

The analytical characterization of naphthenic acids has achieved significant progress, especially as a result of the development of mass spectrometry techniques, which provide detailed information on the molecular weight distribution as well as on the functionalities. When applied to field deposits, these techniques have recently allowed the identification of a very heavy tetra-acid which seems to be involved in the occurrence of such deposits. The next challenge is to be able to identify it in crude oils, where its concentration is extremely low. However, these analytical tools cannot determine any physicochemical information, such as pKa, which is of primary relevance for establishing the phase behavior of the system.

On the laboratory scale, this phase behavior seems to be best investigated by using bottle-test experiments carried out over a range of pH. These experiments provide good predictive information, at least regarding the formation of stable emulsions and the quality of the resolved water. Gel-like phases are also obtained, which are much softer than observed on some fields, and may thus be interpreted as the precursors: in field production, there is an accumulation effect over time which is indeed very difficult to reproduce in the laboratory.

Modeling the partitioning of naphthenic acids and their monovalent salts between oil and water works very well on simple model systems, showing that the main characteristics are correctly captured (pKa and partition coefficient). When dealing with acidic crude oils, the results are also satisfactory using as a first approximation a single partition coefficient and a single average pKa value, although it is recognized that more complete data would be required to describe the behavior of such complex systems. For some crudes, for example, it is found that some naphthenic acids do not turn into salt even at high pH, and thus remain in the oil phase. This modeling, however, provides useful guidelines for the investigation of newly discovered acidic oil fields.

Overall, it appears that very significant progress has been achieved in the development of a methodology to tackle possible issues that arise during the production of acidic crude oils. Its validation is on-going by comparing the results with observations carried out on fields presently under production.

ACKNOWLEDGMENTS

The authors wish to express their appreciation to S. Drouilhet, S. Tallet, B. Abbouab, P. Michaud, and H. Dessinet who carried out most of the experiments; to B. Escoffier, A. Goldszal, J.L. Volle, and H.G. Zhou for fruitful discussions; and to the Total management for permission to publish this work.

REFERENCES

1. RF-Rogaland Research Institute, American Chemical Society. Division of Petroleum Chemistry, 43 (1) 142–145 (1998).
2. Tseng-pu Fann. Energy & Fuels, 5 (3) 371-375 (1991).
3. Koike L., Reboucas L., Marsaioli A., Richnow H., Michaelis W. Organic Geochemistry, 18 (6) 851–860 (1992).
4. Robbins W.K. ACS Petrol Chem Div preprint, 43 (1) 137–140 (1998).
5. Nascimento L., Reboucas L., Koike L., Reis F., Soldan A., Cerqueira J., Marsaioli A. Organic Geochemistry, 30 (9) 1175–1191 (1999).
6. Baugh T., Wolf N.O., Mediaas H., Vindstad J.E. Paper presented at the 228th ACS National Meeting, Div. of Pet. Chem., Philadelphia (Aug. 22–26, 2004).
7. Gallup D., Star J. SPE 87471 presented at the 6th International Symposium on Oilfield Scale (May 2004).
8. Brient J.A., Wessner P.J., Doyle M.N. Kirk Othmer Encyclopaedia, Vol.16, 1017–1029.
9. Seifert W., Herz W., Grisebach H., Kirby G. Progress in the Chemistry of Organic Natural Products, 32, 1–49 (1975).
10. Niyazov A.N., Amanov K.B., Trapeznikova V.F., Chirkova B.V. Khim. Geol. Nauuk, 4, 121–123 (1975).
11. Marquez M.L. AIChE Spring National Meeting Session T6005 (March 1999).
12. Vindstad J.E., Bye A.S., Grande K.V., Hustad B.M., Hutvedt E., Nergard B. SPE 80375 presented at the SPE 5th International Symposium on Oilfield Scale, Aberdeen (Jan. 2003).
13. Hurtevent C., Rousseau G., Zhou H.G. Calcium carbonate and naphthenate mixed scale in deep-offshore fields, SPE 68307 presented at the 3rd International Symposium on Oilfield Scale (Jan. 2001).
14. Horvath-Szabo G., Masliyah J.H., Czarnecki J. J. Coll Interf. Sci., 257, 299 (2003).
15. Horvath-Szabo G., Czarnecki J., Masliyah J.H. J. Coll Interf. Sci., 253, 427 (2003).
16. Dyer S.J., Graham G.M., Arnott C. SPE 80395 presented at the 5th International Symposium on Oilfield Scale, Aberdeen (Jan. 2003).
17. Kane R.D., Cayard M.S. Hydrocarbon Processing, 97–103 (Oct. 1999).
18. Hau J.L., Yépez O., Specht M., Lorenzo R. Paper No. 379, Corrosion/99, NACE International, Houston (1999).
19. Naphthenic Acid Corrosion Review Page, Set Laboratories, Internet site: www.setlaboratories.com/nac.htm.
20. Jenkins G.I. J. Inst. Petrol., 51, 501, 313–322 (1965).
21. Meredith W., Kelland S.J., Jones D.M. Organic Geochemistry, 31 (11) 1059–1073 (2000).
22. Jones D.M., Watson J.S., Meredith W., Chen M., Bennett B. Analytical Chemistry, 73, 703–707 (2001).
23. Mediaas H., Grande K.V., Hustad B.M., Rasch A., Rueslatten H.G., Vindstad J.E. SPE No. 80404 presented at the SPE 5th Int. Symp. On Oilfield Scale, Aberdeen (Jan. 2003).
24. Turnbull A., Slavcheva E., Shone B. Corrosion, 54, 922–930 (1998).
25. Corporate Research, Exxon Research and Engineering Co. Book of Abstracts, 215th ACS National Meeting, Dallas, March 29 to April 2 (1998).
26. Messer B., Tarleton B., Beaton M. Paper N° 04634 Corrosion/2004, NACE International, Houston (2004).

27. St. John W.P., Rughani J., Green S.A., McGinnis G.D. J. Chromatography A, 807, 241–251 (1998).
28. Wong D.C.L., van Compernolle R., Nowlin, J.G., O'Neal D.L., Johnson G.M. Chemosphere, 32, 1669–1679 (1996).
29. Hsu C.S., Dechert G.J., Robbins W.K., Fukuda E.K. Energy & Fuels, 14, 217–223 (2000).
30. Miyabayashi K., Suzuki K., Teranishi T., Naito Y., Tsujimoto K., Miyake M. Chem. Lett., 172–173 (2000).
31. Miyabayashi K., Yasuhide N., Miyake M., Tsujimoto K. Eur. J. Mass Spectrom., 6, 251–258 (2000).
32. Zhan D., Fenn J.B. Int. J. Mass Spectrom., 194, 197–208 (2000).
33. Wilm M., Mann M. Anal. Chem., 68, 1–8 (1996).
34. Barrow M.P., McDonnell L.A., Feng X., Walker J., Derrick P. J. Anal. Chem., 75, 860–866 (2003).
35. Seifert W.K., Howells W.G. Anal. Chem., 41 (4), 554–562 (1969).
36. Rousseau G., Zhou H., Hurtevent C. SPE 68307 presented at the 3rd International Symposium on Oilfield Scale, Aberdeen (Jan. 2001).
37. Goldszal A., Hurtevent C., Rousseau G. SPE 74661 presented at the 4th International Symposium on Oilfield Scale; Aberdeen (Jan. 2002).
38. Verzaro F., Bourrel M., Chambu C. "Solubilization in Microemulsions," 5th Int. Symp. On Surf. In Sol., Bordeaux, KL. Mittal ed., 1137–1157 (1984).
39. Havre T.E., Sjöblom J., Vindstad J.E. J. Disp. Sci. Technol., 24 (6), 789–801 (2003).
40. Bourrel M., Graciaa A., Schechter R.S., Wade W.H. J. Coll. Int. Sci., 72, 161 (1979).
41. Salager J.L., Loaiza-Maldonado I., Miñana-Pérez M., Silva F. J. Disp. Sci. Technol., 3, 279–292 (1982).
42. Goldszal A., Bourrel M. Ind. & Eng. Chem. Research, 39 (8), 2716–2727 (2000).
43. Salager J.L., Morgan J.C., Schechter R.S., Wade W.H., Vasquez E. Soc. Pet. Eng. J., 23, 669 (1983).
44. Winsor P.A. "Solvent Properties of Amphiphilic Compounds", Butterworth, London (1954).
45. Bourrel M., Schechter R.S. "Microemulsions and Related Systems: Formulation, Solvency and Physical Properties", Surfactant Science Series, Vol 30, M. Dekker, New York (1988).
46. Goldszal A., Bourrel M., Hurtevent C., Volle J.L. Paper presented at the Spring A.I.Ch.E Meeting New Orleans (March 2002).
47. Friberg S., Mandell L., Larsson M. J. Coll. Int. Sci., 29, 155 (1969).
48. Havre T.E., Sjöblom J. Coll. Surf. A; Phys. Eng. Aspects, 228, 131 (2003).
49. Lucassen J. J. Phys. Chem., 70, 1824 (1966).
50. Griffin W.C. J. Soc. Cosmetic Chem., 1, 311 (1949) ibid. 5, 249 (1954).
51. Stainsby G., Alexander A.E. Trans Faraday Soc., 45, 585 (1949).
52. Biais J., Bothorel P., Clin B., Lalanne P. J. Disp. Sci. Technol., 1, 67 (1981).
53. Graciaa A., Lachaise J., Sayous J.G., Bourrel M., Schechter R.S., Wade W.H. Proceedings of the Second European Symposium on EOR Techniques, Paris, p. 61 (1982).

14 Formation of Crude Oil Emulsions in Chemical Flooding

Mingyuan Li, Jixiang Guo, Bo Peng, Meiqin Lin, Zhaoxia Dong, and Zhaoliang Wu

CONTENTS

14.1 INTRODUCTION

Alkaline–surfactant–polymer (ASP) flooding is a new technique that has been tested in Da Qing and Sheng Li oil fields in China in the past few years. The technique uses partially hydrolyzed polyacrylamide (HPAM) as the polymer, petroleum sulfonate (ORS-41 or TRS) as the surfactant, and sodium hydroxide (NaOH) or sodium carbonate (Na_2CO_3) as the alkaline component. Compared with waterflood ASP flooding may increase oil recovery efficiency by about 15 to 20%, but the use of the new technique is faced with new problems. The production fluids form stable w/o and o/w emulsions and the separation of crude oil and water becomes more difficult. It is believed that the alkali, surfactant, and polymer used in the chemical flooding causes these problems [1].

The concentration of alkali used for ASP flooding is about 0.8 to 1.2% in water; the alkali reacts with the acidic components (resins and asphaltenes) in crude oil and form interfacially active components that accumulate at the oil–water interface and facilitates the formation of emulsion [1]. The stability of this emulsion depends on the concentration of the reservoir formed alkali–oil surfactant at the interface. This again depends on the concentration of the potential acidic components from crude oil that form interfacially active soap components [1]. However, Da Qing crude oil contains very little asphaltene (paraffinic crude oil), and has lower acid number, when sodium hydroxide used as the alkaline component in the recovery of crude was enhanced and the oil recovered also contained stable water-in-crude-oil emulsion. Some of the studies concerning the stability of water-in-crude-oil emulsions have shown that the surfactant and the polymer are not responsible for enhancing the stability of the water-in-oil emulsion [1]. HPAM enhances mainly the stability of oil-in-water emulsions and makes the water treatment difficult. Our investigations revealed that the resin, asphaltene, and saturated fractions of Da Qing crude oil reacted with sodium hydroxide and for a few days were unable to form stable emulsions. The anomalous behavior of the paraffinic crude oil in enhancing the stability of water-in-crude-oil emulsion in the presence of sodium hydroxide in the oil field led us to carry out a laboratory study regarding the chemical nature of the problem. To understand the formation of water-in-oil emulsions of Da Qing crude oil our focus is to study the long time effect of NaOH with the saturated fraction of the crude.

The concentration of the surfactant used for ASP flooding is about 0.1–0.3% in water. As an effect of surfactant coordinate with alkali the interfacial tension between crude oil and ASP solution may be decreased to 10^{-3} mN m^{-1}. Therefore, the crude and ASP solution can be emulsified in reservoir and pipelines during crude production.

The concentration of the polymer used for ASP flooding is 1000 to 2400 mg/l. After the polymer solution passes through the reservoir for months or years the concentration of the polymer in the production water is about 10 to 300 mg/l. It is believed that the polymer in the water is responsible for enhancing the stability of o/w emulsion [1]. The content of crude oil in the production water after treatment by present technology is about 2300 mg/l that is much higher than the standard value of 30 mg/l in Sheng Li oil field.

In a series of articles Sjöblom and co-workers have investigated the properties of water-in-crude-oil emulsions based on the North Sea crude oils [2–16]. Interests has focused on different aspects of stability, destabilization, and separation of interfacially active components, dielectric properties, and the design of adequate model systems. Furthermore, chemical modifications of the interfacially active components, both artificial and natural aging, have been carried out to understand the effect of molecular properties on the stability of water-in-crude-oil emulsions.

Crude oil is a mixture of aliphatic and aromatic hydrocarbons, and oxygen-, nitrogen-, and sulfur-containing compounds such as resins and asphaltenes. There is no doubt that the interfacially active components come from resins and asphaltenes of the crude oils [17]. The asphaltene fraction from crude oils can be adsorbed on the interface between oil and water, and forms stable film to stabilizing crude emulsions [18]. Resins and asphaltenes are polymeric, containing polyaromatic structures, and possess structural similarities between them [17]. They are differentiated by their solubility in light hydrocarbons such as pentane. This differentiation itself clearly indicates that the resins (soluble in pentane) are smaller on a molecular level compared to asphaltenes (insoluble in pentane). Within each fraction of resins and asphaltenes, there is a range of components that are of different sizes and contain different functional groups. The molecules of asphaltenes contain condensed aromatic rings of different sizes. The asphaltenes contain large condensed aromatic rings compared to resins [2]. The stability of water-in-crude-oil emulsions depends on the total structure of the molecular matrix of the interfacially active components. Size, aromaticity, types of carbonyl functionality, and other functional groups in the bulk play an important role in the total stability of the emulsions.

All these studies have given us insight into the problems relating to stability of water-in-crude-oil emulsions. It is obvious that a study that aims to understand the problems related in the ASP flooding technique should involve the interfacially active components from the crude oil. However, the effect of alkali, surfactant, and polymer reactions with the interfacially active fractions from crude oil on the stability of emulsions has not been fully studied.

In order to study the formation of crude oil emulsions in ASP flooding and the effect of alkali, surfactant, and polymer reactions with the interfacially active fractions from crude oil on the stability of emulsions, the saturated, aromatic, resin, and asphaltene fractions from a crude oil in Da Qing oil field and two Gu Dong crude oils from Sheng Li oil field in China were separated first. Then the acid number, elemental composition, and molecular weight determinations of the fractions were made to evaluate the properties of the fractions. The correlations between the physical properties and molecular parameters such as molecular weight and functional groups were explored. Furthermore, the fractions were then used to prepare model oils with additive-free jet oil. The interfacial tension, interfacial shear viscosity, and zeta potential between the model oils and ASP solution and the stability of the emulsions formed with the model oils and ASP solution were determined.

14.2 EFFECT OF ALKALI

In this study the alkali solution used for Gu Dong model oils is 1.2% Na_2CO_3 solution and for Da Qing model oils is 1.2% NaOH solution. The alkali solution was prepared with double-distilled water.

14.2.1 Experimental

14.2.1.1 Separation of Crude Oil Fractions

The interfacially active fractions of saturated, aromatic, resin, and asphaltene used in these experiments were separated from Da Qing and Gu Dong crude oils. The separation of asphaltene fractions from the crude oil was carried out by pentane precipitation. One hundred grams of crude oil was agitated with 3000 ml of pentane at room temperature for 30 min. The mixture was then left to stand for 15 days. The precipitated asphaltene fraction was filtered and washed with a small portion of pentane and dried. The filtrate from the above was then poured on a column containing Al_2O_3. The saturate fraction was extracted by 8000 ml petroleum ether, the aromatic fraction was extracted by 8000 ml benzene, the resin 1 fraction was extracted by

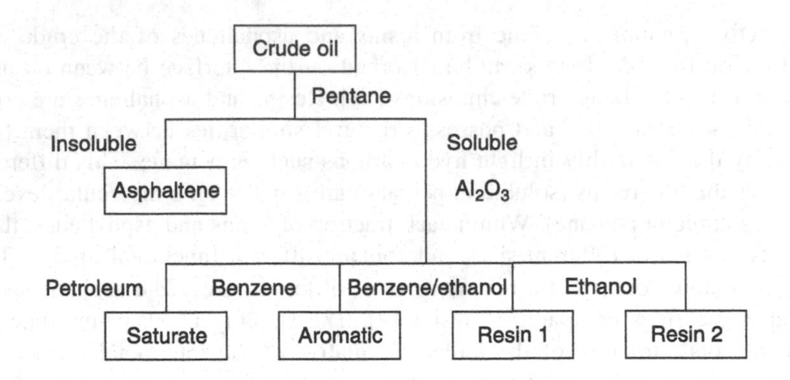

FIGURE 14.1 Separation of crude oil fractions.

TABLE 14.1
Composition of Model Oil (wt%)

	Crude Oil	Saturate	Aromatic	Resin 1	Asphaltene
Gu Dong 1#	10.00	4.59	2.25	1.49	1.44
Gu Dong 4#	10.00	4.02	2.19	1.49	1.14
Da Qing	10.00	6.43	1.52	1.41	0.10

4000 ml benzene/ethanol (benzene/ethanol ratio: 1/1, v/v), and the resin 2 fraction was extracted by 4000 ml ethanol. The separation scheme is presented in Figure 14.1.

14.2.1.2 Elemental Composition and Molecular Weight

The relative molecular weight of the fractions was measured by the VPO method using a molecular weight meter (Knauer, Germany). The elemental composition of the crude oil was determined by Elementar Analysen Systeme, Vario El, Germany.

14.2.1.3 Model Emulsions and Measurement of Interfacial Properties

The water used in the experiments was double-distilled water or alkaline, surfactant, polymer solution. A jet fuel from a refinery without any additives was used as the dispersion media of the model oils. The jet fuel was purified by silica adsorption before the experiments were carried out. The composition of the model oils is shown in Table 14.1. The weight percentage of each fraction is selected to represent the weight percentage of the fraction (around 10%) in the crude oils (Table 14.2). This would allow us to compare the stability of emulsions formed between model oils containing the fractions and the chemical solutions. From Table 14.1 the crude model oil contains 10% of the crude oils in the jet oil and the saturate model oil contains 4.59%, 4.02%, and 6.43% saturated fraction from Gu Dong 1#, Gu Dong 4#, and Da Qing crude oils, respectively.

The model emulsions were prepared by mixing the model oils and alkaline solution (o/w ratio: 2/8, v/v) in a 50 ml cylinder. The emulsification was carried out by shaking the cylinder 50 times at 60 °C. The emulsification was done twice every day and the stability of the emulsions

TABLE 14.2
Composition of Crude Oil Fractions

	Wt %		
Fraction	**Gu Dong 1#**	**Gu Dong 4#**	**Da Qing**
Saturate	47.08	46.54	68.09
Aromatic	23.65	28.18	17.25
Resin 1	14.73	13.81	14.47
Resin 2	0.11	0.12	0.10
Asphaltene	14.43	11.35	0.09

was determined visually by measuring the water separated from the emulsions at 60 °C as a function of time. The emulsion stability experiments were carried out for 84 days. This would allow us to investigate the effect of reaction time of the alkali on the properties of the model emulsions.

The interfacial tension between the model oils and distilled water or the chemical solution was measured by a spinning drop interfacial tension meter (JJ2000A, Shanghai, China) at 45 °C. The interfacial shear viscosities were measured using a SVR·S Interfacial Viscoelastic Meter (Kyowa Kagaku Co. Ltd, Japan) [18] at 25 °C.

The zeta potential on oil droplets was measured by a zeta potential meter (ZPOM, Kyowa Kagaku Co. Ltd, Japan) at 25 °C. There was about 10 mg/l oil in the emulsion (o/w) used for zeta potential measurement.

14.2.2 Results and Discussion

14.2.2.1 Physical Properties of Crude Fractions

The composition of the crude oils is given in Table 14.2; the content of asphaltene fraction in Gu Dong 1# and Gu Dong 4# crude oils is 14.43% and 11.35% but it is only 0.09% in Da Qing crude oil. Da Qing crude contains more saturate fraction than Gu Dong crude oils. Because the resin 2 fraction is too little to do experiments in the work resin 1 fraction is presented as the resin fraction in this chapter.

The elemental composition of the crude oils and their fractions is given in Table 14.3. The value of the H:C ratio decreased in the order of saturated > aromatic ≈ resin > asphaltene for the three crude oils.

Table 14.4 shows that the saturate fraction from Da Qing crude oil contains most oxygen (0.375 g in 100 g crude oil), following with the asphaltene fractions from Gu Dong 1# and Gu Dong 4# crude oils. The asphaltene fraction from Da Qing crude oil contains only 0.001 g oxygen in 100 g crude oil.

The molecular weight and acid number of the fractions are given in Tables 14.5 and 14.6. It is clear that the molecular weight of the fractions is in the order of saturated < aromatic < resin < asphaltene for all of the three crude oils and the molecular weight of the crude oils is close to their saturated fraction, respectively.

The values presented in Table 14.6 clearly shows that the acid number of the fractions is in the order of saturated < aromatic < resin < asphaltene for Gu Dong crude oils. The order is the same as the order of molecular weight of the fractions. The acid number of the fractions from

TABLE 14.3
Elemental Composition of Crude Oils and their Fractions

Crude oil	Fraction	$w(C)/\%$	$w(H)/\%$	$w(O)/\%$	$w(S)/\%$	$w(N)/\%$	$n(H):n(C)$
	Saturate	86.37	13.24	0.25	0.65	1.00	1.84
	Aromatic	87.76	10.38	0.72	0.69	1.20	1.42
Gu Dong 1#	Resin 1	85.06	10.12	1.32	0.72	1.47	1.43
	Asphaltene	85.97	9.38	2.19	0.50	1.47	1.31
	Crude oil	86.70	11.71	0.79	0.38	1.01	1.62
	Saturate	86.12	13.16	0.30	0.59	0.89	1.83
	Aromatic	84.32	10.11	0.75	3.13	1.42	1.44
Gu Dong 4#	Resin 1	84.08	10.31	1.91	2.18	1.48	1.47
	Asphaltene	81.67	8.89	2.79	4.63	1.44	1.31
	Crude oil	84.88	11.35	0.86	0.41	1.10	1.60
	Saturate	85.54	14.26	0.55	0.03	0.73	1.96
	Aromatic	86.92	11.22	0.50	0.34	1.53	1.52
Da Qing	Resin 1	86.25	10.79	1.48	0.45	1.03	1.50
	Asphaltene	86.19	10.53	1.14	0.18	1.09	1.47
	Crude oil	86.53	12.50	0.73	0.13	0.86	1.73

TABLE 14.4
Oxygen in Crude Oil Fractions

	Oxygen in 100 g Crude Oil/g		
Fraction	Gu Dong 1#	Gu Dong 4#	Da Qing
Saturate	0.118	0.140	0.375
Aromatic	0.170	0.211	0.086
Resin 1	0.194	0.264	0.214
Asphaltene	0.316	0.317	0.001

TABLE 14.5
Molecular Weights of Crude Oils and their Fractions

Crude Oil	Saturate	Aromatic	Resin 1	Asphaltene	Crude Oil
Gu Dong 1#	434	601	1025	1499	433
Gu Dong 4#	503	728	1117	1308	427
Da Qing	485	773	1396	2433	480

Da Qing crude oil is in the order of saturated < aromatic < asphaltene < resin. It seems that the resin fraction contains more acidic matters than the asphaltene fraction in Da Qing crude oil.

The above facts indicate that the asphaltene molecules are large, polyaromatic, and contain the highest acidic oxygen components among the fractions for Gu Dong crude oils, and the resin fraction from Da Qing crude oil contains the highest acidic oxygen components among the fractions from the crude oil.

TABLE 14.6
Acid Numbers of Crude Oils and their Fractions

Fraction	Gu Dong 1#	Gu Dong 4#	Da Qing
Saturate	2.397	3.014	0.4760
Aromatic	5.386	5.242	1.039
Resin 1	6.001	8.002	5.101
Asphaltene	16.45	8.378	4.213
Crude oil	3.640	3.217	0.5174

TABLE 14.7
Interfacial Tension Between Model Oils and Aqueous Phases (45 °C)

Crude Oil		Model Oil				
		Saturate	Aromatic	Resin 1	Asphaltene	Crude oil
Gu Dong 1#	w %	4.60	2.53	1.46	1.44	10.00
	$\gamma_{o/w}$/mN m^{-1}	35.70	24.52	22.76	19.38	11.89
	$\gamma_{o/s}$/mN m^{-1}	13.36	5.61	4.63	0.056	0.93
	w %	3.00	3.00	3.00	3.00	3.00
	$\gamma_{o/w}$/mN m^{-1}	33.03	17.90	15.78	12.75	19.27
	$\gamma_{o/s}$/mN m^{-1}	11.22	4.37	3.12	0.0053	0.762
Gu Dong 4#	w %	4.02	2.19	1.38	1.14	10.00
	$\gamma_{o/w}$/mN m^{-1}	33.55	18.85	18.36	16.90	17.21
	$\gamma_{o/s}$/mN m^{-1}	9.12	7.11	4.30	0.86	1.71
	w %	3.00	3.00	3.00	3.00	3.00
	$\gamma_{o/w}$/mN m^{-1}	30.07	14.42	13.33	7.03	21.48
	$\gamma_{o/s}$/mN m^{-1}	10.45	5.32	2.62	0.43	1.35
Da Qing	w %	6.66	1.58	1.45	0.10	10.00
	$\gamma_{o/w}$/mN m^{-1}	39.96	33.57	27.63	33.68	30.46
	$\gamma_{o/s}$/mN m^{-1}	18.31	11.97	8.24	12.81	4.04
	w %	3.00	3.00	3.00	3.00	3.00
	$\gamma_{o/w}$/mN m^{-1}	36.51	30.24	26.34	28.22	29.40
	$\gamma_{o/s}$/mN m^{-1}	14.68	10.30	2.64	4.07	2.86

14.2.2.2 Interfacial Tension

The interfacial tension between the model oils and aqueous phases is shown in Table 14.7.

It shows that the interfacial tension between the model oils and alkali solutions ($\gamma_{o/s}$) is lower than the interfacial tension between the same model oil and distilled water ($\gamma_{o/w}$).

This indicates that all the fractions from the crude oils which reacted with the alkali are able to form interfacially active components, and these components are more interfacially active than the indigenous interfacially active components in the crude oil. Therefore, the interfacial tension was decreased.

It is interesting to note that the interfacial tension between the asphaltene model oils and Na_2CO_3 solution is lower than the interfacial tension between the crude model oils and the alkaline solution for Gu Dong $1^{\#}$ and Gu Dong $4^{\#}$ crude oils, especially the very low interfacial tension between asphaltene model oil and Na_2CO_3 solution for Gu Dong $1^{\#}$ crude oil (0.056 to 0.0053 $mN\,m^{-1}$). It appears that the formation of interfacially active components is very significant in the reaction of the alkaline and asphaltene fraction. It is clear that the asphaltene fraction dominates the interfacial tension between crude model oil and the alkali solution comparing the interfacial tension between other fraction model oils and alkali solution for Gu Dong crude oils. For Da Qing crude oil a resin fraction dominates the interfacial tension between crude model oil and NaOH solution. From Tables 14.4 and 14.6 it is clear that the asphaltene fraction of Gu Dong crude oils contains more oxygen and has a higher acid number, and the resin fraction of Da Qing crude oil contains more oxygen and has a higher acid number, compared with other fractions. It seems that the interfacially active components formed in the reaction of the alkali and the acidic oxygen components in asphaltene or resin fractions contribute to decrease the interfacial tension.

14.2.2.3 Interfacial Shear Viscosity

Figures 14.2 to 14.4 show that the interfacial shear viscosity between asphaltene or resin model oil and alkali solution is higher than the interfacial shear viscosity between the model oils and distilled water. It is clear that the interfacially active components formed in the reaction of the alkali and asphaltene or resin fraction are accumulated at the interface and the interfacial film has higher mechanical strength [19].

Figures 14.5 to 14.7 show that the interfacial shear viscosity between asphaltene model oils and Na_2CO_3 or NaOH solution has the highest value among the model oils and the alkali solutions measured. These indicate that the interfacial film of the interfacially active components formed in the reaction of asphaltene fractions and alkali solutions has the highest mechanical strength and is able to enhance the stability of w/o emulsions. Because the asphaltene fraction has the largest molecular weight it is obvious that the interfacially active components formed in the reaction

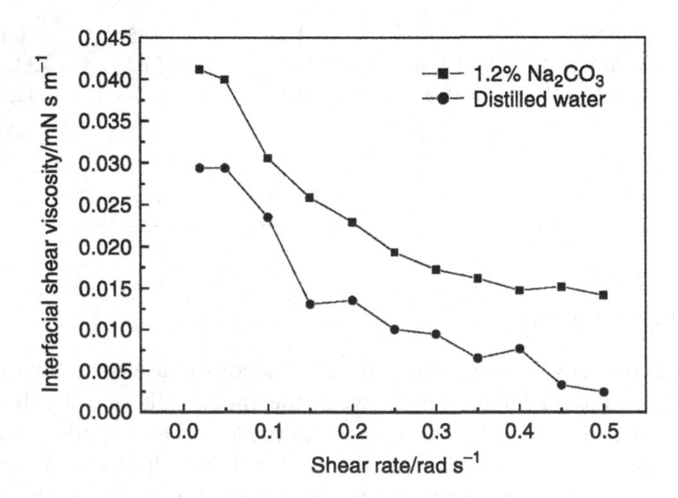

FIGURE 14.2 Interfacial shear viscosity between 2% asphaltene model oil (Gu Dong $1^{\#}$) and distilled water/1.2% Na_2CO_3 solution, 25 °C.

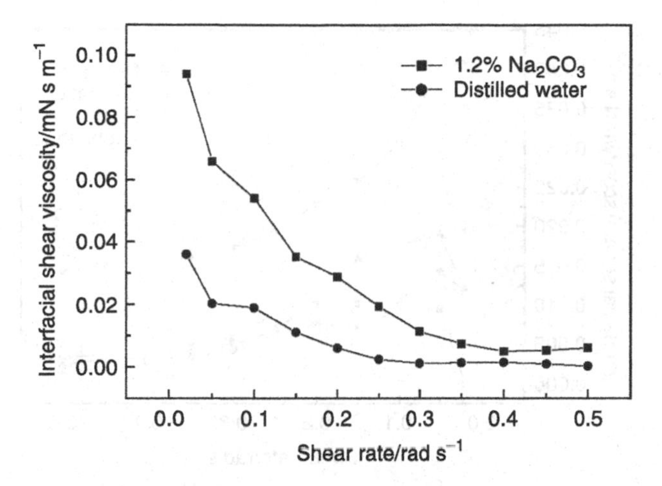

FIGURE 14.3 Interfacial shear viscosity between 2% asphaltene model oil (Gu Dong 4#) and distilled water/1.2% Na_2CO_3 solution, 25 °C.

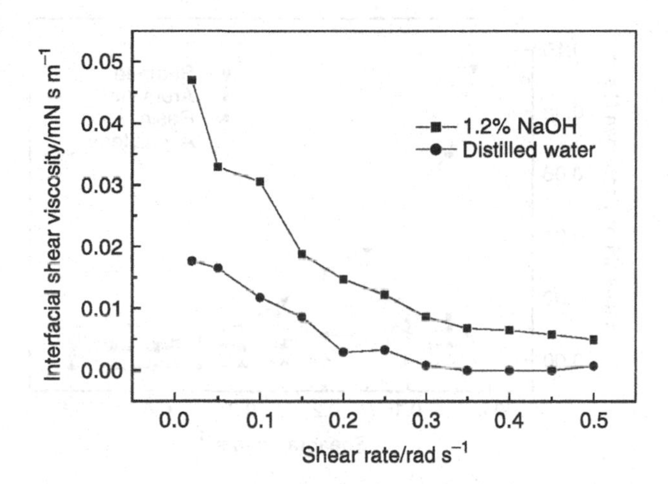

FIGURE 14.4 Interfacial shear viscosity between 2% resin model oil (Da Qing) and distilled water/1.2% NaOH solution, 25 °C.

have a larger molecular weight than that formed in the reaction of the alkali with other fractions. This proves that the interfacially active component which has a larger molecular weight is able to form more stable interfacial film and emulsions. It should also be noted that the interfacial shear viscosity of the film is decreased as the shear rate increases. This phenomenon shows that the film was structured and the structure was broken down as the shear stress was increased.

Figures 14.8 to 14.10 show that the interfacial shear viscosity between the model oils prepared with deasphalted crude oil and alkali solutions is lower than that of crude model oil and the alkali solutions, and the interfacial shear viscosity increases with the concentration of the crude oils in the jet fuel oil. This is a further indication that the asphaltene fraction contributes to the formation of interfacially active components and enhanced the strength of the film.

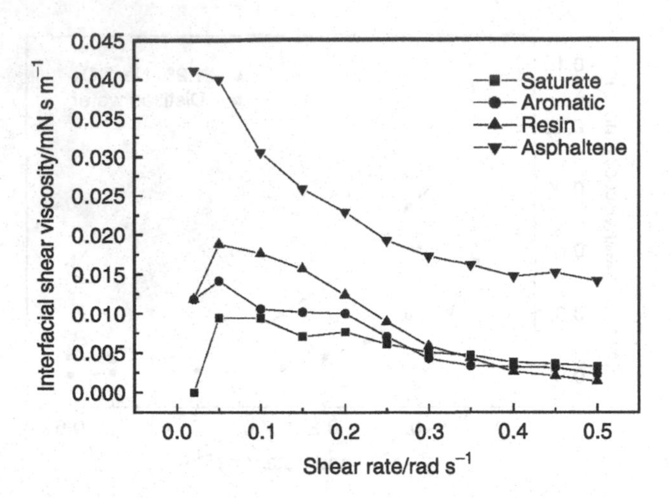

FIGURE 14.5 Interfacial shear viscosity between model oils (Gu Dong 1$^{\#}$) and 1.2% Na_2CO_3 solution, 25 °C.

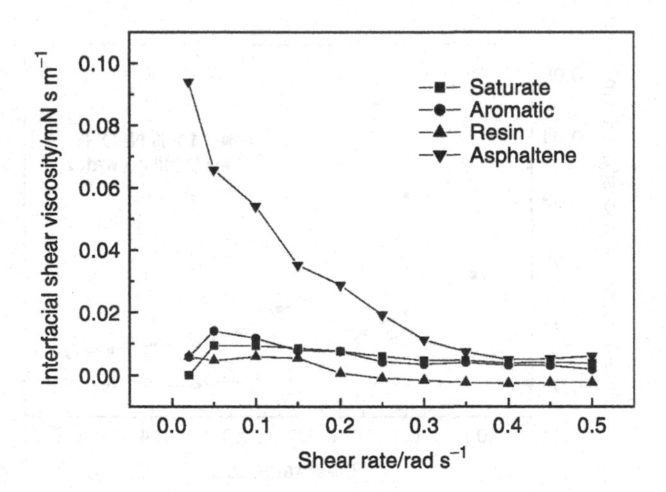

FIGURE 14.6 Interfacial shear viscosity between model oils (Gu Dong 4$^{\#}$) and 1.2% Na_2CO_3 solution, 25 °C.

All these experimental results proved that the interfacial film of the interfacially active components formed in the reaction of asphaltene fractions and alkali solutions has the highest mechanical strength and asphaltene fraction dominates the property of the interfacial film when the alkali solution was as aqueous phase.

14.2.2.4 Zeta Potential

Table 14.8 shows that the absolute value of zeta potential on the asphaltene model oil (Gu Dong 4$^{\#}$ crude oil) droplets was increased with increase of the reaction time of asphaltene and Na_2CO_3. This shows that the interfacially active components formed in the reaction can

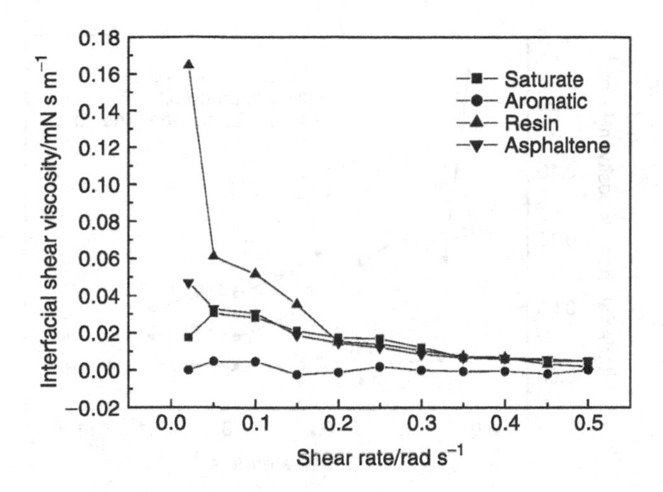

FIGURE 14.7 Interfacial shear viscosity between model oils (Da Qing) and 1.2% NaOH solution, 25 °C.

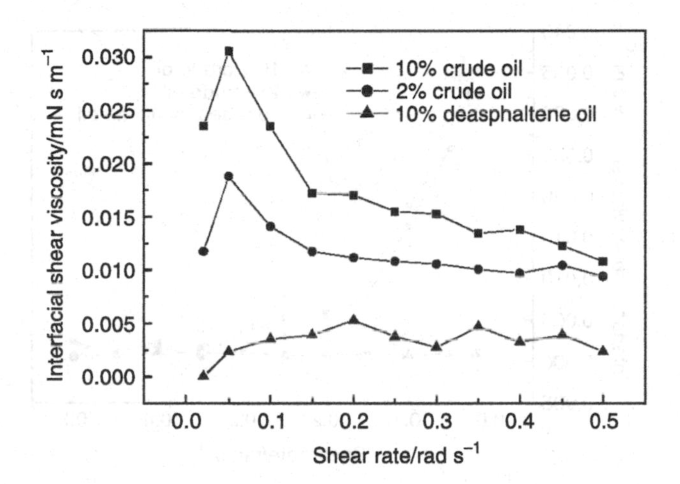

FIGURE 14.8 Interfacial shear viscosity between crude model oils (Gu Dong 1[#]) and 1.2% Na_2CO_3 water solution, 25 °C.

TABLE 14.8
Influence of Reaction Time of Na_2CO_3 Solution and 3% Crude Model Oil on Zeta Potential (mV, 25 °C)

Reaction Time (days)	Zeta Potential (mV)
0	−27.3
1	−45.9
10	−47.5

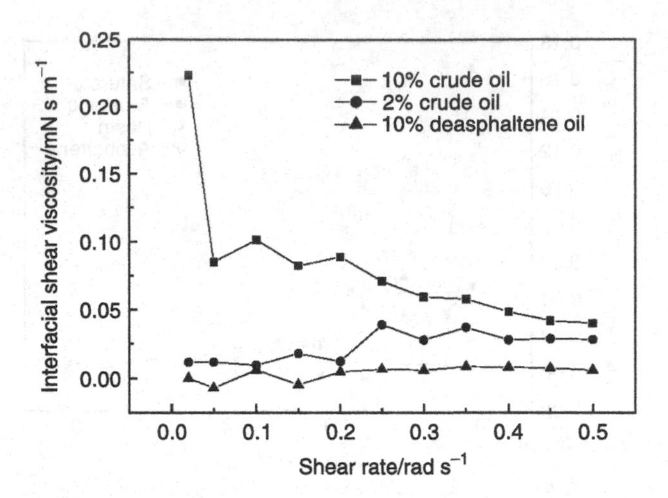

FIGURE 14.9 Interfacial shear viscosity between crude model oils (Gu Dong 4#) and 1.2% Na_2CO_3 water solution, 25 °C.

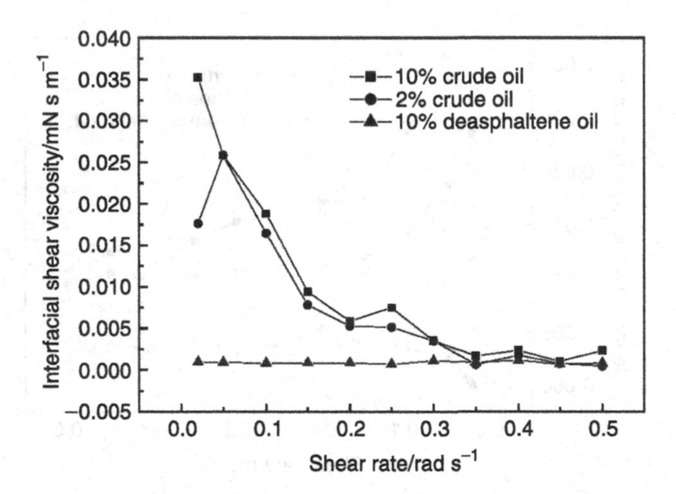

FIGURE 14.10 Interfacial shear viscosity between crude model oils (Da Qing) and 1.2% NaOH water solution, 25 °C.

also change the electrical property of the interfacial film between the oil and water phase, and the effect is more significant as the reaction time is increased. Therefore, the interfacially active components are able to enhance o/w emulsion stability.

14.2.2.5 Stability of Emulsions

The stability of asphaltene model oil emulsions and crude model oil emulsions of Gu Dong 1# crude oil are presented in Figures 14.11 and 14.12. When distilled water was the aqueous phase both the asphaltene model oil and the crude model oil were unable to form stable emulsions as the oils and the water reacted in 54 days. When 1.2% Na_2CO_3 solution was the aqueous phase no water separated from the asphaltene model oil emulsions and crude model oil emulsions after

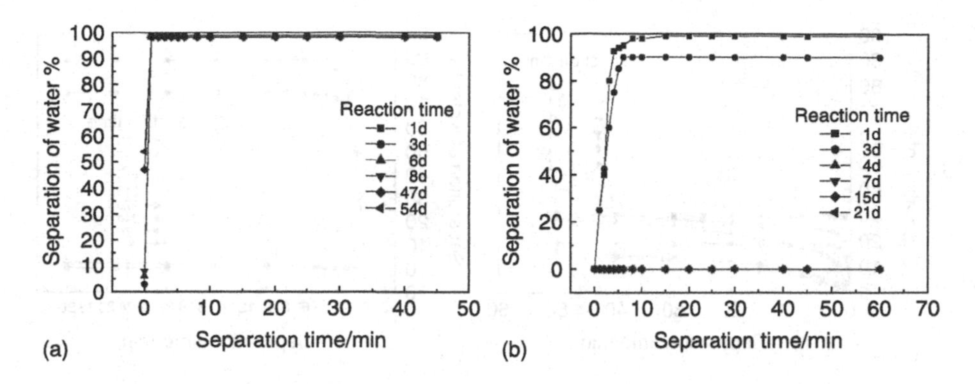

FIGURE 14.11 Stability of the emulsion formed of asphaltene model oil (Gu Dong 1#) and distilled water (a) or 1.2% Na_2CO_3 water solution (b), 60 °C.

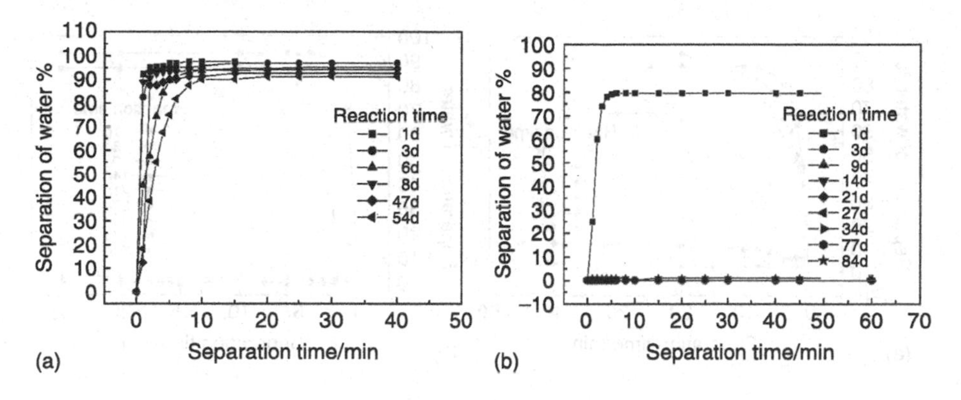

FIGURE 14.12 Stability of the emulsion formed of crude model oil (Gu Dong 1#) and distilled water (a) or 1.2% Na_2CO_3 water solution (b), 60 °C.

a period of 7 or 3 days' reaction of the oil and water. In contrast, the saturated, aromatic, and resin model oils tested in the same manner were not able to form stable emulsions even after 2 months of reaction. It is clear that the asphaltene fraction is responsible for stabilizing Gu Dong 1# crude oil emulsions when Na_2CO_3 was used. This result is supported well by measurement of the interfacial shear viscosity of asphaltene model oil and the alkaline solution discussed above. This result also shows that the reaction of asphaltene and Na_2CO_3 may progress slowly until all the possible acidic oxygen is replaced by sodium atoms.

Figures 14.13 and 14.14 show a similar result to that in Figures 14.11 and 14.12. Also, the asphaltene fraction dominates the stability of Gu Dong 4# crude oil emulsions in the same manner as Gu Dong 1# crude oil emulsions. The difference between Gu Dong 1# and 4# crude oils is that the asphaltene and crude model oil emulsions of Gu Dong 4# crude oil are more stable than the emulsion of Gu Dong 1# crude oil when distilled water was as the aqueous phase.

Comparing the interfacial tension, interfacial shear viscosity, and the physical properties of the fractions from Gu Dong crude oils the study shows clearly that the asphaltene fractions from Gu Dong crude oils have more polar, larger, acidic components than the other fractions. The characteristics can also be generalized to asphaltene from other crude oils. The components

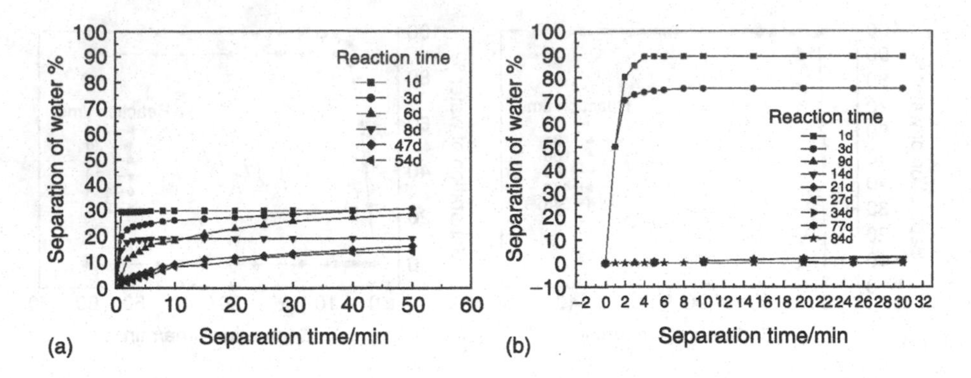

FIGURE 14.13 Stability of the emulsion formed of asphaltene model oil (Gu Dong 4#) and distilled water (a) or 1.2% Na_2CO_3 water solution (b), 60 °C.

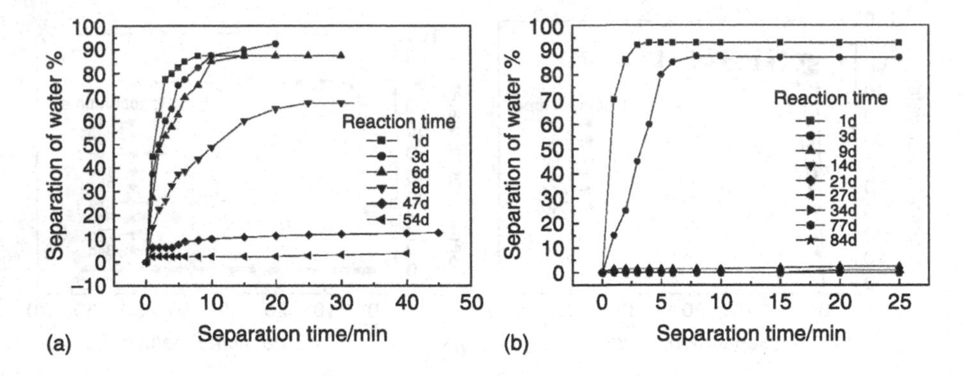

FIGURE 14.14 Stability of the emulsion formed of crude model oil (Gu Dong 4#) and distilled water (a) or 1.2% Na_2CO_3 water solution (b), 60 °C.

in the asphaltene fraction react with Na_2CO_3 solution and form soap-like interfacially active components which accumulate at the oil/water interface and form a rigid film around water droplets. Therefore, the film is able to prevent the coalescence of the droplets. The result also shows that the mechanical strength of the interfacial film can increase with the residence time of the sodium carbonate solution in the reservoir.

Compared with the emulsions of Gu Dong crude oil, the emulsions of Da Qing crude oil have different characteristics. Figures 14.15 and 14.16 show that the saturated and resin fractions reacted with NaOH are able to form more stable emulsions than the asphaltene fraction from Da Qing crude oil. The saturated and resin fractions help to stabilize the crude oil emulsion after reaction with NaOH. Like asphaltene model oil the aromatic model oils tested in the same manner were not able to form stable emulsions after 84 days of reaction with NaOH.

As with the asphaltene fraction from Gu Dong crude oils it is easy to understand why the resin fraction from Da Qing crude oil, when reacted with NaOH, can stabilize crude oil emulsion. The resin fraction has the highest acid number (5.101 in Table 14.6) among the fractions from Da Qing crude oil, and the interfacial tension between resin model oil and NaOH solution has the lowest value (2.64 $mN\,m^{-1}$ in Table 14.7). It is clear that the interfacially active components

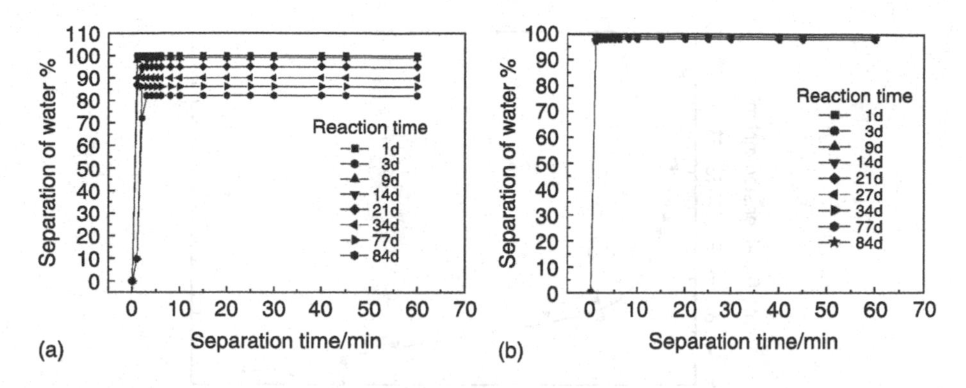

FIGURE 14.15 Stability of the emulsion formed with saturate model oil (a) or asphaltene model oil (b) (Da Qing) and 1.2% NaOH water solution, 60 °C.

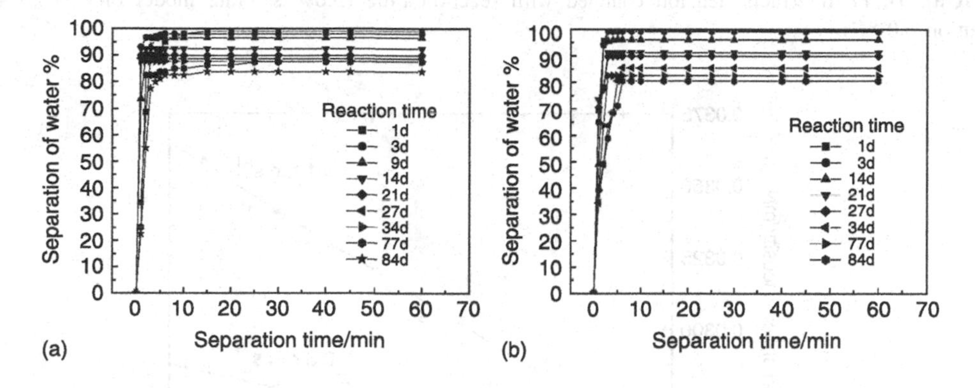

FIGURE 14.16 Stability of the emulsion formed with resin model oil (a) or crude model oil (b) (Da Qing) and 1.2% NaOH water solution, 60 °C.

were formed by the acidic components in the resin fraction reacted with NaOH. However, what are the components in the saturated fraction reacted with NaOH, and how is the emulsion stabilized? As a paraffinic crude oil Da Qing crude oil contains large saturate fraction (68.09%, Table 14.2). The saturate fraction contains the highest amount of oxygen (0.375 g; see Table 14.4) in 100 g crude oil and the lowest acid number (0.476; see Table 14.6). The interfacial tension between the saturated model oil of Da Qing crude oil and NaOH solution is the highest (14.68 to 18.31 $mN\,m^{-1}$; see Table 14.7). It is obvious that not only the acidic components but also some other substances in the saturated fraction reacted with NaOH and the components formed in the reaction are less interfacially active than the components formed in the reaction of resin fraction and NaOH.

To study the properties of the components formed in the reaction of saturate and NaOH the interfacial tension, interfacial shear viscosity during the reaction process, and infrared spectra of saturated fraction and the components formed were measured. The interfacial tension between saturated model oil and NaOH solution changed with reaction time is shown in Figure 14.17. The figure shows that interfacial tension decreased from the first day of the reaction. The interfacial tension decreased rapidly in the first week and slowly during the following 8 weeks.

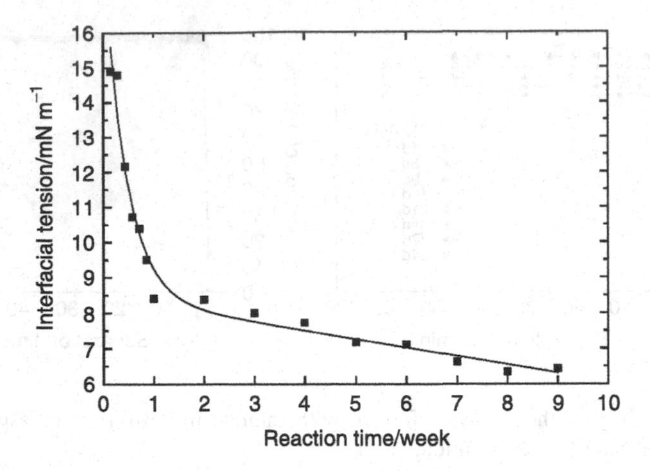

FIGURE 14.17 Interfacial tension changed with reaction time (5.0% saturate model oil, 0.6% NaOH solution, 30 °C).

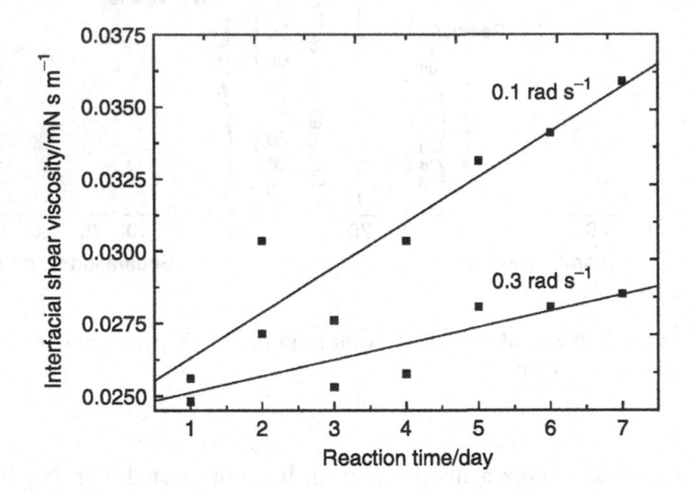

FIGURE 14.18 Interfacial shear viscosity changed with reaction time (3.0% saturate model oil, 0.6% NaOH solution, 25 °C).

The change of interfacial shear viscosity between the saturate model oil and NaOH solution during the first week is shown in Figure 14.18. An increase of the interfacial shear viscosity with reaction time confirms the formation of the interfacially active components again. Furthermore, the interfacial shear viscosity between the model oil and NaOH solution decreased and the increase of shear rate indicated that the interfacially active components form a network at the interface. This proves that the interfacial film formed by the interfacially active components has some strength and is able to stabilize emulsions.

IR spectroscopy (Figure 14.19(b)) shows absorption at 1563 cm^{-1}, which indicates the presence of COO^- groups in the component, and the absorption at 1100 cm^{-1} is due to the C–O stretching vibration of the carboxylic groups. This shows that sodium salts or soaps were formed

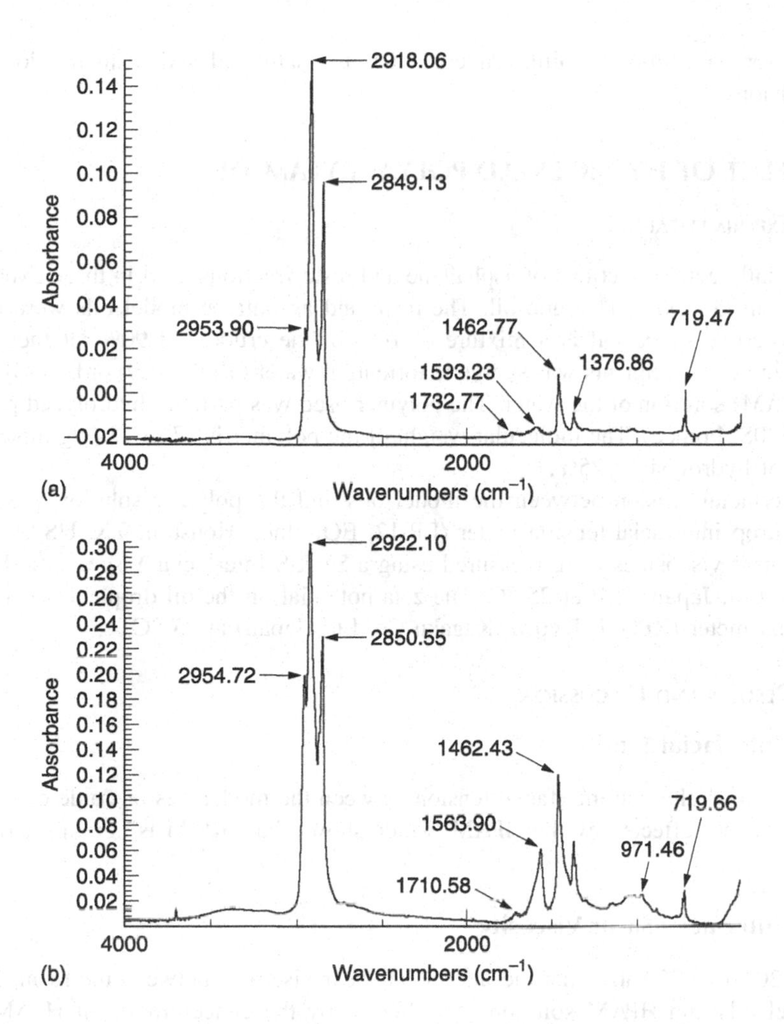

FIGURE 14.19 IR spectroscopy of saturate fraction (a) and the interfacially active components (b) formed in the reaction of saturate and NaOH.

in the reaction of the saturated fraction and NaOH. These salts or soaps may be formed by the reaction of NaOH with the acidic components and ester in the saturate fraction. Because the salts or soaps formed in the reaction have long hydrocarbon chains they are more oil soluble and less interfacially active than the components formed in the reaction of the resin fraction and NaOH.

The study above shows that the ASP flooding process in the reservoir provides sufficient time for the formation of interfacially active components, such as salt or soap, by the reaction of alkali with the acidic or nonacidic components in asphaltene, resin, and saturated fractions. These interfacially active molecules are responsible for providing stability for the water-in-crude-oil emulsion.

It is interesting that by using sodium carbonate solution as the alkaline component in the ASP technique for high acid number Gu Dong crude oils the reaction between the asphaltene fractions and alkali leads to stable emulsions. When sodium hydroxide is used as the alkaline component for paraffinic Da Qing crude oil it is the saturated and resin fractions that lead to stable emulsions.

14.3 EFFECT OF HYDROLYZED POLYACRYLAMIDE

14.3.1 Experimental

The interfacially active fractions of asphaltene and resin fractions used in these experiments were separated from Gu Dong $4^{\#}$ crude oil. The resin and asphaltene model oils were prepared with jet fuel; the crude model oil is a mixture of 10% of the crude and 90% jet fuel. The aqueous phase used in the experiments was synthetic formation water (Table 14.9) or hydrolyzed polyacrylamide (HPAM) solution of the water. The polymer used was partially hydrolyzed polyacrylamide (HPAM, 3530S, France). The molecular weight of the polymer is 17 to 18 (g/g mole $\times$ 10^{-6}) and the degree of hydrolysis is 25%.

The interfacial tension between the model oils and the polymer solution was measured by a spinning drop interfacial tension meter (LP-12, EOR Inc.; Houston, TX, USA) at 24 °C. The interfacial shear viscosities were measured using a SVR·S Interfacial Viscoelastic Meter (Kyowa Kagaku Co. Ltd, Japan) [18] at 25 °C. The zeta potential on the oil droplets was measured by a zeta potential meter (ZPOM, Kyowa Kagaku Co. Ltd, Japan) at 25 °C.

14.3.2 Results and Discussion

14.3.2.1 Interfacial Tension

Table 14.10 shows that the interfacial tension between the model oils or crude oil and the formation water was not affected by the HPAM, which shows that HPAM is not an interfacially active substance.

14.3.2.2 Interfacial Shear Viscosity

Figures 14.20 to 14.22 show that the interfacial shear viscosity between the resin, asphaltene, or crude model oils and HPAM solution was affected by the concentration of HPAM. The higher the concentration of the polymer, the larger the interfacial shear viscosity. It should be noted that the value of the interfacial shear viscosity of the crude model oil is close to that of the asphaltene model oil system and higher than that of the resin model oil system. This shows that the asphaltene fraction dominates the strength of the interfacial film between the crude oil and HPAM solution.

TABLE 14.9
Composition of Synthetic Formation Water

Ion	Concentration (mg l^{-1})
Cl^-	2885.49
HCO_3^-	818.89
$Na^+ + K^+$	2064.55
Mg^{2+}	20.88
Ca^{2+}	66.85

TABLE 14.10
Interfacial Tension Between Model Oil and HPAM Solution (mN m^{-1})

	Model Oil			
HPAM (mg l^{-1})	Jet Fuel	Resin Model Oil	Asphaltene Model Oil	Crude Model Oil
0	52.2	17.2	15.2	23.0
25	52.8	24.2	20.4	30.1
50	55.4	27.7	19.1	29.9
100	54.7	26.2	20.6	29.8
200	46.2	22.8	18.7	27.9
300	48.6	24.4	17.8	27.1
400	49.1	17.1	18.2	31.1

Temperature: 24 °C.

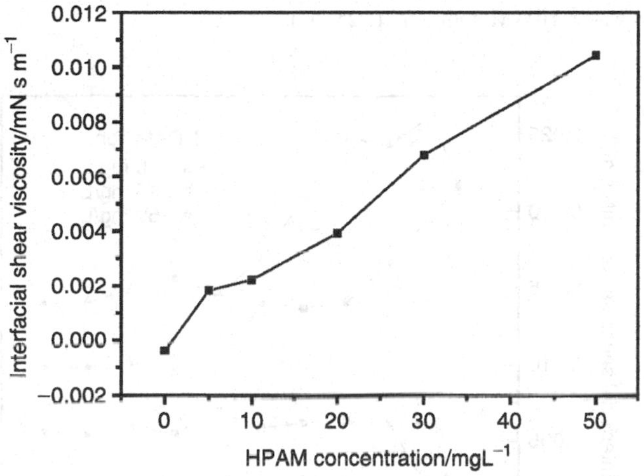

FIGURE 14.20 Interfacial shear viscosity between resin model oil and HPAM solution (oil: 1% resin model oil, water: HPAM solution, T: 25 °C, shear rate: 0.3 rad^{-1}).

This study revealed that HPAM is able to be absorbed at the interface between the model oils and water even when the polymer is not interfacially active and the polymer can enhance the strength of the interfacial film formed by the interfacially active fractions from the crude oil.

14.3.2.3 Zeta Potential

Table 14.11 shows that the absolute value of zeta potential on the resin model oil or asphaltene model oil droplets was increased with the addition of the HPAM. Figure 14.23 shows that the absolute value of zeta potential on the crude oil droplets was also increased as the concentration of HPAM increases. This shows the adsorption of HPAM at the interface between the model oils and water can modify the electrical property of the interface significantly. Therefore, HPAM is able to enhance emulsion stability.

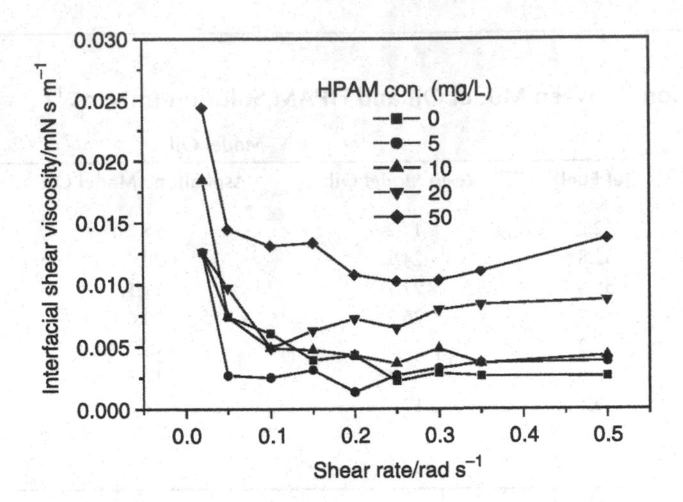

FIGURE 14.21 Interfacial shear viscosity between asphaltene model oil and HPAM solution (oil: 1% asphaltene model oil, water: HPAM solution, T: 25 °C).

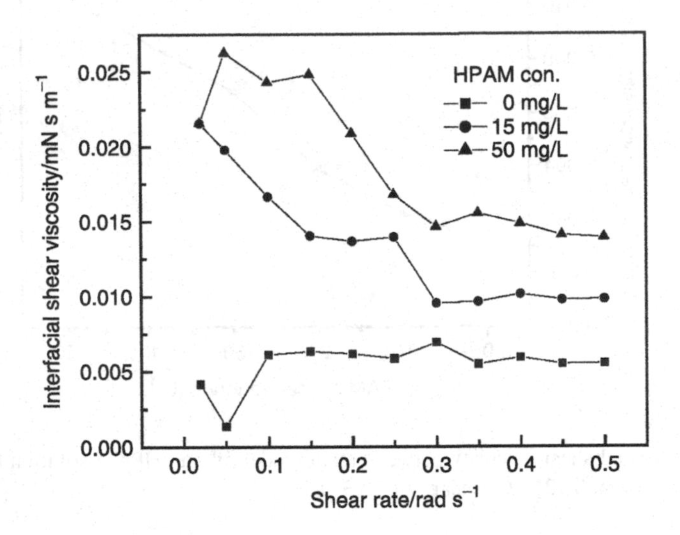

FIGURE 14.22 Interfacial shear viscosity between crude model oil and HPAM solution (oil: 10% crude model oil, water: HPAM solution, T: 25 °C).

14.3.2.4 Stability of O/W Emulsions

Table 14.12 shows that when the concentration of HPAM is in the range 50 to 100 mg/l the amount of oil in the water is much more than when no HPAM has been added to the water. It is clear that the change of electrical properties, caused by HPAM, at the interface between the oil and water contributes to stabilizing the o/w emulsion.

The effect of polymer on the interfacial properties between oil and water is complicated for a simple system and it is much more complex for a crude oil and water system in practice. There are both theoretical and practical problems. From this study it can be said that HPAM can be adsorbed

TABLE 14.11
Zeta Potential on Oil Droplets (mV, 25 °C)

	Oil	
HPAM (mg l^{-1})	**Resin Model Oil (5%)**	**Asphaltene Model Oil (5%)**
0	−11.6	−9.1
50	–	–
100	−15.1	−44.5
200	–	−42.8

TABLE 14.12
The Oil in O/W Emulsion Affected by HPAM (mg l^{-1})

HPAM Solution (mg l^{-1})	5% Resin Model Oil Emulsion	5% Asphaltene Model Oil Emulsion	Crude Emulsion
0	33	40	60
50	100	110	118
100	39	127	99
200	25	49	51

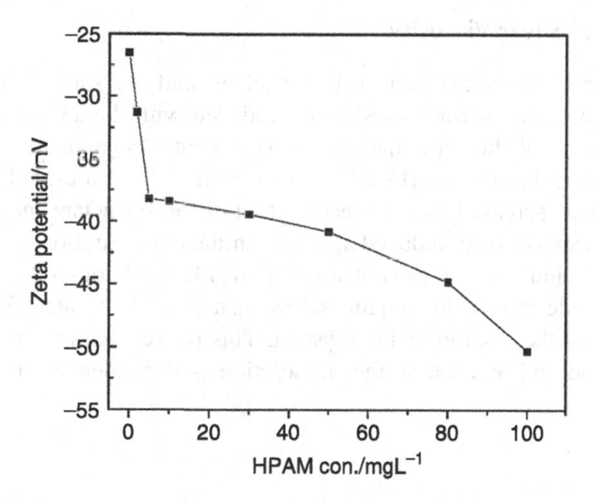

FIGURE 14.23 Influence of HPAM on zeta potential of crude oil droplets (oil: 3% crude model oil, T: 25 °C).

at the interface between the model oils and the synthetic formation water. The adsorption of the polymer increased the interfacial shear viscosity and the strength of the interfacial film between the oil and the water, which enhanced the stability of the emulsion. It is most important that the adsorption of polymer onto the surface of the oil droplets gives the surface a high negative potential. This proved that the oil droplets were not only stabilized by steric stabilization of the polymer but also by electrostatic stabilization. It seems that, in practice, electrostatic stabilization of the polymer dominated the stability of the o/w emulsion.

14.4 EFFECT OF SURFACTANT

14.4.1 Experimental

The interfacially active fractions of asphaltene and resin used in these experiments were separated from Gu Dong 4# crude oil. The resin, asphaltene, and crude model oils were prepared with jet fuel. The surfactant used was a petroleum sulfonate (TRS) produced in Sheng Li oil field in China. The aqueous phase used in the experiments was synthetic formation water (Table 14.9) or TRS solution of the water.

The interfacial tension between the model oils and the surfactant solution was measured by a spinning drop interfacial tension meter (LP-12, EOR Inc.; Houston, TX, USA) at 45 °C. The interfacial shear viscosities were measured using a SVR·S Interfacial Viscoelastic Meter (Kyowa Kagaku Co. Ltd, Japan) [18] at 25 °C. The zeta potential on the oil droplets was measured by a zeta potential meter (ZPOM, Kyowa Kagaku Co. Ltd, Japan) at 25 °C.

14.4.2 Results and Discussion

14.4.2.1 Interfacial Tension

Table 14.13 shows the interfacial tension between the model oils and TRS solution or the synthetic formation water. It is clear that the interfacial tension was decreased dramatically with the addition of the surfactant.

14.4.2.2 Interfacial Shear Viscosity

The interfacial shear viscosity between resin model oil and the surfactant solution is shown in Figure 14.24. The interfacial shear viscosity was reduced with the addition of the surfactant; the higher the concentration of the surfactant the lower the interfacial shear viscosity.

Figure 14.25 shows that the interfacial shear viscosity between asphaltene model oil and the surfactant solution was increased as the concentration of the surfactant increased from 0 to 0.3%, and the interfacial shear viscosity reduced again when the concentration of the surfactant is 0.5%. Figure 14.26 shows a similar result as shown in Figure 14.25, although the value of the interfacial shear viscosity of crude model oil system is less than that of the asphaltene model oil system and higher than that of the resin model oil system. This proves that the strength of the interfacial film between the crude oil and the surfactant solution is dependent on both the asphaltene and resin fractions.

TABLE 14.13
Interfacial Tension Between Model Oils and TRS Solution ($mN\,m^{-1}$), T: 45 °C

Model Oil	TRS (%)			
	0	0.1	0.3	0.5
3% Crude oil	8.5803	0.989	0.0168	0.0099
3% Asphaltene oil	4.3884	0.118	0.0066	0.0012
3% Resin oil	6.8538	0.0212	0.00816	0.0013

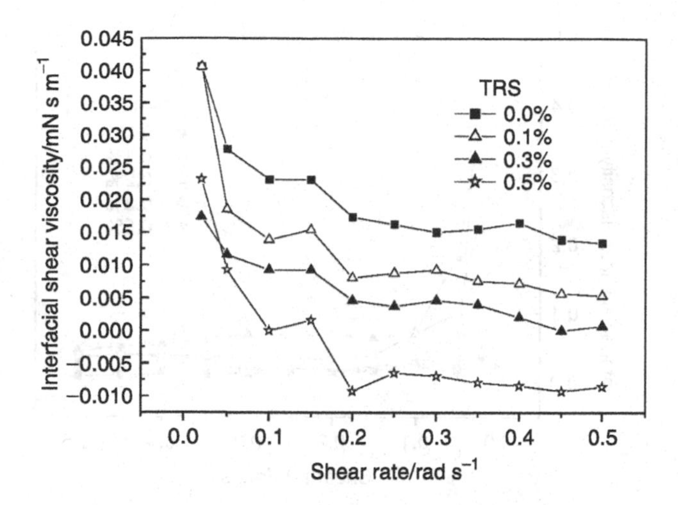

FIGURE 14.24 Interfacial shear viscosity between 3% resin model oil and TRS solution, 25 °C.

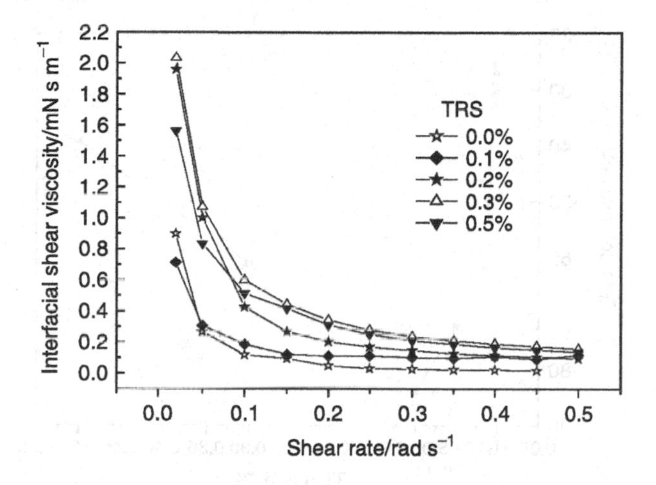

FIGURE 14.25 Interfacial shear viscosity between 3% asphaltene model oil and TRS solution, 25 °C.

14.4.2.3 Zeta Potential

The zeta potential on oil droplets changed with concentration of TRS for resin, asphaltene, and crude model oils is shown in Figures 14.27 to 14.29. The absolute value of zeta potential for resin model oil is increased from 0 to 0.1% of TRS in distilled water. Then the absolute value of zeta potential was reduced when the concentration of TRS is higher than 0.1%.

Figure 14.28 shows the absolute value of zeta potential on asphaltene model oil droplets is increased fast from 0 to 0.05% of TRS in distilled water. Then the absolute value of zeta potential was increased slowly when the concentration of TRS is higher than 0.05%. It is interesting that the change in the absolute value of zeta potential on crude model oil droplets is a combination of the resin and asphaltene model oils (Figure 14.29). It shows that the electrical property of the

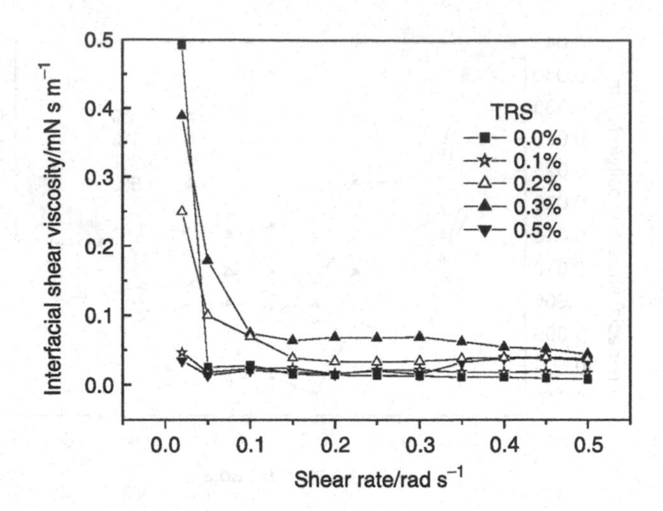

FIGURE 14.26 Interfacial shear viscosity between 3% crude model oil and TRS solution, 25 °C.

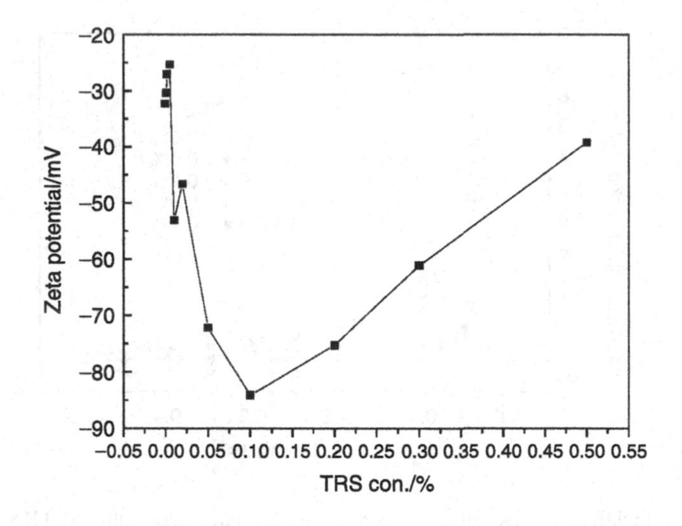

FIGURE 14.27 Influence of TRS on zeta potential (oil: 3% resin model oil, T: 25 °C).

interface between crude model oil and the surfactant solution is also dependent on the asphaltene and resin fractions.

Based on this study it is important to know that the adsorption of the surfactant at the interface between the model oils and distilled water makes the oil droplets negatively charged; the formation of an electrical double layer is able to enhance the stability of o/w emulsions.

14.4.2.4 Stability of Emulsions

The stability of the emulsion studied in this work is evaluated by the value of a volume fraction of $(V_t - V_0)/V_0$. V_t presents the volume of w/o emulsion, V_0 presents the volume of the oil added.

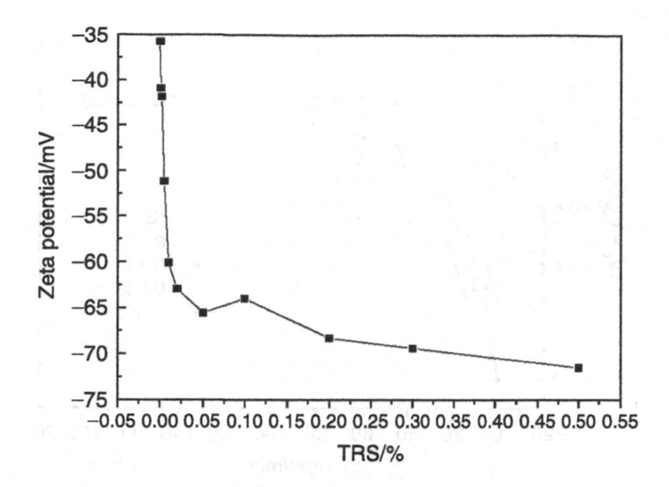

FIGURE 14.28 Influence of TRS on zeta potential (oil: 3% asphaltene model oil, T: 25 °C).

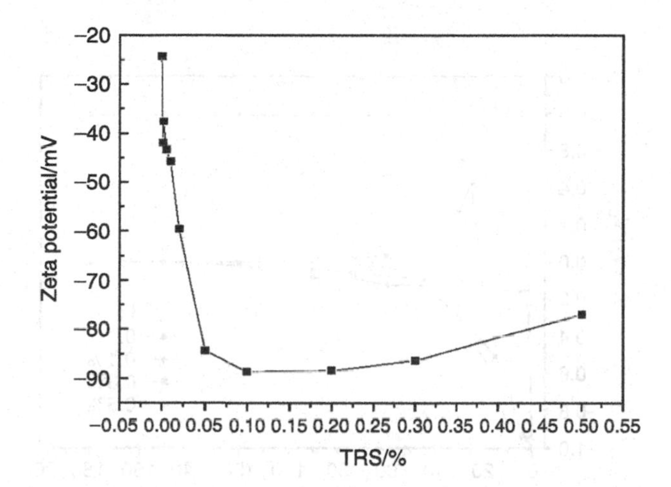

FIGURE 14.29 Influence of TRS on zeta potential (oil: 3% crude model oil, T: 25 °C).

When $(V_t - V_0)/V_0 > 0$, the emulsion is the w/o type. When $(V_t - V_0)/V_0 < 0$, the emulsion is the o/w type, and when $(V_t - V_0)/V_0 = 0$, the emulsion is broken completely.

Figure 14.30 shows the emulsion formed with resin model oil and TRS solution is o/w type, and the time of $(V_t - V_0)/V_0$ changed from −1 to 0 was increased obviously as the concentration of TRS increased. This reveals that the stability of the emulsion was enhanced as the concentration of TRS was increased.

Figure 14.31 shows the emulsion formed with asphaltene model oil and the synthetic formation water is the w/o type. However, the emulsion was changed from the w/o to the o/w type as the concentration of TRS was increased from 0 to 0.5%. As the asphaltene fraction is oil soluble, the interfacially active substance and TRS is a strong water soluble surfactant. This result shows that the emulsion type is dependent on the effect of asphaltene and TRS, and TRS plays an important role when the concentration of the surfactant is high.

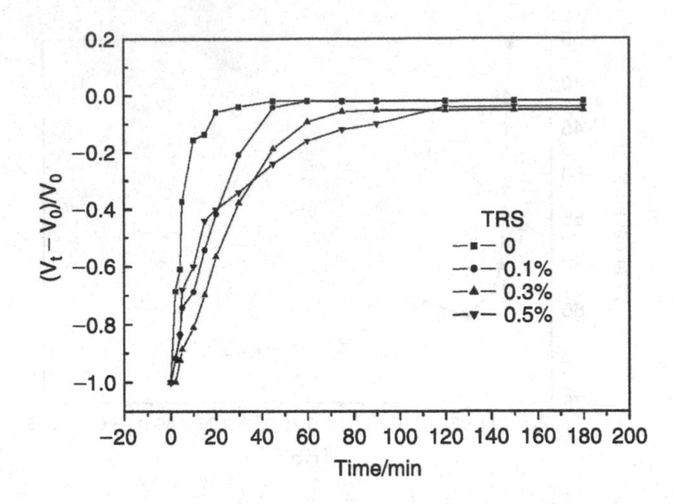

FIGURE 14.30 Stability of emulsion formed with 3% resin model oil and TRS solution, 30 °C.

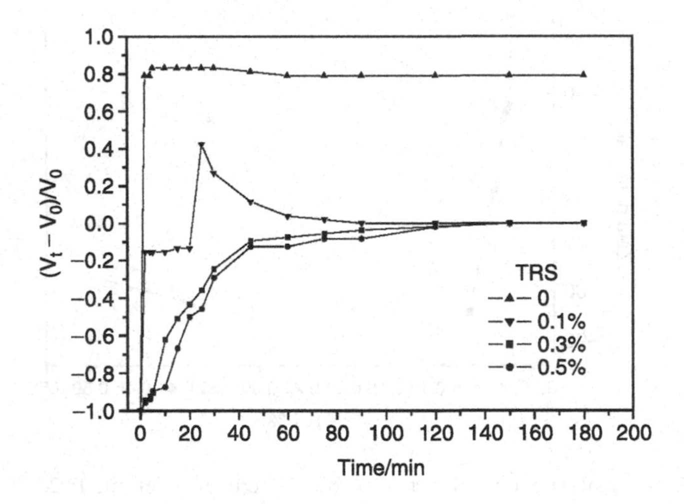

FIGURE 14.31 Stability of emulsion formed with 3% asphaltene model oil and TRS solution, 30 °C.

Figure 14.32 shows the stability of the emulsion formed with crude model oil and TRS solution. The emulsions are the o/w type and the emulsion has a higher stability when the concentration of TRS is 0.3%. It seems that the resin fraction dominates the properties of the emulsion when TRS was added in the water.

14.5 EFFECT OF ALKALINE–SURFACTANT–POLYMER

In this work the experimental methods and condition are the same as described in Section 14.4.1.

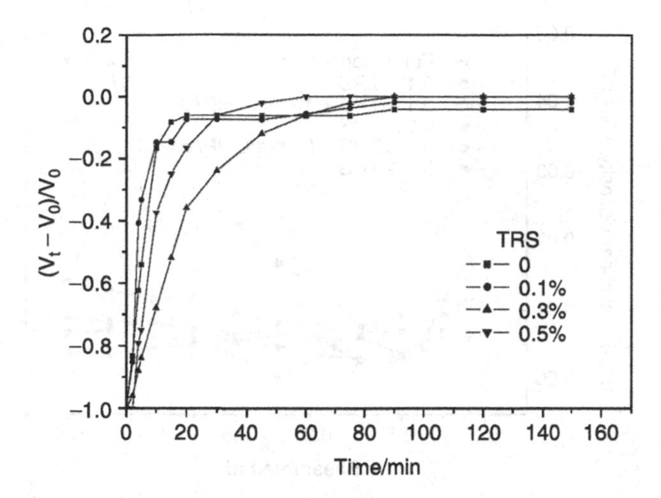

FIGURE 14.32 Stability of emulsion formed with 3% crude model oil and TRS solution, 30 °C.

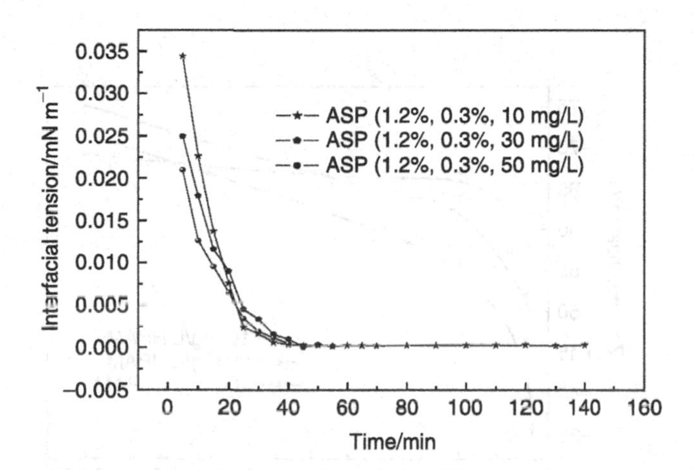

FIGURE 14.33 Interfacial tension between 3% crude model oil and ASP solution (A – Na_2CO_3, S – TRS, P – HPAM, crude Gu Dong 4#, T: 45 °C).

14.5.1 Interfacial Tension

Figure 14.33 shows that when Na_2CO_3 (1.2%), HPAM (10 mg/l), and TRS (0.3%) were added at the same time the interfacial tension between crude model oil (Gu Dong 1#) and the solution is very low. It is obvious that the coordination effect of Na_2CO_3 and TRS make the main contribution to decreasing the interfacial tension.

14.5.2 Interfacial Shear Viscosity

Figure 14.34 shows the interfacial shear viscosity between crude model oil (Gu Dong 4#) and SP solution. Before the measurement the crude model oil had been reacted with 1.2% Na_2CO_3 solution for 10 days to form sufficient interfacially active components. The result shows when

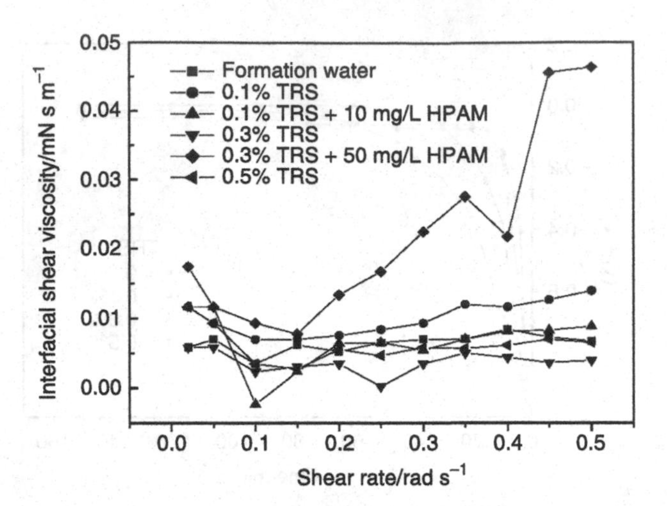

FIGURE 14.34 Interfacial shear viscosity between 0.1% crude model oil (had been reacted with 1.2% Na_2CO_3 solution for 10 days) and SP solution (S – TRS, P – HPAM, crude Gu Dong 4[#], T: 25 °C).

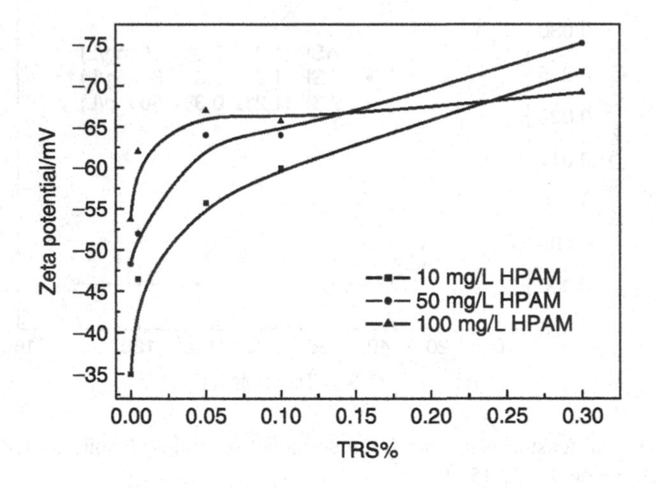

FIGURE 14.35 Influence of ASP on zeta potential (3% crude model oil (Gu Dong 4[#]) had been reacted with 1.2% Na_2CO_3 solution for 1 day, T: 25 °C).

the concentration of HPAM is 50 mg/l the polymer makes a main contribution to interfacial shear viscosity under the experimental condition. It should be noted that the interfacial shear viscosity is increased as the shear rate increases when the aqueous phase contains 0.3% TRS and 50 mg/l HPAM. It seems that there may be some particles formed at the interface [18].

14.5.3 Zeta Potential

Figure 14.35 shows the zeta potential on crude model oil droplets changed with concentration of TRS and HPAM. Before the measurement the crude model oil had been reacted with 1.2%

Na_2CO_3 solution for 1 day to form sufficient interfacially active components. The results show that the zeta potential increased rapidly when the concentration of TRS was below 0.05%, and did not change much when the concentration of TRS was higher than 0.05%. This may be caused by the equilibrium adsorption of TRS on the interface. It can also be seen that the higher the concentration of HPAM in the solution the larger the zeta potential. It is clear that there were more HPAM molecules adsorbed on the interface when the concentration of HPAM was increased.

14.5.4 Stability of Emulsions

The stability of the emulsions formed with crude model oil and SP solution is shown in Figure 14.36. Before determination of the emulsion stability the crude model oil had been reacted with 1.2% Na_2CO_3 solution for 1 or 10 days to form sufficient interfacially active components in part of the experiments. It is very interesting that the emulsion formed with 3% crude model oil and the solution contains 0.1% TRS and 10 mg/l HPAM is the most stable in the experimental condition, and when the concentration of HPAM increased from 10 to 50 mg/l the stability of the emulsion decreased. When crude model oil had been reacted with Na_2CO_3 for 10 days the stability of the emulsion formed with the oil and the solution containing 0.1% TRS and 50 mg/l HPAM is most unstable. The stability of the emulsions formed with the crude model oil that had been reacted with Na_2CO_3 for 1 day and the solution contains 0.1% TRS and 10 to 50 mg/l HPAM is in the middle case. It should be noted that all the emulsions are the o/w type; therefore, electrical double layer stabilization might be one of the main mechanisms for stabilizing the emulsions. When the crude model oil reacted with Na_2CO_3 interfacially active components are formed. Because these interfacially active components are more oil soluble the components might act as demulsifiers to decrease the stability of the o/w emulsions. For longer reaction times, more interfacially active components are formed, and the emulsions are more unstable. The effect of HPAM in these experiments is very complicated. HPAM may be adsorbed at the interface and increase the strength of the interfacial film. The absolute value of the zeta potential

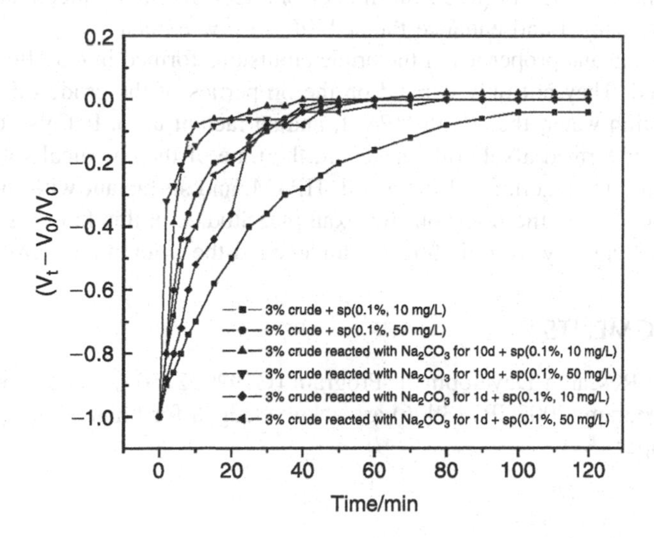

FIGURE 14.36 Stability of emulsion formed with 3% crude model oil and ASP solution (A – Na_2CO_3, S – TRS, P – HPAM, crude Gu Dong 4[#], T: 45 °C).

may be increased, and therefore HPAM is able to enhance the stability of the emulsion. However, HPAM is also able to flocculate the oil drops and decrease the stability of the emulsions. In this experiment the effect of flocculation seems to play an important role, because the stability of the emulsions decreased when the concentration of HPAM was increased.

14.6 CONCLUSION

Based on this study the following conclusions can be drawn:

1. The asphaltene fractions from Gu Dong crude oils of high acid number have more polar, larger, acidic components than the other fractions. The components that react with Na_2CO_3 are able to form interfacially active components that accumulate at the oil/water interface and form a rigid film around water droplets. The asphaltene fraction from Gu Dong crude oils dominates the stability of the emulsions of the crude and Na_2CO_3 solution.
2. It is not only acidic oxygen compounds but also esters in the saturated fractions from paraffinic Da Qing crude oil of low acid number that, when reacted with NaOH, produce components in the reaction which are responsible for stabilizing the emulsions of the crude oil and NaOH solution. The reaction takes place over a long time and the ASP flooding process in the reservoir provides sufficient time for the formation of soap-like components by the reaction of the alkaline and acidic or nonacidic components in the saturation fraction.
3. The effect of HPAM on the formation of the crude oil emulsion in ASP flooding is very complicated. HPAM may be adsorbed at the interface and increase the strength of the interfacial film and the absolute value of zeta potential, therefore enhancing the stability of the emulsion. However, HPAM is also able to flocculate the oil droplets and decrease the stability of o/w emulsions.
4. It is important to know that the adsorption of HPAM, TRS surfactant, and the interfacially active components formed in the reaction of Gu Dong and Da Qing crude oils and the alkali at the interface between the model oils and distilled water makes the oil droplets negatively charged and enhance the stability of o/w emulsions.
5. The formation and properties of the crude emulsions formed in ASP flooding are extremely complicated. They not only depend on the properties of the crude oil, the injection water and formation water, the alkali, HPAM, and surfactant used, but also on the properties of the reservoir formed alkali–oil surfactant, the time of the chemical solution flood through the reservoir, the reactions of the alkali, HPAM, and surfactant with the reservoir, and the solid particles from the reservoir, for example. Studies in this field are just beginning, and we have a long way to go before we understand the scientific nature of the emulsions.

ACKNOWLEDGMENTS

The National Key Research Development Program (G1999022505) and an International Cooperation Research Program (2002CB713906) are acknowledged for financial support and permission to publish this paper.

REFERENCES

1. L. Mingyuan, in S. Pingping, Y. Jiayong (ed.), Basic research for enhancing crude oil recovery, Petroleum industry publisher, 2002.

2. L. Mingyuan, A. A. Christy and J. Sjöblom, in Sjöblom (ed.) Emulsions – A fundamental and practical approach, NATO ASI Ser. Vol. 363, Kluwer Academic Publishers, Dordrecht (1992) 157.
3. E. J. Johansen, I. M. Skjårvø, T. Lund, J. Sjöblom, H. Söderlund and G. Boström, Colloids Surfaces, 34 (1989) 353.
4. J. Sjöblom, H. Söderlund, S. Lindblad, E. J. Johansen and I. M. Skjårvø, Colloid Polym. Sci., 268 (1990) 389.
5. J. Sjöblom, L. Mingyuan, H. Høiland and E. J. Johansen, Colloids Surfaces, 46 (1990) 127.
6. J. Sjöblom, O. Urdahl, H. Høiland, A. Christy and E. J. Johansen, Prog. Colloid Polym. Sci., 82 (1990) 131.
7. K. G. Nordli, J. Sjöblom, J. Kizling and P. Stenius, Colloids Surfaces, 57 (1991) 83.
8. K. G. Nordli, J. Sjöblom and P. Stenius, Colloids Surfaces, 63 (1992) 241.
9. J. Sjöblom, L. Mingyuan, A. A. Christy and T. Gu, Colloids Surfaces, 66 (1992) 55.
10. O. Urdahl, A. E. Møvik and J. Sjöblom, Colloids Surfaces A: Physicochem. Eng. Aspects, 74 (1993) 293.
11. T. Skodvin, J. Sjöblom, J. O. Saeten, O. Urdahl and B. Gestblom, J. Colloid Interface Sci., 166 (1994) 43.
12. K. G. Nordli Børve, Thesis, University of Bergen, Norway, 1991.
13. L. Mingyuan, Thesis, University of Bergen, Norway, 1993.
14. O. Urdahl, Thesis, University of Bergen, Norway, 1993.
15. J. Sjöblom, O. Urdahl, K. G. N. Børve, L. Mingyuan, J. Saeten, A. A. Christy and T. Gu, Adv. Colloid Interface Sci., 41 (1992) 241.
16. A. A. Christy, B. Dahl and O. M. Kvalheim, Fuel, 68 (1989) 431.
17. L. Mingyuan, X. Mingjin, M. Yu, W. Zhaoliang and A. A. Christy, Colloids Surfaces, 197 (2002) 193–201.
18. L. Mingyuan, X. Mingjin, M. Yu, W. Zhaoliang and A. A. Christy, Fuel, 81 (2002) 1847–1853.
19. L. Mingyuan, P. Bo, Z. Xiaoyu, W. Zhaoliang, in Sjöblom (ed.), Encyclopedic handbook of emulsion technology, Marcel Dekker, Inc., New York (2000) 515–523.

15 Electrocoalescence for Oil–Water Separation: Fundamental Aspects

Lars E. Lundgaard, Gunnar Berg, Stian Ingebrigtsen, and Pierre Atten

CONTENTS

15.1 BACKGROUND

On the one hand electrostatic forces are special in that they may act over large distances. Effects such as lightning is one example here. This is contrary to hydrodynamic forces, which are near field forces: one mass element acts on the neighboring elements only. On the other hand, the effects of electrostatic charges and forces are significant down to a molecular level. Electrostatics therefore come into consideration in many different cases.

Electrostatics is a very important aspect of high voltage insulation design [1] where knowledge of the effects of electric fields on materials is required. Electrostatics has also found application in a number of different industrial processes, like gas filtration, separation, spraying, and copying [2,3].

Electrocoalescence is usually the denotation of an industrial separation process used on water-in-oil emulsions, where the electric fields are used to assist merging of small water drops into larger ones that will settle more quickly in a separation tank. This topic has relevance to the petroleum industry, where water-in-oil emulsions are formed by de-pressurization of oil mixed with water and also in the process of removing water deliberately added as fine emulsions in desalinizers in petroleum processing plants. The first patent for an electrocoalescer was filed in 1908 [4]. The electrostatic effects arise from the very different properties of oil and water, water having a dielectric permittivity and conductivity values much higher than those of oil (in practice one often considers water as perfectly conducting like metals).

It is the interfacial effects that stabilize water in crude oil emulsions. A vast literature exists on the chemical aspects of emulsion stability and the use of chemical de-emulsifiers [5]. Electrostatic coalescers have a potential to be an alternative to heating, which is energy consuming, and to the use of chemical emulsion breakers, which may be detrimental from an environmental viewpoint. Different voltages like dc, ac, and even pulsed dc are used for electrocoalescence, and much of the literature has focused on determining the optimum frequencies and electric field levels. Designs vary in respect to whether or not using electrode insulation, and using laminar or turbulent flows, the former one being the most popular.

To a large extent the effects of the electrostatic field are explained by body forces acting on water drops either because they are charged, giving rise to electrophoretic forces, or because they become polarized in divergent fields, resulting in so-called dielectrophoretic forces. Less attention has been paid to the phenomena occurring just when two drops are brought close to each other and eventually merge into one drop.

Without the basic understanding of the final coalescence or merging mechanism, it is not surprising that the efficiency of an electrocoalescer may vary and even disappear depending on the oil source and on the working conditions. How the design and settings should be fitted to a certain well and/or its temporal development is a crucial question. We believe that a closer look into the physical mechanisms may help overcome this type of problem.

Field-enhanced coalescence and movement of dielectric and conductive particles in an insulating environment has been studied within a number of different disciplines, and it may be useful to "take a peek over the fence" to see what is available there. The interaction of water drops in electric dc fields has been extensively studied as an active process in cloud physics and rain formation; movement of dielectric moist cellulose particles in oil has been studied as one process responsible for triggering breakdown in transformer insulation and, finally, the behavior of suspensions of dielectric particles in oil has been investigated for applications like electrorheology.

15.2 INTRODUCTION

During the past 100 years, electrocoalescence has been used in different coalescer designs and various techniques have been promoted based on different voltage shapes, flow characteristics, and electrode shape and coating. Waterman made a survey of different applications and techniques for electrostatic treatment of water-in-oil emulsions [6]. Eow [7,8] has lately given a thorough review of the status and variations of the technique in his two papers.

Experience from oil/water separators on offshore platforms reports that coalescer efficiency often varies between wells and also with time on a given well. It is not obvious how to adapt the electrocoalescence process to the specific emulsion properties for a certain oil well. The crude oil properties will vary with respect to viscosity, conductivity, and water cut, and the interphase layer can have certain surface tension and rheological characteristics that interact with an applied electrical field: Electric forces will act on ions generated by carboxylic acids that precipitate at the interface between water and oil. Knowing that water has a permittivity of around 80 (varying from 87.9 at 0 °C to 55.5 at 100 °C) to be compared with that of crude oil (about 3), and that the conductivity values for water and oil are widely different and may vary over many decades, one cannot expect a universal optimum voltage frequency to exist for ac coalescers. It may be worthwhile to have a closer look at the role played by the components of an emulsion, particularly the influence of materials properties on the field and charge distributions under various conditions: That is to take a closer look at the basic electrostatic processes.

The aim of this chapter is to give an insight into the physical processes and properties that we know or suspect may govern the electrically enhanced and induced coalescence process of water drops in a mineral oil. Even though we do not intend to write a review of the present status of research on electrocoalescence, we will draw attention to examples from our own and other studies in this field, as well as more basic relevant background literature. We will mainly be concerned with electrostatic phenomena in stagnant emulsions and electro-hydrodynamic (EHD) phenomena, and in a limited way only consider hydrodynamic effects in the liquid bulk. Urdahl et al. [9] have given a good review of the role hydrodynamic flows play in electrocoalescence.

This chapter is organized in five parts, going from basic textbook knowledge through simpler cases concerning singular drops or drop pairs towards somewhat more speculative considerations of emulsion behavior and coalescer designs.

In Section 15.3 (Basic electrostatics) we will discuss subjects like forces between charged bodies and polarized bodies, differences between charging and conductance in metallic and dielectric bodies in electric fields, potential distribution in multi-dielectric systems and time constants for charging and charge decay of materials and systems.

In Section 15.4 we look at the forces acting on a single drop, drop pairs, and drops in an emulsion in homogeneous and inhomogeneous electric fields.

Section 15.5 explains how a drop disintegrates when stressed by an electric field and how the roles of voltage frequency and shape, surface tension, and drop size are all consistent with classic theory.

In Section 15.6 (The electrocoalescence mechanism) we discuss the validity of several different hypotheses of what really occurs in the sequence from drop proximity to established continuity in the water phase in the two drops. Weight will be put on explaining an electric field induced surface instability being the basic coalescence mechanism.

Finally, Section 15.7 deals with practical implications and discusses how the present understanding may help in analyzing coalescer performance.

15.3 BASIC ELECTROSTATICS

15.3.1 Introduction

The characteristics and properties of dielectric materials are described with some detail in textbooks on electromagnetism. We recall only some aspects, which are of interest for the water-in-oil emulsions considered here.

15.3.2 Electrostatic Fields and Forces in Vacuum

The theory of electrostatics explains how charges create forces upon other charges and how the electric field and potential are defined.

An electric field is said to exist at a point where a stationary charge experiences a force. The electric field E is defined from the force exerted on a positive unit charge introduced at that point (see Figure 15.1). In practice we have:

$$\vec{E} = \lim_{q' \to 0} \frac{\vec{F}}{q'} \tag{15.1}$$

The limit $q' \to 0$ is required in order for the test charge not to influence the source. The simplest type of electric field, the electrostatic field, is induced by stationary charges. Two point electric charges repel each other when they are of same polarity, and attract each other when they are of opposite polarity. The interaction force follows the law experimentally determined by Coulomb. In vacuum the force between two point charges q and q' with a separation r has an inverse square law variation on r:

$$\vec{F} = \frac{qq'}{4\pi \varepsilon_0 r^2} \hat{e}_r \tag{15.2}$$

where ε_0 is the vacuum permittivity and $\hat{e}_r$ is a unit vector in the r direction. From Equations 15.1 and 15.2 the electrostatic field created by a point charge q at a point lying at the distance r

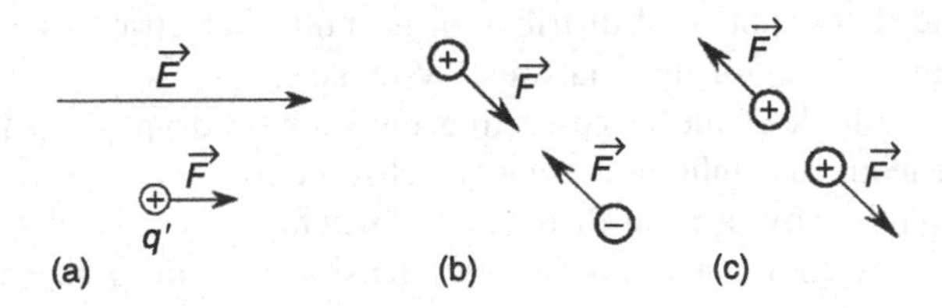

FIGURE 15.1 (a) Force on a point charge and Coulomb force between charges of opposite (b) and equal (c) polarity.

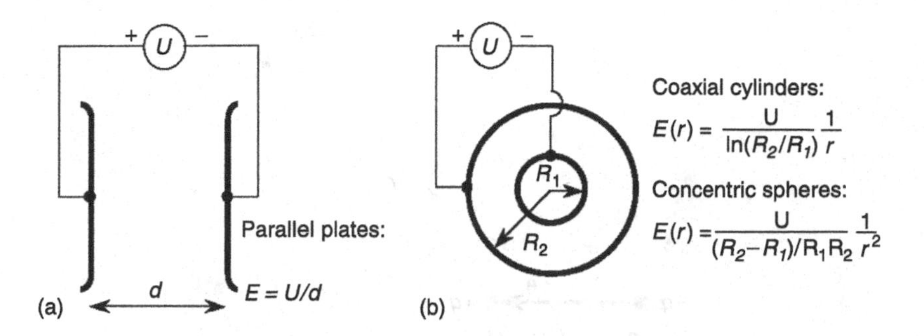

FIGURE 15.2 Examples of homogeneous (a) and inhomogeneous (b) electric field.

from q is given by:

$$\vec{E} = \frac{q}{4\pi\varepsilon_0 r^2}\hat{e}_r \tag{15.3}$$

The concept of electric field is simplest explained by looking at a planar air capacitor with two metallic electrodes of area A separated by a distance d. With a voltage U applied between the two plates one obtains a constant or *homogeneous* electric field E between them: $E = U/d$, as shown in Figure 15.2(a). This field will move free charges towards the electrodes. If one considers a point charge, this charge will induce an electric field around it that is either directed from or towards the charge. The field that decays with distance from the charge is divergent or *inhomogeneous* (Figure 15.2(b)).

The electric potential V is defined from the work done by moving a unit charge between two points in an electric field. The potential difference between two points is obtained by integrating the field along any path between these points. Potential can be defined at points and on surfaces; surfaces of metallic or other conducting materials are equipotential.

If there are many charges distributed over a volume or a surface, then the resulting field can be found by summing up the contributions from each of them. It is then easy to show that the electric field E can be expressed as the gradient of the scalar potential V:

$$\vec{E} = -\nabla V \tag{15.4}$$

The electric potential V at a distance r from a point charge is then given by Equations 15.3 and 15.4:

$$V = \frac{q}{4\pi\varepsilon_0 r} \tag{15.5}$$

If there are several charges, the resulting potential at a point is the scalar sum of the individual potentials. The electric *capacitance* of a conductor is defined as the ratio of the conductor's charge to the applied voltage, $C = Q/V$. The capacitance involves geometrical parameters.

It can further be shown that, at any point in an electrostatic field, the following relation holds:

$$\nabla \cdot \vec{E} = \rho/\varepsilon_0 \tag{15.6}$$

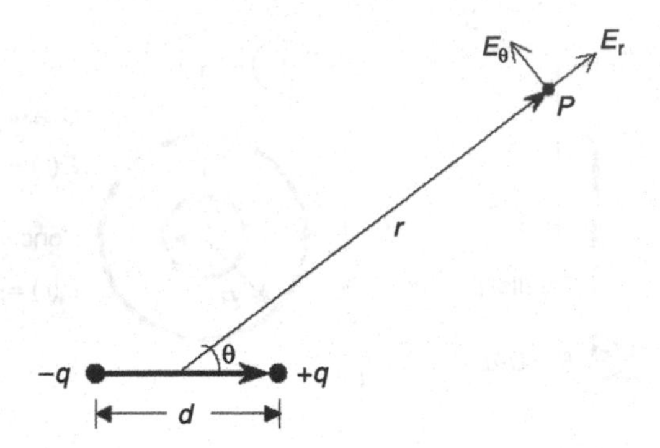

FIGURE 15.3 The electric dipole.

where ρ is the volume charge density. Combining Equations 15.4 and 15.6 leads to the Poisson equation:

$$\nabla^2 V = -\frac{\rho}{\varepsilon_0} \tag{15.7}$$

In a charge-free region ($\rho = 0$) the above expression becomes Laplace's equation. When the electrode- and equipotential surfaces coincide with the surfaces of an orthogonal coordinate system, Laplace's equation will be separable and can be solved analytically. This has been done in Figure 15.2 for rectangular (Cartesian), cylindrical, and spherical coordinates respectively.

Historically, to account for the long distance nature of electromagnetic forces, Faraday, and later Maxwell, considered the region of space occupied by an electromagnetic field to be in a state of stress, where electric (and magnetic) forces are transmitted as tension or compression by elastic lines of forces. Mathematically, this is generalized by Maxwell's stress tensor. For an isotropic medium it is easy to show that the electrostatic energy density is given by:

$$W_{\mathrm{el}} = \frac{1}{2}\varepsilon_0 E^2 \tag{15.8}$$

An electric dipole is an arrangement of two charges $+q$ and $-q$ of opposite polarity separated by a fixed distance d; see Figure 15.3. The resulting potential from both charges at any point is given by Equation 15.5. Assuming $r \gg d$ the potential at point P will be $V = p\cos\theta/\left(4\pi\varepsilon_0 r^2\right)$, where the dipole moment is defined by $p = q \cdot d$. The radial and tangential electric field components of the dipole are found by applying Equation 15.4.

$$\begin{aligned} E_r &= \frac{2p\cos\theta}{4\pi\varepsilon_0 r^3} \\ E_\theta &= \frac{p\sin\theta}{4\pi\varepsilon_0 r^3} \end{aligned} \tag{15.9}$$

15.3.3 Polarization

Most dielectric materials, in particular liquids, are molecular compounds. An applied electric field polarizes the medium, i.e., it creates in every volume Δv a dipolar moment $\vec{p}$ proportional to Δv: $\vec{p} = \vec{P}\Delta v$. The polarization $\vec{P}$ is generally proportional to the electric field: $\vec{P} = \chi\vec{E}$, χ being the electrical susceptibility.

In liquids there are two mechanisms generating $\vec{P}$. The first one, the electronic polarization, is universal and arises from the field induced rearrangement of the electronic clouds in the molecules. The second mechanism, orientation polarization, is specific of the so-called polar materials such that their molecules have a permanent dipole moment (it is the case of water); the field then tends to align the dipoles along its direction.

The polarization makes it necessary to define a second vector, the electric induction $\vec{D}$, in order to describe the electric state of any system with dielectric material ($\vec{D} = \varepsilon_0\vec{E}_0 + \vec{P}$). Most often the susceptibility χ is a constant scalar so that $\vec{D} = \varepsilon\vec{E}_0$, $\varepsilon = \varepsilon_0 + \chi$ being the material permittivity. Inside a dielectric material, the polarization induces an electric field that partly counteracts the field E_0 from which the polarization originates. The dielectric constant $\varepsilon_r = \varepsilon/\varepsilon_0$ gives a measure of the compensation ability.

Two charges q and q' located in a dielectric material with permittivity ε will experience a force according to the Coulomb law $F = qq'/\left(4\pi\varepsilon_0\varepsilon_r r^2\right)$. If a piece of dielectric material is placed in an electric field in vacuum, or in a lower permittivity material, the inner electric field E_i is reduced compared to the background field E_0. For a dielectric sphere this reduction is given by $E_i/E_0 = 3\varepsilon_2/(2\varepsilon_2 + \varepsilon_1)$, where ε_1 is the permittivity of the sphere and ε_2 is the permittivity of the surrounding material.

15.3.4 Conduction

Polarization inherently assumes that the material is a perfect insulator. In practice such materials do not exist. There are always some charge carriers that are free to move once an electric field is applied. Conduction properties must also be taken into account. In a metal, conduction results from movement of free electrons, while in dielectric materials it is movement of ions, electron hopping etc. In a liquid, the conduction mainly takes place through ionic movement. When an electric field E exerts a force upon an ion it will tend to move and gains an average velocity v proportional to the electric field given by its mobility K: $v = KE$. The resulting current density j is found when introducing the number of charge carriers n (assuming univalent charge carriers): $j = ne(K^+ + K^-)E$. The ability of a material to conduct current is described by its conductivity $\gamma = ne(K^+ + K^-)$, which is the inverse of its resistivity, defined by $\rho = E/j$. Ohmic conduction is defined by ρ being independent of E. Resistivity is a material parameter that, combined with geometrical dimensions of the material, gives the resistance $R = U/I$.

The conductivity in a bulk dielectric liquid is mainly governed by the ion concentration. Normally, conductivity is measured with quite low fields. In an insulating liquid (e.g., oil) conduction is ohmic only for low electric fields. Ions appear in the liquid by dissociation of traces of electrolytes dissolved in the liquid. The dissociation constant of a weak electrolyte is a function of the applied electric field according to Onsager's theory [10]. (At room temperature, the dissociation constant will increase by a factor of 10 when the electric field is increased to 5×10^4 V/cm, compared to ohmic conduction.)

Metals are different from dielectrics in that the electrons are free to move around in the conduction band. If a metal is placed in an electric field the electrons will move to the surface of the metal and arrange themselves so that the internal field in the metal is zero.

In many ways one can regard a metal as a dielectric with an infinitely high permittivity or the other way; model a high permittivity dielectric as conducting body. However, disregarding conduction in the dielectric material, even if the two cases look similar from the outside, there is an important difference: The charges on metals are free to move to another object if they get

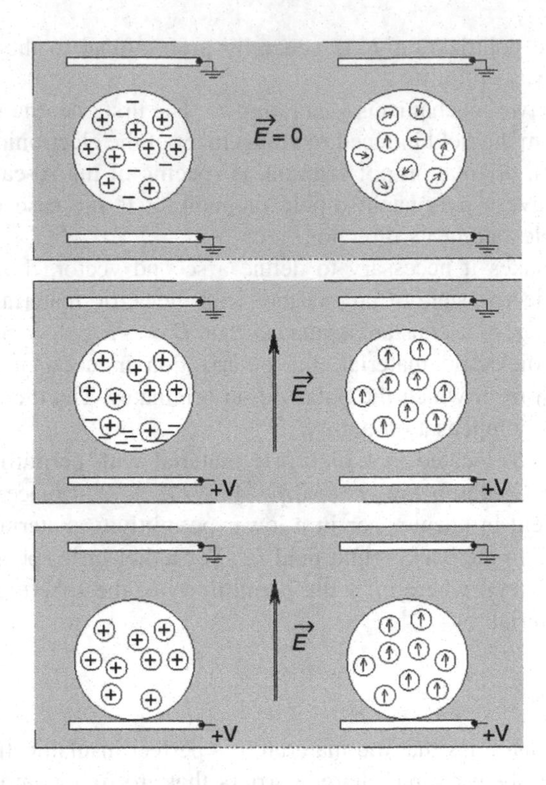

FIGURE 15.4 Polarization of conductive (left) and insulating (right) sphere and interaction with electrode upon contact. *Top*: No electric field. Evenly distributed free charges (left) and randomly oriented dipoles (right). *Middle*: Applied electric field. The conductive sphere is polarized by displacement of the free negative charges. In the dielectric sphere the dipoles try to align with the electric field lines. *Bottom*: Applied electric field, and spheres in contact with the positive electrode. Negative, free charges have moved to the positive electrode, and the conducting sphere is left with positive net charge. The dielectric sphere has no free charges and remains with zero net charge.

into contact as shown in Figure 15.4, while there are no free charges in the dielectric so it will always have zero net charge.

15.3.5 Time Constants

Conductance is the transport of free charges. From the outside it is not a priori easy to see the difference between charge movement (conductance) and charge rearrangement (polarization) within a material. In any case a charge-neutral body in an electric field will behave as a dipole. The main difference is the speed of the processes. For a metal the polarization will, for our practical cases, be instantaneous, while for dielectric bodies it will involve some time. In general, the polarization will be a quick process, compared to redistribution of charges by conduction, which takes more time.

The time constant of charge redistribution or relaxation is a material constant: $\tau = \varepsilon_0 \varepsilon_r / \gamma = \varepsilon_0 \varepsilon_r \rho$. This is equivalent to the time constant for a capacitor, $\tau = RC$, where C is the capacitance and R the internal resistance. This time constant for a material is an important parameter describing whether the material behaves like a conductor or an insulator. For time varying fields changing much faster than the time constant the material will behave like an insulator and vice versa. Another important aspect is that charges placed in a dielectric will leak away at a rate given by this time constant.

15.3.6 Interfaces

At the interfaces between different materials continuity relations must be expressed. For two perfect dielectric materials there is continuity of the electric potential and of the normal component of electric induction:

$$\varepsilon_2(E_2)_n = \varepsilon_1(E_1)_n \tag{15.10}$$

Taking into account the conduction of the materials leads to the general relation

$$\varepsilon_2(E_2)_n - \varepsilon_1(E_1)_n = \sigma_s \tag{15.11}$$

where σ_s is the surface charge density of free charge carriers and depends on the conduction properties. When one of the materials (No. 1 here) is conducting, the field $E_1 = 0$ and the surface charge density σ_s is related to the field E_2:

$$\sigma_s = \varepsilon_2 E_2 \tag{15.12}$$

The action of the field on the surface charge density results in the so-called *electrostatic pressure*:

$$p_{es} = \frac{1}{2}\varepsilon_0 E_{\text{out}}^2 \tag{15.13}$$

15.3.7 Practical Coalescer System

The differences in dielectric properties of emulsions and of materials in electrocoalescers have several important practical implications: Both ac and dc frequencies are utilized, and one should consider the effects of the time constants on both drops in the emulsion and the emulsion itself between the electrodes.

To understand the effect of the applied voltage shape, it is important to be familiar with the terms of *capacitive* and *resistive* voltage distribution. As mentioned earlier in this chapter, the capacitance of a conductor is defined as the ratio of its charge to its potential, $C = Q/V$. As an example, we see from Equation 15.5 that the capacitance of a conducting sphere of radius R is $C_{\text{sphere}} = 4\pi\varepsilon_0\varepsilon_r R$, when the sphere is surrounded by a dielectric of dielectric constant ε_r. A typical capacitor is formed by two parallel conducting plates as shown in Figure 15.5.

The electrodes are separated by a distance d and the space between them is filled by a dielectric. When σ is the surface charge density on the plates according to Equation 15.12, the electric field between the plates is $E = \sigma/\varepsilon = (V_1 - V_2)/d$. Further, if A is the area of the plates, the capacitance of the parallel plates is

$$C = \frac{Q}{U} = \frac{\sigma A}{\Delta V} = \varepsilon_0\varepsilon_r\frac{A}{d} \tag{15.14}$$

Similarly, the resistance R between the plates is given by the conductivity γ of the dielectric, $R = d/(\gamma A)$, and the resulting charge relaxation time is $\tau = RC = \varepsilon_0\varepsilon_r/\gamma$. Under dc step

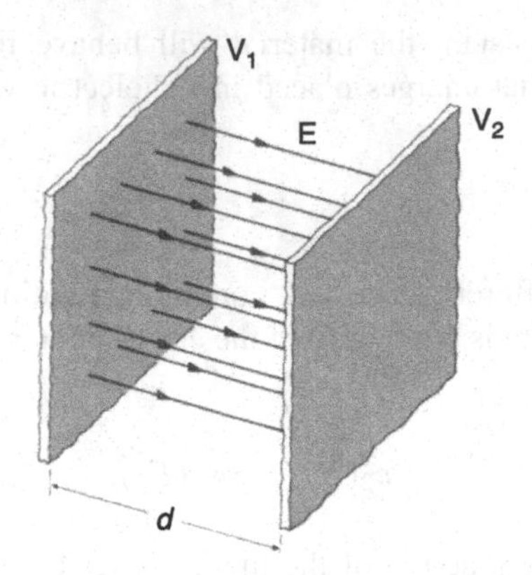

FIGURE 15.5 Parallel plate capacitor.

conditions the total current between the electrodes is the sum of a dc conduction current and a transient displacement current. The current density is expressed by

$$j = j_{\text{conduction}} + j_{\text{displacement}} = \gamma E + \frac{d}{dt}(\varepsilon E) \tag{15.15}$$

Thus, for a constant, or even slowly varying electric field, the current density and the local electric field will be determined by the conductivity of the dielectric material.

Figure 15.6 shows a simplified electrocoalescer with electrodes coated by a solid insulating material. As shown in Table 15.1, the conductivity is much lower for a solid insulating material than for a crude oil emulsion and the time constants will therefore also be very different. If a dc field is applied all ions and charge carriers will quickly move to the insulating barrier. The field distribution is governed by the conductivity of the materials and the surface charge on the solid coating. The effect of this is that the voltage drop between the metallic electrodes mainly takes place across the solid insulation; as a result there is no field in the emulsion that can activate the electrocoalescence process.

TABLE 15.1
List of Material Constants

Material	Conductivity (S/m)	Dielectric Constant	Relaxation Time Constant (s)
Solid insulation	1×10^{-16} to 1×10^{-14}	4 to 6	5.3×10^{3} to 3.5×10^{5}
Pure oil	1×10^{-13} to 1×10^{-12}	2 to 3	27 to 177
Crude oil	1×10^{-9} to 1×10^{-7}	2 to 3	2.7×10^{-4} to 1.8×10^{-2}
Purified water	$> 4 \times 10^{-6}$	20 °C: 80	1.8×10^{-4}
		90 °C: 58	1.3×10^{-4}
Water with	~ 1	20 °C: 80	7.1×10^{-10}
5 wt% NaCl		90 °C: 58	5.1×10^{-10}

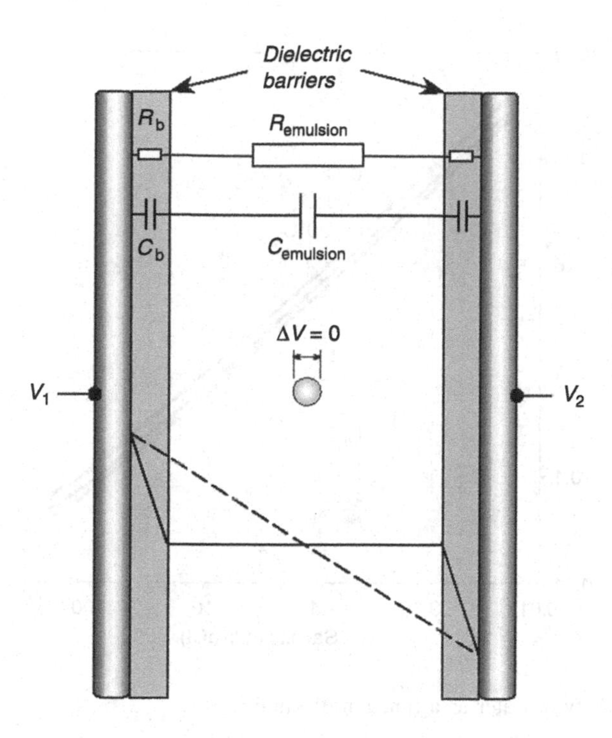

FIGURE 15.6 Example of emulsion between electrically insulated electrodes. Solid line: dc voltage distribution. Dashed line: ac voltage distribution.

If ac is applied at the system then the picture is quite different. Provided the frequency of the applied voltage has a period significantly shorter than the time constant of the emulsion, then the ions and charged particles will not have time to move to the insulating barriers and build up surface charge. It is now the permittivities that govern the electric field distribution; we say we have a capacitive field distribution.

15.3.8 Polarization and Conductivity of Water Drops

Even for highly purified water the electrolytic conductivity will be $\gamma > 4 \times 10^{-6}(\Omega\text{m})^{-1}$. When salt is added to the water the conductivity increases with the salt concentration as shown in Figure 15.7 [11]. Typically, seawater has an average salinity about 35 psu (3.5 wt% salt). The relaxation constant will be less than 1 ns, indicating that one may equal a water drop with a conductive body.

15.3.9 Processes for Charging of Drops

When a conducting body like a water drop hits a metallic plane with an electric field the particle will receive a net charge Q. For a sphere of radius a in a liquid of permittivity ε this is

$$Q = \frac{\pi^2}{6} 4\pi\varepsilon a^2 E_0 \tag{15.16}$$

The more elongated the particle is in the direction of the field the higher the charge [12].

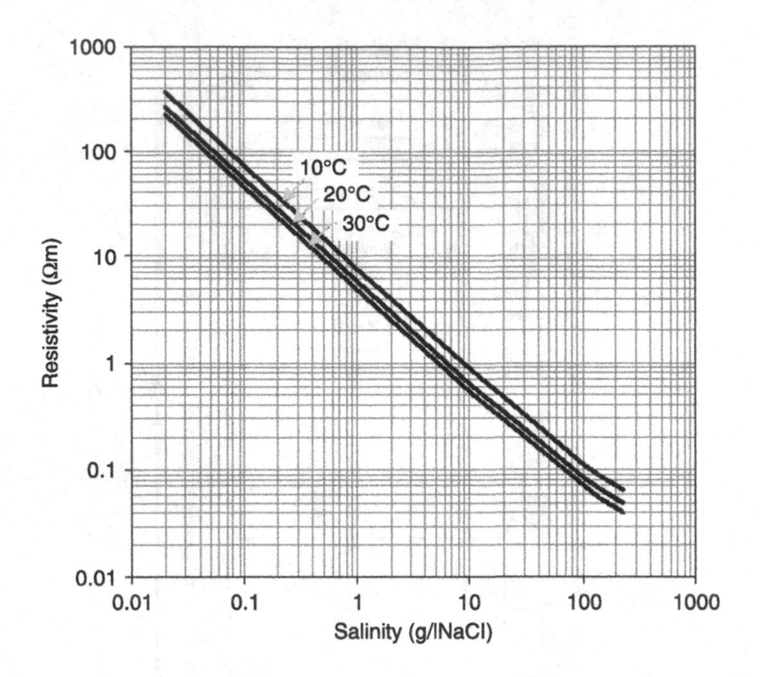

FIGURE 15.7 Resistivity of water as a function of salinity.

Water drops can also be charged in other processes such as charge transfer from other bodies, streaming electrification, adsorption of polar ions on the drop surface (preferential adsorption), and by drop breakup. The latter case is due to polarization of the water drop. As the drop is conductive, positive and negative ions will separate on the drop surfaces and provide a net charge at each drop pole. Consequently, if the drop breaks up, either by collision, turbulence, or a high field induced instability, the resulting droplets will carry net charges.

15.3.10 Chemical Surface Potentials in External Electrical Fields

So far we have considered the interface between the phases (water/oil or oil/electrode) as quite ideal, only described by its interfacial tension and a location where charges may accumulate. In reality the interface has its own microscopic electrostatic environment. Between a metallic surface and liquid there is an electrochemical potential difference; ions from the liquid accumulate at the interface and get bound to it by their electric images in the metal. At thermodynamic equilibrium ($E = 0$) this electric double layer (EDL) has a diffuse part in the liquid. The higher the number of charge carriers available, the faster the field will decrease with distance from the surface. The characteristic distance, called the Debye length, is related to fluid conductivity γ, permittivity ε, and molecular diffusion coefficient D by $\lambda_D = \sqrt{D\varepsilon/\gamma}$. As there are much more ions in water than in oil the Debye length will be much smaller in water than in oil. In aqueous electrolytes the EDL thickness has a typical value of only 1 nm. At room temperature the thermal voltage drop across this layer yields a very high internal electric field $E = (kT/q)/\lambda_D \sim 300$ kV/cm. For low conductivity oils the corresponding higher Debye length and lower internal field are respectively of the order of 5 μm and 50 V/cm [13]. With external fields of 1 kV/cm, which is in the design range for a coalescer, the diffuse layer in the oil will be swept away.

15.4 ELECTROSTATIC FORCES AND MOVEMENT

15.4.1 Electrostatic Forces Acting on a Single Droplet

Depending on the charging and polarization of a body (hereafter denoted the water drop) and the distribution of the external electric field, different forces may act. We distinguish between *electrophoretic* and *dielectrophoretic* forces. We will disregard effects from forces resulting in a change of the drop shape, assuming a fixed spherical shape.

The movement resulting from an electric field acting on a charged drop is denoted electrophoresis. As explained in Section 15.3 the force is $\vec{F} = q\vec{E}$, where the direction of the force depends on polarity of the charge q and the direction of the electric field. In the case where the particle is subjected to a homogeneous ac field the particle will oscillate, but over time no net displacement is observed. For dc fields a net movement will result. These forces do not depend on material properties. The maximum charge a drop may carry is equal to what it may acquire upon contact with an electrode (Equation 15.16). However, this charge may leak away to the surrounding oil. The time constant τ for this is given by the relaxation time constant of the surrounding liquid $\tau = \varepsilon/\gamma$, where ε is its permittivity and γ its conductivity.

Dielectrophoretic forces, on the other hand, do not require net charging, but depend on material properties. This force only arises for inhomogeneous dc or ac fields. For insulating media the direction of the force depends on the dielectric constants of the particle and the medium it is placed in. If the dielectric constant of the particle is higher than that of the surrounding medium the force will pull the particle towards higher fields (positive dielectrophoresis). If the dielectric constant of the particle is lower than for the surrounding medium the particle will be pushed away from the high field region (negative dielectrophoresis). In the case of a drop of water, which is always much more conducting than the suspending medium, the electric field inside the drop is zero and the situation is formally equivalent to a medium of infinite permittivity. Rather simplified, the positive dielectrophoresis can be illustrated by Figures 15.8 and 15.9; the drop is polarized; equal but opposite charge appears on the sides of the drop and the forces on the charge placed in the higher electric field will dominate.

The dipole moment p for a spherical particle is given by

$$p = 4\pi\varepsilon R_0^3 \beta E_0, \quad \beta = \left(\frac{\varepsilon_p - \varepsilon}{\varepsilon_p + 2\varepsilon}\right) \tag{15.17}$$

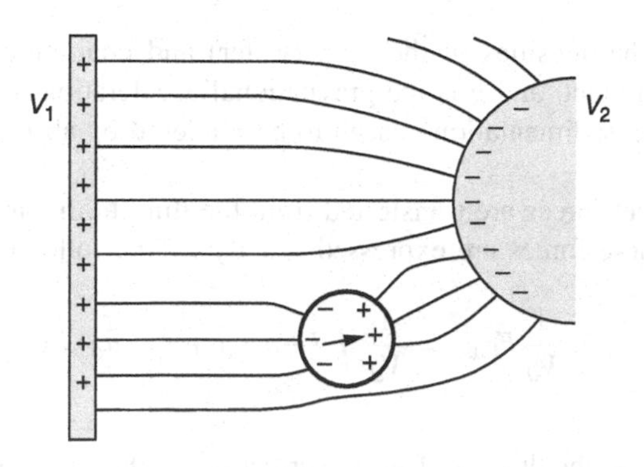

FIGURE 15.8 Dielectrophoretic forces on a water drop in oil.

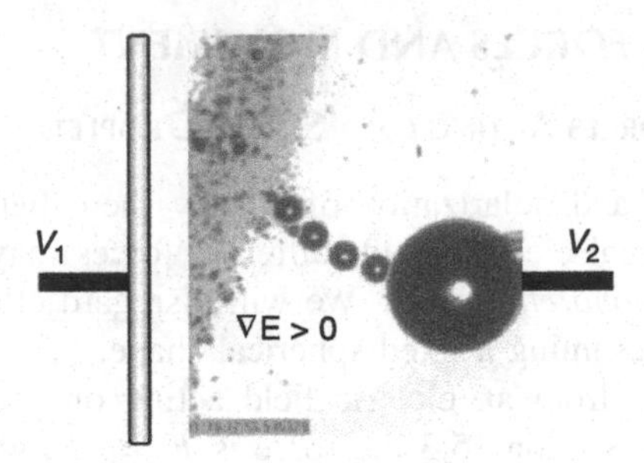

FIGURE 15.9 Experiment showing a sequence where the dielectrophoretic force attracts a water droplet out of an emulsion towards a spherical water electrode (From: A. Pedersen, E. Ilstad, A. Hysveen. Forces and movement of small water droplets in oil due to applied electric field. In: *Conf. Proc. Nordic Insulation Symposium*, NORD-IS, 2003 [14]).

where R_0 is the radius of the particle, ε_p is its permittivity, and ε is the permittivity of the surrounding medium. The dielectrophoretic force is given by the dipole moment of the particle and the gradient of the electric field: $\vec{F} = (p\nabla)\vec{E}$, and the resulting force becomes

$$\vec{F} = 4\pi\varepsilon R_0^3 \beta \vec{E} \nabla \vec{E} \tag{15.18}$$

For a conducting water drop, $\varepsilon_p \to \infty$, Equation 15.18 is simplified to

$$\vec{F} = 4\pi\varepsilon R_0^3 \vec{E} \nabla \vec{E} \tag{15.19}$$

For a homogeneous field $\nabla\vec{E} = 0$, and consequently $\vec{F} = 0$.

15.4.2 Mechanical Forces

In addition to the electric forces there will be forces from gravity, inertia, and from viscous effects. The buoyancy force is given by

$$F_b = (\rho_d - \rho_c)\, g V_d \tag{15.20}$$

where ρ_d and ρ_c are the densities of the drop (water) and continuous phase (oil) respectively, V_d the volume of the droplet, and g is the gravitational acceleration. In emulsions with very fine droplets (i.e., <10 μm), sedimentation is seen to be hindered by absorption of small particles on the drop surface.

For a moving droplet forces are transferred from the fluid to the droplets through friction and pressure difference. These forces are expressed exactly by the following surface integral:

$$\frac{1}{V_d}\vec{F}_{\text{fluid}} = \frac{1}{V_d}\int_{A_d} (-p_s \vec{n}_d + \tau_d \cdot \vec{n}_d)\,dA \tag{15.21}$$

where V_d is the volume of the droplet, A is the surface, p_s is the pressure at the droplet surface, n_d represents the unit outward normal vector, and τ_d is the shear stress tensor at the droplet surface.

The pressure and the friction on the interface are unknown and Equation 15.21 has to be modeled. In the Lagrangian framework the models for the surface integral attempt to provide particular physical meanings.

15.4.2.1 Drag Force

The "steady-state" drag force acts on a droplet in a uniform pressure field when there is no acceleration of the droplet relative to the conveying fluid. The force is given by

$$\vec{F}_d = \frac{1}{2}\rho_c C_d A \left|\vec{u} - \vec{v}\right| (\vec{u} - \vec{v}) \tag{15.22}$$

where $(\vec{u} - \vec{v})$ is the relative velocity of particle and surrounding medium. For a droplet Reynolds number Re_d below 1, the drag coefficient C_d for a rigid sphere is given by:

$$C_d = \frac{24}{\mathrm{Re}_d} \tag{15.23}$$

In fluid spheres, an internal circulation is induced that reduces the viscous part of the drag. For spherical clean bubbles and droplets, the induced internal circulation is accounted for by the Hadamard–Rybczynski equation [15]:

$$C_d = \frac{24}{\mathrm{Re}_d}\frac{\lambda + 2/3}{\lambda + 1} \tag{15.24}$$

where $\lambda = \eta_d/\eta_c$ is the viscosity ratio. In case surfactants are present surface tension gradients might appear. Le Van has suggested a model taking this into account [16,17].

15.4.3 Forces and Movement of Drop Pairs

Again we may distinguish between electrophoretic and dielectrophoretic forces. In both cases the inter-drop forces quickly fall off with distance. When drops are far from each other the forces may be neglected, at intermediate distances simple formulas based on point charges and dipole moments may be used, while when within close range charge distributions and drop geometries must be considered.

The electrophoretic force between two charged drops falls inversely with the square of the distance (Equation 15.2). The maximum charge a water drop may acquire from an electrode in a homogeneous field, disregarding drop breakup due to electric forces, can be found from Equation 15.16. Depending on polarities the force is either attractive or repulsive. These forces will only play a role in the electrocoalescence process when drops are adjacent and carry charges of opposite polarity. Furthermore, charged drops will lose their charge due to conduction through the carrying liquid.

Dielectrophoretic forces between water drops are due to the dipole moment of the drops induced by the external field. Both ac and dc fields are equally relevant. As mentioned in the previous chapter, we assume as a first approximation that the water drops are conductive spheres in an oil continuum, disregarding the conductivity of the oil. When the drops are subjected to an external field there will be induced charges on them, so that the internal field in the drop approaches zero. The charges on each hemisphere will have the same magnitude but opposite polarity. The field at each pole will be higher than the background field by a factor of 3 when the

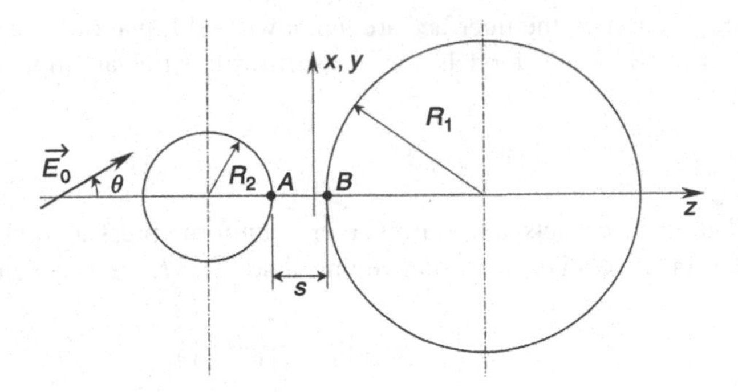

FIGURE 15.10 Two drops of different sizes located with a skew angle θ to a homogeneous electric field E_0.

drops are more than one diameter apart, see Figure 15.11. When two polarized drops get close to each other, the drops experience each other's inhomogeneous field, and they may either attract or repel each other, depending on their relative position in the external field as explained below.

Consider two drops of radii R_1 and R_2 separated by a distance s as shown in Figure 15.10. The axis between the drop centers may have a skew angle θ with respect to the applied field E_0 as indicated in the figure.

15.4.3.1 Dipole–Dipole Interaction

For large droplet distances $s/R \gg 1$ we can approximate the electrostatic interaction between two conducting droplets (the spheres are then equipotential surfaces) as the force between two dipoles located at the sphere centers. This is frequently referred to as the *point-dipole* approximation. The forces in radial direction F_r and tangential direction F_θ are [18]:

$$\begin{aligned} F_r &= 12\pi\varepsilon E_0^2 R_2^3 R_1^3 d^{-4}\left(3K_1\cos^2\theta - 1\right) \\ F_\theta &= -12\pi\varepsilon E_0^2 R_2^3 R_1^3 d^{-4} K_2 \sin 2\theta \end{aligned} \tag{15.25}$$

where d is the distance between drop centers. The coefficients K_1 and K_2 equal 1 in the point dipole approximation for the droplets.

15.4.3.2 The Dipole-Induced Dipole Model

The point-dipole approximation is not valid when the droplets are approaching each other, and the dipole moments are changed due to mutual induction between the spheres. In the literature there are different approaches to find the sphere–sphere forces beyond the point dipole approximation for multiple particles of arbitrary size and position. Clercx and Bossis [19] presented a multi-pole expansion method that gives good results, but the calculation is complex. A more promising model, the *multiple image method*, was presented by Yu et al. [20]. The first two terms in the multiple image method give the dipole-induced dipole model (DID), which is simple and numerically efficient. Siu et al. [21] show that the DID model is in good agreement with the experimental values obtained by Klingenberg et al. [18] for $s/R > 0.1$ with equally sized conductive particles. The DID model is given by Equation 15.25, with coefficients K_1 and K_2

given by

$$K_1 = 1 + \frac{R_1^3 d^5}{\left(d^2 - R_2^2\right)^4} + \frac{R_2^3 d^5}{\left(d^2 - R_1^2\right)^4} + \frac{3R_1^3 R_2^3 \left(3d^2 - R_1^2 - R_2^2\right)}{\left(d^2 - R_1^2 - R_2^2\right)^4}$$
$$K_2 = 1 + \frac{R_1^3 d^3}{2\left(d^2 - R_2^2\right)^3} + \frac{R_2^3 d^3}{2\left(d^2 - R_1^2\right)^3} + \frac{3R_1^3 R_2^3}{\left(d^2 - R_1^2 - R_2^2\right)^3} \quad (15.26)$$

In the limit $d \to \infty$ the coefficients K_1 and K_2 approach unity and we recover the point dipole expression given by Equation 15.25.

15.4.3.3 Analytical Solution

With the geometry defined in Figure 15.10 it is possible to derive analytical solutions for the potential, the electric field, and the force between the two spheres. The problem is analyzed by using bispherical coordinates [22]. Carter and Loh [23] calculated the electric field in a sphere–sphere electrode gap, with equally sized spheres. Later, Davis [24,25] gave a complete solution for the electric potential, field, and mutual forces for two arbitrary sized spheres carrying any net charge.

The maximum electric field, appearing at the drop pole A (see Figure 15.10), is conveniently written

$$E_A = E_0 \cos\theta \cdot E_3 \quad (15.27)$$

where E_0 is the background field. The field coefficient E_3 is an infinite series depending on the drop pair geometry. As shown in Figure 15.11 the electric field between the two drops increases strongly with decreasing drop distance.

The forces are found by integrating the electrostatic pressure over the drop surface. Assuming uncharged spheres the force components on sphere 2 take the simple form:

$$(F_r)_2 = 4\pi\varepsilon R_2^2 E_0^2 \left(F_1 \cos^2\theta + F_2 \sin^2\theta\right)$$
$$(F_\theta)_2 = 4\pi\varepsilon R_2^2 E_0^2 F_3 \sin 2\theta \quad (15.28)$$

where the radial force F_r can be either attractive or repulsive, depending on the angle θ, and F_θ induces a torque tending to align the drop pair with the applied electric field. The force coefficients F_1, F_2, and F_3 are complex series depending on the drop size ratio R_2/R_1 and the drop distance s/R_2. As seen from Figures 15.11 and 15.12 the electric field and the attraction force increases rapidly with decreasing drop distance.

Unfortunately, the computational cost required for calculating F_1, F_2, and F_3 is high in a multi-droplet situation. However, the exact solution is excellent for benchmarking other models in cases with two particles/droplets. The different approximations are compared in Figure 15.13. For large drop separations $s/R_2 \gg 1$ the force coefficients F_1, F_2, and F_3 approach the values of the point-dipole equation (15.25).

For small separations $s/R_2 < 1$, F_2 and F_3 take constant values while F_1 diverges. Atten [26] proposed the following empirical asymptotic expression, valid in the range $10^{-3} \leq s/R_0 \leq 10^{-1}$:

$$F_1 = 1.25/(1 + (R_2/2R_1))^4 \, (R_2/s)^{0.8} \quad (15.29)$$

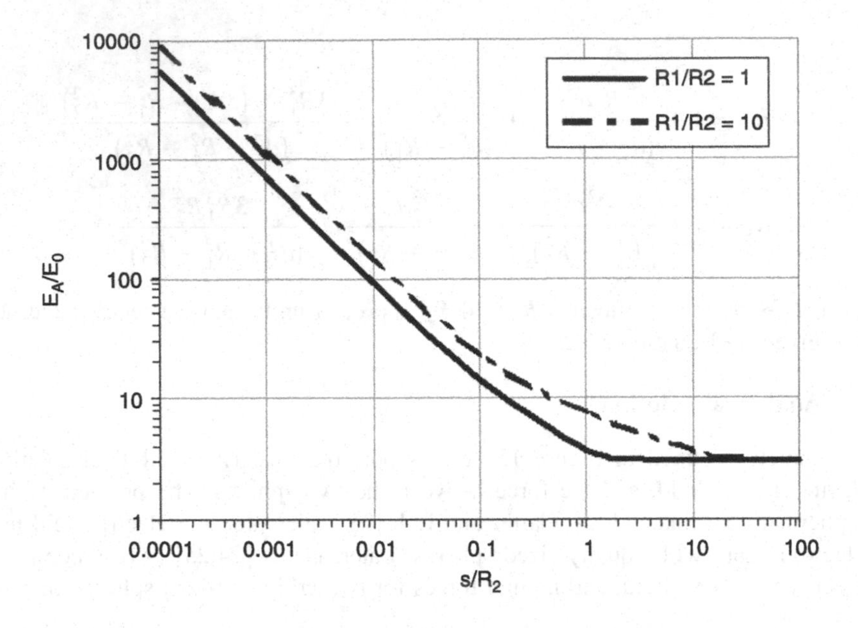

FIGURE 15.11 Electric field enhancement at the pole A as a function of the reduced drop pair distance s/R_2. The electric field E_0 is parallel to the drop axis.

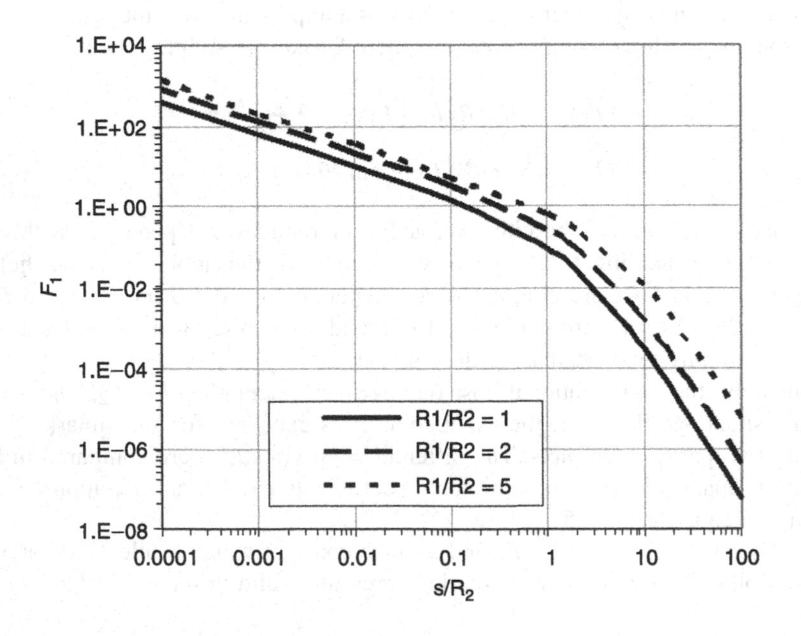

FIGURE 15.12 Force coefficient F_1 as a function of the reduced distance s/R_2 for different drop size ratios.

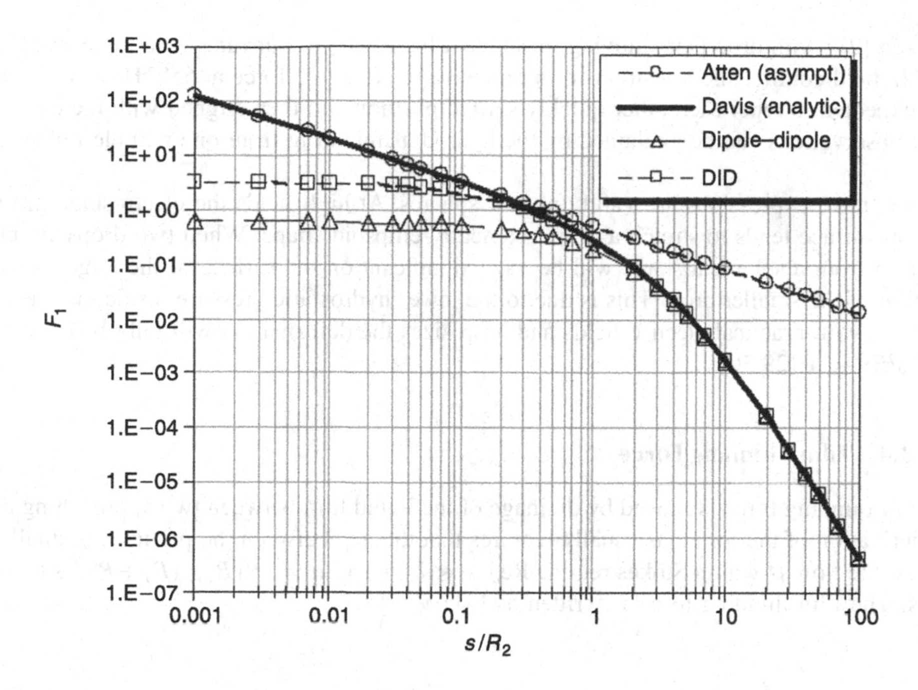

FIGURE 15.13 Comparison of different models for the attraction force between two conducting droplets. The drops are aligned with the electric field (θ = 0). Drop size ratio R_1/R_2 = 2.

In the particular case of spheres of equal radius R_0 aligned with the field ($\theta = 0$), Williams [27] derived asymptotic relations for the potential difference between the two spheres:

$$\Delta V = \frac{2\pi^2}{3} \frac{E_0 R_0}{\log\left(\dfrac{R_0}{s}\right)} \tag{15.30}$$

and the electric field at the facing poles:

$$E_{\text{pole}} \cong \frac{2\pi^2}{3} E_0 \frac{R_0/s}{\log(R_0/s)} \tag{15.31}$$

The attraction force then takes the expression:

$$F_r = \frac{2\pi^5}{9} \varepsilon E_0^2 R_0^2 \frac{R_0/s}{[\log(R_0/s)]^2} \tag{15.32}$$

For drops with a separation of one drop radii or more it is easy to show from Equation 15.25 that the resulting force is repulsive, $F_r < 0$ when $3\cos^2\theta < 1$, that is, when the angle θ between the drop pair and the electric field exceeds 54.7°. Now this does not hold for drops with small separation. In the case of very small separation the angle approaches $\pi/2$, which can be found from Equation 15.28, implying that close drops nearly always attract each other. This analysis is done for a static case, disregarding the influence of hydrodynamic forces. Experiments done by

Pedersen [28] with drop pairs moving in stagnant liquid due to buoyancy and dielectrophoresis in a 50 Hz field confirm the transition from repulsive to attractive force at 55°. However, drops that were expected to repel each other quickly shifted position and then aligned with the electric field. These observations can be explained by the hydrodynamic drag from one particle influencing the other.

The above expressions are valid only for spheres. At high fields the electrostatic pressure on the drop surface tends to stretch a drop to a nearly ellipsoid shape. When two drops are close the deviation from a spherical shape will be most significant on the surface of the larger drop, in the direction of the smaller drop. This is due to the lower hydrostatic pressure inside the largest drop. However, with practical electric fields and drop sizes the deformation will only be of importance when $s/R_2 \ll 1$ [29,30].

15.4.3.4 Film-Thinning Force

The film-thinning force is caused by drainage of the liquid film between two approaching droplets. The derivation of the formulas usually requires that the gap between the particles is small, $s \ll a$, and that the flow is within Stokes regime $\mathrm{Re}_d \cdot s \ll a$, where $a = (R_1 R_2)/(R_1 + R_2)$ is the reduced radius. The film-thinning force is written as [31]

$$\vec{F}_f = \frac{-6\pi \mu_c a^2 (\vec{v}_r \cdot \vec{e}_r)}{s} f^* \vec{e}_r \tag{15.33}$$

where $\vec{v}_r$ is the two drop relative velocity and $\vec{e}_r$ indicates the direction of the relative motion. In the case of rigid sphere the function $f^* = 1$. For very close liquid spheres expressions compensating for proximity effects have been developed by Vinogradova [32] and Barnocky et al. [33].

15.4.4 Electrostatic Models for Multiple Spheres

When more than two spheres are considered, Laplace's equation can only be solved numerically. This is possible with a fast computer code; but with increasing number of spheres the calculation time can be quite considerable. Faster calculations are obtained by using the method of images or multipole expansion [34–41]. These methods are applicable to both dielectric and conductive particles. As mentioned in the previous sections, the electric field of a conducting sphere can be exactly represented by a dipole p or a point charge Q at the sphere center. Two spheres could be modeled, to a first order approximation, by two point charges or two dipoles at a distance s located at the respective sphere centers. However, due to the mutual induction between the charges or dipoles, the corresponding spheres will no longer be equipotential surfaces. This is corrected by recursively adding more charges Q' or dipoles p' at increasing distance from the sphere centers. The major advantage with these methods is that the force exerted on one sphere by the other is simply found by summing the forces between all electric images (Coulomb's law). Further, the polar interactions in a multiparticle system are pairwise additive. Inspired by the work by Jones [37], Bosch [42] developed a simpler method using dipole images to calculate the force between two conducting spheres in an external electric field. The results fit well to the previous results by Davis [25], and according to the author the method can readily be extended to more than two spheres. This method is effective as it uses a simple matrix inversion rather than numerous charge/dipole iterations.

15.4.5 Forces and Movement in Emulsions

For a water-in-oil emulsion the case is more difficult as analytic expressions are not readily available. The two main questions are: What is the local electric field?, and, How can forces between multiple bodies be calculated?

If the distribution of the emulsion is inhomogeneous the background field may be influenced; increasing in the regions with the lower water cut. This may be explained by the fact that when two dielectric materials are put in series in an electric field, the stress increases in the lower permittivity dielectric due to polarization effects. An emulsion with a high water cut will have higher permittivity than one with a low water cut due to the effect from the high permittivity and conductivity of water.

We expect that for emulsions with a water cut below 5% it is possible to use the formulas for fields and forces presented in this section, because only rarely are more than two drops at a time within near-field distance of each other (i.e., within one drop radius distance; see Figure 15.13). This allows simulations based on discrete element method (DEM) code [43].

The topic of multiple particle ($n > 2$) interaction in a flowing dielectric liquid is thoroughly treated in the field of electrorheology, and will not be discussed further here. See for instance Refs. 18–21,34,44–47.

Some experiments were carried out on stagnant emulsions to get a preliminary impression of the behavior of such a complicated system. The amplitude, frequency, and waveform (square/sine) of the applied voltage were varied. The experiments were similar to what was previously reported [48]; a drop of fine water-in-oil emulsion (diameters below 10 μm) was injected into a pure oil phase in a homogeneous electric field between two electrodes.

In the experiments it was impossible to avoid charged drops due to the tribo-electric effects during processing of the emulsion. The bulk fluid had a very low conductivity giving a charge relaxation time of several minutes. The effect of charged droplets seemed to influence the electrocoalescence for all emulsions. Generally, for frequencies below ~100 Hz the movement of the majority of drops was dominated by the electrostatic alternating forces on the droplets being electrically charged. Charged droplets were either resident or obtained from drops touching each other and exchanging charge before separating again. Charges were easily scattered across the emulsion due to the drop–drop interactions. The resulting oscillating movement of the bulk oil lowered the electrocoalescence efficiency. For higher applied frequencies, above ~300 Hz, electrostatic movements of charged drops were locally confined and less influencing the surrounding emulsion. In general, coalescence efficiency decreased with decreasing emulsion average drop size and water cut, and there was no clear observable effect on the efficiency of the coalescence from the voltage shapes. The presence of a larger drop, as shown in Figure 15.14, increased the coalescence rate strongly by "sucking up" small droplets from the emulsion.

Water-in-oil emulsions without chemical stabilizers added, with 10 to 20 μm initial drop size and water cuts up to 20%, showed no field induced coalescence. Chemically stabilized emulsions with even smaller drop sizes (1 to 10 μm diameter) behaved differently. These stabilized emulsions, being vastly denser, showed a highly active electrocoalescence. Furthermore, for the case of the denser chemically stabilized emulsions, we observed an unexpected sudden and continuous expansion of the emulsion volume during some time (and voltage periods) starting at application of the electric field. The phenomenon was promoted by high electric fields, suppressed by an increasing voltage frequency, and most intense using bipolar square voltages. The expansion

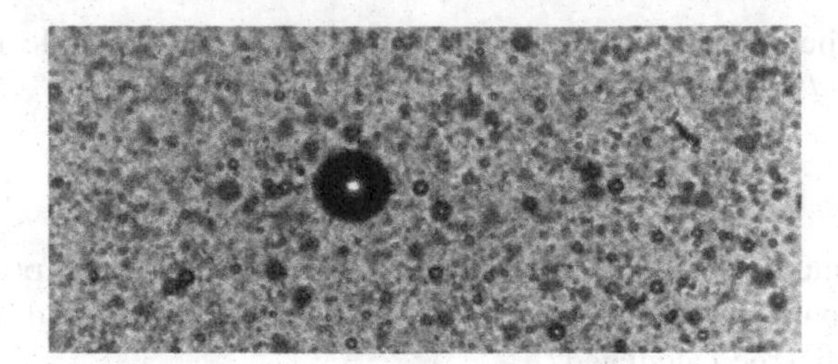

FIGURE 15.14 Picture from emulsion experiments in a horizontal electric field, showing a large drop sucking up small droplets from the emulsion.

FIGURE 15.15 Chain formation in a stagnant liquid. Square wave voltage, 3.5 kV/cm horizontal electric field.

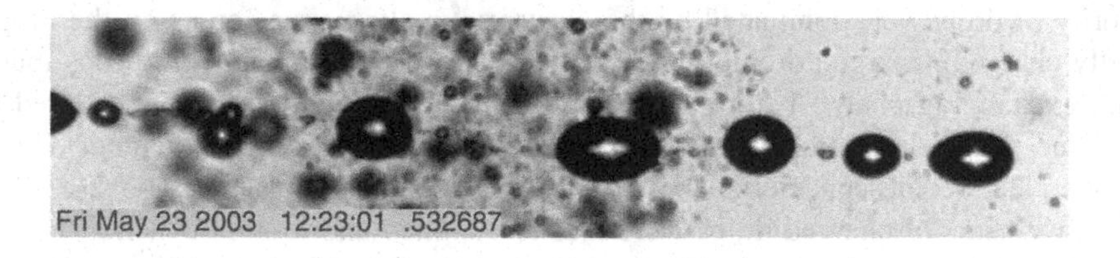

FIGURE 15.16 Chain formation along the horizontal electric field. Small, stable droplets in the chain tend to obstruct coalescence.

continued until the emulsion volume had reached a new and lower density. The higher optimum frequencies preserved emulsion density and therefore coalescence efficiency. For increasing frequencies up to 10 kHz the chain formation became less pronounced as the coalescence efficiency increased.

Very often droplets formed chains that were aligned with the electric field; see Figure 15.15. These drops grew larger in time as chains coalesced internally, and more drops were added to the ends of the chains. Often, the coalescence rate for chains of larger droplets (>100 μm) was lower than theoretically expected, apparently due to intervening particles and very small, stable water droplets as seen in Figure 15.16. The droplet chain formation seemed a natural and significant part of the multi-droplet electrocoalescence.

15.5 DROP INSTABILITIES

So far we have treated the water drops as rigid spheres. However, when an electric field is applied on a soft conductive body like a water drop, it will change its shape, become elongated, and in the end may break up due to electric field forces [49]. This is explained by the classic theory by Taylor [50] developed for water drops in dc fields. We have studied the phenomenon for water drops in oil under various voltage shapes and frequencies, and found good agreement with classic theory. However, classical theory only considers an ideal interface between oil and water only characterized by its interfacial tension. In reality the interface has not a zero thickness, and we expect that for real emulsions surface rheology will play a role introducing new time constants and gradients in the surface characteristics (such as the Marangoni effect known from chemistry).

15.5.1 CLASSICAL THEORY OF DROP INSTABILITIES

In the absence of an electric field the interfacial tension T will keep the water drop spherical with radius r when the gravitational effects are negligible (Bond number $Bo = \Delta \rho g r^2/T \ll 1$). The pressure difference across the interface of a drop of radius r due to surface tension is then $\Delta p = 2T/r$. The shape stability is therefore higher for smaller drops.

When a uniform field is applied, the drop will deform due to the electric stress on the surface, and the drop will elongate in the direction of the electric field. There are two different classical approaches to model the elongation of the drop. Both models assume that the drop takes the shape of a prolate spheroid. Sherwood [51] used an energy argument, where the drop takes a shape that minimizes the total energy, which is the sum of the electrostatic and surface energy. In this model it is possible to vary the permittivity and conductivity of the drop liquid and of the suspending fluid. Taylor [49,50] used a simpler approach, seeking for a static solution such that the pressure difference across the drop interface is the same at any point on the surface. This pressure difference is due to the effect of interfacial tension and electrostatic pressure. Postulating that the shape of the deformed drop is an ellipsoid, it is sufficient to estimate Δp at two points, at the drop pole and at the equator. Balancing the two pressure differences, $\Delta p_{\text{pole}} = \Delta p_{\text{eq}}$, gives:

$$E_0\sqrt{\frac{2r_0\varepsilon}{T}} = 2\cdot\left(\frac{b}{a}\right)^{\frac{4}{3}}\left(2-\frac{b}{a}-\left(\frac{b}{a}\right)^3\right)^{\frac{1}{2}} I_2 \tag{15.34}$$

where

$$I_2 = \frac{1}{2}e^{-3}\ln\left(\frac{1+e}{1-e}\right) - e^{-2} \tag{15.35}$$

where e is the eccentricity defined by $e = \sqrt{1-b^2/a^2}$, and a, b are the semi-axes of the prolate spheroid. The drop elongation as a function of the applied electric field is plotted in Figure 15.17.

The drop elongation increases with the field up to a limiting aspect ratio above which the droplet becomes unstable. The corresponding field is called the critical field E_c. Considering the special case of a fully conductive droplet, Taylor [50] showed that a stationary prolate spheroid

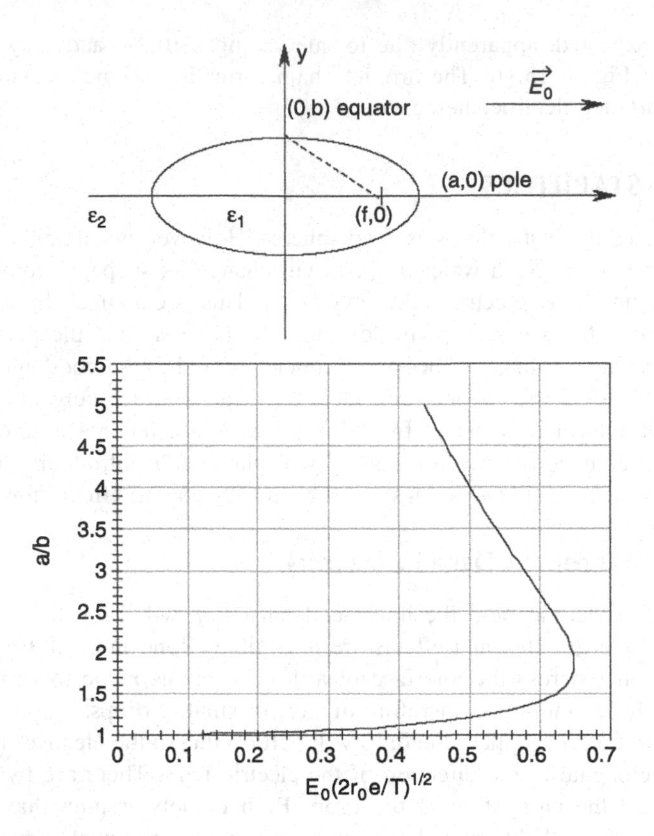

FIGURE 15.17 Theoretical elongation of a water drop as a function of the applied electric field. *a* is the long semi-axis and *b* is the short one.

has a maximum possible aspect ratio $a/b \cong 1.9$, as shown in Figure 15.17. At this maximum value of a/b the applied field E_o equals the critical field E_c, found to be:

$$E_c = 0.648\sqrt{\frac{T}{2\varepsilon r_0}} \tag{15.36}$$

At higher values of the applied field E_o the drop becomes unstable at the poles and breaks up. This model fits well to experimental data for dc and low frequency alternating voltages. More accurate expressions for the critical field can be found in Ref. 52.

15.5.2 Experimental Observations of Disintegration of Water Drops in Oil

A study of instabilities of sub-millimeter water drops in refined naphthenic oil was performed varying the frequency and the shape of the applied ac voltage [53]. The drops were slowly falling by gravity in a horizontal electric field while the voltage was gradually increased. A camera was used to observe the changes in shape and to determine in what conditions the instability occurs.

For dc voltage it was difficult to keep the drop in the field of view as it easily got charged and was pulled toward an electrode. The frequency of applied ac voltage was varied from 2 Hz to 2000 Hz. When increasing the voltage amplitude, the droplets deformed into prolate spheroids

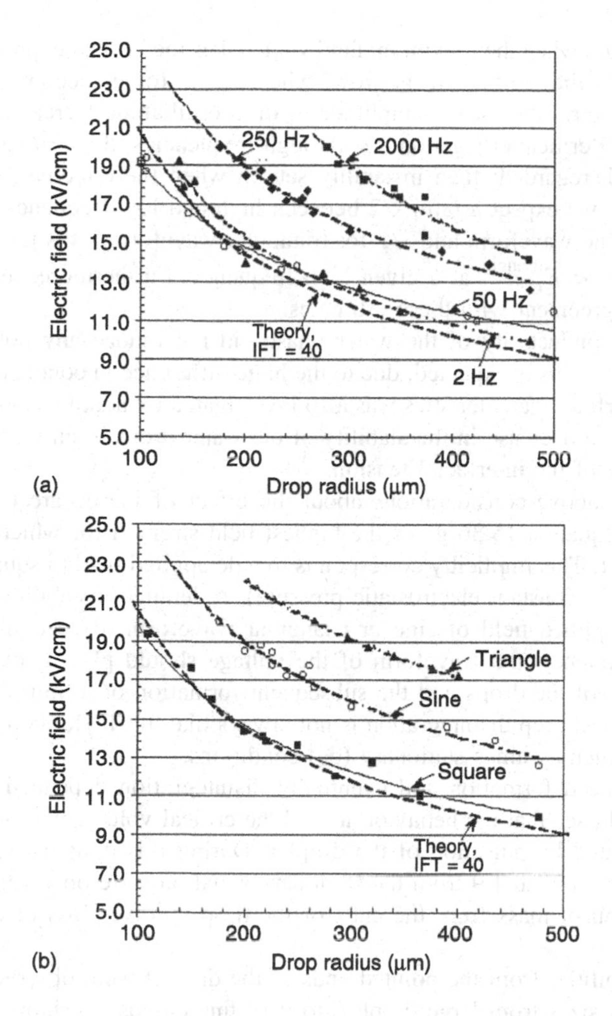

FIGURE 15.18 Drop breakup field strength for (a) varying frequencies of a pure sine voltage and (b) different shapes of applied voltage at 250 Hz.

pulsating at twice the voltage frequency as expected. For low frequencies in the range 2 to 50 Hz the measured critical field for drop disintegration is in very good agreement with the theoretical values obtained from Equation 15.36, independently of the voltage waveform. At higher frequencies, however, the critical field increases with the applied frequency as shown in Figure 15.18(a).

The different voltage waveforms used in this study were triangular, sinusoidal, or square. For frequencies above ~100 Hz the observed critical field was also depending on the voltage waveform when the frequency was kept constant, as shown in Figure 15.18(b).

The disintegration of water droplets is well predicted by theory for low applied frequencies, independently of the voltage waveform. However, at increasing frequencies, a stronger field is required to reach the drop instability. This can be understood taking into account the inertia of the system. For very low frequency, the drop deformation follows the changes in electrostatic pressure so that, despite the oscillating character, at every moment the drop is in quasi-static equilibrium.

Instability then occurs when the maximum field value takes the critical expression (15.36). Raising the frequency makes the inertia effects grow; when inertia forces become of the same order as the capillary and electric forces, the amplitude of drop oscillation decreases and a higher field is required to reach the critical elongation a/b. At high frequencies, the oscillation amplitude is very small and can be disregarded; then instability sets in when the effective field takes the critical value E_C. Therefore we expect a ratio $\sqrt{2}$ between high and low frequency critical values of the effective field for sine waveform and $\sqrt{3}$ for triangular waveform. This justifies the observations that $E_C^{\text{square}} < E_C^{\text{sine}} < E_C^{\text{triangle}}$ at a given high frequency. Furthermore, results of Figure 15.18 are in reasonable agreement with these estimates.

Increasing the conductivity of the water phase did not induce any noticeable effect on the critical field strength. This is expected, due to the huge difference in conductivity of water and oil. The influence of surface characteristics was also investigated by adding asphaltenes. Adding these compounds leads to a decrease in the stability of the water drops in an electric field, presumably due to the lowering of the interfacial tension.

Very likely the above considerations about the effect of inertia are only part of the story. Strictly speaking, Equation 15.36 gives the highest field strength for which a static deformation of the drop can exist. This implicitly corresponds to a dc applied field (a square waveform ac field practically provides a constant electrostatic pressure). A detailed instability analysis is necessary in the case of ac applied field of sine or triangular waveform to account for the influence of frequency. The frequency and waveform of the voltage should play a role in the development of the pointed ends of the drops and the subsequent formation of a thin jet or ejection of small droplets. The observed drop disintegration is not always like the Taylor cone-jet formation shown in Figure 15.19, which assumes stationary (dc) conditions.

At instability the deformation and eventually disintegration depended on the ac frequency and shape. We could see that the behavior around the critical voltage was nonstationary; at lower frequencies dominated by pulsation of the droplet. During this dynamic situation, aspect ratios far exceeding the theoretical 1.9 from the stationary situation were observed. Several factors will play a role: Ejection of mass from the ends of the drop involves loss of charge, viscosity, and drop sizes etc.

Different instabilities from the pointed ends of the droplet were observed, involving ejection of drops of various sizes from large droplets to very fine clouds, as shown in Figure 15.20.

Once the instability has started it will continue until the original drop size is reduced below the critical limit. In our experiments the instability usually ended with a violent final breakup, observed in different modes like separation into a few droplets (sometimes by a sausage instability) or a more explosion-like event.

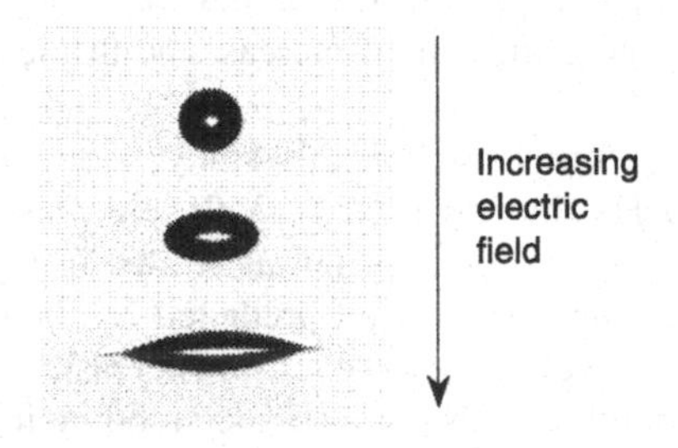

FIGURE 15.19 Drop stretching and Taylor cone formation with thin jet instability during increasing field.

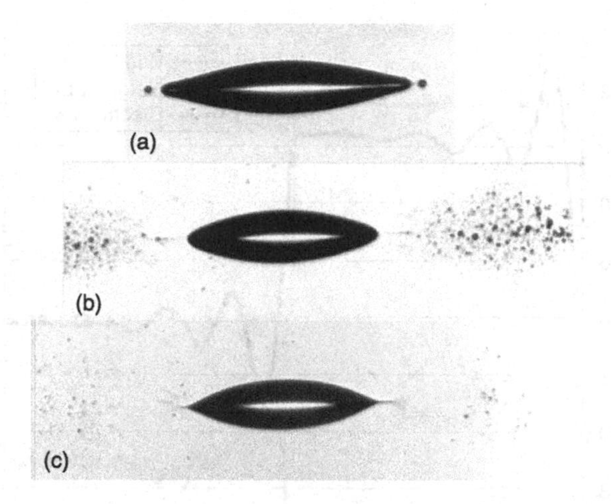

FIGURE 15.20 Observed instability types: (a) ejection of single droplets, (b) ejection of cloud of droplets, and (c) ejection of cloud of fine droplets.

15.5.3 Dynamic Study of Drop Shape for Transient Electric Fields

The resonance modes $n \geq 2$ for a free drop have been calculated theoretically [54] for water drops in air. For a water drop in a dielectric liquid the $n = 2$ mode dominates, as the drop surface dynamics is damped by viscous effects. The transient oscillations of a drop elongating to a steady state spheroid can be studied by applying a short square wave voltage pulse. In the same way, the relaxation of the drop surface can be observed when the voltage is turned off. This can be used to measure the surface elasticity and other dynamical properties of the water/oil interface, such as the time constant of surfactant absorption. The drop surface elasticity e follows the definition given by Gibbs [55,56]

$$e = \frac{dT}{d \ln A} \tag{15.37}$$

where T is the interfacial tension and A is the surface area of the drop. Figure 15.21 shows as a typical example the transient evolution of a water drop in a liquid without surfactants, while Figure 15.22 shows the damping effect when the drop surface is saturated by asphaltenes.

This significantly damped dynamic behavior affects the instability formation, which is slowed down by the less flexible drop surface. Water drops with asphaltene saturated surfaces would frequently exhibit a stationary elongation of more than twice the theoretical break-up value $a/b = 1.9$.

15.6 THE ELECTROCOALESCENCE MECHANISM

From Sections 15.4 and 15.5 we obtain a picture of an applied electric field that induces attractive forces between close drops: it brings pairs of drops closer until they eventually coalesce. This picture applies for stagnant emulsions but not strictly in the case of electrocoalescers in which the suspending oil is in motion. For an emulsion characterized by a laminar or turbulent flow, due to the action of shear, the drops in a volume around a particular drop have a relative motion,

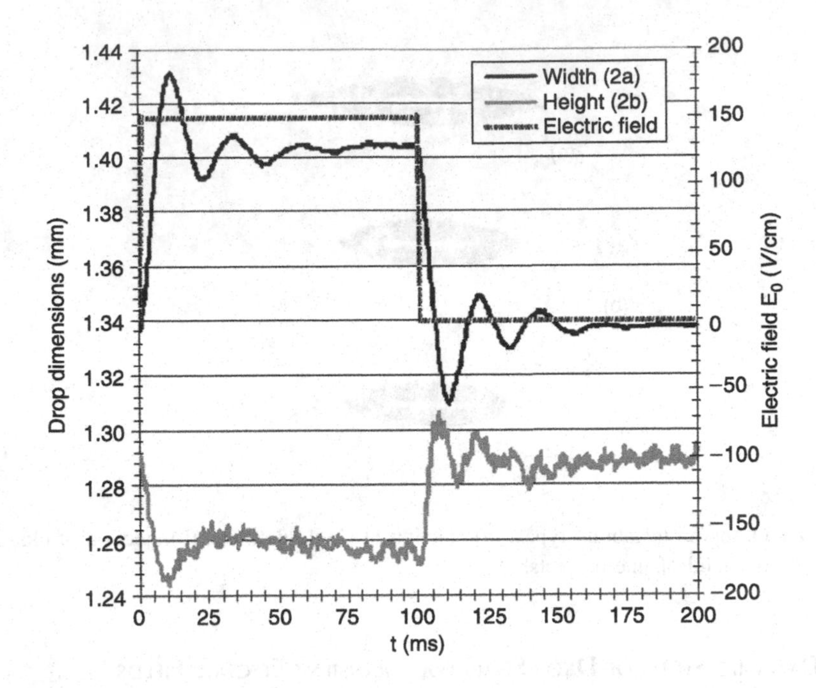

FIGURE 15.21 Transient oscillations of a 1.34 mm diameter water drop in paraffinic oil Exxsol D80.

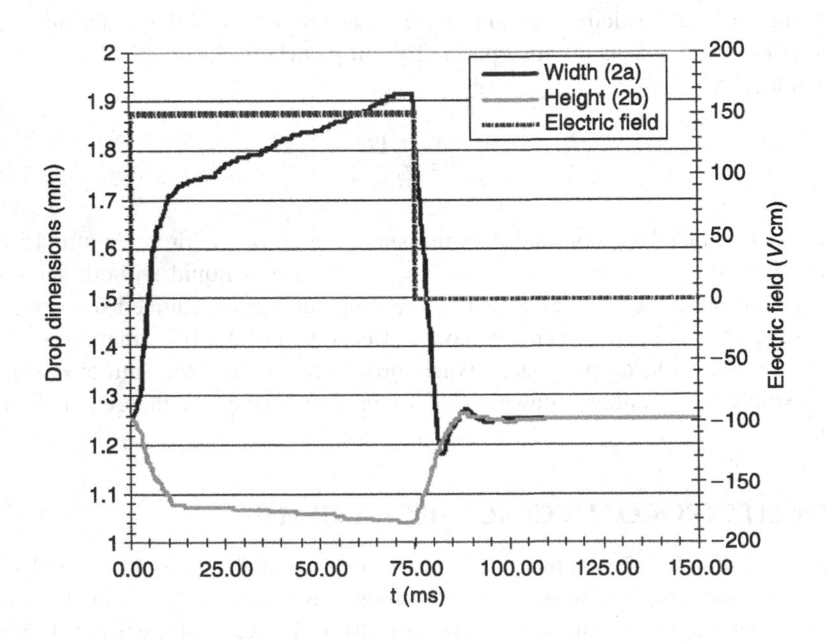

FIGURE 15.22 Transient deformation of a 1.25 mm diameter water drop in naphthenic oil with 100 ppm (wt) asphaltenes dissolved. Measurements after 155 min saturation time.

which can bring some of them in close vicinity at some moments. In the separators not applying electric fields, these quasi-collisions of drops result in pairs of drops rolling off each other during some time. Without an electric field the rate of natural coalescence for adjacent drops is very low. In the presence of an applied electric field, the rate of coalescence is much higher and increases with the field magnitude. We can thus distinguish two stages in the electrocoalescence process, the first one that brings two drops in close vicinity (and which is determined mainly by the flow), and the second one in which the electric field promotes their coalescence [26].

In this section we focus on the second stage, i.e., on the coalescence of two close drops under the action of an electric field. This is a rather complex situation and we consider here the simplest possible case by assuming that:

- The water droplets are electrically neutral (no net charge).
- The water is assumed to be perfectly conducting and the suspending oil perfectly insulating.
- The two droplets are identical and their centers are aligned with the electric field.

Moreover, the following considerations implicitly assume that the drops have a small radius and that the distance between their facing interfaces is smaller than their radius.

15.6.1 Merger of Drop Pairs

The main question that arises concerns the nature of the mechanism resulting in the final merging of two very close drops; this is important for practical applications because a good knowledge of this mechanism could help to augment or hinder electrocoalescence. Three main scenarios may be proposed:

- The field strength between the water drops is high enough to initiate an electric discharge and a shock wave that punctures the oil film.
- The attraction force between the drops brings them closer and closer and coalescence takes place at the end of the process of thinning of the oil film separating them.
- The electrostatic pressure induces a deformation of the water–oil interfaces and coalescence arises for a finite value of the drop separation when at least one of these interfaces becomes unstable.

The possibility of getting partial discharges between water drops was suggested by Lundgaard and colleagues [48]. However, such a mechanism for triggering coalescence appears very unlikely. For drops of diameter exceeding, typically, a millimeter, the electric field should induce their disintegration and not a discharge. For droplets of diameter $2R_0 < 100\ \mu m$ (which is the case for emulsions treated by electrocoalescers), electrical discharges between close droplets appear impossible. Indeed, similarly to the case of gases, there is a minimum required potential difference for electron avalanches in the oil phase. Now from the results given in Section 15.4, the potential difference ΔV between two electrically neutral drops is lower than the product of the field E_0 (applied to the emulsion) and the distance between drop centers. For close identical droplets this gives $\Delta V < 2R_0E_0$. With a background field $E_0 = 1$ kV/mm (being a high field in a coalescer), we obtain $\Delta V < 100$ V, a value lower than the minimum Paschen voltage for self-sustained discharge in the suspending oil. Therefore in electrocoalescers used to augment the size of water droplets of the water-in-oil emulsions, the coalescence cannot be triggered by electrical discharge between droplets.

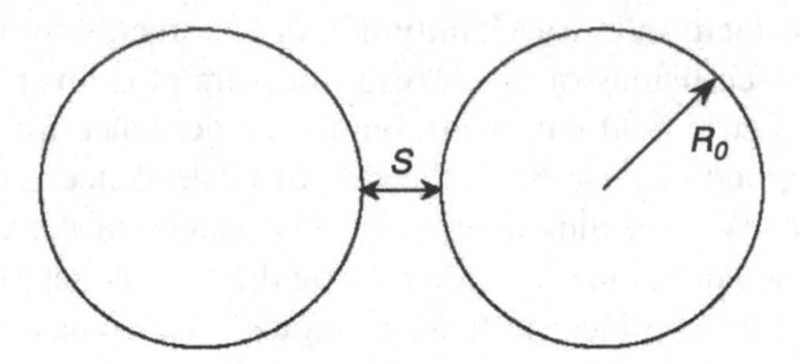

FIGURE 15.23 Two conducting spheres at distance s.

In conventional coalescence models, film drainage between drops plays an important role as a barrier to overcome to get the drops in contact [57]. For two undistorted spheres of radius R_0 at a small distance s (see Figure 15.23) the law for film thinning deduced from Ref. 32, and which is also given by Allan and Mason [58], is:

$$\frac{ds}{dt} = -\frac{2Fs}{3\pi\eta R_0^2} \tag{15.38}$$

where F is the interaction force and η the oil viscosity. Clearly as the attraction force increases as s decreases, the drop separation s decreases to zero in a finite time which gives an upper bound for the coalescence time. Let us consider the simplest case of two drops of radius R_0 between which a potential difference ΔV is maintained. A good approximation of the attraction force is [26]:

$$F = \frac{\pi}{2}\varepsilon(\Delta V)^2 \frac{R_0}{s} \tag{15.39}$$

The velocity of drop approach is then constant:

$$\frac{ds}{dt} = -\frac{1}{3}\frac{\varepsilon(\Delta V)^2}{\eta R_0} \tag{15.40}$$

Defining t_{ec} as the time necessary for the drop separation to pass from $R_0/3$ to zero we obtain:

$$t_{ec} = \frac{\eta R_0^2}{\varepsilon(\Delta V)^2} \tag{15.41}$$

A typical value is $t_{ec} \approx 50$ ms for $\Delta V/R_0 = 1$ kV/cm.

This picture of total film draining holds for rigid spheres of well-defined radius. This apparently excludes the cases where drops upon collision form flat channels, dimples, etc. However, the strong electrostatic pressure in the zone of small distance between droplets tends to keep the facing surfaces convex and very likely prevents the formation of flat channels or dimples. Moreover, the radii of curvature of the facing parts of the interfaces should tend to decrease which should help film draining and, hence, diminish the upper bound of coalescence time.

In the above scenario of film thinning, it was implicitly assumed that the parts of the drop surfaces subjected to strong electric forces keep their smooth shape. If we refer to the qualitatively similar problem of disintegration of a single conducting drop in a uniform field, it is clear that the interfaces will destabilize when the effect of electrostatic pressure, inducing a deformation of the facing interfaces, overcomes the restoring action of surface tension. We can therefore give the following description of the process of electrocoalescence: after a first stage of approach of the two droplets resulting from the action of the shear (or of the electric field in a stagnant emulsion), there is a second stage of film thinning and drop deformation until the interfaces destabilize and lead to the generation of a bridge between the drops in a very short time. The instability development has not yet been investigated and characterized; from the study of single drop disintegration (see Section 15.5), and from recent observations, one can only infer that it is a very rapid process possibly involving the formation of a rather narrow jet. The conditions under which instability and coalescence set in are better delineated.

15.6.2 Critical Conditions

Several studies have focused on this question. Strictly speaking they do not concern the instability conditions but rather the determination of stationary solutions for two identical drops, the shape of which is distorted by the action of the electric forces. In these studies a dc applied field is assumed; the results also apply for ac sinusoidal fields provided that the frequency is very low and that the drops can be considered to be in quasi-static conditions at any time. The critical field $(E_0)_{crit}$ (or critical potential difference ΔV_{crit} between the drops) is determined as the maximum value beyond which a stationary solution no more exists.

The first study was performed in connection with raindrops in thunderstorms. Latham and Roxburgh [59] observed that drops placed not close to each other in an electric field would deform to ellipsoids and eventually disintegrate for high stresses. If the drops are very close, then coalescence would occur instead through an instability resulting in mass transfer from one drop to the other. Their derivation was based on the assumption of the drops taking an ellipsoidal shape and on the formulas of Davis [60] for the field value at the facing poles of spheres. They determined the critical value of the applied field as a function of the relative initial spacing s_0/R_0 of the drops and they found a fairly good agreement with their experimental measurements. A very interesting property can be derived from their results [61]: the critical separation s_{crit} is a fraction of the separation s_0 of the undistorted drops ($s_{crit} \cong 0.63 s_0$). This means that only a limited drop deformation is possible; beyond the critical deformation, the electrostatic pressure is higher than the capillary pressure and there is a dynamic process with the drop locally elongating until it reaches the other drop surface.

Taylor [62] considered the case of two closely spaced drops between which a potential difference ΔV is applied. The drops are anchored on two rings, which prevent them from moving toward each other. His main assumption was to consider that the spherical shape of the drops is distorted only in the zone delimited by the rings. Solving numerically the approximate equation for the deformed interfaces, he obtained predictions in good agreement with his experimental measurements. The most interesting results concern the asymptotic case of very small spacing of drops ($s_0/R_0 \ll 1$) characterized by the following critical conditions:

$$s_{crit} \cong \frac{1}{2} s_0 \qquad \Delta V_{crit} \approx 0.38 s_0 \sqrt{\frac{T}{\varepsilon R_0}} \tag{15.42}$$

Again static solutions exist only for a limited deformation correlated with a limited potential difference. An important point to be noted here is that the critical conditions correspond to a local field value in the oil film between the drops $\Delta V_{crit}/s_{crit}$ independent of the separation s_0.

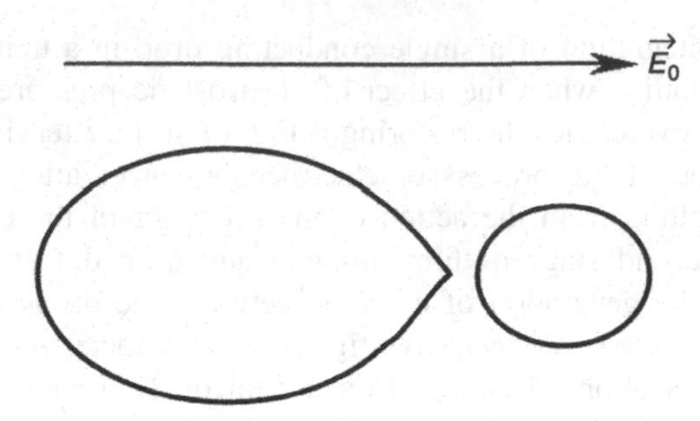

FIGURE 15.24 Deformation of close droplets.

Through a more sophisticated numerical treatment of the drop deformation problem, Brazier-Smith [63] confirmed that for $s_0 \ll R_0$ the surface distortion is mainly restricted to the zone where the drop surfaces are close (see Figure 15.24). By examining the time evolution of the drop shapes, it was shown [64] that for large enough drop separation ($s_0 > 1.2R_0$), like for a single drop, the instability occurs through the cone-jet mode. For smaller separations there is no evidence for a jet-like formation.

In the case of very small values of the distance s_0 between nondeformed drops, which is of interest for electrocoalescer applications, we propose a physical picture [61] which illustrates the phenomenon and justifies the laws obtained by Taylor. The drops are distorted only in the zones where the interfaces are at a short distance and where the electrostatic pressure p_{es} takes high values:

$$p_{es} = \frac{1}{2}\varepsilon E^2 = \frac{1}{2}\varepsilon\frac{\Delta V^2}{z^2} \tag{15.43}$$

By denoting s the separation between drops and by taking the following parabolic expression for the distance z between the interfaces at the radial distance r from the 2 drops axis:

$$z = s + \frac{r^2}{R_0} = s\left[1 + \frac{r^2}{R_0 s}\right] \tag{15.44}$$

we see that the electrostatic pressure distribution has a characteristic scale $(R_0 s)^{1/2}$. The distorted zone of the drops then has a radius $r_{dist} = A(R_0 s)^{1/2}$, A being a numerical constant (outside this region the drop deformation can be neglected). This is an asymptotic solution valid for $s/R_0 < 1/A^2$. Assuming that the centers of the spherical parts of the drops are immobile and that the deformed parts ($r \leq r_{dist}$) are characterized by a radius of curvature $R < R_0$, a simple relation between R and R_0 is deduced [61]:

$$\frac{1}{R} - \frac{1}{R_0} = \frac{1}{A^2}\frac{s_0 - s}{R_0 s} \tag{15.45}$$

where s_0 is the spacing between the undistorted drops. Balancing the capillary pressure difference between pole and spherical part and the electrostatic pressure at the poles leads to [61]:

$$(\Delta V)^2 = \frac{4}{A^2}\frac{T}{\varepsilon R_0}s(s_0 - s) \quad (15.46)$$

The maximum of ΔV gives the critical conditions:

$$\Delta V_{\text{crit}} = \frac{1}{A}s_0\sqrt{\frac{T}{\varepsilon R_0}}, \quad s_{\text{crit}} = \frac{s_0}{2} \quad (15.47)$$

and we recover Equation 15.42 by taking $A \cong 2.6$. In the case of two drops subjected to a uniform field E_0, the main difficulty is to evaluate ΔV; using the asymptotic expression for ΔV valid for spheres (Equation 15.30), the following critical value for the field can be proposed:

$$(E_0)_{\text{crit}} \cong 0.058\frac{s_0}{R_0}\log\left(\frac{2R_0}{s_0}\right)\sqrt{\frac{T}{\varepsilon R_0}} \quad (15.48)$$

This relation can also be used to determine the critical separation s_0 for a given applied field E_0.

All the investigations recalled here considered only two identical drops. In the case of drops of different radii, the biggest drop has the lowest internal pressure (due to surface tension) and its surface is more easily deformed. Therefore it is the interface of the biggest drop that will become unstable first (see Figure 15.24).

Up to now we have examined the case of dc applied field. Using an ac voltage of rectangular shape defines conditions for drop deformation fully equivalent to dc ones. With a sinusoidal applied voltage, the problem is different and it is not possible to take into account the effective applied field only. The electrostatic pressure has a mean value and a sinusoidal component at the doubled frequency. This periodic component induces a periodic modulation of the surface distortion whose amplitude and influence on the interface stability depend on frequency. The example of the disintegration of a single drop (see Section 15.5) illustrates such a role of frequency, which presumably should influence the critical conditions.

In the practical case of demulsification of crude oils, which contain numerous chemical components, the situation is not as ideal as hypothesized in the aforementioned studies. With surfactants the surface tension might vary with the interface deformation and expansion, which should drastically modify the conditions of electrocoalescence (for example naphthenic acids have a stabilizing effect on water-in-oil emulsions). In the case where the drops are covered with a surface film (like asphaltenes, bitumen, and resins) having a certain stiffness, the drop deformation is affected by this stiffness and the thinning of the coating layer might play the major role in controlling the coalescence process. Another phenomenon to take into account is the presence of particulate surface layers which would act as spacers maintaining the surfaces of close drops at a finite distance and thus hindering coalescence. In all these cases it is not possible to use the electrocoalescence conditions deduced from the analysis of the ideal case and, therefore, to design a "universal" electrocoalescer; the basic phenomena must be clearly characterized before attempting to derive critical conditions for electrocoalescence.

15.6.3 Experimental Investigations of Falling Drops

Contrary to many earlier investigations on basic phenomena of drop–drop electrocoalescence we have used ac fields. The influence of the shape, frequency, and magnitude of ac electric fields was studied in an optical bench under homogeneous field conditions described in Ref. 48. Furthermore, the influence of factors like salinity and surface-active substances has been studied.

The greater part of the experiments was done with one smaller drop falling on to a larger drop resting at the lower electrode. The stationary drop was either in direct contact with the electrode surface (non insulated drop), or separated from the electrode by a polypropylene disk (insulated drop). Observations were made with a fast (1000 fps full resolution) video camera. Under sinusoidal ac voltages we typically would see the lower surface (with lower curvature and internal pressure) starting to deform and oscillate with the applied field when the drop approached, and then starting to deform when the falling drop came close. When using square wave ac the oscillations disappeared. As water has a higher density than oil, the drop will fall down and after some time reach a terminal velocity, until it gets close to the surface of the lower drop where it is retarded by the film draining forces as shown in Figure 15.25.

Without an applied field the falling drop would generally roll off the larger drop (Figure 15.26). Applying an electric field will create a force that increases the velocity. For lower fields the falling droplet would now rest for some seconds at the summit of the lower one before coalescing, while for increasing fields a quick coalescence would occur. Even at 10^4 frames per second we could not resolve the final coalescence event. Often, after the collapse, a smaller satellite drop was pinched off from the larger drop, as seen from Figure 15.27(c).

A large number of experiments were done in order to study coalescence efficiency between drops that collide. A set-up like that shown in Figure 15.27(a) was used. The size of the falling

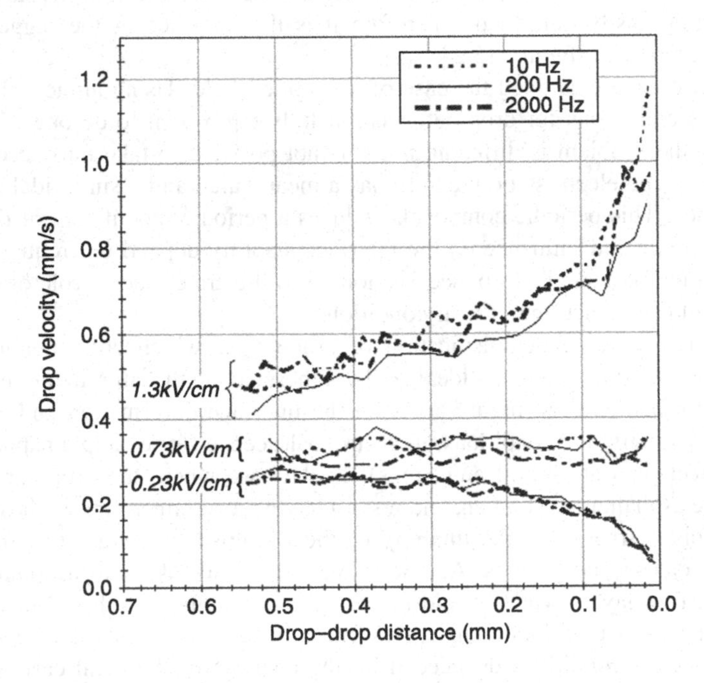

FIGURE 15.25 Velocity of the lower interface of a 200 μm drop as the drop falls on to the larger drop; influence of applied field and frequency. Square wave ac voltage.

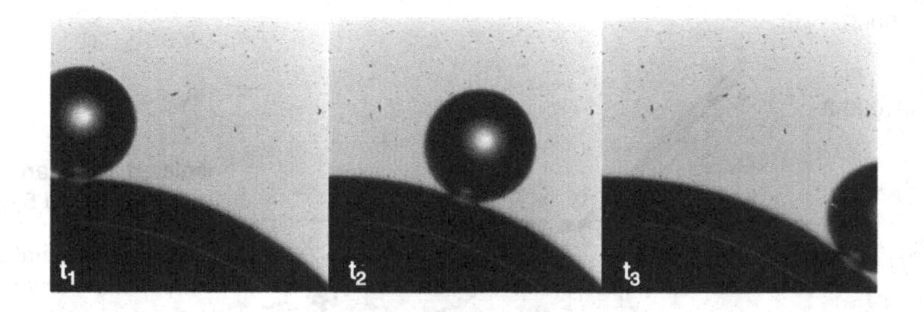

FIGURE 15.26 Drop impact at zero electric field with drop rolling off.

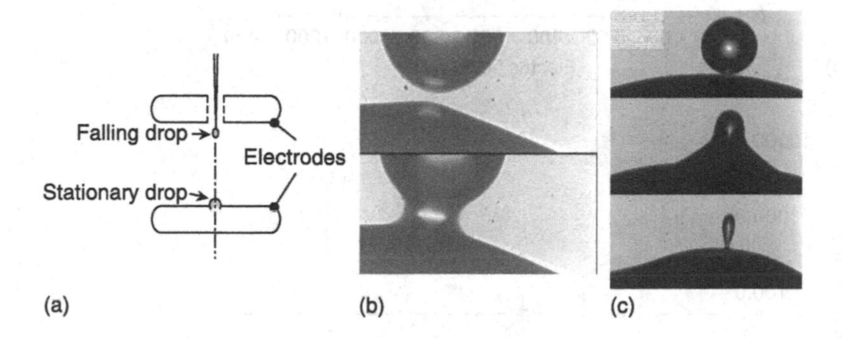

FIGURE 15.27 Experiments with falling drop: (a) Experimental set-up with small drops falling on a large drop at 3 to 4 kV/cm; (b) 90 μm drop falling and deformation of lower drop; and (c) 500 μm drop falling, coalescing, and forming a satellite drop.

drop was in the 200 μm range, and a square ac voltage with varying amplitude was used. The square voltage was used to avoid the more imprecise situation of a time variant force occurring under sine wave conditions. The time for drops to coalesce after they appeared to be in apparent contact was recorded. Then at higher stresses the coalescence would occur instantaneously when the drop hit or came so close that it looked like impact and at even higher stresses the coalescence would occur before the impact. Typical results are shown in Figure 15.28. Figure 15.28(a) suggests that we have two dependencies between time to coalesce: one for times above 0.05 sec where more than one half period is needed for coalescence to occur and one for shorter times where we are in the range where instantaneous coalescence occurs. At the present time we cannot explain the different results between isolated and non-isolated stationary drops.

In another experiment we found indications that for small diameter falling drops we had to go to higher fields, which is perfectly in line with the hypothesis that the instantaneous coalescence is governed by instability formation with a critical field that increases when size of the smaller drop is reduced. However, these results were obtained with sine voltage excitation, and therefore difficult to interpret.

15.6.4 Non-Ideal Surfaces

When the influence of asphaltenes was studied the water drops were kept for 20 min to allow saturation on their surfaces. Typical coalescence times are shown in Figure 15.28(b). The differences between oils with and without asphaltenes can be explained by the reduced interfacial

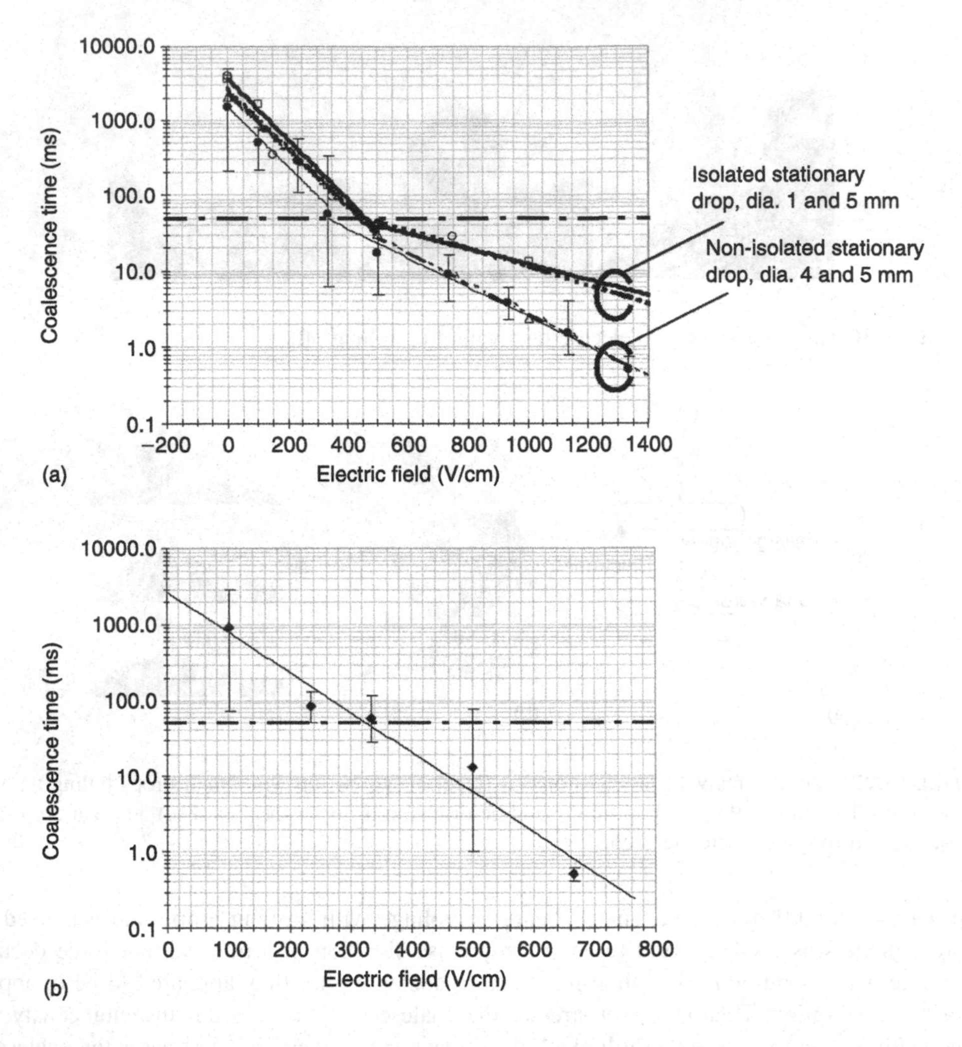

FIGURE 15.28 Coalescence time for water drops added 3.5 wt% NaCl in naphthenic oil under the influence of a 10 Hz square wave electric field. 200 μm falling drop. (a) Pure oil (no additives). A break in the curves occurs at 50 ms, corresponding to half the voltage period. (b) 100 ppm asphaltenes dissolved in oil phase. Saturated drop surfaces. Non-isolated, 4 mm stationary drop.

tension of the asphaltene oil. For oils with asphaltenes added the interfacial tension varied with time starting at 35 mN/m and falling towards 5 mN/m over a 20 min period, compared to the pure oil having a surface tension of 40 mN/m. One important difference between the pure oil and oil containing asphaltenes was that the voltage to get a coalescence time of 0.5 msec was reduced from 1350 V/cm to 700 V/cm. Another important difference was the time needed to drain the upper water drop once puncture of the surfaces was established. For the pure oil the drop collapsed within one millisecond, while for asphaltene saturated water drops the draining could take more than 1 min (see Figure 15.29). In the latter case no satellite was formed. This clearly shows the influence of the stiff surface layer. How such a stiff layer would influence

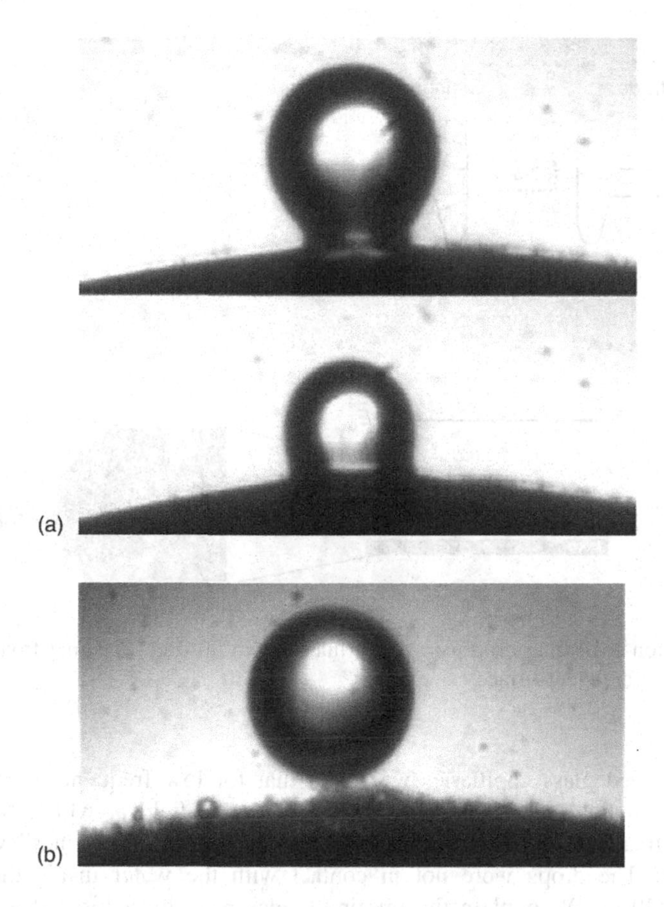

FIGURE 15.29 Surface covering. *Top* (a): Slow draining (>0.5 sec) of asphaltene covered coalesced drop. *Bottom* (b): Particulate maters preventing coalescence of a 200 μm droplet.

the formation of the instability cone remains unknown. However, our results indicate that the timescales for the instability formation are sub-millisecond in all cases.

For the experiments with asphaltenes the surface of the larger drop often got contaminated with small particles. These would tend to concentrate at the summit of the larger drop and could prevent coalescence, as shown in Figure 15.29.

Experiments in oils with asphaltenes added gave the same results for 10, 200, and 2000 Hz square wave voltage. This is in line with our theory that it is the level of the instantaneous field, and the dipole moment this field induces, that in the case of uncharged droplets govern the coalescence process.

15.6.5 Experiments with Supported Drops

As it is troublesome to do visual observations with moving drops, several experiments were done with drops fixed in various ways; hanging on capillaries or resting on PTFE surfaces as shown in Figure 15.30. These experiments often produced effects that could not be reproduced with free drops.

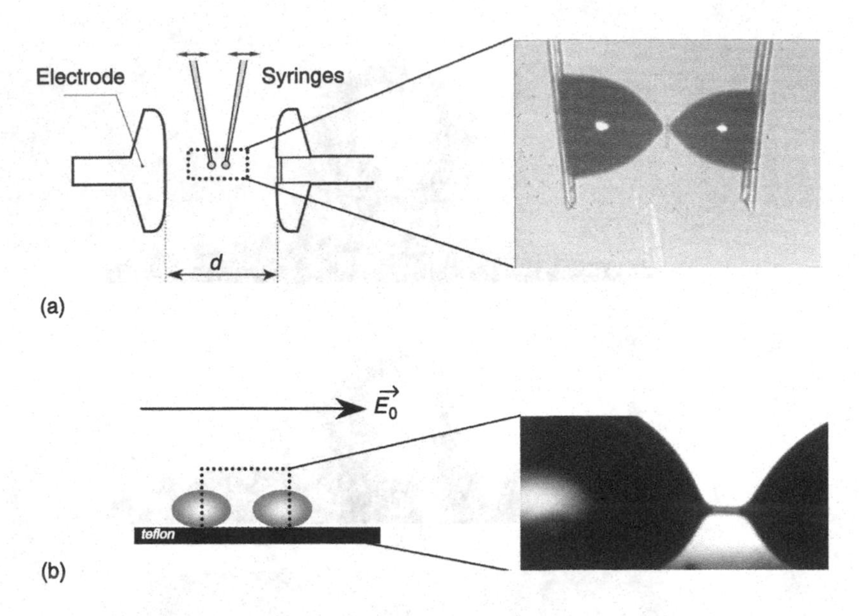

FIGURE 15.30 Non-coalescing drop pairs. Experiments with (a) drops hanging from glass capillaries and (b) drops resting on a Teflon surface.

With water filled glass capillaries we found that for low frequencies coalescence occurred at low background fields. With increased frequencies the field at which coalescence occurred increased until finally at 10 kHz we could get stable "instability cones" without coalescence occurring at all. The drops were not in contact with the water inside the capillary as seen from Figure 15.30(a). We explain the missing coalescence by a force due to the good capacitive coupling at higher frequencies between the water drop and the water column inside the capillary that "glued" the drops to the capillaries, and thereby preventing them to freely coalesce. Furthermore, the water jets from the capillaries hit each other with opposing direction.

Similar experiments were performed with drops resting on a horizontal Teflon surface. Small drop pairs became unstable and coalesced as expected, independent of the applied frequency in the range 10 Hz to 10 kHz. However, with large droplets (radius $\sim$ 1 mm) we could observe a stable channel forming between the two drops as shown in Figure 15.30(b). Apparently, again this was due to forces adhering the drop to the solid surface, thus preventing their mass centers to move.

Attempts to reproduce these effects for free drops and emulsions failed, and we conclude that one should be careful to extrapolate experiences with supported drops to real emulsions.

15.7 SOME CONSIDERATIONS REGARDING PRACTICAL ELECTROCOALESCERS

Possibly the largest problem for an oil/water separator, and also for an electrocoalescer, is to treat oils with a low water cut and small drops: Small drops gravitate slowly towards the bottom of a sedimentation tank, and the distances between drops will be considerable. The overall coalescence efficiency is given by the drop collision rate and the probability that the drops will coalesce once they are in close proximity to each other.

This far we have mainly looked at electrostatic forces on drops and drop pairs under stagnant or laminar flow conditions. This is relevant for the typical older generation electrocoalescers in large sedimentation tanks. However, during the last decade a new concept has arrived, where shear movement in the emulsion is applied to create more frequent drop collisions [9,26,65]. From an analysis of movement of drops suspended in liquids one may calculate the collision frequency. With this we do not mean direct impacts between drops, but situations where the drops gets close enough to allow the short range electrostatic forces to give a large enough force to pull them together. So "impact" is based on a criterion determined by when the drops come within a certain capture radius where the electrostatic attractive forces exceeds the viscous drag. Increasing turbulence and shear flow can increase this collision rate. The upper allowed limit of the turbulence is where it rips apart drops that get in proximity of each other, or even more extreme; results in drop fragmentation. An *efficient collision cross section* can be developed depending on drop size, electric field, and liquid shear movement.

Once the drops are within each other's efficient cross section, the achieved time to coalescence is reduced with increasing background fields as described in Figure 15.28. At a certain critical voltage and drop distance combination, instantaneous coalescence will occur. The coalescence efficiency can be increased by raising the field towards the limit where the larger drops in the emulsion start to become unstable.

The designer has, on the one hand, to balance shear flow (or turbulence) between high collision frequency between drops and ripping apart drop pairs or existing large drops. On the other hand, he has to balance the electric field between the need of getting large collision cross sections and high coalescence efficiency, and the danger of destroying large drops by initiating instabilities.

Our analysis points at electrostatic attractive forces and surface instability as being the governing factors for electrocoalescence. From this point of view a high dc field, where the driving field is continuously high, seems favorable. However, if covered electrodes are chosen in order to reduce the risk of breakdown from water plugs, this may result in a resistive voltage distribution. Then the solid, highly resistive dielectric barrier may take all the voltage leaving the emulsion as a low field region. The use of so-called pulsed dc, which has been suggested by several authors [66], is an attempt to solve this problem. During the voltage steps, the pulsed dc redistributes the field from a resistive to a capacitive distribution. Though after some time, which is given by the time constant of the emulsion, the distribution will fall back to the resistive mode, and a new voltage transient is needed. Transients therefore need to be applied with short intervals. However, if they become too short, the system will not get time to relax back to its initial state and the effect of the steps will be reduced. A better alternative is to use high frequency ac fields that distribute voltages capacitively all the time, giving increased fields in the emulsion compared to the low frequency or dc case. It is the net conductivity of the emulsion that determines how high a frequency is required, as described in Section 15.3.

For a drop pair, an ac field will, in general, result in a lower coalescence efficiency than a dc field, because during parts of the period the field will have fallen below the critical level; the duty cycle, being the percentage of time where the field is above the critical value, will be less than 100%. The duty cycle of a dc voltage is 100%. However, as explained above, dc can result in problems in coalescers with covered electrodes. To overcome this, one can use an ac square voltage. The forces will act equally efficiently as under dc field conditions, with a duty cycle of 100%. Increasing the field will also increase the duty cycle, but also stress the insulation system, which often is sensitive to the peak stress. The downside of using square ac voltages is that it will invoke some form of power electronics, which can be costly, while sine voltages are easily produced using a transformer.

There have been discussions about optimum frequencies for ac coalescers. It is difficult to imagine any "universal optimum." We would suggest that in the case of covered electrodes, as explained above, the main criterion is that the duration of one half period of the voltage should be short compared to the time constant of the oil, as explained in Section 15.3. For uncovered electrodes where the drops are charged at the electrode surfaces, under laminar flow or stagnant fluid condition, the drift velocity and the separation of the drops will be determining factors. When coalescence occurs between singular droplets and not at the electrode surfaces, it will mainly occur between oppositely charged drops. The drops must therefore be allowed to keep their acquired charge long enough to meet drops of opposite polarity. As drops lose charge with time as they propagate away from one electrode towards the opposite it could be advantageous to divide the electrode gaps into shorter gaps where less charge is lost. Additionally, it would be advantageous to have an inhomogeneous field that allows uncharged drops to be attracted to an electrode and become charged. One should also consider that in addition to the attractive or repulsive forces between charged drop pairs, the applied field will also act on each charged drop. As we observed in our emulsion experiments, the interaction with the applied field (Equation 15.1) can be stronger than between the drops (Equation 15.2).

We have this far mainly considered emulsions with a low water cut. The coalescer design will have to be adapted to the conditions of the particular oil reservoir as discussed by Tsabek [52], for example. For systems with high water cuts the drop–drop collisions become more frequent and chain formation more likely to occur. Forces that keep drop chains together are described in studies dealing with particle/oil suspensions used for rheological purposes [44].

In the production of crude oil, it is not only the emulsified water that is the problem; there are gases under varying pressures, and there may be sand particles and salts that must be removed by gas scrubbers, cyclones, sedimentation tanks etc. The sequence in which this equipment is assembled is not indifferent to the final result; e.g., if one at an early stage can remove smaller particles then particle stabilizing of the drop surfaces is less likely to occur in the electrocoalescer. This far the electrocoalescers have usually been combined with and built into the sedimentation tanks. It seems advantageous to apply a more compact unit in front of the sedimentation tank, taking advantage of the liquid mixing in the turbulent flow in pipelines before the sedimentation tank with a more stagnant condition. If drop sizes at the inlet can be increased, then the volume of the sedimentation tanks can be reduced. The effect of an experimental compact coalescer is shown in Figure 15.31.

Finally, one must not forget a possible influence of streaming electrification and triboelectric effects that may start to play a role at higher liquid velocities. Most scientific studies are done either as optical studies in stagnant emulsions or looking at drop size distributions for model systems with liquid flows. In these studies there is little support for hypotheses or conclusions on the significance of this aspect. However, charging of droplets occurred frequently in our experiments that mainly concerned forces on neutral, polarized drops and was a nuisance throughout our investigations. Most likely, charging will also occur in practical coalescers, but as time constants of the oil are small one would expect the effects to be limited. Still, this is an open question that should be further investigated.

ACKNOWLEDGMENTS

This work would not have been undertaken without the initial support from Statoil and ABB Offshore Systems (now Vetco Aibel) and the later substantial support from the Norwegian Research Council. We are most grateful to Pål Jahre Nilsen from ABB Offshore Systems in helping to start this project. We also acknowledge the assistance of Professor Johan Sjöblom in dealing

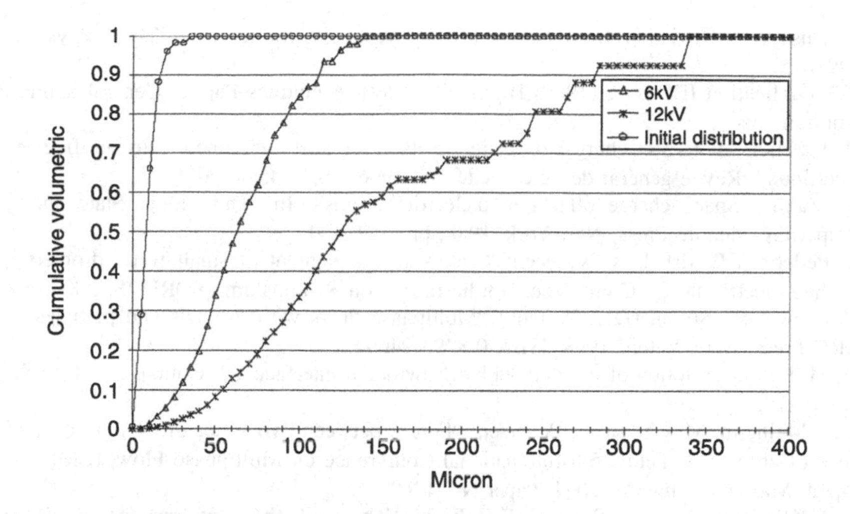

FIGURE 15.31 Cumulative water droplet size distribution in an emulsion treated with an electrostatic ac field. The emulsion flow has a Reynolds number of 20,000, and a residence time of 1 sec in the electric field. (Courtesy of Vetco Aibel.)

with the surface chemistry problems. The authors thank the International Electrotechnical Commission (IEC) for permission to reproduce figure 6 (page 40) from its International Standard IEC 60060-1(1989-11) 2nd edition — High-voltage test techniques. Part 1: General definitions and test requirements. All such extracts are copyright of IEC, Geneva, Switzerland. All rights reserved. Further information on the IEC is available from www.iec.ch. IEC has no responsibility for the placement and context in which the extracts and contents are reproduced by the author; nor is IEC in any way responsible for the other content or accuracy therein.

REFERENCES

1. E. Kuffel, W. S. Zaengl: "High Voltage Engineering", Pergamon Press, Oxford, 1988.
2. "Handbook of Electrostatics", Ed. J. Chang et al., Marcel Dekker, New York, 1995.
3. J. Cross: "Electrostatics: Principles, Problems and Application", Adam Hilger, Bristol, 1987.
4. F. G. Cottrell: "Breaking and separating oil water emulsions", US patent specification 895 729, 1908.
5. J. Sjöblom, N. Aske, I. H. Auflem, Ø. Brandal, T. E. Havre, Ø. Sæther, A. Westvik, E. E. Johnsen, H. Kallevik: "Our current understanding of water-in-crude oil emulsions. Recent characterization techniques and high pressure performance", Adv Colloid Interface Sci, 100–102 (Special Issue: A collection of invited papers in honour of Professor J. Th. G. Overbeek on the occasion of his 90th birthday), 2003, pp. 399–473.
6. L. C. Waterman: "Electrical coalescers", Chemical Engineering Progress, 61 (10), 1965, pp. 51–57.
7. J. S. Eow, M. Ghadiri, A. O. Sharif, T. Williams: "Electrostatic enhancement of coalescence of water droplets in oil: a review of the current understanding", Chem. Eng. J., 84 (3), 2001, pp. 173–192.
8. J. S. Eow, M. Ghadiri: "Electrostatic enhancement of coalescence of water droplets in oil: a review of the technology", Chem. Eng. J., 85, 2002, pp. 357–368.
9. O. Urdahl, N. J. Wayth, H. Førdedal, T. J. Williams, A. G. Bailey: "Compact Electrostatic Coalescer Technology", In: Encyclopedic Handbook of Emulsion Technology (Ed. J. Sjöblom), Marcel Dekker, 2001, pp. 679–694.

10. L. Onsager: "Deviation from Ohm's law in weak electrolytes", J. Chem. Phys., 2, 1934, pp. 599–615.
11. IEC publication IEC 60-1 (1989) High-voltage test techniques Part 1: General definitions and test requirements.
12. N. J. Felici: "Forces et charges de petits objets en contact avec une électrode affectée d'un champ électrique", Revue général de l'électricité, 75, 1966, pp. 1145–1160.
13. M. Zahn: "Space charge effects in dielectric liquids". In: The Liquid State and its Electrical Properties, Plenum Press, New York, 1988, pp. 367–430.
14. A. Pedersen, E. Ilstad, A. Nysveen: "Forces and movement of small water droplets in oil due to applied electric field", Conf. Proc. Nordic Insulation Symposium, NORD-IS, 2003, pp. 127–133.
15. C. Crowe, M. Sommerfeld, Y. Tsuji: "Multiphase flows with droplets and particles", 2nd edition, CRC Press, Boca Raton, 1998, ISBN 0-8493-94694.
16. D. M. LeVan: "Motion of droplets with a Newtonian interface", J. Colloid Interface Sci., 83, 1981, pp. 11–17.
17. J. A. Melheim, M. Chiesa, A. Pedersen: "Forces between two water droplets in oil under the influence of an electric field", 5th International Conference on Multiphase Flow, ICMF'04, Yokohama, Japan, May 30 to June 4, 2004, Paper No. 126.
18. D. J. Klingenberg, F. van Swol, C. F. Zukoski: "The small shear rate response of electrorheological suspensions. II. Extensions beyond the point-dipole limit", J. Chem. Phys., 94 (9), 1991, pp. 6170–6178.
19. H. J. H. Clercx, G. Bossis: "Many-body electrostatic interactions in electrorheological fluids", Phys. Rev. E, 48 (4), 1993, pp. 2721–2738.
20. K. W. Yu, T. K. Wan Jones: "Interparticle forces in polydisperse electrorheological fluids", Comput. Phys. Commun., 129, 2000, pp. 177–184.
21. Y. L. Siu, T. K. Wan Jones, K. W. Yu: "Interparticle force in polydisperse electrorheological fluid: Beyond the dipole approximation", Comput. Phys. Commun., 142, 2001, pp. 446–452.
22. P. M. Morse, H. Feshbach: "Methods of Theoretical Physics, Part II", McGraw-Hill, New York, 1953.
23. G. W. Carter, S. C. Loh: "The Calculation of the Electric Field in a Sphere-gap by Means of Dipolar Co-ordinates", IEE Monograph, No. 325 M, pp. 108–111, 1959.
24. M. H. Davis: "The Forces Between Conducting Spheres in a Uniform Electric Field", RM-2707-1-PR, Rand Corporation, 1962.
25. M. H. Davis: "Two Charged Spherical Conductors in a Uniform Electric Field: Forces and Field Strength", RM-3860-PR, Rand Corporation, 1964.
26. P. Atten: "Electrocoalescence of water droplets in an insulating liquid", Journal of Electrostatics, 30, 1993, pp. 259–270.
27. T. J. Williams, "The resolution of water-in-oil emulsions by the application of an external electric field", PhD Thesis, Univ. Southampton, UK, 1989.
28. A. Pedersen, E. Ilstad, A. Nysveen: "Forces and movement of water droplets in oil caused by applied electric field", IEEE Conference on Electrical Insulation and Dielectric Phenomena, Colorado, USA, 2004.
29. K. Adamiak: "Force of attraction between two conducting droplets in electric field", Proceedings of the 1999 IEEE Industry Applications Society Annual Meeting, Phoenix, Arizona, Vol. 3, 1999, pp. 1795–1800.
30. K. Adamiak: "Interaction of two dielectric or conducting droplets aligned in the uniform electric field", Journal of Electrostatics, 51–52, 2001, pp. 578–584.
31. R. H. Davis, J. A. Schonberg, J. M. Rallison: "The lubrication force between two viscous drops", Phys. Fluids A, 1 (1), 1989, pp. 77–81.
32. O. I. Vinogradova: "Drainage of a thin liquid film confined between hydrophobic surfaces", Langmuir, 11, 1995, pp. 2213–2220.
33. G. Barnocky, R. H. Davis: "The lubrication force between spherical drops, bubbles and rigid particles in a viscous fluid", Int. J. Multiphase Flow, 15, 1989, pp. 627–638.
34. D. J. Klingenberg, F. van Swol, C. F. Zukoski: "Dynamic simulation of electrorheological suspensions", J. Chem. Phys., 91, 1989, pp. 7888–7895.

35. I. V. Lindell, G. Dassios, K. I. Nikonskinen: "Electrostatic image theory for the conducting prolate spheroid", J. Phys. D: Appl. Phys., 34, 2001, pp. 2302–2307.
36. R. T. Bonnecaze, J. F. Brady: "Dynamic simulation of an electrorheological fluid", J. Chem. Phys., 96, 1992, pp. 2183–2202, 1992.
37. T. B. Jones: "Forces and torques on conducting particle chains", Journal of Electrostatics, 21, 1988, pp. 121–134.
38. T. B. Jones: "Dipole moments of conducting particle chains", J. Appl. Phys., 60, 1986, pp. 2226–2230.
39. T. B. Jones: "Effective dipole moment of intersecting conducting spheres", J. Appl. Phys., 62, 1987, pp. 362–365.
40. I. V. Lindell, J. C. E. Sten, K. I. Nikonskinen: "Electrostatic image method for the interaction of two dielectric spheres", Radio Science, 28, 1993, pp. 319–329.
41. J. C. E. Sten, K. I. Nikoskinen: "Image polarization and dipole moment of a cluster of two similar conducting spheres", Journal of Electrostatics, 35, 1995, pp. 267–277.
42. E. v. d. Bosch: "Electrostatic coalescence and light scattering", PhD Thesis, TU Eindhoven, Netherlands, 1996. ISBN 90-386-0179-4.
43. J. A. Melheim: "Cluster integration method for Lagrangian particle dynamics", *Comput. Phys. Comun.*, 171(3), 2005, pp. 155–161.
44. P. Atten, P. Gonon, J.-N. Foulc, C. Boissy: "Estimate of the electric stress in the liquid lying between particles of an electrorheological fluid", 1999 IEEE Conf. El. Ins. and Diel. Phenomena, pp. 333–336.
45. P. Atten, C. Boissy, J.-N. Foulc: "Slightly conducting spheres immersed in a dielectric liquid subjected to a DC field: attraction force, electric field on the liquid and partial discharges", Proc. 13th Intl. Conf. Diel. Liquids, ICDL, Japan, 1999, p. 111.
46. R. T. Bonnecaze, J. F. Brady: "Dynamic simulation of an electrorheological fluid", J. Chem. Phys., 96, 1992, pp. 2183–2202.
47. L. C. Davis: "Polarization forces and conductivity effects in electrorheological fluids", J. Appl. Phys., 72, 1992, pp. 1334–1340.
48. L. Lundgaard, G. Berg, A. Pedersen, P. J. Nielsen: "Electrocoalescence of water drop pairs in oil", Proc. 14th Int. Conf. on Dielectric Liquids, ICDL, Austria, 2002, pp. 215–219. ISBN 0-7803-7350-2.
49. C. T. R. Wilson, G. I. Taylor: Proc. Camb. Phil. Soc. 22, 1925, p. 728.
50. G. I. Taylor: "Disintegration of water droplets in an electric field", Proc. R. Soc., London, A 146, 1964, pp. 383–397.
51. J. D. Sherwood: "Breakup of fluid droplets in electric and magnetic fields", J. Fluid Mech., 188, 1988, pp. 133–146.
52. L. K. Tsabek, G. M. Panchenkov, V. V. Papco: "Theoretical basis of operation of equipment for electrical dehydration and electrical desalting of oil emulsions", Special paper no. 7, 8th Petroleum World Congress, Moscow, 1971, pp. 423–430.
53. G. Berg, L. Lundgaard, M. Becidan, R. S. Sigmond: "Instability of electrically stressed water drops in oil", IEEE ICDL 2002, Conference Record, Graz, 2004, pp. 220–224. ISBN 0-7803-7350-2.
54. S. B. Sample, B. Raghupaty, C. D. Hendricks: "Quiscent distortion and resonant oscillations of a liquid drop in an electric field", Int. J. Engn. Sci., 8, 1970, pp. 97–109.
55. R. Myrvold and F. K. Hansen: "Surface elasticity and viscosity from oscillating bubbles measured by automatic axisymmetric drop shape analysis", Journal of Colloid and Interface Science, 207, 1998, pp. 97–105.
56. T. Frømyr, F. K. Hansen, A. Kotzev, A. Laschewsky: "Adsorption and surface elastic properties of corresponding fluorinated and nonfluorinated cationic polymer films measured by drop shape analysis", Langmuir, 17, 2001, pp. 5256–5264.
57. C. W. Angle: "Chemical demulsification of stable crude oils and bitumen emulsions in petroleum recovery – a review". In: Encyclopedic Handbook of Emulsion Technology, Ed J. Sjöblom, Marcel Dekker, NY, 2001, pp. 541–594.
58. R. S. Allan, S. G. Mason: "Particle motions in sheared suspensions. XIV. Coalescence of liquid drops in electric and shear fields", J. Colloid Sci., 17, 1962, p. 383–408.

59. J. Latham, I. W. Roxburgh: "Disintegration of pairs of water drops in an electric field", Proc. Royal Soc., London, A295, 1966, pp. 84–97.
60. M. H. Davis: "Two charged spherical conductors in a uniform electric field: Forces and field strength", Quart. J. Mech. Appl. Math., 17, 1964, pp. 499–510.
61. P. Atten, L. Lundgaard, G. Berg: "A simplified model of electrocoalescence of two close water droplets in oil", Proceed. 5th Intern. EHD Workshop, Poitiers, 2004, pp. 119–123.
62. G. I. Taylor: "The coalescence of closely spaced drops when they are at different potentials", Proc Royal Soc. A, London, 306, 1968, pp. 423–434.
63. P. R. Brazier-Smith: "Stability and shape of isolated and pairs of water drops in an electric field", Physics of Fluids, 14 (1), 1971, pp. 1–6.
64. P. R. Brazier-Smith, S. G. Jennings, J. Latham: "An investigation of the behavior of drops and drop pairs subjected to strong electrical forces", Proc. Royal Soc. A, London, 325, 1971, pp. 363–376.
65. O. Urdahl, T. J. Williams, A. G. Bailey, M. T. Thew: "Electrostatic destabilisation of water-in-oil emulsions under conditions of turbulent flow, IChemE, 74 (part A), 1996, pp. 158–165.
66. P. J. Bailes, S. K. L. Larkai: "Liquid phase separation in pulsed dc fields", Trans IChemE, 60, 1982, pp. 115–121.

16 Numerical Simulation of Fluid Mechanisms and Separation Behavior in Offshore Gravity Separators

Ernst W.M. Hansen and Geir J. Rørtveit

CONTENTS

Abstract

The offshore activities to produce oil and gas have grown rapidly during the last few decades. The equipment used for bulk-separation operations is of considerable size and weight. The most widely used concept to separate bulk flows of water/oil/gas/sand on platforms are vessels in which gravity settling due to the density differences of the fluids takes place. The operation of the separator must provide sufficient time to allow the phases to separate by gravity. Both the operation in the field and the internal devices should make separators efficient and with low maintenance. Computational fluid dynamics (CFD) can provide a valuable insight, and the fluid flow behavior in the liquid bulk zone inside the separator may such be analyzed. Water-in-oil forms unstable emulsion in processing, and the drop-growth in the complex fluid will influence the separation efficiency. Many practical, design, and redesign applications may be performed by CFD modeling and simulations. Examples are (1) the development of the vessel inlet configurations that improve the uniformity of the gas and liquid flows, (2) sensitivity of a separator design to changes in operating conditions, (3) influence of internal equipment on separation performance, and (4) coalescence and breaking of emulsion.

16.1 INTRODUCTION

Separation during petroleum production takes place in a process composed of a number of equipment units. In many fields the process is also composed of a number of separation steps, where the pressure is reduced stepwise and gas flashed off. All oil/gas/water/sand separation is based on differences in density, and most separators are gravity separators, i.e., they utilize the common acceleration of gravity in huge pressurized vessels. Later development has brought compact, reliable high-g equipment to the market, particularly hydro-cyclones and centrifuges for oil and water polishing. The fluid flow behavior of a three-phase separator shows that different physical and chemical phenomena are important in different zones.

Separator sizing must satisfy several criteria for good operation during the lifetime of the producing field:

- Provide sufficient time to allow the immiscible gas, oil, and water phases to separate by gravity
- Provide sufficient time to allow for the coalescence and breaking of emulsion droplets at the oil–water interface
- Provide sufficient volume in the gas space to accommodate rises in the liquid level that result from the surge in the liquid flow rate
- Provide for the removal of solids that settle to the bottom of the separator
- Allow for variation in the flow rates of gas, oil, and water into the separator without adversely affecting separation efficiency.

The simple design methods for gravity separators like those found in Refs. 1 and 2 do not match the complicated multiphase fluid flow behavior through such equipment in any great detail. The design of the separators is not accurate, and the design may be too conservative. This frequently gives oversized vessels, both in volume and weight, and low separation efficiency [2]. Development of design and simulation tools is under way [3–12]. These are based on knowledge of basic hydrodynamic and physicochemical principles, and possibly a combination of these effects. Even at the present stage these tools represent a significant enhancement of the design and redesign of jobs for gravity separators offshore [13,14]. Computational fluid dynamics (CFD) applied to single-phase flow has become more and more important for modeling and design of industrial processes and its components [15–19]. The extension of single-phase CFD techniques to two-phase or multi-phase flow calculations is not easy. Some of the problems and the difficulties have been reviewed by Jayanti and Hewitt [20].

This chapter discusses the characterization of the mixture of oil and water and the distribution of the multi-phase fluid flow in a horizontal gravity separator. CFD can provide valuable insight to develop and demonstrate the technology required for the realization of efficient, compact, low maintenance gravity separators. Water-in-oil will form unstable emulsions in processing and the drop-growth in the complex fluid will influence the separation efficiency. A phenomenological model for drop growth in batch separation is described and a simulation performed. The fluid flow behavior in the liquid bulk zone inside a separator is analyzed. The use of CFD demonstrates how the flow field is related to inlet flow conditions and the performance of internals.

16.2 FLOW BEHAVIOR IN HORIZONTAL GRAVITY SEPARATORS

Each separator operates at a fixed pressure. The multiphase fluid flow enters the vessel through the inlet nozzle as a high momentum jet hitting an inlet arrangement such as a momentum

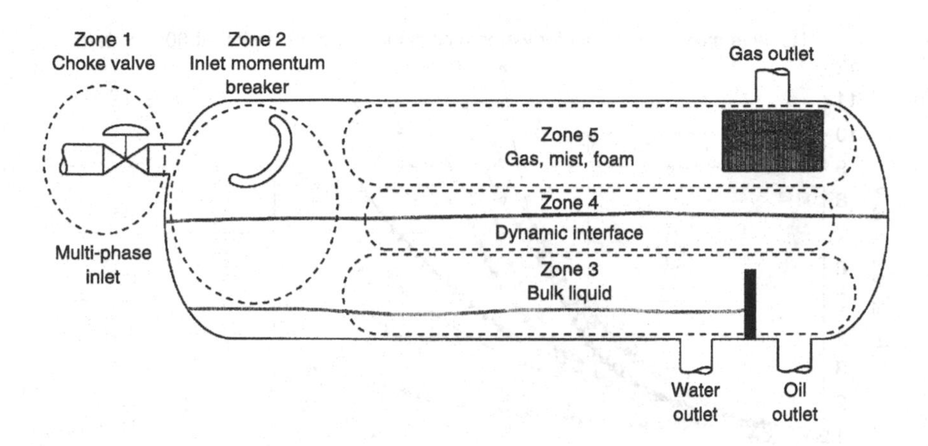

FIGURE 16.1 Typical flow zones in a horizontal gravity separator.

breaker device or cyclones. The liquid is diverted into the liquid pool in the lower part of the vessel. Inside the liquid pool, near the inlet, the multiphase fluid flows as a dispersion with low horizontal velocity. The low-density gas and mist rises, and the oil, water, and emulsion will separate by gravity on the way to the outlet. In the upper part of the vessel, the oil and water particles together with condensed gas will fall down to the liquid interface as the multiphase fluid flows to the outlet. For the modeling and simulation of fluid flow and phase separation behavior inside a gravity separator, it is helpful to characterize the different flow regimes or zones; see Figure 16.1.

16.3 FLUID SYSTEM AND MECHANISMS OF SEPARATION

The mixture of oil, gas, water, and particles that is the inlet stream to a separator train exhibits large variations in both composition and physicochemical properties. The thermodynamic properties, which determine the gas/oil ratio through the various steps of the process, are well known, and will form a well-described system. The properties of the liquid phase, represented by bulk viscosity, emulsification, coalescence, and settling properties, are less well understood, and consequently insufficiently described in terms of realistic mathematical models.

The separation process is dominated by two factors: (1) The emulsification process, which takes place in the choke and other equipment components with high shear, and (2) the coalescence and settling effects, where drops grow and settle or cream to its homo-phase.

16.3.1 Dispersed Mixtures of Oil and Water

Normally, the gas is released quickly in the inlet arrangement of the separator, and the water-in-oil or oil-in-water will form an emulsion in between the oil and water. The amount of water, the water cut, and the stability of the emulsion will create the resistance in the flow field. The bulk viscosity is an important parameter in the fluid dynamics and different correlations for apparent viscosity versus water cut for dispersions have been derived and presented in the literature. Some of the models suggested in the literature [21–23] are reviewed and plotted in Figure 16.2 for an oil/water mixture. The plot shows the relative viscosity versus water cut. The relative viscosity for an emulsion is the viscosity for the emulsion divided by the viscosity of the continuous phase.

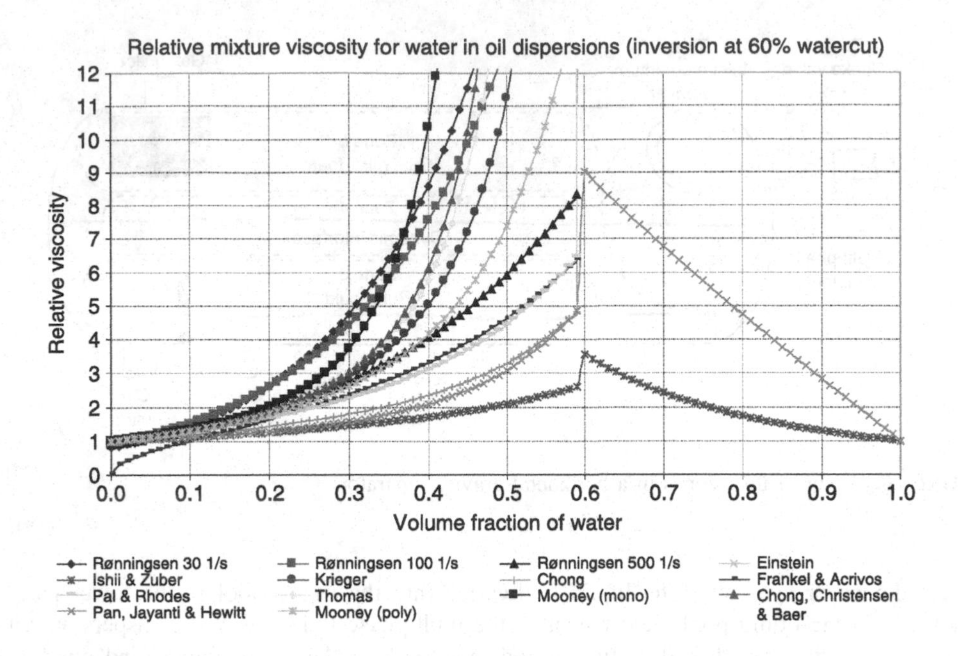

FIGURE 16.2 Relative viscosity correlations for dispersed oil/water mixtures.

All the models agree well for a dilute dispersion (water cut below 10%), but for higher water cuts the models give quite a scattered result. Thus, the viscosity model used for fluid dynamic calculation in a gravity separator should rely on the emulsion behavior investigated.

16.3.2 The Bottle Test (Batch Separation)

A common method of determining relative emulsion stability of water-in-oil or oil-in-water is a simple bottle test (batch separation test), as described by Schramm [22]. A batch separation test involves agitating a mixture of oil and water to form a uniform dispersion, then removing the agitation to allow the dispersion to settle naturally. These batch sedimentation and coalescence experiments are an attractive method for studying the separation behavior of emulsions because of simple, repeatable, and inexpensive experiments. When the percentage of water in the mixture exceeds about 20% by volume, two "interfaces" become visible as the mixture settles. One is a "sedimentation" interface between the settling dispersion and the bulk oil phase (in which a small amount of dispersed water may remain). The other is a "coalescence" interface between the dispersion and the bulk water phase (which contains a small part of dispersed oil). As time goes on, the thickness of the dispersion layer gets smaller and the two interfaces approach each other.

The mixture in a batch settler behaves as a quasi-homogeneous flow and may be described by an advanced mixture model, such as the drift-flux model [9,24–27]. The drift-flux model is one of the multi-phase flow models in the commercial CFD program FLOW-3D [28].

The settling of drops initiates an upward flow in the container and this dynamic behavior is well defined within the drift-flux model. The model regards the whole mixture as a single flowing continuum, by solving both the volume continuity and the momentum equations for the whole mixture. This mixture has macroscopic properties like the bulk viscosity, previously discussed. The model describes the relative flow of the immiscible fluids with different densities. One of the important variables in focus is the local, instantaneous volume fraction, called the water cut. Furthermore, a drop-system or a drop-size distribution is the resulting macroscopic value called the water cut, in a polydispersed water-in-oil emulsion.

In the separation of the phases, the process is dominated by the coalescence and the settling rates of the dispersed phase. The coalescence rate depends on a number of parameters: drop-size distribution, drop density, shear/turbulence, and interfacial properties. The settling rates are determined by the density difference, and the viscosity of the dispersion, computed by Stokes' law or models for hindered settling. The binary coalescence rate is the product of collision frequency and coalescence efficiency, i.e., the number of collisions that lead to a successful merging of drops, which gives the resulting drop growth in the mixture system. The driving forces in the collision frequency are: the settling of drops, the shear flow, and the turbulence in the flow field.

16.3.3 Modeling and Simulation of Batch Separation

The mathematical modeling of coalescence in unstable oil-in-water emulsions is thoroughly described in the literature [29–32] and by the more fundamental theories in Refs. 33 and 34. Numerical simulations for batch settling of a poly-dispersed water-in-oil emulsion are performed and the results are presented in Figure 16.3. The water cut in the emulsion was 20%, the density difference of the two fluids is 200 kg/m^3, and the height of the settling container was 0.3 m. The drop-size distribution (Figure 16.4) was initially configured in ten different drop classes. The Sautern mean diameter in the distribution is 160 μm. Settling in the batch was simulated with and without a coalescence model, and as seen in Figure 16.3 the separation run faster with the drop-growth model (coalescence model) included. The separation time for the present batch system is 430 sec, calculated by Stokes' unhindered settling equation.

The separation time is 1300 sec, calculated by Kumar and Hartland's hindered settling equation (see Ref. 35). The simulated result with a constant Sautern mean drop-size 160 μm (original drift-flux model) gives a long settling time, about 1800 sec. The simulated result with a drop-growth model (coalescence model) shows the approach of the two interfaces after 1150 sec, which

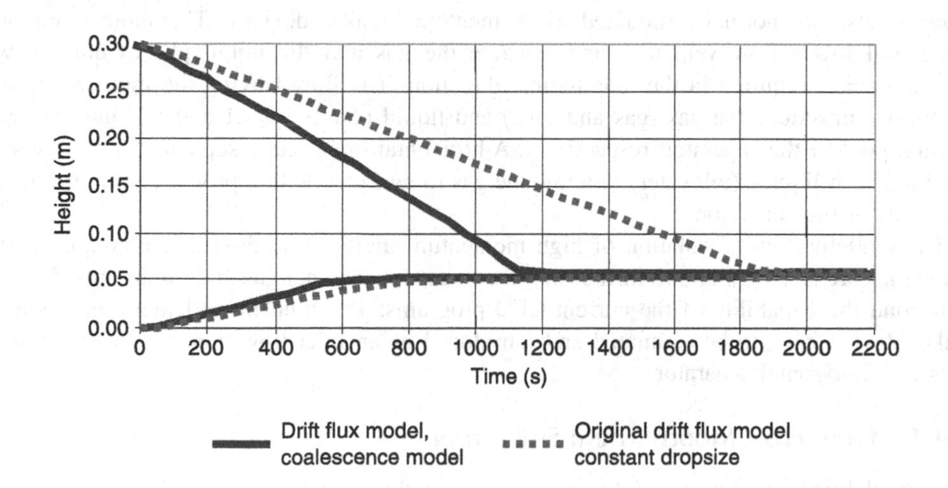

FIGURE 16.3 Simulated results of separation interfaces in batch settling. (From EMU-SIM. Oil/water separation offshore: optimized flow- and separation behavior. JIP, 1999–2002.)

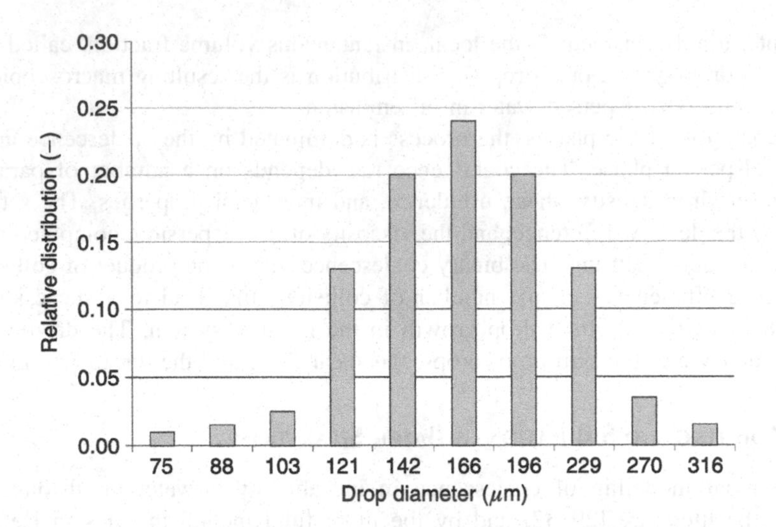

FIGURE 16.4 Drop-size distribution for an oil/water mixture.

is in relatively good agreement with the hindered settling equation. The different simulated results are presented in Figure 16.3.

16.4 HYDRAULIC BEHAVIOR IN A HORIZONTAL SEPARATOR

The gas/liquid and liquid/liquid systems will result in two relatively separated flow systems in pipelines or separators, due to the density differences. Between the two fluids a dynamic interface will form, stratified or annular in pipelines and stratified in separators. The characteristics of such flow systems may be developing flows, entrance flows, co-current flow transition, etc. and the full dynamic interaction between the phases is important.

The gas and liquid flows with rather high velocities enter through pipe and inlet devices to the separator. Inlet devices, such as cup-shaped plates, turbine-vane arrangements, or cyclone arrangements, are normally mounted as momentum breaker devices. The momentum breaker device will lower flow velocities and separate the gas and the liquid phases quickly with a minimum space required in the longitudinal direction. The flow leaving the momentum breaker will further introduce the gas (gas and mist) and liquid phases (water and oil) into gas volume or liquid pool of the separator, respectively. A horizontal three-phase separator is normally about half-filled with liquids (oil/water) to allow the gas to evacuate with a proper retention time in the upper part of the separator.

The modeling and simulation of high momentum multi-phase flow in the complicated inlet momentum breaker region, and further into the three-dimensional gas and liquid bulk flow zones, are beyond the capability of the current CFD programs. The fluid flows leaving the momentum breaker device have to be estimated and introduced as an inlet flow condition to the bulk flow zones in a horizontal separator.

16.4.1 Fluid Flow Modeling and Simulation

CFD simulations are performed to study the fluid flow behavior in the liquid bulk flows in a horizontal separator. The test separator is a conventional horizontal gravity separator with a diffuser and a cascade tray as the inlet arrangement (Figure 16.5). The inlet arrangement, a diffuser and a cascade tray, is mounted on the inlet riser, which is located centrally at the end.

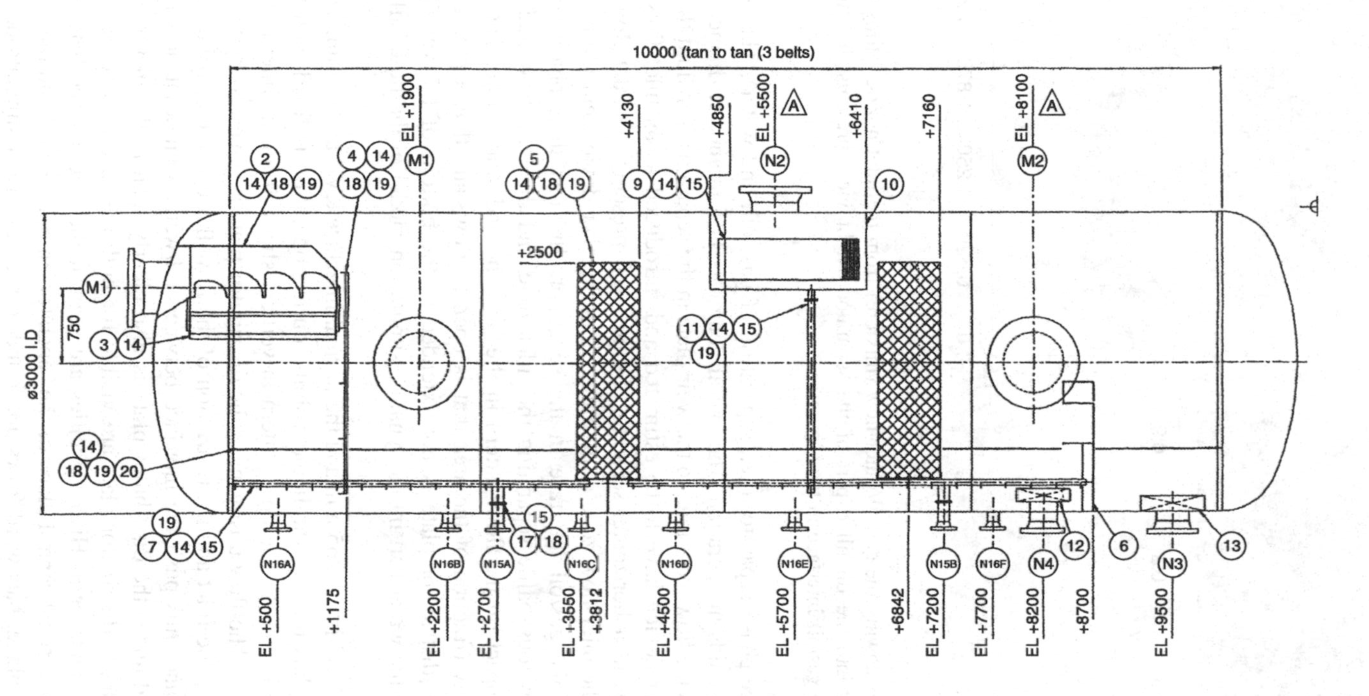

FIGURE 16.5 Sketch of the test separator. (2) and (3) are the inlet diffuser and the cascade tray, respectively. (4) is the distributor plate just downstream of the inlet and (6) is the fixed weir between the water and the oil outlets. A horizontal baffle (20) and the sand wash assembly (7) are in the lower part of the separator. Mellapak, structured packing, is shown as the crosshatched areas in the separator.

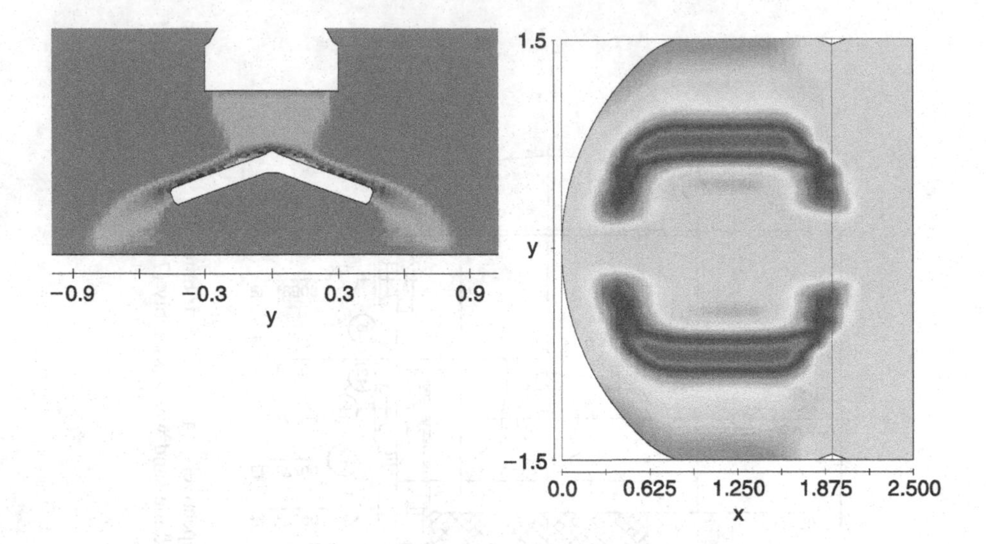

FIGURE 16.6 Left view showing the gas/oil/water-flow (mixture) from the inlet nozzle hitting the cascade tray and flowing further into the oil filled part of the separator. Right view presents the mixture flow, oil/water, diving into the gas/oil interface in the separator.

The separator is a three-phase separator (i.e., separation of gas, oil, and water). The separator is about half filled with liquids in operation. The overall length is 11.5 m and the diameter is 3.0 m. The flow model covers the flow from inlet to the weir plate in the separator. The fluid flows leaving the momentum breaker device have to be estimated and introduced as an inlet flow condition to the bulk flow zones in a horizontal separator. Figure 16.6 presents the inlet flow condition in a cross section in the inlet flow zone at the momentum breaker device and the mixture flow, oil/water, diving into the gas/oil interface in the separator. In most three-phase separators a distributor plate, the porous baffle separating the inlet zone and the bulk flow and separation zones, has a strong impact on the flow pattern in the oil zone and water zone. The distributor plate has a low porosity (fraction of the area that is open to flow) and thus a pressure loss for the flow through. The distributor plate is not extended to the bottom of the vessel and it is important to introduce the water part into a dynamic simulation model of the liquid filled part of the separator.

The normal liquid level is at 1265 mm and the normal interface level at 942 mm. The oil zone thickness is 323 mm; thus most of the liquid volume during operation is filled with water. The water cut in this study is 29.3%. The flow pattern in vertical slices, from the inlet to the weir, is presented in Figure 16.7. The flow upstream of the distributor plate is governed by the flow from the inlet. The flow from the inlet and the restriction of the flow through the distributor plate will increase the pressure and thus push the interface between oil/water somewhat down. The flow of water that is pushed under the distributor plate is also clearly seen in the picture. Relatively high velocities through the distributor plate are visible for both the forward and backward flow through the distributor plate area. High velocities are seen at the free surface between oil and water, especially in the vertical central plane of the separator. In both the oil and the water zones, away from the central plane, backward flows are found in the numerical simulation.

Figure 16.8 gives the velocity fields in horizontal slices, from the inlet to the weir (along the main flow direction). In the oil zone some re-circulation zones are seen just downstream of the distributor plate. In the water zone the re-circulation patterns are more pronounced. The figure

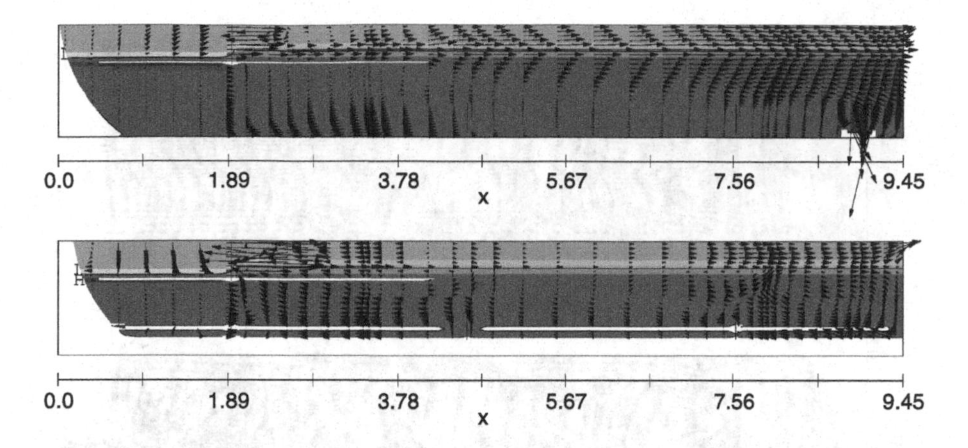

FIGURE 16.7 Numerical simulation of liquid flow in the test separator. Flow patterns are seen in vertical slices, from inlet to weir, along flow direction. Maximum velocity is different in the two different views. Top view is along central plane and max velocity is 59.4 cm/sec. Lower view is about midway between central and wall plane and max velocity is 32.8 cm/sec.

shows both a shorter re-circulation downstream of the distributor plate, and a longer re-circulation pattern from the outlet area back to the distributor plate along the separator wall. The high velocities under the distributor plate govern the re-circulation pattern just downstream of the plate. The Mellapak support plates also govern the re-circulated water in the separator.

Figure 16.9 presents the velocity field in vertical slices, perpendicular to the flow direction. For the vertical motion in the oil zone the fluid is in general slowing down toward the weir. In the water zone it is demonstrated that the horizontal baffles, above the sand wash assembly, create a vertical circulation flow pattern just below the oil/water interface.

16.4.2 Tracer Response and Hydraulic Efficiency Indicators in Liquid Zones

16.4.2.1 Tracer Response in Oil Zone

The numbers from the tracer simulations performed in the separator are summarized and presented in Table 16.1. A theoretical retention time of 169.7 sec was expected for the oil. The calculated median, or "average," retention time from the tracer response curve is 145.9 sec. The total short-circuiting number is 1.0 for ideal plug flow. For Case 1, the total short-circuiting number is 0.86. The value is reasonably close to unity, indicating little stagnant space in the oil flow. The tail of the tracer response curve indicates the mixing in the oil flow. The hydraulic indicator for the "worst" short-circuiting states that values as low as 0.05 clearly indicate serious short-circuiting. The test separator has an indicator of 0.33. The dispersion index is between 2.5 and 3.0, which is the indicator for mixing in the oil flow. From literature, a "good settling basin" should have a dispersion index of less than 2.0.

16.4.2.2 Tracer Response in Water Zone

The water flow through the separator has a theoretical retention time of 835.1 sec. The hydraulic indicators for the water presents non-favorable separation conditions for oil drops (and small sand particles) in the separator. This number, as well as the re-circulating pattern in the flow field, indicates a high degree of mixing. The most serious short-circuiting clearly indicated a value

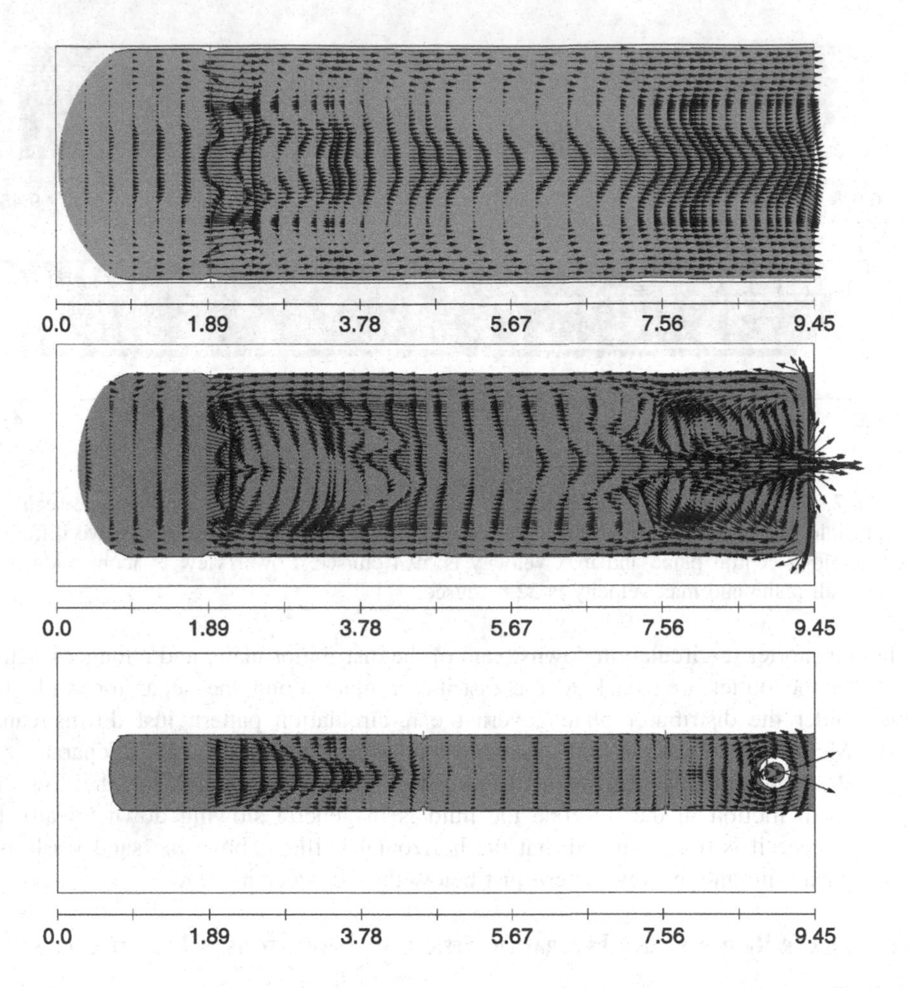

FIGURE 16.8 Case 1 numerical simulation of liquid flow in Test separator. Flow pattern seen in horizontal slices, from inlet to weir, along the main flow direction. Maximum velocity is different in the three different views. Top view is 12.6 cm below oil/gas interface (NLL) and max velocity is 33.6 cm/sec. Middle view is midway between oil/water interface (NIL) and max velocity is 6.98 cm/sec. Lower view is just below distributor plate and max velocity is 17.1 cm/sec.

as low as 0.07, resulting that some flow is passing through the separator much faster than the theoretical retention time.

16.5 CONCLUSIONS

Offshore separators are huge pressurized vessels, in which oil/water/gas/sand are separated. The operation of the separator is to provide sufficient time to allow the phases to separate by gravity. The operation in the field and the internal devices should make efficient and low maintenance separators.

Advanced multiphase flow modeling by the CFD program FLOW-3D can simulate the fluid dynamic effects in a gravity separator. The modeling of flow in separators is based on dividing

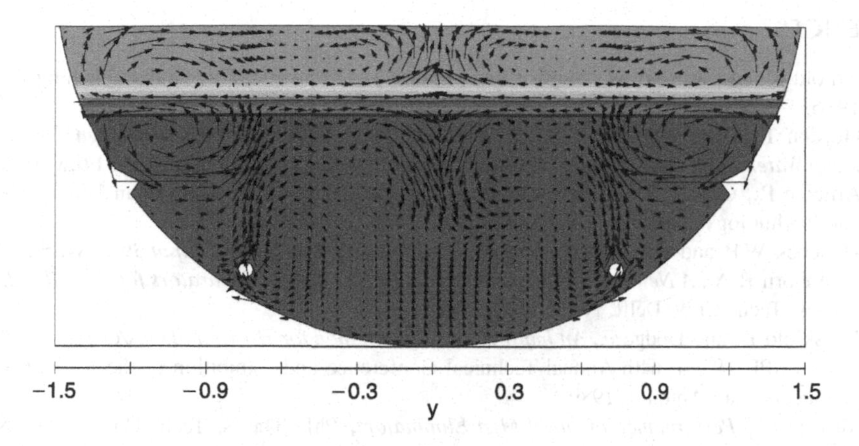

FIGURE 16.9 Numerical simulation of liquid flow in the test separator. Flow pattern is seen in vertical slices, perpendicular to flow direction. The plane view is midway between distributor plate and weir and max velocity is 0.99 cm/sec.

TABLE 16.1
Theoretical Retention Times in Oil and Water Flow Volumes for the Test Separator, Tracer Response Characteristics, and Hydraulic Efficiency Indicators

Flow Volume	Prod. Rates	Theoret. Retention Time, Tr	Init. Appear., Ti	10% Tracer, T10	90% Tracer, T90	Total Short-Circ.	Most Serious Short-Circ.	Disper. Index
	M3/d	s	S	s	s			
Oil	3572	169.7	55.4	88.5	262.0	0.86	0.33	2.96
Water	1479	835.1	62.5	153.6	1909.6*	0.67	0.07	12.43

*T88: 88% of tracer particles counted at outlet.

the separator in different zones, and limiting the study to two phases at a time; the entire separator can be analyzed. FLOW-3D has been used in a number of studies and engineering jobs.

The rheology of oil/water emulsions is described mathematically. The bulk viscosity models generally lack validation towards relevant field or pilot scale model experiments. The drop-growth (coalescence rate) modeling implemented in a drift-flux flow model gives realistic separation and the behavior of the dispersion interfaces. Simulations of oil/water separation in a typical batch settler are performed.

The coalescence can be modeled with semi-empirical relations, but the relation between the coalescence rate and the mixing/turbulence in bulk flow zones in a separator is not yet sufficiently described.

Computational fluid dynamics and numerical simulations represent a significant enhancement of design jobs and the great challenge is to combine the chemical and the fluid dynamic effects.

Many practical, design, and redesign applications may be performed by CFD modeling and simulations. Examples are (1) development of vessel inlet configurations that improve the uniformity of the gas and liquid flows, (2) sensitivity of a separator design to changes in operating conditions, (3) influence of internal equipment on separation performance, and (4) coalescence and breaking of emulsion.

REFERENCES

1. Arnold K. and Stewart M., *Surface Production Operations – Vol. 1*, Gulf Publishing Co., Houston, 1986.
2. Gordon I.C. and Fairhurst C.P., *Multi-Phase Pipeline and Equipment Design for Marginal and Deep Water Field Development*, 3rd International Conference on Multi-Phase Flow, Hague, 1987.
3. Arntzen R., Gravity separator revamping, Dr.ing.thesis 2001:16, Norwegian University of Science and Technology, Trondheim, Norway, 2001.
4. Hancock W.P. and Hagen D.A., *Fine Tuning Increases Statfjord B Output 39%*, World Oil, 1986.
5. Swanborn R.A., *A New Approach to the Design of Gas–Liquid Separators for the Oil Industry*, Ph.D. Thesis, Tech. Univ. Delft, Netherlands, 1988.
6. Hafskjold B. and Dodge F., *An Improved Design Method for Horizontal Gas/Oil and Oil/Water Separators*, SPE 19704, 64th Annual Technical Conference and Exhibition of the Society of Petroleum Engineers, San Antonio, 1989.
7. Verlaan C., *Performance of Novel Mist Eliminators*, PhD Thesis, Tech. Univ. Delft, Netherlands, 1990.
8. Hansen E.W.M., Heitmann H., Laksâ B., Ellingsen A., Østby O., Morrow T.B. and Dodge F.T., *Fluid Flow Modelling of Gravity Separators*, 5th International Conference on Multi-Phase Production, Cannes, France, 1991.
9. Hansen E.W.M., *Modelling and Simulation of Separation Effects and Fluid Flow Behavior in Process-Units*, SIMS '93 – 35th Simulation Conference, Kongsberg, Norway, 1993.
10. Hansen E.W.M, Celius H.K. and Hafskjold B., *Fluid Flow and Separation Mechanisms in Offshore Separation Equipment*, Two-Phase Flow Modeling and Experimentation, Rome, Italy, 1995.
11. EMU-SIM, *Oil/Water Separation Offshore: Optimized Flow- and Separation Behavior*, JIP, 1999–2002.
12. Hansen E.W.M., *Phenomenological Modeling and Simulation of Fluid Flow and Separation Behavior in Offshore Gravity Separators*, PVP-Vol. 431, ASME 2001.
13. Hansen E.W.M., Heitmann H., Laksâ B. and Løes M., *Numerical Simulation of Fluid Flow Behavior Inside, and Redesign of a Field Separator*, 6th International Conference on Multiphase Production, Cannes, France, 1993.
14. Iversen I.R., *The Use of CFD to Solve Topside Process Problems*, Production Separation Systems, Stavanger, Norway, 1999.
15. Boris J. P., *New Directions in Computational Fluid Dynamics*, Annu. Rev. Fluid Mech., 21, 1989. Industries, Trans IChemE, Vol. 68, Part A, 1990.
16. Colenbrander G.W., *CFD in Research for the Petrochemical Industry*, Appl. Scientific Res., 48, 1991.
17. Foumeny E.A. and Benyahia F., *Can CFD Improve the Handling of Air, Gas and Gas–Liquid Mixtures?* Chem. Eng. Prog., Feb. 1993.
18. Hunt J.C.R., *Industrial and Environmental Fluid Mechanics*, Annu. Rev. Fluid Mech., 23, 1991.
19. Sharratt P.N., *Computational Fluid Dynamics and Its Application in the Process Industries*, Trans IChemE, Vol. 68, Part A, 1990.
20. Jayanti S. and Hewitt G.F., *Review of Literature on Dispersed Two-Phase Flow with a View to CFD Modelling*, AEA-APS-0099, AEA Petroleum Services, AEA Technology, 1991.
21. Pal R. and Rhodes E., *A Novel Viscosity Correlation for Non-Newtonian Concentrated Emulsions*, J. Coll. and Interf. Sci., Vol. 107, No. 2, 1985.
22. Schramm L.L., *Emulsions: Fundamentals and Applications in the Petroleum Industry*, American Chemical Society, Washington, DC, 1992.
23. Pan L., Jayanti S. and Hewitt G.F., *Flow Patterns, Phase Inversion and Pressure Gradient in Air–Oil–Water Flow in a Horizontal Pipe*, The 2nd International Conference on Multiphase Flow '95-Kyoto, Kyoto, Japan, 1995.
24. Hirt C.W., Romero N.C., Torrey M.D. and Travis J.R., *SOLA-DF: A Solution Algorithm for Nonequilibrium Two-Phase Flow*, Los Alamos Scientific Laboratory Report, NUREG/CR-0690, LA-7725-MS, 1979.

25. Ishii M., *Thermo-Fluid Dynamic Theory of Two-Phase Flow*, Eyrolles, Paris, 1975.
26. Ungarish M., *Hydrodynamics of Suspensions*, Springer-Verlag, Berlin, Heidelberg, 1993.
27. Zuber N. and Findley J.A., *Average Volumetric Concentration in Two-Phase Flow Systems*, J. Heat Trans., 87, 1965.
28. FLOW-3D, *Excellence in Flow Modeling Software*, Flow Science, Santa Fe, US, 1983–2004.
29. Davis J.T., *Drop Sizes of Emulsions Related to Turbulent Energy Dissipation Rates*, Chem. Eng. Science, Vol. 4, No. 5, 1985.
30. Hafskjold B., Hansen E.W.M., Singstad P. and Celius H.K., *E-FLOSS; Interim Technical Report no. 1*, IKU-Report 91.053, 1991.
31. Hafskjold B., Celius H.K. and Aamo O.M., *A New Mathematical Model For Oil/Water Separation in Pipes and Tanks*, SPE 38796, San Antonio, Oct. 1997.
32. Meijis F.H. and Mitchell R.W., *Studies on the Improvement of Coalescence Conditions of Oilfield Emulsions*, JPT, May 1974.
33. Coualoglou C.A. and Tavlarides L.L., *Description of Interaction Processes in Agitated Liquid–Liquid Dispersions*, Chem. Eng. Sci., 32, 1977.
34. Randolph A.D. and Larson M.A., *Theory of Particulate Processes – 2nd edition*, Academic Press, Inc., San Diego, 1988.
35. Panoussopoulos K., *Separation of Crude Oil–Water Emulsions: Experimental Techniques and Models*, Diss. ETH No.12516, 1998.

17 Design Criteria and Methodology for Modern Oil–Water Separation Systems

John D. Friedemann

CONTENTS

Abstract

The development, verification, and qualification of gravity separator design guidelines should have the highest focus in sub-sea projects as the design of sub-sea processing presents challenges not normally experienced by the process development team. Despite the fact that footprint real estate is easily obtained, new limitations are imposed because sub-sea installation traditionally occurs through the moonpool of a drilling rig, or with vessels with limited lifting capacity. Although large separators are found on platform processes, seabed processes should be as compact as possible. The purpose of this chapter is to present an overview of current understanding in how oil/water separator efficiency may be influenced by various parameters. The ability to predict separator performance based on fluid characterization is discussed. Additionally, the chapter includes a comparison between theoretical models for separation performance and laboratory experience. The main conclusions to be drawn are:

- Traditional design rules for topside separators are based on over-simplified models, and are not sufficient for sub-sea use.
- More comprehensive approaches exist, based on more data describing the fluids involved and their behavior. This will involve testing with high pressure test cells. These approaches give a better understanding and a more correct separator design, but there is still a need to establish the missing link between pressurized test cell behavior (laboratory scale) and large scale dynamic behavior.
- The mixing of water continuous wells with oil continuous wells normally gives complex emulsions that are hard to separate in a gravity separator. No reliable method to account for these effects is identified in this study and should be established experimentally.

17.1 INTRODUCTION

The gravity separation process is a key part of almost any realistic sub-sea processing scheme. In the major part of the potential sub-sea processing projects seen to date the gravity separator is well suited for doing the job alone. The performance of the gravity-settling sections of the cyclonic separators will still be crucial to the total performance of the system.

This chapter will demonstrate that the complexity of the fluids and production systems makes it difficult to calculate the exact performance of a separator, and that the design guidelines in use today are normally over-simplified. This is why the development, verification, and qualification of gravity separator design guidelines should have the highest focus in the sub-sea projects.

Conceptual solutions for sub-sea production and processing have been on the drawing board for decades, yet the concepts have been slow to gain acceptance. This is related to a combination of risk aversion within the oil industry and risk ignorance amongst suppliers. Both parties see significant market advantages with qualified solutions for sub-sea systems yet active effort towards pushing the products in the direction of a risk reduced production train has been difficult.

One potential cause of this conflict is related to operational problems observed at surface production facilities. Typical facilities have production start-up challenges that are very difficult to analyze even with surface access. These can range from mechanical to chemical sourced problems such as:

- Accumulation of well cleanup products in the separation train
- Mechanical failures in rotary equipment caused, indirectly, by misinterpreted reservoir fluid analyses
- Incorrect process flow sheets
- Under-instrumentation

Typically the solution to these types of problem on surface facilities has been to form a multiparty task force involving the original suppliers, the relevant operator staff, and one or more specialist consulting firms. These are then given the opportunity to perform various activities designed to solve the operator's problem.

The reservoir production environment has a different set of rules for its problem solving activities. These are very relevant to sub-sea-processing as they include the sub-sea wellhead solution set. Facilities for well diagnostics have to be requested long in advance and typically have a 1-year life cycle. In fact, the increased expense and complexity of sub-sea well diagnostics has resulted in a controversy as to the effect of such solutions on ultimate field recovery.

This chapter reviews the current state of the art on separator performance modeling. The paper focuses on two-phase oil–water gravity separation. However, gas is an important factor as it will influence the inlet conditions and, additionally, it is most likely that there will be a gas pocket in the separator. This is the case even when an upstream degasser is applied or when the wellhead pressure is above the saturation pressure.

17.1.1 The Importance of Characterization

The interaction between heavy and solid crude oil constituents, such as asphaltenes, resins, waxes, sand, and salt, for example, forms the basis for the stabilization of crude oil emulsions. In fact, many of the heavy organics are present in a state not unlike solid particles. Resins and the other, lighter, polar components contribute by stabilizing the dispersions and inhibiting coagulation. (There is a trade-off as these systems show less tendency towards well plugging by adhesion or sedimentation because of these naturally occurring dispersants.)

This condition is strongly altered by the presence of water. In the flowing system, fresh interfacial areas are created and polar molecules, such as resins, are more strongly attracted to these. Hence these diffuse quickly to the surfaces and, in turn, form a protective coating on the droplet surfaces. Decisive factors determining the final attachment positions of the resins are (1) the hydrophilic/lipophilic balance of these molecules and (2) the corresponding properties of the solid surface. It can be imagined that a very hydrophobic particle surface and a very polar water–oil interface would extract different types of resin for different activities. However, highly interfacially active resins will show preference for the water–oil interface not only over less hydrophobic resin molecules but also over asphaltenes.

As a consequence of the transport process, the resin concentration in the bulk fluid is reduced. This, in turn, changes the solubility conditions of the resulting precipitation of asphaltene particles which then tend to accumulate at the droplet surfaces. The second layer further enhances both droplet rigidity and emulsion stability. Central to this stabilization process are mechanisms related to steric hindrance.

Depending upon the surface chemistry of the particles, other solids, including sand, clays, and even crystallized salts, can be attracted to the surfaces. These inorganic solids further enhance the droplet protection mechanisms. Changes in the interfacial conditions due to ageing of the water–oil emulsion should also be borne in mind.

17.1.1.1 Coalescence

Coalescence is defined as the combination of two or several droplets to form a larger drop. When these droplets approach each other a thin film of the continuous phase will therefore be trapped between the droplets, and it is obvious that the properties of this film will determine the stability of the emulsion (Brown, 1968). In the Brown model, the mechanism of coalescence occurs in two stages: film thinning and film rupture. In order to have film thinning there must be a flow of fluid in the film and a pressure gradient. It is obvious that the rate of film thinning is affected by:

- Phase rheology
- Interfacial tension
- Dilation properties (viscosity and elasticity)
- Droplet size
- Fluid chemistry

Many authors have provided notable research into the modeling of these systems (de Vries, 1958; Frank and Mysels, 1960; Lang, 1962; Liem and Woods, 1974; Lin and Slattery, 1982; Malhorta, 1984; Zapryanov et al., 1983).

Attempts at predicting critical film thicknesses for the final rupture phase have resulted in three conclusions. The first was given by Schedluko and Manner (1968) who developed an expression for the critical film thickness for surface fluctuations with the assumption that the system is described only by van der Waals forces:

$$d_C = \left(\frac{A\pi}{32K^2\gamma_0}\right)^{0.25} \tag{17.1}$$

where A is the Hamaker constant, γ_0 is the interfacial tension, and K is the wave number of the surface fluctuations. The second and third conclusions were given by Vrij (1966) who derived two alternative expressions for d_C and large and small thickness models. For a large thickness

$$d_C = 0.268\left(\frac{A^2R^2}{\gamma_0\pi f}\right)^{0.14} \tag{17.2}$$

where R is the droplet radius and f is $f(d)$. For a small thickness

$$d_C = 0.22\left(\frac{AR^2}{\gamma_0 f}\right)^{0.25} \tag{17.3}$$

Like many of the models for droplet behavior, the thick film model shows nonphysical behavior at extremes of interfacial tension and/or droplet diameter.

17.1.2 Indigenous Stabilizing Components

17.1.2.1 Saturates and Aromatics

The saturates (or aliphatics) are the nonpolar compounds containing no double bonds and include both the alkanes and the cycloalkanes. Wax is a sub-class of the saturates. The aromatics consist of all compounds with one or more benzene rings. These ring systems may be linked up with naphthene rings and/or aliphatic side chains.

17.1.2.2 Resins

This fraction is comprised of polar molecules often containing heteroatoms such as nitrogen, oxygen, or sulfur. This fraction is operationally defined, and one common definition of resins is as the fraction soluble in light alkanes such as pentane and heptane, but insoluble in liquid propane. Naphthenic acids are a part of this fraction.

17.1.2.3 Asphaltenes

Asphaltenes are polar molecules that can be regarded as similar to the resins, but with higher molecular weight, typically 500 to 1500 g/mole. The asphaltene fraction, like the resins, is defined as a solubility class, namely the fraction of the crude oil precipitating in light alkanes such as pentane, hexane, or heptane. The precipitate is soluble in aromatic solvents such as toluene and benzene. The asphaltene fraction contains the largest percentage of heteroatoms (O, S, N) and organometallic constituents (Ni, V, Fe) in the crude oil. The structure of asphaltene molecules is believed to consist of polycyclic aromatic clusters, substituted with varying alkyl side chains. The molecular weight of asphaltene molecules has been difficult to measure due to the tendency of asphaltenes to self-aggregate, but molecular weights in the range 500 to 2000 g/mole are believed to be reasonable.

17.2 SEPARATOR PERFORMANCE

17.2.1 Production Related Challenges

As they rise from the reservoir and undergo differential separation process in the production tubing, the reservoir fluids undergo significant changes in properties. This is indicated in Figure 17.1. Heavy components become less soluble and become apparent as asphaltenes, waxes, scale, for example. These small particles may agglomerate to form much larger particles, which can accumulate on the production train. Otherwise, these provide the basis for particle stabilization in emulsion or foaming systems.

Ideally, the separation process should be simplest at "nearest to reservoir" conditions to achieve maximal benefit from:

- Maximal density differences between the oil and water phases
- Minimal effects of carbonate scale deposition by removing the water
- Minimal emulsion effects on flowline capacity

Separation curves are shown in Figure 17.2.

Determining the source of the stabilizing mechanism is very important. Different mechanisms require different production chemicals or production solutions. Efforts to identify water production, asphaltene deposition, and, ultimately, the correct solution to a problem in production

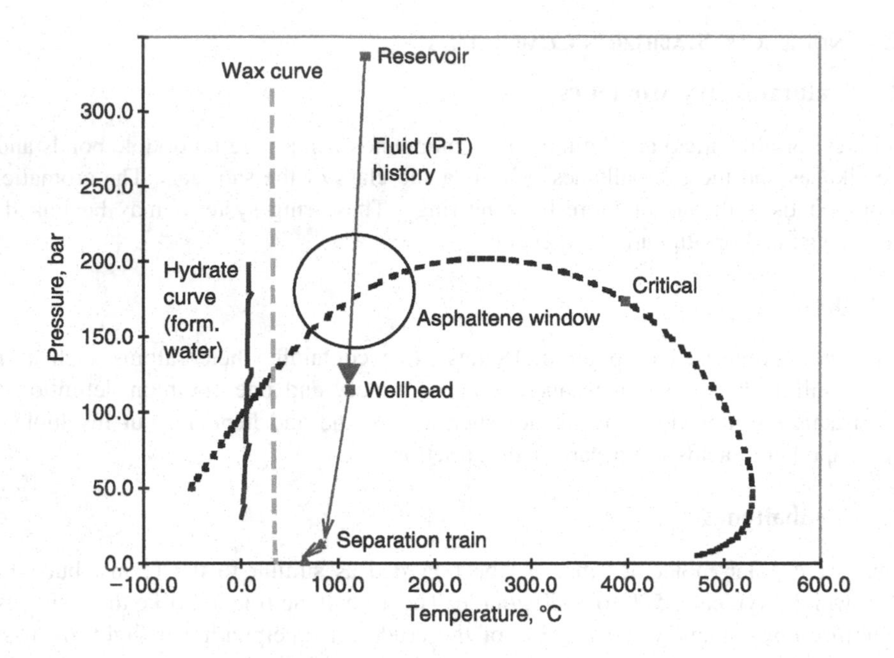

FIGURE 17.1 Phase envelope for a typical emulsion-forming oil.

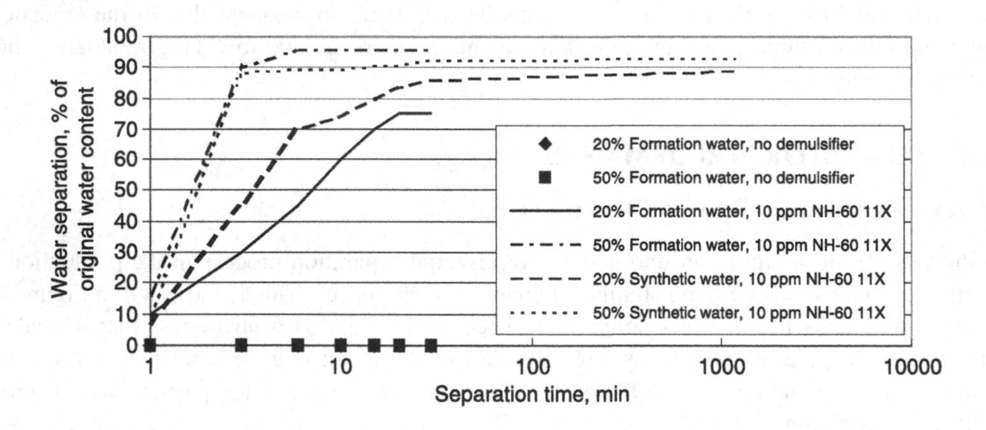

FIGURE 17.2 Separation curves for a particle (salt, scale, and asphaltene) emulsion system.

chemistry can be expensive. Published material is available which describes the efforts to locate problem wells in networked fields with as many as 150 wells (with as many as five completion zones per well) producing through a common export line from individual fields.

The system described in Figure 17.3 is then quite simple. The driving force can be dramatic as Figure 17.4 indicates. The operator risked incurring a multi-million dollar loss when the salt content exceeded refinery tolerances (in the middle of Figure 17.4).

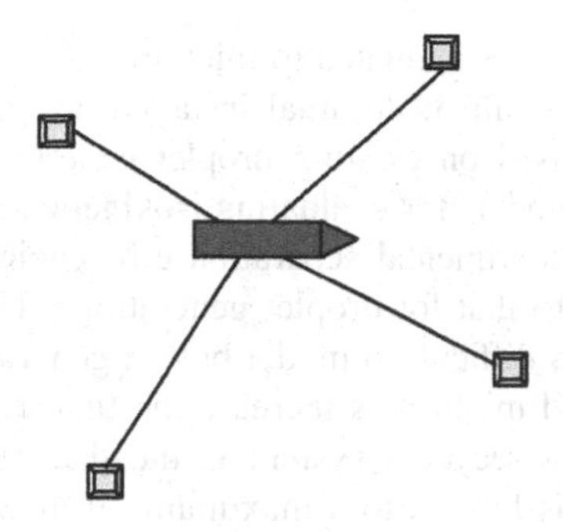

FIGURE 17.3 Typical seabed arrangement.

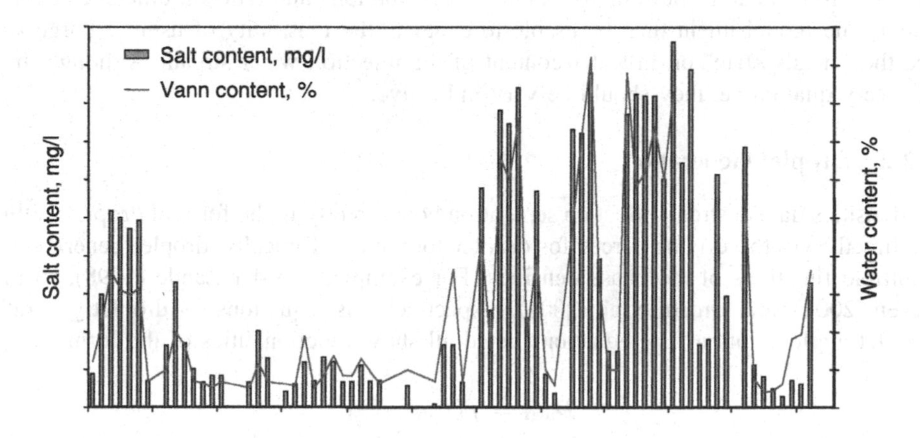

FIGURE 17.4 Final result after initial water breakthrough.

17.2.2 Inlet Conditions

17.2.2.1 Critical Droplet Size Evaluation

In summary, the design of a conventional gravity separator is very much controlled by four factors:

- The insecurity in fluid properties (which is controllable)
- The insecurity in droplet sizes which the separator is exposed to (can be predicted approximately)
- The insecurity in phase continuity and the presence of complex emulsions
- The insecurity in inlet device performance with respect to free gas carry under and/or the ability to release dissolved gas to form small bubbles

The first of the uncertainties can be controlled by reviewing the reservoir data as well as conventional separation tests. The influence of fluid chemistry on separation problems has been evaluated in several joint industry projects. As is the nature of such projects much of the data and experience remains unpublished, but indications from the open literature (e.g., Friedemann, 1997; Osa and Holt, 2001) are that methods exist for the reliable onshore evaluation of production chemicals.

The second of the uncertainties is related to inlet droplet size. This is the result of flow energy and, as such, very much the result of the final installed design. A system can still be evaluated from a sensitivity analysis based on existing droplet generation models (see Section 17.2.2.2). Experience with gas droplet models for evaluating isokinetic sampling systems indicates that this is a relevant method. (The experimental separation efficiencies derived from field data indicate a trend relationship identical to that for droplet generation.) The influence of cyclones and other inlet devices on droplet size is difficult to model but, in general, cyclones will reduce the droplet size (see Section 17.2.2.3) and might thus increase the time required for oil–water separation.

A potential problem for a recycled system is the danger of accumulating droplets in the discharge water. The system is limited to a maximum oil-in-water content. The oil limit is very much determined by the rate of accumulation of very small oil droplets in the recycle loop. In this case a study based on a determination of critical droplet sizes within the system can be proposed. A model can be built up using cut-off distributions and removal efficiencies for each of the units. The model might then be usable to evaluate the feasibility of using a purge stream to reduce the "steady-state" oil-in-water content of the injection water stream. Although the results will be very qualitative, they should be very indicative.

17.2.2.2 Droplet Generation

Phase densities have a strong effect on separation both directly in the form of droplet settling rates and indirectly via the driving forces for droplet formation. Typically, droplet generation models are multiple functions of the phase densities. For example, van der Zande (1998), Arntzen and Andresen (2001), and Urdahl et al. (2001), respectively, list equations for droplet generation and for droplet agglomeration. The equations listed all show functionalities of the form:

$$D_{\max} = f(\tilde{E}, \rho_\chi, \xi, \sigma) \tag{17.4}$$

where $\tilde{E}$, ρ_χ, ξ, and σ represent applied energy, continuous phase density, droplet concentration, and interfacial tension. Chokes, pumps, and shear stresses, for example, can supply energy. A simple and understandable example is given for surfactant-stabilized droplets in an impeller-based mixer (Skelland and Slaymaker, 1992):

$$\frac{d_{32}}{d_I} = 0.05C_s(1 + 2.316\varphi)\left(\frac{d_I}{T}\right)^{-0.75} N_{\mathrm{Fr}}^{-0.13} N_{\mathrm{We}}^{-0.6} \tag{17.5}$$

where

C_s = correction factor
d_{32} = Sauter mean droplet diameter
φ = dispersed phase volume fraction
d_I = impeller diameter
T = tank diameter
N_{Fr}, N_{We} = Froude number and Weber number, respectively

The work by Skelland and Slaymaker goes on to relate these values to time dependent droplet diameters. This work provides a functionality that is important to consider in systems with recycle streams.

17.2.2.3 Effect of Upstream and Downstream Equipment

As indicated, droplet sizes in two-phase flow are very much a result of flow shear. For example, the reason for using cyclonic devices for separating solids from liquids is because they directly attack the gravity term in Stokes' law:

$$v_t = \frac{2R_{\text{droplet}}^2 g\left(\rho_w - \rho_O\right)}{9\mu} \tag{17.6}$$

Here v is the velocity, R the droplet radius, and μ the viscosity. The functionality for this is described by the relationship:

$$g = \frac{v_{\text{tangential}}^2}{R_{\text{cyclone}}} \tag{17.7}$$

Unfortunately, the relationship for droplet size as a function of flow shear is:

$$R_{\text{droplet}} \propto v_{\text{tangential}}^{-1.2} \tag{17.8}$$

Combined, the settling velocity (in the gravity separator) for such cyclonic devices used for degassing and degassing the feed of a gravity will be described as:

$$v_t \propto \frac{R_{\text{droplet}}^2 \left(\rho_w - \rho_O\right)}{v_{\text{tangential}}^{2.4} \mu} \tag{17.9}$$

where R_{droplet} is the droplet radius upstream the process. Using upstream cyclones (e.g., degasser, Figure 17.5) the coalescer device will have to compensate completely for these cyclones.

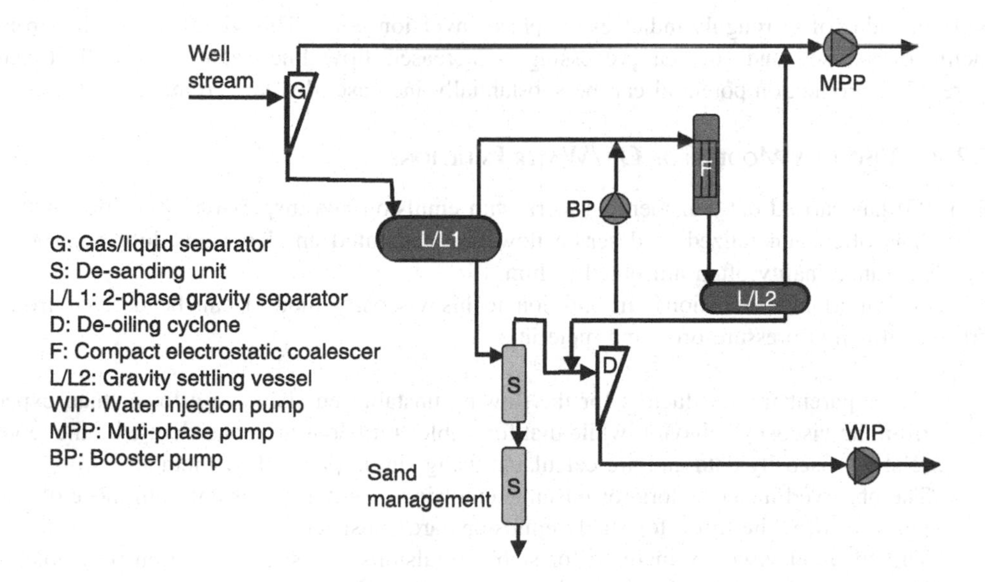

FIGURE 17.5 Simplified SPC flow process.

17.2.3 Density Dependent Variables

It should be noted then that the functions used to estimate the applied energy (flow related shear) and interfacial tension in Equation 17.1 are typically of the form:

$$\sigma = [P\,(\rho_I - \rho_v)]^4 \tag{17.10}$$

and

$$\rho_{\text{gas,rel}} = \text{polynomial}(\rho, T, Z\ldots)$$
$$\rho_{\text{liq,rel}} = \text{function}(P_c, T_c, Z_c\ldots)$$

where the viscosity polynomials are all strongly density dependent. Note that the practice is to extend the liquid polynomial models for use with both phases. The resulting model has a weak (direct) density functionality, and may show strong inaccuracies as the components approach their freezing points (Reid et al., 1988; Calsep, 1991).

The benefit of downhole and sub-sea separation processes is then clearly dependent upon the relative behavior of the oil phase and water phase density and interfacial tensions as the oil degases. Without taking into account the surface chemistry effects on particle stabilization, relatively dead heavy oils would, in theory, show little benefit from downhole and sub-sea separation, as the water–oil phase densities are similar.

Early removal of water shows other benefits, which are directly related to the viscosity of the oil phase. This is often handled using a simple hyperbolic model, such as that proposed by the Taylor-based model of Pal:

$$\mu_{\text{rel}} \propto k_0\left(1 + \frac{wc}{k_3 + wc}\right)^{2.5} \tag{17.11}$$

where the value of k_3 roughly indicates the phase inversion point. This yields one of the strongest benefits of sub-sea and sub-sea processing – increased flow line capacity. As illustrated in Figure 17.6, production potential can be substantially increased by early removal of water.

17.2.4 Viscosity Models for Oil/Water Emulsions

Pal (1993) has carried out considerable work with emulsion and dispersion viscosities. Although his work is often generalized to describe flow line generated emulsions, it does not show the very clear functionality often attributed to him.

Three important conclusions (in addition to his viscosity model) can be derived from his work into flowline pressure-drop and metering:

- The apparent friction factors for the flowing unstable emulsions are lower than expected from the viscosity behavior, while that for stable emulsions are consistent with the reported relative viscosity data and are calculable using single-phase flow methods.
- The observed meter-factors for unstable emulsions are not consistent with those of single-phase flow, while those for stable emulsions are consistent.
- The apparent viscosity increase for stable emulsions is a strong function of droplet size. Viscosities increase radically with decreasing droplet size and the fluids show strong non-Newtonian behavior.

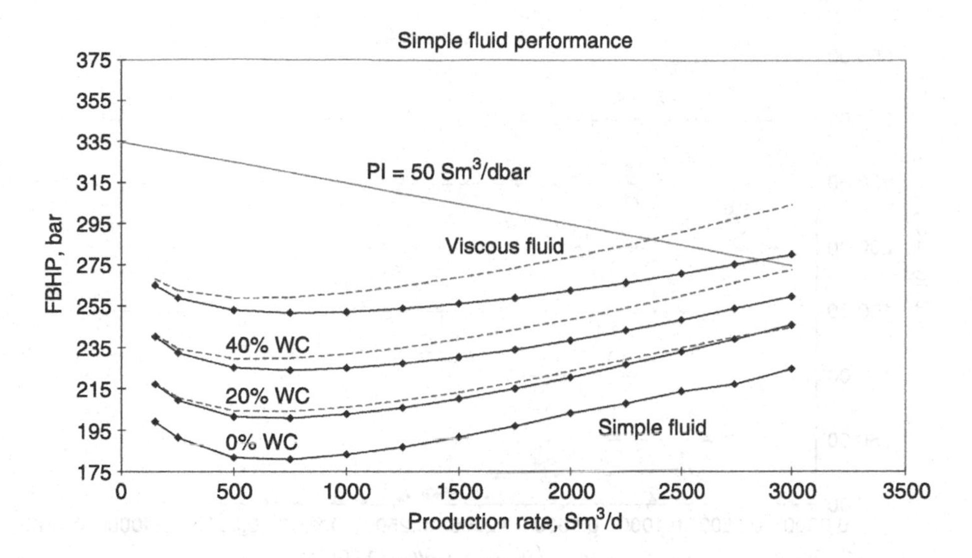

FIGURE 17.6 Effect on production potential of emulsion viscosities. Calculations are performed on a 15 km sub-sea tieback.

Pal (1993), Krawczyk et al. (1991), and Skelland and Slaymaker (1992) relate the droplet and viscosity behavior to chemical surfactant behavior. The simple experiments of Skelland and Slaymaker show a very strong relationship between steady-state droplet diameters and surfactant concentrations.

The last observation may explain the poor results experienced for processes with strong recycle streams. These may be accumulating large numbers of very small droplets, which, in combination with the increase in stabilizing compounds, occur with these recycles. These, in turn, contribute to a viscous dispersion band in the separator. Such a system may be so "stiff" that flow short-circuiting occurs within the separator.

These models are also utilized to describe other systems such as waxy crudes, magma flows, etc. Studies of these systems (e.g., Wardaugh and Boger, 1991) indicate that transitions from high shear to low shear for systems with stable droplets can give extremely high apparent viscosities. This effect is also seen in work with monodisperse particles.

All of this is quite important because (as illustrated in Figures 17.6 to 17.9) the design equations for the system are very sensitive to viscosity data (which is very difficult to predict using an equation of state) and also to surface active chemicals (such as asphaltenes and saponifiers). It is difficult to predict the behavior of the latter if geochemical information is not available.

17.2.5 Liquid–Liquid Separation Models

17.2.5.1 Traditional Sizing

The traditional method for estimating separator size is based on simplistic assumptions. An example can be taken from the exploration well test separator. The design method used to size these simple devices is based on Stokes' law separation for all of the phases and an assumed droplet size of 100 microns. The droplet generation functions illustrated previously show strong

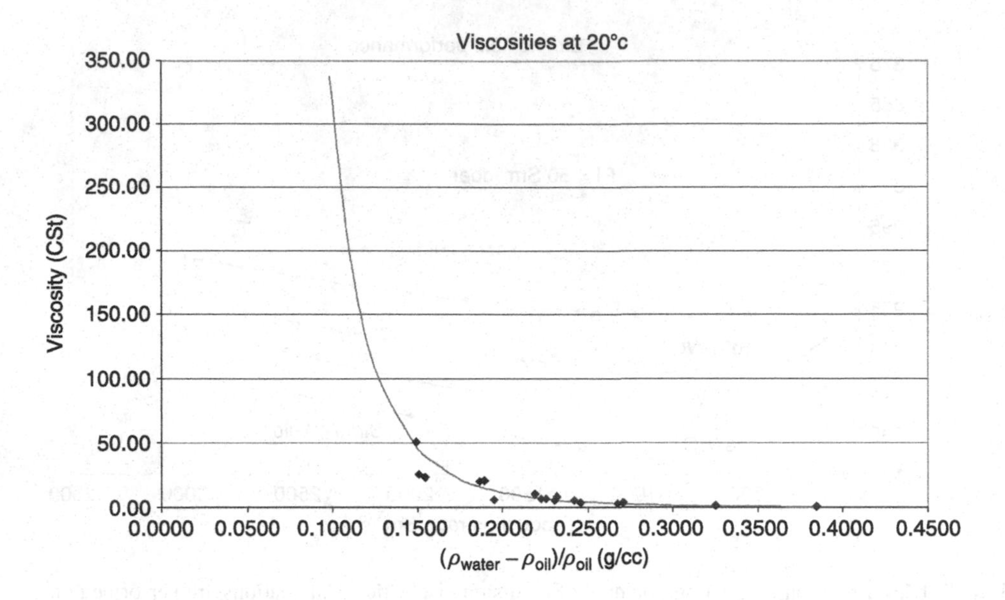

FIGURE 17.7 Example of correlation between stock tank oil density and viscosity.

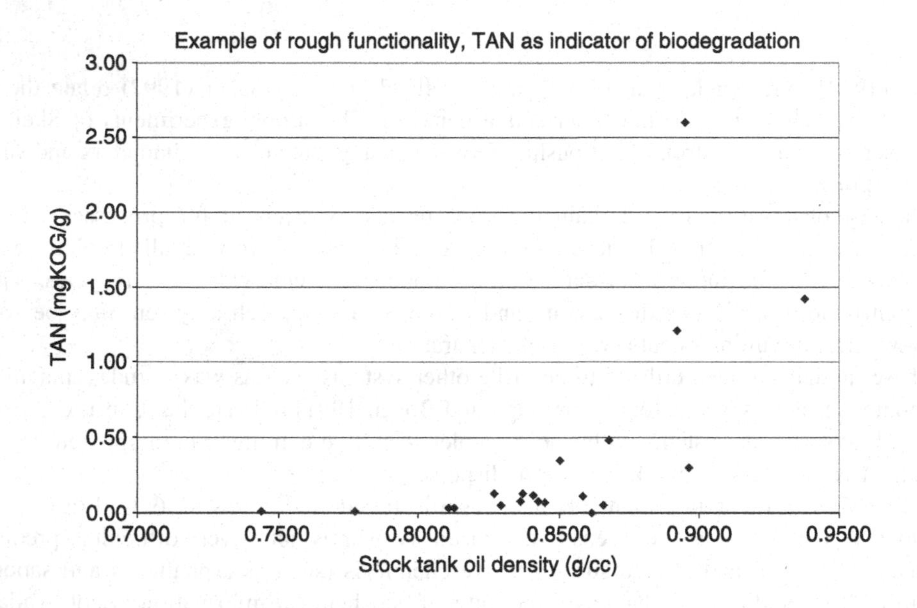

FIGURE 17.8 Relationship between the density of a heavy oil and its degree of biodegradation. The illustration is based on TAN as an indicator of biodegradation.

rate dependence as well as strong functionality with pressure and density. Experience with these separation devices showed strong fall-off of efficiency with increasing rate (a factor that has been a driving force for the use of isokinetic sampling techniques during gas well tests). The resulting sizing equations are little better than the American Petroleum Institute 12J separation time guidelines used previously (Table 17.1). Alternatives to these exist, such as those suggested

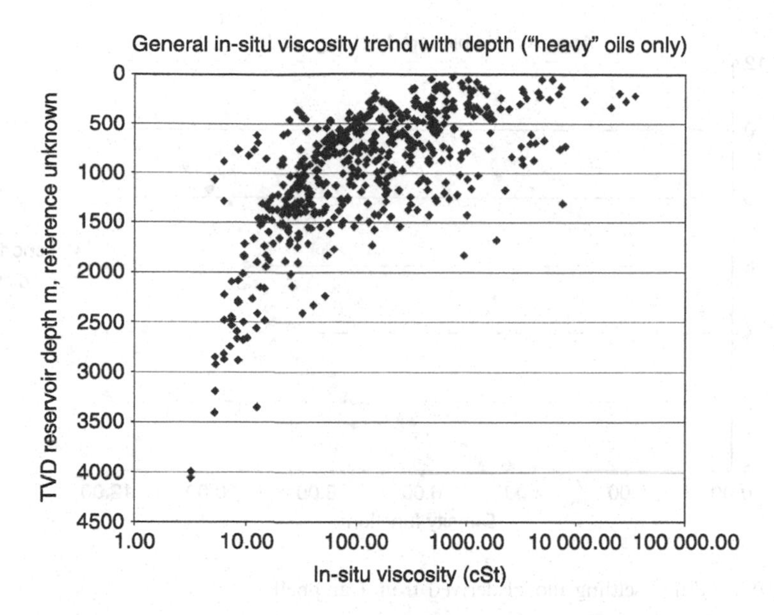

FIGURE 17.9 Consistent relationship for viscosity, illustrating the relationship between reservoir depth (and temperature) and *in situ* viscosity. (Source, US Department of Energy.)

TABLE 17.1
American Petroleum Industry Separation Time Guidelines

Oil Density (g/cc)	Residence Time (min)
<0.85	3
>0.85	5 to 10

in the Campbell gas processing series (Campbell, 1988), where the settling time is given as:

$$\alpha\mu_O/(\rho_w - \rho_O)/C(wc, \Delta_O) \quad (17.12)$$

where the function $C(wc, \Delta\rho_O)$ has a functionality similar to that of hindered droplet settling:

$$C(wc, \Delta\rho_O) = \left\{a + b\Big/\left[(\rho_w - \rho_O)\,\rho_O \times \left(1 - \frac{wc}{100}\right)^{1/4}\right]\right\}^4 \quad (17.13)$$

Figure 17.10 shows the relationship between density function and settling rate factor, as derived from Campbell.

17.2.5.2 The Shell Model

Shell has presented a new model for the design of separators (Polderman et al., 1997) which looks at the flux of the dispersed phase through an interface. The limiting flux can be established

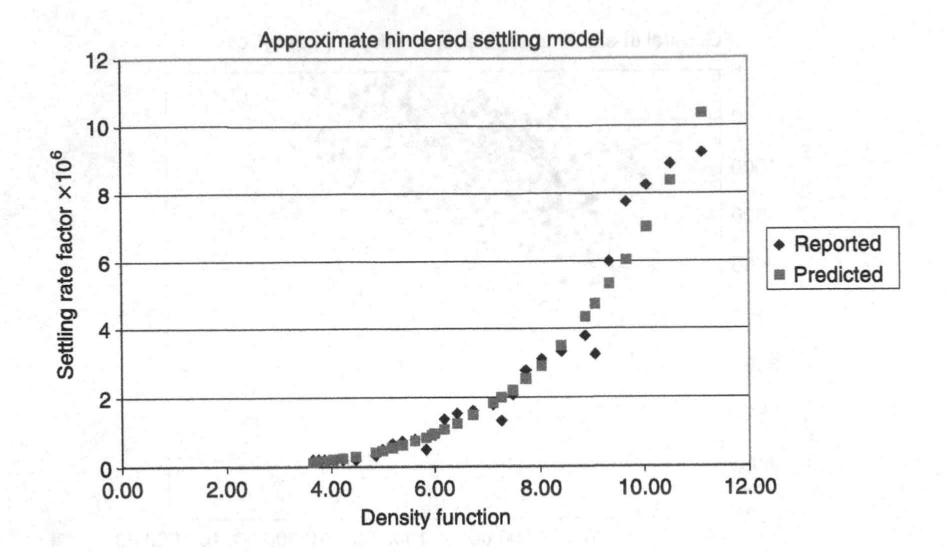

FIGURE 17.10 Modified settling model derived from Campbell.

by a batch test in a droplet characterization rig (DCR) or in the field.

$$\frac{Q_l}{A_{\text{interface}}} = \frac{h_d}{a + bh_d} \tag{17.14}$$

$$\frac{dh_d}{dt} = \frac{Q_l}{A_{\text{interface}}} \tag{17.15}$$

$$\frac{h_d}{dh_d/dt} = a + bh_d \tag{17.16}$$

where

Q_l = liquid flux of dispersed phase
h_d = height of emulsion band
$A_{\text{interface}}$ = horizontal cross-sectional area of the vessel
a and b = empirical constants

Generalized charts for separator loading can be calculated with this method as given by Polderman et al. in Figure 17.11. As can be seen, the proposed allowable flux span is 1:3, indicating the uncertainty of the method.

17.2.5.3 The Hartland Model

A recent model that has been shown to be very applicable to describing oilfield separation processes is that of Jeelani and Hartland. The settling velocity is given by:

$$v_0 = \frac{12\mu_C(1-\varepsilon_0)}{0.53\rho_C\varphi_0}\left[-1 + \sqrt{1 + \frac{0.53\rho_C\Delta_\rho g\varphi_0^3(1-\varepsilon_0)}{108\mu_C^2\left(1+4.56\varepsilon_0^{0.73}\right)}}\right] \tag{17.17}$$

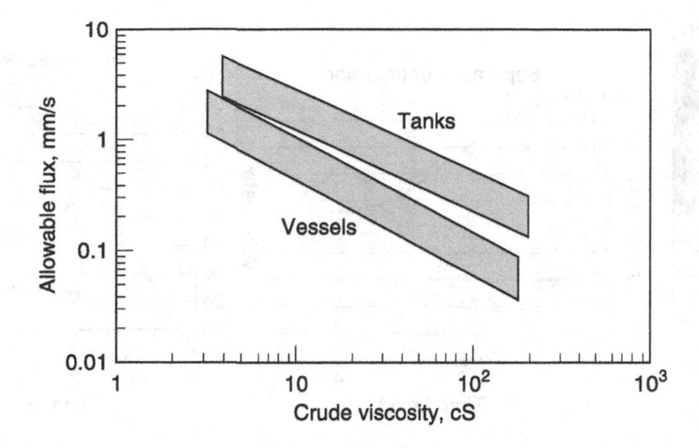

FIGURE 17.11 Dispersed phase flux chart. (From Polderman et al., 1997.)

The rate of coalescence, the drainage from the dense packed layer, can be found by:

$$\Psi_i = \frac{\varepsilon_P \left[2H_0 \left(1 - \varepsilon_0\right) - v_0 t_i\right]}{\left(1 - \varepsilon_P\right)\left(3\tau_0 + t_i\right)} \tag{17.18}$$

where

μ_C = viscosity of the continuous phase
ε_0 = the holdup fraction of the dispersed phase at $t = 0$
φ_0 = initial average droplet size
ρ_C = density of the continuous phase
$\Delta\rho$ = density difference between continuous and dispersed phase
g = gravitational force
ε_P = the holdup fraction of the dispersed phase in the dense packed layer at t_i
H_0 = total height of the emulsion at $t = 0$
τ_0 = coalescing time at $t = 0$

This model accounts for both settling and coalescence in the dense packed layer (see Figure 17.12). The nice feature of this model is the inclusion of emulsion stability due to various chemical components and mechanisms through the τ_0, coalescing time. Hindered settling is modeled based on a representative droplet diameter. These two parameters can be experimentally tuned in a DCR and the model has been shown to give good description of separator performance in simple flow conditions, i.e., with single well, no complex emulsions. Still, some work remains to link the chemical characterization data to the coalescence time in the model, i.e., to reliably predict the parameter based on fluid characterization.

Neither of the models presented above take into account droplet size distribution. The effect of the dispersion bed is also unclear as the majority of the models utilized the continuous-phase viscosity as the reference viscosity.

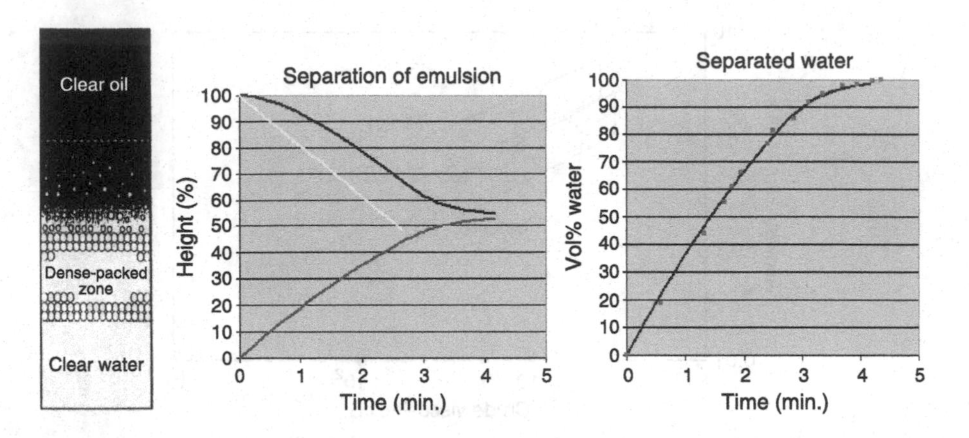

FIGURE 17.12 Hartland model versus experimental data.

17.2.5.4 The Hafskjold Model

Hafskjold et al. (1999) developed an empirical model for oil/water separation in pipes and tanks that accounts for droplet-size distribution. The model accounts for settling using a model for hindered settling that is dependent on the droplet diameter. Coalescence is calculated as the product between number of collisions and the probability of coalescence given a collision. Then these are summarized over all droplet classes. In general, the model by Hafskjold et al. is based on the following volume balance:

$$\frac{\Delta V_{i,n}}{\Delta t} = u_i \frac{\Delta V_{i,(n-1)\text{ or }(n+1)}}{\Delta z} - u_i \frac{\Delta V_{i,n}}{\Delta z} + \text{source} \tag{17.19}$$

where the left-hand side represents change in volume of drops of class i in control volume n during time Δt, the two first terms on the right-hand side represents the volume of drops of class i flowing in from control volume $n-1$ or $n+1$ and out of control volume n. The source term is given by:

$$\text{source} = v_i^c - v_i^a \tag{17.20}$$

where v_i^c is the volume of droplets of size class i created by coalescence, and v_i^a is the volume of droplets of size class i destroyed by coalescence.

There are five adjustable variables in this empirical model. These are related to the coalescence rate and the fluid. The model can be applied to analysis and improvement of existing separator processes, as well as design of new equipment.

17.3 EXPERIMENTAL VERIFICATION OF SEPARATOR PERFORMANCE

In order to reduce risk in separator design, experiments have been proposed in the following steps:

- Fluid characterization:
  1. American Petroleum Institute, reservoir dept. characterization

2. Bottle tests
3. DCR 1, simple
4. DCR 2, twin/triple (two to three wells)
5. DCR 3, high pressure; repeat steps 3 and 4

- Qualification/fluid characterization:
  6a. LP flow test with model oils
  6b. Flow test in HP/HT flow loop (Porsgrunn type)

The experimental steps are described below. Note that the combination of experiments and tests are used to scale up from sample to real conditions. The experimental work should also be combined with modeling.

17.3.1 Sampling, PVT, and Assay Study Results

It is essential that the oil sample be collected without air contact. To ensure realistic inlet conditions for the test separator detailed assessment of fluid flow and droplet size distribution must be carried out. Samples and studies designed to emulate multi-well flow conditions should be included to provide for a worst-case assessment. In this case, a phased approach should be considered. The crude oil assay contains important information for the evaluation. Particular concerns are:

- High total acid number or iodine values
- Problems with excess water
- High values for S, Vd, and Ni
- Asphaltene contents over 1%

(See also Section 17.4.)

17.3.2 Bottle Tests

Bottle tests are standard in order to find properties of the fluid. These are important for simple screening, but because of the simple methodology, the test process does not accurately emulate the production process. Droplet characterization rigs provide an improved model of the production process.

17.3.3 Droplet Characterization Rigs 1 to 3

Droplet characterization rigs are classified DCR1, 2, and 3. DCR1 is just an advanced bottle test that includes a certain approach to inlet shear. The fluids are pumped through a valve and into a container or PVT cell. After filling the cell, the change in free water level is monitored. The thickness of the emulsion layer may also be tracked with additional equipment.

The second type of droplet characterization rig, DCR2, is similar to DCR1 but also provides a certain insight into the effect of mixing wells by allowing for mixing of more than one emulsion stream before the mixture flows in the capture cell. DCR3 is the third type of rig. It is similar to DCR1 and DCR2, except that it operates at high pressure.

There are many low-pressure rigs (DCR1 and DCR2), and they are cheap to build and operate. On the other hand, fewer high-pressure units (DCR3) exist, and they are also expensive to build.

17.3.4 Dynamic Flow-Loop Tests

The best-known way to validate separator performance prior to installation is to carry out a scaled test with real fluids at real operating pressure and temperature. However, prior to such high-pressure tests, low-pressure tests should be carried out to optimize the design.

The purpose of a flow-loop test is to:

- Investigate the effect of free gas and released gas
- Investigate the dynamic behavior of the gravity separator
- Verify the effect of mixing wells with different water cuts/complex emulsions
- Verify the retention time and the geometry
- Verify the pressures and water cuts
- Investigate the impact of continuous sand flushing
- Investigate the impact of one selected demulsifier
- Investigate the impact of level settings
- Uncover unknown scale effects

17.4 DATA SOURCES

As is indicated in the preceding text, methods exist to allow the operators and vendors to design and deliver systems more appropriate to the field's fluid behavior. Typically, the supplier is often restricted by the information (or lack of it) which the operator supplies. To make a proper system design the vendor should be supplied with:

- A compositional analysis for all fluids, detailing:
    1. Trace elements
    2. Reservoir conditions
    3. Surface and stabilized fluid properties
    4. Formation water description including scaling potential
- A copy of the relevant reservoir fluid studies (including any special studies):
    1. Viscosity studies
    2. Differential liberation, miscibility, etc.
    3. Wax
    4. Asphaltene
- A copy of a relevant crude-oil assay data:
    1. Metals
    2. Sulfur
    3. Iodine number
    4. Acid number
    5. Viscosities
- Emulsion separation studies if performed

These give the vendor access to the majority of the fluid data required to meet the operator's needs for information about emulsion forming tendencies. Proper equipment sizing can then be based on "rules of thumb" and experience with the various stabilizing mechanisms for emulsions.

TABLE 17.2
Typical Data Provided to Process Supplier

Component	Mol %
Nitrogen	0.00
Carbon dioxide	45.07
Hydrogen sulfide	5.37
Methane	5.44
Ethane	0.98
Propane	2.85
i-Butane	1.24
n-Butane	1.80
Hexane+	36.41
Total	100.00
Pour point, °C	12
Wax content (wt%)	Not supplied
Density at 15 °C (kg/m^3)	862
Kinematic viscosity at 15 °C (cSt)	7.60
Sulfur (wt%)	Not supplied
Hydrogen sulfide (ppm wt)	Not supplied

TABLE 17.3
Results from a Reservoir Fluid Study

Pressure (MPa)	R_s(Sm^3/Sm^3)	B_o(m^3/Sm^3)	u_o(cP)
36	130	1.366	0.79
33	130	1.373	0.74
31	130	1.378	0.70
28	130	1.386	0.65
26.1*	130	1.391	0.62
23	114	1.349	0.67
21	104	1.321	0.71
19	93	1.294	0.76
16	78	1.254	0.84
14	68	1.227	0.91
11	53	1.186	1.05
8	36	1.145	1.24
6	26	1.118	1.40
3	10	1.078	1.76
1	0	1.050	2.54

Note: Asterix denotes the bubble point for the liquid.

The data presented in Tables 17.2 and 17.3 are normally available to the reservoir engineering department but not released to the process engineers. Indeed, the well test engineers are the first to properly test the fluids for any unusual behavior. In one relevant case (Friedemann, 1997) the crude assay samples were the only ones in which water was observed. A later review of the well-test pressure behavior indicated that the well tests produced at 50% water cuts. Because the observed salinity was at or near saturation, this form of stability represented a threat of significant price penalties from both the tanker owners as well as the refineries.

The data in Table 17.3 are of very great importance for sub-sea and downhole processing. This is because the information gives the system designers fluid property data at the correct operating condition. These data are also important for tuning the fluid property models of the process simulators.

17.5 ALTERNATIVE SEPARATOR DESIGN

Conventional gravity separators are devices designed to provide for phase disengagement by encouraging stratified flow. This is done by using large diameter devices with typical length to diameter ratios (L/D) of 3 to 5. The result is a device designed to optimize footprint, but not necessarily weight or installability.

From a multiphase flow viewpoint the long slim device gives advantages in that droplet rising/settling distances are much shorter. The challenge is then to hold the flow rates so that the system is within the stratified flow region. Once that flow regime is achieved, the liquid levels are a result of the same physics, which yields holdup.

This puts a challenge on the control system, as the sensitivity of the level sensing system will be important, as indicated in Section 17.6.2. The advantage is that long slim separators may offer more efficient separation volumes than traditional gravity separators. Therefore an investigation of the pros and cons for installation of long thin vessels should be performed.

17.6 CONTROL SYSTEMS

The challenges of remote and sub-sea field operations are made more difficult by a lack of direct access to operational data. Insufficient planning for problem-solving solutions often challenges traditional topside facilities. Typically, operating companies may be challenged with crash-course diagnostic activities to determine the causes of repeated equipment failures. Although the topside systems are easily accessible, the diagnostic activities are made even more difficult. This places an emphasis on control system design.

Traditionally, the system buffer for the level control system is the gas in the gravity separator. Two factors that influence the need and use for separator control are:

- Instability of the inlet flow (slugging)
- Quality of the separation

Both of these are strongly influenced by fluid property models and (in the case of oil–water systems) the rheology of production enduced emulsions.

17.6.1 Slugging

Gas and liquid flow can operate within different flow regimes (annular, slug, stratified, bubbly) depending on the flow rates. Slug flow has a special characteristic, as it is an unstable regime where liquid slugs are followed by gas pockets. The length (and resultant volumes) of hydrodynamic slugs can vary dramatically from a few to very many pipe diameters and can include internal gas as well. Gravity and cyclonic separators will have to handle such hydrodynamic slugs.

A more severe case can occur with so-called terrain slugs. These are created by liquid collecting in the lower parts of a flowline, and then transported through the flow line. Note that flow regime maps cannot predict this type of slug.

Many different methods exist for estimating slug lengths. These vary from the use of simple tubing diameter based methods and include complex dynamic simulations. It is the experience of the author that the latter method is to be preferred as it allows for the inclusion of nonsteady state analyses because the result of a start-up, shutdown, restart sequence can often give surprising effects on the slugging behavior of a system. The slug period of terrain slugs will vary from minutes to hours, while hydrodynamic slugs typically have a period of a few seconds.

The control system required for a specific field depends on the flow line profile and diameter, but also on the separator design. Smaller equipment or cyclones are more sensitive to slugs, and therefore more dependent on the control system.

As mentioned, to get terrain slugs, the terrain must be hilly with dips, and the flow line must have a certain length, much longer than the slug length. By placing the sub-sea separator next to the wellhead one will avoid terrain slugs. Furthermore, in a real application, a transient computer simulation should be performed in order to estimate the slug sizes.

17.6.2 Level Control

Recall the Hartland model presented in Section 17.2.5.3. The model predicts how the quality of the fluid, and the inter-phase levels, change with time. A level sensor should indicate the region of dense packed zone, but also the profile of the water cut.

The required accuracy on the water cut profile depends on specifications. There are several qualified level sensors to measure the oil–water interface inside a sub-sea separator. The two different level monitoring systems installed in Troll Pilot are also capable of giving the concentration profile. However, these profilers can not reliably distinguish 1% oil-in-water from 0% oil-in-water, or 1% water-in-oil from 0% water-in-oil. Nevertheless, accounting for knowledge of the fluids and their separation behavior compared to the level profile readouts, likely oil and water qualities can be deduced.

A level sensor operating in a liquid-filled vessel with a sensitivity of 10 cm will give an acceptable resolution when placed in a 3 m diameter tank. The same device mounted in a tank of 1 m diameter will need a higher resolution.

17.7 CONCLUSIONS

Sub-sea separation calls for highly reliable separator equipment. To ensure the best possible performance for the entire life of a field, a detailed characterization of the oil–water emulsion should be carried out at proper pressure and temperature. Chemicals to be used should be tested with the oil in the same manner.

Traditional design rules for topside separators are based on over-simplified models, and are not sufficient for sub-sea use, while the more comprehensive approach (see Section 17.2.5.3) gives a better understanding and a more correct separator design. However, there is still a need to establish the missing link between pressurized test cell behavior (laboratory scale) and large-scale dynamic behavior.

The complex emulsions formed by commingling different well streams with different water cuts are hard to predict in terms of separability. To ensure no disastrous effects of mixing well streams, a proper high pressure test should be carried out to explore the separator efficiency under the most unfavorable conditions. In particular, the inlet devices should be carefully examined to ensure efficient performance throughout the field lifetime.

REFERENCES

Arntzen, R. and Andresen, P.A.K. (2001). Three-phase wellstream gravity separation. In: *Encyclopedic Handbook of Emulsion Technology*, J. Sjøblom (ed.). New York: Marcel Dekker.

Aske, N., Kallevik, H. and Sjöblom, J. (2001). Determination of saturate, aromatic, resin and asphaltenic (SARA) components in crude oils by means of infrared and near infrared spectroscopy. *Energy Fuels* **15**(5), 1304–1312.

Brown, A.H. (1968). *Chem. Ind.* **30**, 990.

Buenrostro-Gonzales, E., Groenzin, H., Lira-Galeana, C. and Mullins, O.C. (2001). *Energy & Fuels* **15**, 972–978.

Calsep, PVT. (2005) Sim technical reference.

Campbell. (2003) Gas Processing Series, Vol II.

de Vries, A.J. (1958). Part IV: Kinetics and activation. Energy of film rupture. *Recl. Trav. Chim.* **77**, 383, 441.

Frank, S.P. and Mysels, K.J. (1960). On the 'dimpling' during the approach of two surfaces. *J. Phys. Chem.* **66**, 190.

Friedemann, J.D. (1997). Guest Lecture Texas A&M and New Mexico Institute of Technology. Depts of Petroleum Engineering, April 1997.

Hafskjold, B. et al. (1999). A new mathematical model for oil/water separation in pipes and tanks. *SPE Prod. & Facilities* **14**(1), 30–36.

Hartman, M. et al. (1994). Free settling of non-spherical particles. *IEChem. Res.* **33**(8).

Jeelani, S. A. K. and Hartland, S. (1998). Effect of dispersion properties on the separation of batch liquid–liquid dispersions. *Ind. Eng. Chem. Res.* **37**, 547–554.

Krawczyk, M.A. et al. (1991). Chemical demulsification of petroleum emulsions using oil-soluble demulsifiers. *IEChem. Res.* **30**(2).

Lang, S.B. (1962). A hydrodynamic mechanism for the coalescence of liquid drops. PhD thesis, University of California.

Liem, A.J.S. and Woods, D.R. (1974). Review of coalescence phenomena. *AIChE Symp. Ser.* **70**(8).

Lin, C.Y. and Slattery, J.C. (1982). *AIChE J.* **28**(5), 786.

Malhorta, A.K. (1984). PhD thesis, Illinois Institute of Technology, Chicago.

Osa, K. and Holt, Ø. (2001). *Testing of Emulsion Breakers on Live Crude using MTU*. Oilfield Chemistry Symposium, April 2001.

Pal, R. (1993). Flow of oil-in-water emulsions through orifice and venturi meters. *IEChem. Res.* **32**(4).

Polderman, H.G. et al. (1997). *Design Rules for Dehydration Tanks and Separator Vessels*. SPE 38816. SPE Annual Technical Conference.

Reid, Prausnitz and Sherwood. (1988). *Properties of Gases and Liquid*.

Schedluko, A. and Manner, E. (1968). Critical thickness of rupture of chlorbenzene and aniline films. *Trans. Faraday Soc.* **64**, 1123.

Sheu, E.Y. and Mullins, O.C. (1995). *Asphaltenes: Fundamentals and Applications*. New York: Plenum Press.

Skelland, A.H.P. and Slaymaker, E.A. (1992). Effects of surface-active agents on droplet sizes in liquid–liquid systems. *IEChem. Res.* **29**(3).

Urdahl, O. et al. (2001). Compact electrostatic coalescer technology. In: *Encyclopedic Handbook of Emulsion Technology*, J. Sjøblom (ed.). New York: Marcel Dekker.

van der Zande, M.J. and vand den Broek, W.M.G.T. (1998). *Emulsification of Production Fluids in the Choke Valve*. SPE 49173. SPE Annual Technical Conference.

Vrij, A. (1966). Possible mechanism for the spontaneous rupture of thin, free liquid films. *Disc. Faraday Soc.* **42**, 23.

Wardaugh, L.T. and Boger, D.V. (1991). Flow characteristics of waxy crude oils: application to pipeline design. *AIChem. J.* **37**(6).

Zapryanov, Z., Malhorta, A.K., Aderangi, N. and Wasan, D.T. (1983). An analysis of the effects of bulk and interfacial properties on film mobility and drainage rate. *Int. J. Multiphase Flow* **9**(2), 105.

FURTHER READING

Ali, M.A. and Nofal, W.A. (1994). *Fuel Sci, Technol. Intern.* **12**(1), 21–33.

Andersen, S.I. and Speight, J.G. (2001). *Petroleum Sci. Technol.* **19**(1&2), 1–34.

ASTM (1991). **D 4124-91**.

ASTM (1993). **D 2007-93**.

Bollet, C., Escalier, J.-C., Souteyrand, C., Caude, M. and Rosset, R. (1981). *J. Chromatog.* **206**, 289–300.

Chatzi, E.G. et al. (1989). Generalized model for prediction of steady-state droplet size distributions in batch stirred vessels. *IEChem. Res.* **28**(11).

Dark, W.A. (1982). *J. Liq. Chromatog.* **5**(9), 1645–1652.

Gramme, P. et al. (1999). *MTU – The Mobile Test Unit for Investigating Offshore Separation Problems and Optimizing the Gas/OilWater/Separation Process.* SPE 56847. SPE Annual Technical Conference, Houston, Texas.

Grizzle, P.L. and Sablotny, D.M. (1986). *Anal. Chem.* **58**, 2389–2396.

Groenzin, H. and Mullins, O.C. (2000). *Energy & Fuels* **14**(3), 677–684.

Leontaritis, K.J. (1997). In: *SPE International Symposium on Oilfield Chemistry,* Houston, Texas. pp. 421–440.

Lundanes, E. and Greibrokk, T. (1994). *J. High Resolution Chromatog.* **17**, 197–202.

Radke, M., Willsch, H. and Welte, D.H. (1980). *J. Am. Chem. Soc.* **52**(3), 406–411.

Reynolds, A. (1886). On the theory of lubrication and its application to Mr. Beauchamps Tower's experiments. *Phil. Trans. R. Soc.* **A177**. 157.

Sheu, E.Y. (2002). *Energy & Fuels* **16**(1), 74–82.

Sonntag, H. and Strenge, K. (1969). *Coagulation and Stability of Disperse Systems.* New York: Halstead-Wiley.

Speight, J.G. (1999). *The Chemistry and Technology of Petroleum.* New York: Marcel Dekker.

Sterling, C.V. and Scriven, L.E. (1959). *AIChem. J.* **5**, 514.

Suatoni, J.C. and Swab, R.E. (1975). *J. Chromatographic Sci.* **13**, 361–366.

Wang, J. (2000). *Predicting Asphaltene Flocculation in Crude Oils.* New Mexico Institute of Mining & Technology, Socorro, New Mexico.

18 Droplet Size Distributions of Oil-in-Water Emulsions under High Pressures by Video Microscopy

Pål V. Hemmingsen, Inge H. Auflem, Øystein Sæther, and Arild Westvik

CONTENTS

18.1 INTRODUCTION

In crude oil production, gas, oil, and co-produced water are mixed due to pressure drop and high shear forces as the fluid passes the wellhead and various choke valves from the reservoir to the separation facilities. The primary task at the oil producing facility is to separate gas from liquid and oil from water. The oil and water mixture is typically processed through a separation train of two or three separators, where water is removed. The produced water contains dispersed oil droplets, which need to be removed in order to meet environmentally based, governmental demands on water quality before discharge to sea. The produced water is therefore cleaned through additional separators, cyclones, etc. before it is disposed to sea or re-injected into the reservoir.

One of the major factors influencing performance in both hydrocyclones and traditional flotation cells/plate separators is the droplet size distribution of the dispersed phase. The separation of smaller droplets is slower and more difficult. In the design and evaluation of oil–water separators in the oil industry, information concerning the sizes of droplet is critical [1]. Droplet size is an important parameter in equations describing droplet movement, such as Stokes' law for gravity induced sedimentation/creaming [2], the Stokes–Einstein equation for diffusion [3], and in equations describing sedimentation and coalescence profiles [4–6].

Samples of the produced water stream are often analyzed in a particle/droplet analyzer at the oil production facility. Such particle/droplet analyzers operate at atmospheric conditions. As the pressure of the sampled water is reduced to atmospheric pressure, dissolved gas in both the oil droplets and water may cause coalescence or re-agitation of the dispersed droplets and thereby influence the droplet size distribution. Also time will be important as the oil-in-water emulsion will not be a stable system, but will continue to separate during the time from the water sampling to the time of analysis. All these factors imply that the best way to measure the droplet size distribution is to measure it *in situ* under real conditions.

At Statoil R&D Center, development of laboratory scale high pressure separation equipment makes it possible to mix water and oil under realistic process conditions [7–9]. Connecting a video microscopy system [10] to a high pressure separation rig makes it possible to investigate the droplet distribution of emulsions at high pressures [11].

Many different techniques are used for droplet and particle size analysis: optical microscopy (photomicrography) [12,13], video microscopy [10,14–29], light scattering [30–32], electrozone sensing [1,33–35], photon correlation spectroscopy [30,36,37], turbidimetry [38–41], pulse field gradient nuclear magnetic resonance (PFG-NMR) [42–47], and ultrasonic spectroscopy [48]. Light scattering, the electrical zone sensing method, and photon correlation spectroscopy require dilution of the samples.

The different methods have different size regions which they can measure [49,50]. Not all the methods are easy to implement for measuring at high pressures. Some of the methods also need longer time for the measurement, like the PFG-NMR, which takes about 15 to 30 min, although recent advances make it possible to perform faster analyses [44]. Many of the techniques for measuring particle sizes are typically based on various forms of light scattering. The results from the light scattering techniques have to be fitted to a scattering function, where in some cases one has to assume the particle shape and also the type of size distribution (Gaussian, log–normal, etc.). These methods do not discriminate between droplets and particles.

Although optical microscopy has a long history in the determination of particle and droplet size, there has been an increase in the use of image analysis for particle and shape analysis, due to the emergence of new imaging systems and improved software. Video microscopy is a technique that combines the magnification power of a microscope with the image-acquisition capability of a video camera. The resulting data matrix, from which information about the sample can be extracted, is an image or a series of images. Current image-analysis software provides a wide range of analytical capabilities for acquiring, enhancing, and analyzing images. Video microscopy is a direct method that measures the actual size of the discrete objects. This means that multimodal size distributions will be detected properly. It also gives the shape of objects, whether or not you have multiple emulsions, and if the sample is flocculated. All the above parameters are central to the understanding of emulsion behavior and emulsion stability. Another advantage with optical microscopy, which should not be overlooked, is the fact that, as a visual image is obtained, experimental defects such as dirt covering the flow cell windows or entrapped gas bubbles can easily be detected.

18.1.1 Limitations in Video Microscopy

First, the sample must have certain optical properties, since the technique relies on reflection, refraction, scattering, and absorption of visible light. This means that the sample must be transparent and that the continuous liquid and the droplets must have different refractive indices or different colors, i.e., properties that make them optically distinguishable. Second, the resolution limit, and hence the operational size domain, is governed by the wavelength of the light. This feature is known as the Rayleigh limit and results in a physical limit of about half the illumination wavelength, i.e., 0.2 μm in the visible area.

The Rayleigh criterion [51] states that:

$$R = \frac{0.61\lambda}{\text{NA}} \tag{18.1}$$

where λ is the wavelength of the light source, and NA is the numerical aperture. The numerical aperture of a microscope objective is a measure of its ability to gather light and resolve fine specimen detail at a fixed object distance. The practical limit tends to be slightly higher, 1 to 2 μm, due to rapidly increasing measurement errors with decreasing object dimensions. This is caused by diffraction: the image of an object is actually a diffraction pattern, and the overlapping patterns of closely spaced objects result in image blurring. When it comes to magnification, there is no theoretical upper limit, although increasing the magnification only renders larger, blurred images of the objects. Still, innovations in the optics have proved that the diffraction-imposed barrier is not absolute, and may be an important tool in the near future of microscopy.

For size measurements there is also a limit of how concentrated the emulsion can be, due to overlap of the droplets. A maximum droplet concentration of 1 vol% is often referred to. This is in the order of 10,000 ppm, which, in the oil production, is considered as highly polluted water, and which by no means could be disposed of without thorough cleaning. Normally, at the outlet of the first stage production separator, the amount of oil in the water varies from above 1000 ppm down to 100 ppm depending on the field conditions and the stability of the oil-in-water emulsion. The separated water then needs to be cleaned further until it meets governmental demands before it is disposed of. In the Norwegian sector of the North Sea, the maximum allowed oil content in produced water disposed to the sea is 40 ppm (in 2004), but it is proposed that the maximum is reduced to 30 ppm from the end of 2005. In this region, video microscopy should be ideal for measuring the oil droplets, with no need for dilution. Video microscopy attached to a high-pressure flow cell could then be used for online monitoring of the quality of the produced water at various stages in the water treatment process.

In this chapter we present and discuss a video microscopy system for measuring the droplet size distributions in separated water from mixed crude oil/water emulsions at realistic process conditions, i.e., at elevated temperatures and pressures.

18.2 THEORY

18.2.1 Droplet Size Distribution

Due to the high industrial importance of measuring sizes of particles as part of the manufacturing of powders, most of the theory of size measurements is based on the measuring of particles and not of droplets. However, what applies to particles with regard to size measurements and size distributions will also apply to droplets. In many cases, the term "diameter" would be well representative of the size of spherical droplets, while particles tend to be more irregular in shape and thus the meaning of "diameter" may be interpreted differently. However, although single droplets dispersed in a diluted emulsion would be spherical, there might be some degree of flocculation, which would give clusters of droplets with irregular shape and different diameters depending on what direction it is measured.

The reporting of droplet size and droplet size distribution can be done in various forms and with various techniques [52] and it is often related to the property (number, surface area, or volume) important for the given emulsion. If the different amount of the particles/droplets is important, then the number mean diameter and number distribution should be used. This might be used in a process where one of the quality parameters is that the size of the particles should be between 10 and 20 μm, for example.

In other cases, like in the investigation of a catalyst or the amount of surfactants at the droplet interface in an emulsion, parameters describing the surface area are important. In produced water in the oil industry, the total amount of the oil in the water is important, so one should relate the diameters to the mass or volume of the dispersed oil droplets.

18.2.2 Average Diameters

The purpose of an average is to represent a group of individual values in a simple and concise manner in order to obtain an understanding of the group. It is therefore important that the average should be representative of the group. All averages are a measure of central tendency. The mean, $\overline{D}$, describes the center of gravity of the distribution [53]:

$$\overline{D} = \frac{\sum D_i \mathrm{d}\phi}{\sum \mathrm{d}\phi} \tag{18.2}$$

where ϕ is a given property.

For a number distribution $\mathrm{d}\phi = \mathrm{d}N$, giving the mean diameter as:

$$D\,(1,0) = \frac{\sum D_i \mathrm{d}N}{\sum \mathrm{d}N} \tag{18.3}$$

For a mass or volume distribution $\mathrm{d}\phi = \mathrm{dV} = D_i^3 \mathrm{d}N$, giving the volume moment mean diameter:

$$D\,(4,3) = \frac{\sum D_i^4 \mathrm{d}N}{\sum D_i^3 \mathrm{d}N} \tag{18.4}$$

18.2.3 Droplet Shape

For particles, the shape is a fundamental powder property affecting powder packing and properties like bulk density, porosity, and so on. For image analysis of droplets, the shape parameters can be used to discriminate objects that differ too much compared to a spherical shape. It can also be used to investigate the degree of flocculation, as flocculated clusters of droplets will have a different shape to that of a spherical droplet. The shape of a droplet can be described in many ways; the three used in this work are roundness, aspect, and radius ratio. The roundness is given by:

$$\text{roundness} = \frac{P^2}{4\pi A} \tag{18.5}$$

where P is the perimeter of the droplet, and A its area. The aspect shape factor reports the ratio between the major and the minor axis of the ellipse equivalent of the object (i.e., an ellipse with the same area). The radius ratio shape factor describes the ratio between the maximum and minimum radius of each object. For a perfectly round sphere, all these parameters are equal to 1.

Combining these shape parameters, we can define an overall droplet shape factor, SF:

$$\text{SF} = \sqrt{\frac{\text{roundness}^2 + \text{aspect}^2 + \text{radius ratio}^2}{3}} \tag{18.6}$$

For a perfect sphere, the droplet shape factor defined in Equation 18.6 is equal to 1.

18.3 EXPERIMENTAL

18.3.1 Sample Preparation

Live crude oil samples and produced water were received from different North Sea oil fields. The crude oil and produced water samples were typically sampled from the outlet of the inlet production separator. The fluids were kept under pressure from field to laboratory. In order to simulate the separation in the inlet production separator, the fluids should meet the temperature and pressure conditions at the choke valve before the inlet production separator. In order to meet these fluid conditions, the crude oil samples were first heated to the elevated temperature. Any free water left in the sample was removed and the crude oil was recombined with a natural gas mixture to obtain the pressure condition at the oil field prior to the inlet production separator. The produced water samples were only heated to the experimental temperature, but not recombined with the natural gas mixture, as water has very limited solubility of natural gases.

18.3.2 High-Pressure Separation Rig

The separation rig was used to mix water and oil at elevated temperature and pressure and to monitor the subsequent separation of the phases. Figure 18.1 shows an outline of the whole rig.

The principle of the rig is that water and oil meet at elevated pressure in a pipeline and flow through a choke valve into the separation cell. The rig has three choke valves (P10, P12, and P13), giving the possibility to mix the oil and water far from the separator, in order to achieve coalescence effects in the pipeline to the separator, or to mix oil and water close to the separator. It is also possible to study the mixing of two different fluid systems by mixing each system on both sides (using choke valves P10 and P13) and then combine them at choke valve P12. The low-pressure side of choke valve P12 equals the separator pressure. The pressure drop over the choke valves provides the shear forces necessary for the mixing of water and oil and depending on the water cut one obtains a water-in-oil or oil-in-water emulsion. At the same time the light end of the oil undergoes a phase transition from liquid to gas. The gas evolved may cause foam and may influence the sedimentation and coalescence of the dispersed droplets. The separation cell is filled with a constant liquid flow for a given time. The amount of the different phases (foam, oil, emulsion layer, and water) is recorded as a function of time.

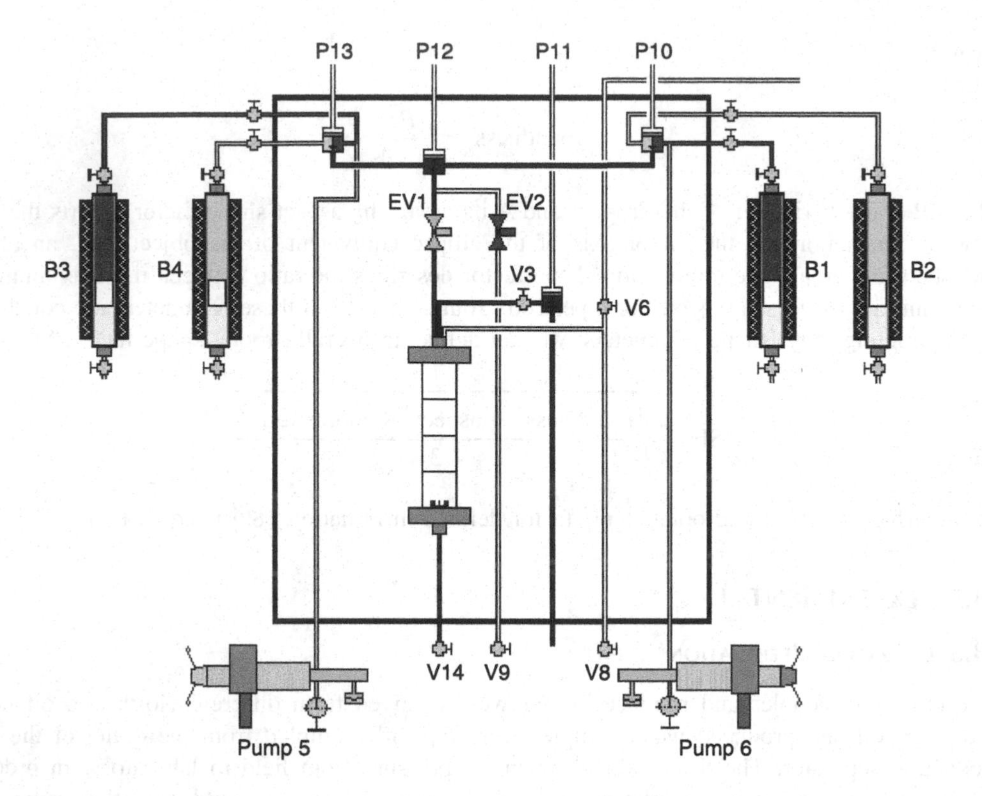

FIGURE 18.1 Schematic lay-out of high pressure separation rig.

The rig consists of four 600-cm^3 high-pressure sample cylinders, marked B1 to B4 in Figure 18.1. A selection of cylinders is filled with water or oil; with the aid of up to four independent motor driven high capacity piston pumps, the water and oils were pumped and mixed through choke valves. The pressure drop through the choke valves is backpressure controlled. The choke valve P12 is connected to the separation cell by a 50 cm 1/4" tube. The pressure in the separation cell is regulated by a backpressure-controlled valve (P11 in Figure 18.1). The separation cell is made of sapphire, assuring full visibility of the separation process. The maximum pressure in the cell is 200 bars. The cell is pressurized with natural gas to the experimental pressure before filling of the cell. Two pumps (5 and 6) were available for on-line chemicals injection. In the bottom of the cell there is a stirrer that is used at low rpm in order to give a small shear to the system during separation.

18.3.3 High-Pressure Video Microscopy

The video microscopy equipment is shown schematically in Figure 18.2. The system consists of a monochrome CCD camera (Pulnix TM-1300) with a Nikon 10X micro objective. A stroboscope (Polytec) with adjustable light intensity was used as light source. The stroboscope works by sending a brief, intense pulse of electric current through an inert gas, in this case xenon, which then emits a brilliant burst of white light. The stroboscope and the camera are synchronized so that while the camera is acquiring an image, the stroboscope emits a very short flash of bright

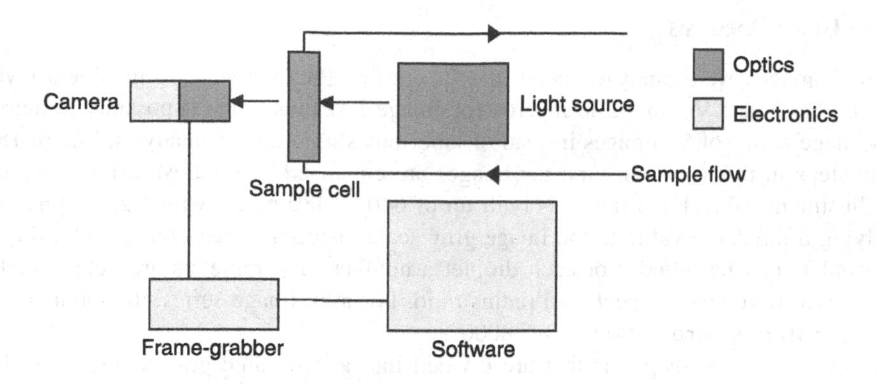

FIGURE 18.2 Schematic illustration of the high pressure video microscopy set-up.

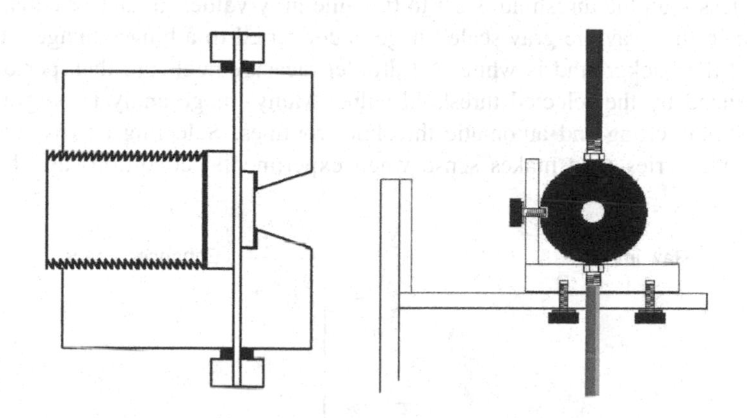

FIGURE 18.3 Sketch of high pressure cell and attachment.

light through the sample cell and the flowing emulsion. The effect on the grabbed images is that the droplets seem "frozen" in its movement.

The high pressure flow cell shown in Figure 18.3 was connected to the bottom outlet of the high pressure separation cell. The high pressure cell consists of a sapphire window in front and Plexiglas window in the rear. The cell was tested to withstand a pressure of 650 bars. A needle valve was placed after the high pressure flow cell, ensuring that the pressure in the flow cell was approximately the same as in the separation cell. Keeping the pressure constant in the separation cell and the flow cell is important as any release of dissolved gas from the dispersed oil droplets (due to pressure drop) would influence the oil droplet distribution. It would also be hard to distinguish between oil droplets and gas bubbles.

After a given separation time, the needle valve was opened and the separated water phase flowed through the video microscopy flow cell at a rate of approximately 1 to 2 ml/min. During this time an image series of 50 images was acquired. The video camera and the stroboscope were connected to a frame grabber PC card (Matrix Vision SDIG). The Image-Pro Plus version 4.2 (Media Cybernetics) software was used both for acquiring and analysis of the images. In order to obtain the correct droplet sizes, the image system was spatially calibrated using a graticule.

18.3.4 Image Analysis

The digital images were analyzed using the Image-Pro Plus software from Media Cybernetics, version 4.2. In-house Visual Basic macros for Image-Pro Plus made it possible to automatically process image series of 50 images instead of laborious single images analysis. Figure 18.4 shows the main steps in this analysis. First the images are enhanced by contrast, brightness, and sharpening adjustments [54]. Each image is built up of 640 × 480 pixels with a 256-value gray scale. By applying a threshold value to the image gray scale histogram (see Figure 18.5), the boundary of the droplets can be found. For each droplet a number of parameters are determined, like the diameter, area, roundness, aspect, and radius ratio. For a 50 image series, the number of droplets found varied from around 2000 to over 8000.

The images consist of pixels that are divided into a 256-value gray scale, where 0 is black and 255 is white. The droplets are darker than the background due to higher absorbance of light by the oil, and due to light scattering. In order to distinguish the droplets from the background, a threshold value is set on the basis of the gray scale pixel value characteristic for the droplets. Gray values less than the threshold is set to 0, while gray values greater or equal to the threshold is set to 255. In this way the gray scale image is converted to a binary image where the droplets are black and the background is white. All droplet measurements are then performed, but are of course influenced by the selected threshold value. Many image analysis programs include both manual threshold setting and automatic threshold routines. Selecting an overall threshold value for every image series only makes sense when experimental conditions are the same, like the

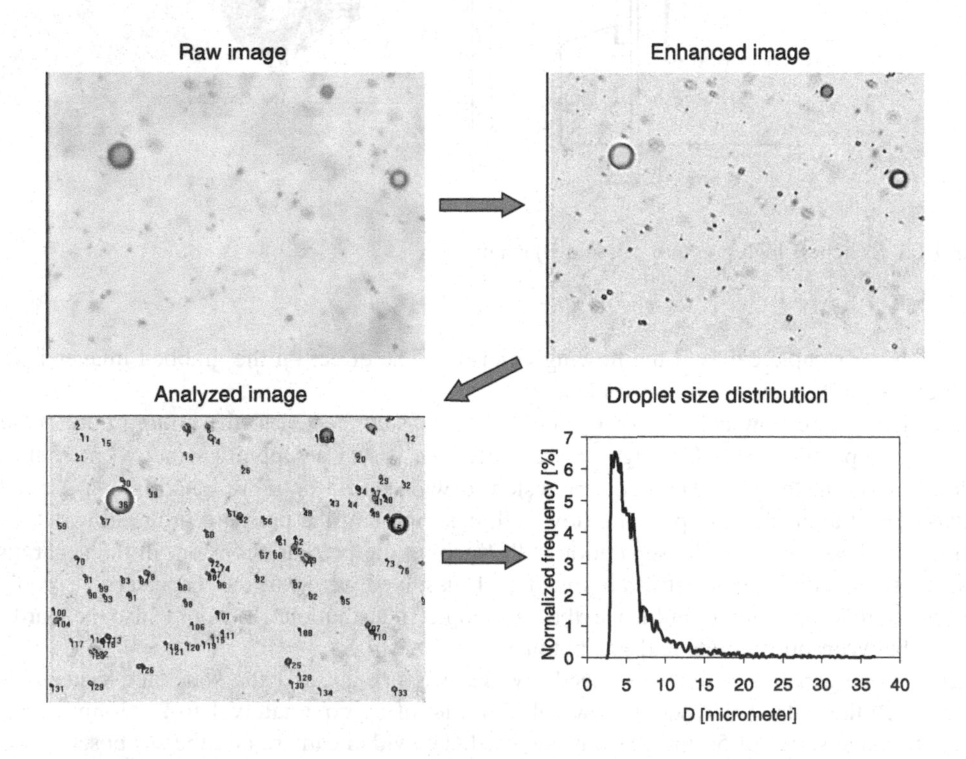

FIGURE 18.4 Image analysis. The details in the raw image are first enhanced, and then the boundary of the droplets is found by applying a threshold to the image gray scale histogram. The droplets are counted and droplet diameters are converted to droplet size distributions.

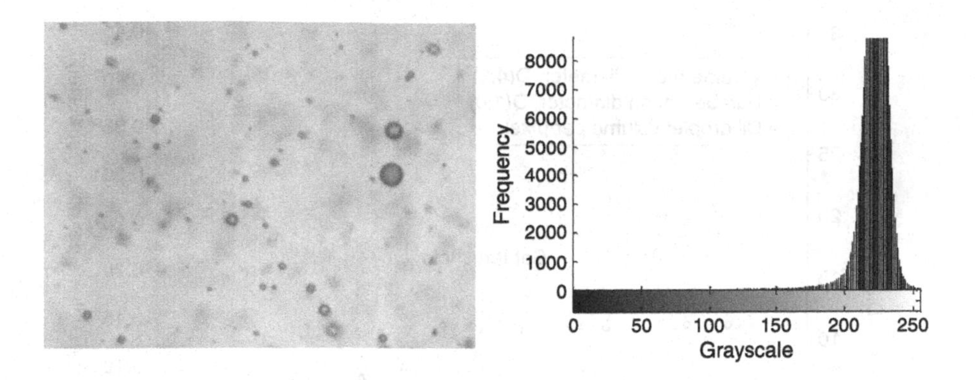

FIGURE 18.5 Image and its corresponding gray scale histogram.

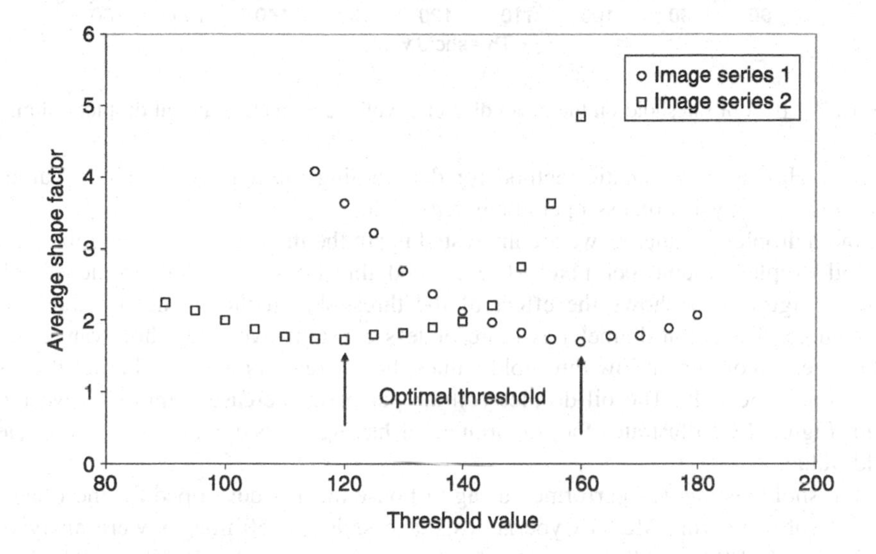

FIGURE 18.6 The average shape factor defined in Equation 18.6 as a function of the threshold for two image series. The two image series have different experimental settings for the intensity of the light source.

light source intensity, the focusing of the optics, and the overall optical density of the sample. The optical density will definitely change during the experiments as an oil-in-water emulsion containing many large oil droplets will absorb more light (and therefore the images will be darker) than an emulsion with few small droplets. It will also be difficult to obtain the same light intensity and focusing of optics during extended periods of time. All these concerns force one to have a method for setting the threshold.

One way to perform a test on the threshold value is to test how the average shape factor (defined in Equation 18.6) for all the droplets changes as one changes the threshold (see Figure 18.6). A threshold that is too high will add part of the background to the droplets and the average shape factor will increase. A threshold that is too low will only include a part of the droplets, and likewise the shape factor will increase. The optimal threshold value should then be the threshold value giving the global minimum for the average shape factor, as indicated in

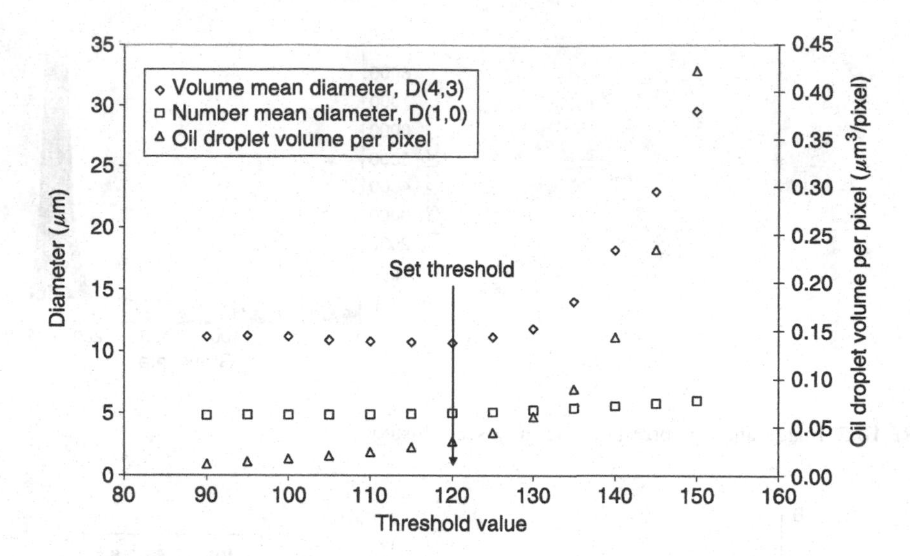

FIGURE 18.7 Effect of threshold on the mean diameter, volume diameter, and oil droplet volume per pixel.

Figure 18.6. Having a systematic method for determining the applied threshold value will also make the image analysis process operator independent.

The main droplet parameters we are interested in are the mean diameters (number and volume) and the oil droplet volume per pixel. The applied threshold value will influence all of these parameters. Figure 18.7 shows the effect of the threshold on the diameter values and the oil droplet volume. The mean diameter is more or less constant over the whole range; the volume mean diameter is constant at low threshold values, but increases rapidly at higher thresholds than the set optimal threshold. The oil droplet volume per pixel increases rapidly as we increase the threshold. Figure 18.7 illustrates the importance of having a systematic method for selecting the threshold value.

The threshold testing was performed using in-house macros developed for the Image-Pro Plus version 4.2 software from Media Cybernetics. Each series of 50 images were analyzed starting from a threshold of 90 to 180, increasing the threshold stepwise by 5. All the calculated droplet parameters were dumped to Excel (Microsoft) and the optimum threshold value was selected for a given image series. After finding the optimum threshold value, the different droplet sizes and volume were compared based on the experimental conditions in the high-pressure separation rig. In order to discriminate between spherical droplets and irregular shaped particles, all objects having a shape factor above 2.3 were rejected from the count. On average about 10% of the objects were removed in this process. The oil droplet volume per pixel is found from the summation of the volume of all droplets divided by the image resolution (640×480) and the number of images in each image series.

18.4 RESULTS

In the high pressure separation rig, oil and water are mixed through a choke valve and into the separation cell. The subsequent emulsion then starts to separate as oil droplets coalesce and cream upwards (for water continuous emulsions). Figure 18.8 shows a snapshot of the separation cell after a given separation time. Observing the height of the different phases (water, emulsion layer, oil, and foam) in the separation cell, a plot of the separation rate can be drawn (see Figure 18.9).

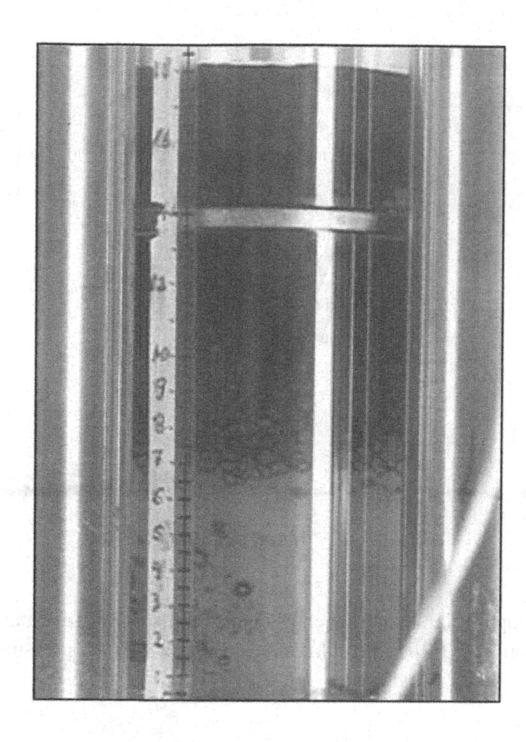

FIGURE 18.8 High pressure separation cell.

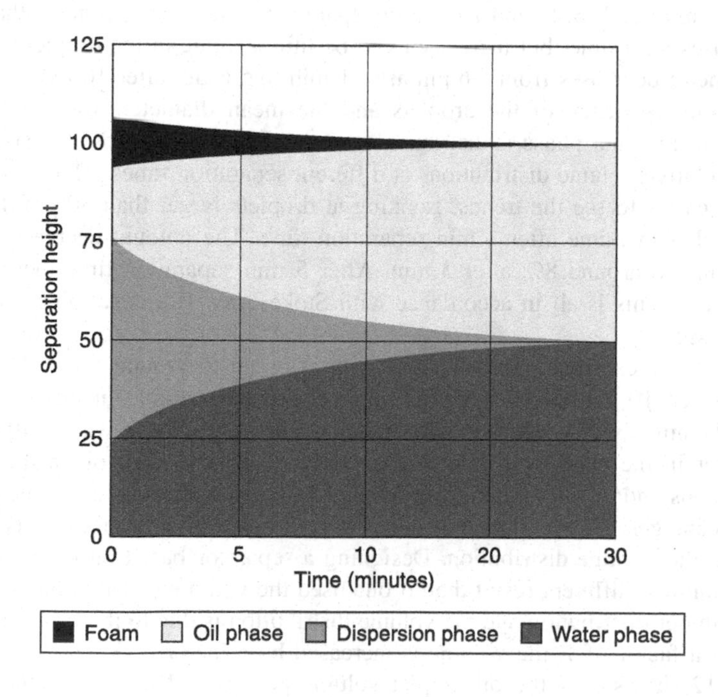

FIGURE 18.9 Separation curves.

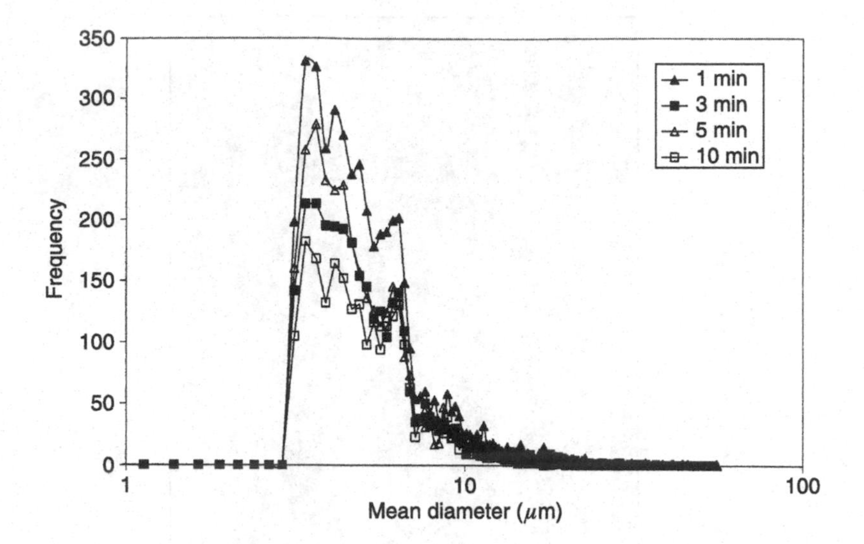

FIGURE 18.10 Mean droplet diameter frequency distributions of oil-in-water emulsions for varying retention time in the separation cell. Experimental conditions: T = 80 °C, pressure before choke valve: 65 bar, pressure drop over choke valve: 5 bar, water cut: 80%.

Both the amount of oil droplets and the droplet size distribution will change as the separation evolves.

Figure 18.10 shows the number distributions of the dispersed oil droplets in the water phase for a given system after 1, 3, 5, and 10 min of separation. The general trend is that the amount of droplets decreases with time, but there seems to be little change in the droplet size distributions. The mean diameter decreases from 6.6 μm after 1 min to 5.6 μm after 10 min. However, instead of looking at the frequency of the droplets and the mean diameter, we can look at how the volume of the oil-in-water phase is distributed across the different oil droplet sizes. Figure 18.11 shows the cumulative volume distributions at different separation times. The effect is much larger and it is much easier to see the trends. Looking at droplets larger than 30 μm, they account for 30% of the total oil volume after 1 min separation time. The amount decreases to around 12% after 1.5 min and to around 8% after 3 min. After 5 min separation time there are no droplets larger than 30 μm. This is all in accordance with Stokes' law [2], describing sedimentation and creaming processes.

Looking at the mean diameters, droplets with sizes up to around 6 μm do not account for more than between 10 and 20 vol% of the total oil droplet volume. In oil processing the main goal is to get the amount of water in export oil and oil in the produced water within certain limits. Too much water in the export oil reduces the price, while too much oil in the produced water implies regulations and/or fines from governmental authorities. In both cases the amount is based on weight percentages. This means that the correct droplet size distribution is not the number distribution but the volume distribution. Designing a separator based on the number distribution would lead to a much different result than if one used the volume distribution. The big difference between the number distribution and the volume distribution is due to the fact that as the diameter is increased by a factor of 1, the volume is increased by a factor of 3.

Figure 18.12 shows how the oil droplet volume per pixel decreases as the separation time increases. The change is largest in the beginning; the oil droplet volume is halved from 1 min

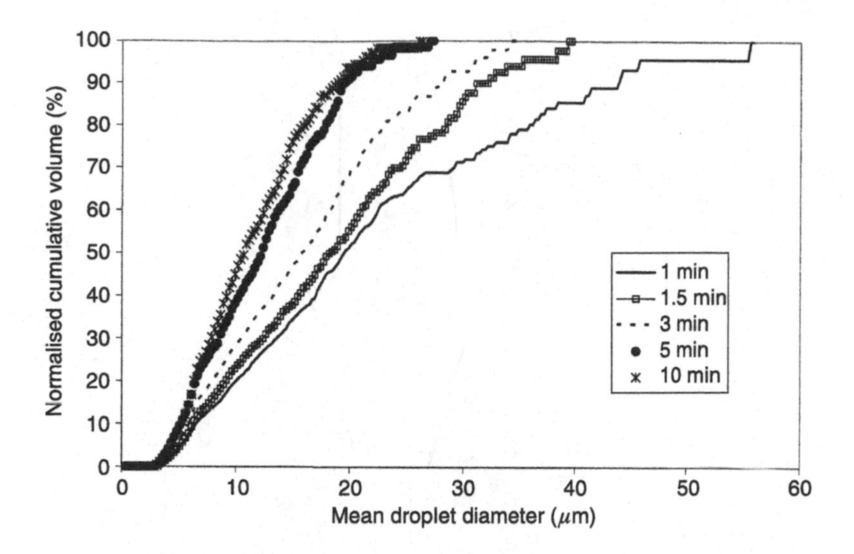

FIGURE 18.11 Cumulative volume distributions. Same experimental conditions as in Figure 18.10.

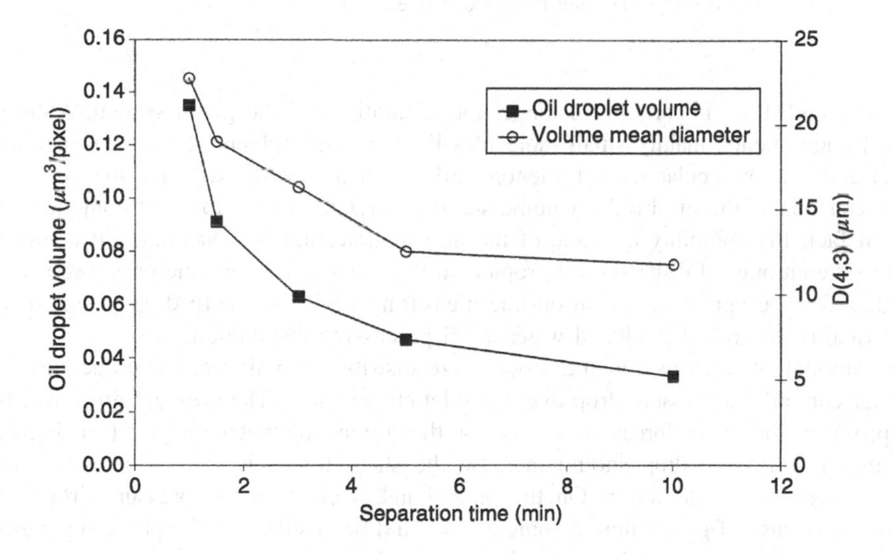

FIGURE 18.12 Changes in the oil droplet volume per pixel and the volume mean diameter, $D(4,3)$, as a function of the separation. Same experimental conditions as in Figure 18.10.

to 3 min separation time, while the next halving occurs after 10 min. For the volume mean diameter the same trend follows; it decreases rapidly in the beginning, while from 5 min to 10 min separation time, there is not much change. In the separation cell, there are two processes which affect the volume mean diameter. The creaming of the droplets removes the largest droplets, while flocculation and coalescence increases the droplet sizes. From Figure 18.12 one can qualitatively conclude that the creaming process is dominant in the beginning while the coalescence starts

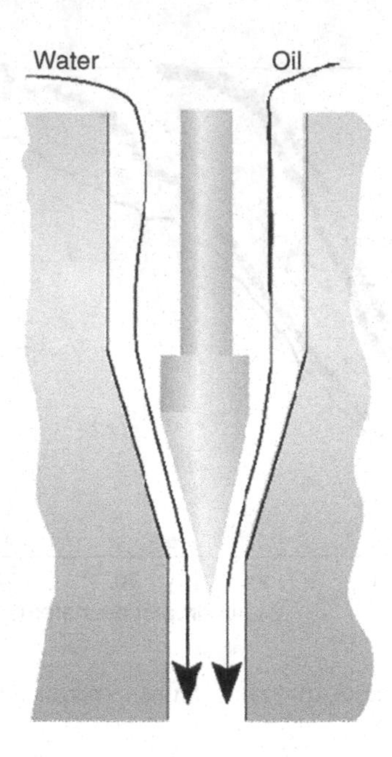

FIGURE 18.13 High pressure separation rig choke valve.

to be more and more important the longer the separation for the given system. Although some crude oil components, mainly small molecules like benzene, toluene, ethyl benzene, and xylene (BTEX) and low molecular weight phenols and naphthenic acids, will dissolve to some degree in the water phase, the oil droplet volume should be related to the total oil content in the water phase. In fact, the solubility of some of the higher molecular weight crude oil components are related to the amount of dispersed oil droplets in the water [55]. The video microscopy technique could therefore be applicable as an on-line measuring device for both droplet size distributions and for quality control of produced water at oil processing installations.

The amount of droplets and the droplet size distribution will also be influenced by factors like water cut and the pressure drop over the inlet choke valve. The pressure drop over the choke valve provides the shear forces necessary for the mixing of water and oil (see Figure 18.13). Increasing the pressure drop should increase the shear forces in the choke valve, leading to higher mixing of oil and water. On the other hand, increasing the pressure drop would also increase the release of gas, which in some cases could be positive for the phase separation, as the released gas bubbles can adsorb at the oil droplet surface and thereby increase the drag upwards. Increasing the pressure drop from 5 to 15 bars increases the amount of oil in the water phase (see Figure 18.14), due to the increased mixing in the choke valve. The effect is greater at 80% water cut, than for 95% water cut, for this particular system. Looking at the volume mean diameter in Figure 18.15, the increase in pressure drop gives an increase for $D(4,3)$ for 80% water cut, while there is no or little effect for 95% water cut.

The video microscopy is also capable of identifying inversion points for emulsions. Figure 18.16 shows how the amount of dispersed oil and the volume mean diameter of the oil droplets changes as we change the water cut for a given system. From the figure we can identify the following regions: at 50% water, the choke valve produces an oil continuous phase. Water sediments and only small oil droplets are drawn into the water phase. This leads to both

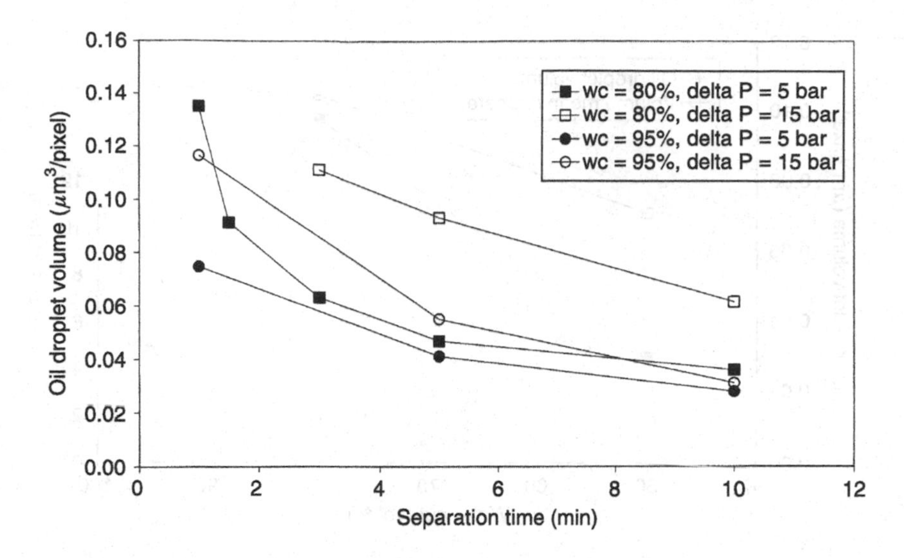

FIGURE 18.14 Effect of pressure drop on the oil droplet volume per pixel. Experimental conditions: T = 80 °C, pressure before choke valve: 65 bar.

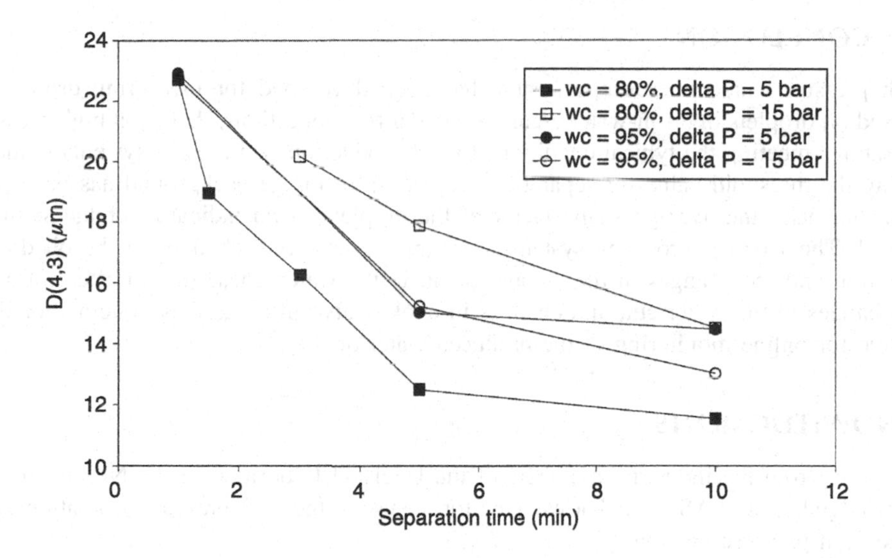

FIGURE 18.15 Effect of pressure drop on the volume mean diameter. Same experimental conditions as in Figure 18.14.

a low volume mean diameter and a low content of oil in the water phase. Moving to 80% water cut gives a huge increase in the oil content in the water phase, and the volume mean diameter is also increasing. At 80% water cut we have most likely a water continuous emulsion. We might also be close to the inversion point. At 95% water cut we have a water continuous emulsion and from 80% to 95% water cut we would expect a decrease in the amount of oil simply because we mix less oil into the water phase.

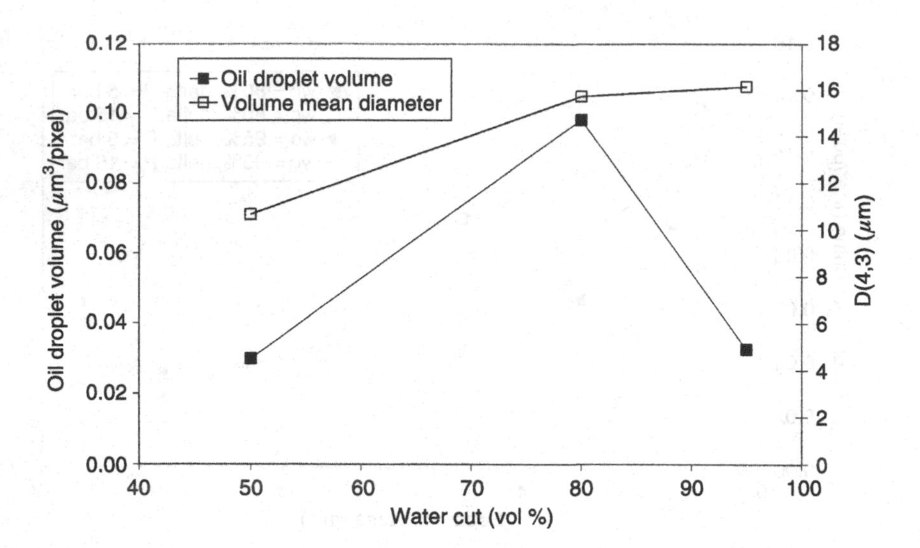

FIGURE 18.16 Effect of water cut. T = 65 °C, pressure before choke valve: 65 bar, pressure drop over choke valve: 5 bar.

18.5 CONCLUSIONS

A high pressure video microscopy system has been developed for measuring droplet sizes of dispersed oil droplets in oil-in-water emulsions under real conditions, that is, at high pressures and temperatures meeting the typical conditions for oil production facilities. A systematic method for selecting the threshold value for separating droplets from image background has been proposed. The method uses the average shape factor of the droplets as an indicator for the setting of the threshold. The video microscopy system is capable of detecting changes in the oil droplet size distribution and the changes in the amount of oil in the water phase due to effect of separation time, changes in the water cut, and changes in choke valve mixing. This system should be very attractive for online monitoring of the produced water quality.

ACKNOWLEDGMENTS

The support from the industrial members of the Ugelstad Laboratory at NTNU in Trondheim is acknowledged. Statoil ASA is acknowledged for the use of the high pressure separation equipment and the high pressure laboratory.

REFERENCES

1. Flanigan, D.A.; Stolhand, J.E.; Scribner, M.E.; Shimoda, E. Droplet Size Analysis: A New Tool for Improving Oilfield Separations. *63rd Ann. Tech. Conf. Exhib. Soc. Petr. Eng., Houston* **1988**, *SPE 18204*.
2. Geankoplis, C.J. *Transport Processes and Unit Operations*. Prentice Hall: New Jersey, 1983.
3. Cussler, E.L. *Diffusion. Mass Transfer in Fluid Systems. 2nd edition*. Cambridge University Press: Cambridge, 1997.
4. Jeelani, S.A.K.; Hartland, S. Effect of dispersion properties on the separation of batch liquid–liquid dispersions. *Ind. Eng. Chem. Res.* **1998**, *37*, 547–554.

5. Kumar, A.; Hartland, S. Gravity settling in liquid–liquid dispersions. *Can. J. Chem. Eng.* **1985**, *63*, 368–376.
6. Yu, G.Z.; Mao, Z.S. Sedimentation and coalescence profiles in liquid–liquid batch settling experiments. *Chem. Eng. Tech.* **2004**, *27*, 407–413.
7. Auflem, I.H.; Westvik, A.; Sjöblom, J. Destabilization of water-in-crude oil emulsions based on recombined oil samples at various pressures. *J. Dispers. Sci. Technol.* **2003**, *24*, 103–112.
8. Auflem, I.H.; Kallevik, H.; Westvik, A.; Sjöblom, J. Influence of pressure and solvency on the separation of water-in-crude-oil emulsions from the North Sea. *J. Petrol. Sci. Eng.* **2001**, *31*, 1–12.
9. Sjöblom, J.; Johnsen, E.E.; Westvik, A.; Ese, M.-H.; Djuve, J.; Auflem, I.H.; Kallevik, H. Demulsifiers in the Oil Industry. In *Encyclopedic Handbook of Emulsion Technology*. Sjöblom, J.; Ed.; Marcel Dekker: New York, 2001, pp 595–619.
10. Sæther, Ø. Video-enhanced Microscopy Investigation of Emulsion Droplets and Size Distributions. In *Encyclopedic Handbook of Emulsion Technology*. Sjöblom, J.; Ed.; Marcel Dekker: New York, 2001, pp 349–360.
11. Sjöblom, J.; Aske, N.; Auflem, I.H.; Brandal, Ø.; Havre, T.E.; Sæther, Ø.; Westvik, A.; Johnsen, E.E.; Kallevik, H. Our current understanding of water-in-crude oil emulsions. Recent characterization techniques and high pressure performance. *Adv. Colloid Interface Sci.* **2003**, *100*, 399–473.
12. Bhardwaj, A.; Hartland, S. Kinetics of coalescence of water droplets in water-in-crude oil-emulsions. *J. Dispers. Sci. Technol.* **1994**, *15*, 133–146.
13. Mason, S.L.; May, K.; Hartland, S. Drop size and concentration profile determination in petroleum emulsion separation. *Colloid Surf. A* **1995**, *96*, 85–92.
14. Holt, Ø.; Sæther, Ø.; Sjöblom, J.; Dukhin, S.S.; Mishchuk, N.A. Video enhanced microscopic investigation of reversible Brownian coagulation in dilute oil-in-water emulsions. *Colloid Surf. A* **1997**, *123*, 195–207.
15. Holt, Ø.; Sæther, Ø.; Sjöblom, J.; Dukhin, S.S.; Mishchuk, N.A. Investigation of reversible Brownian flocculation and intradoublet coalescence in o/w emulsions by means of video enhanced microscopy. *Colloid Surf. A* **1998**, *141*, 269–278.
16. Jokela, P.; Fletcher, P.D.I.; Aveyard, R.; Lu, J.R. The use of computerized microscopic image-analysis to determine emulsion droplet size distributions. *J. Colloid Interface Sci.* **1990**, *134*, 417–426.
17. Mishchuk, N.A.; Verbich, S.V.; Dukhin, S.S.; Holt, Ø.; Sjöblom, J. Rapid Brownian coagulation in dilute polydisperse emulsions. *J. Dispers. Sci. Technol.* **1997**, *18*, 517–537.
18. Pacek, A.W.; Moore, I.P.T.; Calabrese, R.V.; Nienow, A.W. Evolution of drop size distributions and average drop diameter in liquid–liquid dispersions before and after phase inversion. *Chem. Eng. Res. & Des.* **1993**, *71*, 340–341.
19. Pacek, A.W.; Moore, I.P.T.; Nienow, A.W.; Calabrese, R.V. Video technique for measuring dynamics of liquid–liquid dispersion during phase inversion. *Aiche J.* **1994**, *40*, 1940–1949.
20. Pacek, A.W.; Nienow, A.W. Measurement of drop size distribution in concentrated liquid–liquid dispersions – video and capillary techniques. *Chem. Eng. Res. & Des.* **1995**, *73*, 512–518.
21. Pacek, A.W.; Nienow, A.W.; Moore, I.P.T. On the structure of turbulent liquid–liquid dispersed flows in an agitated vessel. *Chem. Eng. Sci.* **1994**, *49*, 3485–3498.
22. Sæther, Ø.; Sjöblom, J.; Verbich, S.V.; Mishchuk, N.A.; Dukhin, S.S. Video-microscopic investigation of the coupling of reversible flocculation and coalescence. *Colloid Surf. A* **1998**, *142*, 189–200.
23. Sæther, Ø.; Dukhin, S.S.; Sjöblom, J.; Holt, Ø. The lifetime of a doublet of miniemulsion droplets and its measurement with the use of video microscopy. *Colloid J.* **1995**, *57*, 793–799.
24. Wang, Y.Y.; Bian, S.Z.; Wu, D. A simplified method for measuring dilute emulsion stability. *Pestic. Sci.* **1995**, *44*, 201–203.
25. Lashmar, U.T.; Richardson, J.P.; Erbod, A. Correlation of physical parameters of an oil-in-water emulsion with manufacturing procedures and stability. *Int. J. Pharm.* **1995**, *125*, 315–325.
26. Pather, S.I.; Neau, S.H.; Pather, S. A comparison of 2 quality assessment methods for emulsions. *J. Pharm. Biomed. Anal.* **1995**, *13*, 1283–1289.

27. Kamel, A.H.; Akashah, S.A.; Leeri, F.A.; Fahim, M.A. Particle-size distribution in oil–water dispersions using image-processing. *Comp. & Chem. Eng.* **1987**, *11*, 435–439.
28. Hazlett, R.D.; Schechter, R.S.; Aggarwal, J.K. Image-processing techniques for the estimation of drop size distributions. *Ind. Eng. Chem. Fundam.* **1985**, *24*, 101–105.
29. Scott, T.C.; Sisson, W.G. Droplet size characteristics and energy input requirements of emulsions formed using high-intensity-pulsed electric-fields. *Sep. Sci. Tech.* **1988**, *23*, 1541–1550.
30. Yan, Y.D.; Clarke, J.H.R. Insitu determination of particle-size distributions in colloids. *Adv. Colloid Interface Sci.* **1989**, *29*, 277–318.
31. Glatter, O.; Hofer, M. Interpretation of elastic light-scattering data. 3. Determination of size distributions of polydisperse systems. *J. Colloid Interface Sci.* **1988**, *122*, 496–506.
32. Lindner, H.; Fritz, G.; Glatter, O. Measurements on concentrated oil in water emulsions using static light scattering. *J. Colloid Interface Sci.* **2001**, *242*, 239–246.
33. Washington, C.; Sizer, T. Stability of tpn mixtures compounded from lipofundin-s and aminoplex amino-acid solutions – comparison of laser diffraction and coulter-counter droplet size analysis. *Int. J. Pharm.* **1992**, *83*, 227–231.
34. Eberth, K.; Merry, J. A comparative-study of emulsions prepared by ultrasound and by a conventional method – droplet size measurements by means of a coulter-counter and microscopy. *Int. J. Pharm.* **1983**, *14*, 349–353.
35. Sontum, P.C.; Kolderup, E.M.; Veldt, D.I. Coulter counting and light diffraction analysis applied to characterisation of oil–water emulsions. *J. Pharm. Biomed. Anal.* **1997**, *15*, 1641–1646.
36. Yan, Y.D.; Clarke, J.H.R. Dynamic light-scattering from concentrated water-in-oil microemulsions: The coupling of optical and size polydispersity. *J. Chem. Phys.* **1990**, *93*, 4501–4509.
37. Finsy, R. Particle sizing by quasi-elastic light-scattering. *Adv. Colloid Interface Sci.* **1994**, *52*, 79–143.
38. Crawley, G.; Cournil, M.; DiBenedetto, D. Size analysis of fine particle suspensions by spectral turbidimetry: Potential and limits. *Powder Technol.* **1997**, *91*, 197–208.
39. Raphael, M.; Rohani, S. On-line estimation of solids concentrations and mean particle size using a turbidimetry method. *Powder Technol.* **1996**, *89*, 157–163.
40. Li, M.Z.; Wilkinson, D. Particle size distribution determination from spectral extinction using evolutionary programming. *Chem. Eng. Sci.* **2001**, *56*, 3045–3052.
41. Roland, I.; Piel, G.; Delattre, L.; Evrard, B. Systematic characterization of oil-in-water emulsions for formulation design. *Int. J. Pharm.* **2003**, *263*, 85–94.
42. Balinov, B.; Urdahl, O.; Söderman, O.; Sjöblom, J. Characterization of water-in-crude-oil emulsions by the NMR self-diffusion technique. *Colloid Surf. A* **1994**, *82*, 173–181.
43. Balinov, B.; Soderman, O.; Warnheim, T. Determination of water droplet size in margarines and low-calorie spreads by nuclear-magnetic-resonance self-diffusion. *J. Am. Oil Chem. Soc.* **1994**, *71*, 513–518.
44. Hollingsworth, K.G.; Sederman, A.J.; Buckley, C.; Gladden, L.F.; Johns, M.L. Fast emulsion droplet sizing using NMR self-diffusion measurements. *J. Colloid Interface Sci.* **2004**, *274*, 244–250.
45. Vandenenden, J.C.; Waddington, D.; Vanaalst, H.; Vankralingen, C.G.; Packer, K.J. Rapid-determination of water droplet size distributions by pfg-nmr. *J. Colloid Interface Sci.* **1990**, *140*, 105–113.
46. Packer, K.J.; Rees, C. Pulsed nmr studies of restricted diffusion. 1. Droplet size distributions in emulsions. *J. Colloid Interface Sci.* **1972**, *40*, 206.
47. Li, X.Y.; Cox, J.C.; Flumerfelt, R.W. Determination of emulsion size distribution by NMR restricted diffusion measurement. *AICHE J.* **1992**, *38*, 1671–1674.
48. Alba, F.; Crawley, G.M.; Fatkin, J.; Higgs, D.M.J.; Kippax, P.G. Acoustic spectroscopy as a technique for the particle sizing of high concentration colloids, emulsions and suspensions. *Colloid Surf. A* **1999**, *153*, 495–502.
49. Haskell, R.J. Characterization of submicron systems via optical methods. *J. Pharm. Sci.* **1998**, *87*, 125–129.
50. Goodwin, J. *Colloids and Interfaces with Surfactants and Polymers*. Wiley: Chichester, UK, 2004.
51. Murphy, D.B. *Fundamentals of Light Microscopy and Electronic Imaging*. Wiley-Liss: New York, 2001.

52. Jillavenkatesa, A.; Dapkunas, S.J.; Lum, L.-S.H. In *Particle Size Characterization*. National Institute of Standards and Technology, 2001; Special Publication.
53. Allen, T. *Particle Size Measurement. Volume 1. Powder Sampling and Particle Size Measurement*. Chapman & Hall: London, 1997.
54. Russ, J.C. *The Image Processing Handbook, 4th Edition*. CRC Press: Boca Raton, 2002.
55. Grini, P.G.; Hjelsvold, M.; Johnson, S. Choosing Produced Water Treatment Technologies Based on Environmental Impact Reduction. *Paper presented at SPE International Conference on Health, Safety and Environment in Oil and Gas Exploration and Production, Kuala Lumpur, Malaysia* **2002**, *SPE 74002*.

19 Monitoring the Demulsification of Crude Oil Emulsions by Using Conductivity Measurements

Jan van Dijk, Trond E. Havre, and Hans-Jörg Oschmann

CONTENTS

19.1 INTRODUCTION

The selection and performance evaluation for demulsifiers is commonly based on the bottle test, which, to date, is the most common and well-accepted method in the crude oil production industry worldwide. The method relies on the visual observation of phase separation of fresh crude oil emulsions in standard graded bottles. Interpretation is based on the speed of separation as well as on visual appearance. It is a hands-on method that is time consuming and an experienced engineer is required to perform the test in order to interpret the results and select an effective demulsifier.

The development of the method described in this chapter was driven by the idea to find an electronic, measurable signal, which stands quantitatively and qualitatively in relation with the demulsification process. With this background an automated demulsification monitoring method was developed.

In the following sections, first a short review of methods for monitoring of demulsification is given. We then present the method for monitoring the demulsification process by means of conductivity measurements. The principle, test equipment, and test procedure are described. Experimental results, where the demulsification monitoring by the new method is compared to the bottles test, are reported and discussed.

19.2 EXPERIMENTAL TECHNIQUES FOR DEMULSIFICATION MONITORING

Crude oil emulsions have been known since the start of crude oil production. The use of demulsifiers for separating the emulsified water and its development started in the beginning of the twentieth century.

The classical way for selecting an effective demulsifier has always been the bottle test, a method with a long tradition, which is still the most common way for demulsifier selection. The bottle test is described by several authors, for example by Lissant [1] and Becker [2]. The disadvantage with the bottle test is that it is time consuming and it relies on the subjective reading of the engineer.

As a result of attempts to remove subjective interpretation in the bottle tests, various methods for the investigation of the demulsification process have been presented in the literature, many of which use highly sophisticated techniques.

Many of the techniques presented to determine the separation speed of emulsions are based on light transmission [3]. One of these is a laser dispersant tester (Turbiscan), which can be used for the monitoring of emulsion separation [4,5]. During the test, a near infrared laser scanner moves along the height of the test tube and analyzes transmission and back scattering of light over the whole sample length at given time intervals. As water separates, the transmission increases at the lower part of the sample and the separation can be monitored. The advantage of this technique is that it provides an objective and accurate measure of the separation process. The instrument is also rather small (20 kg) and can be transported to remote oilfield locations [6]. However, the relative small diameter of the test tubes ($\sim$1 cm) can be of influence on the demulsification process. This laser dispersant tester is also limited to perform only one test at a time.

An alternative technique to the Turbiscan is presented under the commercial name Lumifuge [7,8]. This instrument is based on the same principle, with the additional advantage of accelerated tests through centrifugation. It is capable of performing up to eight tests at the same time, but with a weight of 38 kg is certainly not what the traveling service engineer would like to carry during a field visit!

It has been demonstrated that the water content in a crude oil emulsion can be determined by the use of an electroacoustic method [9,10]. In this method an ultrasonic field is applied to the emulsion and the electric field is measured. The decrease in water as the emulsion separates can then be monitored.

Above is presented techniques that are based on measuring how fast water is separated from an emulsion by means of light or electrical measurements. The separation process is the same as for the bottle test; it is the monitoring method that makes them different. There is also a range of other techniques that more indirectly assess the emulsion stability, and which can be used to determine the effectiveness of demulsifiers. Examples of such are methods based on critical electric field [11], high frequency spectroscopy [12], correlation to microemulsion phase behavior [13], or studies of Langmuir films [14]; just to mention a few.

Although many of these methods have provided valuable insights in the demulsification mechanism, they are costly, and few of them can be adapted for use in the field and even fewer would survive the trips to the remote locations, which have become the center of attention in our industry.

Another frequent disadvantage of the methods mentioned above is that they are commonly only capable of dealing with one or very few samples at the same time.

19.3 A NEW CONCEPT: THE DEMCOM METHOD

The target of the development of the new method was to find an electronic, measurable signal that stands quantitatively and qualitatively in relation with the demulsification process. In the method development, the criteria and the benefits of the traditional bottle test should be respected.

The solution was found in measuring conductivity of the separated water by a novel approach. In this section, the principle, apparatus, and test procedure of the new method are described. In the next section the experiments performed with one measuring cell are reported. This new method may lead to future advantages in demulsifier selection.

19.3.1 Principle of the Method

The method is based on the idea that the conductivity of the emulsified water is a constant and the same for the first as well as for the last separated drop of water. However, if the separated water is continuously mixed with a known amount of water of a different, known conductivity, the conductivity of the mixture will change proportionally with the amount of separated water. The same principle has been used for many years in crude oil washing processes where the amount of salt separated from the crude oil can be monitored by measuring the increase in conductivity of the washing water [15,16].

Measuring the conductivity of the water phase continuously as a function of time, and by using data logging, a suitable program, and a personal computer, the demulsification process can be registered automatically. Objective data of how fast the water separates are obtained and can be used to determine the performance of demulsifiers.

We have named the method DEMCOM, which is an acronym for "***de***mulsification ***m***onitoring by ***co***nductivity ***m***easurement."

19.3.2 Test Equipment

The DEMCOM apparatus consists of standard equipment. The set-up is presented schematically in Figure 19.1. The apparatus consists of a vessel (~70 ml) with a thermostatic jacket, magnetic stirrer, and an electrode placed in the water phase. The probe must be placed such that the oil phase does not come in contact with the measuring electrode. The electrode is connected to a conductivity meter, which again is connected to a device that registers the signal as a function of time, such as a personal computer.

19.3.3 Test Procedure

In order to generate a measurable change in the conductivity versus time, the difference in conductivity of the mixing water and the emulsified water should be large. Typically, the emulsified water has a relatively high conductivity and demineralized water will be suitable for mixing. In those cases where the conductivity of the separated water is relatively low, mixing water with a high conductivity may be used.

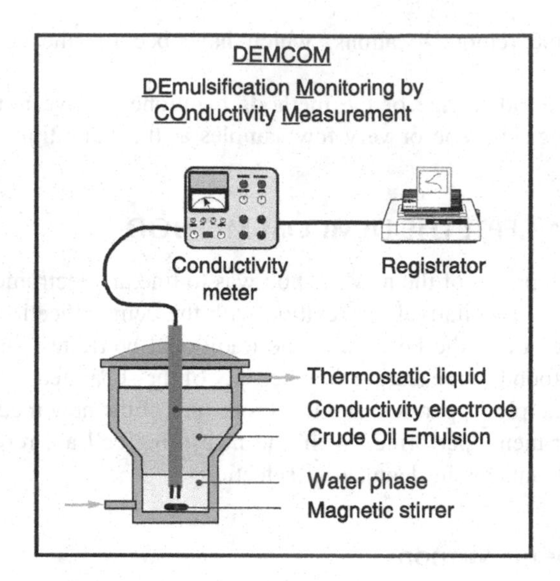

FIGURE 19.1 Scheme of the DEMCOM apparatus.

19.3.3.1 Calibration

In order to be able to convert the measured conductivity into the amount of separated water it is necessary to know the relationship between the measured signal and the ratio of each aqueous phase.

Measuring the conductivity of prepared ratios of each water quality allows a calibration curve to be drawn. For the resulting curve, a linear trend line and its formula can be determined. The formula is then used for calculating the separated amount of water. In Equation 19.1 the conductivity, Y, is shown as a linear function of the fraction X, in percent, of separated water in the water bulk phase.

$$Y = \alpha \cdot X + C \tag{19.1}$$

where α and C are constants determined by the curve fitting. Based on the measured conductivity and the calibration function, the fraction X of separated water in the water bulk phase is determined. However, the parameter of interest is the amount of separated water in percent of the initially emulsified water, Z. The parameter Z can be calculated using Equation 19.2 where $M_{emul,i}$ is the amount of initially emulsified water and $M_{bulk,i}$ is the amount of bulk water before the separation starts.

$$Z = \frac{X}{(100 - X)} \cdot \frac{M_{bulk,i}}{M_{emul,i}} \cdot 100 \tag{19.2}$$

As will be shown later in this chapter it may be sufficient to measure the conductivity of each of the water phases and then assume a linear calibration line.

19.3.3.2 Performing the Test

The procedure for adding the demulsifier is the same as for the bottle test. The crude oil emulsion is heated to the test temperature and the demulsifier under investigation is added and mixed into the emulsion by shaking.

The test cell (Figure 19.1) is filled with a measured quantity of water for mixing and thermostated to the test temperature. Once the conductivity electrode is installed in the water phase, the crude oil emulsion is poured carefully on top of this layer. In order to avoid disturbing of the reliability of the measuring signal, the electrode should not be contacted by the oil phase.

The magnetic stirrer is started at a moderate speed to obtain a fast and reliable signal response. Care has to be taken to ensure that the preconditioned water will not function as wash water for the crude. Formation of a vortex has to be avoided. Moderate agitation also occurs in practice where the crude oil emulsion in a separator always is under a slight movement.

The measuring starts from time zero and will continue as long as the monitored and logged signal indicates that the separation is still in progress.

19.4 EXPERIMENTAL RESULTS AND DISCUSSION

In the following we show experimental results obtained when using the DEMCOM method. We first determine the reproducibility of the calibration needed to convert the measured conductivity into separated water. The effectiveness of demulsifiers is then studied and the DEMCOM results are compared to results obtained using the traditional bottle test.

19.4.1 Calibration Results

Demineralized water was used for mixing with the separated water. The water separated from eight samples of one crude oil emulsions where used to make eight different calibration curves. The average values, with standard deviation, of the constants α and C in Equation 19.1 were determined to be 0.106 ± 0.006 and 0.23 ± 0.06, respectively.

By making the calibration formula based only on the measurement of the conductivity of each of the water phases, α and C are determined to be 0.108 and 0.2, respectively. This is in agreement with the values obtained by making various blending ratios. Since produced water normally primarily consists of strong electrolytes, the conductivity will be linearly dependent on the mixing ratio and it would be sufficient to determine the conductivity of each water phase. Figure 19.2 shows the experimental points used for the calibration. The line is drawn between the measuring points for each of the two water phases.

19.4.2 Demulsification Experiments

For the investigation in the research of this method, crude oil emulsions from one origin were used. The demulsifiers used were selected based on bottle test results and represent both good and bad performing products for this crude oil.

The amount of deionized water used for mixing with the separated water was 10 g. Twenty-eight grams of crude oil emulsion was added on top of this water. The water cut in the prepared emulsions varied from 38 to 54%. All experiments were performed at 40 °C.

Products A, B, C, D, E1, E2, and E3 was dosed at 30 ppm to the crude oil emulsion. The separation was studied by the bottle test and the DEMCOM method and the results are presented in Figures 19.3 to 19.7 where the amount of separated water from the emulsion is plotted as a

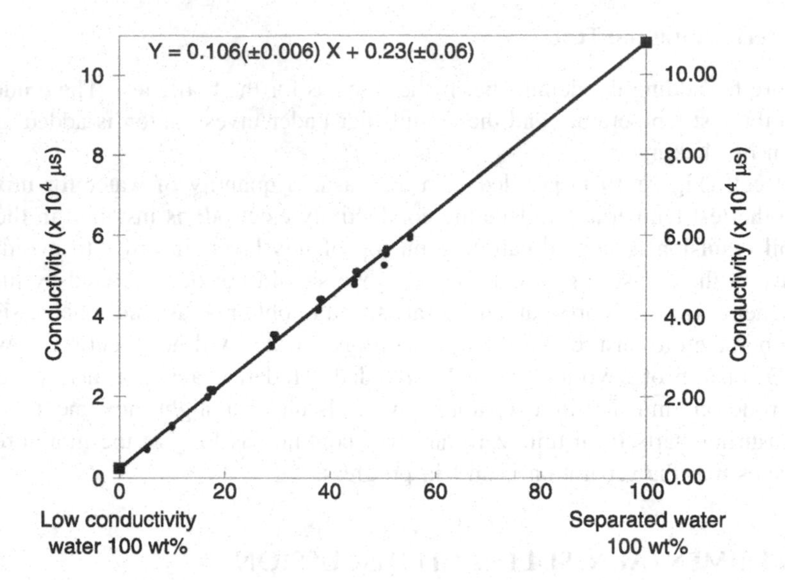

FIGURE 19.2 Conductivity calibration data for mixtures of separated water and low conductivity water. The line is drawn between measuring points for each of the two water phases. The points are determined conductivities for various mixing ratios.

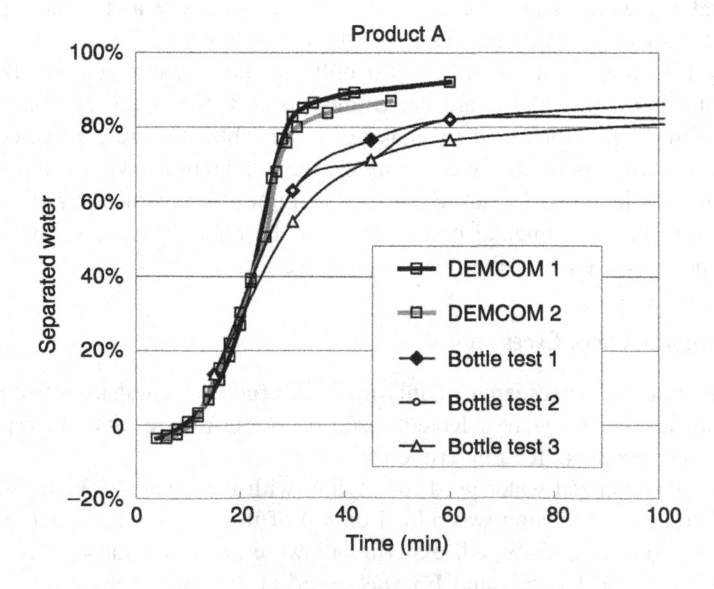

FIGURE 19.3 Amount of separated water versus time, as determined with the bottle test and the DEMCOM method, for a crude oil emulsion treated with 30 ppm of product A.

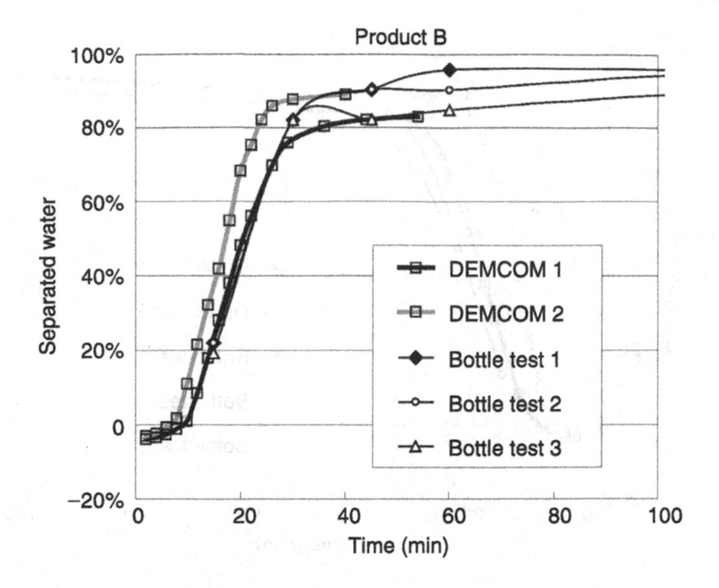

FIGURE 19.4 Amount of separated water versus time, as determined with the bottle test and the DEMCOM method, for a crude oil emulsion treated with 30 ppm of product B.

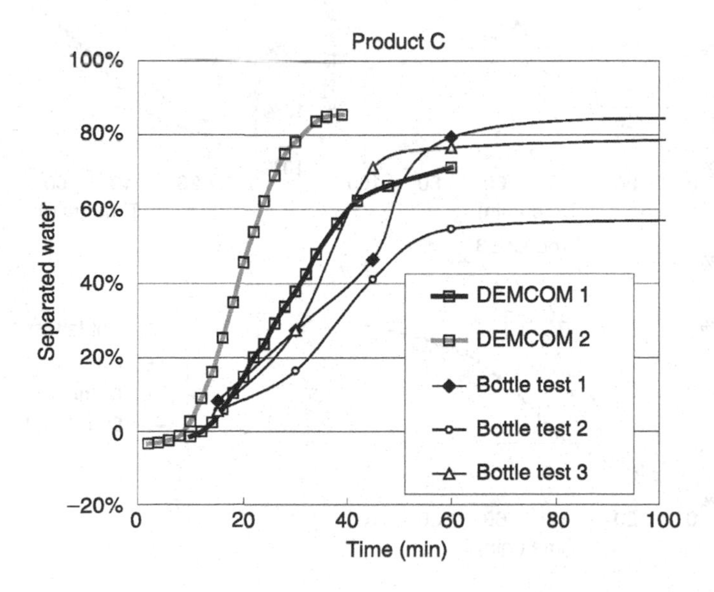

FIGURE 19.5 Amount of separated water versus time, as determined with the bottle test and the DEMCOM method, for a crude oil emulsion treated with 30 ppm of product C.

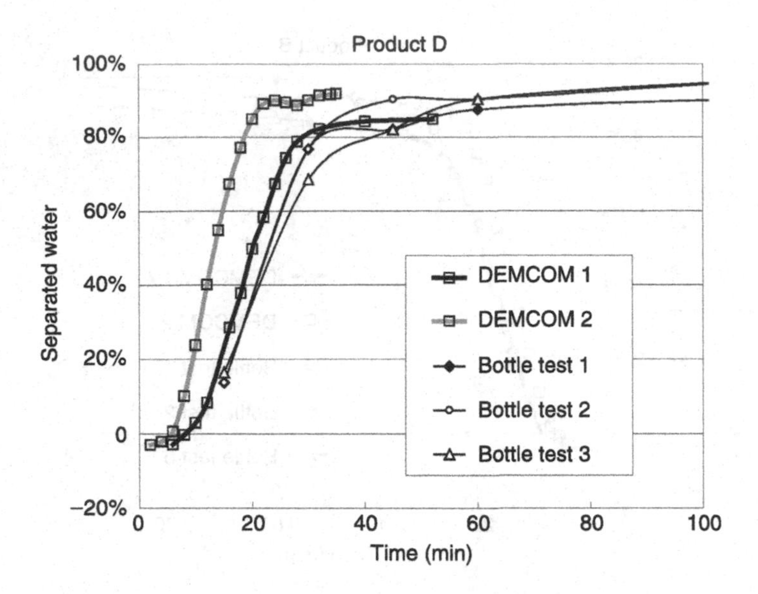

FIGURE 19.6 Amount of separated water versus time, as determined with the bottle test and the DEMCOM method, for a crude oil emulsion treated with 30 ppm of product D.

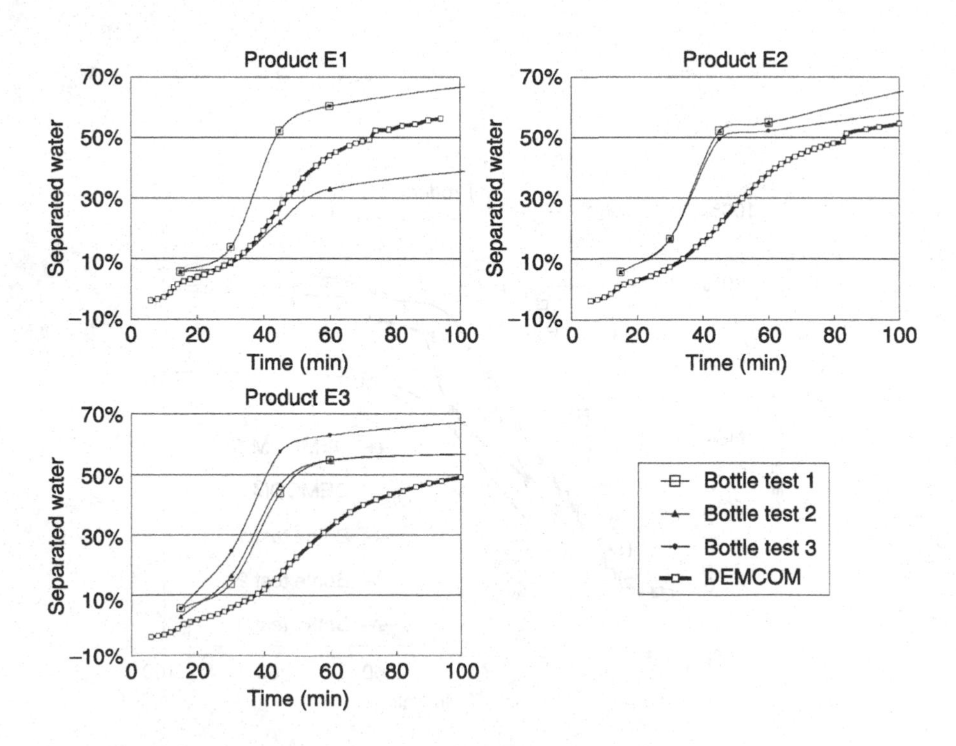

FIGURE 19.7 Amount of separated water versus time, as determined with the bottle test and the DEMCOM method, for a crude oil emulsion treated with 30 ppm of products E1, E2, or E3.

function of time. Bottle tests were performed in triplicate for each system and the DEMCOM tests were performed singly or repeated.

When comparing the bottle test with the DEMCOM test there is no consistent trend in which method that shows the higher amount of water separated. In the tests where products A and C are used as demulsifiers, the DEMCOM method indicates slightly higher amount of water separated than the bottle test (Figures 19.3 and 5). When products B and D are used, the amount of separated water at a given time is the same for the two methods (Figures 19.4 and 6). However, for products E1 to E3 the bottle tests show a higher amount of separated water than the DEMCOM tests (Figure 19.7).

Some factors can contribute to a difference in the results for the two methods. In the DEMCOM method, an agitated water phase is present from the start of the experiment. This may have an influence on the separation speed when compared to the bottle test. The interface between crude oil emulsion and water may be difficult to read with the naked eye, and this will lead to a random or systematic error in the estimated amount of water separated for the bottle test.

The results obtained by the DEMCOM method are not always exactly the same as the results obtained by the bottle test. Nevertheless, it is clear that the DEMCOM test generates the same trend of the effectiveness of the demulsifiers as the bottle test. The products found to be effective in the bottle test were also good in the DEMCOM method. Looking at the results presented here, it is seen that products A, B, and D are highly effective in destabilizing the current crude oil emulsion, product C is moderately effective, and products E1 to E3 show limited performance. Neither with the bottle test nor with the DEMCOM test was it possible to perform a more detailed ranking based on these results; i.e., to get an exact ranking of the products E1 to E3 or A, B, and D in terms of performance.

The shape of the lines is representative for the demulsifying characteristic; this is the same in both tests. However, the resolution of each curve is much higher with DEMCOM and the separation profile becomes clearer.

An idea of the reproducibility for each of the test methods can be found by looking at the duplicate runs for the DEMCOM method and comparing with the triplicate for the bottle test. It is clear that the reproducibility of the DEMCOM test is at least as good as for the bottle test. The experiments reported here were performed with the aim of proving that the DEMCOM principle can be used for demulsification monitoring. It is believed that further work with the experimental set-up and test procedure can result in improved reproducibility.

During the performance of the DEMCOM method no unexpected difficulties occurred. Once the operator is familiar with the procedure it is an effective and efficient method.

19.5 ADVANTAGES AND DISADVANTAGES WITH THE DEMCOM METHOD

The main incentive for developing the DEMCOM method has been to find an alternative for demulsification monitoring, especially when selecting demulsifier products at field locations. Today this is more or less always done by bottle testing. In the following we discuss advantages and disadvantages that can be identified when comparing the DEMCOM method with the traditional bottle test.

Obviously, a major advantage of the bottle test is that the equipment needed is limited to graded bottles. For the DEMCOM method a more advance instrument set-up is needed. However, the DEMCOM apparatus consists of low cost parts and it would be possible to design it so that it can be easily transported to field locations. This makes the DEMCOM method unique when compared to other automated demulsification monitoring techniques. In order to work accurately with the DEMCOM method, a balance with an accuracy of ±0.01 g is necessary, an apparatus which could also be very helpful for the bottle test.

In our experiments we have used one single measurement cell. For the DEMCOM method to work as an alternative to the bottle test in demulsifier screening it will be absolutely necessary to have multi-cell equipment available. Further developments have to be made before the DEMCOM method will have a portable form for experiments in the field. Once this is done, the DEMCOM method will be able to compete with the bottle test.

The DEMCOM method needs some extra preparation in respect to the generating of the calibration curve. However, for a bottle test it is also necessary to check the salt content of the crude oil emulsion. The cleaning of the DEMCOM apparatus after the test will be somewhat more laborious than for the bottle test, while the preparation of the emulsion is the same for both methods.

One of the main advantages with the DEMCOM method is that when the experiment is running, the demulsification process is registered automatically and accurately without any interference by the operator. While for the bottle test a series of tests has to be started at the same time, with the DEMCOM method each test can be performed on its own. During the course of the test it can be decided to continue or to stop. The latter can be done if the extrapolation of the curve is not predicting a good result. After terminating the test, a new experiment can immediately be prepared. For the bottle test it will only be practical to start a series of experiments at the same time.

The amount of crude oil emulsion for one DEMCOM test is about 30 ml, which is less than for the bottle test. This can be an advantage when a crude oil emulsion sample is limited in amount or difficult to obtain.

It is of a great advantage to generate a demulsifier characteristic and to have it directly available in an electronic data system. Comparing a great number of results in a demulsifier selection process is much easier this way. The generated data gives accurate information regarding the demulsifier performance, the initial activity, the demulsification rate, and what can be expected in respect to the end separation effect. While for the bottle test there will be limitations in how often a measuring point can be recorded, the DEMCOM apparatus can be set to collect data points as often as the operator decides.

Given the measuring principle and its accuracy it is expected that a rest water content determination of the separated crude will not be necessary for the DEMCOM method.

To summarize the comparison of the bottle test with the DEMCOM test it is clear that the great advantage of the bottle test is the simple equipment needed. Less work is also required for preparation of the separation vessels. On the other hand, the advantages of the DEMCOM method are its automatic and objective collection of separation data. It is our belief that the DEMCOM method can be developed so that it will be a good alternative to the bottle test for demulsifier screening.

19.6 CONCLUSIONS AND RECOMMENDATIONS

The name chosen for the described method is "The DEMCOM method," which is an acronym for "***de***mulsification ***m***onitoring by ***co***nductivity ***m***easurement."

The data generated with the DEMCOM method can be considered as the demulsifying characteristic of a particular demulsifier for a particular crude. This characteristic describes information such as the initial activity, the progress in separating water, and the effectiveness in respect to rest water and rest salt in the crude. Having the data directly available in an electronic data system makes the comparison of a great number of results much easier.

The precision of the method is good and is the same for the first as well as for the last separated drop of water; therefore the determination of the rest water or rest salt in the crude oil after finishing the separation process is no longer required.

Crude oil emulsions containing just a few percent of water, which are almost impossible to test in the classical bottle test, can be easily tested by this method. This will be of interest for refineries and for monitoring the wash process for the crude.

The DEMCOM method can be used for quality control of supplied demulsifier batches to crude oil producers.

The first experimental results with this newly developed DEMCOM method are very promising. The method is based on a well-known principle and is simple to perform. It is expected that once a multi-cell unit is developed with suitable data logger and software a very accurate method will be available for field use.

ACKNOWLEDGMENT

The authors would like to thank Mr Marcel Huis in't Veld posthumously for his support in performing the tests, his critique on interpretation, and in discussing the results.

REFERENCES

1. Lissant, K. J., Demulsification, industrial applications. Surfactant Science Series. Vol. 13. 1983, New York: Marcel Dekker Inc.
2. Becker, J. R., Crude oil waxes, emulsions, and asphaltenes. 1997, Tulsa: PennWell Publishing Company.
3. Miller, D. J., Coalescence in crude oil emulsions investigated by a light transmission method, Colloid Polym Sci 265(4) (1987) 342–346.
4. Mengual, O., Meunier, G., Cayre, I., Puech, K. and Snabre, P., TURBISCAN MA 2000: multiple light scattering measurement for concentrated emulsion and suspension instability analysis, Talanta 50 (1999) 445–456.
5. Dalmazzone, C. and Noik, C., Development of new "green" demulsifiers for oil production. In: SPE International Symposium on Oilfield Chemistry. 2001. Houston, USA.
6. Oschmann, H.-J., New methods for the selection of asphaltene inhibitors in the field. In: Chemistry in the Oil Industry VII: Performance in a Challenging Environment, H. A. Craddock, H. Frampton, J. Dunlop and G. Payne, Editors. 2002, Royal Society of Chemistry. p. 254.
7. Lerche, D., Sobisch, T. and Küchler, S., Stability analyser LUMiFuge 116 for rapid evaluation of emulsion stability and demulsifier selection. In: Third World Congress on Emulsions. 2002. Lyon, France.
8. Sobisch, T., Lehnberger, C. and Lerche, D., Efficient measuring system for objective characterization and optimization of separation processes, Wasser, Luft und Boden 42(10) (1998) 38–40.
9. Babchin, A. J., Chow, R. S. and Sawatzky, R. P., Electrokinetic measurements by electroacoustical methods, Adv Colloid Interface Sci 30 (1989) 111–151.
10. Isaacs, E. E., Huang, H., Chow, R. S. and Babchin, A. J., Electroacoustic method for monitoring the coalescence of water-in-oil emulsions, Colloids Surf 46 (1990) 177–192.
11. Sjöblom, J., Aske, N., Auflem, I. H., Brandal, Ø., Havre, T. E., Sæther, Ø., Westvik, A., Johnsen, E.E., and Kallevik, H., Our current understanding of water-in-crude oil emulsions, recent characterization, techniques and high pressure performance, Adv Colloid Interface Sci 100–102 (2003) 399–473.
12. Kageshima, K., Takel, T. and Sugitani, Y., Monitoring the stability of an emulsion by high-frequency spectroscopy, Analytical Sciences 19 (2003) 757–759.

13. Goldszal, A. and Bourrel, M., Demulsification of crude oil emulsions: correlation to microemulsion phase behavior, Ind Eng Chem Res 39 (2000) 2746–2751.
14. Ese, M.-H., Galet, L., Clausse, D. and Sjöblom, J., Properties of Langmuir surface and interfacial films built up by asphaltenes and resins: influence of chemical demulsifiers, J Colloid Interface Sci 220 (1999) 293–301.
15. Magaril, R. Z., Zevako, V. K. and Zablotskii, A. G., Continuous automatic determination of salt in desalted crude oil, Neftepererabotka 4 (1958) 32–33.
16. Romanov, I. S., Apparatus for determining salt in desalted petroleum, Neftepererabotka i Neftekhim 6 (1961) 39–40.

Index

Printed in the United States
by Baker & Taylor Publisher Services

Printed in the United States
by Baker & Taylor Publisher Services